D1163038

Properties and Processes of Earth's Lower Crust

Geophysical Monograph Series

Including

IUGG Volumes
Maurice Ewing Volumes
Mineral Physics Volumes

GEOPHYSICAL MONOGRAPH SERIES

Geophysical Monograph Volumes

1 **Antarctica in the International Geophysical Year** *A. P. Crary, L. M. Gould, E. O. Hulburt, Hugh Odishaw, and Waldo E. Smith (Eds.)*
2 **Geophysics and the IGY** *Hugh Odishaw and Stanley Ruttenberg (Eds.)*
3 **Atmospheric Chemistry of Chlorine and Sulfur Compounds** *James P. Lodge, Jr. (Ed.)*
4 **Contemporary Geodesy** *Charles A. Whitten and Kenneth H. Drummond (Eds.)*
5 **Physics of Precipitation** *Helmut Weickmann (Ed.)*
6 **The Crust of the Pacific Basin** *Gordon A. Macdonald and Hisashi Kuno (Eds.)*
7 **Antarctica Research: The Matthew Fontaine Maury Memorial Symposium** *H. Wexler, M. J. Rubin, and J. E. Caskey, Jr. (Eds.)*
8 **Terrestrial Heat Flow** *William H. K. Lee (Ed.)*
9 **Gravity Anomalies: Unsurveyed Areas** *Hyman Orlin (Ed.)*
10 **The Earth Beneath the Continents: A Volume of Geophysical Studies in Honor of Merle A. Tuve** *John S. Steinhart and T. Jefferson Smith (Eds.)*
11 **Isotope Techniques in the Hydrologic Cycle** *Glenn E. Stout (Ed.)*
12 **The Crust and Upper Mantle of the Pacific Area** *Leon Knopoff, Charles L. Drake, and Pembroke J. Hart (Eds.)*
13 **The Earth's Crust and Upper Mantle** *Pembroke J. Hart (Ed.)*
14 **The Structure and Physical Properties of the Earth's Crust** *John G. Heacock (Ed.)*
15 **The Use of Artificial Satellites for Geodesy** *Soren W. Henricksen, Armando Mancini, and Bernard H. Chovitz (Eds.)*
16 **Flow and Fracture of Rocks** *H. C. Heard, I. Y. Borg, N. L. Carter, and C. B. Raleigh (Eds.)*
17 **Man-Made Lakes: Their Problems and Environmental Effects** *William C. Ackermann, Gilbert F. White, and E. B. Worthington (Eds.)*
18 **The Upper Atmosphere in Motion: A Selection of Papers With Annotation** *C. O. Hines and Colleagues*
19 **The Geophysics of the Pacific Ocean Basin and Its Margin: A Volume in Honor of George P. Woollard** *George H. Sutton, Murli H. Manghnani, and Ralph Moberly (Eds.)*
20 **The Earth's Crust: Its Nature and Physical Properties** *John G. Heacock (Ed.)*
21 **Quantitative Modeling of Magnetospheric Processes** *W. P. Olson (Ed.)*
22 **Derivation, Meaning, and Use of Geomagnetic Indices** *P. N. Mayaud*
23 **The Tectonic and Geologic Evolution of Southeast Asian Seas and Islands** *Dennis E. Hayes (Ed.)*
24 **Mechanical Behavior of Crustal Rocks: The Handin Volume** *N. L. Carter, M. Friedman, J. M. Logan, and D. W. Stearns (Eds.)*
25 **Physics of Auroral Arc Formation** *S.-I. Akasofu and J. R. Kan (Eds.)*
26 **Heterogeneous Atmospheric Chemistry** *David R. Schryer (Ed.)*
27 **The Tectonic and Geologic Evolution of Southeast Asian Seas and Islands: Part 2** *Dennis E. Hayes (Ed.)*
28 **Magnetospheric Currents** *Thomas A. Potemra (Ed.)*
29 **Climate Processes and Climate Sensitivity (Maurice Ewing Volume 5)** *James E. Hansen and Taro Takahashi (Eds.)*
30 **Magnetic Reconnection in Space and Laboratory Plasmas** *Edward W. Hones, Jr. (Ed.)*
31 **Point Defects in Minerals (Mineral Physics Volume 1)** *Robert N. Schock (Ed.)*
32 **The Carbon Cycle and Atmospheric CO_2: Natural Variations Archean to Present** *E. T. Sundquist and W. S. Broecker (Eds.)*
33 **Greenland Ice Core: Geophysics, Geochemistry, and the Environment** *C. C. Langway, Jr., H. Oeschger, and W. Dansgaard (Eds.)*
34 **Collisionless Shocks in the Heliosphere: A Tutorial Review** *Robert G. Stone and Bruce T. Tsurutani (Eds.)*
35 **Collisionless Shocks in the Heliosphere: Reviews of Current Research** *Bruce T. Tsurutani and Robert G. Stone (Eds.)*
36 **Mineral and Rock Deformation: Laboratory Studies—The Paterson Volume** *B. E. Hobbs and H. C. Heard (Eds.)*
37 **Earthquake Source Mechanics (Maurice Ewing Volume 6)** *Shamita Das, John Boatwright, and Christopher H. Scholz (Eds.)*
38 **Ion Acceleration in the Magnetosphere and Ionosphere** *Tom Chang (Ed.)*
39 **High Pressure Research in Mineral Physics (Mineral Physics Volume 2)** *Murli H. Manghnani and Yasuhiko Syono (Eds.)*
40 **Gondwana Six: Structure, Tectonics, and Geophysics** *Gary D. McKenzie (Ed.)*

41 **Gondwana Six: Stratigraphy, Sedimentology, and Paleontoloty** *Garry D. McKenzie (Ed.)*
42 **Flow and Transport Through Unsaturated Fractured Rock** *Daniel D. Evans and Thomas J. Nicholson (Eds.)*
43 **Seamounts, Islands, and Atolls** *Barbara H. Keating, Patricia Fryer, Rodey Batiza, and George W. Boehlert (Eds.)*
44 **Modeling Magnetospheric Plasma** *T. E. Moore, J. H. Waite, Jr. (Eds.)*
45 **Perovskite: A Structure of Great Interest to Geophysics and Materials Science** *Alexandra Navrotsky and Donald J. Weidner (Eds.)*
46 **Structure and Dynamics of Earth's Deep Interior (IUGG Volume 1)** *D. E. Smylie and Raymond Hide (Eds.)*
47 **Hydrogeological Regimes and Their Subsurface Thermal Effects (IUGG Volume 2)** *Alan E. Beck, Grant Garvin and Lajos Stegena (Eds.)*
48 **Origin and Evolution of Sedimentary Basins and Their Energy and Mineral Resources (IUGG Volume 3)** *Raymond A. Price (Ed.)*
49 **Slow Deformation and Transmission of Stress in the Earth (IUGG Volume 4)** *Steven C. Cohen and Petr Vaníček (Eds.)*
50 **Deep Structure and Past Kinematics of Accreted Terranes (IUGG Volume 5)** *John W. Hillhouse (Ed.)*

IUGG Volumes

1 **Structure and Dynamics of Earth's Deep Interior** *D. E. Smylie and Raymond Hide (Eds.)*
2 **Hydrogeological Regimes and Their Subsurface Thermal Effects** *Alan E. Beck, Grant Garvin and Lajos Stegena (Eds.)*
3 **Origin and Evolution of Sedimentary Basins and Their Energy and Mineral Resources** *Raymond A. Price (Ed.)*
4 **Slow Deformation and Transmission of Stress in the Earth** *Steven C. Cohen and Petr Vaníček (Eds.)*
5 **Deep Structure and Past Kinematics of Accreted Terranes** *John W. Hillhouse (Ed.)*

Maurice Ewing Volumes

1 **Island Arcs, Deep Sea Trenches, and Back-Arc Basins** *Manik Talwani and Walter C. Pitman III (Eds.)*
2 **Deep Drilling Results in the Atlantic Ocean: Ocean Crust** *Manik Talwani, Christopher G. Harrison, and Dennis E. Hayes (Eds.)*
3 **Deep Drilling Results in the Atlantic Ocean: Continental Margins and Paleoenvironment** *Manik Talwani, William Hay, and William B. F. Ryan (Eds.)*
4 **Earthquake Prediction—An International Review** *David W. Simpson and Paul G. Richards (Eds.)*
5 **Climate Processes and Climate Sensitivity** *James E. Hansen and Taro Takahashi (Eds.)*
6 **Earthquake Source Mechanics** *Shamita Das, John Boatwright, and Christopher H. Scholz (Eds.)*

Mineral Physics Volumes

1 **Point Defects in Minerals** *Robert N. Schock (Ed.)*
2 **High Pressure Research in Mineral Physics** *Murli H. Manghnani and Yasuhiko Syono (Eds.)*

Geophysical Monograph 51
IUGG Volume 6

Properties and Processes of Earth's Lower Crust

Robert F. Mereu
Stephan Mueller
David M. Fountain
Editors

American Geophysical Union
International Union of Geodesy and Geophysics

Geophysical Monograph/IUGG Series

Library of Congress Cataloging-in-Publication Data

Properties and processes of earth's lower crust / edited by Robert F. Mereu, Stephan Mueller, David M. Fountain.
p. cm.—(Geophysical monograph ; 51 /IUGG series ; 6)
"Collection of papers which were presented at the IUGG Symposium U7 (Lower Crust: Properties and Processes) held in Vancouver, Canada in August, 1987"—Pref.
ISBN 0-87590-456-4 (American Geophysical Union)
1. Earth—Crust—Congresses. 2. Geophysics—Congresses. I. Mereu, R. F. (Robert Frank), 1930- . II. Mueller, Stephan, 1930- . III. Fountain, David. IV. International Union of Geodesy and Geophysics. V. American Geophysical Union. VI. IUGG Symposium U7—Properties and Processes of Earth's Lower Crust (1987 : Vancouver, B.C.) VII. Series.
QE511.P78 1989 89-2144
551.1'3—dc20

Printed in the United States of America.

CONTENTS

Preface
R. F. Mereu, St. Mueller and D. M. Fountain xi

I. THE LOWER CRUST FROM RESULTS OF NEAR-VERTICAL REFLECTION EXPERIMENTS

1. **Large-Scale Lenticles in the Lower Crust Under an Intra-Continental Basin in Eastern Australia**
   D. M. Finlayson, J. H. Leven, and K. D. Wake-Dyster 3
2. **Dating Lower Crustal Features in France and Adjacent Areas from Deep Seismic Profiles**
   C. Bois, B. Pinet and F. Roure 17
3. **Styles of Deformation Observed on Deep Seismic Reflection Profiles of the Appalachian-Caledonide System**
   J. Hall, G. Quinlan, J. Wright, C. Keen, and F. Marillier 33
4. **Laterally Persistent Seismic Characteristics of the Lower Crust: Examples from the Northern Appalachians**
   Francois Marillier, Charlotte E. Keen, and Glen S. Stockmal 45
5. **Contrasting Types of Lower Crust**
   Scott B. Smithson 53
6. **A "Glimpce" of the Deep Crust Beneath the Great Lakes**
   Alan G. Green, W. F. Cannon, B. Milkereit, D. R. Hutchinson, A. Davidson, J. C. Behrendt, C. Spencer, M. W. Lee, P. Morel-à-l'Huissier, and W. F. Agena 65
7. **Crustal Extension in the Midcontinent Rift System—Results from Glimpce Deep Seismic Reflection Profiles over Lakes Superior and Michigan**
   J. C. Behrendt, A. G. Green, M. W. Lee, D. R. Hutchinson, W. F. Cannon, B. Milkereit, W. F. Agena, and C. Spencer 81

II. THE LOWER CRUST FROM RESULTS OF REFRACTION/WIDE-ANGLE REFLECTION EXPERIMENTS

8. **The Glimpce Seismic Experiment: Onshore Refraction and Wide-Angle Reflection Observations from a Fan Line over the Lake Superior Midcontinent Rift System**
   Duryodhan Epili and Robert F. Mereu 93
9. **The Complexity of the Continental Lower Crust and Moho from PmP Data: Results from Cocrust Experiments**
   R. F. Mereu, J. Baerg, and J. Wu 103
10. **A Petrological Model of the Laminated Lower Crust in Southwest Germany Based on Wide-Angle P- and S-Wave Seismic Data**
    W. Steven Holbrook 121
11. **DSS Studies over Deccan Traps Along the Thuadara-Sendhwa-Sindad Profile, Across Narmada-Son Lineament, India**
    K. L. Kaila, I. B. P. Rao, P. Koteswara Rao, N. Madhava Rao, V. G. Krishna, and A. R. Sridhar 127

12. **Synthetic Seismogram Modeling of Crustal Seismic Record Sections from the Koyna DSS Profiles in the Western India**
V. G. Krishna, K. L. Kaila, and P. R. Reddy 143
13. **Study of the Lower Crust and Upper Mantle Using Ocean Bottom Seismographs**
Yuri P. Neprochnov 159
14. **A Seismic Refraction Study of the Crustal Structure of the South Kenya Rift**
W. Henry, J. Mechie, P. K. H. Maguire, J. Patel, G. R. Keller, C. Prodehl, M. A. Khan 169

III. THE LOWER CRUST FROM RESULTS OF SEISMICITY STUDIES

15. **Seismicity near Lake Bogoria in the Kenya Rift Valley**
Philippa Cooke, Peter Maguire, Russ Evans, and Nicholas Laffoley 175
16. **Anelastic Properties of the Crust in the Mediterranean Area**
A. Craglietto, G. F. Panza, B. J. Mitchell, and G. Costa 179
17. **Earthquakes and Temperatures in the Lower Crust Below the Northern Alpine Foreland of Switzerland**
N. Deichmann and L. Rybach 197
18. **Heat Flow, Electrical Conductivity and Seismicity in the Black Forest Crust, SW Germany**
H. Wilhelm, A. Berktold, K.-P. Bonjer, K. Jäger, A. Stiefel, and K.-M. Strack 215

IV. THE LOWER CRUST FROM RESULTS OF OTHER GEOPHYSICAL STUDIES

19. **On the Vertical Distribution of Radiogenic Heat Production in the Continental Crust and the Estimated Moho Heat Flow**
Vladimír Čermák and Louise Bodri 235
20. **Fluids in the Lower Crust Inferred from Electromagnetic Data**
Leonid Vanyan and Andrej Shilovski 243
21. **The Magnetisation of the Lower Continental Crust**
Albrecht G. Hahn and Hans A. Roeser 247
22. **Generalized Inversion of Scalar Magnetic Anomalies: Magnetization of the Crust off the East Coast of Canada**
J. Arkani-Hamed and J. Verhoef 255
23. **Inversion of Magnetic and Gravity Data in the Indian Region**
B. P. Singh, Mita Rajaram, and N. Basavaiah 271
24. **Gravity Field, Deep Seismic Sounding and Nature of Continental Crust Underneath NW Himalayas**
R. K. Verma and K. A. V. L. Prasad 279

V. THE LOWER CRUST FROM STUDIES OF GEOLOGICAL PROCESSES

25. **Growth and Modification of Lower Continental Crust in Extended Terrains: The Role of Extension and Magmatic Underplating**
David M. Fountain 287
26. **Granulite Terranes and the Lower Crust of the Superior Province**
John A. Percival 301
27. **Phase Transformations in the Lower Continental Crust and its Seismic Structure**
Stephen V. Sobolev and Andrey Yu. Babeyko 311
28. **Geophysical Processes Influencing the Lower Continental Crust**
D. L. Turcotte 321
29. **Geochemistry of Rb, Sr and REE in Niutoushan Basalts in the Coastal Area of Fujian Province, China**
Yu Xueyuan 331

PREFACE

This AGU volume is a collection of papers which were presented at the IUGG Symposium U7 (Lower Crust: Properties and Processes) held in Vancouver, Canada in August, 1987. The principle objective of this symposium was to update our understanding of the properties of the lower crust and to review the physical and chemical processes which may have taken place and are still going on in the deep crust. Recent high resolution seismic experiments have thrown new light on the structure of the lower crust and crust-mantle transition which shows a much greater degree of complexity than had previously been assumed. In this symposium the complexity problem was examined from a multidisciplinary approach with papers being presented by earth scientists from the seismological, geological, geochemical, geothermal, geoelectric, geomagnetic and tectonophysics communities. A total of 38 oral and 33 poster papers were presented at the meetings. Eighteen of these papers were invited. Twenty nine of the participants have their work presented in this volume.

The major highlight of the symposium was the presentation of numerous new seismic images of the earth's interior from recent data sets obtained by ACORP (Australia), BIRPS (Britain), COCORP (USA), DEKORP (Germany), ECORS (France). LITHOPROBE (Canada), NFP20 (Switzerland), and USGS (USA) research groups. The state-of-the-art of reflection and refraction imaging methods of the crust has clearly been advanced particularly by the use of airgun sources which are recorded by both marine hydrophone streamers and fixed land stations. Interpretational methods with a greater emphasis on S wave observations, anisotropic and anelastic effects also received considerable attention.

The problem of the origin of the complex reflective zones within the crust dominated several of the papers and subsequent multidisciplinary debates which took place after many of the presentations. A number of papers discussed the origin and characteristics of the laminated layer just above the Moho in terms of various underplating mechanisms and how these play a role in the evolution of the continental crust during both continental collision and extension processes. New rheological and magnetotelluric studies tended to support the seismic data that the lower crust, in general, has different physical properties from the upper crust. Some evidence, based on magnetotelluric observations, was also presented that showed the presence of a highly conductive layer, probably caused by fluids in the mid-crust while the lower crust is considered to be "dry". The symposium closed with an account of the history of the "Conrad discontinuity" and how, in the light of our present knowledge about intracrustal discontinuities, numerous early seismic data sets were erroneously interpreted.

Editors

R. F. Mereu
Department of Geophysics
University of Western Ontario
London, Ontario, Canada, N6A 5B7

St. Mueller
Institute of Geophysics
ETH- Hönggerberg
CH- 8093 Zurich
Switzerland

D. M. Fountain
Department of Geology and Geophysics
P.O. Box 3006
University of Wyoming
Laramie, Wyoming, 82071
U.S.A.

Properties and Processes of Earth's Lower Crust

Section I

THE LOWER CRUST FROM RESULTS OF NEAR-VERTICAL REFLECTION EXPERIMENTS

LARGE-SCALE LENTICLES IN THE LOWER CRUST UNDER AN INTRA-CONTINENTAL BASIN IN EASTERN AUSTRALIA.

D. M. Finlayson, J. H. Leven, and K. D. Wake-Dyster.

Bureau of Mineral Resources, Geology & Geophysics,
Canberra, Australia.

Abstract. Under the central Eromanga Basin region in eastern Australia seismic reflection events at two-way time greater than 7 seconds outline large-scale lenticles or pod-shaped zones in the lower crust which are possible indicators of processes at such depths. The three-dimensional shape of these lenticles has been approximately outlined along a network of deep reflection traverses across the Barcoo Trough. The lenticles are a feature of the lower crust under areas of significant Devonian deposition and deformation, and appear to be attenuated, altered or absent from the lower crust under adjacent basement highs. The basement highs appear to have different deep structural features and velocities compared with the intervening basins, pointing to a different evolutionary history.

The velocity in the lenticles is greater than in the overlying upper crustal rocks of the Thomson Fold Belt. This, together with the persistence of lower crustal sub-horizontal reflections within the lenticles, suggests some form of dense high-velocity fraction from the upper mantle has been added to the lower crust early in the history of the Devonian basins. A discontinuity observed in the reflection data at mid-crustal depths between the lower crust and the overlying deformed Thomson Fold Belt suggests that the processes in the lower crust are younger than those in the upper crustal basement. Reflections at 13-14 seconds two-way time, interpreted as being at the Moho based on coincident refraction data, appear to cut dipping lower crustal reflectors near the basin margins, suggesting that the Moho has re-equilibrated after lower crustal features were formed.

Introduction

The recording of reflections from deep within the Earth's crust along seismic profiling traverses in Australia, North America and Europe has now been a feature of lithospheric transect studies for a number of years [Barazangi and Brown, 1986a,b]. In some places seismic images feature prominent reflections throughout the whole crustal section with another common observation being the decrease in amplitude and continuity of reflections below the interpreted Moho. In cases where upper crustal reflections are present it is sometimes possible to correlate them with observed surface features. Lower crustal reflections, however, often have no direct association with surface features making their interpretation speculative. Moreover, in some other areas the upper crust appears to be non-reflective or "transparent" to seismic waves, with very few observed events in the first 6-8 s two-way time (TWT). In such places there is commonly an increase in amplitude and continuity of reflections in the lower crust.

If the interpretation of deep crustal reflections is to be advanced, then other attributes of the reflecting segments must be considered and more attention paid to the 3-dimensional imaging of deep reflection events. Matthews and Cheadle [1986] have discussed the possible origins of deep reflections and pointed out that, although they must be derived from formations with scale lengths of at least 300 m, they are most likely to arise from constructive interference of wavelets in thin rock layers or sills. There are essentially two strong contenders as possible reflection sources: 1) acoustic impedance contrasts generated in layered dry rocks at depth, and 2) changes in bulk physical properties due to fluid filled cracks. Any attempts to interpret deep reflections must therefore take these possibilities into account when considering the processes by which such reflections have been formed. Hurich and Smithson [1987] tend to favour deep reflections being the result of constructive interference of reflected seismic waves in layered sequences each 35-80 m thick. In the situation addressed in the crust of the central Eromanga Basin we must bear in mind that the processes apply to deep reflection events evident over hundreds of kilometers across the basin at depths of 20-40 kilometres. In this paper we consider the envelope of these deep crustal reflection zones.

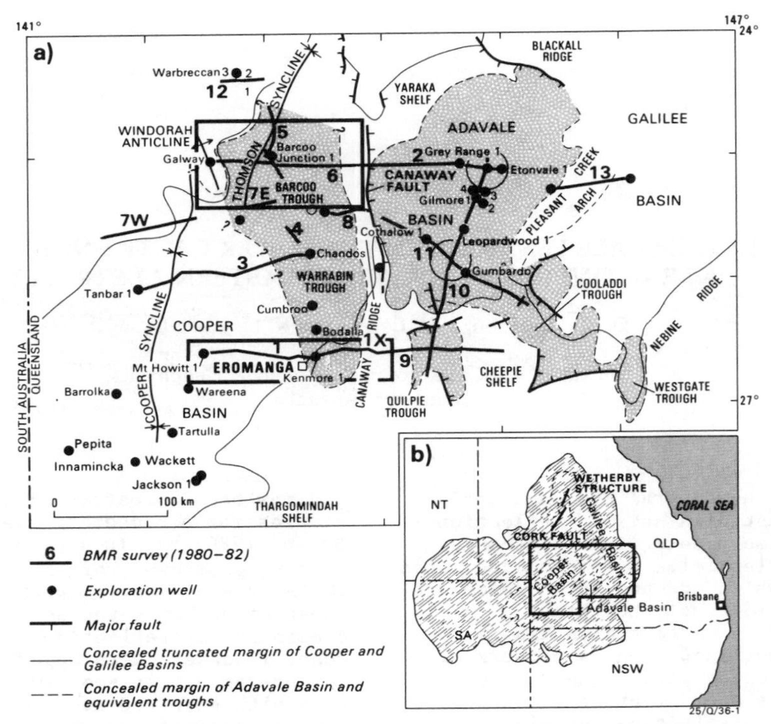

Fig.1. Central Eromanga Basin study region, eastern Australia, and the location of BMR deep seismic reflection traverses (numbered 1 to 13). Boxes indicate the study areas in the Barcoo and Warrabin Troughs; circles indicate traverse crossover areas for data used in Figure 3. Shaded area in a) indicates areas of known Devonian sedimentary sequences. In b) the shaded area is the outline of the Jurassic-Cretaceous Eromanga Basin sequence.

Seismic investigations of the deep crustal features under the central Eromanga Basin in eastern Australia were conducted in the period 1980-82 and have been described by a number of authors [Finlayson et al.,1984; Wake-Dyster et al.,1983, Mathur, 1983]. These investigations provided a network of deep seismic reflection profiling traverses (20 s data) and, in some places, coincident wide-angle reflection and refraction traverses. Generally, the reflection profiling data provided information on the structures within the crust, and the wide-angle reflection and refraction data provided information on the velocities within the crust and upper mantle. The locations of the various traverses shot by the Bureau of Mineral Resources (BMR) are shown in Fig.1. The region also has an extensive network of industry exploration seismic traverses.

This paper is mainly concerned with the deep structure under concealed Devonian basins, the Barcoo and Warrabin Troughs beneath the central Eromanga Basin region of eastern Australia. Features seen in the deep seismic reflection data at two-way times (TWT) of 7 to 13 seconds along a network of traverses provide an indication of gross three-dimensional structures in the lower crust under an intra-continental basin and its neighbouring basement highs. The associated wide-angle reflection and refraction profiling demonstrate the high velocities in these lower crustal zones compared with the upper crustal basement.

Geological Summary

The area now occupied by the epicontinental Jurassic to Cretaceous Eromanga Basin in eastern Australia developed during four major episodes on the rifted eastern margin of the Australian Precambrian craton [Finlayson & Leven, 1987]. A marginal sea is postulated for the region in the early Palaeozoic and it is possible that fragments of the Precambrian craton occur to the east of the present-day Precambrian boundary [Harrington, 1974; Veevers & Powell, 1984]. This episode ended with the early Silurian deformation and metamorphism which formed the upper crustal basement, the Thomson Fold Belt [Murray & Kirkegaard, 1978]. Rocks obtained from drillholes penetrating into this basement include a variety of volcanic, metamorphic and granitic rocks. This continental basement was, however, not yet a stable craton and the second major basin-forming episode occurred during Devonian time with the formation of the transitional tectonic Adavale Basin and its associated troughs, the Barcoo, Warrabin, Westgate, Quilpie and Cooladdi Troughs.

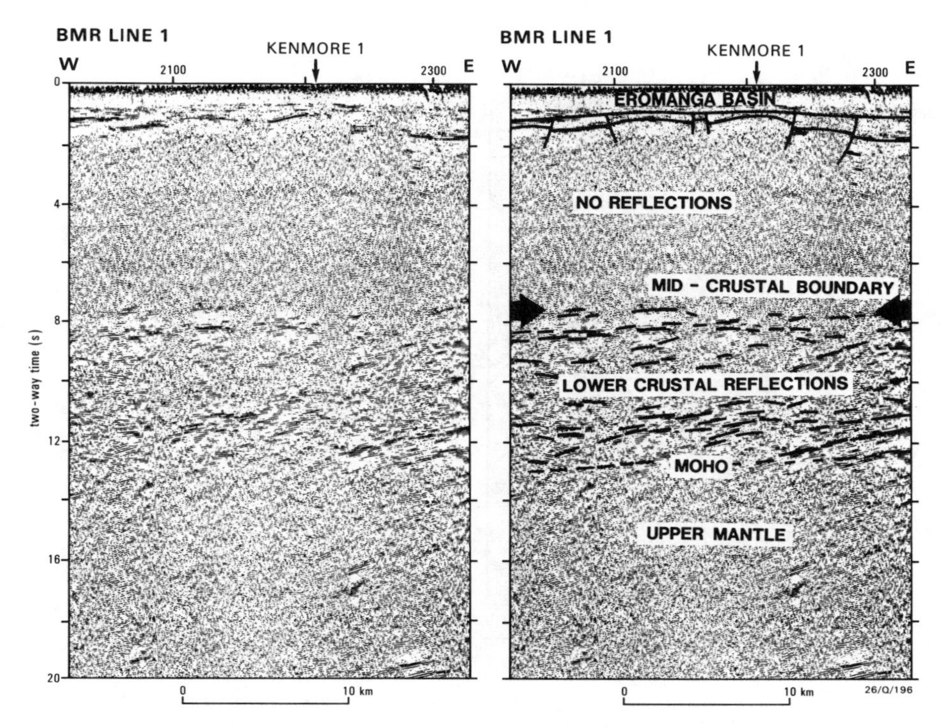

Fig.2. A 20 s reflection profile from BMR Line 1 west of the Warrabin Trough illustrating the characteristics of deep records from the central Eromanga Basin region. The upper crustal basement between about 2 and 8 s two-way time (TWT) is largely non-reflective or "transparent". There is a discontinuity at mid-crustal depths and many reflections from the lower crust. Below the interpreted Moho at 13 s there is a marked decrease in the amplitude and continuity of reflections.

Veevers & Powell [1984] envisaged a process having analogues in the Cainozoic episodes of the Basin and Range province in western USA. The Devonian sedimentary rocks in the Eromanga region were extensively deformed during two Carboniferous events and eroded during the early Permian [Leven & Finlayson, 1987].

The third episode comprises the Permo-Triassic deposition in the Cooper Basin to the southwest of the present study region, but which included the deposition of a thin platformal sedimentary sequence beneath much of the central Eromanga Basin region. Kuang [1985] identified mid-Permian and Triassic deformation events in the Cooper Basin but these are not evident in the study region for this paper. There was a depositional hiatus during the Late Triassic and then the whole region underwent a major episode of Jurassic-Cretaceous platformal deposition which blanketed an area of about one million square kilometres with up to 2 km of sediment [Senior et al., 1978]. This Jurassic-Cretaceous Eromanga Basin sequence was deformed during the mid-Tertiary by relatively mild northeast-southwest compression, with the added possibility of strike-slip movement in some places [Finlayson et al., 1987a]. This ended the fourth major episode in the formation of the crust in the area of the largest intracontinental basin in eastern Australia.

Some data considered in this paper are from the Barcoo Trough and were recorded along BMR deep seismic reflection Lines 5, 6 and 7E (Fig.1). The Barcoo Trough is a comparatively shallow undeformed remnant of the more widespread Devonian Adavale Basin, underlying a thin platformal Permo-Triassic Cooper Basin sequence and a thick platformal (1.5-2.5 km) Jurassic-Cretaceous Eromanga Basin sequence. It can be regarded as a comparatively minor infra-basin with relatively little structural deformation. On the western side of the Barcoo Trough is the Windorah Anticline, a basement high formed along a series of northwest-southeast trending faults [Finlayson et al.,1987a]. On the northwest boundary of the trough is the Warbreccan Dome. To the east is the north-south trending Canaway Ridge, regarded as being a major structural element in the tectonics of the region [Pinchin & Anfiloff,1982].

South of the Barcoo Trough is the Warrabin Trough [Pinchin and Senior, 1982], another structural remnant of the formerly widespread Devonian Adavale Basin. It is bounded on the east by the Canaway Ridge (Fig.1) and contains up to 3000 m of sedimentary rocks which are folded and

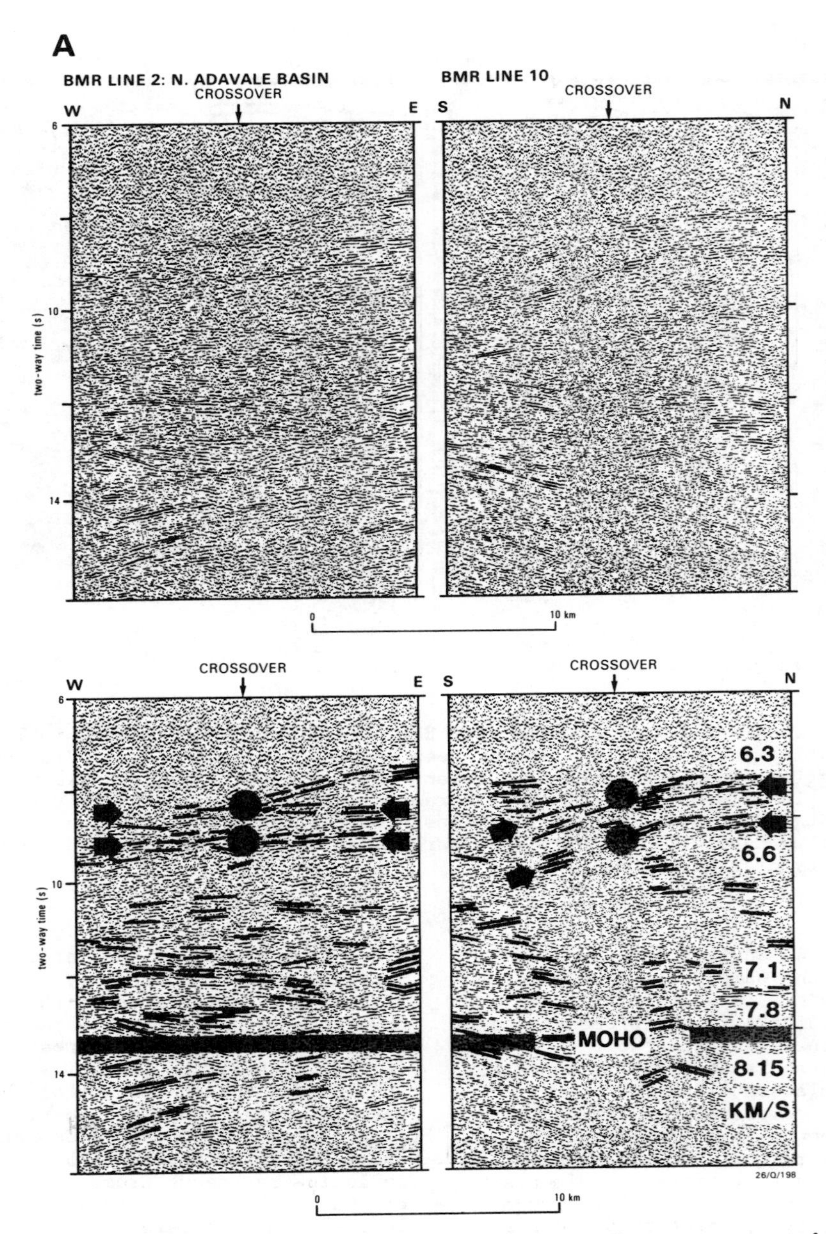

Fig.3. Seismic reflections at mid-crustal depths at the crossover points between BMR Line 10 and a) BMR Line 2, and b) BMR Line 11 showing substantial agreement of mid-crustal reflection features on the pairs of lines. The velocities in the middle-to-lower crust and upper mantle determined from wide-angle reflection and refraction profiling are indicated in the lower panels of a) and b).

displaced by high-angle reverse faults. The Devonian sedimentary sequences of the Warrabin Trough thin to the west and north, there being only a thin sequence joining it to the Barcoo Trough in the north. BMR Line 1 extends across the southern part of the Trough (Fig.1).

Seismic Data Acquisition and Processing

The data acquisition and processing sequences for BMR deep seismic data from the region of the Eromanga Basin were as follows. Continuous reflection profiling data were acquired using a

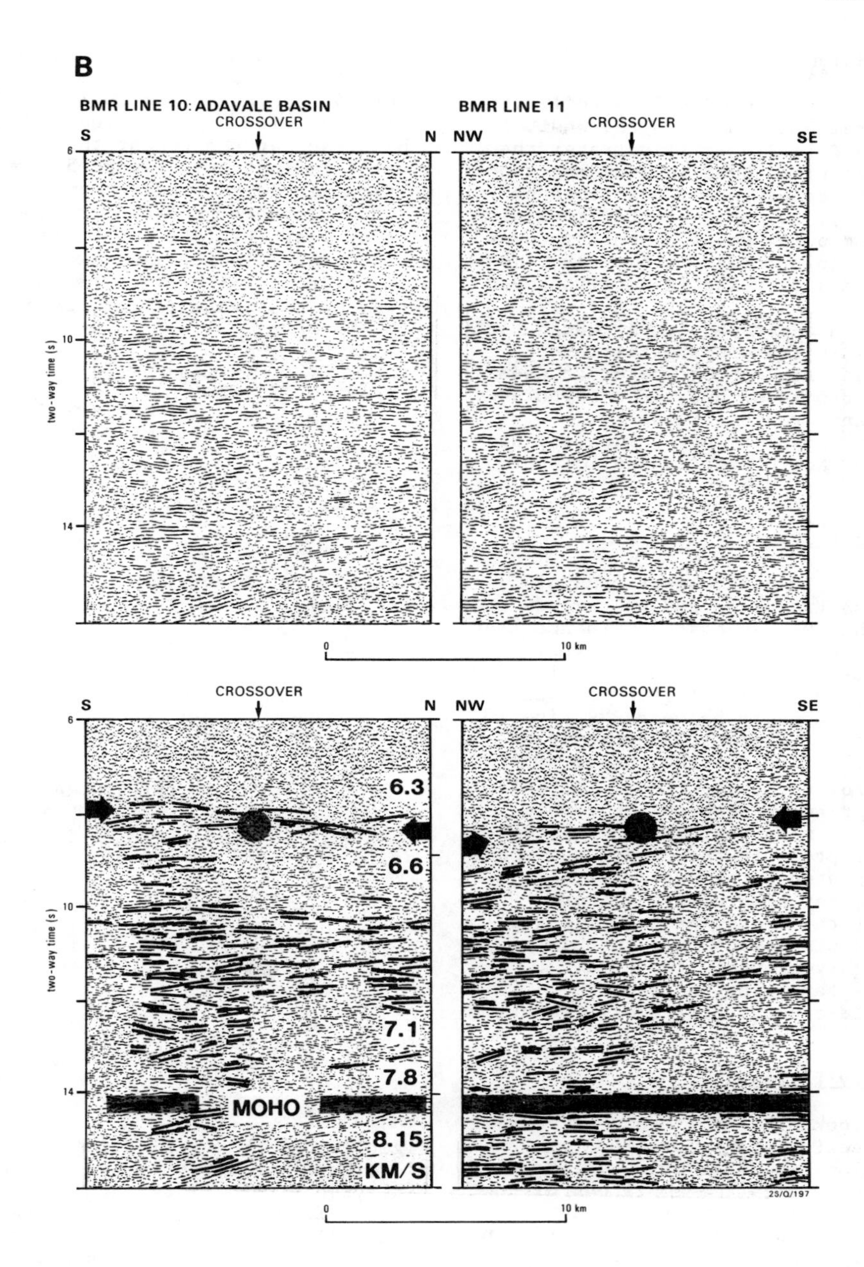

Texas Instruments DFS-IV recording system with 6-fold common mid-point (CMP) subsurface coverage. Geophone group spacing was 83.3 m and 48 channels were recorded. 8 Hz geophones were used , 16 per channel. Shotholes were drilled at 333 m intervals to a depth of 40 m so that explosive charges were below the base of surface weathering. Charge sizes varied from 7 to 20 kg [Finlayson et al., 1987b].

Data were acquired at 2 ms sampling rate on 20 second records. Static corrections for elevation differences were made using uphole recordings with a replacement velocity of 2000 m/s.

Processing was conducted by industry contract and initially concentrated on the first 4 seconds of record to achieve high quality record sections for the sedimentary sequences, considered a prerequisite to good imaging of the deeper data. To obtain the deeper record sections the data were resampled at 4 ms intervals, with deeper data (greater than 4 s TWT) filtered in the bandpass 10-30 Hz, followed by time variant equalisation with 1 s gate lengths, and display. The 20 s display was improved by use of a routine to enhance the display of events with lateral continuity, Digistack, a proprietary program of Digicon (Australia) Pty. Ltd.

Deep Seismic Traverses

Figure 2 shows an example of a 20 s reflection profiling record from BMR Line 1 to the west of the Warrabin Trough (Fig.1). It illustrates the essential characteristics found in many locations throughout the Eromanga Basin and described by Mathur [1983], namely, an upper crustal zone with very few reflections below the sedimentary cover rocks, a reflective lower crust at two-way times of 8-13 seconds, and a decrease in amplitude and continuity of reflections at greater two-way times from within the upper mantle. Although the number of deep seismic traverses is limited, these seismic characteristics of the crust seem to be particularly prominent under the Devonian basins in the Eromanga region. These characteristics are, however, not unique and are seen elsewhere under North America and Europe.

Traverse Cross-over Points

It is important to identify lower crustal reflections on cross-traverses and on records at large offsets so that we can have confidence that we are dealing with primary reflections from major crustal features. Figure 3 shows unmigrated seismic sections of the onset of lower crustal reflections at the crossover between BMR Line 10 and BMR Lines 2 and 11. There is substantial agreement between the times of the shallowest reflection zones from the lower crust. There is also substantial agreement between the double zones shown in Fig.3b near the top of the lower crust. Dipping reflectors in the cross-over zones move updip when migrated but the general conclusion that the reflectors have a source approximately under the crossover point is still valid. Migration is discussed later in this paper. It seems highly probable that the cross-traverses are targeting the same closely-related mid-crustal geological features.

Mid-crustal Velocity Increase

Another way of looking at the features causing lower crustal reflections is to examine wide-angle reflections at large offset distances. At velocity discontinuities the amplitude of events at angles of incidence near that for critical point reflections will be large and can be recognised easily. In the region of the central Eromanga Basin there are a number of locations where there are coincident reflection profiles and wide-angle reflection/refraction profiles. These are on BMR Line 1 where there were large shots 37.5 km west of Mt. Howitt No.1 well (at Terebooka Bore), and on the western side of the Warrabin Trough at Tallyabra, 11.5 km east of Kenmore No.1 well (Fig.1). On BMR Line 10 there were large shots near the crossover with BMR Line 11 (Fig.1) to the south of Adavale. At all these locations a velocity increase at 22-24 km depth is interpreted with velocities going from 6.3 km/s above this depth to 6.6 - 6.9 km/s below [Finlayson et al., 1984; Finlayson & Collins, 1986]. The large-amplitude wide-angle reflections from BMR Line 1 shots at offset distances out to over 120 km are illustrated in Figure 4. Finlayson and Collins [1986] have shown that these wide-angle reflections are from the sub-surface at the top of the zone of deep crustal reflections. Figure 5 shows the onset of vertical-incidence reflections about 40 km west of the Tallyabra shot and the location of the zone of near-critical point reflections from the mid-crustal velocity increase interpreted from wide-angle seismic data. The same target is imaged by the two different modes of recording.

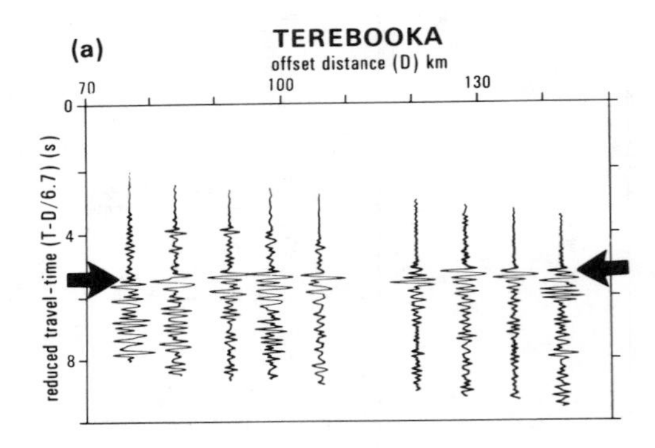

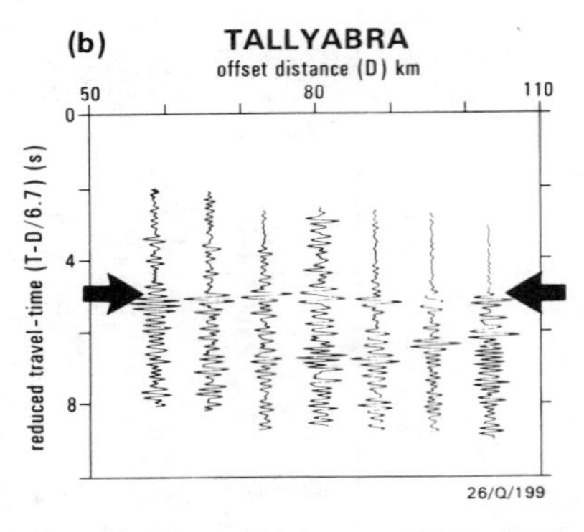

Fig.4. Wide-angle reflections (arrowed) from the mid-crustal velocity increase determined by Finlayson et al. [1984] at 22.5 to 24.0 km depth. a) from Terebooka shot 37.5 km west of Mt. Howitt No.1 well recorded eastward along BMR Line 1; and b) Tallyabra shot in the Warrabin Trough recorded westward along BMR Line 1. Model travel-time branches are virtually coincident with the arrowed events.

The seismic features outlined above from the central Eromanga Basin region provide some confidence that the deep crustal reflections are derived from quite specific structures in the lower crust which can be imaged in 3-dimensions where cross-traverses are available, and that their onset at mid-crustal levels is identified with velocity increases of 0.3-0.6 km/s interpreted from the wide-angle data. It is

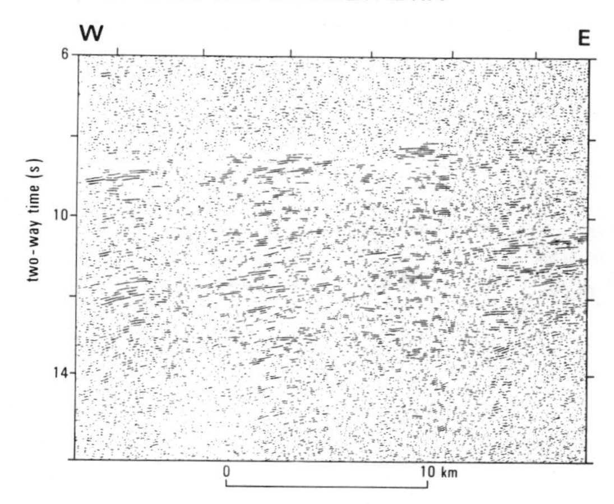

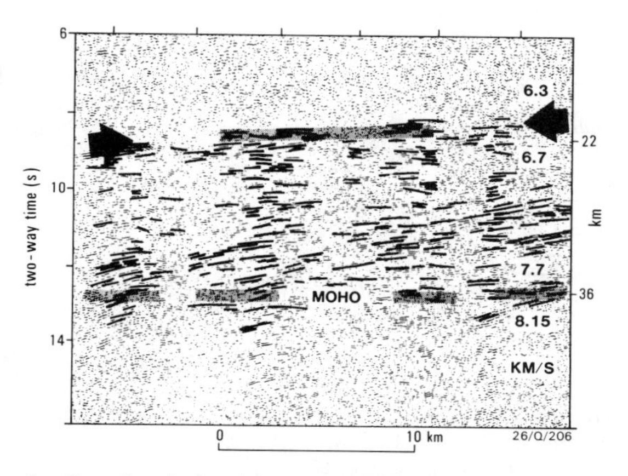

Fig.5. Vertical-incidence reflections from BMR Line 1 corresponding to the subsurface location (arrowed) of wide-angle reflections shown in Figure 4b.

appropriate to examine the regional form of these structures in the lower crust.

Barcoo Trough

In the Barcoo Trough (Fig.1) a number of BMR traverses can be used to give an indication of the 3-dimensional nature of structures in the lower crust (Fig.6). The structures within the sedimentary sequences can be interpreted from the extensive company seismic profiling in the area, only some of which is shown in Figure 6. Finlayson et al. [1987a,b] have mapped the major structures, and Figure 6b gives some indication of the structural features evident within the sedimentary rocks along BMR Line 6. Within the upper crustal basement there are generally no reflections between about 2.6 s and 7 s two-way time.

Between 7 and 14 s, however, there are prominent seismic features under the Barcoo Trough which can be used as indicators of structure. Figure 7 shows the deep reflections from BMR Line 7E. At the western end of BMR Traverse 7E, adjacent to the Windorah Anticline, deep crustal reflections are only evident at two-way times (TWT) greater than 9 s. Reflections at two-way times greater than 13 s decrease in amplitude and continuity. This is interpreted as indicating the Moho where velocities increase to an upper mantle value of 8.15 km/s [Finlayson et al.,1984]. Dipping reflection events between 13 and 14 s TWT tend to be cross-cutting, indicating some structure on the lower boundary of the deep reflection zone. The most striking deep event on Traverse 7(east) is the arc-like reflector decreasing in depth from 10.5 s TWT in the west to 8 s in the east. This reflector seems to emerge from a band of reflectors at 11-13 s TWT and form part of a lenticle or "pod" structure in the lower crust.

BMR Line 6 is located 30-40 km to the north of Line 7E (Fig.1) and it too has the Windorah Anticline at its western end; at its eastern end is the Canaway Ridge. Figure 8a,b shows the deep reflections at the western and eastern limits of the Barcoo Trough. Along the line the deepest reflections appear to concentrate in a band at 10.5-13 s TWT with a tendency to have a saucer shape. Some of the features of this concave-upwards horizon can be attributed to velocity pull-down under the deepest sedimentary rocks. Some eastward-dipping reflectors below 13 s TWT migrate into the lower crust and don't alter the general impression of the lower boundary of the lenticle structure. At shallower levels between 7 and 10.5 s TWT there is a quite distinct reflector which is convex upwards and forms the upper boundary of a lower crustal lenticle. The depth extent of the reflections forming the upper boundary of the lenticle is quite small compared to the deeper (10.5-13 s) reflection band, but it is quite clearly seen at its ends where it converges with the deeper band. The upper boundary of the lenticle is shallowest at 7.0-7.5 s TWT and it can be traced along the length of the seismic line. /BMR Line 5 generally trends north-south across Line 6 (Fig.6) and deep crustal reflections from the southern part of this traverse are shown in Fig.9. The character of the lower crustal reflections is confirmed by this intersecting traverse and at the intersection point the depth of the upper and lower boundaries of the lower crustal lenticle are substantiated. At its southern end BMR Line 5 confirms that the shallowest upper boundary of the lenticle is at 7.0-7.5 s TWT and the band of deeper reflectors forming the lower boundary is at 11-13 s TWT. At its southern end Line 5 is only 18 km from BMR Line 7E where we have identified a lower crustal lenticle with upper boundary at 8 s TWT and a lower boundary at 11-13 s TWT.

Hence, we have identified a large-scale lenticle of lower crustal material under the Barcoo Trough. It has a lower boundary at about 13 s TWT everywhere. Its upper boundary is identified by an upwardly convex series of reflectors which is shallowest at 7.0-7.5 s TWT under the basin but plunges towards the lower

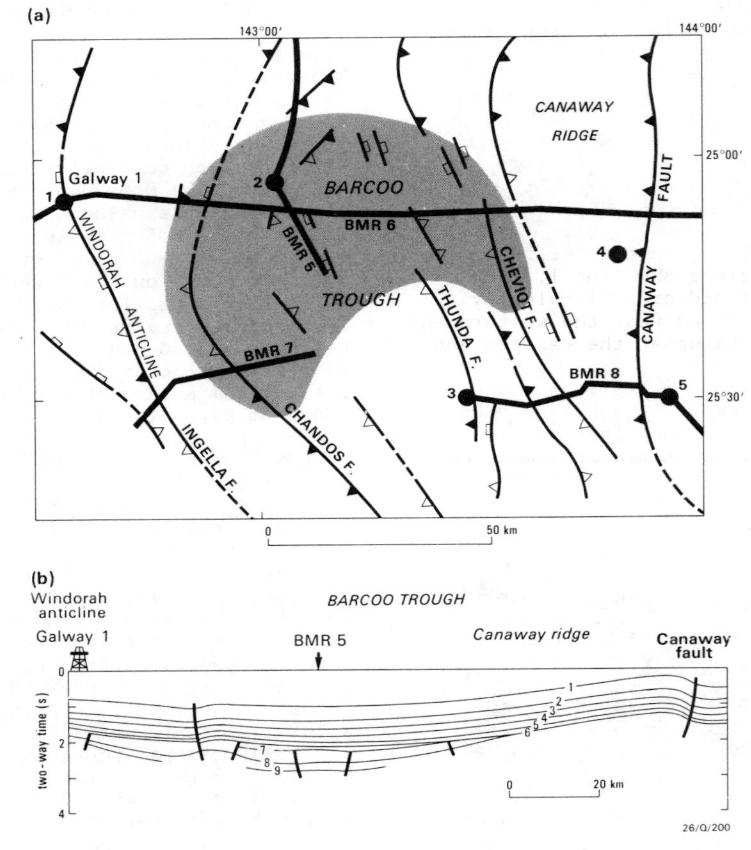

Fig.6. a) Basement structures and the location of the lower crustal lenticle or "pod" within the Barcoo Trough between the Windorah Anticline and the Canaway Ridge. Reverse faults are indicated as follows; throws greater than 100 ms two-way time - closed triangles, less than 100 ms - open triangles, and monoclines - open squares. BMR deep seismic traverses - thick lines. The shaded area indicates where the lenticle is at a two-way time of less than 10 s.
b) Sketch of the structural features across the Barcoo Trough at depths less than 5 km derived from BMR seismic reflection Traverse 6. Seismic horizons labelled 1-4 are within the Jurassic-Cretaceous Eromanga Basin; horizon 5 is at the top of the Permo-Triassic Cooper Basin sediments; horizon 6 is the top-Devonian unconformity; horizons 7-9 are within the Devonian Barcoo Trough.

boundary as the basement highs are approached to the west and north, and loses definite shape in the east against the Canaway Ridge. The outline of the lower crustal lenticle where it is shallower than 10 s two-way time is shown in Figure 6a. Throughout the interpretation we were mindful of other seismological explanations for curved and dipping reflectors such as diffractions and reflections from outside the plane of the seismic line. However, the evidence that we have from locations where cross-traverses are available suggests that features on the seismic records are real in the geological sense.

BMR Line 1

Prominent upwardly convex reflectors are not confined to the Barcoo Trough. Fig.10a shows a digitized line diagram of the deep crustal reflections from BMR Line 1 where it runs eastward from the Mt.Howitt Anticline to the Canaway Ridge (about 160 km). One striking feature of the deep crust at the Mt.Howitt end of the line is the arc-shaped reflector ascending from 11 s TWT to 9 s TWT. The deepest reflectors appear to form a lower boundary in a band at 11-13 s TWT.

Along much of the length of BMR Line 1 there is a lenticle in the lower crust with an upper boundary at 8.0-8.5 s TWT and a lower boundary at 12.5-13.0 s TWT. The section shown in Figure 2 is taken from this line in an area just west of the Warrabin Trough. The general impression of a lower crustal lenticle is again gained from BMR Line 1, but, unfortunately, with no cross-traverse confirmation. In Figure 10b the digitized line diagram has been migrated; this generally moves dipping reflectors from below the Moho to a position in the lower crust, emphasizing further the lenticle nature in the lower crustal process. Some earlier impressions of westward-dipping reflectors within the lower crust gained from unmigrated data [Moss & Mathur, 1986] may have to be revised. At the ends of the line, under the adjacent basement highs, the upper boundary of lower crustal reflections deteriorates and largely disappears.

Finlayson et al. [1984] have interpreted velocity information along BMR Line 1. They identified a prominent mid-crustal velocity horizon at a depth of 24 km near the western end of Line 1 and at 22.5 km under the eastern end. At this velocity horizon the P-wave velocity increases from 6.3 km/s to 6.7-6.9 km/s and the velocity in the lower crust is markedly higher than in the upper crust. The crust/mantle boundary is interpreted as a transitional feature at 36-37 km depth where the velocity increases from 7.7 km/s to the upper mantle velocity of 8.15 km/s.

Migration

Migration of any dipping reflection events in the deep crust has proved difficult [Warner, 1987] and the resulting sections are often a poor representation of the unmigrated data. The prominent reflection events in the deep data from the central Eromanga Basin region have been digitized to form line sections and the data at the boundaries of the Devonian basins have been migrated. Figure 10b is an example of migrated digitized reflections on BMR Line 1. The crustal velocity model used in the migration process was that of Finlayson et al. [1984] for the Warrabin Trough region. The migration of the dipping events at the boundaries of lenticles under BMR Lines 1, 6 and 7 move the events up to 15 km basinward with a corresponding increase in dip. However, The basic "lenticular" form of the envelope of deep crustal events remains the same. We have assumed that the boundary reflections are from within the vertical plane under the seismic line and that the lines are positioned approximately normal to strike.

Sub-Moho Reflections

The interpretation of reflections at two-way times greater than 13 seconds i.e., from below the normal incidence Moho depth, is not easily resolved with only single isolated traverses. As mentioned earlier these reflections tend to be cross-cutting and migrate to a higher position in the section. This is the case with quite a few of the so-called "upper mantle" reflections under BMR Line 1. In all cases they have migrated to a two-way time less than 13 seconds, i.e., within the lower crust. Despite the sometimes irregular nature of reflections near 13 seconds two-way time there is always a persistent series of sub-horizontal events at about 13 seconds TWT which we interpret as indicating the transition

BMR LINE 7E: W. BARCOO TROUGH

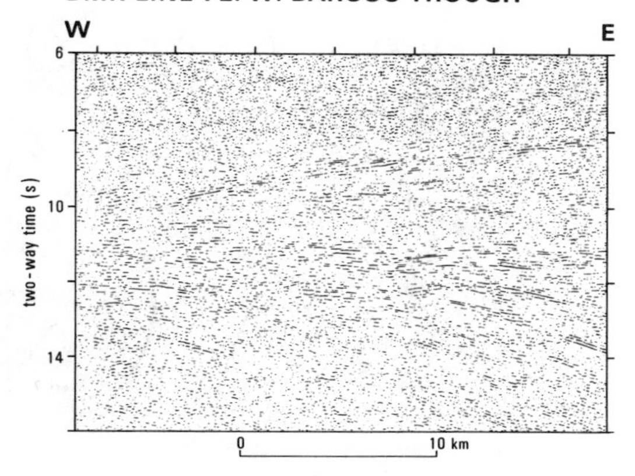

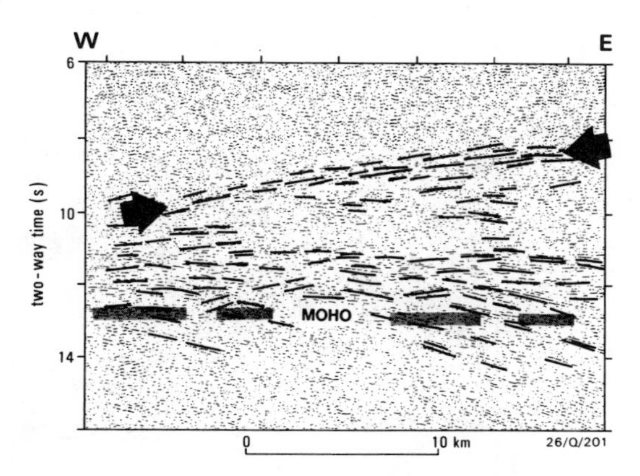

Fig.7. Mid-crustal reflections from BMR Line 7E, illustrating the prominent top boundary of the lower crustal lenticle dipping westward under the Windorah Anticline.

zone from the lower crust to the upper mantle. This transition zone is consistent with the coincident wide-angle reflection and refraction profiling interpretation of the Moho. From migrated seismic data there is an impression that dipping reflectors at the edges of the Devonian basins meet the sub-horizontal Moho reflections at an acute angle and are truncated. One valid interpretation of this is that the Moho has re-equilibrated since the formation of the lower crustal reflectors, and hence its present morphology is younger than the lower crustal reflections.

Although we are convinced that most dipping reflection events at greater than 13 seconds TWT have their source in the lower crust rather than the upper mantle, there are isolated instances of deeper sub-horizontal events. We cannot rule out the possibility of their being primary reflections from within the upper mantle, in some way connected with structuring of the lithosphere. These deeper events are not, however, the subject of this paper.

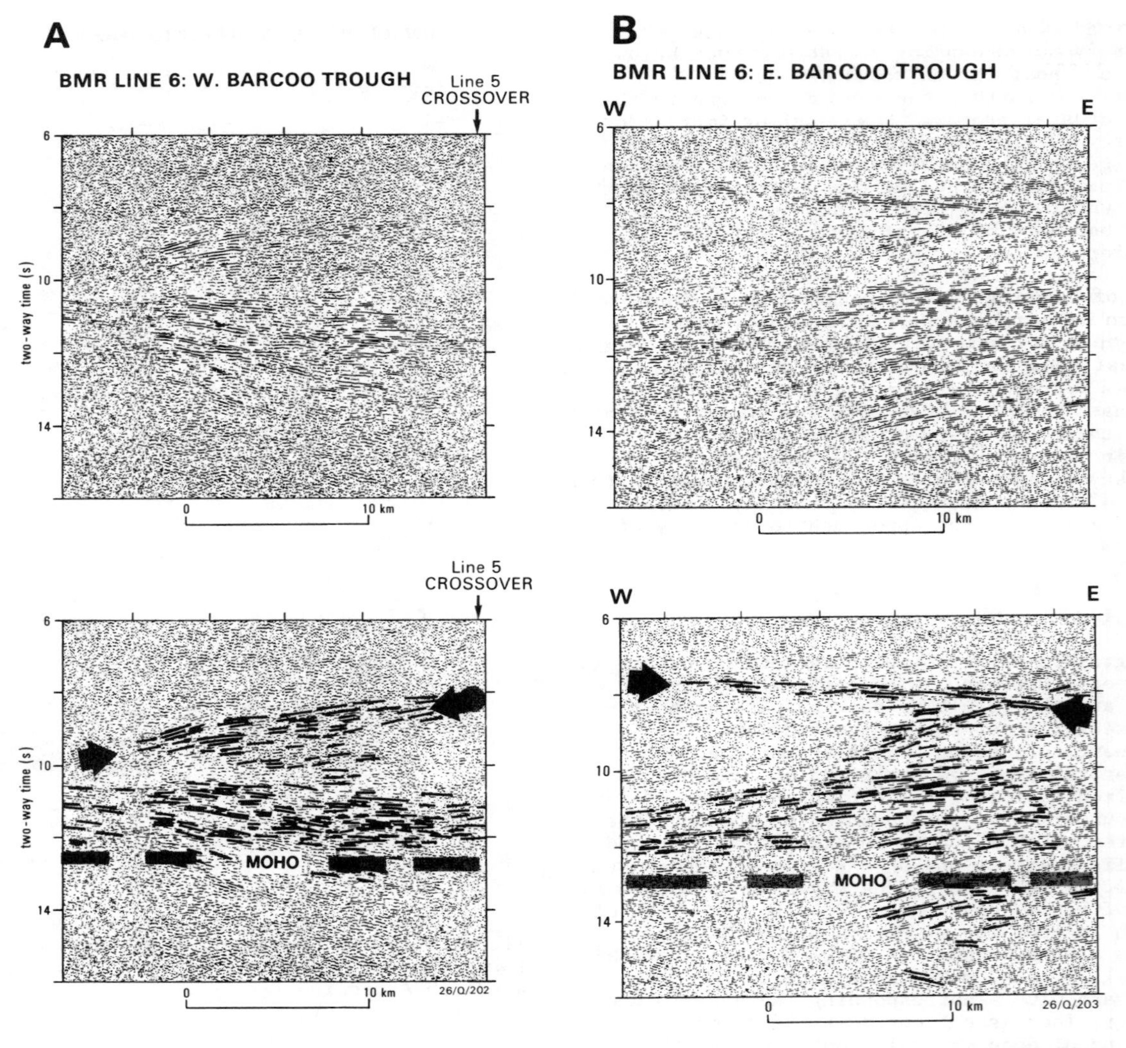

Fig.8. Mid-crustal reflections from a) the western end of BMR Line 6 where it crosses onto the Windorah Anticline, and b) the eastern side of the Barcoo Trough where BMR Line 6 crosses onto the Canaway Ridge. Arrows indicate the prominent mid-crustal reflection feature interpreted as the upper surface of the lower crustal lenticle.

Lenticular Lower Crustal Zone

From the Eromanga region data, we believe that the lenticular shape of the lower crustal reflection zone is well-defined, and that the P-wave velocity within it is a strong indication that it has a denser, more mafic composition than the upper crust. Finlayson et al. [1984] also showed that the velocity structure under the Canaway Ridge was markedly different from that under the basin to the west, there being no prominent velocity increase at mid-crustal depths and the Moho being shallower. These results were confirmed by subsequent interpretations by Finlayson and Collins [1986, 1987] in the same region. The basement highs are not only different in their lower crustal structure evident from reflection events but also in their lower crustal velocity structure, pointing strongly towards an evolutionary history different from that under the areas of deepest Devonian sedimentation in the intervening basins.

Discussion

There is much interest in the structure of sedimentary basins and the processes by which they are formed. Some summaries have been made by Ziegler[1982], Kent et al.[1982], DeRito et al.[1983], and Houseman and England [1986]. Those models of particular interest in this paper are those involving thinning of the crust during an episode of lithospheric extension [McKenzie,1978; Dewey,1982]. Such models envisage subsidence in a region affected by a thermal event resulting from stretching of the lithosphere and emplacement of upper mantle material at shallower levels. DeRito et al.[1983] proposed that, in basins where multiple depositional cycles can be identified, the model includes the emplacement of a mass excess in the lower crust, and that subsequent thermal events reactivate subsidence until the crust reaches isostatic equilibrium. Under the area of Devonian sedimentation in the Eromanga region we have detected large-scale lenticles of high-velocity material which we tentatively suggest could be identified with such a mechanism.

Other indications of possible processes under intra-continental basins are evident in areas of younger crustal extension. Allmendinger et al. [1987] considered four models when discussing deep seismic data and intra-continental basin formation across the Basin and Range Province in western USA. They indicated that a symmetric horst and graben model was unlikely, most of the structures being asymmetric like those found in eastern Australia [Finlayson et al., 1987a]. More likely, the Basin and Range Province is compatible with a mid-crustal sub-horizontal decoupling model or a similar anastomosing shear model for the lower crust.

The reflection features described in this paper have been seen elsewhere on deep seismic profiling records. Nelson et al. [1985] observed a zone of dipping reflections under the COCORP Georgia Line 14 at two-way times greater than 6 s. They pointed out the acute angle between the dipping reflectors in the lower crust and the relatively flat-lying Moho reflections at 12.5 s TWT, implying that the lower crustal features predate the Moho fabric. In the Basin and Range Province of western USA where there are basin-bounding faults with throws of 3-4 km, Hauge et al.[1987] have indicated that the crust/mantle boundary is a relatively smoothly varying feature and took its present shape at or later than the Cainozoic extension. On the boundary of the Colorado Plateau, Allmendinger et al [1987] have described the change in character of deep reflections across the boundary.

Elsewhere in western U.S.A. Cheadle et al. [1986] have describe arc-like deep crustal reflections at 6-10 s TWT on the boundary between the Mojave Block and the Basin and Range province. At deep levels across the province boundary there is an arcuate zone of numerous discontinuous reflections which terminate at a series of horizontal Moho-depth reflections. Roy-Chowdhury et al. [1983] interpreted a mid-crustal velocity increase from 6.1 to 6.4 km/s around 20 km depth, and describe a series of complex reflectors dipping at 35 degrees above a near-horizontal Moho. Prodehl [1979] also interprets a velocity increase at mid-crustal depths.

The difficulties in interpreting reflections from the lower crust have been highlighted by Barton [1986] and Klemperer et al. [1986] among others. Strain fabrics, igneous layering, free fluids and thermal effects are all listed as possible causes. Free fluids in the crust should be evident as electrical conductivity anomalies. Spence and Finlayson [1983] have shown from magnetotelluric observations that pronounced high conductivity layers within basement above 20 km are unlikely in the Eromanga Basin. Conductivity values begin to increase in the lower crust but do not rise significantly until depths greater than 40 km. Although the cause of any conductivity increase may be fluids in the lower crust as suggested by Gough [1986], it is difficult to reconcile the significant increase in velocity in the lower crust with the presence of fluids alone. Changes in rheology resulting in a change of strain fabric at mid-crustal depths is another possible cause of reflections [Ranalli and Murphy, 1987; Reston and Blundell, 1987] but does not explain the increased velocity in the lower crust.

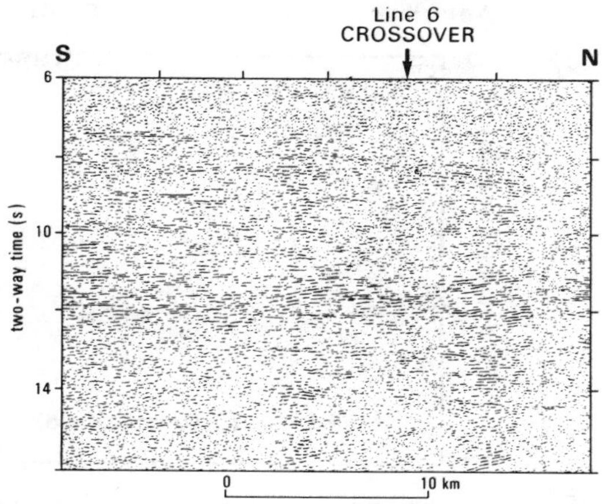

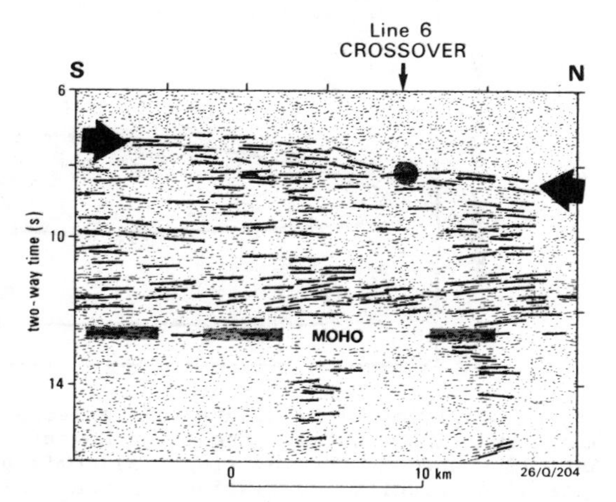

Fig.9. Mid-crustal reflections from BMR Line 5 where it crosses BMR Line 6, illustrating the agreement in depth to the upper boundary of the lower crustal lenticle.

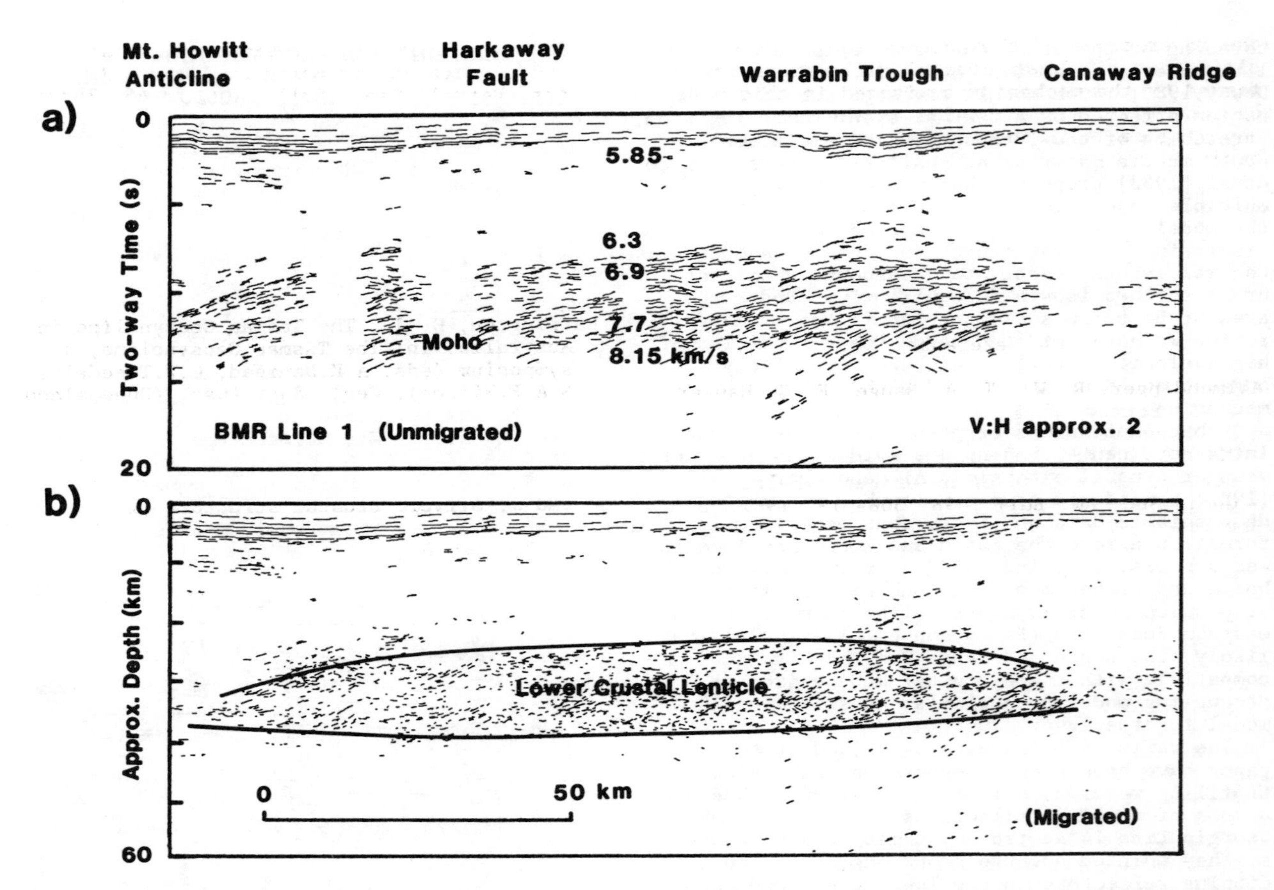

Fig.10. Digitized line diagram of reflectors along BMR Line 1 between the Mount Howitt Anticline and the Canaway Ridge: a) unmigrated data with velocity information derived from coincident seismic refraction data; b) migrated data converted to depth using an average crustal velocity of 6.0 km/s. The migrated data emphasize the reflections in a lower crustal lenticle.

Drummond and Collins [1986] indicate that, in Australia, the geochemical and physical properties of lower crustal nodules, when considered with the interpreted seismic velocities, imply a lower crust composed of large volumes of mafic and ultramafic rocks, the lower crust being formed by multiple underplating episodes. We feel that a thermal mechanism would be the main driving force. Such a mechanism relates composition of the lower crust with heat flow regimes at province boundaries, upward movement of partial melts and magma, increased velocities compared with the upper crust, the observation of lower crustal reflections and their three-dimensional expression at province boundaries, and the subsidence of basins, in a way not evident by other mechanisms alone.

Conclusions

In this paper we have identified deep crustal features correlated across a network of traverses at depths of 20 to 40 km under the Barcoo and Warrabin Troughs of eastern Australia. The three-dimensional shape of features adjacent to basement highs has been tentatively established and we believe we have outlined the general shape of lenticles or pod-shaped zones in the lower crust which have a velocity markedly greater than that in the overlying upper crustal rocks. The edges of the lenticles have boundaries rising upwards from near the crust/mantle transition at 11-13 seconds two-way time to a minimum of about 7 seconds two-way time. The maximum thickness of the lenticles shown in this paper is about 16 km and lateral dimensions are at least 70 km. The persistant onset of sub-horizontal reflection features at the mid-crustal boundary are interpreted as a discontinuity formed in the process of Devonian basin formation. The lower crustal features, therefore, are younger than the deformed basement and upper crustal non-reflective or "transparent" zone, the Early Palaeozoic Thomson Fold Belt.

The main conclusion of this paper is that we have related specific reflection features in the lower crust with specific increases in velocity and outlined the shape of a lenticular feature in the lower crust of the Barcoo Trough. We do not claim that all lower crustal reflections are formed by the mechanism envisaged in this paper, or even that this mechanism is the only one applying early in the history of the Devonian basin in the Eromanga region. We do suggest, however, that it is a major contributing factor.

Acknowledgements. This paper is published with the permission of the Director, Bureau of Mineral Resources, Geology and Geophysics, Canberra.

References

Allmendinger, R. W., T. A. Hauge, E. C. Hauser, C. J. Potter, S. L. Klemperer, K. D. Nelson, P. Knuepfer, and J. Oliver, Overview of the COCORP 40 deg. N transect, western United States: the fabric of an orogenic belt. Geol. Soc. Am. Bull., 98, 308-319, 1987.

Barazangi, M., and L. Brown,(editors), Reflection seismology: a global perspective. Am. Geophys. Un., Geodynamic Series Vol.13, 311 pp, 1986a.

Barazangi, M., and L. Brown,(editors), Reflection seismology: the continental crust. Am. Geophys. Un., Geodynamics Series Vol.14, 339 pp, 1986b.

Barton, P., Deep reflections on the Moho. Nature, 323,392-393, 1986.

Cheadle, M. J., B. L. Czuchra, T. Byrne, C. J. Ando, J. E. Oliver, L. D. Brown, and S. Kaufman, The deep crustal structure of the Mojave Desert, California, from COCORP seismic reflection data. Tectonics, 5, 293-320, 1986.

DeRito, R. F., F. A. Cozzorelli, and D. S. Hodge, Mechanism of subsidence of ancient cratonic rift basins. Tectonophysics, 94, 141-168, 1983.

Dewey, J. F., Plate tectonics and the evolution of the British Isles. J. Geol. Soc. Lond., 139, 371-412, 1982.

Drummond, B. J. and C. D. N. Collins, Seismic evidence for underplating of the lower continental crust of Australia. Earth and Planet. Sci. Lett., 79, 361-372, 1986.

Finlayson, D. M., C. D. N. Collins, and J. Lock, P-wave velocity features of the lithosphere under the Eromanga Basin, eastern Australia, including a prominent mid-crustal (Conrad?) discontinuity. Tectonophysics, 101, 267-291, 1984.

Finlayson, D. M. , and C. D. N. Collins, Lithospheric velocity beneath the Adavale Basin, Queensland, and the character of deep crustal reflections. BMR J. Aust. Geol. & Geophys., 10, 23-37, 1986.

Finlayson, D. M. , and C. D. N. Collins, Crustal differences between the Nebine Ridge and the central Eromanga Basin from seismic data. Aust. J. Earth Sci., 34, 251-259, 1987.

Finlayson, D. M., and J. H. Leven, Lithospheric structures and processes in Phanerozoic eastern Australia from deep seismic investigations. Tectonophysics, 133, 199-215, 1987.

Finlayson, D. M., J. H. Leven, and M. A. Etheridge, Structural styles in the central Eromanga Basin region, eastern Australia and constraints on basin evolution. Am. Ass. Petrol. Geol. Bull., 70, 33-48, 1987a.

Finlayson, D. M., J. H. Leven, S. P. Mathur, and C. D. N. Collins, Geophysical abstracts and seismic profiles from the central Eromanga Basin region, eastern Australia. Bur. Miner. Resour., Geol. & Geophys., Aust., Report 278, 1987b.

Gough, D. I., Seismic reflectors, conductivity, water and stress in the continental crust. Nature, 323, 143-144, 1986.

Harrington, H. J., The Tasman Geosyncline in Australia. In: The Tasman Geosyncline, a symposium (eds. A.K.Denmead, G.W.Tweedale, & A.F.Wilson), Geol. Soc. Aust. (Queensland Div.), 383-407, 1974.

Hauge, T. A., R. W., Allmendinger, C. Caruso, E. C. Hauser, S. L. Klemperer, S. Opdyke, C. J. Potter, W. Sanford, L. Brown, S. Kaufman, and J. Oliver, Crustal structure of western Nevada from COCORP deep seismic reflection data. Geol. Soc. Am. Bull., 98, 320-329, 1987.

Houseman, G., and P. England, A dynamic model of lithospheric extension in sedimentary basin formation. J. Geophys. Res., 91, 719-729, 1986.

Hurich, C. A. and S. B. Smithson, Compositional variation and the origin of deep crustal reflections. Earth & Plan. Sc. Letts., 85, 416-426, 1987.

Kent, P., M. H. P. Bott, D. P. McKenzie, and C. A. Williams,(editors), The evolution of sedimentary basins. Phil. Trans. Roy. Soc. Lond., Series A, 305, 1489, 338 pp, 1982.

Klemperer, S. and the BIRPS group, Reflectivity of the crystalline crust: hypotheses and tests. Geophys. J. Roy. Astron. Soc., 89, 217-222, 1987.

Kuang, K. S., History and style of Cooper and Eromanga Basin structures. Bull. Aust. Soc. Expl. Geophys., 16, 245-248, 1985.

Leven, J. H., & Finlayson, D. M., Lower crustal involvement in upper crustal thrusting. Geophys. J. Roy. Astron. Soc., 89, 415-422, 1987.

Mathur, S. P., Preliminary deep crustal reflection results in the central Eromanga Basin, Queensland, Australia. Tectonophysics, 100, 163-173, 1983.

Matthews, D. H. and M. J. Cheadle, Deep reflections from the Caladonides and Variscides west of Britain and comparison with the Himalayas. In: Reflection Seismology: a Global Perspective, M. Barazangi & L. Brown (Eds.), Am. Geophys. Un. Geodynamics Series Vol. 13, 5-19, 1986.

McKenzie, D., Some remarks on the development of sedimentary basins. Earth Planet. Sci. Lett., 40, 25-32, 1978.

Moss, F. J., and S. P. Mathur, A review of continental reflection profiling in Australia. In: Reflection Seismology: a Global Perspective, M. Barazangi & L. Brown (Eds.), Am. Geophys. Un. Geodynamics Series Vol. 13, 67-76, 1986.

Murray, C. C., and A. G. Kirkegaard, The Thomson

Orogen of the Tasman Orogenic Zone. Tectonophysics, 48, 299-326, 1978.

Nelson, K. D., J. A. Arnow, J. H. McBride, J. H. Willemin, J. Huang, L. Zheng, J. E. Oliver, L. D. Brown, and S. Kaufman, New COCORP profiling in the southeastern United States. Part 1: Late Paleozoic suture and Mesozoic rift basin. Geology, 13, 714-718, 1985.

Pinchin, J., and V. Anfiloff, The Canaway Fault and its effects on the Eromanga Basin. In: The Eromanga Basin (eds. P.S. Moore & T.J.Mount), Petrol. Expl. Soc. Aust., & Geol. Soc. Aust. (South Australian Divs.), 161-170, 1982.

Pinchin, J. and B. R. Senior, The Warrabin Trough, western Adavale Basin, Queensland. J. Geol. Soc. Aust., 29, 413-424, 1982.

Prodehl, C., Crustal structure of the western United States. U.S. Geol. Survey Professional Paper 1034, 1979.

Ranalli, G., and D. C. Murphy, Rheological stratification of the lithosphere. Tectonophysics, 132, 281-295, 1987.

Reston, T. J. and D. J. Blundell, Possible mid-crustal shears at the edge of the London Platform. Geophy. J. Roy. Astron. Soc., 89, 251-258.

Roy-Chowdhury, K., A. M. Suteau, and R. A. Phinney, Crustal structure and velocity estimation for Mojave CERP data. EOS, 64, 259, 1983.

Senior, B. R., A. Mond, and P. L. Harrison, Geology of the Eromanga Basin. Bur. Miner. Resour., Geol. & Geophys., Aust. Bull. 167, 1978.

Spence, A. G. and D. M. Finlayson, The resistivity structure of the crust and upper mantle in the central Eromanga Basin, Queensland, using magnetotelluric techniques. J. Geol. Soc. Aust., 30, 1-16, 1983.

Veevers, J. J., and C. McA. Powell, Uluru and Adelaidian regimes. In: Phanerozoic Earth History of Australia (ed. J.J.Veevers), Clarendon Press, Oxford, 270-350, 1984.

Wake-Dyster, K. D., F. J. Moss, and M. J. Sexton, New seismic reflection results in the central Eromanga Basin, Queensland, Australia: the key to understanding its tectonic evolution. Tectonophysics, 100, 247-162, 1983.

Warner, M., Migration - why doesn't it work for deep continental data? Geophys. J. Roy. Astron. Soc., 89, 21-26, 1987.

Ziegler, P. A., Thoughts on mechanisms of basin subsidence. In: Geological Atlas of Western and Central Europe, Shell International Petroleum, 100-106, 1982.

DATING LOWER CRUSTAL FEATURES IN FRANCE AND ADJACENT AREAS FROM DEEP SEISMIC PROFILES

C. Bois, B. Pinet and F. Roure

Institut Français du Pétrole
B.P.311, 92506 Rueil-Malmaison Cedex, France

Abstract. France's geological history may be broadly subdivided into three main phases. The first one corresponds to the consolidation of Western Europe during the Variscan orogeny. The second one is related to the formation of large cratonic basins such as the Paris and Aquitaine ones and the opening of new oceanic domains such as the Tethys ocean and the North Atlantic. The third phase corresponds to the Tertiary closure of the Tethys and the formation of the Pyrenees and Alps collisional ranges.

The deep seismic profiles shot in SW Britain and northern France across the Variscan orogenic belt show a comparatively flat Moho and a highly reflective lower crust displaying prominent layering over 2 to 4.5 s TWT. On the profiles shot on the Aquitaine shelf (Parentis basin) and across the Pyrenees and Alps ranges, the Moho displays definite deformation and the layered lower crust shows local changes in thickness, especially beneath the Parentis Basin. The crust is partly or entirely cut by dipping seismic events or vertical discontinuities that are interpreted as thrust or strike-slip faults related to the last major orogenic event having occurred in the area.

The Paleozoic thrusts and faults seem to have been obliterated by the lower crustal layering which, on the other hand, is involved in the Tertiary deformation. A study of the layered crust beneath various sedimentary basins suggests that the layering was emplaced either during the late Carboniferous-Permian, or the late Triassic-early-middle Jurassic. Both these periods correspond to major geodynamic events associated with regional increases of the thermal flow: reequilibration of the crust following the Variscan orogeny and opening of the Tethys ocean. To assign one of these events to the layering formation depends on the model of formation chosen for the SW Britain sedimentary basins.

Two geological causes may be proposed for the emplacement of the layering in the lower crust: metamorphism and mafic magmatic intrusion and/or cracks filled by fluids associated with moderate extensional shearing.

The seismic Moho was formed at the same time as the layering in the lower crust. Its emplacement across former crustal features suggests that rocks presently beneath the Moho may have been a part of the crust which was largely metamorphosed into higher velocity rocks that cannot be differentiated from those of the original mantle by the usual seismic methods. Some relics of these rocks show the former roots of the Variscan mountain range. Such a metamorphic process might be in progress in the deepest part of the crust beneath the Pyrenees and the Alps.

Introduction

Time is a major parameter in the geologist's reasoning. While surface rocks and bore-hole samples can be easily dated through the observation of fossils, magnetism and/or radioactivity, dating the rocks becomes more difficult when going deeper and farther from any reliable reference. The interpretation of deep seismic sections in terms of geology and geodynamics is typically faced with this kind of difficulty. However, the many structural events which occurred during geological history have left some imprints within the whole continental crust. Seismic interpretation attempts to relate the observed markers to these structural events and to give them at least a relative chronology. The purpose of this paper is to review some of the data which may lead to dating the seismic features observed in the deep crust of France and adjacent areas, taking advantage of the most advanced stage of geological knowledge in these regions. Geological models will be proposed as a consequence of this dating. They result from observations in the area under study and are not intended to explain crustal features in regions which underwent a quite different structural history.

Geological Framework

The geological history in France and neighboring area may be broadly subdivided into three main geodynamic phases. The first one corresponds to the consolidation of the crust in Western Europe during the Variscan orogeny (380-300 Ma). The overall crustal shortening resulted from the North-South impingement of Africa against the European continent and a number of intervening terranes (Aquitaine, Central Armorica, English Channel (Fig. 1)). Former oceanic areas were sutured in the Devonian in northern Aquitaine and in the late Carboniferous in southern England (Lizard). The northern front of deformation of this orogeny is well evidenced from southern Germany to southern Ireland, while south vergent thrusts are known in the Massif Central. The axial part of the Variscan orogeny is assumed to extend from the North of the Massif Central to the Armorican Massif where it was strongly obliterated by late-Variscan strike-slip faulting (Matte, 1986).

The second phase is related to the subsidence of the Variscan basement and the breaking up of the Variscan continent during the Permian and the Mesozoic. This led to the formation of large cratonic basins such as the Paris and Aquitaine basins and the opening of new oceanic domains in the Mesozoic: the Alpine Tethys ocean in the late Jurassic and the North Atlantic and Bay of Biscay in the middle Cretaceous. The third phase of the geological history corresponds to another plate convergence and collision during the late Cretaceous and the Tertiary following the closure of the Tethys ocean. During this period, in the Oligocene, large grabens were formed in the eastern part of France (Rhine and Rhone valleys) and the Western Mediterranean margin was rifted.

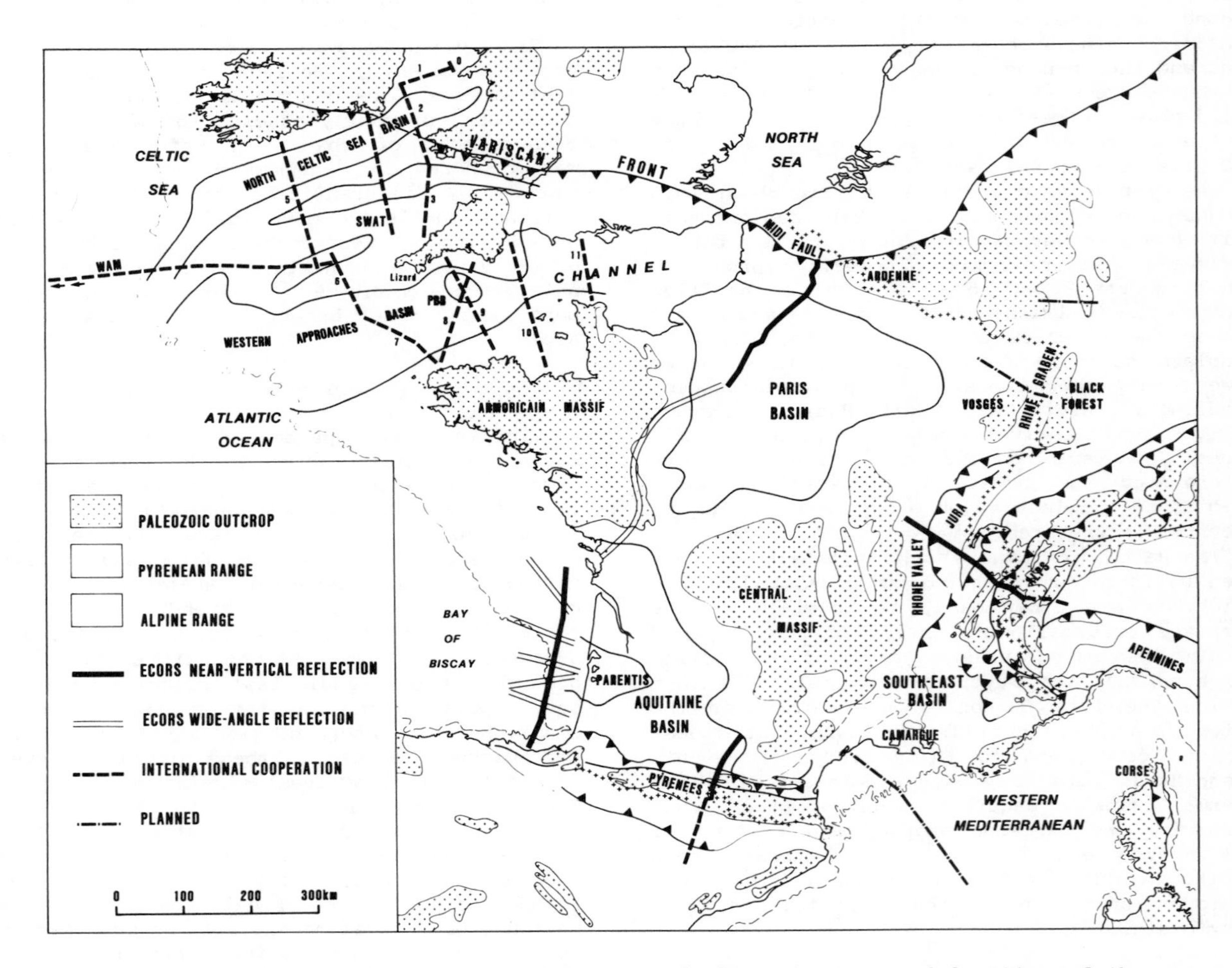

Fig. 1. Main structural features of France and adjacent areas and location of the ECORS and BIRPS-ECORS deep seismic profiles. Portions of the ECORS profiles have been completed thanks to international cooperation. PBB = Plymouth Bay basin.

For many years, sparse geological and geophysical data have been collected on the deep crust in France (Bois et al., 1988). Since 1983, the ECORS deep seismic project has provided a comprehensive picture of the deep crust and some regional phenomena could be tentatively interpreted. The ECORS profiles together with the BIRPS-ECORS ones in SW Britain (SWAT and WAM projects) have imaged major structural features that are related to the different phases of the regional geological history (Fig. 1).

Deep Structures of the Crust Imaged by the Seismic Profiles

The Layering of the Lower Crust and the Moho

Fig. 2a shows the line-drawings of deep seismic profiles shot from southern Ireland and Britain to southern France and northern Spain and Italy. The main structures resulting from their interpretation appear in Fig. 2b. The most striking feature occuring on these line-drawings is the layering that is largely displayed between 7 and 12 s TWT. This layering forms a highly reflective zone, 2 to 4.5 s TWT thick, which is located in the lower part of the crust. The bottom of this zone, which often corresponds to a band of slightly higher amplitude reflections, was identified as the Moho through ECORS wide-angle and refraction experiments in northern France (Cazes et al., 1986; Bois et al., 1986), the Bay of Biscay (Pinet et al., 1987) and the Alps (Bayer et al., 1987) and former experiments in the Pyrenees (Hirn et al., 1980) and the Celtic Sea (Blundell, 1981). The average P-wave velocity in the layered zone is 6.7 $km.s^{-1}$ while it is only 5.9 $km.s^{-1}$ in the upper crust.

The layering consists of straight, flat and subparallel reflections a few to ten kilometers in length. The overall character of this seismic layering is fairly constant in all the area under study. The layered lower crust can easily be identified in seismic profiles shot in regions as different as SW Britain, northern and southern France, northern Spain and northern Italy. All these regions, however, have in common a basement which underwent strong Variscan deformation. Some changes observed in vertical distribution of the layered reflections cannot be related to any known regional geological event (McGeary, 1987). The layered zone is however fairly constant in thickness and character beneath most of the sedimentary basins of the Celtic Sea (SWAT 4) and the English Channel (SWAT 9) (BIRPS and ECORS, 1986). The layered zone extends northward beyond the Variscan front in the southern Irish Sea (Brewer et al., 1983) and the Celtic Sea (SWAT 4). North of the ECORS northern France profile, the layered zone fades and the Moho itself cannot be observed anymore beneath the Variscan front (Bois et al., 1986). The crust is similarly transparent farther North in Belgium (Boukhaert et al., 1988). A wide-angle seismic reflection experiment carried out in northern France however shows a deeper Moho in that area (Hirn et al., 1987) (Fig. 6). This latter one was only imaged by wide-angle seismic reflection due to the use of lower frequency geophones and subcritical ray paths (Hirn et al., 1987).

The Moho and the layered zone are comparatively flat on the time sections of the profiles shot in the northern part of the area under study: Celtic Sea, English Channel and northern France. The depth sections should however show some Moho uplift beneath the deep sedimentary basins of the Celtic Sea and the English Channel (Warner, 1987). Assuming an average velocity of 4-5 $km.s^{-1}$ in the sediments, this uplift should be around 50-20% of the sediment thickness, i.e. only a few kilometers beneath the deepest basins.

On the profiles shot in the southern part of the area under study (Bay of Biscay, Pyrenees and Alps), the Moho displays definite deformation and we can also observe major thickness changes in the overlying layered zone. This is especially striking in the Bay of Biscay (Fig. 2) where the layered zone of the lower crust has nearly completely disappeared beneath the Parentis basin while the Moho was uplifted 2 s TWT, i.e. 15 km (Pinet at al., 1987). This interpretation is supported by ESPs and gravity modeling. The Parentis basin was initiated in the late Jurassic-early Cretaceous as a half-graben that was super-imposed on the Triassic-Jurassic Aquitaine basin. The attenuation of both the upper and lower crusts is readily related to the rifting of the Bay of Biscay's margin and the opening of an oceanic domain in the early Albian. Similar results were observed on the WAM profile shot across the continental margin of the Atlantic Ocean (Fig. 1) (Pinet et al., 1988).

On the Pyrenees profile (ECORS Pyrenees Team, 1988), the Iberian crust shows a gentle northward flexure toward the European crust. The layered lower crust, especially prominent along most of the profile, is attenuated on the southern edge of the European plate beneath the north Pyrenean zone. This attenuation may be related to the strong subsidence which occurred in the Albian-late Cretaceous in this area (Curnelle et al., 1982). The seismic section also shows a striking deterioration of the layered reflections and the Moho deeper than 15 s TWT.

On the Alps profile (Bayer et al., 1987), the European crust shows a gentle eastward flexure, well outlined by the Moho and the layered zone. This latter one is however thicker than usual. Farther east, the layered zone and the Moho cannot easily be picked on the seismic section. The Moho could however be traced thanks to a

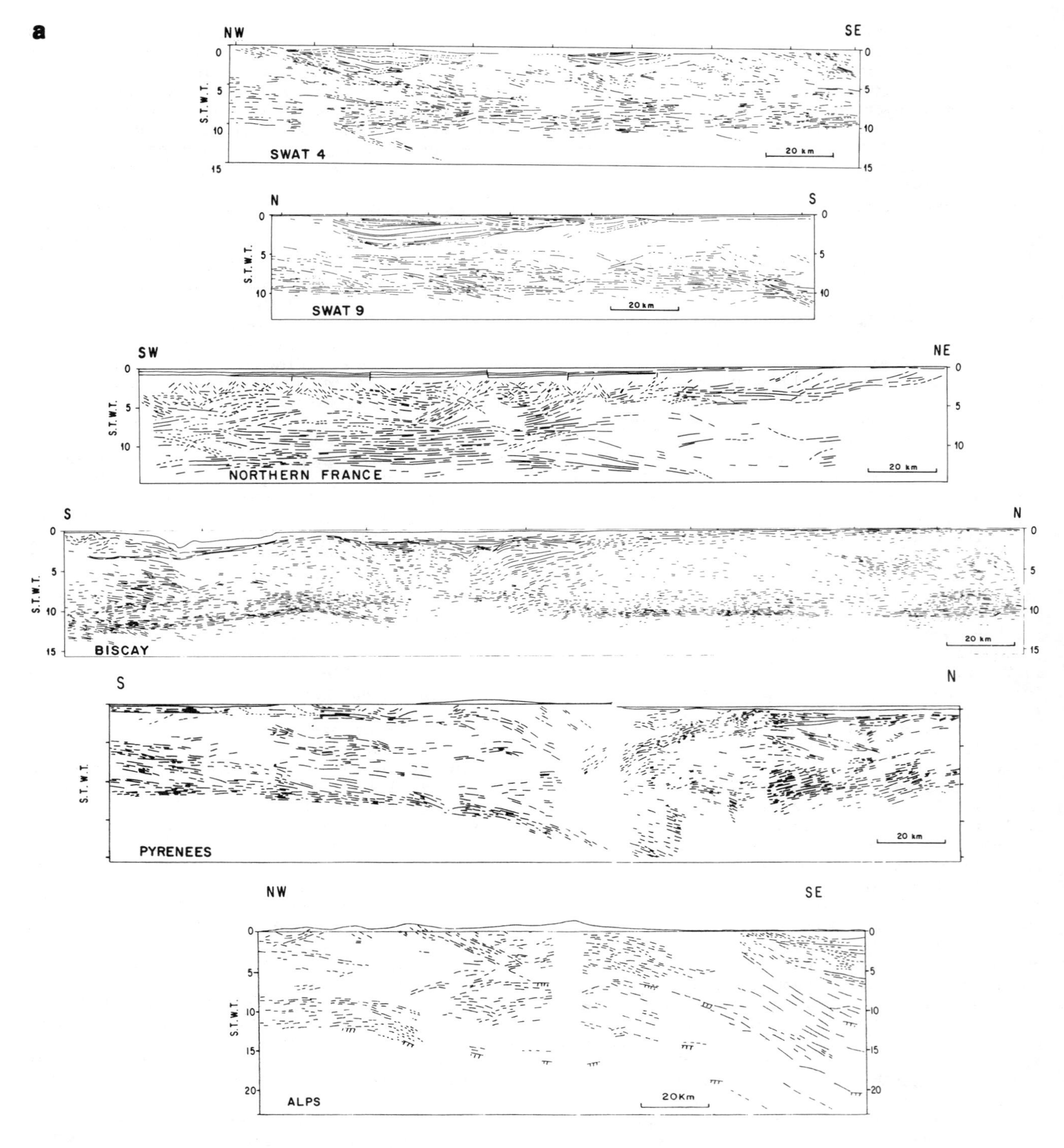

Fig. 2. SWAT 4, SWAT 9, northern France, Bay of Biscay, Pyrenees (French-Spanish cooperation) and Alps (French-Italian cooperation) deep seismic profiles. a = line-drawings of the unmigrated sections, b = main structures displayed on the seismic sections. AB = Aquitaine Basin, AF = Alpine Front, AZ = Axial Pyrenean Zone, BF = Bray Fault, EB = Ebro Basin, ECM = External Crystalline Massif, IL = Insubrian Line, LH = Landes High, PB = Parentis Basin, PF = Penninic Front, PFD =

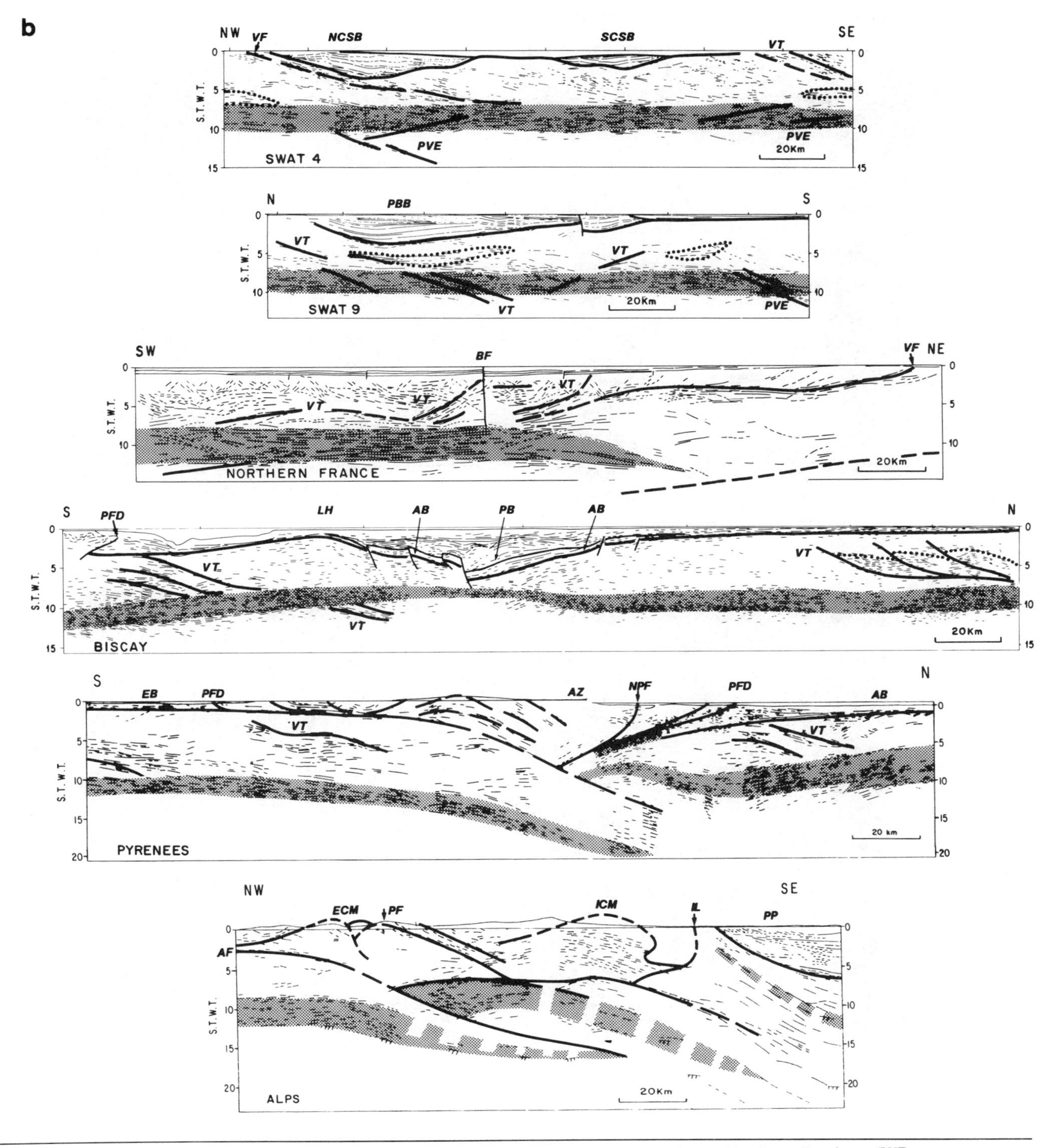

Pyrenean front of deformation, PP = Po Plain, PBB = Plymouth Bay Basin, PVE = Pre-Variscan event, NCSB = North Celtic Sea Basin, NPF = North Pyrenean Fault, SCSB = South Celtic Sea Basin, VF = Variscan Front, VT = Variscan Thrust. Dotted lines outline shallow crustal layering. Stippled areas = layered lower crust. On the Alps profile, the short markers underlined by diagonal ruling correspond to the Moho from a preliminary wide-angle experiment.

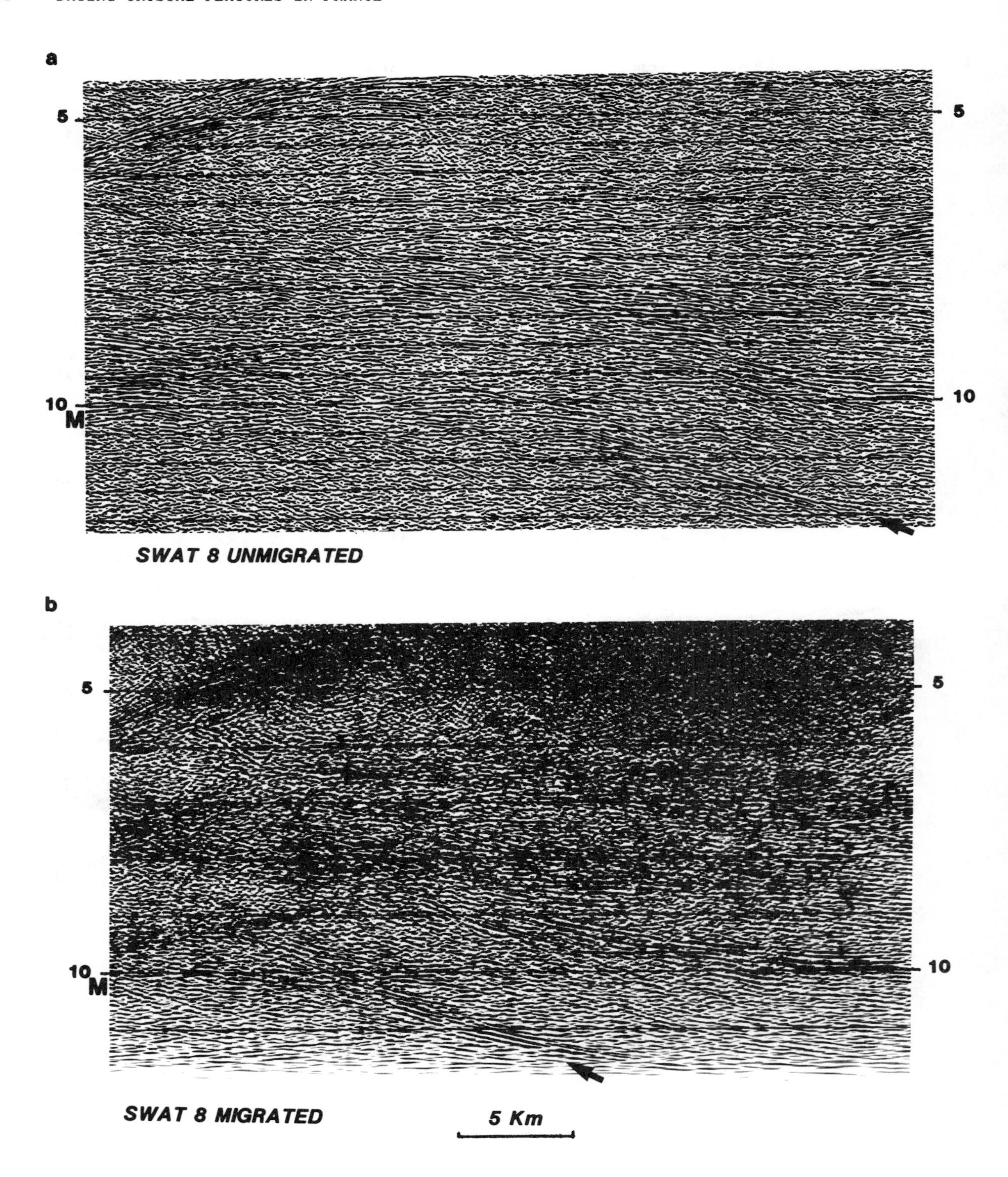

Fig. 3. Portion of the SWAT 8 section. a = unmigrated, b = migrated. M = Moho. The arrow shows a dipping event which cut across the Moho down to the upper mantle on both unmigrated and migrated sections. This event projects northward to the surface in the area of Lizard ophiolitic suture.

preliminary wide-angle seismic reflection experiment.

The geological cause of the crustal layering will be discussed in a further section. Let us merely recall that this layering was observed on seismic sections recorded by quite different methods and equipment both offshore and onshore, and it was also found in a number of other regions (Barazangi and Brown, 1986; Matthews and Smith, 1987). In the deep crust, the most continuous horizontal reflections with the highest amplitudes have been recorded in the European Variscan domain by the ECORS, BIRPS and German groups. Therefore, we shall assume that the layered reflections correspond to a real property of the lower crust.

Dipping Seismic Events Within the Continental Crust

Besides the horizontal layering of the lower crust, more or less energetic dipping events can be observed in the crust (Fig. 2). The profiles were generally shot parallel to the main direction of tectonic transport as it was established by a lot of previous studies of surface geology, bore-holes and conventional seismic surveys. Therefore we may reasonably assume that most of those dipping reflections do not come from lateral geological features. The sections displayed in Fig. 2 are not migrated, but wave equation migration was completed for most of the profiles. An example of this migration is shown in Fig. 3. Even though the deep events are somewhat disrupted as is frequent at these depths, the major dipping event seen on the unmigrated section can also be identified on the migrated section. In both cases, it is found to cross the Moho located around 10 s TWT and to extend into the upper mantle.

On the Pyrenees profile, the main dipping events are the north and south vergent thrusts of the Axial zone related to the Pyrenean compression. Most of them do not intersect the layered lower crust and the Moho which have however been deformed and offset. The deep seismic profile clearly images the structural relationships between the upper, middle and lower crust (Fig. 2), leading to definitely assigning the deformation and offset of the Moho and layered lower crust to the Pyrenean compression (ECORS Pyrenees team, 1988). At the deep crust level, this deformation is concentrated beneath the Axial Zone. North-dipping reflections can, however, be observed in the crust on both sides of this zone. They are truncated by the Mesozoic unconformity which was not deformed by the Pyrenean compression. Therefore, these reflectors are regarded as south-vergent Variscan thrusts associated with a well-known tectonic event in the region (Matte, 1986). The Alps profile displays a similar picture. The main Alpine dipping features are the Alpine front beneath the External Crystalline Massif, the Penninic front and dipping markers related to inner areas (Bayer et al., 1987).

In contrast to the Pyrenees and Alps profiles, the profiles shot across the northern part of the Variscan orogen (SWAT and northern France profiles) show a number of rather straight dipping events which cut across the flat layered lower crust and sometimes go down to the upper mantle such as on the SWAT 4 and 9 profiles (Fig. 2). Among the most prominent ones are the south-dipping events that can be observed north of the English Channel and are labelled VT on Profile 9. They are related to the Lizard thrust, a Variscan feature associated with an outcropping ophiolitic suture. Where they cut across the Moho and the layered lower crust, these latter features do not seem to have been offset (Fig. 3), suggesting that they were formed after the last compressional phase which built the Variscan belt.

A number of vertical faults of major significance in the sedimentary basins were crossed by the profiles, e.g. the mid-Channel fault system in the English Channel and the Bray fault in the Paris basin (Fig. 2). All these faults are considered as old features that were later reactivated by Mesozoic extension and occasionally inverted by Tertiary compression. The crustal displacements of such faults are extremely difficult to assess. The only data available concern the Bray fault which affected at least the upper crust in the late Carboniferous as suggested by the lack of correlation among the seismic features across the fault and the observation of a ductile dextral slip in a basement core coming from a nearby borehole (Matte et al., 1986). The lower crustal layering is interrupted on the near-vertical seismic reflection (Figs. 2 and 4) and was first regarded as offset by the fault (Bois et al., 1986). An explosive shot recorded by a large spread straddling the fault however shows a fair continuity of the layering across the fault (Hirn et al., 1987; Damotte, 1988). The big difference in the results obtained by the two methods probably comes from the shallow unfavourable conditions near the fault that the wide-angle rays keep clear of. Either the Bray fault never displaced the lower crust or the layering was emplaced after its major strike-slip displacement in the late Carboniferous. The Mesozoic and Tertiary reactivation of the fault was comparatively small and should not have led to any crustal offset visible on the section. The conclusion is that the Bray fault does not provide any reliable evidence for the dating of the lower crust layering.

Other Seismic Events

The top of the layered zone of the lower crust is generally located around 7 to 8 s TWT

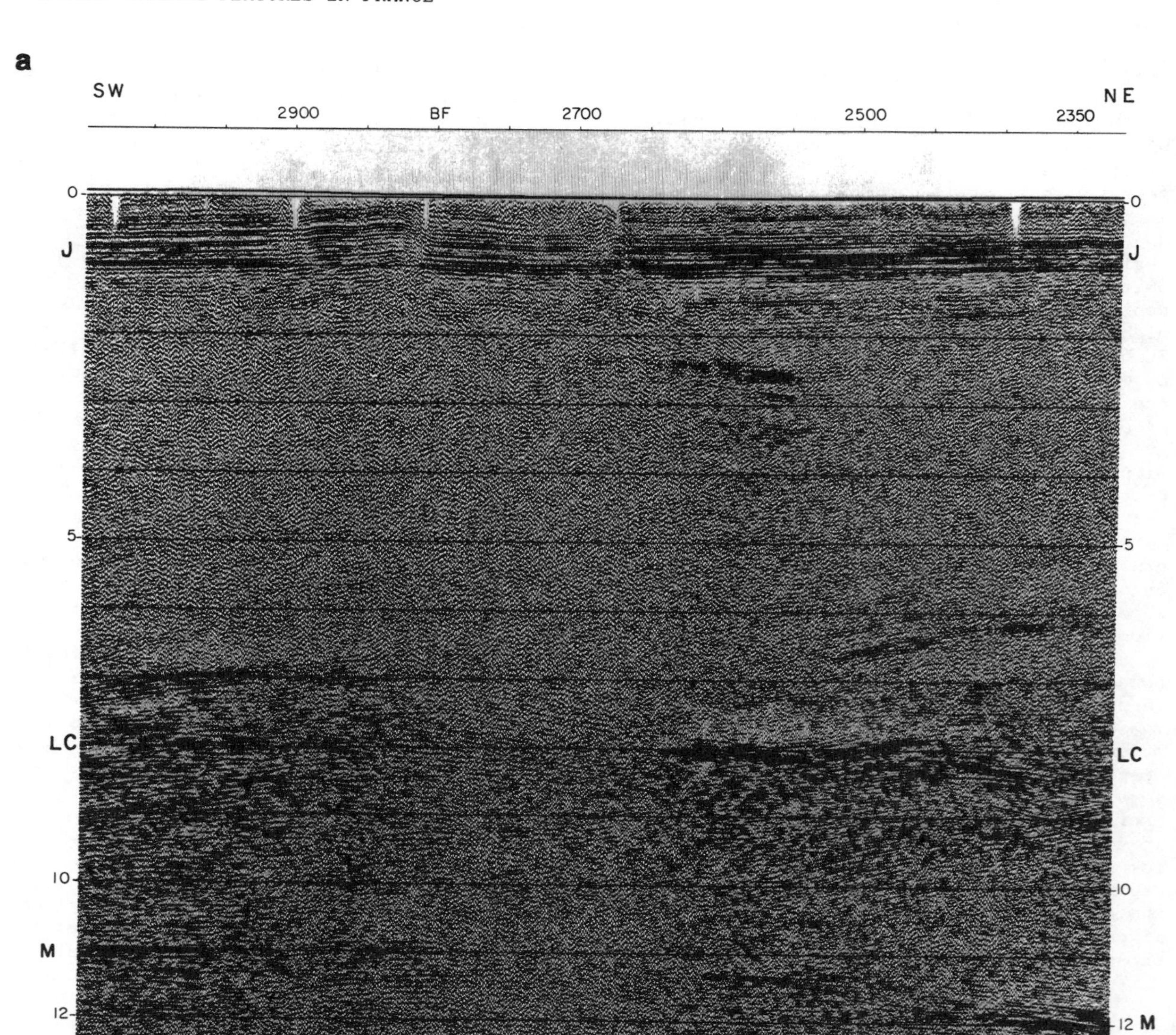

Fig. 4. The Bray Fault area. a = portion of the vibroseismic section of the northern France profile (Cazes et al., 1986; Bois et al., 1986): the Bray fault area corresponds to a blank almost 10 km wide; the picture suggests the northern block was offset downward. b = record from a single dynamite shot on the northern France profile (Hirn et al., 1987; Damotte, 1988): the lower crust layering (8-12 s TWT) shows a fair continuity across the fault. BF = Bray Fault in surface, J = bottom Jurassic, LC = top lower crust, M = Moho, S = explosive shot.

on most of the profiles (Fig. 2). However, layered flat reflections that display a character similar in many respects to the lower crust can occasionally be observed between 2 and 7 s TWT. They form packets irregularly distributed along the profiles. They are especially prominent in the profiles across the English Channel and in the northern end of the Bay of Biscay profile (Fig. 2). Unfortunately, the velocity of these packets has never been measured. Hence there are very few constraints on their geological interpretation. Possible explanations for this shallow layering are: (1) horizontal shearing, (2) thrust sheet or nappe of reflective material, either sediments or basement, (3) petrologic differentiation related

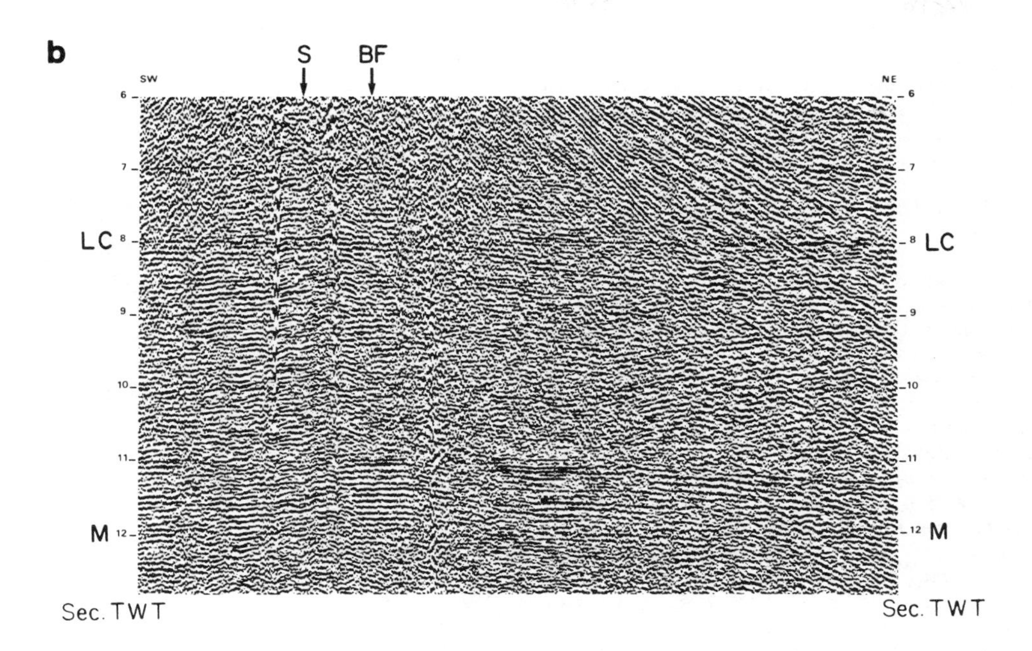

to the emplacement of a granite batholith (Maguire et al., 1982; Matthews, 1987) or (4) magmatic sills.

Hypothesis 2 was preferred for the northern Bay of Biscay because of the northern Aquitaine collision zone, just to the north, associated with nappes and a large magnetic anomaly (Pinet et al., 1987). Hypothesis 3 may be advanced for some areas in the southern English Channel. Hypothesis 4 may be proposed for the reflections that occur beneath the Plymouth Bay basin, around 5 s TWT, because of the presence of volcanics in the Exeter basin and probably in the Plymouth Bay basin itself.

Timing and Causes of the Emplacement of the Layering in the Lower Crust

Constraints on the Timing

The layering and the compressive structures. On the Pyrenees and Alps profiles, the Moho and the layered zone of the lower crust were deformed and offset by the Alpine compression (Fig. 2). Near the subduction zones, the upper mantle together with the lower crust are involved in imbricated thrusts. The layering was then emplaced before the Alpine deformation in the lower crust.

The structural studies carried out for years in the Variscan belt led to the conclusion that this range followed a structural model similar to the Alpine one (Matte, 1986). However, the SW Britain and northern France profiles show that the Moho and the layered lower crust are flat and cut across dipping seismic events related to Variscan thrusts and subduction zones such as the Lizard one. These events even go into the upper mantle (Fig. 3). The emplacement of the layering then occurred after the last Variscan orogenic events. There is neither any geophysical nor geological reason to substantiate any major changes in the age of the layering emplacement within the comparatively limited area under study. The simplest interpretation consists in assuming this process occurred between the late Carboniferous and the late Cretaceous in the whole area. This discussion does not preclude the existence of a pre-Variscan layered lower crust involved in the Variscan orogeny and subsequently obliterated by the late Carboniferous-Mesozoic layering. Relics of such an old layering may well be present in the northern part of the Bay of Biscay profile, as suggested above.

The layering and the sedimentary basins. The relationships between the layered zone of the lower crust and the sedimentary basins developed between the late Carboniferous and the late Cretaceous provide more time constraints on the layering emplacement.

The basins imaged on the sections in Fig. 2 were formed by several subsidence cycles. The first one occurred in the Permian and the Triassic. Stress relaxation and thermal subsidence followed the Variscan range building, crustal thickening and granite intrusions (Mascle and Cazes, 1987). The following cycles are related to the rifting of three major features: the Tethys ocean, the North Sea graben system and the Bay of Biscay. The rifting of the Tethys margin started in the early Jurassic and led to the opening of an oceanic domain in the late Jurassic in the Alps and the central

Atlantic (Lemoine et al., 1986). The rifting in the North Sea, late Jurassic in age, resulted in the formation of an extensive graben system which never reached the oceanic stage. The Bay of Biscay rifting was initiated at the end of the late Jurassic and the continental breakup was completed in the late Aptian-early Albian (Pinet et al., 1987). Each rifting was associated with extension along a specific direction. The various basins are more or less influenced by one or another cycle of subsidence according to their locations with respect to the above features.

We can observe in Fig. 2 that the layered zone of the lower crust was strongly attenuated beneath the Parentis basin. This attenuation was related to the basin rifting that occurred in the early Cretaceous and the coeval opening of the oceanic domain of the North Atlantic and the Bay of Biscay. The emplacement of the layering in the lower crust was therefore earlier. The attenuation of the layered lower crust during the late Jurassic rifting of the Viking graben in the North Sea (Beach, 1986; Pinet et al., 1987) provides a time limit that is still slightly earlier.

Beneath the deep basins of the Celtic Sea and the English Channel the upper crust was attenuated by about 50% (Fig. 2). The small number of normal faults controlling the subsidence of the basins and or the moderate displacements across these faults which were active from the Permo-Triassic to the late Cretaceous make it difficult to identify a rifting stage followed by a thermal subsidence stage. Anyway, the largest portion of the tectonic subsidence was completed by the end of the Triassic (Buchet, 1986; Dyment, 1987): 75% of the whole tectonic subsidence in the North Celtic Sea basin and 90% in the Plymouth Bay basin (Fig. 5). However, whatever the actual shape of the Moho may be, the layered zone of the lower crust does not show any change in thickness and character beneath the basins (Fig. 2). Only the upper crust is strongly attenuated as a result of the basin's emplacement.

Given the interpretation that the relative timing of the lower crustal layering is determined by the geometry described, there appear to be two possible time intervals for its development. As a first hypothesis, we may propose that the emplacement of the layering post-dated the basins' formation. Assuming that the layering was emplaced in the lower crust at the same time in all the area under study, we find this emplacement period to be early-middle Jurassic. Taking into account that the subsidence was slightly earlier in the Plymouth Bay basin, we might also contemplate an earlier emplacement in the southern part of the area under study, i.e. late Triassic. As a second hypothesis, we shall consider that the layering may have a different age beneath and outside the

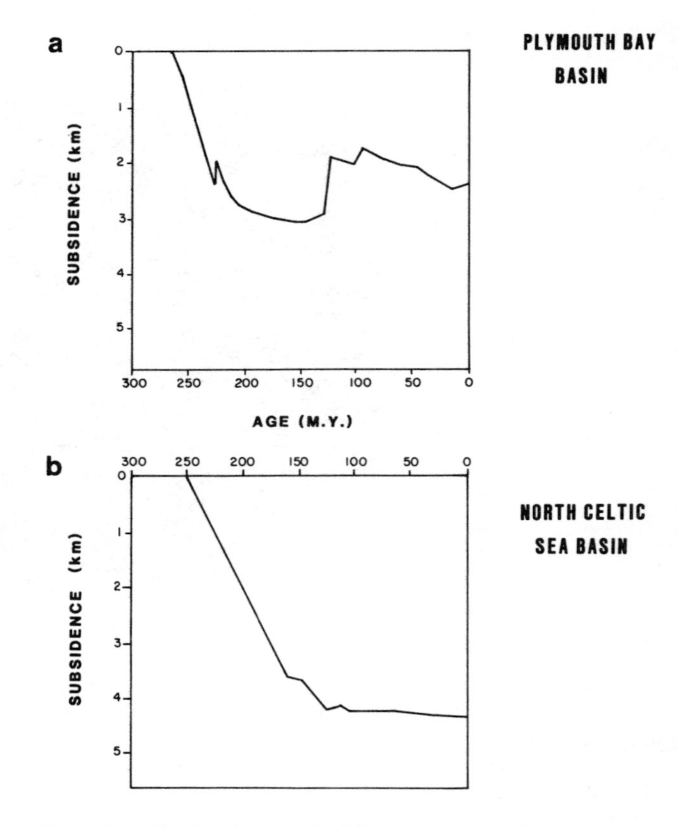

Fig. 5. Tectonic subsidence of the crust beneath sedimentary basins (local isostatic compensation). a = Plymouth Bay Basin, SWAT 9, point 1600 (Buchet, 1986). b = North Celtic Sea Basin, SWAT 5, point 800 (Dyment, 1987).

basins. A widespread layering may have been emplaced in the late Carboniferous-Permian, then locally affected by the Permo-Triassic rifting and crustal attenuation beneath the basins and eventually restored during and after this period.

Both the above conclusions imply that the basin's formation was associated with crustal attenuation beneath the basins. Other models of basin may be contemplated in which the lower crust attenuation is either small or laterally offset. The extensional reactivation of a former compressional detachment may result in a large offset of the lower crustal thinning with respect to the basin (Wernicke, 1981). This model was proposed by Gibbs (1987) for the North Celtic Sea basin, the detachment responsible of the basin's subsidence being the Variscan front. However, the data do not support this model. The basin and the Variscan front show quite different trends (Fig. 1) and the depocenters shift southwards with the subsidence when the model predicts a northward shift (Bois and ECORS, 1988). Moreover, there is no evidence on the seismic profiles of any lower crustal attenuation which would be offset with respect to the subsidence areas.

Permo-Triassic basins might also have been created by mechanisms that the current extensional models do not take into account. An accumulation of thick sedimentary piles may be explained by the collapse and the subsidence of large areas of the Variscan hinterland related to the thermal contraction of a hot lithosphere and the availability of large masses of sediments through erosion. This process followed the relaxation of tectonic stresses and the equilibration of a thickened crust (Ziegler, 1982; Mascle and Cazes, 1987). The emplacement of the layering in the lower crust might then be late Carboniferous-Permian.

Causes of the Layering in the Lower Crust

Re-equilibration of the crust after a major compressional event. We first proposed that the layering in the lower crust could be primarily caused by the Variscan compression (Bois et al., 1988). This layering would result from crust-mantle delamination and shearing during the compression and subsequent melting, intrusion and metamorphism in the deep part of the thickened crust. Such a hypothesis does not fit with the above conclusions. Moreover, the correlation between the areal distributions of the Variscan belt and the layering's occurence is not very good. Layering extends beyond the Variscan front in the Celtic Sea (SWAT 4) and in the southern Irish Sea (Brewer et al., 1983). On the other hand, it fades in the inner part of the Variscan range in southern Germany (DEKORP, 1987).

On the seismic sections across the Celtic Sea, the English Channel and northern France, the layered lower crust and the Moho are comparatively flat, and without being offset they intersect a number of features assigned to Variscan structures, e.g. the Lizard thrust and possibly the Bray fault (Figs. 3 and 4). Such a setting for the layered lower crust and the Moho then suggests that they result from a late equilibration of the crust-mantle transition. Such an equilibration may have occurred as early as the late Carboniferous-Permian and have been associated with crustal melting, metamorphism and major magmatism (late Variscan granite). This is the period of major erosion, widespread peneplanation and the disappearance of the Variscan mountain roots either by mechanical processes or subcrustal metamorphism. The Alps and Pyrenees profiles do not show any evidence of incipient layering in the deep crust as might be expected after the Tertiary compression, unless this reequilibration is possibly underway and not yet completely done.

The slight surface evidence of Western European lower crust in the Pyrenees (northern Pyrenean massifs), the Alps (Ivrea zone) and from the xenoliths brought to the surface by Neogene and Quaternary volcanoes (Massif Central) supports the above hypothesis. In France most granulitic rocks show a thermal reequilibration which is late Variscan in age (315-285 Ma) (see references in Bois et al., 1988). This major thermal event may account for the formation of the layering in the lower crust.

Mechanical transformation of the crust related to a moderate extension. Another hypothesis suggests that the layering was later superimposed on a lower crust that was already equilibrated after the Variscan orogeny but subject to the profound disturbances which led to the formation of the Permo-Mesozoic sedimentary basins. This layering would have been formed during the late Triassic-early-middle Jurassic, after the first phase of basin subsidence of large sedimentary basins in France (Paris, Aquitaine) and the rifting of the Alpine Tethys ocean. The asthenospheric uplift should have been associated with a regional heat flow increase obliterating part of the previous features and reequilibrating the crust-mantle transition. This model is somewhat supported by an inverse correlation between the present thermal flow and the depths of the Moho and the top of the lower crust (Klemperer, 1987; Meissner et al., 1987). Such a thermal event may be associated with metamorphism and mafic magmatism as evidenced by Triassic and Jurassic magmatism reported in Western Europe in areas which underwent major subsequent rifting (Ziegler, 1982). Widespread thermal events during the same period were reported in France and adjacent areas (Bonhomme et al., 1987; Merceron et al., 1987). However, this latter hypothesis is not supported by the age of the rocks samples in the lower crust.

To account for an early-middle Jurassic age of the lower crustal layering, we should look for a purely mechanical transformation of the existing material. It has been proposed that the layering was the trace of a major extensional shearing in the ductile lower crust while brittle extension controlled the formation of the basins and margins (Matthews and BIRPS, 1987; Klemperer and BIRPS, 1987). Extension occurred along many normal faults in the upper crust, but the extension rates are very low (Tremolières, 1981; Chenet et al., 1983; Rudkiewicz, 1988) suggesting that this mechanical factor was subordinate. On the other hand, if the layering was formed by major extension, the dipping seismic events which cross the lower crust should have been strongly deformed and even destroyed in most cases. Therefore we propose that the layering was not related to the lower crustal granulites themselves but to a mechanical transformation of them. This might be a moderate extensional shearing leading to the formation of cracks possibly filled by fluids. This hypothesis fits well with the discovery of a low-resistivity layer more or less coincident with the layered lower crust in northern France (Cazes et al., 1986).

Significance of the Seismic Moho and its Relation with the Upper Mantle

All the seismic data available in the area under study show that the Moho is coincident with the bottom of the layered zone in the lower crust. This raises the question of the emplacement of the Moho and the relationship between the lower crust and the upper mantle.

Several seismic events can be observed below the present Moho in the upper mantle. They are dipping reflections in the SWAT profiles and flat reflections in the northern part of the Bay of Biscay and northern France profiles (Figs. 2, 3 and 6). A spectacular deep dipping reflection was observed in Scotland down to 30 s TWT (Brewer and Smythe, 1983, Warner and McGeary, 1987). According to the above models, the present Moho should result from the reequilibration of a crust-mantle boundary which may have had a complicated pattern. The rocks presently beneath the Moho reflection may have locally been part of the crust which was largely metamorphosed into higher velocity rocks under the temperature and the high pressure prevailing deeper than 30-35 km. Such rocks can hardly be differentiated from those of the original mantle by the usual seismic methods. Some relics of former crustal features, especially traces of dipping Moho, thrusts and sutures, have however been preserved.

This view is in good agreement with the flat shape of the Moho reflection in a large part of the area under study and its geometric relationship with the Variscan dipping events. It may also explain the Moho overlap observed in the northern part of the northern France profile (Fig. 6): the late Moho (Mv) seems to have been overprinted on the southern edge of the foreland where a former deeper Moho (Mwa), only found by wide-angle seismic reflection, has been locally preserved. This latter Moho dips southward and probably images a foreland flexure beneath the frontal nappes which is quite similar to the one observed on the northwestern edge of the Alps profile and on the southern part of the Pyrenees profile (Fig. 2). The geometrical relationships between the two Mohos make it extremely difficult to assume they result from the Variscan thrusting of a formerly single feature. The layered lower crust is not present on the near-vertical seismic reflections sections across the foreland Brabant Massif (Cazes et al., 1986; Boukhaert et al., 1988). This suggests that the areal distribution of the layering might be somewhat controlled by the petrographic composition and/or the structure of the crust on which it was overprinted. Allmendinger et al. (1987) proposed that the lack of layering and Moho reflection was characteristics of the foreland.

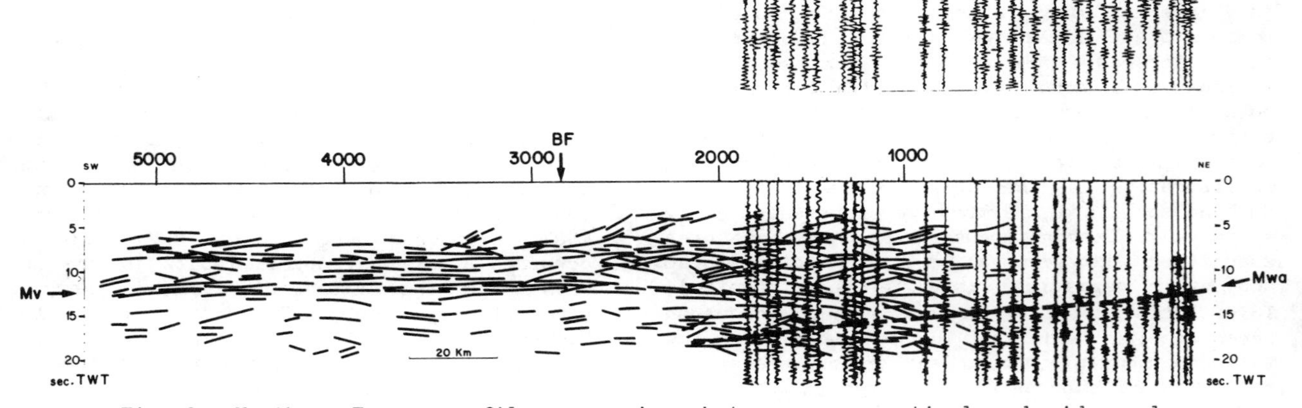

Fig. 6. Northern France profile: comparison between near-vertical and wide-angle seismic reflection (Damotte, 1988). The line-drawing corresponds to the near-vertical seismic reflection recorded from explosive shots with offsets from 0 to 45 km. Note the flat reflections occurring beneath the Moho in the northern portion of the profile. The seismic traces correspond to the wide-angle seismic reflection from low-frequency self-recording stations. BF = Bray Fault, Mv = Moho from near-vertical reflection, Mwa = Moho from wide-angle reflection.

This hypothesis does not fit with the presence of a well developped layered lower crust in the Variscan foreland in the northern Celtic Sea and southern Ireland Sea (BIRPS and ECORS, 1986; Brewer et al., 1983).

The northern France section also suggests that metamorphic processes should have played an important part in the disappearance of the Variscan mountain roots. This disappearance is generally explained by mechanical processes such as surface erosion and crustal uplift through isostatic balance, crustal extension and/or lower crustal creeping (Meissner et al., 1987). However, gabbros present in the deep crust should have been transformed into eclogites beneath 30-35 km in areas of normal heat flow. These rocks have the same seismic properties as the mantle. Such a process might be in progress beneath the Pyrenees and the Alps where the sections show a definite fading of the lower crust reflections and the Moho below 15 s TWT. The present geometry of the Moho is well imaged by only wide-angle seismic reflection. However, we cannot dismiss the idea that this fading could also be caused by seismic noise and/or energy absorption due to shallow structural complication.

Conclusion

The deep seismic profiles carried out in France and adjacent areas have provided major data on the deep crust of this part of Western Europe. Variscan and Tertiary orogenic belts have been imaged as well as Permo-Mesozoic basins of various ages. The set of profiles presented in this paper shows that the layered lower crust is almost flat and uniform in the areas which only underwent the Variscan orogeny. In contrast, the layered lower crust is readily deformed and/or attenuated in the areas which experienced late Jurassic-Cretaceous rifting and Alpine orogeny. This observation indicates that the observed layering was emplaced between the late Carboniferous and the early-middle Jurassic in the lower crust. Two major geodynamic events occurring during this period may also be accounted for the layering's emplacement: (1) re-equilibration of a thickened crust following the Variscan orogeny and (2) rifting of the Tethys ocean and subsidence of large cratonic basins. Both events were associated with a large regional heat flow increase and moderate extension. To assign one of these events to the formation of the layering depends on the model of formation of the Celtic Sea and English Channel basins. Two geological causes may be proposed for the emplacement of the layering in the lower crust: metamorphism and mafic magmatism and/or cracks filled by fluids associated with moderate extensional shearing.

The seismic Moho emplaced at the bottom of the layered lower crust may have notably changed through geological time. Metamorphosed crustal rocks that cannot be differentiated from the original mantle may locally exist beneath the present Moho and form relics of the former mountain roots.

The puzzling problem of the reflective layering of the lower crust obviously requires further investigation. Better knowledge of the areal extent of this layering could for instance provide important clues to the formation and the geological composition of this enigmatic feature.

Acknowledgments. The ECORS project is managed and founded by Institut Français du Pétrole, Institut National des Sciences de l'Univers (Conseil National de la Recherche Scientifique), Elf Aquitaine and Institut Français pour la Recherche et l'Exploitation de la Mer. National and international cooperation was achieved on the Bay of Biscay, Pyrenees and Alps profiles. The SWAT and WAM projects were operated by the BIRPS group with the financing and technical participation of the ECORS project.

The authors are deeply indebted to the members of the ECORS scientific teams for the many discussions from which they greatly benefited. They also thank L. Brown and F. Cook for their fine reviews of the manuscript and the pertinent questions they have raised.

References

Allmendinger R.W., Nelson K.D., Potter C.J., Barazangi M., and Oliver J.E., Deep seismic reflection characteristics of the continental crust, Geology, 15, 304-310, 1987.

Barazangi M. and Brown L. (eds.), Reflection Seismology: A Global Perspective, A.G.U. Geodynamic series, 13 and 14, 1986.

Bayer R., Cazes M., Dal Piaz G.V., Damotte B., Elter J., Gosso G., Hirn A., Lanza R., Lombardo B., Mugnier J.L., Nicolas A., Nicolich R., Polino R., Roure F., Sacchi R., Scarascia S., Tabacco I., Tapponnier P., Tardy M., Taylor M., Thouvenot F., Torreilles G., Villien A., First results of a deep seismic profile through the western Alps (ECORS-CROP Program), Comptes Rendus Acad. Sci., Paris,105, série II, 1461-1470, 1987.

Beach A., A deep seismic reflection profile across the northern North Sea, Nature, 323, 53-55, 1986.

BIRPS and ECORS, Deep Seismic reflection profiling between England, France and Ireland, J. geol. Soc. London, 143, 45-52, 1986.

Blundell D., The Nature of the Continental Crust beneath Britain, in Petroleum Geology of the Continental Shelf of North-West Europe, Illing. I.V. and Hobson G.D. (eds.), Institute of Petroleum, London, 58-64, 1981.

Bois C., Cazes M., Damotte B., Galdeano A., Hirn A., Mascle A., Matte P., Raoult J.F. and

Torreilles G., Deep Seismic Profiling of the Crust in Northern France: The ECORS Project, in Reflection Seismology: A Global Perspective, Barazangi M. and Brown L. (eds.), A.G.U. Geodynamic Series, 13, 21-29, 1986.

Bois C., Cazes M., Hirn A., Mascle A., Matte P., Montadert L., and Pinet B., Contribution of Deep Seismic Profiling to the Knowledge of the Lower Crust in France and Neighbouring Areas, Tectonophysics, 145, 253-275, 1986.

Bois C. and ECORS Scientific party, Major crustal features disclosed by the ECORS deep seismic profiles, Ann. Soc. Belge Geol., in press, 1988.

Bonhomme M.G., Baubron J.C and Jebrak M., Minéralogie, Géochimie, Terres rares et âge K-Ar des argiles associés aux minéralisations filonniennes, Chemical Geology, (Isotope Geosc. Sect.) 65, 321-339, 1987.

Boukhaert J., Fock W. and Vandenberghe N., First results of the Belgium Geotraverse, (BELCORP), Ann.Soc. Belge de Géol., in press, 1988

Brewer J.A., Matthews D.H., Warner M.R., Hall J., Smythe D.K. and Whittington R.J.,BIRPS deep seismic reflection studies of the British Caledonides, Nature, 305, 206-210, 1983.

Brewer J.A. and Smythe D.K., MOIST and the continuity of crustal reflectors geometry along the Caledonian-Appalachian orogen, J. geol. Soc. London, 141, 105-120,1984.

Buchet B., Mémoire de Diplôme d'Etude Approfondie en Géologie-Géophysique, Univ. Paris-Orsay, 118 p, 1986.

Cazes M., Mascle A. and Scientific Party, Large Variscan overthrusts below the Paris Basin, Nature, 323, 144-147, 1986.

Chenet P.Y., Montadert L., Gairand H. and Roberts D., Extension ratio measurements on the Galicia, Portugal and northern Biscay continental margin: Implications for evolutionary models of passive continental margins, in Studies in Continental Margin Geology, Watkins J. (ed.), Am. Assoc. Pet. Geol., Mem. 34, 703-715, 1982.

Curnelle R., Dubois P. and Seguin J.C., The Mesozoic-Tertiary evolution of the Aquitaine Basin, Phil. Trans. R. Soc. London, A 305, 63-84, 1982.

Damotte B., Le profil Nord de la France. Traitement et interprétation de la sismique réflexion à l'explosif, in Etude de la croûte terrestre en France par méthode sismique. Profil Nord de la France, Cazes M., Torreilles G. and Raoult J.F. (eds), Technip, Paris, 1988.

DEKORP Research Group, Results of deep reflection seismic profiling in the Oberpfalz (Bavaria), Geophys. J. of R. astr. Soc., 89, 353-363, 1987.

Dyment J.,Etude des bassins sédimentaires celtiques et de leurs relations avec la croûte sous-jacente à l'aide des profils de sismique-réflexion profonde SWAT, Diplôme Ecole de Physique du Globe, Strasbourg, 187 p,1987.

ECORS Pyrenees Team, The ECORS deep reflection seismic survey across the Pyrenees, Nature, 331, 508-511, 1988.

Gibbs A.D.,Basin development, examples from the United Kingdom and comments on hydrocarbon prospectivity, Tectonophysics, 133,189-198, 1987.

Hirn A., Daignières M., Gallard J. and Vadell M.,Explosion seismic sounding of throws and dips in the continental Moho, Geophys. Res. Lett., 7, 4, 263-266, 1980.

Hirn A., Damotte B. and Torreilles G., Crustal reflection seismics: the contribution of oblique low frequency and shear wave illumination, Geophys. J. of R. Astr. Soc., 89, 1, 287-295, 1987.

Klemperer S.L. and the BIRPS group, Reflectivity of the crystalline crust: hypotheses and tests, Geophys. J. of R. Astr. Soc., 89, 217-222, 1987.

Lemoine M., Bas T., Arnaud-Vanneau A., Arnaud H., Dumont T., Gidon M., Bourbon M., Graciansky P.C. de, Rudkiewicz J.L., Megard-Galli J. and Tricart P., The continental margin of the Mesozoic Tethys in the Western Alps, Mar. Petr. Geol., 3, 179-199, 1986.

Maguire P.K.H., Andrew E.M., Arter G., Chadwick R.A., Greenwood P., Hill I.A., Kenolty N. and Khan M.A., A deep seismic reflection profile over a Caledonian granite in Central England, Nature, 297, 671-673, 1982.

Mascle A., and Cazes M., Geodynamic setting of Permian basins in Northwestern Europe, 4th E.U.G., Terra Cognita, 7, 2-3, 178, 1987.

Mathur S.P., Deep Crustal Reflections Results from the Central Eromanga, Australia, Tectonophysics, 100, 163-173, 1983.

Matte P., Respaut J.P., Maluski H., Lancelot J.R., and Brunel M., La faille NW-SE du Pays de Bray, un décrochement ductile dextre hercynien: déformation à 320 Ma d'un granite à 570 Ma dans le sondage Pays de Bray 201, Bull. Soc. Geol. France, 8, II, 1, 69-77, 1986.

Matte P., Tectonics and Plate Tectonics Model for the Variscan Belt of Europe, Tectonophysics, 126, 329-374, 1986.

Matthews D. and Smith C. (eds.), Deep seismic reflection profiling of the continental lithosphere, Geophys. J. of the R. Astr. Soc., Sp. issue, 89, 1, 1-477, 1987.

Matthews D.H., Can we see granites on seismic reflection profiles? Ann. Geophys. Terrest. Planet. Phys., 5, B 4, 353-356, 1987.

Matthews D.H. and BIRPS, Some unsolved BIRPS problems, Geophys. J. of R. Astr. Soc., 89, 209-215, 1987.

McGeary S., Non typical BIRPS on the margin of the northern North Sea: the SHET survey, Geophys. J. of R. Astr. Soc., 89, 231-237, 1987.

Meissner R., Wever T. and Flüh E.R., The Moho in Europe - Implications for crustal development, Ann. Geophys. Terrest. Planet. Phys., 5, B 4, 357-364, 1987.

Merceron T., Bonhomme M.G., Fouillac A.M., Vivier G. and Meunier A., Pétrologie des altérations hydrothermales du sondage GPF Echassières n° 1, in Géologie Profonde de la France, le forage scientifique d'Echassières (Allier), Cuney M. and Autran A. (eds), Géologie de la France, 2-3, 259-269, 1987.

Pinet B., Wannesson J., Whitfield M. and Vially R., Coincident Deep Seismic Reflection-Refraction profiles beneath the Viking Graben (North Sea), EOS Trans. A.G.U., 68, 44, 1480,1987.

Pinet B., Montadert L. and ECORS Scientific Party, Deep Seismic Reflection and Refraction Profiling along the Aquitaine Shelf (Bay of Biscay), Geophys. J. of R. Astr. Soc., 89, 305-312, 1987.

Pinet B., Montadert L., Curnelle R., Cazes M., Marillier F., Rolet J., Tomassino A., Galdeano A., Patriat P., Brunet M.F., Olivet J.L., Schaming M., Lefort J.P., Arrieta A. and Riaza C., Crustal thinning on the Aquitaine shelf, Bay of Biscay, from deep seismic data, Nature, 325, 513-516, 1987.

Rudkiewicz J.L., Quantitative Subsidence of the Tethyan Western Margin during Lias and Dogger, on the Grenoble Briançon Transect, Mem. Soc. Geol. France, in press, 1988.

Trémolières P., Mécanismes de la déformation en zones de plate-forme : méthode et applications au Bassin de Paris. Rev. Institut Franç. Pétr., 36, 4, 395-428, 5, 579-592, 1981.

Warner M.R., Seismic reflection from the Moho - the effect of isostasy. Geophys. J. of R. Astr. Soc., 88, 425-435, 1987.

Warner M. and McGeary S.,Seismic reflection coefficients from mantle fault zones, Geophys. J. of R. Astr. Soc., 89, 223-230, 1987.

Wernicke B., Low-angle normal faults in the Basin and Range Province: nappe tectonics in an extended orogen, Nature, 291, 645-647, 1981.

Ziegler P.A., Geological Atlas of Western and Central Europe, Shell Internationale Petroleum Maatshapij B.V., The Hague, 130 p., 40 encl., 1982.

STYLES OF DEFORMATION OBSERVED ON DEEP SEISMIC REFLECTION PROFILES OF THE APPALACHIAN-CALEDONIDE SYSTEM

J. Hall[1], G. Quinlan[1], J. Wright[1], C. Keen[2] and F. Marillier[2]

Abstract. Images of the deep crust and upper mantle from BIRPS, Lithoprobe and AGC seismic reflection profiles show a variety of discontinuities interpreted as faults or shear zones. Much of the observed diversity can be attributed to variation in depth of detachment and the position of the structure within the mobile belt. Reactivation of old structures may also be involved, and in places may occur well into the continental 'foreland' marginal to the mobile belt.

Introduction

The Appalachians in eastern North America and the Caledonides in northwest Europe and Greenland were formed during the opening and closure of the early Paleozoic Iapetus Ocean [Wilson, 1966; Dewey, 1969; Stockmal et al., 1987]. The crustal structures so formed were subsequently disrupted during the Mesozoic and Tertiary opening of the present day Atlantic Ocean and the contemporaneous development of its marginal basins [Keen et al., 1987]. The complex tectonic history introduced by superposed orogenic cycles is further complicated by along-strike variability in the timing of both Paleozoic collision events [Phillips et al., 1976; Colman-Sadd, 1982] and Mesozoic rifting [Srivastava, 1978]. While such complexity makes it difficult to propose unique interpretations of seismic profiles, it also provides a rich diversity of lithospheric response from which generalisations may be made about the ways that structures evolve in both convergent and divergent tectonic environments.

In this paper, we review briefly some similarities and contrasts exhibited in deep seismic profiles across the Appalachian - Caledonide belt. The locations of profiles used are shown on the base map of Figure 1, which restores the British Isles and Canada to their approximate relative positions prior to Atlantic opening. Around Britain, use is made of profiles shot for the British Institutions Reflection Profiling Syndicate (BIRPS); in eastern Canada, we use marine seismic data recorded for the Atlantic Geoscience Centre (AGC) of the Geological Survey of Canada as part of their contribution to Lithoprobe and the Frontier Geoscience Program. Parts of lines AGC 84.1,2 [Keen et al., 1986] have been reprocessed on the NSERC/Petro-Canada CONVEX C1 at Memorial University using the Merlin SKS software. A more detailed discussion of the effects of reprocessing on the interpretation of terrane boundaries around Newfoundland will be published elsewhere.

Comparative Geology of Newfoundland and the British Isles

The various structures and terranes described here are shown in Figure 1. The trans-Atlantic correlations follow Williams [1978]. From northwest to southeast, the principal terranes are as follows: (i) foreland; this is of Grenville age in Canada, Archaean in NW Britain (but with some later pre-Paleozoic cover); (ii) foreland capped by early Paleozoic passive margin sequences, over-ridden by thrust stacks of these rocks and some of oceanic affinity (Humber terrane in Newfoundland, northwestern parts of the Highlands in Scotland); (iii) a central mobile belt, bounded by the dashed lines of Figure 1, and which includes a diversity of rocks including arc and back arc complexes (Dunnage, Midland Valley, Southern Uplands) and a complexly deformed continental margin sequence (Gander, Lake District); (iv) a continental platform (Avalon, Anglesey), stable in early Paleozoic time.

An increasing awareness of the significance of strike-slip faults in juxtaposing unrelated 'suspect' terranes [e.g. Soper and Hutton, 1984] should promote caution in attempts to extrapolate deep structures across terrane boundaries in cross sections. With this in mind, generalised sections of both areas on Figure 1 indicate the possible relationships. The specific structures

[1]Centre for Earth Resources Research, Department of Earth Sciences, Memorial University of Newfoundland, St. John's, Newfoundland, Canada A1B 3X7.

[2]Atlantic Geoscience Centre, Geological Survey of Canada, Bedford Institute of Oceanography, Dartmouth, Nova Scotia, Canada B2Y 4A2.

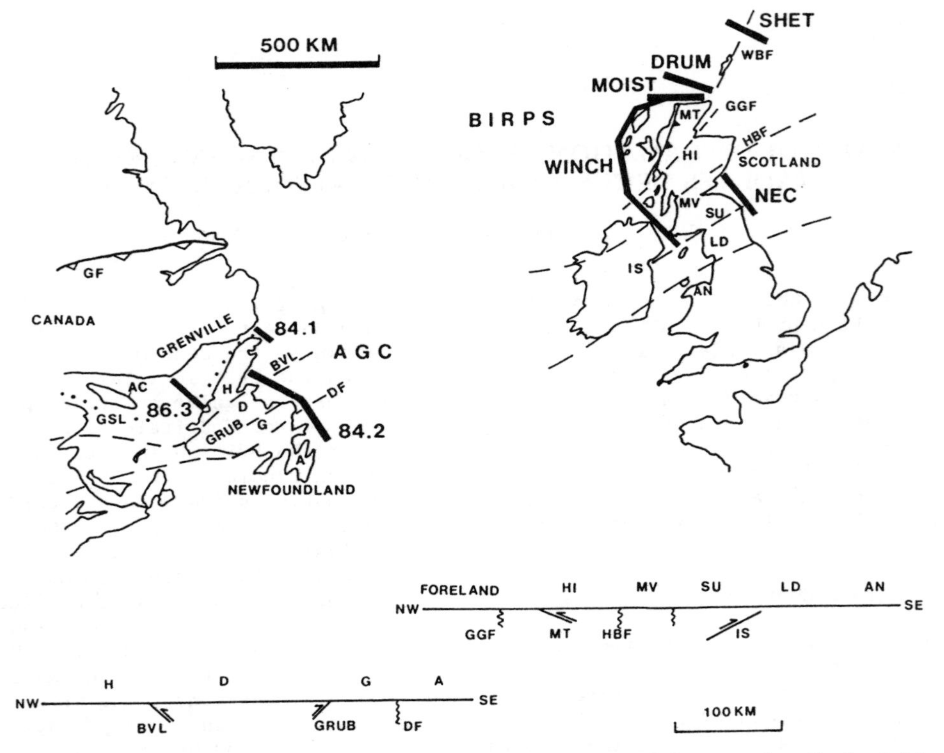

Fig. 1. Map showing locations of seismic reflection profiles used. Dashed lines indicate central mobile belt of the Appalachian/Caledonide system. Dotted line shows northern limit of Appalachian deformation in eastern Canada. Europe and N. America in approximate relative position prior to Atlantic opening. Sections below are generalised views of structural sequence in the two areas considered, from the foreland of the Laurasian continent in the northwest to the edge of the European (Baltic) continent in the southeast. Around Newfoundland, terranes are abbreviated as follows: H = Humber, D = Dunnage, G = Gander, A = Avalon; other features: AC = Anticosti Island, BVL = Baie Verte lineament, DF = Dover fault, GF = Grenville front, GRUB = Gander River Ultrabasic Belt, GSL = Gulf of St Lawrence. Around Britain, terranes are: HI = Highlands, MV = Midland Valley, SU = Southern Uplands, LD = Lake District, AN = Anglesey; other features: GGF = Great Glen fault, HBF = Highland Boundary fault, IS = Iapetus suture, MT = Moine thrust, WBF = Walls Boundary fault.

examined below may be viewed in this broad framework. Two kinds of structure are examined: deep structure associated with strike-slip faults and dipping zones which may be interpreted as dip-slip structures, sometimes associated with known thrusts. Strike-slip faults discussed are the Great Glen fault system in Scotland and the Dover fault in Newfoundland. The linked Great Glen and Walls Boundary faults [Kennedy, 1946; McQuillin et al., 1982] lie within the northwest continental foreland where it is covered by marginal sequences telescoped by compressional structures such as the Moine thrust [Peach et al., 1907]. This fault system cuts southwestwards obliquely across the structural grain in northern Scotland and, immediately to the west of Scotland, reaches the exposed foreland traditionally defined as basement rocks west of the Moine thrust (MT on Fig. 1). The Dover fault [O'Brien et al., 1983] forms the southeast boundary of the mobile belt in Newfoundland, separating Gander and Avalon terranes. The lack of association of these two terranes suggests that the Dover fault is a strike-slip fault with considerable displacement.

In northern Britain the boundary of the northwest continental margin and the arc terranes to the southeast is defined by the Highland Boundary fault [Bluck, 1984]. The Baie Verte lineament [Williams and St-Julien, 1982] is the equivalent structure in Newfoundland. In both areas, ophiolitic rocks representing former oceanic crustal remnants are found north of these two marginal structures, demonstrating the original separation of the arc terranes from the continental forelands. These two bounding structures are steeply dipping at surface and of complex form with repeated movement. The Highland Boundary fault is not imaged on deep seismic profiles, and the deep crust in its vicinity is not imaged well either. The Baie Verte line can be associated with deep structure, and is discussed further below.

The site of the former Paleozoic ocean (Iapetus) is now represented by the Iapetus suture [Phillips et al., 1976], interpreted as separating Southern Uplands and Lake District terranes in Britain, and as the Gander River Ultrabasic Belt (GRUB line) separating Gander and Dunnage terranes in Newfoundland [Jenness, 1958]. It should be noted that neither of these structures can be unambiguously traced along strike. Both these structures are examined below.

The timing of deformation is not directly determinable from the seismic images. Surface mapping defines two broad ages of Appalachian deformation in Newfoundland - Taconic (Ordovician) and Acadian (Silurian-Devonian). The earlier deformation is thought to represent arc-continent collision with the late orogenic event related to the final closure of Iapetus. The equivalent deformational episodes in Britain are described as the Grampian and the Caledonian [Lambert and McKerrow, 1976].

Seismic Evidence for Faults and Shear Zones

Many features seen on deep seismic sections are interpreted as shear zones although the reasons for doing so are not uniformly compelling. Geometrical reasons include (i) coplanarity, or at least parallelism, with known surface structures; (ii) offset of other reflection packages; and (iii) abrupt lateral termination or change of reflectivity. Other arguments are based on mechanical requirements that shear displacement does not die out suddenly and that rocks tend to more ductile behaviour with increasing temperature. Thus, zones of sub-horizontal reflections, some occupying the entire lower crust, are commonly interpreted as detachments because known surface shears appear to sole into them and do not reappear below them. Such interpretations have field analogs in the shallow sub-horizontal detachments found below parts of the extensional Basin and Range terrain [Wernicke, 1985], and the basal thrusts found lying within incompetent strata in convergent structures such as the Moine thrust [Peach et al., 1907].

We use a range of such arguments in this paper, but recognise that these are interpretations and not well substantiated. We also recognise that deep seismic data constrain current structural relationships without necessarily providing much insight into the age of the shears or the sense of displacement across them. A further problem in the interpretation of deep structure is that it is not possible to view the third dimension in an isolated seismic profile. For these reasons there is always a danger that interpretation of deep seismic data is over-simplistic. Nevertheless the approach is useful in providing hypotheses for future testing.

The line drawings which are used here to illustrate deep seismic profiles are deliberately simpler than those originally published and referenced in the appropriate section. The aim for this paper is to illustrate the overall form and dimensions of structures on different profiles, using a common line drawing criterion of amplitude and lateral continuity, and publishing with common scale. Heavier lines overlaid on the line drawings outline interpreted shear zones with the sense of early- to mid-Paleozoic motion. In attempting to synthesise the deformation in the two areas, some differences may be noted from earlier interpretations. The alternatives may not necessarily be discriminable from the data: the same series of reflectors may be identified, but the sequence of movements is open to debate.

The following sections address three observed patterns in deep seismic data: those related to known strike-slip faults; those apparently associated with compressional tectonics; and those which are not simply explained by commonly-used analogs.

Seismic Images across Strike-Slip Faults

Strike-slip faults often have rather straight courses in plan view, appear to dip steeply, and often have large offsets across them. These characteristics might suggest that such faults cut completely through the lithosphere. This is not universally the case as has been argued by Hall [1986] on seismic grounds and by Beaumont et al. [in press] on the grounds of the flexural consequences implied by such lateral decoupling of the lithosphere.

As shown on Figure 1, the Great Glen fault system crosses mainland Scotland. The main (Paleozoic) displacement on the fault has been variously estimated at, for example, 100 km sinistral [Kennedy, 1946], 2000 km sinistral [Van der Voo and Scotese, 1982], 100 km dextral [Garson and Plant, 1972]. Post-Paleozoic movements involving oblique slip are relatively modest [McQuillin et al., 1982]. The fault system has quite a different aspect on seismic profiles across its marine extensions to the southwest and the northeast. Figure 2 shows a line drawing of the WINCH profile, off the west coast of Scotland [Brewer et al., 1983; Hall et al., 1984]. The fault is seen here in the Sea of the Hebrides west of the Moine thrust, that is, in the foreland. The fault is still conspicuous in its effects near surface, where variable post-Caledonide reactivation is implied by the fault zone defining a Mesozoic graben overlain by

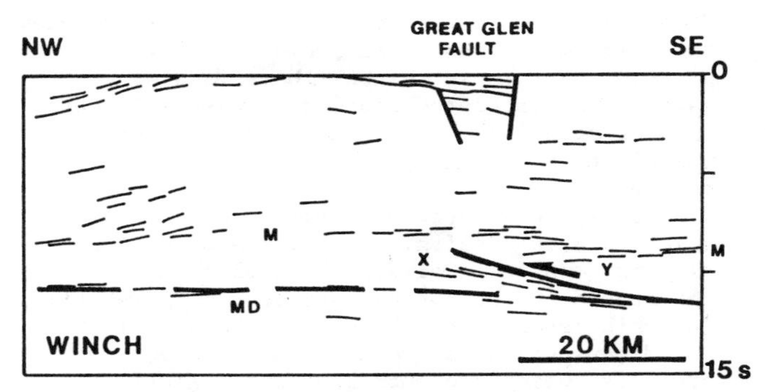

Fig. 2. Line drawing (thin lines) and interpretation (thick lines) of part of the migrated WINCH seismic profile across the Great Glen fault based on Hall [1986]. The Great Glen fault zone though conspicuous at surface does not appear to affect the Moho (M), below which a mantle detachment (MD) and rising thrust (between X and Y) are unaffected by the fault. This suggests that the movement on the fault here is not large and what there is dies out downwards so as to have no significant effect below the crust.

a Tertiary half-graben. However at depth the fault does not appear to affect sub-crustal reflectors. 500 km to the northeast, the fault zone is observed as the marine extension of the Walls Boundary fault in the Shetland Isles, imaged on the SHET profile [McGeary, 1987; and Fig. 3]. High amplitude diffraction patterns at Moho level but directly below the surface position of the fault may be taken to imply that this portion of the fault extends at least through the crust. Such along-strike variability of the fault system led Hall [1986] to suggest that it dies out to the southwest, as the foreland is approached. The variable amount of displacement may be taken up by reactivation in oblique movement on older, cross-cutting mid-crustal thrusts, such as the Moine thrust which lies between the two sections cited here.

Significant strike-slip faults are also recognized in the Appalachians in Newfoundland. The most compelling example seen on seismic profiles is the Dover fault which separates Gander and Avalon terranes at the eastern side of the orogen [O'Brien et al., 1983]. Extrapolation of this feature a short distance offshore places it immediately above a pronounced change in the seismic character of the lower crust [Keen et al., 1986], suggesting that the Dover fault penetrates the entire crust. Occasionally, lateral change in deep seismic character can be attributed to artifacts of the shallow section [Hirn et al., 1987]: this is not the case here because the reflection amplitude envelope has quite different character, and there is nothing in the near-surface section which indicates lateral change. This profile does not resolve the issue of whether the Dover fault penetrates the entire lithosphere. Reprocessing has emphasised the contrast between terranes quite dramatically (Fig. 4). The lower crust below the Avalon terrane has a broad zone of diffuse layering between 10 and 13 seconds two-way-time (TWT), or approximately 30 - 40 km depth. In contrast the lower crust below the Gander terrane is characterized by a narrow zone of reflectivity lying between 10 and 11 seconds TWT with one particularly strong reflector being dominant. Although this character has elsewhere been considered typical of young extensional terranes [Nelson et al., this volume], such an interpretation is untenable here. This is because the latest documented extensional event is associated with Mesozoic rifting of the Avalon terrane lithosphere to form the Atlantic ocean. It is difficult to see how this extension could have affected the Moho of the adjacent Gander terrane but not Avalon Moho.

The following conclusions may be drawn:
(i) Strike-slip faults can cut through the entire crust (exemplified by the Walls Boundary fault and the Dover fault).
(ii) Strike-slip faults do not necessarily cut through the entire crust (exemplified by the Great Glen fault, which appears to die out both laterally and with depth).
(iii) Although strike-slip faults may possibly cut the entire lithosphere, there is no direct seismic reflection evidence for this.
(iv) Well-defined, narrow bands of horizontal reflectors from Moho depths need not imply rejuvenation of the Moho by the most regionally important extensional event.

Seismic Images in Overthrust Terranes

The Appalachian-Caledonide belt includes a number of well-documented thrusts with large

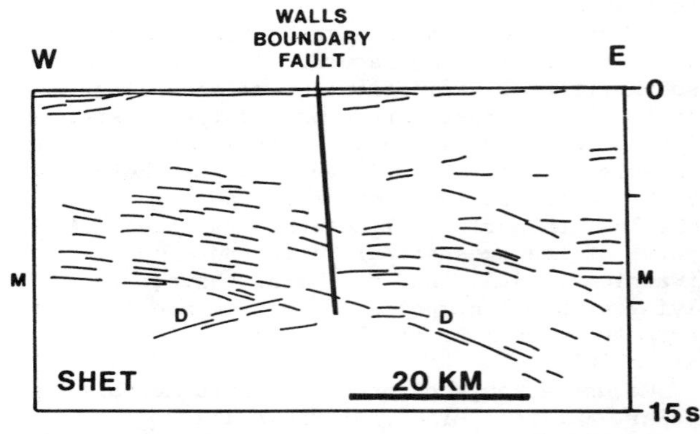

Fig. 3. Line drawing and interpretation of the unmigrated SHET seismic profile across the Walls Boundary fault. Diffraction patterns (D) suggest that the fault zone cuts the Moho (M) as a sharp feature, vertically below the surface position of the fault.

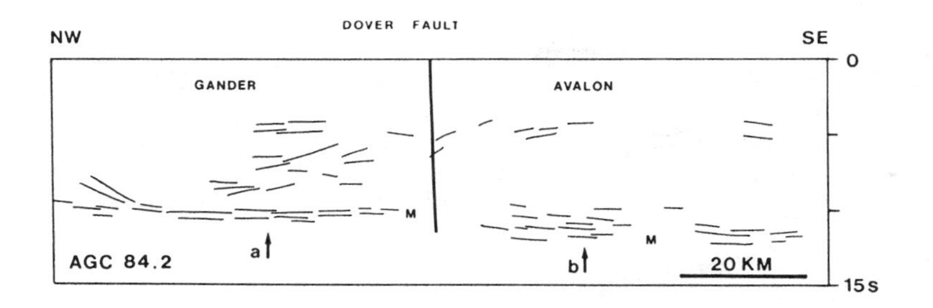

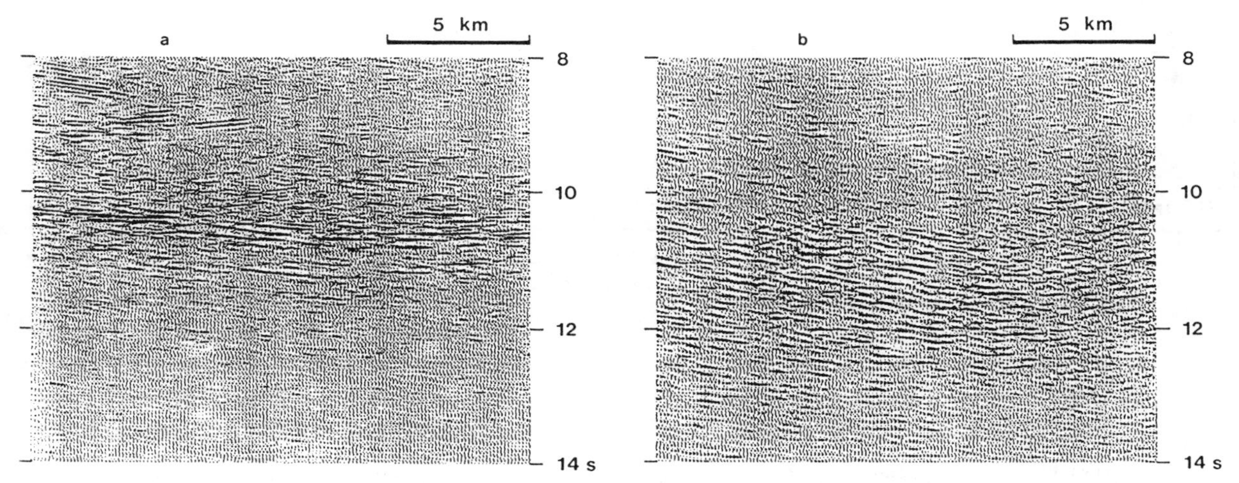

Fig. 4. Line drawing of part of the unmigrated seismic profile AGC 84.2 across the Dover fault which separates Gander and Avalon terranes. The fault appears to separate quite different deep crustal signatures in the two terranes. Base of crust = M. Two segments of migrated deep seismic data, one from each of the two sides of the fault, show the contrast in seismic character of the two terranes.

displacements. Deep seismic sections (Figs. 5-8) show that a variety of deep structures may be associated with thrusts mapped at the surface. Other geometrically related structures also exist and might also be thrusts.

In northern Britain, the Moine thrust has traditionally been viewed as the western limit of Caledonian deformation. Recent work, particularly relating isotopic signatures to structure within the mobile belt, shows that the

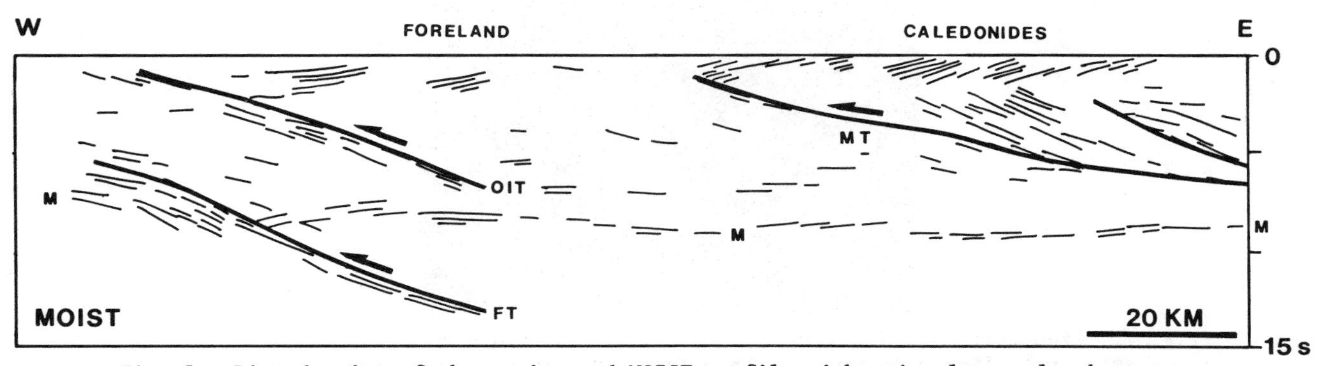

Fig. 5. Line drawing of the unmigrated MOIST profile with major low angle shears identified. MT = Moine thrust; OIT = Outer Isles fault; FT = Flannan fault. M = base of crust. Note westerly dipping half graben in near surface formed by late Paleozoic and Mesozoic reactivation in extension of deep shears. Half arrows indicate interpreted Caledonide movement.

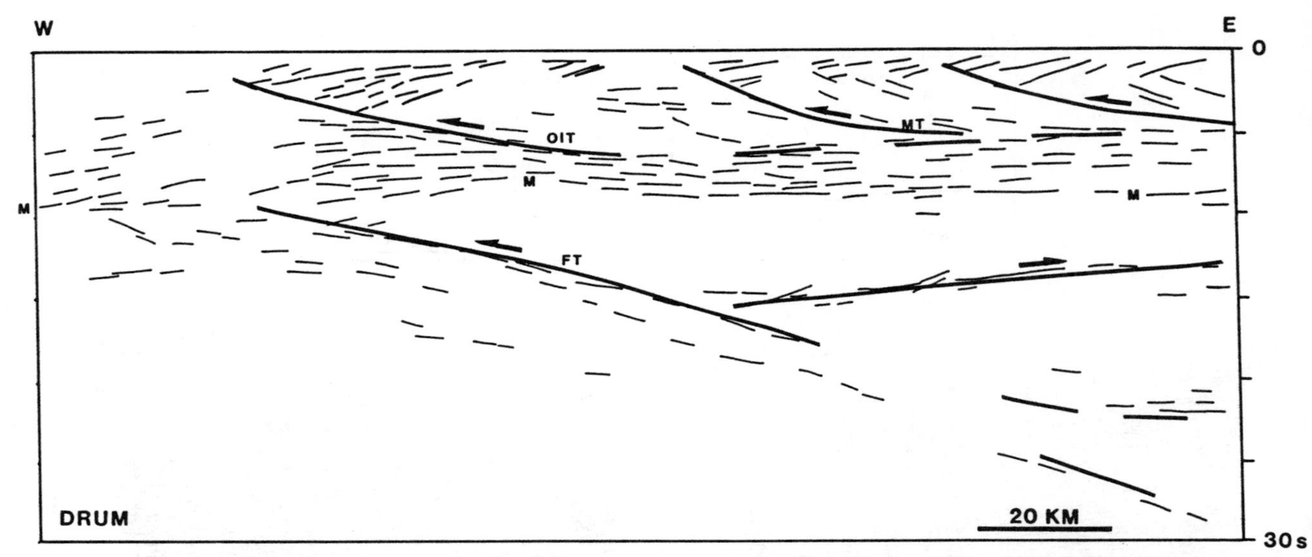

Fig. 6. Line drawing of the unmigrated DRUM seismic profile, with major dislocations noted. OIT = Outer Isles fault; FT = Flannan fault; M = Moho. Half arrows indicate interpreted Caledonide movement.

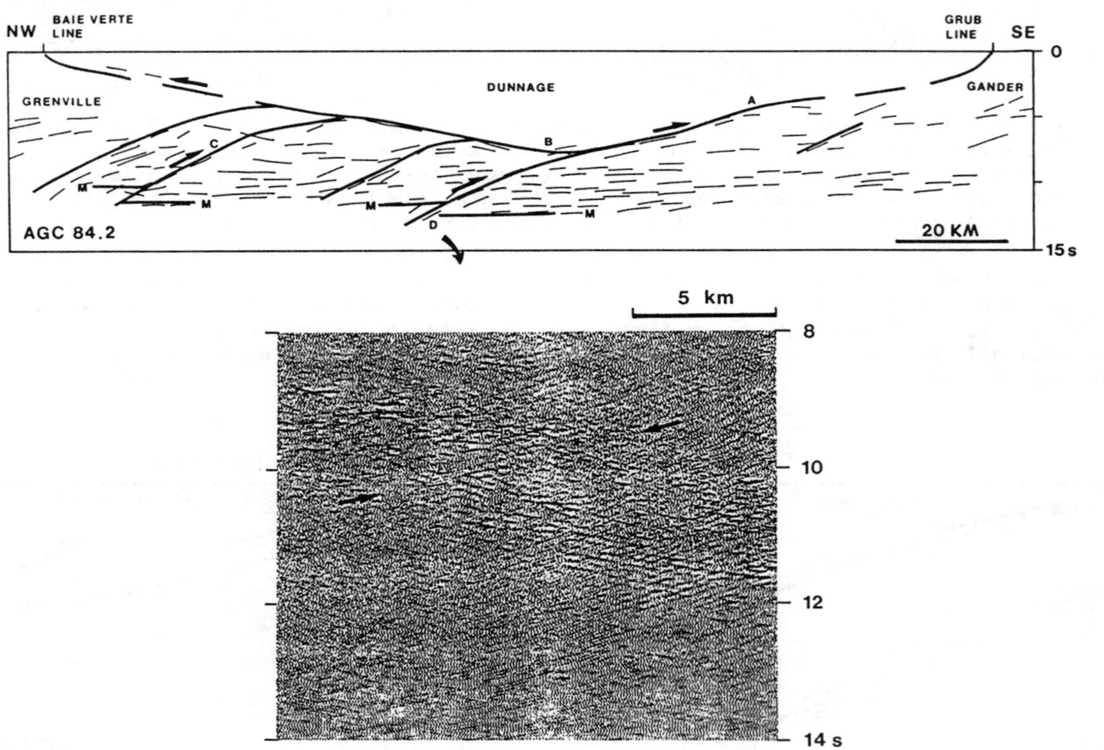

Fig. 7. Line drawing and interpretation of the unmigrated profile AGC 84.2. GRUB line appears to link a continuous shear from surface down to mantle. Duplex structure in lower crust appears to retain reversed sense of displacement of base of crust (M). Seismic section below gives view of migrated data showing this (arrows indicate position of shear). Baie Verte line may be older and be partly reactivated during GRUB line movement. A-D describe features referred to in text. Half arrows indicate interpreted Appalachian movement.

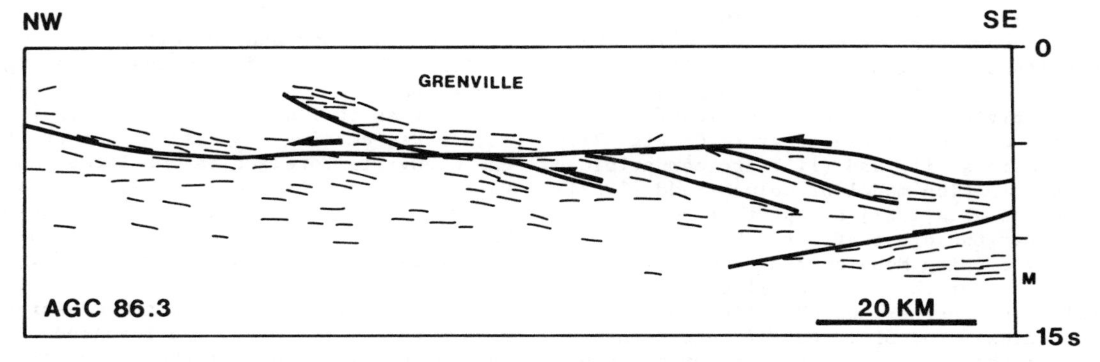

Fig. 8. Part of the migrated seismic profile AGC 86.3, showing apparent mid-crustal detachment with thrusts rising off it into the upper crust. Near surface, the Anticosti basin is too poorly imaged to show whether it is affected by these structures. At southeast end of section westerly dipping structure cutting Moho may be equivalent of similar structures on line 84.2, and is shown in more detail in Figure 10. Half arrows indicate interpreted Appalachian movement.

rocks immediately to the east of the thrust preserve some pre-Caledonian ages rather than being wholly reworked in the Grampian orogeny [van Breemen et al., 1978]. It is only many tens of kilometres into the belt that the rocks become universally overprinted by Caledonian effects. The seismic sections of Figures 5 and 6 show that the Moine thrust may be only one of a set of similarly-oriented structures [Smythe et al., 1982; McGeary and Warner, 1985; Brewer and Smythe, 1986]; others, such as the Outer Isles fault and the Flannan fault, lie to the west of the supposed limit of Caledonian deformation. There is no conclusive evidence for the earliest movements on these structures, which may have been reactivated during Caledonian convergence and subsequently during late Paleozoic and/or Mesozoic extension. For example the Outer Isles fault is considered to have an early Proterozoic history from intersection relationships with the Lewisian rocks it affects where exposed [Fettes and Mendum, 1987]. What is clear from this is that the traditional view of a sharp boundary to Caledonide reworking and deformation is no longer tenable. Westward of the zone of wholesale reworking (a few km west of the Great Glen fault), strain becomes more inhomogeneous with increasing concentration into shear zones and then thrusts, so that the final effects may be represented by isolated faults within what was traditionally regarded as the undeformed foreland (Fig. 6). That nothing has been previously observed in surface geology to tie with such foreland deformation may be attributed to its concentration on discrete structures interpreted as older, and the modesty of its effect on crustal thickening, which precluded obvious metamorphic re-equilibration. The horizontal range over which this variation of strain pattern may occur is a few kilometres short of the distance from the Great Glen fault to the location of the Outer Isles fault, which is over 200 km.

In eastern Canada, the Appalachian belt follows a sinuous course from the mainland across the Gulf of St Lawrence and then across Newfoundland to the Atlantic margin. The relationship of the Newfoundland Appalachians to equivalent structures to the southwest is examined elsewhere [Stockmal et al., 1987; Marillier et al., this volume, and in press]. Here we restrict attention to structures seen immediately around Newfoundland and which can be more easily related with structures mapped there. The two principal convergent structures here are the Baie Verte lineament [Williams and St-Julien, 1982] and the Gander River Ultrabasic Belt (GRUB line), [Jenness, 1958], both complex structures formed during the closure of Iapetus and the juxtaposition of its borderlands.

Deep seismic reflection profile 84.2 [Keen et al., 1986] from northeast of Newfoundland (Fig. 7) indicates that neither the Baie Verte lineament nor the GRUB line can be directly linked with deep structure. The upper section of the seismic data is very noisy and there are few coherent reflectors. The few that there are can be used to suggest that the two structures converge on shallowly dipping detachments which isolate the Dunnage terrane as an upper-crustal allochthon sitting on Grenville and Gander (eastern central block, sensu Stockmal et al. [1987]) basements. Although definitive links between surface and deep structure are lacking because of severe water-bottom multiple problems, a number of interpretations based on this weak correlation is possible. (i) A weakly defined single reflector rising from the top of a deep crustal layered structure may connect to the GRUB line (feature A of Fig. 7). (ii) The GRUB line may level out in the mid-crust to form the top of a detachment (feature B, Fig. 7) which then rises towards the west. This detachment may continue in the mid-crust far to the north-west. This may be a very old structure extending into the Grenville foreland and having a long history of repeated movement (see below). The same mid-crustal shear zone may be one into which the Baie Verte lineament also detaches. (iii) The GRUB line may extend into a complex duplex structure involving the whole of the lower crust and some of the upper mantle (feature C, Fig. 7).

Interpretations (ii) and (iii) are compatible with the delamination model of Colman-Sadd [1982] and Stockmal et al. [1987] except that here we do not assert that the crust affected by the northwesterly dipping Moho-cutting shears is necessarily part of the old Grenville foreland.

Regardless of whether the GRUB line merges with this duplex structure, the reflectors that define the north-west dipping slices cut through the sub-horizontal lower crustal layering and the base of the crust. The most obvious of these, feature D on Figure 7, preserves a sense of reverse displacement of the Moho (assumed to coincide with the base of the lower crustal layering). This seems a clear example of a situation in which more recent extension, associated with Mesozoic opening of the Atlantic, has not undone the effects of earlier Paleozoic convergence.

It should be mentioned here that Figure 7 represents a line drawing of an unmigrated section, so that dipping reflectors are not in their true position relative to horizontal ones. Migration of this section shows that the displacement of lower crustal layering remains and the dipping shear is moved somewhat up-dip: the interpretation remains unchanged.

The extension of the Baie Verte line to depth is less clear. A patchy zone of west dipping segments is taken to mark the Baie Verte line as the top of the lower crustal duplex described above, but the connection with the surface structure is still obscure. There is a near-surface dipping reflector in the correct position to connect with the Baie Verte line, but this may just be the base of a known Carboniferous basin. If that is so, the basin could have formed in the hanging wall of a later extension reactivating the Baie Verte structure, in a manner analogous to the formation of basins by Mesozoic extension of the Moine thrust in NW Britain [Smythe et al., 1982]. Reprocessing on this section continues in attempts to clarify this problem. If the Baie Verte line does descend into this part of the crust, it is likely from Figure 8 that the structure into which the thrust soles continues at mid-crustal level much further north-west, as noted above. Such continuation is both compatible with the presence of Taconic or Acadian thrusting in the Humber terrane [Williams and Cawood, 1986] and with further evidence of mid- to upper-crustal detachment seen in the Grenville elsewhere, especially on AGC 86.3 (Fig. 8). Here a sub-horizontal mid-crustal reflective zone spawns reflection packets rising into the upper crust. These look very much like thrust horses, and are interpreted as such. Their age is uncertain.

Thus a variety of shear zones combines to slice up the Appalachian belt below AGC lines 84.1,2. The times of displacement along the shear zones can be assessed from the seismic evidence as follows: (i) The NW dipping structures cutting the base of the crust below the Grenville foreland may be later than the Baie Verte line, and its possible mid-crustal relations in the Grenville. This conclusion is based on the possible continuity of the GRUB line structure down through the crust and into the mantle. (ii) On AGC line 86.3, the south-east dipping thrusts rising from the mid-crustal detachment in the Grenville foreland would affect the sedimentary rocks of the Anticosti basin near surface, if the thrusts have Paleozoic movement. These rocks are not well imaged on line 86.3, but show no evidence of major thrusting here or elsewhere. These thrusts in the Grenville would thus appear to be older structures than those further south-east, for instance the north-west dipping ones which cut the Moho (see Fig. 8 for example), but may have been reactivated to some extent later, at least as far north-west as the Baie Verte line and probably into the Humber terrane, but not necessarily into the Grenville below the Gulf of St Lawrence.

Lower Crustal 'Wedges' and the Iapetus Suture

Around Britain, there are two examples of thrust-like structures in the lower crust which appear not to cut the Moho (Fig. 9). In each case the Moho appears to continue beyond the toe of the wedge formed by it and the thrust. In both these cases [Brewer et al., 1983; Klemperer and Matthews, 1987], the structure is likely to be associated with the Iapetus suture, though the reflection character of the lower crustal layering changes from one example to the other. The shear zone associated with a major plate boundary would be expected to cut through the Moho into the mantle. No evidence of this is present here, though a possibly associated conducting layer may continue down into the mantle [Beamish and Smythe, 1986]. The lack of evidence of Moho displacement may be the result of later oppositely-directed shearing balancing out the earlier thrust displacement, circumstantial evidence for this existing in the presence of the Peel basin in the hanging wall of the Iapetus suture on WINCH [Hall et al., 1984]. Other possibilities are discussed below.

On line AGC 84.2, the northwesterly dipping zone linking the GRUB line with the lower crustal duplex may be the Newfoundland equivalent of the suture. However in this case the Moho is definitely cut by the shears. A wedge-like structure appearing on the southeastern edge of AGC line 86.3 (Fig. 10) could be another such structure but the origin and interpretation of it are even more unclear. Possible links between the two AGC lines 84.2 and 86.3 may be clarified

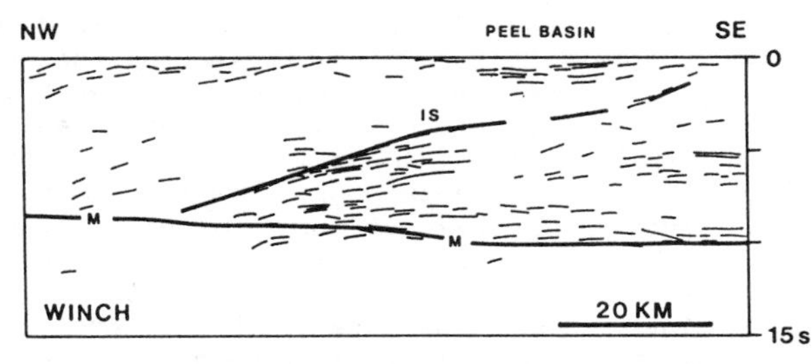

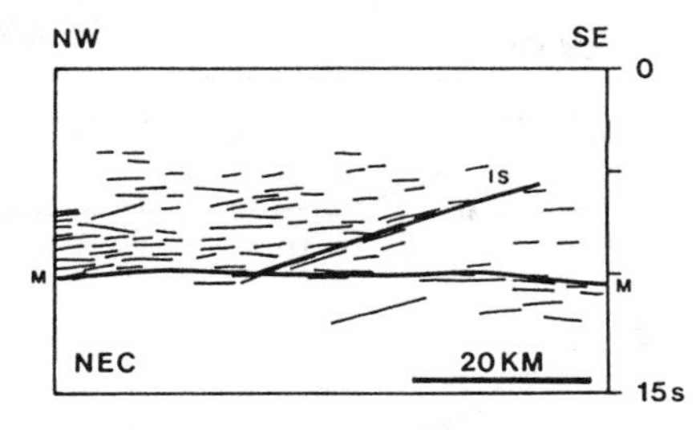

Fig. 9. Examples of lower crustal wedges associated with the Iapetus suture (IS) around Britain. Figure 9a shows an unmigrated part of the WINCH profile, in which the suture appears to be truncated by the base of the crust (M). Updip the suture contains the Carboniferous Peel basin in its hanging wall. This may indicate later reactivation as an extensional shear. Figure 9b shows an unmigrated part of the NEC line, along strike from WINCH. Here the suture juxtaposes reflective lower crust above less reflective lower crust (in contrast with the case on WINCH).

in 1989 by the shooting of the LITHOPROBE East line on Newfoundland.

Elsewhere [e.g. Hauser et al., 1987] dipping features in the lower crust truncated by a level Moho have been taken to indicate the relative youth of the Moho. We see no circumstantial evidence for this here. We intend to model these features structurally in order to clarify some of the interpretative problems presented by them. For example, the possibility exists that the shear zones turn to follow the Moho, which might be expected to be a zone of stress concentration. A change of dip with a radius of curvature similar to depth could produce an apparently sharp change of dip on an unmigrated profile. This would be one means of explaining the wedges without requiring post-shear Moho rejuvenation.

Discussion

A diversity of deep structure is observed on profiles across the Appalachian and Caledonide belts in north-west Britain and around Newfoundland. Some general conclusions can be drawn from this.

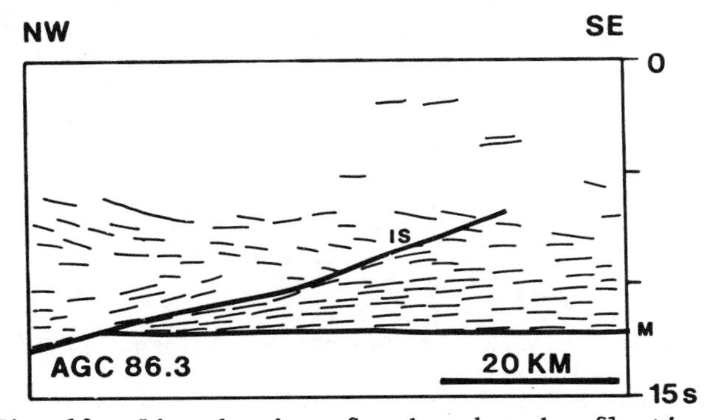

Fig. 10. Line drawing of wedge shaped reflective zone in the lower crust, seen on the migrated profile AGC 86.3. In this case the dipping structure (IS) appears to truncate the Moho (M) to the east and continues downdip to the west, where it may run below or along the base of the crust.

Strike-slip faults in the upper crust do not necessarily cut through the crust and into the mantle without some change of style. The Great Glen fault appears to die out downwards and outwards from the mobile belt as it is traced towards the foreland; displacement may be taken up by reactivation of earlier thrusts. Diffraction patterns observed at Moho level below the Walls Boundary fault indicate a sudden truncation of layering at the edge of the fault zone. The zone must have sharp edges (a few hundred metres across only) and a width of a few kilometres: this is very similar to the surface expression of the Great Glen fault. The Dover fault also appears to descend directly into the mantle; no diffraction patterns are observed, but the width of the shear zone at Moho level is not greater than 5 km.

Shear zones attributed to convergence detach at various depths. In the Grenville and below the equivalent Moines of the northern Highlands in Scotland, we suggest that thrusts may detach in the upper or middle crust. Some structures (Outer Isles fault, Flannan fault) may detach in the lower crust. By contrast, a possible lower crustal extension of the GRUB line may appear to have a duplex form in the lower crust with the sole (unobserved) in the mantle. This variability of behaviour could result from differences in temperature or composition, both of which may radically affect rheology [Kusznir and Park, 1986]. One possibility of explaining the deep structure of the GRUB line is to hypothesise that the lower crust here is part of the old Grenville passive margin and therefore may have been subjected to underplating during the rifting stage. This would have the effect of changing the overall composition of the lower crust to something much more mafic, so that the behaviour of quartz would no longer dictate the rheology: more brittle behaviour would then

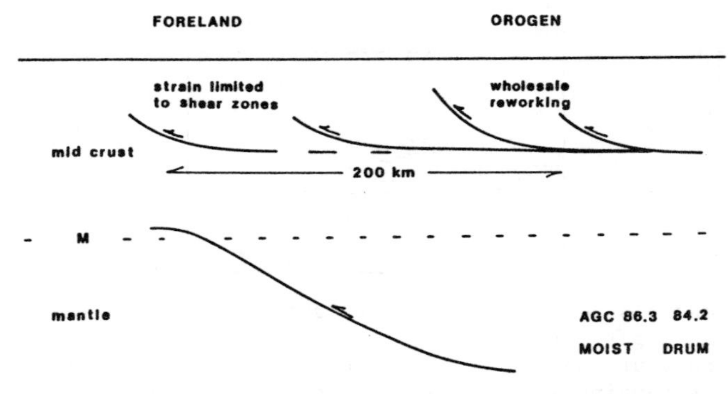

Fig. 11. Cartoon illustrating the possible extent of deformational advance by thrusting ahead of the mobile belt. Shearing may advance both along a detachment in the crust and also by displacement in the mantle.

result, i.e. there would be a lesser tendency to detach at this level.

Reactivation of old structure is seen repeatedly in these data. The Moine thrust is suspected of having an early history prior to the main late Caledonian (Acadian?) event; later Mesozoic extension has reversed the sense of displacement on the thrust. In the Newfoundland area, the Grenville foreland is characterised by a mid-crustal detachment, probably older than the Anticosti foreland basin, and possibly linking with the Grenville Front. But the later Baie Verte line and other thrusts to the west of it may well sole into that older structure. The GRUB line is probably younger still and may be deflected into the old mid-crustal shear which, for a short distance forms the top of a lower crustal duplex which carries the GRUB line displacement deeper into the lithosphere.

The most significant conclusion from this is that the margin of the mobile belt can no longer be readily associated with just one dominating structure. An orogenic 'front' may be diffused across a wide zone, in which the strain appears to become less homogeneous as the foreland is approached, with a number of related thrusts reaching forward into the foreland by distances of many kilometres beyond the traditionally-held limit of mobile belt deformation (Fig. 11).

If shearing in a reverse fault sense propagates far into the foreland, then crustal thickening must do likewise. A consequence of this is that the crustal thickening also may not terminate abruptly, but extend with gradual diminution into the foreland. The absence of an obvious foreland basin in some parts of the Caledonides, e.g. in north-west Britain, may be a result of this: even modest elevation may be enough to drive sediments further across the foreland or, equally likely, along strike to areas where less foreland propagation has occurred.

There is no evidence of post-Paleozoic extensional structure or reactivation in the data we have considered around Newfoundland, despite the large extensions witnessed by the basins further out on the Grand Banks. The change of crustal signature across the Dover fault is likely to be at least as old as Paleozoic. Reverse displacement is preserved in shears cutting the base of the crust associated with the GRUB line. Only one kind of structure shows any sign of Moho readjustment: this is the wedge of lower crust associated with the Iapetus suture. Either the earlier reverse movement has been just about exactly matched by later extensional movement, or the shear turns suddenly to run along the base of the crust [Klemperer and Matthews, 1987], or the Moho has been rejuvenated. No answer to this choice can be made from these data at present.

Acknowledgments. Copies of BIRPS sections were provided by the Natural Environment Research Council at cost vie the British Geological Survey; earlier work on them by Hall was assisted by a grant from Shell UK Expro. The seismic processing centre at Memorial University is funded by NSERC and Petro-Canada.

This paper is LITHOPROBE contribution number 49, and Geological Survey of Canada contribution number 18288.

References

Beamish, D., and Smythe, D.K.,Geophysical images of the deep crust: the Iapetus suture, J. Geol. Soc., 143, 489-497, 1976.

Beaumont, C., Quinlan, G., and Stockmal,G., The evolution of the western interior basin: causes, consequences and unsolved problems, Spec. Paper Geol. Ass. Can., (in press), 1988.

Bluck, B.J., Pre-Carboniferous history of the Midland Valley of Scotland, Trans. R. Soc. Edinb.: Earth Sci., 75, 275- 295, 1984.

Brewer, J.A., Matthews, D.H., Warner, M.R., Hall, J., Smythe, D.K., and Whittington, R.J., BIRPS deep seismic reflection studies of the British Caledonides, Nature, 305, 206-210, 1983.

Brewer, J.A., and Smythe, D.K., Deep structure of the foreland to the Caledonian orogen, NW Scotland: results of the BIRPS WINCH profile, Tectonics, 5, 171-194, 1986.

Colman-Sadd, S.P., Two stage continental collision and plate driving forces, Tectonophys., 90, 263-282, 1982.

Dewey, J., Evolution of the Appalachian/Caledonian orogen, Nature, 222, 124-129, 1969.

Fettes, D.J., and Mendum, J.R., The evolution of the Lewisian complex in the Outer Hebrides, In Park, R.G. and Tarney,J. (eds), Evolution of the Lewisian and comparable Precambrian high grade terrains, Geol. Soc. Spec. Publn. No. 27, 27-44, 1987.

Garson, M.S., and Plant, J.A., Possible dextral movements on Great Glen and Minch faults in Scotland, Nature Phys. Sci., 240, 31-35, 1972.

Hall, J., Brewer, J.A., Matthews, D.H. and Warner, M.R., Crustal structure across the Caledonides from the WINCH seismic reflection profile: influences on the evolution of the Midland Valley of Scotland, Trans. R. Soc. Edinb. Earth Sci., 75, 97-109, 1984.

Hauser, E., Potter, C., Hauge, T., Burgess, S., Burtch, S., Mutschler, J., Allmendinger, R., Brown, L., Kaufman, S. and Oliver, J., Crustal structure of eastern Nevada from COCORP deep seismic reflection data, Bull. Geol. Soc. Am., 99, 833-844, 1987.

Hirn, A., Damotte, B., Torreilles, G. and ECORS Scientific Party, Crustal reflection seismics: the contributions of oblique, low frequency and shear wave illuminations, Geophys. J. R. Astron. Soc., 89, 287-296, 1987.

Jenness, S.E., Geology of the Gander River ultrabasic belt, Newfoundland, Geological Survey of Newfoundland, Report 11, 1958.

Keen, C.E., Keen, M.J., Nichols, B., Reid, I., Stockmal, G.S., Colman-Sadd, S.P., O'Brien, S.J., Miller, H., Quinlan, G., Williams, H., and Wright, J., A deep seismic reflection profile across the northern Appalachians, Geology, 14, 141-145, 1986.

Keen, C.E., Stockmal, G.S., Welsink, H., Quinlan, G., and Mudford, B., Deep crustal structure and evolution of the rifted margin northeast of Newfoundland: results from Lithoprobe East, Can. J. Earth Sci., 24, 1537-1549, 1987.

Kennedy, W.Q., The Great Glen fault, Q. Jl. Geol. Soc. Lond., 102, 41-72, 1946.

Klemperer, S.L., and Matthews, D.H., Iapetus suture located beneath the North Sea by BIRPS deep seismic reflection profiling, Geology, 15, 195-198, 1987.

Kusznir, N.J., and Park, R.G., Continental lithosphere strength: the critical role of lower crustal deformation, In Dawson, J.B., Carswell, D.A., Hall, J. and Wedepohl, K. (eds), The nature of the lower continental crust, Spec. Publn. Geol. Soc., 24, 79-93, 1986.

Lambert, R. St.J., and McKerrow, W.S., The Grampian orogeny, Scott. J. Geol., 12, 271-292, 1976.

Marillier, F., Keen, C.E., Stockmal, G.S., Quinlan, G., Williams, H., Colman-Sadd, S.P., and O'Brien, S.J., Crustal structure and surface zonation of the Canadian Appalachians: implications of deep seismic reflection data, Can. J. Earth Sci., in press, 1988.

McGeary, S., Non-typical BIRPS on the margin of the northern North Sea: the SHET survey, Geophys. J. R. Astron. Soc., 89, 231-238,1987.

McGeary, S., and Warner, M.R., Seismic profiling the continental lithosphere, Nature, 317, 795-797, 1985.

McQuillin, R., Donato, J.A and Tulstrup, J., Development of basins in the Inner Moray Firth and the North Sea by crustal extension and dextral displacement of the Great Glen Fault, Earth Planet. Sci. Lett., 60, 127-139, 1982.

O'Brien, S.J., Wardle, R.J., and King, A.F., The Avalon zone: a pan-African terrane in the Appalachian orogen in Canada, Geol. J., 18, 185-222, 1983.

Peach, B.N., Horne, J., Gunn, W., Clough, C.T., Hinxman, L.W., and Teall, J.J.H., The geological structure of the NW Highlands of Scotland, Mem. Geol. Surv. U.K., 1907

Phillips, W.E.A., Stillman, C.J. and Murphy, T., A Caledonian plate tectonic model, J. Geol. Soc., 132, 579-609, 1976.

Smythe, D.K., Dobinson, A., McQuillin, R., Brewer, J.A., Matthews, D.H., Blundell, D.J. and Kelk, B., Deep structure of the Scottish Caledonides revealed by the MOIST reflection profile, Nature, 299, 338-340, 1982.

Srivastava, S.P., Evolution of the Labrador Sea and its bearing on the early evolution of the North Atlantic, Geophys. J. R. Astron. Soc., 52, 313-357, 1978.

Stockmal, G.S., Colman-Sadd, S.P., Keen, C.E., O'Brien, S.J., and Quinlan, G., Collision along an irregular margin: a regional plate tectonic interpretation of the Canadian Appalachians, Can. J. Earth Sci., 24, 1098-1107, 1987.

Van Breemen, O., Halliday, A.N., Johnson, M.R.W., and Bowes, D.R., Crustal additions in late Precambrian times, In Bowes, D.R., and Leake, B.E. (eds), Crustal evolution in northwestern Britain and adjacent regions, Spec. Issue Geol.J., No. 10, 81-102, 1978.

Van der Voo, R., and Scotese, C., Palaeomagnetic evidence for a large (2000 km) sinistral offset along the Great Glen fault during Carboniferous times, Geology, 9, 583-589, 1982.

Wernicke, B., Uniform sense simple shear of the continental lithosphere, Can. J. Earth Sci., 22, 108-125, 1985.

Williams, H., Geological development of the northern Appalachians: its bearing on the evolution of the British Isles, In Bowes, D.R., and Leake, B.E. (eds), Crustal evolution in northwestern Britain and adjacent regions, Spec. Issue Geol. J., No. 10, 1-22, 1978.

Williams, H., and Cawood, P.A., Relationships along the eastern margin of the Humber Arm allochthon between Georges Lake and Corner Brook, western Newfoundland, Current Res. Geol. Surv. Can., Paper 86-1A, 759-765, 1986.

Williams, H. and St-Julien, P., The Baie Verte - Brompton line: early Paleozoic continent-ocean interface in the Canadian Appalachians, In St-Julien, P., and Beland, J. (eds), Major structural zones and faults in the northern Appalachians, Geol. Ass. Can. Spec. Paper, 24, 177-207, 1982.

Wilson, J.T., Did the Atlantic close and then reopen, Nature, 211, 676-681, 1966.

LATERALLY PERSISTENT SEISMIC CHARACTERISTICS OF THE LOWER CRUST: EXAMPLES FROM THE NORTHERN APPALACHIANS

Francois Marillier, Charlotte E. Keen, and Glen S. Stockmal

Geological Survey of Canada, Atlantic Geoscience Centre, Bedford Institute of Oceanography, P.O. Box 1006, Dartmouth, Nova Scotia, B2Y 4A2, Canada

Abstract Over 1600 km of deep seismic data were recently acquired to probe the crust and the upper mantle of the Appalachians in Canada. Although it is usually difficult to relate the surface expression of tectonic provinces to characteristic seismic signatures in the crust, our data show consistent seismic patterns in the lower crust, over distances of several hundreds of kilometres. These patterns enable us to identify three lower crustal blocks which underlie the orogen. In some places, boundaries between crustal blocks are identified by consistent differences in the seismic character of the Moho reflections. We suggest that one of these block boundaries (between the Central and Avalon crustal blocks) corresponds to a fault at the surface which has been identified as the Gander-Avalon terrane boundary.

High reflectivity of the lower crust and/or bright elongated Moho reflections are observed along some profiles where velocities higher than 7.0 km/s have been measured by earlier refraction studies. Although similar observations have been associated with extensional terranes by many authors, our data indicate that they can also be found in convergent environments.

Introduction

In 1984 and 1986 a total of 1620 km of deep multichannel seismic reflection data were gathered northeast of Newfoundland and in the Gulf of St. Lawrence area, as part of the Canadian LITHOPROBE and Frontier Geoscience Programs. The seismic profiles (Fig. 1) sample the eastern edge of the Grenville Craton, as well as all the major tectonostratigraphic zones of the Northern Appalachians: Humber; Dunnage; Gander; Avalon; and Meguma [Williams, 1979; Williams and Hatcher, 1983].

Long range refraction profiles collected in the offshore Canadian Appalachians in the 60's [Dainty et al., 1966], provide information on the distribution of the average crustal seismic velocities, as well as crustal thickness, but it is difficult to relate these results to the near-surface geology. The deep reflection seismic data enables us to address this problem more effectively.

The first interpretations of the reflection data [Keen et al., 1986; Marillier et al., in press] led to a model of the crust in which three lower crustal blocks underlie the Canadian Appalachians. They were referred to as Grenville, Central and Avalon. These blocks and their boundaries were identified by consistent seismic patterns in the lower crust, as well as consistent differences in the seismic character of the reflection Moho. In this paper we review these features and address the question of lateral persistence of seismic characteristics in more detail.

The Lower Crustal Blocks

The Grenville Block

Four profiles sample the Grenville zone and cross the Appalachian deformation front: 86-2, 86-3, 86-4 and 84-1 (Fig. 1). West-dipping reflections are observed along profiles 84-1 and 84-2; they are labeled A in Fig. 2. Clusters of these reflections, which in some places extend from 5 to 13 seconds two-way time (TWT), intersect zones of horizontal reflections (B). Similar west-dipping reflections (A) are observed along profile 86-3, although in this case they are generally deeper than 10 seconds TWT and extend below Moho (B) (Fig. 3). A particular case is a major reflection, in the eastern half of the profile, which extends from about 9 seconds to the upper mantle at 15 seconds, along a horizontal distance of about 75 km (A_1). Along this profile, the west-dipping reflections in the lower crust appear to offset the Moho in a normal sense. No dipping reflections are

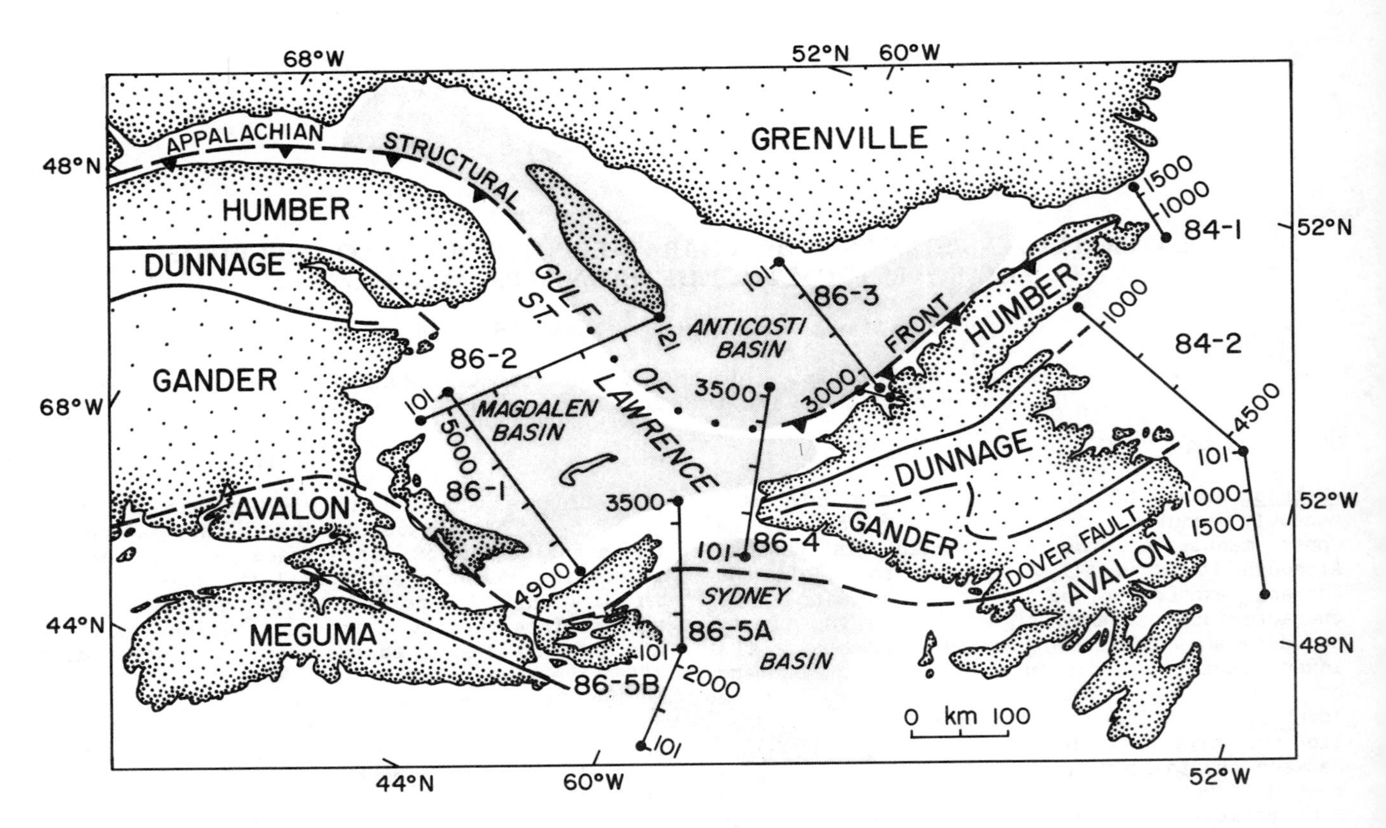

Fig. 1 Location of 1986 and 1984 seismic reflection profiles, tectonostratigraphic zone boundaries in the northern Appalachians [after Williams and Hatcher, 1983], and major basins.

observed in the lower crust along 86-2 and 86-4; this may be caused by the orientation of these profiles, approximately north-south, along which west-dipping reflections are not likely to be imaged.

Because reflectors A and B show a continuous pattern along profile 84-1 and the western half of 84-2, Keen et al. [1986] suggested that they belong to a single lower crustal block. The comparable characteristics of the lower crust

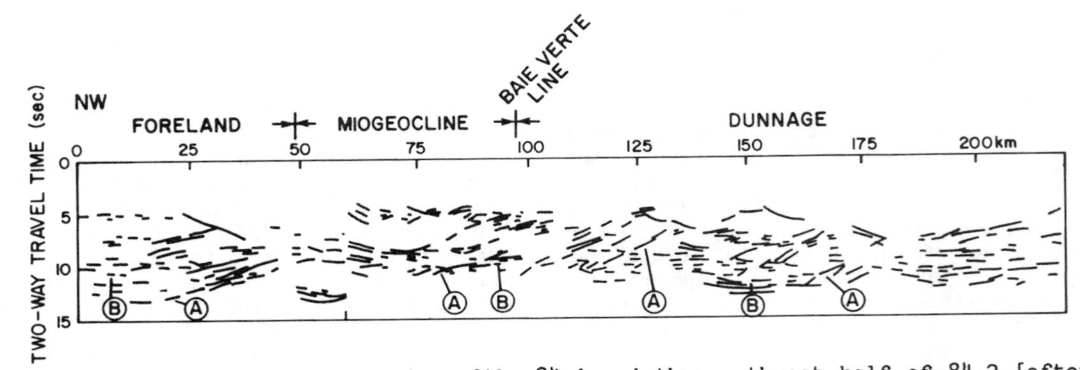

Fig. 2 Composite line-drawing of profile 84-1 and the northwest half of 84-2 [after Keen et al., 1986] (see Figure 1 for location). The vertical to horizontal scale ratio is 1:1 at 5 km/s, as well as for all the seismic sections in following figures. The west-dipping reflections (A) are identified as characteristic signatures of the Grenville lower crustal block. B shows interpreted position of the reflection Moho. The line-drawings and data sections shown in this paper are migrated.

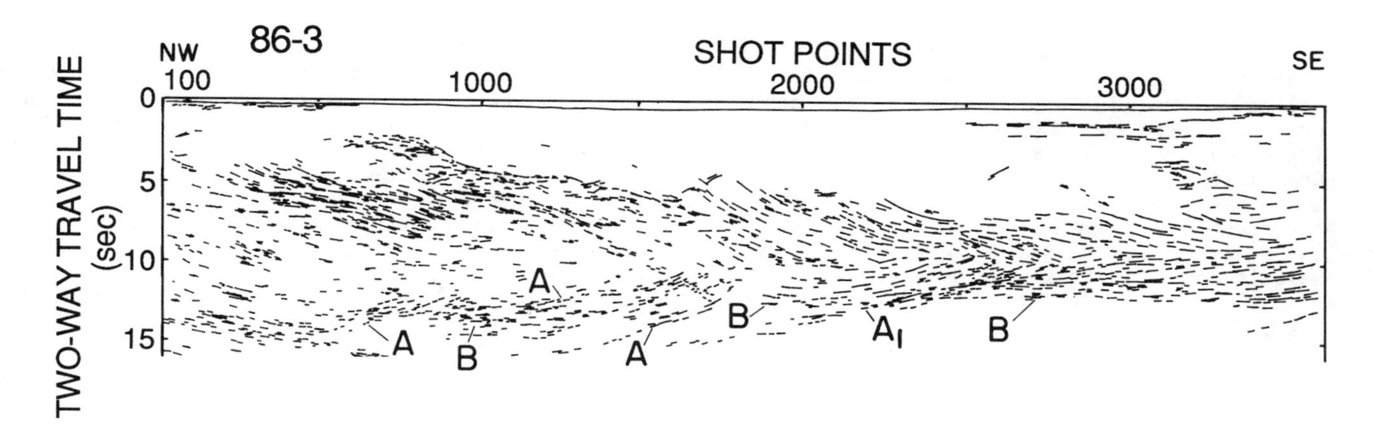

Fig. 3 Line-drawing of profile 86-3. Note that the same reflection patterns as in Figure 1 (A- west-dipping reflections, B- reflection Moho) are present on this profile.

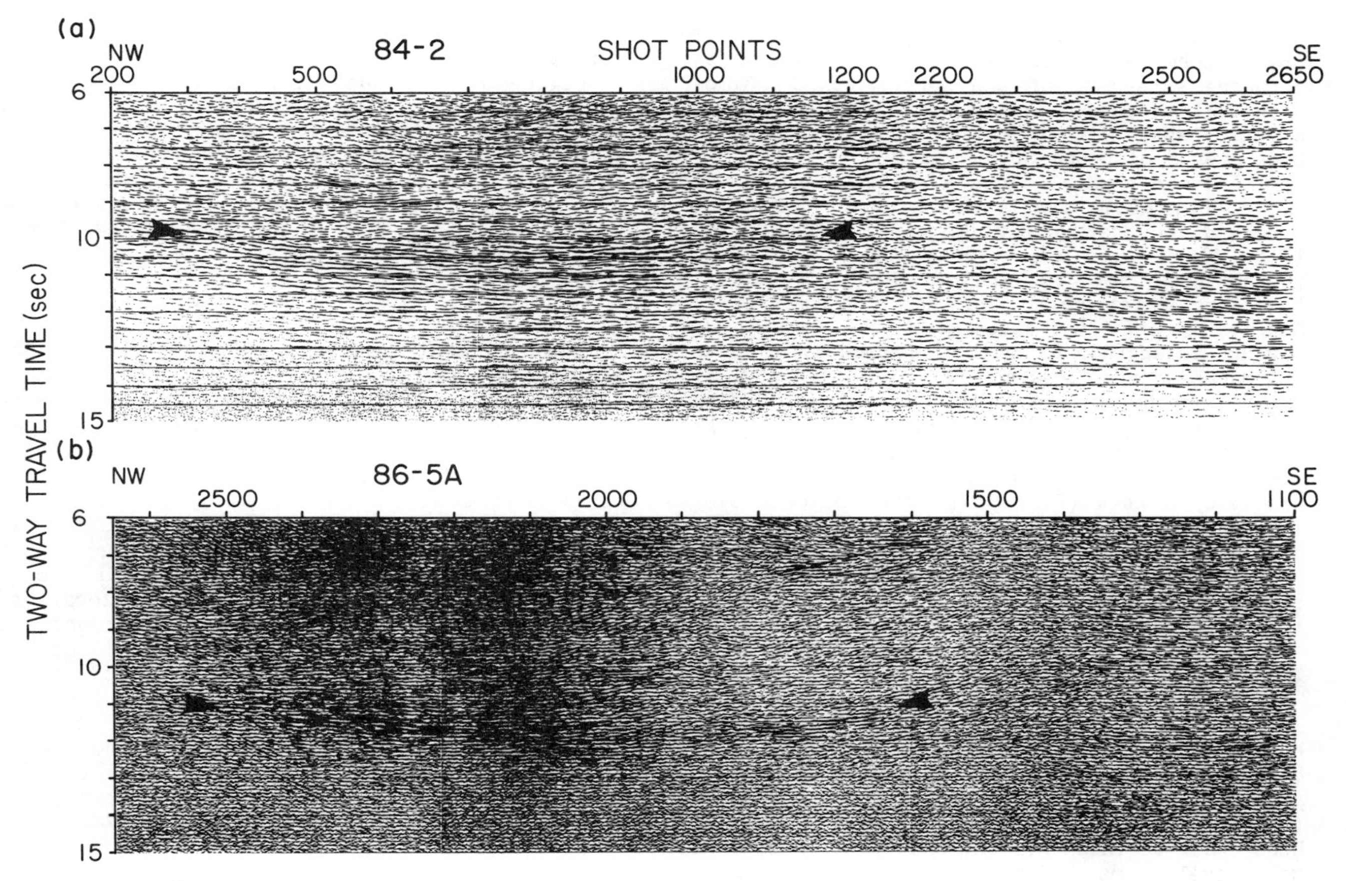

Fig. 4 Lower portion of segments of profiles 84-2 (a) and 86-5A (b). Note the similarity in the bright and elongated Moho reflections between the arrows, from 9.5 to 10 seconds in (a), and from 11 to 12 seconds in (b). To the SE of shotpoints 1050 in 84-2, and 1500 in 86-5A, the reflections have a completely different character; this change in the reflection patterns is interpreted as the boundary between the Central lower crustal block to the NW and the Avalon block to the SE.

along profile 86-3 suggest that this profile is underlain by the same lower crustal block, called "Grenville" by Marillier et al. [in press]. These authors made use of several geologic observations, such as the exposure of Grenville age craton in Quebec and Labrador immediately west of line 84-1 and 86-3, to interpret the Grenville crustal block as underlying the orogen eastward to a major change in seismic character at km 180 on Figure 2.

The Central Block

Beneath the Gander zone on profile 84-2, a band of bright and elongated reflections is observed between 10 and 11 seconds (shotpoints 200 to 1050 in Fig. 4a) which is interpreted as the reflection Moho. When superimposed on the Moho reflections observed in profile 86-5A (Fig. 4b), the two bands match almost perfectly, provided the reflective band of 84-2 is shifted downwards by 0.5 seconds. According to a revised geologic zonation of Cape Breton Island [Barr and Raeside, 1986], the portion of profile 86-5A where this band occurs, corresponds to the offshore extension of the Gander zone. Almost identical reflection patterns are thus found to be associated with the same tectonostratigraphic zone over a distance of 600 km.

The lower crustal block identified by Keen et al. [1986] along profile 84-2 under the Gander zone and the eastern portion of the Dunnage zone (Fig. 1), is located between the Grenville and Avalon (see below) lower crustal blocks, and therefore referred to as the "Central" block. Surface geology indicates that the Gander zone may be allochthonous over the underlying crust [Colman-Sadd and Swinden, 1984].

The northwest limit of this lower crustal block under 84-2 is defined as a wedge of bright reflections whose sharp upper boundary dips to the west (Fig. 5a); a very similar feature is observed in 86-4 (Fig. 5b). The apparent similarity of these wedges, and the position of line 86-4 relative to the offshore extension of the Gander zone, suggest that the wedge on profile 86-4 also marks the northwest limit of the Central lower crustal block.

The Avalon Block

The elongated band of bright reflections corresponding to the reflection Moho in profiles 84-2 and 86-5A, and which were described in the previous section, end abruptly to the south (shotpoints 1050 and 1500 in Figs. 4a and b). Less well-defined reflections then follow at Moho level, some of them dipping slightly to the east. The seismic data are compatible with a nearly vertical boundary through the crust on both profiles 84-2 and 86-5A. In profile 84-2, the change of seismic reflection patterns is located beneath the offshore extension of the Dover fault [O'Brien et al., 1983] which corresponds to the Gander-Avalon zone boundary. South of Newfoundland, the offshore extension of this zone boundary is not known. However, major northeast-southwest magnetic trends across the Sydney basin [B. Loncarevic, pers. comm.] suggest a link between the Avalon-Gander boundary in Newfoundland and in Cape Breton, where Barr and Raeside [1986] recently defined it. Its offshore extension would fall near the change in seismic pattern on profile 86-5A at shotpoint 1500.

Beneath the Avalon crustal block, immediately southeast of the Central-Avalon block boundary, the Moho reflections are diffuse on both the 86-5A and 84-2 profiles. Further to the SE, some northwest-dipping reflections are observed between 11 and 12 seconds, but it is hard to tell whether they belong to Moho; they are associated with diffractions on the unmigrated section. The contrasting Moho reflections observed on both sides of the inferred boundary between the Central and the Avalon blocks, suggests that the crust-mantle boundary is of different nature beneath these two units.

Figure 6 gives the distribution of the lower crustal blocks in the Canadian Appalachians, based on the interpolation between all the profiles. The boundary between the Grenville and the Central blocks has the same shape as the Appalachian structural front. The width of the Central block reaches a minimum in the Sydney basin.

The Intermediate Velocity Layer

Velocity models derived from offshore refraction data [Dainty et al., 1966], show that rather high velocities are found in the upper and in the lower crust of the Canadian Appalachians. In particular, an intermediate velocity layer at the base of the crust, with velocities ranging from 7.05 to 7.52 km/s, has been detected northeast and west of Newfoundland, as well as along an east-west profile across the Magdalen basin (Fig. 7). The refraction profiles sample areas which include the locations where bright and flat lying Moho reflections are observed in the reflection profiles 86-1 and 86-3. The Central crustal block was interpreted by Marillier et al. [in press] to occur in the same areas, except for the profile running to the west of Newfoundland. It is tempting to relate the intermediate velocity layer to the Central crustal block, although more detailed refraction work is needed to establish this relationship more firmly.

Discussion

The west-dipping reflections observed in the Grenville block can either be related to extension of a continental margin or to compression

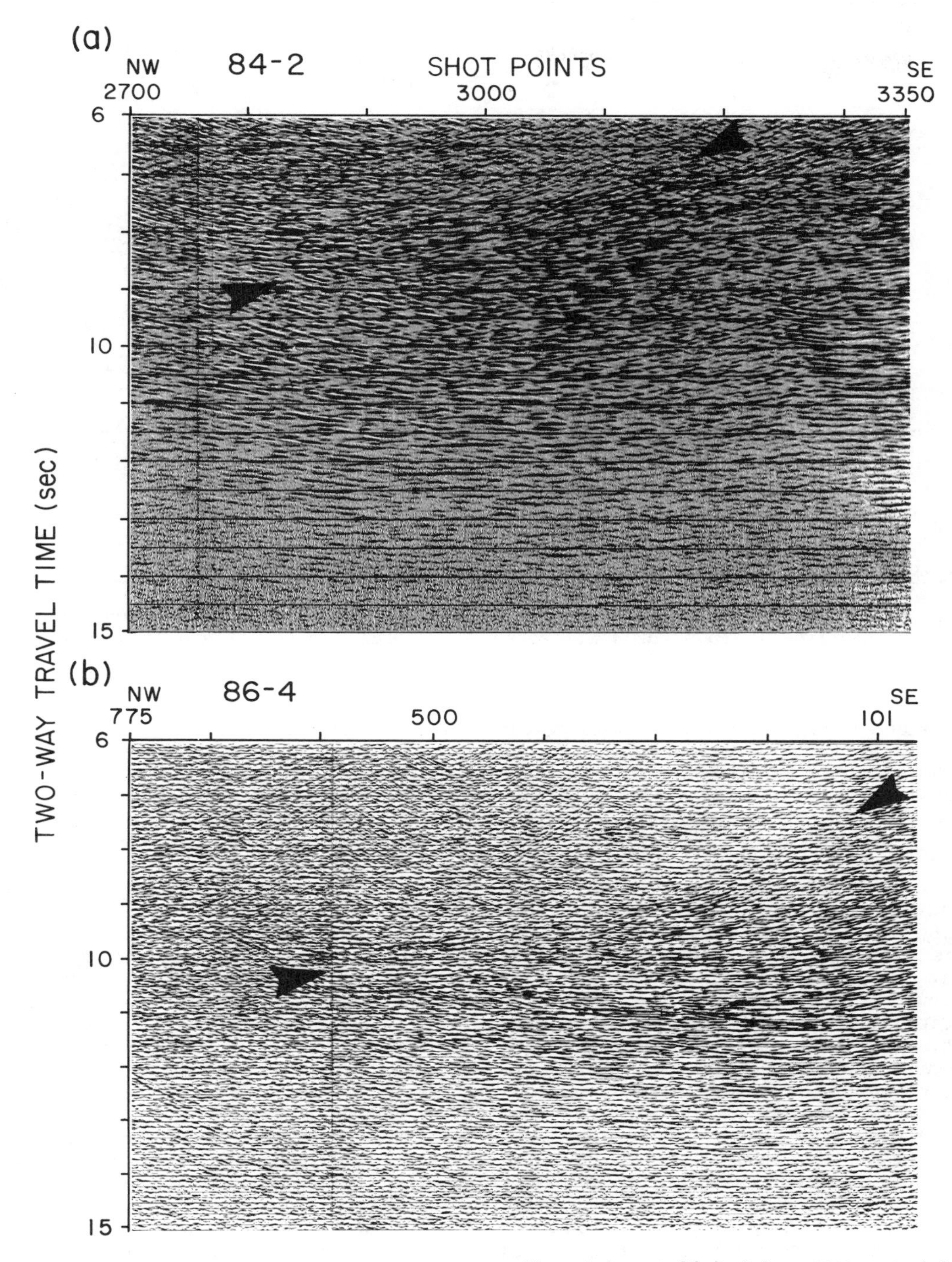

Fig. 5 Lower portion of segments of profiles 84-2 (a) and 86-4 (b). Although (a) is plagued by sea-bottom multiples, both sections display similar wedges of bright reflections in the lower crust, the upper limit of which are indicated by arrows. We interpret the NW end of the wedges as the boundary between the Grenville and Central lower crustal blocks.

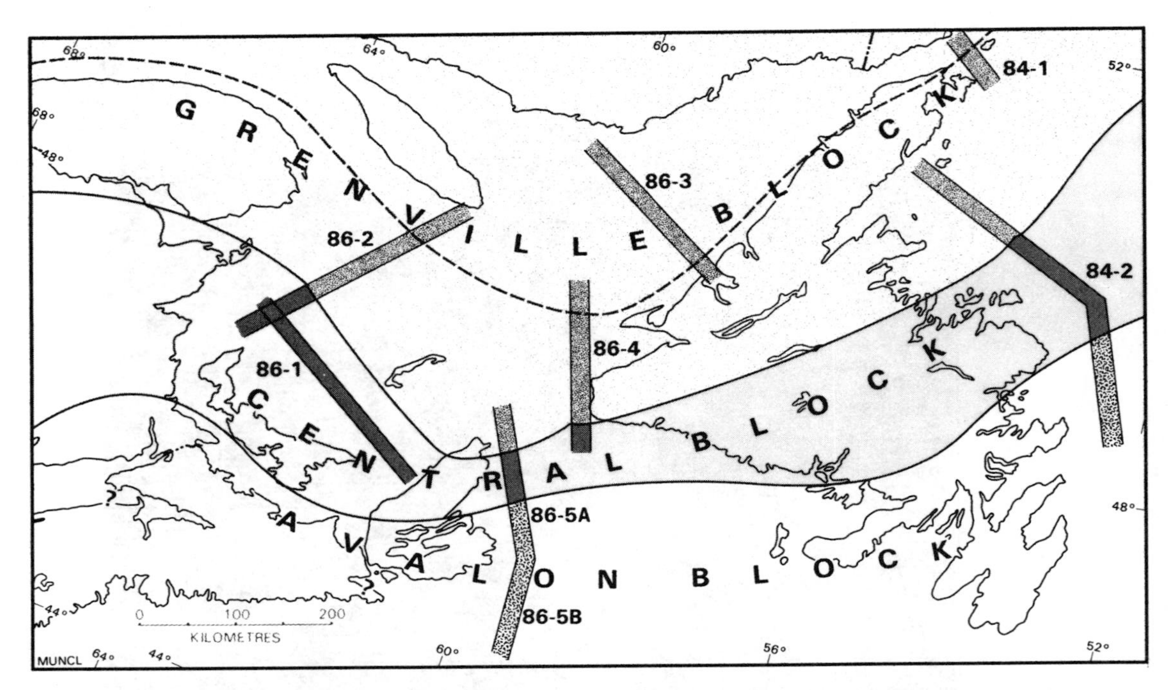

Fig. 6 Major crustal blocks of the Canadian Appalachians. Boundaries defined on seismic sections at lower crustal depths with total extent extrapolated between sections [after Marillier et al., in press].

during collision. The latter has been advocated by Stockmal et al. [1987] who propose a model involving delamination of the Grenville lithosphere during closure of the Iapetus ocean. Reprocessing of the 1984 data provided further evidence for compression in some places along profile 84-2, where the west-dipping reflections clearly indicate a sense of reversed displacement of the Moho reflections [Hall et al., this volume]. By contrast, normal displacement of the Moho reflections is dominant along profile 86-3, and it is associated with thinning of the Grenville crust. Thus our data are compatible with west-dipping shear zones or faults in the lower crust, which accommodated thinning of the Grenville crust over large distances during late Precambrian rifting, and which later were reactivated as thrust faults in the area of the Appalachian orogen. Whatever the reason, our data suggest that the Moho has retained its topography since at least Paleozoic time. This is a much longer time span than the 10 to 100 Myr suggested by Kusznir [1987] for Moho topographic decay, when the crust is overthickened by orogenesis.

The well-defined Moho offset by dipping reflections, as observed in the Grenville block (profile 86-3), do not match the seismic characteristics usually observed in the crust of Precambrian cratons [e.g. Allmendinger et al., 1987]. The reason may either be that our marine data are of better quality than the land profiles obtained so far in this kind of environment, or that the crust is more reflective in this area for other reasons, for example the late Precambrian extension of the Grenville crust suggested above may have changed the characteristics of the lower crust. The relative paucity of profiles in Precambrian cratons makes generalized comparisons difficult.

The association of zones of crustal reflectivity with crustal velocities higher than 7.0 km/s has been observed in many places, and Mooney and Brocher [1987] suggested that it is mainly found in extensional terranes. Bright and elongated Moho reflections are observed in reflection profile 86-1 which runs across the Carboniferous Magdalen basin, beneath which Dainty et al. [1966] detected a lower crustal layer with a velocity of 7.35 km/s. Thus the seismic characteristics beneath the Magdalen basin, which is believed to be a pull-apart basin [Bradley, 1982], correspond to those of extensional terranes.

In the Central crustal block, northeast of Newfoundland, a band of high reflectivity at Moho level is also observed in association with high velocities in the lower crust [Dainty et al., 1966]. In this case, there is no evidence of extension related to these features. The question then arises whether they can be related to a thermal event associated with extension which occurred before this block was accreted, or whether they have an origin which is not

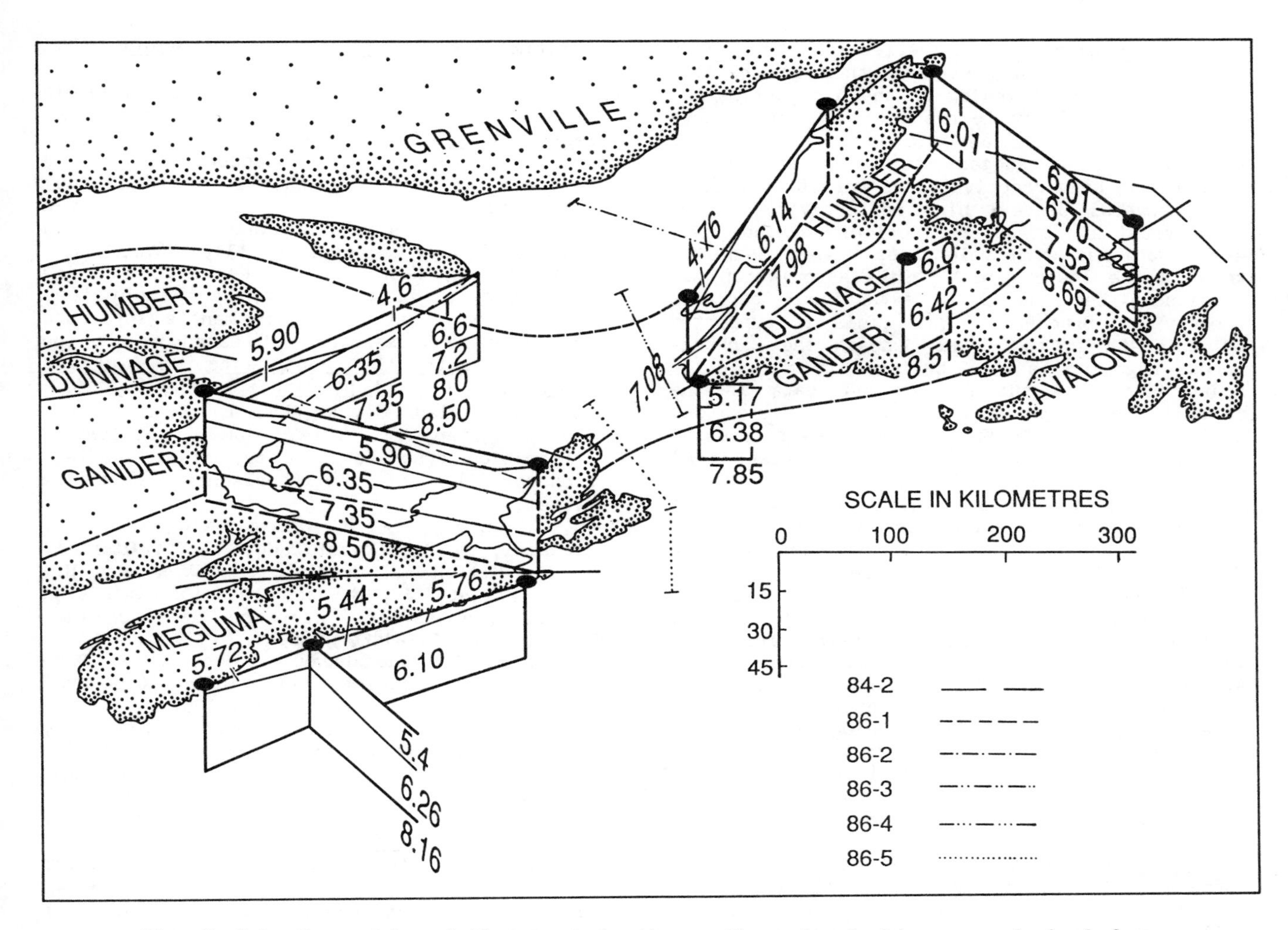

Fig. 7 Velocity models of the crust in the northern Appalachians, as derived from refraction experiments [after Dainty et al., 1966]. The black dots indicate the locations of the recording stations. Profiles with two dots have been reversed. Note the high velocity layers in the lower crust, along reversed refraction profiles located northeast and west of Newfoundland, and in the Magdalen basin. These profiles include the locations where bright and elongated Moho reflections are observed. The locations of the deep seismic profiles are also shown.

related to extension. Similar characteristics are observed of the lower crust in the reflection profiles 84-2 and 86-5A, which are about 600 km apart. If these reflection patterns have a common origin, then a lateral extent of 600 km suggests that it was a major regional tectonic event. A possible candidate is the Late Paleozoic strike-slip movement believed to have displaced the northeastern units of the Canadian Appalachians relative to the continent, by more than a hundred kilometres [e.g., Arthaud and Matte, 1977]; this would imply that this event has rejuvenated the Moho which may correspond to a detachment between the crust and the lower lithosphere in this area.

Conclusion

We have been able to identify seismic characteristics in the lower crust using deep reflection and refraction data, which led us to define three major lower crustal blocks underlying the northern Appalachians. Because these characteristics remain the same over along-strike distances of several hundreds of kilometres, we believe that they result from phenomena important enough to affect the crustal blocks on a regional basis.

West-dipping reflections in the lower crust of the Grenville craton are interpreted as faults or shear zones which offset the Moho.

Where they displace the reflection Moho in a normal sense, we suggest that they are related to extension and thinning of the old continental margin of the Grenville craton. In places where these faults are associated with reverse displacement, we suggest that the extensional faults have been reactivated as thrust faults.

In several places, high reflectivity of the lower crust and bright reflections at Moho level are associated with velocities higher than 7.0 km/s in the lower crust. This association has previously been noted in extensional terranes [Mooney and Brocher, 1987]. This is supported by our data in the crust beneath the Carboniferous Magdalen basin - believed to be a pull-apart feature - but not by our data in the Central lower crustal block. This block exhibits high reflectivity and lower crustal velocities, but it is not an extensional terrane.

Attempts to relate different tectonic provinces with variations in the character of deep seismic data have not always been successful, as for example in the area around Great Britain where much deep seismic data have been collected [McGeary and Matthews, 1987]. The quality of the data may be responsible for being able or not to clearly image the lower crust. Although we generally cannot find a direct relation between the seismic characteristics in the lower crust and the surface geology (except for the Central-Avalon block boundary for which we suggest a near vertical fault cutting through the entire crust), our interpretation is at least consistent in the lower crust.

Acknowledgments. We thank B. de Voogd and B. Loncarevic for their review of this paper, as well as two anonymous reviewers for their detailed comments. The Lithoprobe-East Group has helped in the interpretation of the seismic data. P. Durling has helped us to prepare the figures.

This paper is Lithoprobe contribution number 46, and Geological Survey of Canada contribution number 24488.

References

Allmendinger, R. W., Nelson K. D., Potter, C. J., Barazangi, M., Brown, L. D., and Oliver, J. E., Deep seismic reflection characteristics of the continental crust. Geology, 15, 304-310, 1987.

Arthaud, F., and Matte, Ph., Late Paleozoic strike-slip faulting in Southern Europe and Northern Africa: result of a right lateral shear zone between the Appalachians and the Urals. Geological Society of America Bulletin, 88, 1305-1320, 1977.

Barr S. M., and Raeside, R. P., Pre-Carboniferous tectono-stratigraphic subdivisions of Cape Breton Island. Nova Scotia. Maritime Sediments and Atlantic Geology, 22, 252-263, 1986.

Bradley, D. C., Subsidence in late Paleozoic basins in the northern Appalachians. Tectonics, 1, 107-123, 1982.

Colman-Sadd, S. P. and Swinden, H. S., A tectonic window in central Newfoundland? Geological evidence that the Appalachian Dunnage Zone is allochthonous. Canadian Journal of Earth Sciences, 21, 1349-1367., 1984.

Dainty, A. M., Keen, C. E., Keen. M. J., and Blanchard J. E., Review of geophysical evidence on crust and upper mantle structure on the eastern seaboard of Canada. In The earth beneath the continents. American Geophysical Union Monograph, 10, 349-369, 1966.

Hall, J., Quinlan, G., Wright, J., Keen, C. E., and Marillier, F., This volume. Styles of deformation observed on deep seismic reflection profiles of the Appalachian-Caledonide system.

Keen, C. E., Keen, M. J., Nichols, B., Reid, I., Stockmal, G. S., Colman-Sadd, S. P., O'Brien, S. J., Miller H., Quinlan, G., Williams, H., and Wright, J., A deep seismic reflection profile across the northern Appalachians. Geology, 14, 141-145, 1986.

Kusznir, N. J., The decay of continental Moho topography. Abstract U7-32, IUGG XIX General Assembly, Vancouver, Aug. 9-27, 1987.

Marillier, F., Keen, C. E., Stockmal, G. S., Quinlan, G., Williams, H., Colman-Sadd, S. P., and O'Brien, S. J., Crustal structure and surface zonation of the Canadian Appalachians: implications of deep seismic reflection data. Canadian Journal of Earth Sciences, In press.

McGeary, S., and Matthews, D. H., Classification of lower crustal reflectivity: nontypical BIRPS. Abstract U7-12, IUGG XIX General Assembly, Vancouver, Aug. 9-27, 1987.

Mooney, W. D., and Brocher, T. M., Coincident seismic reflection/refraction studies of the continental lithosphere: a global review. Geophysical Journal of the Royal Astronomical Society, 89, 1-6, 1987.

O'Brien, S. J., Wardle, R. J., and King, A. F., The Avalon zone: a Pan-African terrane in the Appalachian orogen in Canada. Geological Journal, 18, 185-222, 1983.

Stockmal, G. S., Colman-Sadd, S. P., Keen, C. E., O'Brien, S. J., and Quinlan, G., Collision along an irregular margin: a regional plate tectonic interpretation of the Canadian Appalachians. Canadian Journal of Earth Sciences, 24, 1098-1107, 1987.

Williams, H., Appalachian orogen in Canada. Canadian Journal of Earth Sciences, 16, 792-807, 1979.

Williams, H., and Hatcher, R. D., Jr., Appalachian suspect terranes. In Hatcher, R. D., Jr., Williams, H., and Zietz, I., eds., Contributions to the tectonics and geophysics of mountain chains. Geological Society of America Memoir, 158, 33-53, 1983.

CONTRASTING TYPES OF LOWER CRUST

Scott B. Smithson

Program for Crustal Studies, Dept. of Geology and Geophysics
University of Wyoming, Laramie

Abstract. The deep crust of the Basin and Range of NE Nevada and the Minnesota Archean represent strongly contrasting tectonic features. The lower crust in the Basin and Range of NE Nevada is horizontally layered, probably as a result of extensional plastic flow during the Cenozoic, representing a major metamorphic event generating granulites in the deep crust. The lower crust is compositionally layered and predominantly mafic underlain by a more mafic 3-km zone of cumulate or residual material related to magmatic history. The Moho is a sharp compositional boundary marked by strong interlayering of mantle and crustal rocks to produce unusually high reflectivity, and the crustal rocks may be partially molten to explain the reflectivity. About 50% of the crust is mafic material, and probably a maximum of 25% of the crust represents mafic additions and underplating during Cenozoic extension. In the Minnesota-early Archean crust, reflections become weaker and less continuous in the lower crust, and absent in the lowermost crust and Moho zone. Here most of the deep crust seems to have been thickened by stacking of nappes, which are "floating" in small bodies of later anatectic granite. The lowermost Archean crust is a distinctly different tectonic pattern possibly formed by subhorizontal compressional shearing that may have erased larger scale structures or by later gabbroic underplating that thickened the early Archean crust. The Moho in Minnesota must be strikingly different from the Basin and Range. It is either gradational or a relatively small step-function change in chemical composition such as garnet pyroxene granulite to peridotite; it is probably not layered unless the velocity changes across the layering are very small. If it is gradational, then it is probably formed by the gabbro-eclogite phase change, and the gabbroic zone could have been underplated.

Introduction

Crustal reflection profiling has revealed different reflection patterns in different tectonic regimes related to the burning questions about the nature of the lower crust and Moho. In general, lower crust has been affected by a series of events and processes and is polygenetic. When vertical-incidence reflection profiling is complemented with wide-angle reflection data, more reliable velocity models can be determined and related to lithologies and ultimately crustal genesis. Petrologic and geothermal information together with reflection data are used to interpret contrasting lower crustal styles in the young hot extensional Basin and Range and the cold, early Archean of Minnesota, areas of very different age and tectonic history that emphasize how dissimilar the lower crust and Moho may be in different tectonic provinces and the different considerations for interpreting lower crust.

Nevada

The crust in NE Nevada, which has been the site of extensive COCORP crustal reflection profiling and detailed profiling by the University of Wyoming, contains some of the strongest subhorizontal reflections found, and these are very common at lower crustal levels. Most events are short (1 - 2 km) and discontinuous, and the most continuous events (5 - 10 km) are found in the mid-crust and lower crust. Probably the fewest events occur between about 1 to 3 s, although the relationship may be reversed with more events in the upper crust and few at depth in some places. Some strong, west-dipping events are found at mid-crustal levels (5 - 7 s) on COCORP Nevada Line 4 (Fig. 1). The most consistent events, which cover about one third of the sections, are a multicyclic event or series of events between 9 and 11 s with a duration of about 0.5 s (Fig. 1). This has been identified as the reflection from the Moho (Klemperer et al., 1986); this event is consistent in its reappearance within a certain time range and with a multicyclic character. In some places, the event is distorted by near-surface velocity contrasts and in others it appears to dip. This event may be accompanied by another event about 1 s higher that has been called a second Moho. This event, named the "X" reflection (Klemperer, 1986) is weaker, but displays a similar character. The lowermost event

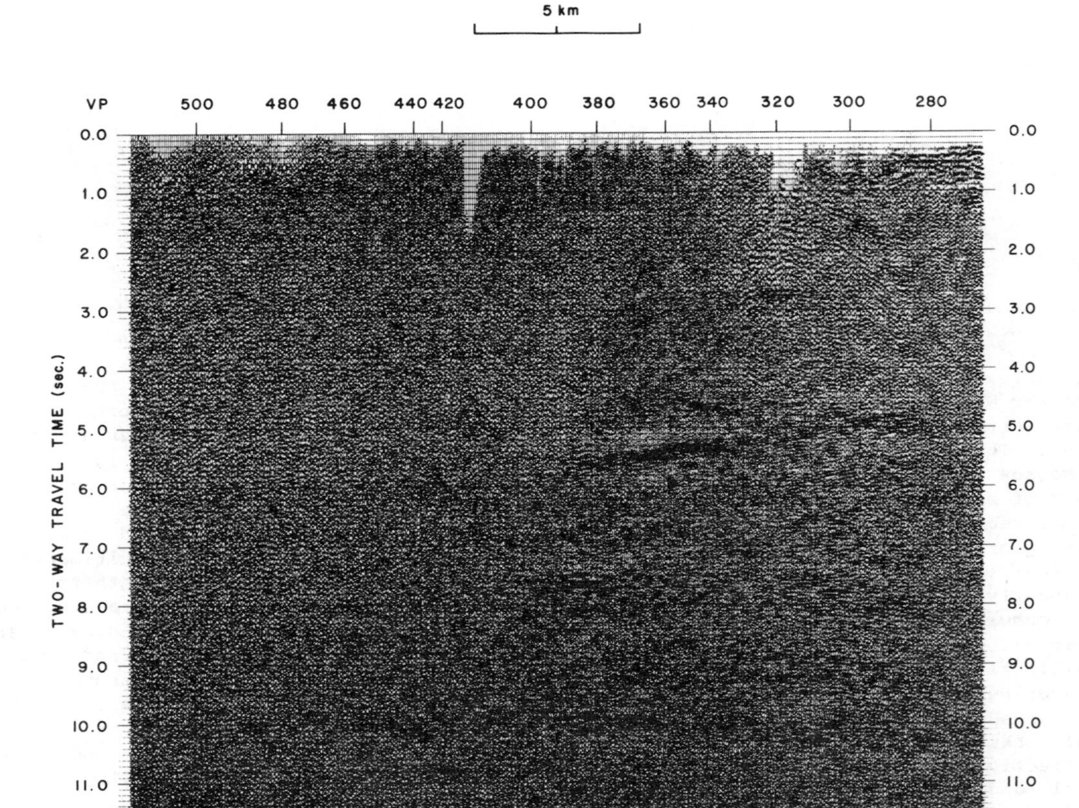

Fig. 1. COCORP Line 4 from Nevada across the Cherry Creek Range. Reprocessed at the University of Wyoming. Note strong reflections at 5 s and 8 to 10 s passing laterally in an almost totally transparent section.

around 10 s is best interpreted as a Moho reflection; the event about 1 s above it is most likely a lower crustal reflection (Valasek et al., 1987).

Metamorphic core complexes typify the eastern Basin and Range (Crittendon et al., 1980; Snoke and Lush, 1984) and appear to be the site of strongest crustal reflections in this area of abundant reflections (Hurich et al., 1985). Several seismic reflection lines together with wide-angle data have been recorded across in the Ruby Mountains core complex by the University of Wyoming 70 km north of the COCORP (Hauser et al., 1987) profile. A CDP profile across the plunge of the core complex images the mylonite zone at shallow depths, reflections through the entire crust, and particularly strong events at 2.8, 4, and 6 s. This profile reveals Basin-and-Range faults cutting the mylonite zone and, as in the Kettle dome (Hurich et al., 1985), shows the mylonite zone to be a good reflector. Other strong upper- to middle-crustal reflectors may also be mylonite zones formed by crustal-scale

TABLE 1.

Discontinuity	Depth	Velocity	T-corrected
	0.0		
		3.80	
	1.5		
		5.50	
A	3.0		
		6.10	6.20
B	12.5		
		6.35	6.4-6.5
C	19.5		
		6.6	6.7-6.8
X	28.0		
		7.2-7.5	7.4-7.8
M	31.0		
		7.8	7.9-8.1

Velocity model matching travel times and moveouts for 5 bands of reflections identified on a 27.6 km offset shot gather. Velocity for the surface layer is an average for a multilayered zone. Right-hand column gives temperature-corrected velocities.

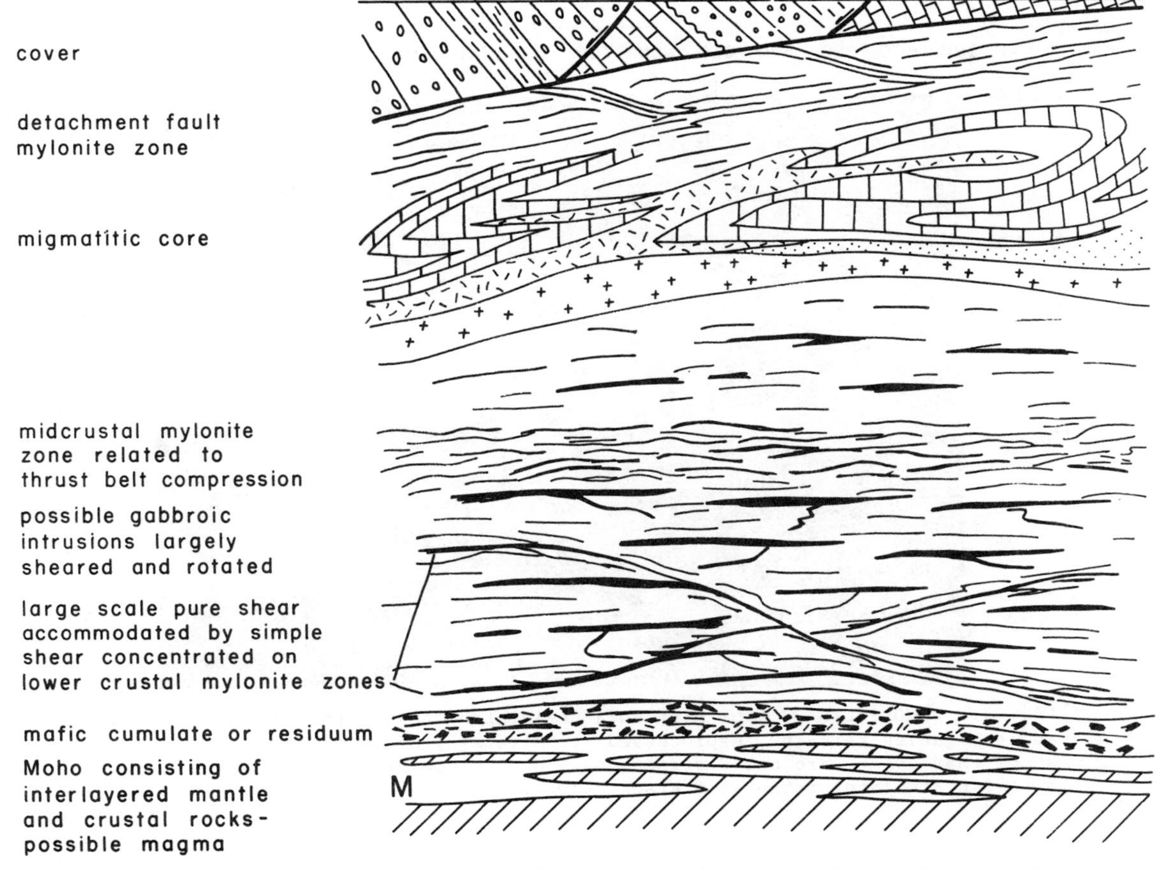

Fig. 2. Interpretative cross-section of crust in the northern Ruby Mountains of the Basin and Range. Upper plate overlies detachment fault and mylonite zone. Subhorizontal nappes of the migmatitic core (infrastructure) are underlain by granitic intrusions and mylonite zones (concentrated shear). These pass downward into mafic rocks, possible intrusions, and sheared lower crust and Moho. Mid-crustal mylonite zone accommodates crustal shortening that formed foreland thrust belt. Simple shear in lower crustal mylonites accommodates pure shear on a crustal scale. Layered Moho is accentuated by shear accompanying extension and/or intrusion. Shallow structure from A.W. Snoke (personal communication).

extension. The seismic section offers a strong contrast with the rather transparent COCORP section to the south and shows the upper crust as well as the lower crust to be highly reflective.

Wide-angle recording at offsets of 18 to 50 km provides moveouts large enough to estimate velocities to the base of the crust (Valasek et al., 1987) and will be used to develop a more detailed model of the lower crust than would otherwise be possible. Several crossing wide-angle profiles were incorporated into a seismic model that matches the travel time and moveout of the various reflections although it does not attempt to match the detailed layering implied by the multicyclic character within the bands (Table 1). This velocity model was tested against reflections on four other shot gathers and proved consistent for different offsets (Valasek et al., 1987).

The strong layering demonstrated by seismic sections suggests that most of the crust is strongly anisotropic where the amount and type of anisotropy depend on mineralogy. This anisotropy has not been measured in the field but can be deduced directly from the presence of large-scale layering because layered crystalline rocks are invariably anisotropic (Birch, 1960; Christensen, 1982; Christensen and Szymanski, 1988). This anisotropy must be considered when interpreting velocities. Because of the strong multicyclic reflections (Fig. 1) most of the various interval velocities (Table 1) must be averages of

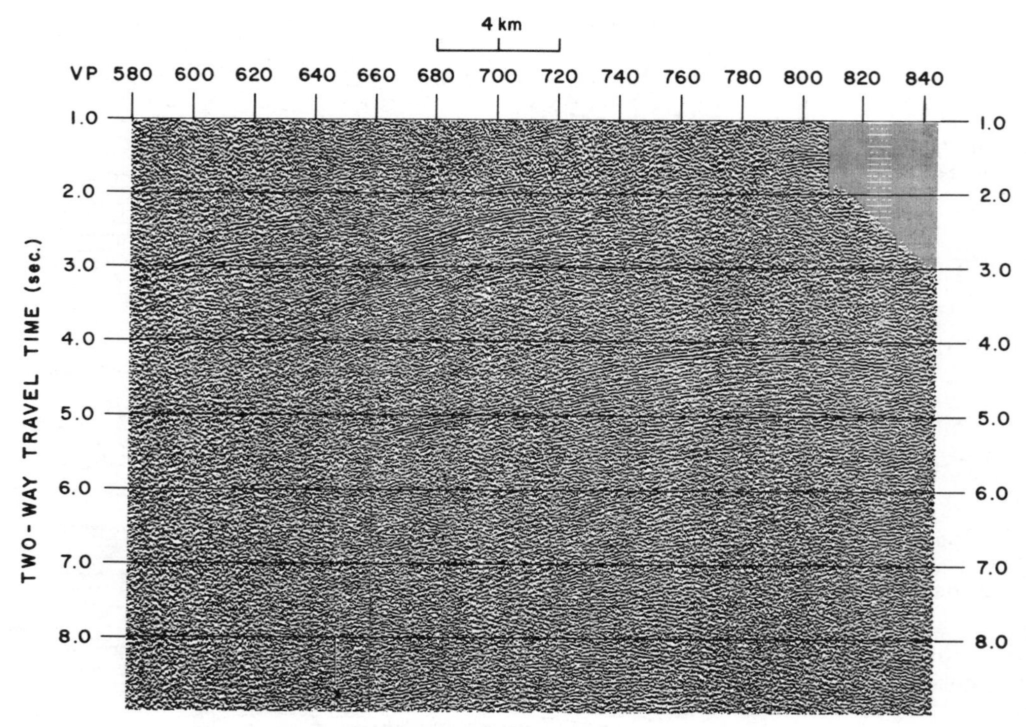

Fig. 3. COCORP Line 3 from Minnesota unmigrated showing reflections in ancient gneiss terrain. Reprocessed at University of Wyoming. Migration shows that events are not diffractions. Gentle convex upward events from 1.5 s to 9 s are interpreted as a stack nappes.

interlayered high- and low-velocities. Most of the reflections can be interpreted in terms of compositional differences; however, anisotropy probably plays a role in contributing to reflectivity (Christensen and Szymanski, 1988). Other causes of reflections such as fluids, while possible, are not necessary to explain reflections except perhaps at the Moho. Interpretation of rock types on the basis of velocity is not unique but velocities near 6 km/s are generally associated with quartzo-feldspathic rocks and those near 7 dm/s with mafic rocks (Christensen, 1982). Based on laboratory velocity determinations, (Christensen, 1982) velocities of 6.1 - 6.2 km/s correspond to quartzofeldspathic rocks; this agrees with observations of interlayered quartzites, schists, marbles, and layered to lenticular granites in the upper crust. Velocities of 6.4 to 6.5 km/s correspond to slightly more mafic, but not intermediate (andesitic) compositions. These rocks probably are mostly quartzofeldspathic (deformed granites and metasediments) in which about one-third of the material is mafic, possibly originally intrusive. The next zone from a depth of 20 to 28 km has model velocities of 6.7 to 6.8 km/s. If it is in granulite facies, which seems likely (Fig. 6 and Sandiford and Powell, 1986), then it is considerably more felsic than gabbro, possibly about one-third granitic material. This is the zone that would have to be the primary locus for underplating accompanied by partial melting of the overlying more felsic rocks to produce the granitic intrusions in the upper crust. The lowermost 3 km in the crust (the zone between the "X" reflection and the Moho) has a relatively high velocity of 7.4 to 7.8 km/s that lies between normal velocities for lower crust and upper mantle, representing layered cumulates or a mafic residuum. Garnet-pyroxene granulites correspond to the lower part of this velocity range (Christensen, 1982) but more mafic than normal gabbroic composition is more likely because of the near-mantle velocity.

An interpretive cross-section (Fig. 2) is based on available seismic reflection data and surface geology projected down plunge, and shows a near surface mylonite zone representing Cenozoic extension above nappes of the infrastructure formed during Mesozoic compression and denuded during the later extension. This is underlain by a zone of concentrated shear (anastomozing mylonites) above a zone of intrusion and pure shear on a large scale accommodated by simple shear on a smaller scale. The generally reflective crust, the subhorizontal reflections,

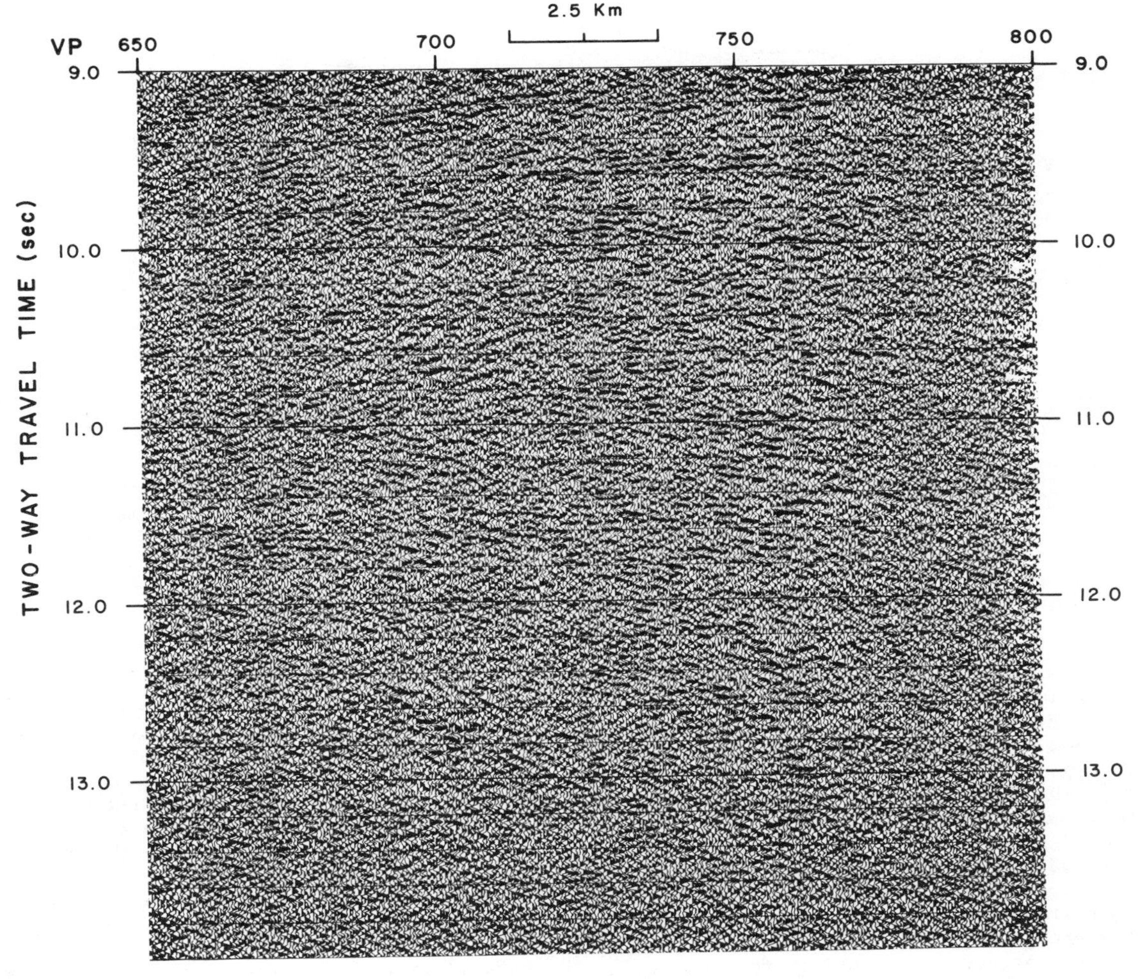

Fig. 4. COCORP Line 3 from Minnesota showing weak to transparent reflective character of lower crust. No visible Moho reflection.

and the strong mid-crustal, lower crustal, and Moho reflections are attributed to plastic flow accompanying extension which has generated sharp subhorizontal layering into which intrusions are drawn out. The seismic profiles thus show the effect of extensional regional metamorphism on the deep crust.

Minnesota

Some of the oldest Archean crust in North America is found in the Precambrian of Minnesota, which has dates of 3.6-3.8 b.y. (Goldich and Hedge, 1974). Here the ancient Minnesota Valley gneiss terrain is bounded on the north by the Great Lakes tectonic zone (Sims et al., 1980) and a series of late Archean greenstone belts where COCORP recorded a number of seismic reflection profiles (Gibbs et al., 1984).

The gneiss terrain which is correlated with the ancient Minnesota River valley gneisses, at the south end of COCORP Minnesota Line 3 yields some of the most interesting reflection results because of the implications for early crustal genesis. In the ancient gneiss terrain, a distinctive reflection pattern consists of broad asymmetrical arcs (Fig. 3). Seismic reflection data from the southern end of COCORP line 3 show packets of reflections that extend almost continuously from about 2 km to 30 km (0.8-9.5 s) in depth (Fig. 3). The reflections come from a series of layered zones (apparently discontinuous) that are up to 15 km long and 3 to 4.5 km (1-1.7 s) thick, separated by transparent zones 1 - 3 km (0.3 - 1.5 s) thick. Short (1 - 2 km) discontinuous subhorizontal reflections (Fig. 4) are found from 30 km to about 35 km (9 - 11 s). No distinct Moho reflection can be seen.

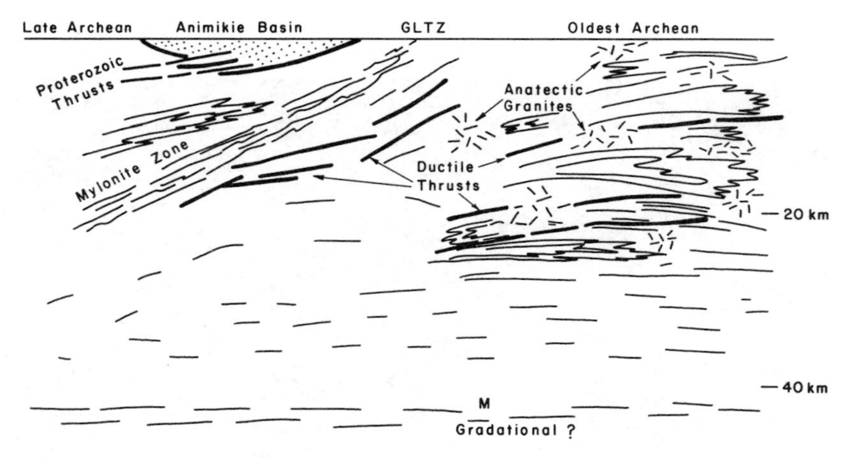

Fig. 5. Interpretative cross section of Archean crustal structure in Minnesota showing suture, Animikie basin and ancient gneiss terrain. Archean basement is remobilized along discrete thrusts to deform supracrustal rocks in Animikie basin. Complex, moderately dipping suture zone (GLTZ = Great Lakes Tectonic Zone) can be followed geophysically to about 20 km. Oldest Archean crust passes from a thick (approx. 30 km) stack of nappes interspersed with anatectic granites into a subhorizontal tectonic regime in lowermost crust about a gradational Moho.

Based on metamorphic facies, the present depth of exposure is about 20 km; therefore, the upper part of the early Archean lower crust is exposed. Of particular importance is the fact that the reflection pattern indicates essentially the same gross structure from the surface down to a depth of 30 km with strong heterogeneity. Comparison with surface geology and seismic modeling suggests that the rocks are recumbently folded and thrust into nappes, probably during a 3.55 Ba event (Bauer, 1980). The parallelism of reflections that approach possible fold hinges in places is an expression of early recumbent folding found in most old crystalline terrains. The layering is almost certainly the result of transposition so that layering could represent any combination of igneous and sedimentary precursors in what is now a metamorphic sequence.

The reflections are best interpreted as a stack of nappes of highly variable composition, probably originally an igneous and sedimentary mix that extends from near the present surface to a depth of 30 km (Fig. 5). The reflection pattern almost certainly indicates a horizontal (compressive) tectonic regime to this depth. The folded layered rock sequences most plausibly represent a large-scale migmatite, in part a restite, from which the ubiquitous granites were "sweated out." Such reflections would be best generated from alternating felsic and mafic layers (rocks with maximum contrast in reflectivity), but the effect of differentiation through partial melting can result in some unusual compositions and physical properties; e.g., mineralogies rich in garnet and aluminum silicate (Winkler, 1972) so that high-density, high-velocity rocks occur within mafic layers, probably increasing the velocity contrast.

Short, subhorizontal reflections below 30 km (9 s) indicate the presence of a different tectonic regime (Fig. 4), which may be caused by effects of crustal underplating, or these reflections could indicate horizontal tectonic movements under increased plasticity compared with the rocks above. Early Archean crust may have been thin because of high geothermal gradients (Fyfe, 1974) so the possibility of magmatic underplating of this crust must be considered. If so, the underplating took place in only the lowermost 5-10 km of the Archean crust, implying that the deformed early Archean crust was about 50 km thick before underplating although one can postulate earlier injections that became interfolded with the nappes to thicken the crust before the 3.55 Ba event.

Moho reflections are not identified. Generally, the deepest events are short discontinuous events between 9 and 12 s (Fig. 4) particularly underneath the ancient gneiss terrain (VPS 650-800 on Line 3). Deep reflections simply die out after these travel times so that no distinct Moho reflection is seen at vertical incidence. The lack of a clear vertical-incidence Moho reflection in the presence of other deep events is peculiar and must be telling us omething about the nature of the ancient Moho, which has been detected at about 42 km from refraction data (Greenhalgh, 1981).

Discussion

We compare the lower crustal structure of these two areas starting with the Basin-and-Range crust

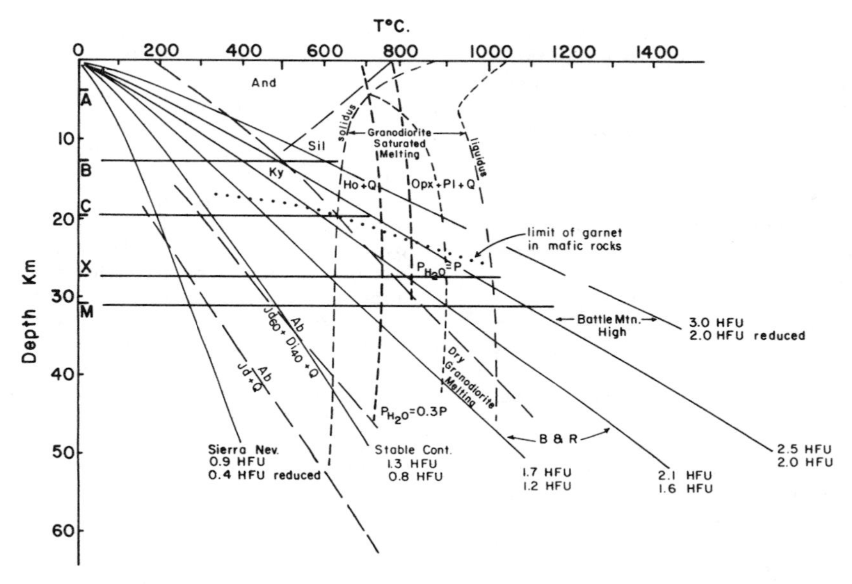

Fig. 6. Geotherms calculated for Basin and Range after Lachenbruch and Sass (1978) and superposed petrogenetic grid. Garnet stability in mafic rocks after Newton (1983). Hornblende stability after Binns (1969). A, B, C, X and M represent position of reflectors for which interval velocities were calculated for Basin and Range. And = andalusite, Ky=kyanite, Sil = sillimanite, Ho = hornblende, Q = quartz, Opx = orthopyroxene, Pl = plagioclase, Di = diopside, Ab = albite, Jd = jadeite.

of the Ruby Mountains core complex. Here the crust has been tectonically denuded rapidly during the mid-Cenozoic to expose sillimanite-grade rocks indicating a depth of exposure of at least 15 km so that what was the middle crust during mid-Cenozoic is now exposed. The exposed rocks including mid-Cenozoic granitic intrusions are metamorphosed and strongly layered and foliated indicating strong deformation after Oligocene magmatism (Snoke and Lush, 1984). Reflection profiles show that the entire crust is subhorizontally layered and that if anything, lower crustal layering is even more pronounced. This crustal layering subparallels known Cenozoic plastic-deformation zones (mylonites) and is most plausibly a Cenozoic feature caused by plastic extension. This also indicates that the entire crust must be distinctly anisotropic because layered crystalline rocks are invariably anisotropic. For velocity, this anisotropy can range from 5 to 15% (Birch, 1960; Christensen, 1982); the amount of anisotropy expected will be controlled by minerals present of which micas and amphiboles will show the greatest effects (Alexandrov and Rhyzova, 1961; Christensen, 1982).

Geothermal profiles taken from Lachenbruch and Sass (1978) are presented in order to aid interpretation of velocity structure and petrologic history. Temperature profiles are presented for Sierra Nevada (low heat flow), stable craton (normal heat flow), characteristic Basin and Range, and anomalous Basin and Range (Fig. 6) superposed on a petrogenetic grid. This will be used to predict both mineral stability with depth and melting relations. The characteristic Basin and Range geothermal profiles will be used for estimates of present temperatures; for the magmatic peak during the Oligocene, the higher temperatures shown by the Battle-Mountain-high profiles will be used as an upper-temperature limit because they encompass the temperature ranges calculated by Lachenbruch and Sass (1978, p. 233, 234 and 241) for combinations of extending, intruded and underplated Basin-and-Range lithosphere.

The petrogenetic grid (Fig. 6) indicates that, for the highest geotherm, hornblende would be unstable at about 20 km depth and garnet would appear in mafic rocks at about 23 km. Granodiorite would begin to melt at about 16 km and be completely molten at about 25 km. These depths will not correlate exactly with present crustal depths because of the dynamic interaction of extension, intrusion and uplift but rather present some rough estimates. The disappearance of hornblende at intermediate crustal depths suggests that extreme anisotropy (15%) associated with amphibolites (Christensen, 1982) will not be present in the more mafic (V=6.7-6.8 km/s) zone in the lower crust but rather the relatively small anisotropy (3-6%) found in pyroxene granulites. In addition, the probable occurrence of garnet in pyroxene granulite (Fig. 6) means that the velocity will be higher by several tenths of a km/s for a given mafic composition or that the composition inferred from velocity is somewhat more felsic when garnet is present.

The velocity profile is related to the geotherm and petrogenetic grid in order to better infer rock types and their amounts at depth. In the 6.2 km/s zone, mica will be stable and will tend to lower the velocity slightly for a given composition because of its orientation parallel or somewhat oblique to layering. Amphibolites are probably scattered throughout the quartzofeldspathic succession as shown by exposures. The 6.5 km/s zone is about one third mafic rocks that could range from amphibolite to pyroxene granulite interlayered with quartzofeldspathic rocks. The interlayering of extreme compositions rather than more uniform diorites is inferred from the abundant multicyclic reflections (Fig. 1). If we assign a velocity of 7.0 km/s for normal gabbroic rocks at lower crustal depths, the 6.8 km/s zone in the lower crust consists of about one third granitic rocks and two thirds mafic rocks. This is also consistent with the high reflectivity of the lower crust. Widespread garnet in mafic rocks, which seems likely, would increase velocity and decrease the amount of mafic rocks estimated from velocity. The 3-km-thick, high-velocity (7.4-7.8 km/s) zone at the base of the crust indicates a composition intermediate between mafic and ultramafic such as pyroxene- and possibly garnet-rich rocks. This zone, which is also layered, but without much continuity for the individual layers, could consist of a residuum from partial melting or a cumulate zone from crustal underplating.

The amount of mafic material in the crust is an important parameter because numerous models for Basin and Range development rely on mafic intrusion and/or underplating (Lachenbruch and Sass, 1978; Klemperer et al., 1986; Thompson and McCarthy, 1986). The above estimates indicate that at the most 50% of the present crust consists of mafic material. Another way of looking at the problem, though certainly not independent, is by means of mean crustal velocity (Smithson and others, 1981). A mean crustal velocity of 6.6 km/s also suggests that the crust is about 50% felsic and 50% mafic. If thick Sierran-type crust were developed during the Mesozoic (Coney and Harms, 1984), it would have had a thickness of about 50 km and almost exactly the same proportion of felsic to mafic rocks (Bateman and Eaton, 1967) as this part of the Basin and Range. Therefore, homogeneous extension would result in essentially the same amount of mafic material in the crust after it was thinned to 31 km. This leaves little room for mafic intrusion or crustal underplating. If we allow for differences in the two data sets and for variability in geologic assumptions, we postulate that an absolute maximum of about 25% of the Basin-and-Range crust might have been added through intrusion and/or underplating; i.e., the pre-intrusion crust at the beginning of the Cenozoic was only about 25% mafic.

Deformed Oligocene granites in the present upper crust have a crustal isotopic signature (Kistler et al., 1981). These were generated in the deep crust as the temperature rose due to intrusion and lithospheric thinning and separated from a solid residuum; therefore, some of the layered material in the 6.8 km/s zone is probably a garnet-pyroxene residuum (Winkler, 1972) and some is injected mafic material (now garnet-pyroxene granulite). As modest amounts of mafic material were added to the deep crust, granites separated out and moved upward to form much of the 6.2 km/s zone. Brodie and Rutter (1987) suggest that lower crust of the Ivrea zone extended along listric granulite-facies mylonite zones during the Mesozoic and conclude that simple shear was an important process. Similarily, in the Basin-and-Range crust, simple shear may have operated on a small scale to accommodate pure shear on a crustal scale so that granitic and gabbroic intrusions together with country rocks were rotated and drawn out into subhorizontal layers and metamorphosed during the strong Cenozoic extension (Fig. 2).

The amplitude of Moho reflections (Klemperer and others, 1986; and Fig. 7) appears to be unique for the Basin and Range. The multi-cyclic character of Moho reflections (Fig. 1) demonstrates that the Moho is a layered zone, and the duration of these events (0.3-0.4 s) indicates that the zone is 1.2 to 1.5 km thick. These values for Moho thickness are somewhat arbitrary because individual reflections are not continuous but appear and disappear at slightly different levels to give a continuous zone of Moho reflections. This indicates a laterally discontinuous layering or large-scale lensing (possibly boudinage) marking the Moho zone. The amplitude of the reflection indicates a large velocity contrast such as interlayering of

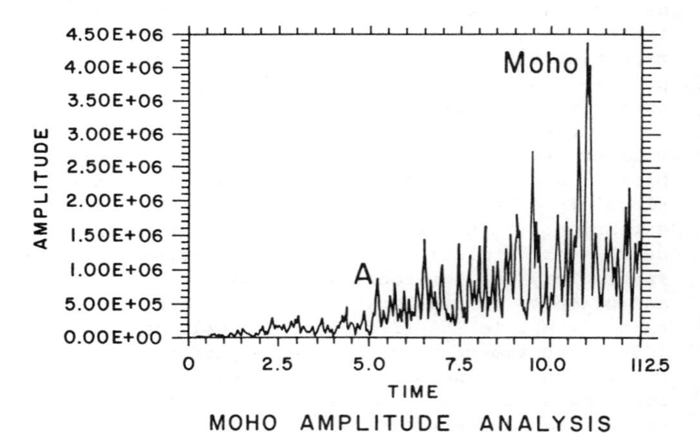

Fig. 7. True-amplitude plot of trace 34 from the 26.3 km wide-angle gather from Ruby Mountains. Based on summing 10 adjacent traces and correcting for spherical divergence with the crustal velocity function, the large-amplitude Moho event at 11.0 s is 13db above the basement reflection A at 5.4 s.

peridotite with more felsic material. Thus gneissic rocks ranging from granite to diorite in composition might be required to generate such large amplitude reflections; however, the petrogenetic grid (Fig. 6) shows that rocks as felsic as granite would be melted at today's probable range of P-T conditions, or another possibility is that basaltic magma might be ponded as discontinuous sills along the crust mantle boundary. This suggests that the high-amplitude Moho reflection is either caused by felsic-to-intermediate rocks (granites to diorites compositionally) interlayered with peridotite and possibly the presence of some melt in the more felsic (lower velocity) layers or by basaltic magma. The Moho zone is strongly layered and discontinuous because of intershearing of mantle and crustal rocks or because of intrusion of basaltic melt from the mantle along the mantle-crust transition.

The deep crust and Moho in Minnesota provide a sharp contrast with that of the Basin and Range. The broadly arched, gently inclined discontinuous reflections that extend from near the surface to 9 s depth (about 30 km) in part of the ancient gneiss terrain are most plausibly interpreted as a stack of nappes. Because surface exposures and isotope studies suggest periods of anatexis and migmatite generation, I interpret transparent zones as "sweated-out" granitic accumulations and the layering shown by the reflection pattern as indicating a combination of migmatites passing into restites at deeper levels. This structure indicative of early Archean compression extends to almost 30 km - an astonishing vertical extent. The 6-to-8-km-thick zone of short horizontal reflections beneath this may indicate greater horizontal shearing that wiped out the large-scale structures or may possibly indicate underplating. In any case it represents a different structural style and passes into a transparent zone (no reflections) from 38 to approximate Moho depth of 42 km as determined from a refraction study (Greenhalgh, 1981).

Compilations of heat flow from cratons (Drury, 1988) indicate that values are the lowest and average about 0.9 HFU. Because these values are similar to that of the Sierra Nevada, the Sierran geotherm of Lachenbruch and Sass (1978) is used to estimate present temperatures (Fig. 6). This geotherm is similar to those calculated for the craton by Blackwell (1971) and Smithson and Decker (1974) based on different assumptions. In any case, we shall see that kinetics rather than the details of the Archean geotherm will be the critical factor. Reflections down to 11 s indicate that adequate seismic energy should be arriving at Moho depth so that lack of a Moho reflection suggests that the Moho produces either a weak reflection or none at all. If the Moho were a step-function compositional change, particularly if it went from garnet-pyroxene granulite (V=7.2-7.5 km/s) to peridotite (V=8.0-8.2 km/s), this single velocity contrast would produce a much smaller reflection than the reflections from layering in the crust. This result is shown by a 1-D synthetic seismogram comparing a 7.0-8.2 km/s Moho velocity contrast with reflections from typical layered crustal rocks (Smithson et al., 1977; Fig. 1, p. 260) which shows that crustal reflections from layering might be 3 times stronger than a single Moho reflection.

The other possibility that no reflection is generated suggests that the Moho is gradational over seismic-reflection wavelengths. The most likely explanation for this is that the Moho is a phase change, and the appropriate phase change is the gabbro-eclogite transition because this has been documented experimentally (Green and Ringwood, 1967) and because these rocks yield the appropriate velocities (Christensen, 1982). A phase change is unlikely to be represented by a sharp boundary but more likely is spread out over some distance because of kinetic effects. The effect would be to smear out the velocity contrast so that the transitional zone would be a poor reflector at vertical incidence but still mark a seismic Moho for long-wavelength refracted waves and wide-angle diving waves. In this case, the crust-mantle transition would take place within a rock of constant composition, gabbro, and this would be an example of Griffin and O'Reilly's suggestion (1987) that the seismic and petrologic Moho are different.

The petrogenetic grid (Fig. 6) indicates that the Minnesota ancient gneiss terrain would be in the eclogite facies stability range at about 15-20 km depth; however, recent sudies by Wood (1988) suggest that the low-temperature extrapolation of Green and Ringwood's (1967) field boundary is in error and eclogites don't become stable until about 40 km depth. However, crustal rocks of cratons could have passed through eclogite-facies conditions as they cooled but do not preserve any record of achieving this mineralogy. This is because of kinetic factors that seem to commonly leave eclogite facies unattained during decreasing temperatures. Austreheim and Griffin (1985) demonstrated that conversion of mafic granulites to eclogites only occurred locally in response to shearing and fluids. The ancient gneiss terrain has undergone a series of events in Archean and early Proterozoic time, and one of the events may have catalyzed an eclogitic recrystallization in the deep crust, or temperatures at 40-50 km depth may have been sufficient for an eclogitic recrystallization.

If the Moho in this area does represent the gabbro-eclogite phase change, this has important implications for lower crustal composition. Examples of quartzofeldspathic gneisses, which are interlayered with scattered eclogites, are found with relics of eclogite mineralogy in the classic eclogite terrain of Norway (Krogh, 1980). The disappearance of plagioclase in favor of jadeite can greatly increase the density and velocity; e.g., graywacke recrystallized in eclogite facies may exhibit a velocity increase from 6.1 to 6.7 km/s. This means that quartzofeldspathic rocks in the eclogite facies exhibit typical lower crustal velocities, but velocities still somewhat lower than gabbro. Thus if the Moho in Minnesota is formed by the gabbro-eclogite phase change, lower crustal rocks above the Moho might be significantly quartzofeldspathic and still exhibit lower crustal velocities.

Conclusions

The deep crust of the Basin and Range and the Minnesota Archean represent strongly contrasting tectonic features. The lower crust in the Basin and Range is strongly horizontally layered, probably as a result of extensional plastic flow during the Cenozoic, representing a major metamorphic event that generated lower crustal granulites during extension. Any folding present is represented by intrafolial (isoclinal folds) and the strong parallelism results from this horizontal flow. The lower crust in the Basin and Range is compositionally layered and predominantly mafic underlain by a more mafic 3-km zone of cumulate or residual material related to the magmatic history (either underplating or partial melting). The Moho is a sharp compositional boundary marked by strong interlayering of mantle and crustal rocks to produce unusually high reflectivity, and the crustal rocks may be partially molten to explain the reflectivity. Moho interlayering may be caused by shearing and/or intrusion. Emplacement of gabbroic magma in the lower crust resulted in melting and separation of granitic material that intruded the upper crust followed by plastic deformation during extension. About 50% of the crust is mafic material, and probably an absolute maximum of 25% (8 km) of the crust represents mafic additions and underplating during Cenozoic extension. About 10% (3 km) might be a more realistic figure. A crustal addition to thicken the crust and compensate for exhumation and doming of the Ruby Mountains core complex (Thompson and McCarthy, 1986) does not seem to be present, but detailed velocity profiles adjacent to the domed areas may be necessary in order to recognize such additions.

In the Minnesota-early Archean crust reflections became weaker and less continuous in the lower crust, and absent in the lowermost crust and Moho zone (based on the Moho depth from a single refraction interpretation). The early Archean crust was about 50 km thick and developing during compressive tectonic regime leaving 10-12 km of the lowermost crust that might have been underplated. Here most of the deep crust seems to have been thickened by stacking of nappes, which are "floating" in small bodies of later anatectic granite. In contrast, the granite melts in the Basin and Range have been sheared out into the layering. The lowermost Archean crust is a distinctly different tectonic regime, possibly formed by subhorizontal compressional shearing that may have erased larger scale structures or by later gabbroic underplating that thickened the early Archean crust. The Moho in Minnesota, which may be typical for cratons (Smithson et al., 1987) must be strikingly different from the Basin and Range. It is either gradational or a relatively small step-function change in chemical composition such as garnet pyroxene granulite to peridotite; it is probably not layered unless the velocity changes across the layering are very small. If it is gradational, then it is probably formed by the gabbro-eclogite phase change along a gabbroic zone, and this gabbroic zone could have been underplated.

If the gabbro-eclogite transition has taken place in the lowermost Archean crust of Minnesota, it has important consequences for lower crust of the craton. This may explain the apparently poor Moho reflectivity underneath the craton. If eclogite is formed at lower crustal P-T conditions at some stage in the history of cratonic crust, then gabbroic material will have a velocity of 8 km/s and appear as mantle to seismic investigations. Similarly if quartzofeldspathic rocks recrystallize in eclogite facies, they will have velocities approaching those of gabbro. The effect would truly be to underestimate crustal thickness as too small and to overestimate the amount of mafic rocks in the crust if a distinction is made between seismically determined crust (the original definition) and compositionally determined crust (Griffin and O'Reilly, 1987). In any case, the nature of the Moho must be very different between the Basin and Range and Minnesota Archean, and we should expect to find different compositions and structures for lower crust and Moho in different tectonic provinces.

Acknowledgments. Support for this research was received from NSF Grants EAR-8511919, EAR-8512083, EAR-8618633, EAR-8519153 and EAR-8618633. Processing was carried out on the DISCO VAX 11/780 of the Program for Crustal Studies. I am indebted to Professor B.R. Frost for advice on phase petrology and to Professor A.W. Snoke for advice on structure.

References

Alexandrov, K.S. and T.V. Rhyzhova, Elastic properties of rock forming minerals 2, Layered silicates, Bull. Acad. Sci., USSR, Geophys. Ser., 9, 1165-1168, 1961.

Austrheim, H., and W.L. Griffin, Shear deformation and eclogite formation within granulite-facies anorthosites of the Bergen arcs, western Norway, Chemical Geology, 50, 267-281, 1985.

Bateman, P.C., and J.P. Eaton, Sierra Nevada batholith, Science, 158, 1407-1417, 1967.

Bauer, R.L., Multiphase deformation in the Granite Falls-Montevide area, Minnesota River Valley, Geol. Soc. Amer. Spec. Paper 182, 1-17, 1980.

Binns, R.A., Hydrothermal investigations of the amphibole-granulite facies boundary, Geol Soc.

Aust. Spec. Publ. No. 2, edited by D.A. Brown, 341-344, 1969.

Birch, F., The velocity of compressional waves in rocks to 10 kilobars, 1, Journ. Geophys. Res., 65, 1083-1102, 1960.

Blackwell, D.D., The thermal structure of the continental crust, in The Structure and Physical Properties of the Earth's Crust, Geophys. Monogr. Ser., 14, edited by J. G. Heacock, p. 169, AGU, Washington, D.C., 1971

Brodie, K.H., and E.H. Rutter,Deep crustal extensional faulting in the Ivrea Zone of Northern Italy, Tectonophysics, 140,193-212, 1987.

Christensen, N.I., Seismic velocities, in Handbook of Physical Properties of Rocks, II, edited by R.S. Carmichael, CRC Press, 2-228, 1982.

Christensen, N.I. and D.L. Szymanski, Origin of Reflections from the Brevard fault zone, Journ. Geophys. Res., 93, 1087-1073, 1988.

Coney, P.J. and T.A. Harms, Cordilleran metamorphic core complexes: Cenozoic extensional relics of Mesozoic compression, Geology, 12, 550-555, 1984.

Crittenden, M.D., Jr., P.J. Coney, F.H. Davis, editors, Cordilleran metamorphic core complexes, Geol. Soc. Am. Mem., 153, 490 p., 1980.

Drury, M.J., Tectonothermics of the North American Great Plains basement Tectonophysics, 148, 299-208, 1988.

Fyfe, W.F., Archean tectonics, Nature, 249, 338-340, 1974.

Gibbs, A.K., B. Payne, T. Letzer, L.D. Brown, J.E. Oliver, and S. Kaufman, Seismic reflection study of the Precambrian crust of central Minnesota, Geol. Soc. of Amer. Bull., 95, 280-294, 1984.

Goldich, S.S. and C.E. Hedge, 3,800 myr granitic gneiss in southwestern Minnesota, Nature, 252, 467-468, 1974.

Green, D.H. and A.E. Ringwood, An experimental investigation of the gabbro to eclogite transformation and its petrological applications: Geochim et Cosmochim. Acta 31, 767-833, 1967.

Greenhalgh, S.A., Seismic investigations of crustal structure in east central Minnesota, Phys. Earth Planet Interiors, 25, 372-389, 1981.

Griffin, W.L., and S.Y. O'Reilly, Is the continental Moho the crust-mantle boundary?, Geology, 15, 241-244, 1987.

Hauser, E., C. Potter, T. Hague, S. Burgess, S. Burtch, J. Mutschler, R. Allmendinger, L. Brown, S. Kaufman, and J. Oliver, Crustal structure of eastern Nevada from COCORP deep seismic reflection data, GSA Bulletin, 99, 1987.

Hurich, C.A., S.B. Smithson, D.M. Fountain, and M.C. Humphreys, Seismic evidence of mylonite reflectivity and deep structure in the Kettle Dome metamorphic core complex, Washington, Geology, 13, 577-580, 1985.

Kistler, R.W., E.D. Ghent, and J.R. O'Neil, Petrogenesis of garnet two-mica granites in the Ruby Mountains, Nevada, JGR, 86, 10,591-10,606, 1981.

Klemperer, S.L., T.A. Hauge, E.C. Hauser, J.E. Oliver, and C.J. Potter, The Moho in the northern Basin and Range province, Nevada, along the COCORP 40½N seismic-reflection transect, Geol. Soc. Amer. Bull., 97, 603-618, 1986.

Krogh, E.J., Compatible P-T conditions for eclogites and surrounding gneisses in the Kristiansund area, western Norway, Contrib. Min. Pet., 75, 387-393, 1980.

Lachenbruch, A.H., and J.H. Sass, Models of an extending lithosphere and heat flow in the Basin and Range province, GSA Mem 152, 209-250, 1977.

Newton, R.C., and E.C. Hansen, The origin of Proterozoic and late Archean charnockites -- evidence from field relations and experimental petrology, GSA Mem. 161, 167-179, 1983.

Sandiford, M. and R. Powell, Deep crustal metamorphism during continental extension: modern and ancient examples, Earth Planet. Sci. Lett., 79, 151-158, 1986.

Sims, P.K., K.D. Card, G.B. Morry, and Z.E. Peterman, The Great Lakes tectonic zone - A major crustal feature in North America, Geol. Soc. Amer. Bull., 91, 690-698, 1980.

Smithson, S.B. and E.R. Decker, A continental crust model and its geothermal implications, Earth Planet. Sci. Lett., 22, 215, 1974.

Smithson, S.B. and P.N. Shive, Seismic velocity, reflections, and structure of the crystalline crust, Geophys. Mono. 20, AGU, 254-271, 1977.

Smithson, S.B., R.A. Johnson, Y.K. Wong, Mean crustal velocity: a critical parameter for interpreting crustal structure and crustal growth, EPSL, 53, 323-332, 1981.

Smithson, S.B., R.A. Johnson, C.A. Hurich, P.A. Valasek, and C. Branch, Deep crustal structure and genesis from contrasting reflection patterns: an integrated approach, Geophysical Journal of R. astr. Soc., 89, in press, 1987.

Snoke, A.W., and A.P. Lush, Polyphase Mesozoic-Cenozoic deformational history of the northern Ruby Mountains-East Humboldt Range, Nevada, in Western Geological Excursions (GSA annual meeting guidebook, edited by J. Lintz, Mackay School of Mines, Department of Geological Sciences, Reno, Nevada, 232-260, 1984.

Thompson, G. and J. McCarthy, Geophysical evidence for igneous inflation of the crust in highly extended terraines (abs.) EOS Trans AGU, 67, 1184, 1986.

Valasek, P.A., R.B. Hawman, R.A. Johnson, and S.B. Smithson, Nature of the lower crust and Moho in eastern Nevada from "Wide-Angle" reflection measurements, GRL, 14, 1111-1114, 1987.

Winkler, H.G., Petrogenesis of Metamorphic Rocks, Springer-Verlag, New York, 334, p., 1972.

Wood, B.J., Phase relations and seismic properties of mafic rocks under lower crustal conditions, Italy--U.S. Workshop on the Nature of the Lower Continental Crustal, Programme - Abstracts, Field Trip Guide, Universita degli Studi di Milano, 77-79, 1988.

A "GLIMPCE" OF THE DEEP CRUST BENEATH THE GREAT LAKES

Alan G. Green[1], W. F. Cannon[2], B. Milkereit[1], D. R. Hutchinson[3], A. Davidson[1], J. C. Behrendt[4], C. Spencer[1], M. W. Lee[4], P. Morel-à-l'Huissier[1], W. F. Agena[4],

Abstract. Approximately 1350 km of multichannel seismic reflection data and an equivalent amount of seismic refraction data have been collected across the North American Great Lakes. The seismic surveys, sponsored by the Great Lakes International Multidisciplinary Program on Crustal Evolution (GLIMPCE), were designed to resolve the deep crustal structure of the 1.89-1.82 Ga Penokean orogen, the 1.3-1.0 Ga Grenville orogen and the 1.11-1.09 Ga Midcontinent rift system. In north-central Lake Huron a band of gently east-dipping reflections at about 18-20 km depth separates a complex and highly reflective lower crustal layer from a markedly less reflective upper crustal layer. The lower layer is interpreted as attenuated crust of the Superior cratonic margin and the upper layer is assumed to be composed of continental margin deposits of the Huron supergroup, an exotic mass that collided with the cratonic margin during the Penokean orogeny, and younger granites and rhyolites. The intervening band of gently east-dipping reflections may delineate a master décollement zone, active during the Penokean orogeny. All structures in north-central Lake Huron are truncated abruptly at the western end of Georgian Bay by the Grenville Front, represented by the westernmost event of a spectacular series of moderately east-dipping reflections. These reflections, which extend from the surface to about 55 km depth, are interpreted as discontinuities between gneissic and migmatitic rocks of varying lithology. Where exposed in the Grenville front tectonic zone, the different rock units are commonly juxtaposed along mylonite zones with a northwest thrust sense. Moho may deepen by 5-10 km along a narrow region just southeast of the Grenville front.

Seismic reflection data from Lake Superior reveal an extraordinary thickness of Keweenawan mafic lavas and sedimentary rocks deposited in a number of discrete grabens/half-grabens of the Midcontinent (Keweenawan) rift system. Total vertical thickness of layered Keweenawan strata exceeds 30 km beneath some parts of the lake. The geometry of the central basin changes significantly along the axis of the Midcontinent rift system. In western Lake Superior the Keweenawan lavas thicken northward toward a south-dipping growth fault, in the center of the lake they thicken southward toward a north-dipping growth fault, and in the eastern part of the lake the structures are relatively symmetric. These mega-grabens and mega-half-grabens with differing asymmetries are separated by transfer faults or accommodation zones recognized on potential field maps. The segmented character of the MRS is similar to that observed along some passive continental margins and the East African rift system.

[1] Geological Survey of Canada, Ottawa, Ontario, K1A 0Y3,
[2] United States Geological Survey, Reston, Virginia, 22092,
[3] United States Geological Survey, Woods Hole, MA 02543,
[4] United States Geological Survey, Denver, CO 80225

Introduction to GLIMPCE

Proterozoic tectonism and magmatism in the Great Lakes region had a major influence on the evolution of the Canadian Shield [Hoffman, 1988]. A number of the more important Proterozoic structures on the North American continent occur there (Figure 1), and although rocks under the Great Lakes have been sampled only rarely, major structures can be projected beneath the lakes on the basis of potential field data and other geoscientific information.

To help resolve the history and three-dimensional geometry of the crust in this region, the Great Lakes International Multidisciplinary Program on Crustal Evolution (GLIMPCE) was initiated in the fall of 1985 by a consortium of more than sixty scientists with a common interest in the geology of the midcontinent. Under the auspices of this program approximately 1350 km of multichannel seismic reflection data and an equivalent amount of high quality seismic refraction data were collected in Lakes Huron, Michigan and Superior, and new aeromagnetic surveys were flown across Lakes Huron and Superior. The surveys were designed to resolve the deep structure of the Huronian continental margin and Grenville front in Lake Huron, the Wisconsin magmatic terrane and Niagara fault in Lake Michigan, and the Midcontinent (Keweenawan) rift system (MRS) in Lake Superior and northern Lake Michigan. Proposed future projects include new gravity surveys in several of the lakes, enhanced geological, geochemical and geochronological studies on the surrounding land, and eventually deep drilling at key locations.

In this paper we (i) outline briefly the Proterozoic geology of the Great Lakes region, (ii) present the results of seismic reflection profiling across the Grenville front and adjacent terranes in Lake Huron and (iii) describe the structure of the MRS deduced from seismic reflection data recorded in Lake Superior. In companion papers Behrendt et al. (this volume) show unmigrated seismic reflection sections from Lakes Superior and Michigan and review processes that might have created the massive volumes of magma emplaced along the MRS, and Epili and Mereu (this volume) present in-line and broad-side seismic refraction data recorded in Lake Superior.

Geologic and Tectonic Setting

Penokean Orogen

Several periods of Proterozoic extension and compression are recorded in rocks assigned to the Penokean orogen (Figure 1). Soon

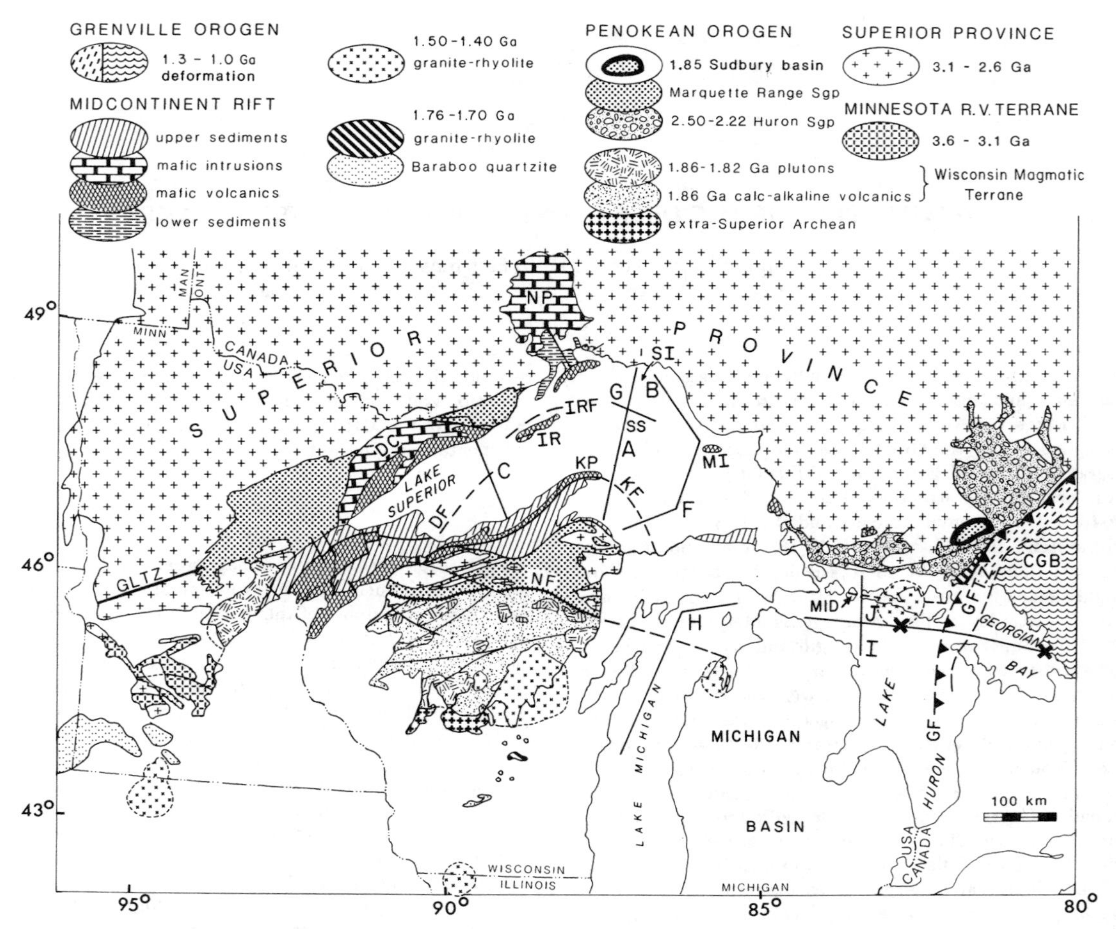

Fig. 1. Simplified Precambrian geology map of the Great Lakes region (modified from Hoffman, 1988) showing the locations of GLIMPCE deep seismic reflection profiles A to J. The crosses on line J delineate the region covered by the seismic section shown in Figure 2. CGB - Central Gneiss belt; DC - Duluth complex; DF - Douglas fault; GF - Grenville front; GFTZ - Grenville front tectonic zone; GLTZ - Great Lakes tectonic zone; IR - Isle Royale; IRF - Isle Royale fault; KF - Keweenaw fault; KP - Keweenaw Peninsula; MI - Michipicoten Island; MID - Manitoulin Island discontinuity; NF - Niagara fault; NP - Nipigon plate; Sgp - supergroup; SI - Slate Islands; SS - Superior Shoals. Manitou Island (not shown) lies between profile A and Keweenaw Peninsula. The uniform dot pattern refers to the Marquette Range supergroup and equivalent rock units. Aeromagnetic and gravity gradient data have been used to extrapolate southwards the Grenville front tectonic zone from its exposure along the shores of Georgian Bay. The Central Gneiss belt in the interior of the Grenville orogen has been sub- divided into a number of microterranes by Davidson (1984a, b), one of which (Britt domain) is crossed by seismic profile J on its eastern end.

after the close of the Archean, at about 2.50 Ga [Krogh et al., 1984], crustal attenuation and rupturing of the southern Superior craton in the vicinity of Lake Huron resulted in the creation of a south-facing continental margin on which rift and passive margin rocks of the Huron supergroup were deposited [Card et al., 1984; Zolnai et al., 1984]. At about 2.33 Ga the supracrustal rocks were folded [Zolnai et al., 1984] and at 2.22 Ga [Corfu and Andrews, 1986] there may have been a second phase of extension when Nipissing diabase widely intruded all levels of the marginal sequence. A later episode of crustal rupturing, which occurred predominantly to the west of previous Proterozoic extension, led to the formation of a south-facing continental margin in the region of Lake Superior and deposition of the lower units of the Marquette Range supergroup and equivalent rocks [Van Schmus, 1976; Medaris, 1983; Sims et al., 1987]. Although there may be some temporal overlap between the lower parts of the Marquette Range supergroup and the upper parts of the Huron supergroup, any spatial overlap is hidden beneath Lakes Michigan and Huron.

Between 1.89 and 1.82 Ga the entire southern edge of the Superior craton from north of Georgian Bay to west of Lake Superior was affected by one or more collisional events that resulted in the Penokean orogeny [Bickford et al., 1986]. South of Lake Superior, the Niagara fault zone juxtaposes the exotic Wisconsin magmatic terrane, a complex assemblage of Proterozoic island arc and structurally dismembered ophiolitic rocks, against deformed sediments of the Marquette Range supergroup. The Wisconsin magmatic terrane and Niagara fault zone can be traced into Lake Michigan on potential field maps, but farther to the east in Lake Huron the location of equivalent structures is problematic. Van Schmus et al. [1975] delineated an east-west trending discontinuity on Manitoulin Island (MID in Figure 1) that separates highly deformed Huronian sequences to the north from a poorly understood southern terrane buried entirely beneath Paleozoic sediments and the waters of Lake Huron. Referred to as Manitoulin terrane by Green et al. [1988], this composite terrane likely includes slices of the Huronian continental margin sequence, vestiges of a northward transported exotic mass that collided with the margin during the Penokean orogeny [Zolnai et al., 1984], and younger granites and rhyolites (see below). Whether or not Manitoulin terrane contains rock assemblages equivalent to those of the Wisconsin magmatic terrane has yet to be determined; gneissic rocks of suspect origin from Manitoulin Island have yielded poor quality, presumably reset, Rb/Sr ages of 1.8-1.6 Ga [Van Schmus et al., 1975].

Granite-Rhyolite Magmatism

Subsequent to the Penokean orogeny two periods of granite-rhyolite magmatism affected the Great Lakes area. Between about 1.76 and 1.70 Ga, approximately coeval with the Central Plains orogeny in the midcontinent [Sims and Peterman, 1986], the Killarney felsic volcano-plutonic complex and related granites were emplaced along the southeastern edge of the Penokean orogen [van Breemen and Davidson, 1988], and within the Wisconsin magmatic terrane there were rhylotic ash flows and intrusion of "anorogenic" granites [Van Schmus and Bickford, 1981]. The second phase of magmatism is represented by the widespread emplacement of "anorogenic" granites and rhyolites dated at 1.50-1.40 Ga [Van Schmus and Bickford, 1981; Davis and Sutcliffe, 1985; Bickford et al., 1986; Van Schmus et al., 1987; van Breemen and Davidson, 1988]. Both suites of granite-rhyolite are expected to occur in the region of Lake Huron crossed by our seismic profiles [Green et al., 1988].

Grenville Orogen

Along the eastern edge of the Superior craton and Penokean orogen lies the Grenville front, one of the continent's great structural boundaries [Wynne-Edwards, 1972]. For most of its nearly 4000-km length it marks the junction of Archean and Proterozoic rocks largely unaffected by Grenvillian orogeny with uplifted lower crustal rocks of the Middle Proterozoic Grenville orogen. On aeromagnetic and gravity gradient maps the trend of distinctive anomalies associated with the Grenville front changes from a general southwesterly direction across the exposed Shield to a more southerly trend beneath Lake Huron (Figure 1; Geological Survey of Canada, 1987; Sharpton et al., 1987].

In southern Ontario the Grenville orogen includes the parautochthonous Grenville front tectonic zone (GFTZ), and a variety of allochthonous micro-terranes that have no obvious affinity with the Archean/Early Proterozoic foreland to the northwest [Davidson, 1984a,b, 1986a,b; Moore, 1986; Dickin and McNutt, 1988]. The GFTZ is composed of a broad band of rocks that were thrust toward the foreland along southeast-dipping ductile shear and mylonite zones under high pressure/high temperature conditions (800-1000 MPa, 600-800°C) [Anovitz, 1987]. On the northern shores of Georgian Bay the boundaries of the GFTZ are the Grenville front to the northwest and a recently discovered major mylonite zone to the southeast [Davidson and Bethune, 1988]. The Grenville front at this location is a broad, southeast-dipping zone of mylonite across which metamorphic grade increases from greenschist facies or less in the foreland to at least middle amphibolite facies in the GFTZ. Although the Grenville front delineates abrupt changes in structural orientation, metamorphic grade and magnetic anomaly pattern, rocks on opposite sides of it seem to be lithologically equivalent. For example, Killarney and 1.50-1.40 Ga granites mapped northwest of the front appear within the GFTZ in the form of orthogneiss, layered gneiss and migmatite. General equivalence of rock assemblages on the two sides of the Grenville front is supported by direct comparisons in the field [Davidson and Bethune, 1988], chemical analyses of mafic dykes that intrude both regions [Bethune and Davidson, 1988], and the results of U/Pb and Sm/Nd studies [van Breemen and Davidson, 1988; Dickin and McNutt, 1988].

Allochthonous terranes of the southern Grenville orogen, which are distinguished by differences in their lithology, internal structure, metamorphic grade and potential field signature, are bounded by ductile shear zones (not shown in Figure 1; see Davidson, 1984a,b). Like the GFTZ these ductile shear zones contain mylonites and kinematic indicators showing a dominant southeast over northwest sense of transport [Davidson, 1984a,b, 1986a; Hanmer and Ciesielski, 1984]. Quantitative estimates of displacement are lacking, but the contrasting geology of some adjacent terranes and the high states of strain in the ductile shear zones are compatible with large displacements.

The pre-metamorphic U/Pb ages of the allochthonous terranes fall into two groups. Plutonic rocks from terranes in the high grade Central Gneiss belt [Wynne-Edwards, 1972] have yielded ages between 1.80 and 1.30 Ga [Krogh and Davies, 1973; van Breemen et al., 1986], and the majority of igneous rocks in the lower grade Central Metasedimentary belt to the southeast of our map area have ages younger than 1.30 Ga [summarized in Davidson, 1986a]. Recent Sm/Nd studies [Dickin and McNutt, 1988] show that rock units southeast of the GFTZ have not been significantly contaminated by material from the foreland. Most terranes within the southern Grenville orogen were subjected to high grades of metamorphism during the various phases of Grenvillian orogeny between about 1.30 and 1.00 Ga. Grenvillian-style tectonism was initiated close to the Grenville front before 1.24 Ga Sudbury dikes were intruded, and the age of later phases of tectonic transport under ductile conditions ranges from 1.15 Ga near the front to 1.06 Ga at the highly sheared boundary between the Central Gneiss belt and the Central Metasedimentary belt [Krogh and Wardle, 1984; van Breemen and Hanmer, 1986; van Breemen et al, 1986].

Midcontinent (Keweenawan) Rift System (MRS)

While the compressional Grenvillian orogeny was active along the eastern margin of the Archean/Early Proterozoic craton, crustal extension was taking place a short distance to the west along the MRS [Wold and Hinze, 1982; Green, 1983; Van Schmus and Hinze, 1985; Dickas, 1986]. The MRS is the source of some of the largest potential field anomalies in the interior of the continent, extending along a 2000-km arc from Kansas to southeastern Michigan via Lake Superior. Only around the margin of Lake Superior are rift-related rocks of the Keweenawan supergroup exposed (Figure 1) [Morey and Green, 1982]; at other locations the MRS is covered by younger sediments. There are few obvious relationships between the trend of the MRS and pre-existing major structures [Klasner et al., 1982]. It transects the Central Plains and Penokean orogens and cuts an arc into the Superior craton. A possible third arm of the MRS extends in a northerly direction across the craton toward Lake Nipigon (NP in Figure 1) [Davis and Sutcliffe, 1985].

Sparsely exposed sandstones and quartzites deposited in one or more broad basins are the lowermost units of the Keweenawan supergroup [Ojakangas and Morey, 1982a]. The principal period of extension and subsidence was confined to a more limited area and was accompanied by extrusion of voluminous flood basalts [Green, 1983], deposition of minor interflow sediments, and intrusion of ig-

TABLE 1. Processing sequence applied to the GLIMPCE data

Sequence No.	Processing Step
1.	Demultiplex
2.	Recover gain
3.	Geometry definition
4.	Trace edit
5.	Filter and resample to 8 ms
6.	F-K filter in shot domain
7.	Common reflection point (CRP) sort (24 or 30 fold)
8,18,21.	Automatic gain control
9,11.	Velocity analysis
10.	Preliminary stack
12.	Multiple suppression in the CRP domain
13.	Normal moveout correction
14.	First break suppression (mute)
15.	Deconvolution
16.	CRP stack
17.	Time varying bandpass frequency filter
19.	Predictive deconvolution
20.	Vertical sum (4:1) to give 50 m CRP spacing (optional)
22.	Two dimensional amplitude smoothing
23.	Display CRP stack
24.	Bandpass frequency filter (optional)
25.	F-K filter (optional)
26.	Constant velocity F-K migration tests
27.	F-K migration
28.	Display F-K migrated section
29.	Vertical sum (2:1) to give 100 m CRP spacing (optional)
30.	Compute and apply coherency filter (optional)
31.	Display coherency enhanced data (optional)

neous bodies, including the huge Duluth layered complex [Weiblen, 1982]. Precise U/Pb dating reveals that the bulk of volcanism happened between 1.11 and 1.09 Ga [Davis and Sutcliffe, 1985; Palmer and Davis, 1987; Paces and Davis, 1988].

Following the main phase of volcanism, the upper Keweenawan Oronto group of sedimentary rocks was deposited in broad subsiding basins [Daniels, 1982; Morey and Ojakangas, 1982; Ojakangas and Morey, 1982b]. The lower unit of the Oronto group, the Copper Harbor conglomerate, rests conformably on lavas and consists of a thick sequence of coarse conglomerate and sandstone derived largely from Keweenawan igneous rocks. Near its base, lava flows are intercalated with the sediments. Conformably overlying the Copper Harbor conglomerate is the hydrocarbon-rich shale of the Nonesuch formation and this in turn is overlain by sandstones of the Freda formation. During and after a poorly understood episode of compression there was additional broad-scale sagging of the crust accompanied by deposition of Jacobsville and equivalent sandstones unconformably on the older rocks [Kalliokoski, 1982; Morey and Ojakangas, 1982].

Several major faults have been mapped onshore, or identified offshore on the basis of geophysical data (Figure 1). In outcrop these faults bring Keweenawan lavas in the hanging wall against younger Keweenawan sedimentary strata in the footwall, demonstrating at one time they were the site of reverse movements. Whether or not these faults played a role in the extensional history of the MRS was one of the principal questions addressed by GLIMPCE.

Seismic Program - Data Acquisition and Processing

Although deep lakes are an impediment to geological studies they are well suited for cost effective seismic surveying. Seismic reflection data can be collected more quickly (ten times as fast) and more cheaply (about 15-20% of the cost) than equivalent surveys on land, and coupling of the source and receivers is uniformly good in deep water, typically resulting in superior depth penetration. Locations of the GLIMPCE seismic reflection profiles are shown in Figure 1. Profile I in Lake Huron crosses part of the Huronian continental margin and its presumed junction (MID) with Manitoulin terrane, and profile J extends from Manitoulin terrane across the Grenville front to Britt domain, one of the microterranes identified in the Central Gneiss belt of the Grenville orogen [Davidson, 1984a,b; Davidson and Bethune, 1988]. The southern part of profile H in Lake Michigan traverses the Wisconsin magmatic terrane and the bounding Niagara fault. Profiles A, B, C, F and G in Lake Superior and the northern part of profile H in Lake Michigan cross various structures associated with the MRS. We shall be concerned primarily with information contained on migrated seismic sections from the eastern portion of profile J in Lake Huron and profiles A and C in Lake Superior.

GLIMPCE seismic reflection data were collected by a contractor using acquisition parameters similar to those used by BIRPS around the British Isles and LITHOPROBE off the east and west coasts of Canada. A 3024-m streamer with 25-m long hydrophone arrays and 27 sensors per array was towed at depths of 10-14 m. The energy source consisted of an 80-m wide tuned airgun array of 60 active guns with a total capacity of 127 litres and dominant frequency range of about 6-57 Hz. The 20 s of data were recorded on a 120-channel acquisition system with a 4-ms sample interval. Airgun "pops" every 50 or 62.5 m yielded 30- or 24-fold coverage. Navigation was achieved by continuous Loran-C and doppler-sonar with frequent calibration by GEONAV satellite positioning. A Global Positioning System (GPS) was available for back-up.

Table 1 shows a typical processing sequence applied to the seismic reflection data. The data were processed to common mid-point (CMP) stack at the Denver center of the USGS [Lee et al., 1988] and post-stack processing was completed at the Ottawa center of the GSC [Milkereit et al., 1988].

Results From the Grenville Front and Adjacent Terranes - Profile J

A line diagram and interpretation sketch based on data collected along the eastern part of profile J is presented in Figure 2, and an example of the F-K migrated seismic data is displayed in Figure 3 [Green et al., 1988]. Tectonic boundaries shown along the top of the figures were positioned using new potential field maps of the region [Geological Survey of Canada, 1987, and unpublished]; well defined anomalies on aeromagnetic and gravity gradient maps allow the GFTZ to be confidently extrapolated from outcrop along the north coast of Georgian Bay southward to the vicinity of seismic profile J (Figure 1). The seismic section partitions naturally into three units on the basis of characteristic reflection patterns. Units defined by the seismic data relate one-for-one with major tectonic units projected from the exposed shield. These are Manitoulin terrane in the west, the GFTZ in the center, and combined terranes of the Central Gneiss belt in the east.

The region underlying Manitoulin terrane is distinguished by a complex and highly reflective lower crustal layer and a markedly less reflective upper crustal layer. The two layers are separated by a prominent band of strong reflections that dip gently eastward from 4 s at position C to 5 s at B and 6 s (18-20 km depth] at A (Figure 2a). Reflection Moho, defined as the base of significant crustal reflections

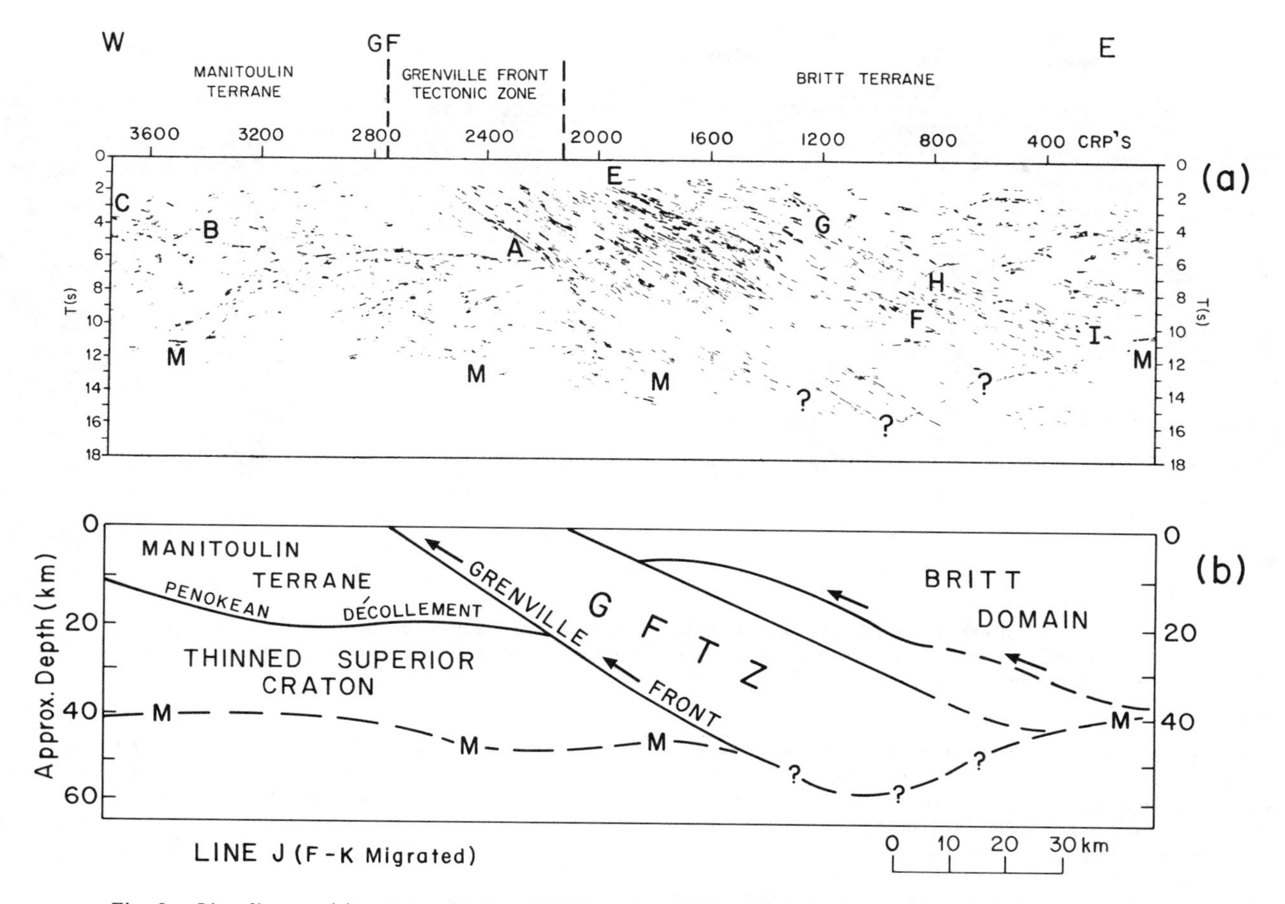

Fig. 2. Line diagram (a) and simplified model (b) based on F-K migrated seismic reflection data collected along the eastern part of profile J across the Grenville front in Lake Huron/Georgian Bay. Vertical scale for 2a is travel time in seconds and for 2b is approximate depth in kilometers. M is reflection Moho determined from amplitude decay studies of groups of shot gathers; CRP - common reflection points. Other letters in 2a are referred to in the text. An example of the F-K migrated seismic reflection data centered on the Grenville front tectonic zone is shown in Figure 3, and an example of an average amplitude decay curve is presented in Figure 4. The location of this segment of profile J is outlined by crosses in Figure 1.

on the stacked seismic section [Klemperer et al., 1986], is relatively flat in this region at 11-13 s travel time (40-46 km depth; Figure 2b). Supporting estimates of the depth to Moho, shown by the letter M, represent peaks and/or notable gradient changes in average amplitude decay curves derived from groups of shot gathers (see the example amplitude decay curve in Figure 4).

In our cross-sectional model (Figure 2b) the reflective lower layer is shown as thin crust of the Superior craton margin, which was attenuated and ruptured during the 2.5 Ga extensional event and upon which continental shelf and rise assemblages of the Huron supergroup were initially deposited. The A-B-C band of reflections is interpreted as the 1.89-1.82 Ga Penokean décollement, active when an exotic mass from the south was thrust over the Superior continental margin causing the shelf and rise sediments to be driven northward toward the stable craton [Zolnai et al., 1984; Green et al., 1988]. Manitoulin terrane constitutes the upper crustal layer. It is interpreted as a composite unit containing displaced Huronian strata, the transported exotic mass, and younger granites and rhyolites. The proposed interpretation is not unique, but is consistent with: (i) the tectonic environment deduced for this region from geological and geochronological studies of rocks exposed north of Lake Huron [Card et al., 1984; Zolnai et al., 1984; Davidson, 1986b]; (ii) minimum estimates for the total shortening of marginal sediments [Zolnai et al., 1984], which lead to the conclusion that Huronian strata once extended well to the south of seismic line J [Green et al., 1988]; (iii) the high amplitudes of the A-B-C reflections; (iv) the very different nature of the seismic section above and below the A-B-C band of reflections; (v) the 18-22km (6-7 s travel time) thickness of the lower layer, which is similar to the thickness of crust beneath modern passive continental margins; (vi) the complex reflectivity of the lower crustal layer, a feature often observed in regions affected by crustal extension. Our model for this region is similar in many respects to that described recently for the Quebec Appalachians, where a Taconic décollement delineated by strong reflections separates highly reflective continental margin basement from an overlying less reflective magmatic arc complex and exotic continental mass [Spencer et al., 1987].

Manitoulin terrane and underlying structures are sharply truncated by a spectacular 32-km-wide zone of apparent east-dipping re-

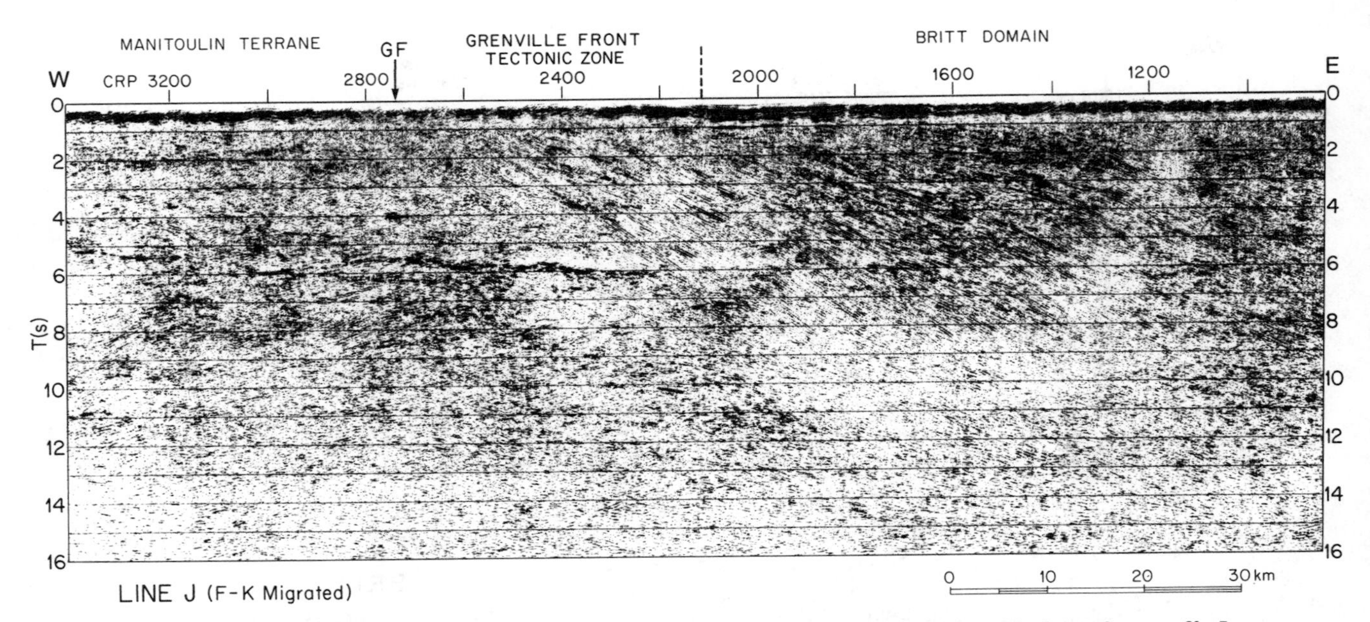

Fig. 3. A portion of F-K migrated seismic reflection data recorded across the Grenville front along profile J. Vertical scale is travel time in seconds. Only every eighth trace is plotted in the section shown. CRP - common reflection points.

flections associated with the GFTZ. High amplitude reflections along the western margin of the GFTZ delineate the Grenville front mylonite zone at depth and a marked change in the strength and density of apparent east-dipping reflections, EF in Figure 2a, coincides with the mylonite zone along its southeastern boundary. Within the GFTZ the apparent dip of events decreases systematically from about 35° at the Grenville front to about 25° along EF, in general agreement with the changing attitude of equivalent structures mapped onshore [Davidson, 1986a,b]. Strong colinear reflections extend to 9 s travel time and weaker but still discernible events can be traced to about 15.5 s on the migrated sections. If these later events originate within the crust beneath the survey line, then reflection Moho would be quite deep (about 55 km) at this location. A narrow zone of thickened crust would be consistent with the results of seismic refraction surveys conducted across three areas of the exposed shield in Quebec [Mereu and Jobidon, 1971; Berry and Fuchs, 1973; Mereu et al., 1986]. All three refraction data sets require a 5-10 km increase in crustal thickness near the GFTZ.

An explanation for the high reflectivity of the GFTZ is found in rock units mapped along the north shore of Georgian Bay [Davidson and Bethune, 1988]. Pronounced layering, tectonically induced under ductile conditions, is ubiquitous there. Most rock units have been intensely flattened and aligned parallel or sub-parallel to the Grenville front; northeast-trending lenticular units of orthogneiss, layered gneiss and migmatite abut along shear and mylonite zones that dip moderately southeast. Mylonite zones vary from tens to hundreds of metres wide. Reflections likely occur at highly strained contacts between the gneissic and migmatitic rocks of varying lithologies, with particularly strong reflections originating at anisotropic zones of mylonite and cataclasite. Rock units seem to be so deformed that layering created by extreme attenuation and plastic flow has yielded a seismic image resembling a well-stratified sedimentary unit.

In the Central Gneiss belt immediately to the east of the GFTZ, reflections continue to dip in an apparent easterly direction as far as the events marked GHI in Figure 2a. Farther to the east the orientations of reflections are more variable with a predominant subhorizontal fabric. We interpret the region bounded by east-dipping reflections EF and EGHI as a buried microterrane, and the region to the east as part of Britt domain (Figure 2b). Along the coast

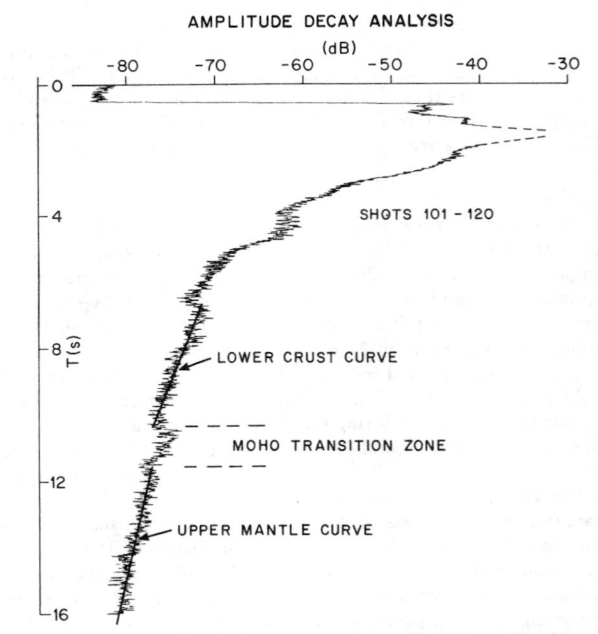

Fig. 4. Average amplitude decay curve determined from 20 adjacent shot gathers near the eastern end of seismic profile J.

of Georgian Bay, the transition from the GFTZ to Britt domain is marked by a change from northeast to northwest in the orientations of regional foliation, gneissic layering and disposition of mapped units. Within Britt domain there are foliated, sheet-like units of migmatitic orthogneiss and layered gneiss folded about gently southeast plunging axes. Because the seismic profile crosses the southern extension of Britt domain at a highly acute angle it is difficult to interpret individual reflections or reflection packages in terms of specific geological structures. However, we suspect that many of the planar and curvilinear shallow-dipping reflections recorded in this region are related to folded structures similar to those mapped onshore. Reflection Moho beneath the eastern end of seismic profile J is well defined at 10-11 s travel time (Figures 3 and 4).

Results from recent geological mapping and geochronological and geobarometric/geothermometric studies in the exposed southern Grenville orogen provide important constraints on the evolution of structures imaged in Figures 2 and 3. Between 1.8 and 1.3 Ga, granites and rhyolites west of the Grenville front and equivalent rocks within the GFTZ were emplaced at shallow crustal levels near the eastern margin of the amalgamated Superior craton and Manitoulin terrane [Davidson, 1986a,b; van Breemen and Davidson, 1988; Davidson and Bethune, 1988]. During an early stage of Grenvillian orogeny, perhaps coeval with 1.30-1.25 Ga island arc magmatism in the Central Metasedimentary belt to the southeast, the outer margin was overridden by northwesterly transported microterranes. At this time rock units in the GFTZ were depressed to lower crustal levels and subjected to intense compression (flattening) and metamorphism under high pressure/high temperature conditions. Two of the transported microterranes are represented in the seismic section (Figure 2).

A later episode of northwest directed thrusting cut deeply into the crust of the buried margin, ramping GFTZ rocks back up to the near surface. This phase of Grenvillian orogeny occurred after the intrusion of 1.24 Ga Sudbury dikes [Krogh et al., 1987] and may have been responsible for 1.15 Ga pegmatoids in the GFTZ [Krogh and Wardle, 1984] and 1.16 Ga syntectonic pegmatites in shear zones along the eastern boundary of the Britt domain [van Breemen et al., 1986]. Although a major component of displacement at this time was concentrated at the Grenville front, where the greatest changes in metamorphic grade are recorded, the large number of mylonite zones mapped along the shores of Georgian Bay and imaged on the seismic reflection section are testament to the considerable differential movements that must have taken place within the GFTZ. A particularly important result of the GLIMPCE seismic reflection survey is the discovery that ductile imbrication penetrated the entire crust.

Results from the Midcontinent Rift System (MRS) - Profiles A and C

Seismic reflection profiles C, A and F cross or almost cross the width of the MRS in western, central and eastern Lake Superior respectively. Figures 5 and 8 show interpreted cross-sections drawn from profiles C and A, and Figures 6, 7 and 9 to 11 display key portions of the stacked and migrated data. Profile F has been presented by Behrendt et al. [1988, and this volume]. Our interpretations of the seismic sections are constrained by the geology at the ends of the lines, potential field data (projected aeromagnetic and gravity profiles are shown above the interpreted cross-sections), and crustal velocities and Moho depths determined from coincident, intersecting, and neighboring seismic refraction surveys [Berry and West, 1966; Smith et al., 1966; O'Brien, 1968; Halls and West, 1971; Ocola and Meyer, 1973; Luetgert and Meyer, 1982; Halls, 1982; Luetgert, 1982]. The interpretation of profile A is also constrained by the geology mapped on nearby Manitou Island, Keweenaw Peninsula, Superior Shoal and the Slate Islands (Figure 1).

There are similarities and some notable differences among the various GLIMPCE seismic sections recorded in Lake Superior. An outstanding feature on all profiles is the great thickness of layered reflections contained in the central rift basin, representing an enormous volume of lavas and sedimentary rocks deposited during Keweenawan extension (Figures 5-11) [Behrendt et al., 1988 and this volume]. Individual seismic events are laterally continuous for distances up to 70 km and layered reflections extend from near the bottom of the lake to 10-12 s travel time, corresponding to 30-36 km depth. Such a thickness of rift strata under Lake Superior far exceeds earlier estimates based on interpretations of potential field and seismic refraction data [see papers in Wold and Hinze, 1982], and is much greater than the thickness of deposits determined along other parts of the MRS [Serpa et al., 1984; Zhu and Brown, 1986; Chandler et al., 1988]. Indeed, the basin beneath Lake Superior may contain the thickest section of continental rift deposits on the planet [Behrendt et al., 1988].

We interpret the zone of high reflectivity in the middle and lower portions of the central rift basin as a thick pile of Keweenawan (MRS) lavas, the overlying zone of intermediate reflectivity as sedimentary rocks and minor lavas of the Oronto group, and the zone of low reflectivity in the uppermost section, largely hidden by water bottom multiples, as monotonous sandstones of the youngest Keweenawan formations (the various post-magmatic sedimentary units are not differentiated in Figures 5 and 8). The strong, continuous reflections from the thick pile of lavas originate at contacts between lavas and interflow sediments [Zhu and Brown, 1986], and at lithological discontinuities and porosity changes within and between individual lava flows [Behrendt et al., 1988; Lippus, 1988].

The major differences among the GLIMPCE seismic sections recorded in Lake Superior are related to lateral variations in (i) the geometry of the central rift basin, (ii) the attenuation of the original Archean/Early Proterozoic crust during MRS tectonism, and (iii) the depth to Moho. Behrendt et al. [1988, and this volume] discuss the latter two aspects of our data. Here, we shall be concerned primarily with the varying geometry of the central rift basin and adjacent structures. Profile F, described by Behrendt et al. [1988, and this volume], reveals a relatively symmetric graben beneath eastern Lake Superior, whereas profiles C and A cross markedly asymmetric half-grabens that have opposite dips (polarities) in the western and central parts of the lake. In the following we present a broad overview of the most important structures imaged by the GLIMPCE data from Lake Superior. Cannon et al [1988] discuss the more detailed aspects of the interpretation.

Geometry of Structures Beneath Profile C - A North-Dipping Mega-Half-Graben

Sedimentary rocks and lavas of the MRS occur onshore south of profile C, and lavas and intrusive rocks outcrop to the north (Figure 1). The general form of structures inferred to the south of our seismic section (Figure 5) is consistent with the surface geology and with information from a nearby industry seismic reflection survey [Hinze et al., 1988]. The central asymmetric rift basin is bounded on the south by the Keweenaw fault and on the north by the northeast continuation of the Douglas fault [Cannon et al., 1988]. Both faults are relatively steep at the surface, becoming listric at depth. Where exposed the faults have experienced considerable reverse displacements, presumably as a consequence of the late Keweenawan compressional event. However, our seismic reflection data demonstrate that they played a much more important role during the earlier extensional history of the MRS. In western Lake Superior the plane of the Keweenaw fault was originally a rotating deposition surface hinged at shallow depths, and the Douglas fault was a major normal growth fault.

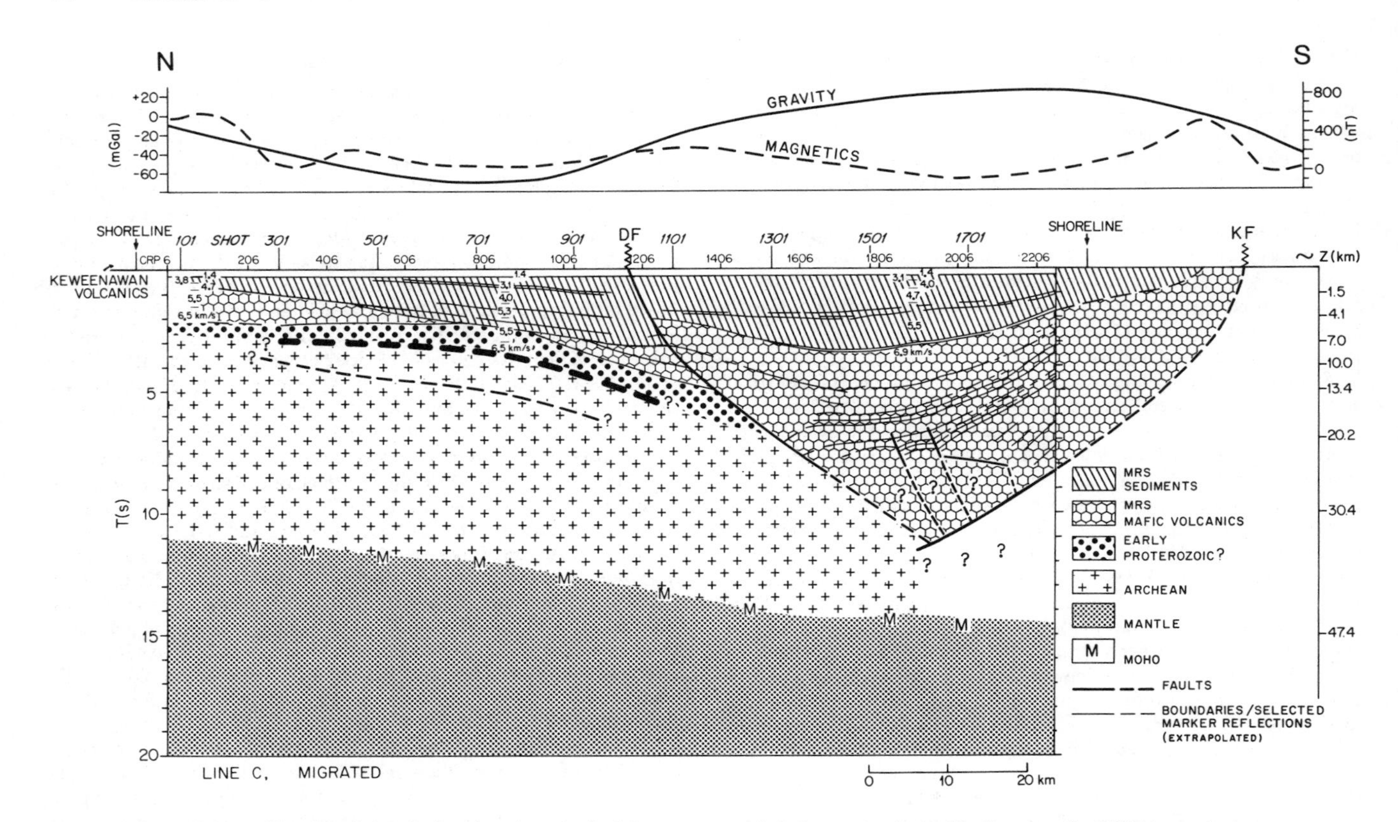

Fig. 5. Simplified sketch showing the principal features recorded along seismic profile C across the MRS in western Lake Superior. A selection of some of the most continuous reflections is shown together with interpreted faults and other structures. See the text for a discussion of the reflection zone marked by the heavy dashed line. DF - Douglas fault as extrapolated eastwards from its outcrop in Wisconsin and Minnesota; CRP - common reflection points; KF - Keweenaw fault; M - Moho transition zone; MRS - Midcontinent rift system. Velocity depth profiles near CRP's 850 and 1900 are based on an interpretation of coincident seismic refraction data by Luetgert and Meyer (1982), and that near CRP 120 is from an interpretation of a nearby seismic refraction profile by Halls and West (1971). Gravity and magnetic profiles shown along the top of the section were taken from maps in Wold and Hinze (1982). The geometry of the Keweenaw fault and sediment-volcanic boundary to the south of profile C is consistent with a nearby seismic reflection profile presented by Hinze et al. (1988). The location of profile C is shown in Figure 1, and examples of the seismic reflection data are shown in Figures 6 and 7.

Except for about 3 km of late reverse displacement on the Douglas fault, post-magmatic sedimentary rocks (Figure 5) are relatively continuous across western Lake Superior, having been deposited in a broad flexural basin centered at about common reflection point (CRP) 1800. Below the sedimentary rocks the Douglas fault divides the section into two distinct structural/depositional regimes. To the south, a very large volume of Keweenawan lavas was deposited in a north-dipping mega-half-graben. The plane of the Keweenaw fault is parallel to the deeper reflections in the half-graben (Figures 6a and 6b), suggesting it was the basement-lava contact early in the history of the MRS. Only minor normal faulting would have occurred on the Keweenaw fault at this location [Hinze et al., 1988]. During extension, the basement would have rotated about a southern pivot, guided by listric normal faulting along the Douglas fault and subordinate crustal flexing. The conspicuous northward fanning of reflections (Figures 5 and 6) and the roll-over as they approach the Douglas fault are compelling evidence for substantial syndepositional normal faulting [Cannon et al., 1988]. On the seismic section the Douglas fault is delineated by a zone of disrupted reflections, by truncations of strong reflections, and by reflections from the fault plane itself. Normal slip on the Douglas fault probably exceeded 20 km [Cannon et al., 1988].

The unusual curved events that interrupt the linearity of reflections in the lower rift basin are tentatively interpreted as Keweenawan lavas and sediments arched or draped over rotated blocks of earlier Keweenawan strata or basement. Analogous normal faulted structures have been identified within the buried MRS in Kansas by Serpa et al. [1984]. North-dipping reflections in the deep central basin truncate the south-dipping events, indicating that reverse displacements along or close to the basement-lava contact (ie. the Keweenaw fault) postdated the last movements on the Douglas and other faults (Figures 5 and 6).

In contrast to the thick sequence of highly reflective lavas south of the Douglas fault, the seismic data to the north reveal only two thin wedges of high reflectivity that may be associated with lava flows. The northern wedge can be tied to exposures of volcanic rock onshore, but we have few controls on the interpretation of the southern wedge; the two wedges pinch-out against an intervening "basement" high (Figure 5 and 7). Both wedges are underlain by a weakly reflective region of probable Early Proterozoic sediments equivalent to those of the Marquette Range supergroup; Early Proterozoic sediments north and south of Lake Superior (Figure 1) are presumed to have been continuous at one time [Sims et al., 1987].

Underlying the weakly reflective region is a conspicuous curvilinear zone of south-dipping reflections (represented by the thick dashed line in Figure 5 and outlined by the upper set of question marks in Figure 7). There are at least six explanations for this complex feature. It could be (1) an Archean structure, (2) an Early Proterozoic structure, (3) the contact between Early Proterozoic sediments and Archean basement, (4) a major shear or mylonite zone associated with Keweenawan extension, (5) a Keweenawan igneous intrusion, or (6) some combination of 1 to 5. It occurs less than 25 km south of troctolitic and gabbroic rocks of the Duluth layered intrusion [Weiblen, 1982], and its complex layered geometry is very similar to that of a large intrusion seismically imaged beneath the western Atlantic continental margin [Joppen et al., 1987]. The distinctive zone of reflections may therefore represent the base of Early Proterozoic sedimentary rocks intruded by a large sill-like body and/or sheared during Keweenawan tectonism. The deeper boundary marked by question marks has a similar range of possible explanations (see below).

Geometry of Structures Beneath Profile A - A South-Dipping Mega-Half-Graben

South of profile A there is a thin layer of Jacobsville sandstone overlying Archean basement, and to the north of the profile lies the Archean Superior craton (Figure 1). Along its length the profile passes close to exposed Keweenawan lavas and sedimentary rocks on Manitou Island, Keweenaw Peninsula, Superior Shoal and the Slate Islands. Like profile C, the spectacular central rift basin is bounded by major faults that now juxtapose lavas against younger sedimentary rocks, and which had earlier larger movements during the extensional phase of the MRS (Figures 8-11). On the south side of the central rift basin there is the Keweenaw fault and associated splays, and on the north side there is the eastward continuation of the Isle Royale fault [Cannon et al., 1988]. Our data indicate that the Keweenaw fault in central Lake Superior was at one time the principal normal growth fault, and that the plane of the Isle Royale fault may include the original basement-lava contact. Both fault systems are now steeply-dipping near the surface and are delineated by truncations of strong reflections and zones of disrupted reflectivity. The faults divide the section into three structural/depositional regimes.

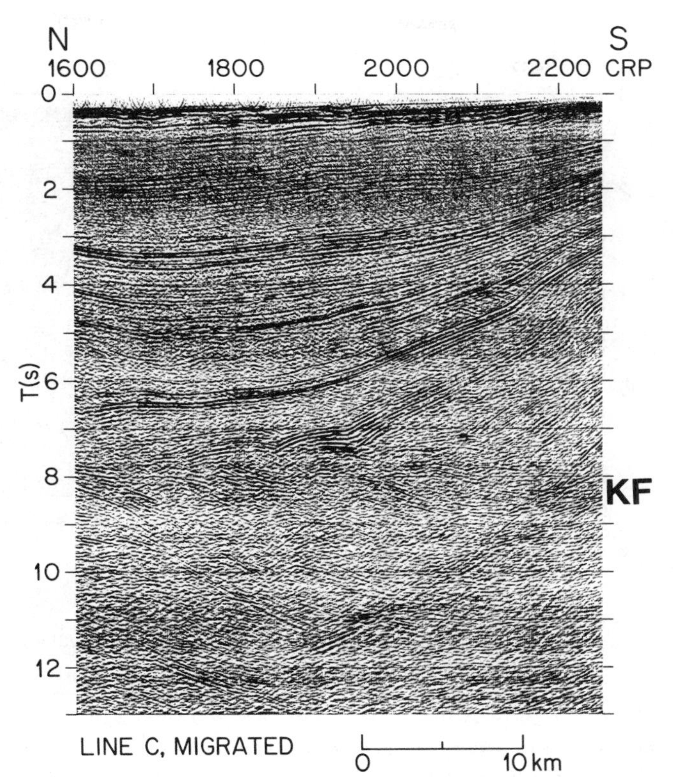

Fig. 6b. Migrated version of Figure 6a.

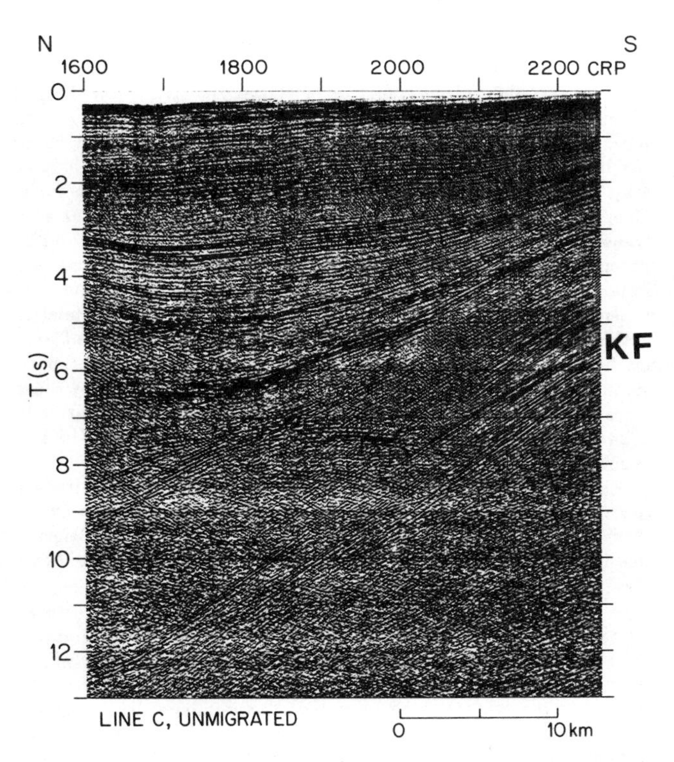

Fig. 6a. Non-migrated seismic reflection data collected along the southern segment of profile C. CRP - common reflection points; KF - Keweenaw fault.

In the area south of the Keweenaw fault, slightly attenuated Archean/Early Proterozoic basement is overlain by a northward thickening wedge of Jacobsville sandstone (Figures 8 and 9). Separating the sediments and basement is a band of strong reflections that may be evidence for a thin layer of intervening lavas. Within the basement is a curved discontinuity that dips northward from 5 s near the southern shoreline to about 9 s north of the Keweenaw fault. It is defined by a subtle line of reflections (outlined by arrows in Figure 9) and by noticeable truncations in both the hanging wall and footwall; no significant reflections cross the discontinuity. In Figure 8 the discontinuity is displayed as a deep detachment zone that merges in the lower crust with the Keweenaw and other faults.

The central mega-half-graben between the Keweenaw and Isle Royale faults contains the huge thickness of Keweenawan strata. The

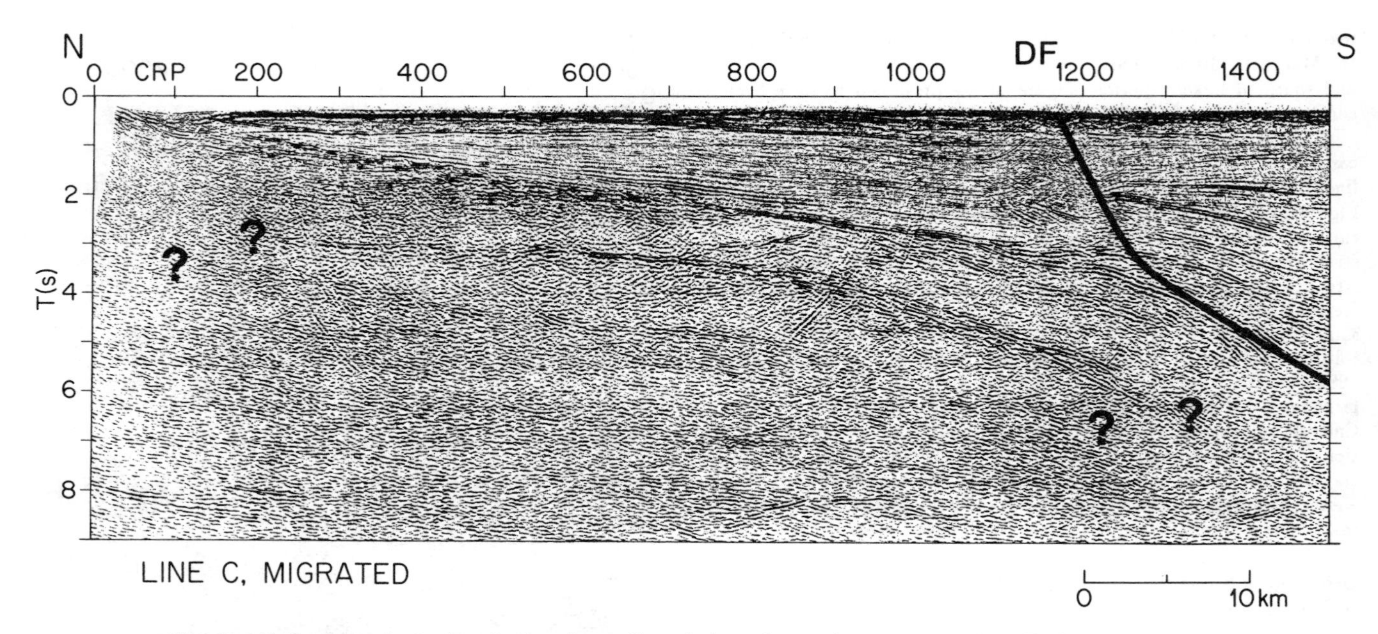

Fig. 7. F-K migrated seismic reflection data collected along the northern segment of profile C. CRP - common reflection points; DF - Douglas fault.

structure is very similar to that observed along profile C, except for its sense of asymmetry. Beneath profile C the growth fault occurs on the north side of the half-graben and is south-dipping, the basement-lava contact dips to the north from the vicinity of the southern boundary (Keweenaw) fault, and Keweenawan strata thicken northward. In contrast, the principal growth fault beneath profile A is the north-dipping Keweenaw fault on the south side of the half-graben, the basement-lava contact dips to the south from the vicinity of the northern boundary (Isle Royale) fault, and Keweenawan strata thicken southward. The half-graben under central Lake Superior is therefore a mirror image of that under the western part of the lake. Even the detachment identified on line A has a mirror counterpart - the reflective boundary highlighted by the lower set of question marks in figures 5 and 7. The two reflective boundaries have the same form and attitude, and occur at approximately the same depth relative to their respective growth faults. An additional feature on profile A is a prominent hinge line (Figure 10), across which the dip and fanning of Keweenawan strata increase markedly. Our interpretation of the curved reflections at the base of the half-graben is similar to that described for profile C.

In the third region of the seismic section (Figures 8 and 11), to the north of the Isle Royale fault, a relatively thin sequence of Keweenawan strata unconformably overlies a deeper basin, inferred to contain Early Proterozoic sedimentary rocks [Cannon et al., 1988]. This portion of profile A crosses the eastern margin of a major elliptical structure identified on magnetic and gravity anomaly maps [Halls, 1972]. Our data support the proposal by Halls [1972] that the source of the prominent potential field anomalies is a large basin of Keweenawan lava and younger clastic rocks. Based on a comparative review of the potential field and seismic reflection data we judge that the thickness of lava at the center of the basin must be substantially greater than that recorded under profile A, and is likely much greater than that inferred by Halls [1972].

Segmentation of the MRS

It is clear that somewhere between profiles C and A the polarity of the central half-graben reverses. The early role of the Keweenaw fault changes along strike from that of a basement-lava contact hinged near the surface in western Lake Superior, to that of a major growth fault in the central region of the lake. Similar changes occur along the northern boundary fault, with normal faulting on the Douglas fault and a rotating basement-lava contact near the Isle Royale fault. Cannon et al. [1988] suggest the reversal in polarity happens at a transfer fault or accommodation zone, delineated by truncations of potential field anomalies a little to the west of Isle Royale. An equally important change in the character of the central rift basin occurs between profiles A and F. Under profile F the main graben is symmetric with poorly resolved normal growth faults at both ends. The transfer from asymmetric half-graben beneath central Lake Superior to symmetric graben under the eastern part of the lake probably takes place 20-30 km to the east of profile A along the Thiel fault zone [Cannon et al., 1988], identified on the basis of potential field and seismic refraction data [Hinze et al., 1982].

Each area sampled in Lake Superior seems to have its own distinctive geometry. Recent seismic reflection surveys across other parts of the MRS have revealed an equally diverse set of structures. Beneath the Michigan basin, a portion of the southeastern limb of the MRS surveyed by COCORP appears to be a relatively symmetric trough [Zhu and Brown, 1986], whereas along the southwestern limb there are symmetric grabens and asymmetric half- grabens with alternating polarities [Serpa et al., 1984; Chandler et al, 1988]. Together, these new data establish, for the first time, that the MRS system developed by movements along a number of semi-independent segments in much the same manner as that proposed for some modern passive margins [Gibbs, 1984; Lister et al., 1986; Klitgord et al., 1988] and the East African rift system [Rosendahl, 1987; Bosworth, 1987].

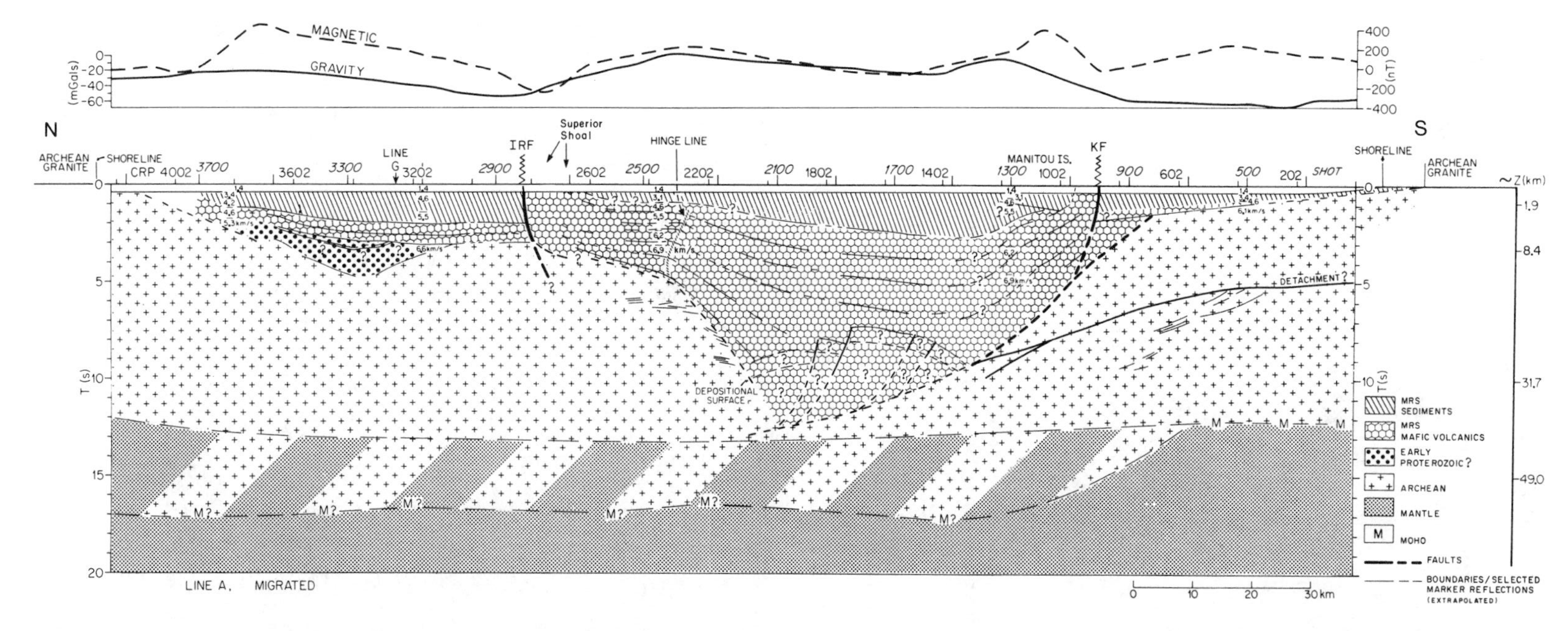

Fig. 8. Simplified sketch showing the principal features recorded along seismic profile A across the MRS in central Lake Superior. A selection of some of the most continuous reflections is shown together with interpreted faults and other structures. CRP - common reflection points; IRF - Isle Royale fault as extrapolated eastwards from its inferred location north of Isle Royale; KF - Keweenaw fault; M - Moho transition zone; MRS - Midcontinent rift system. The location of Superior Shoals is indicated on Figure 1 and Manitou Island lies between seismic profile A and Keweenaw Peninsula. Velocity depth profiles and approximate depth scale are taken from interpretations of intersecting and nearby seismic refraction profiles by Halls and West [1971] and Luetgert [1982]. Gravity and magnetic profiles shown along the top of the section were taken from maps in Wold and Hinze [1982]. The stripe pattern between Archean basement and mantle represents the anomalous lower crust discussed by Behrendt et al. [1988]. Location of the profile is shown in Figure 1, and examples of the F-K migrated seismic reflection data are presented in Figures 9 to 11.

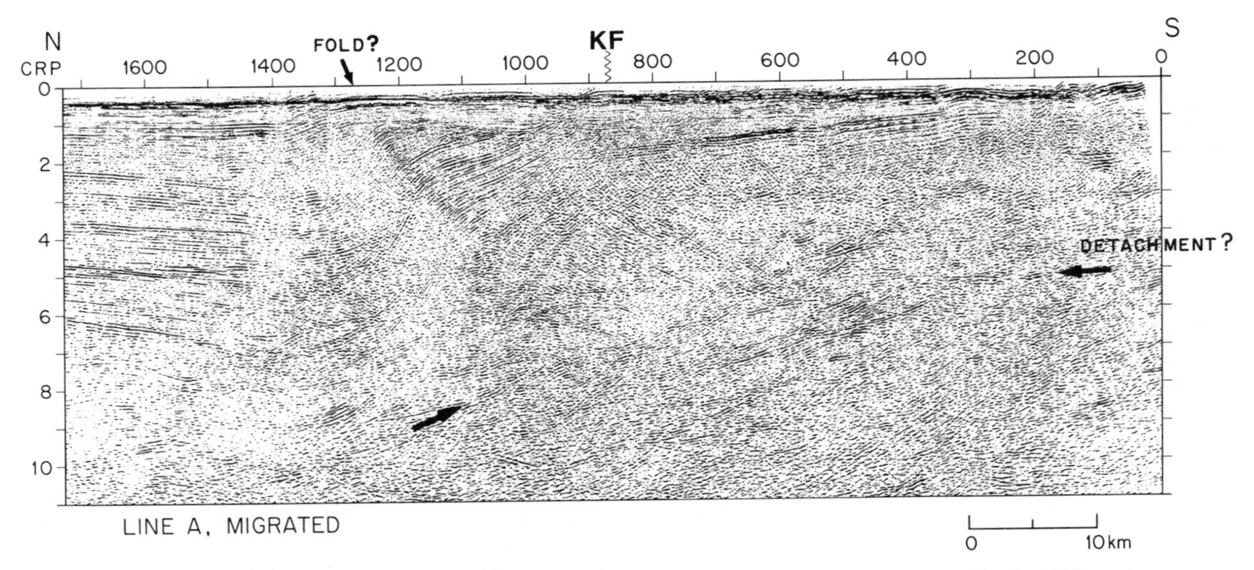

Fig. 9. F-K migrated seismic reflection data collected along the southern segment of profile A. CRP - common reflection points; KF - location of Keweenaw fault projected from its outcrop on Keweenaw Peninsula using aeromagnetic data.

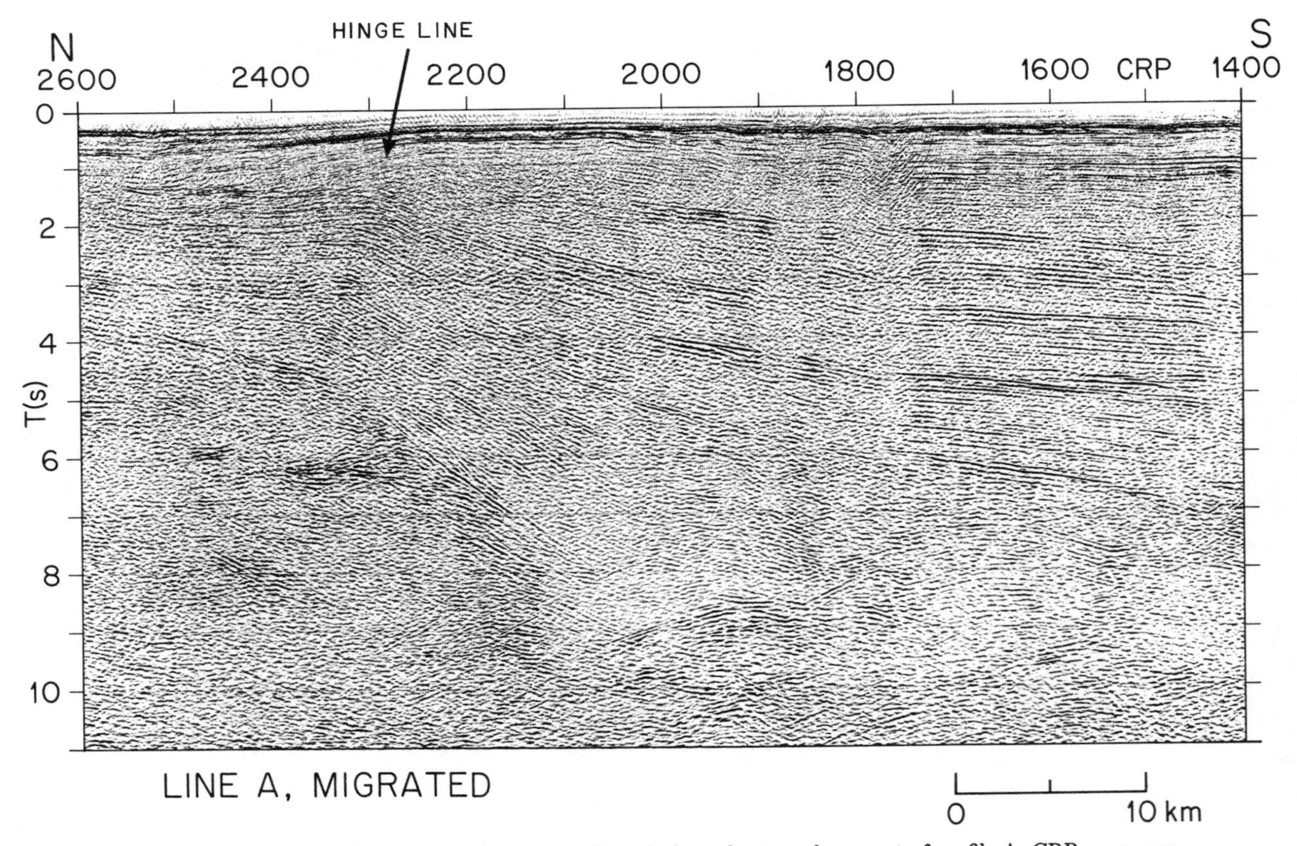

Fig. 10. F-K migrated seismic reflection data collected along the central segment of profile A. CRP - common reflection points.

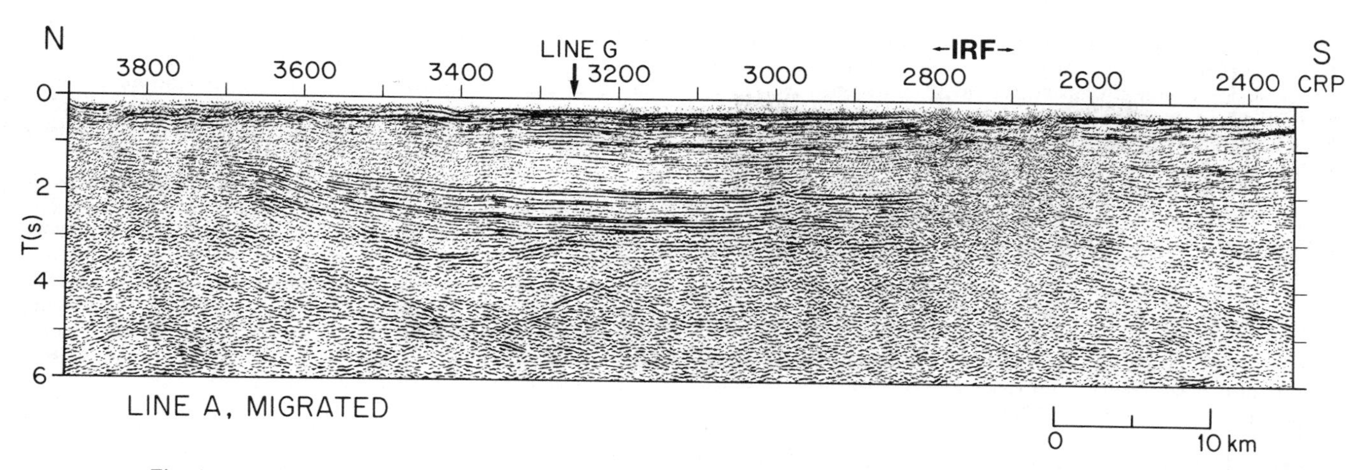

Fig. 11. F-K migrated seismic reflection data collected along the northern segment of profile A. CRP - common reflection points; IRF - location of Isle Royale fault projected from its inferred location north of Isle Royale using aeromagnetic data.

The Keweenawan Compressional Event

The GLIMPCE seismic reflection data provide important structural information pertaining to the late Keweenawan compressional event. Late reverse displacements are recorded on all boundary faults (Figures 5-11). Along the southern margin of the central rift basin, in both western and central Lake Superior, reverse faulting on the Keweenaw fault was responsible for southward thrusting of lavas against younger sediments. Reverse faulting focussed at the basement-lava contact in the western part of the lake and at the marginal normal fault in the central region. Under profile A a short distance to the north of the Keweenaw fault there is also a hint that the lavas and sedimentary rocks were folded or buckled (Figure 9). The stacked seismic section [Behrendt et al., this volume] reveals numerous diffractions in this region (many collapse upon migration, leaving a void on the migrated section), suggesting the geometry of structures is more complicated than that depicted in the simplified sketch of Figure 8.

At the northern margin of the central rift basin our data require late reverse displacements of about 3 km on the Douglas fault and greater than 5-6 km on the Isle Royale fault. At both locations the lavas have been transported northward over the younger sediments. Like the southern margin, thrusting was concentrated along the older normal growth fault and at the basement-lava contact. The effects of the late compressional event have also been observed along the buried southeastern and southwestern limbs of the MRS [Zhu and Brown, 1986; Chandler et al., 1988].

Conclusions

Some of the most important conclusions from this preliminary interpretation of GLIMPCE deep seismic data and related information are:

(1) The region underlying Manitoulin terrane in northern Lake Huron is characterized by a highly reflective lower crustal layer and a distinctly less reflective upper crustal layer that is separated by a major discontinuity delineated by a band of gently east-dipping reflections at about 18-20 km depth. Based on a variety of data we interpret (i) the lower layer as attenuated crust of the southern Superior cratonic margin, (ii) the upper layer as a composite terrane consisting of displaced Huronian strata, the remains of an exotic mass that collided with the Superior cratonic margin during the 1.89-1.82 Ga Penokean orogeny, and younger granites and rhyolites, and (iii) the intervening discontinuity as the master Penokean décollement.

(2) Seismic reflection data across the 1.3-1.0 Ga Grenville orogen are dominated by an extraordinary 32-km-wide band of moderately east-dipping reflections associated with the Grenville front tectonic zone (GFTZ). Its westernmost reflection corresponds to the Grenville front at depth, abruptly truncating all structures to the west. A change in strength and density of east-dipping reflections coincides with the eastern margin mylonite zone of the GFTZ. The east-dipping reflections of the GFTZ, which seem to extend from the surface to the crust-mantle boundary, probably originate at highly strained contacts between gneissic and migmatitic rocks of varying lithology. Shear and mylonite zones are the likely source of the very strong reflections. Moho may be 5-10 km deeper beneath the GFTZ compared to adjacent terranes. Rocks now exposed in the Grenville orogen were buried and exhumed during various phases of middle Proterozoic tectonism that involved northwestward thrust transport of Grenvillian microterranes.

(3) The 1.11-1.09 Ga Midcontinent rift system (MRS) under Lake Superior contains a massive volume of Keweenawan volcanic and sedimentary rock, perhaps representing the greatest thickness of continental rift deposits on Earth. The structure of the central rift basin is highly variable along the length of the MRS, having evolved as a series of discrete grabens and asymmetric half-grabens with opposing polarities. For example, in western Lake Superior the half-graben formed by northward rotation of basement with normal faulting on the south-dipping Douglas fault, whereas in central Lake Superior the central half-graben formed by southward rotation of basement with normal faulting on the north-dipping Keweenaw fault. Under eastern Lake Superior the central rift basin appears to be a relatively symmetric structure. Significant flexing of the crust also occurred at some locations. The different segments of the MRS are probably separated by transfer faults or accommodation zones identified by truncations of MRS potential field anomalies. This development of the MRS by asymmetric graben formation along discrete segments is similar to that recognized along some passive continental margins and along the East African rift system. The last Precambrian tectonic event to affect the MRS caused substantial reverse faulting, by reactivating older normal faults and by creating new faults along or near the basement-lava contact.

Acknowledgements. GLIMPCE seismic reflection profiling in the Great Lakes was funded jointly by the U.S. Geological Survey and the Geological Survey of Canada through its contribution to LITHOPROBE. The Canadian contributors to this paper thank Ray Price, John McGlynn, Mike Berry and members of the LITHOPROBE Scientific Committee for support. We thank Paul Hoffman, Klaus Schulz and Terry Offield for critical reviews of the first draft of the manuscript. The data were collected by Geophysical Services Inc. and processed by us. Contribution of the Geological Survey of Canada number 26288.

References

Anovitz, L.M., Pressure-temperature-time constraints on the metamorphism of the Grenville Province, *Ph.D. thesis*, Univ. of Michigan, Ann Arbor, Michigan, 479 pp., 1987.

Behrendt, J.C., A.G. Green, W.F. Cannon, D.R. Hutchinson, M.W. Lee, B. Milkereit, W.F. Agena, and C. Spencer, Crustal structure of the Mid-continent Rift System: Results from GLIMPCE deep seismic reflection profiles, *Geology, 16*, 81-85, 1988.

Behrendt, J.C., A.G. Green, M.W. Lee, D.R. Hutchinson, W.F. Cannon, B. Milkereit, W.F. Agena, and C. Spencer, Crustal extension in the Mid-continent Rift system - results from GLIMPCE deep seismic reflection profiles over Lakes Superior and Michigan, *this volume*.

Berry, M.J., and K. Fuchs, Crustal structure of the Superior and Grenville provinces of the northeastern Canadian Shield, *Bull. Seismol. Soc. Am., 63*, 1393-1432, 1973.

Bethune, K.M., and A. Davidson, Diabase dykes and the Grenville front southwest of Sudbury, Ontario, *Geol. Surv. Can. Pap., 88-1A*, 151-159, 1988.

Berry, M.J., and G.F. West, A time-term interpretation of the first arrival data of the 1963 Lake Superior experiment, in *The Earth Beneath the Continents*, edited by J.S. Steinhart, and T.J. Smith, *Am. Geophys. Un. Mono., 10*, 166-180, 1966.

Bickford, M.E., W.R. Van Schmus, and I. Zietz, Proterozoic history of the midcontinent region of North America, *Geology, 14*, 492-496, 1986.

Bosworth, W., Off-axis volcanism in the Gregory rift, East Africa: Implications for models of continental rifting, *Geology, 15*, 397-400, 1987.

Cannon, W.F., A.G. Green, D.R. Hutchinson, M.W. Lee, B. Milkereit, J.C. Behrendt, H.C. Halls, J.C. Green, A.B. Dickas, G.B. Morey, R. Sutcliffe, and C. Spencer, The Midcontinent rift beneath Lake Superior from GLIMPCE seismic reflection profiling, *Tectonics, in press*, 1988.

Card, K.D., V.K. Gupta, P.H. McGrath, and F.S. Grant, The Sudbury structure: Its regional geological and geophysical setting, in *The Geology and Ore Deposits of the Sudbury Structure*, edited by E.G. Pye, A.J. Naldrett, and P.E. Giblin, *Ont. Geol. Surv. Sp. Vol., 1*, 25-43, 1984.

Chandler, V.W., P.L. McSwiggen, G.B. Morey, W.J. Hinze, and R.L. Anderson, Interpretation of seismic reflection, gravity and magnetic data across the Keweenawan midcontinent rift system in western Wisconsin, eastern Minnesota, and central Iowa, *Am. Ass. Petrol. Geol. Bull., in press*, 1988.

Corfu, R., and A.J. Andrews, A U-Pb age for mineralized Nipissing diabase, Gowganda, Ontario, *Can. J. Earth Sci., 23*, 107-109, 1986.

Daniels, P.A., Upper Precambrian sedimentary rocks: Oronto group, Michigan-Wisconsin, in *Geology and Tectonics of the Lake Superior Basin*, edited by R.J. Wold, and W.J. Hinze, *Geol. Soc. Am. Mem., 156*, 107-134, 1982.

Davidson, A., Tectonic boundaries within the Grenville Province of the Canadian Shield, *J. Geodyn., 1*, 433-444, 1984a.

Davidson, A., Identification of ductile shear zones in the southwestern Grenville Province of the Canadian Shield, in *Precambrian Tectonics Illustrated*, edited by A. Kroner, and R. Greiling, E. Schweizerbart'sche Verlagsbuchhandlung, Stuttgart, 263-279, 1984b.

Davidson, A., New interpretations in the southwestern Grenville Province, in *The Grenville Province*, edited by J.M. Moore, A. Davidson, and A.J. Baer, *Geol. Ass. Can. Spec. Pap., 31*, 61-74, 1986a.

Davidson, A., Grenville front relationships near Killarney, Ontario, in *The Grenville Province*, edited by J.M. Moore, A. Davidson, and A.J. Baer, *Geol. Ass. Can. Spec. Pap., 31*, 107-117, 1986b.

Davidson, A., and K.M. Bethune, Geology of the north shore of Georgian Bay, Grenville Province of Ontario, *Geol. Surv. Can. Pap., 88-1A*, 135-144, 1988.

Davis, D.W., and R.H. Sutcliffe, U-Pb ages from the Nipigon plate and northern Lake Superior, *Geol. Soc. Am. Bull., 96*, 1572-1579, 1985.

Dickas, A.B., Comparative Precambrian stratigraphy and structure along the Midcontinent rift, *Am. Ass. Petrol. Geol. Bull., 70*, 225-238, 1986.

Dickin, A.P. and R.H. McNutt, Nd model age mapping of the SE margin of the Archean foreland in the Grenville Province of Ontario, *Geology, in press*, 1988.

Epili, D., and R.F. Mereu, The GLIMPCE seismic experiment: Onshore refraction and wide-angle reflection observations from a fan line over the Lake Superior rift system, *this volume*.

Gibbs, A.D., Structural evolution of extensional basin margins, *J. Geol. Soc. London, 141*, 609-620, 1984.

Green, A.G., B. Milkereit, A. Davidson, C. Spencer, D.R. Hutchinson, W. F. Cannon, M.W. Lee, W.F. Agena, J.C. Behrendt, and W.J. Hinze, Crustal structure of the Grenville front and adjacent terranes, *Geology, in press*, 1988.

Green, J.C., Geological and geochemical evidence for the nature and development of the Middle Proterozoic (Keweenawan) Midcontinent Rift of North America, *Tectonophysics, 94*, 413-437, 1983.

Geological Survey of Canada, Total field magnetic maps NL16/17 and NK17, Geol. Surv. Can., Ottawa, Ontario, 1987.

Halls, H.C., Magnetic studies in northern Lake Superior, *Can. J. Earth Sci., 9*, 1349-1367, 1972.

Halls, H.C., Crustal thickness in the Lake Superior region, in *Geology and Tectonics of the Lake Superior Basin*, edited by R.J. Wold, and W.J. Hinze, *Geol. Soc. Am. Mem., 156*, 239-244, 1982.

Halls,, H.C., and G.F. West, A seismic refraction survey in Lake Superior, *Can. J. Earth Sci., 8*, 610-630, 1971.

Hanmer, S. and A. Ciesielski, Structural reconnaissance of the northwest boundary of the Central Metasedimentary belt, Grenville Province, Ontario and Quebec, *Geol. Surv. Can. Pap., 84-1B*, 121-131, 1984.

Hinze, W.J., L.W. Braile, and V.W. Chandler, The southern margin of the Midcontinent rift system in western Lake Superior, *Geology, in press*, 1988.

Hinze, W.J., R.J. Wold, and N.W. O'Hara, Gravity and magnetic anomaly studies of Lake Superior, in *Geology and Tectonics of the Lake Superior Basin*, edited by R.J. Wold, and W.J. Hinze, *Geol. Soc. Am. Mem., 156*, 203-222, 1982.

Hoffman, P.F., United Plates of America, the birth of a craton: Early Proterozoic assembly and growth of Laurentia, *Ann. Rev. Earth and Planet. Sci., 16*, 543-603, 1988.

Joppen, M., R.S. White, G.D. Spence, and G.K. Westbrook, The seismic structure of Rockall trough (abstract), *EOS Trans. Am. Geophys. Un., 68*, 1372, 1987.

Kalliokoski, J., Jacobsville sandstone, in *Geology and Tectonics of the Lake Superior Basin*, edited by R.J. Wold, and W.J. Hinze, *Geol. Soc. Am. Mem., 156*, 147-156, 1982.

Klasner, J.S., W.F. Cannon, and W.R. Van Schmus, The pre-Keweenawan tectonic history of southern Canadian Shield and its influence on formation of the Midcontinent Rift, in *Geology and Tectonics of the Lake Superior Basin*, edited by R.J. Wold, and W.J. Hinze, *Geol. Soc. Am. Mem., 156*, 27-46, 1982.

Klemperer, S.L., T.A. Hauge, E.C. Hauser, J.E. Oliver, and C.J. Potter, The Moho in northern Basin and Range province, Nevada, along COCORP 40°N seismic reflection transect, *Geol. Soc. Am. Bull., 97*, 603-618, 1986.

Klitgord, K.D., D.R. Hutchinson, and H. Schouten, U.S. continental margin: Structure and tectonic framework, in *The Atlantic Continental Margin: U.S.*, edited by R.E. Sheridan, and J.A. Grow, *Geol. Soc. Am., The Geology of North America, I-2*, 19-56, 1988.

Krogh, T.E. and G.L. Davis, The effect of regional metamorphism on U-Pb systems in zircon and a comparison with Rb/Sr systems in the same whole rock and its constituent minerals, *Carnegie Inst. Washington Yearbook, 72*, 601-610, 1973.

Krogh, T.E., and R. Wardle, U-Pb isotopic ages along the Grenville front, *Geol. Ass. Can. Prog. with Abstr., 9*, 80, 1984.

Krogh, T.E., F. Corfu, D.W. Davis, G.R. Dunning, L.M. Heaman, S.L. Kamo, N. Machado, J.D. Greenough, and E. Nakamura, Precise U-Pb isotope ages of diabase dykes and mafic to ultramafic rocks using trace amounts of baddeleyite and zircon, in *Diabase Dyke Swarms*, edited by H.C. Halls, and W. F. Farig, *Geol. Ass. Can. Spec. Pap., 34*, 147-152, 1987.

Krogh, T.E., D.W. Davis, and F. Corfu, Precise U-Pb zircon and baddeleyite ages for the Sudbury area, in *The Geology and Ore Deposits of the Sudbury Structure*, edited by E.G. Pye, A.J. Naldrett, and P.E. Giblin, *Ont. Geol. Surv. Sp. Vol., 1*, 431-446, 1984.

Lee, M.W., W.F. Agena, & D.R. Hutchinson, Processing of the GLIMPCE multichannel seismic reflection data, *U.S. Geol. Surv. Open-File Rept., 88-225*, 46 pp., 1988.

Lister, G.S., M.A. Etheridge, and P.A. Symmonds, Detachment faulting and the evolution of passive continental margins, *Geology, 14*, 246-250, 1986.

Lippus, C.S., The seismic properties of mafic volcanic rocks of the Keweenawan supergroup and their implications, *M.Sc. thesis*, Purdue Univ., Lafayette, Indiana, 117 pp., 1988.

Luetgert, J.H., The Earth's crust beneath Lake Superior - an interpretation of cross-structure seismic refraction profiles, *Ph.D. thesis*, Univ. of Wisconsin, Madison, Wisconsin, 86 pp., 1982.

Luetgert, J.H., and R.P. Meyer, Structure of the western basin of Lake Superior from cross structure refraction profiles, in *Geology and Tectonics of the Lake Superior Basin*, edited by R.J. Wold, and W.J. Hinze, *Geol. Soc. Am. Mem., 156*, 245-256, 1982.

Medaris, L.G., *editor, Early Proterozoic Geology of the Great Lakes Region, Geol. Soc. Am. Mem., 160*, 141 pp., 1983.

Mereu, R.F., and G. Jobidon, A seismic investigation of the crust and Moho on a line perpendicular to the Grenville front, *Can. J. Earth Sci., 8*, 1553-1583, 1971.

Mereu, R.F., D. Wang, O. Kuhn, D.A. Forsyth, A.G. Green, P. Morel, G.G.R. Buchbinder, D. Crossley, E. Schwarz, R. duBerger, C. Brooks, and R. Clowes, The 1982 COCRUST seismic experiment across the Ottawa-Bonnechere graben and Grenville front in Ontario and Quebec, *Geophys. J. R. Astr. Soc., 84*, 491-514, 1986.

Milkereit, B., A.G. Green, P. Morel-à-l'Huissier, M.W. Lee. and W.F. Agena, 1986 Great Lakes seismic reflection survey migrated data, *Geol. Surv. Can. Open File, 1592*, 33 pp. with seismic sections, 1988.

Moore, J.M., Introduction: The "Grenville problem" then and now, in *The Grenville Province*, edited by J.M. Moore, A. Davidson, and A.J. Baer, *Geol. Ass. Can. Spec. Pap., 31*, 1-11, 1986.

Morey, G.B., and J.C. Green, Status of the Keweenawan as a stratigraphic unit in the Lake Superior region, in *Geology and Tectonics of the Lake Superior Basin*, edited by R.J. Wold, and W.J. Hinze, *Geol. Soc. Am. Mem., 156*, 15-26, 1982.

Morey, G.B., and R.W. Ojakangas, Keweenawan sedimentary rocks of eastern Minnesota and northwestern Wisconsin, in *Geology and Tectonics of the Lake Superior Basin*, edited by R.J. Wold, and W.J. Hinze, *Geol. Soc. Am. Mem., 156*, 135-146, 1982.

O'Brien, P.N.S., Lake Superior crustal structure - a reinterpretation of the 1963 experiment, *J. Geophys. Res., 73*, 2669-2689, 1968.

Ocola, L.C., and R.P. Meyer, Central North American rift system 1: Structure of the axial zone from seismic and gravimetric data, *J. Geophys. Res., 78*, 5173-5194, 1973.

Ojakangas, R.W., and G.B. Morey, Keweenawan pre-volcanic quartz sandstones and related rocks of the Lake Superior region, in *Geology and Tectonics of the Lake Superior Basin*, edited by R.J. Wold, and W.J. Hinze, *Geol. Soc. Am. Mem., 156*, 85-96, 1982a.

Ojakangas, R.W., and G.B. Morey, Keweenawan sedimentary rocks of the Lake Superior region: A summary, in *Geology and Tectonics of the Lake Superior Basin*, edited by R.J. Wold, and W.J. Hinze, *Geol. Soc. Am. Mem., 156*, 157-164, 1982b.

Paces, J.B., and D.W. Davis, Implications of high precision U-Pb age dates on zircons from Portage Lake volcanic basalts on Midcontinent rift subsidence rates, lava flow repose periods and magma production rates, *Thirty-fourth Ann. Meet. Inst. Lake Superior Geology, Proc. and Abstr., 34*, 85-86, 1988.

Palmer, H.C., and D.W. Davis, Paleomagnetism and U-Pb geochronology of volcanic rocks from Michipicoten Island, Lake Superior, Canada: Precise calibration of the Keweenawan polar wander track, *Precambrian Res., 37*, 157-171, 1987.

Rosendahl, B.R., Architecture of continental rifts with special reference to East Africa, *Ann. Rev. Earth and Planet. Sci., 15*, 445-503, 1987.

Serpa, L., T. Setzer, H. Farmer, L. Brown, J. Oliver, S. Kaufman, and J. Sharp, Structure of the Keweenawan rift from COCORP surveys across the midcontinent geophysical anomaly in northeastern Kansas, *Tectonics, 3*, 367-384, 1984.

Sharpton, V.L., R.A.F. Grieve, M.D. Thomas, and J.F. Halpenny, Horizontal gravity gradient: An aid to the definition of crustal structure in North America, *Geophys. Res. Let., 14*, 808-811, 1987.

Sims, P.K., E.B. Kisvarsavyi, and G.B. Morey, Geology and metallogeny of Archean and Proterozoic basement terranes in the northern midcontinent, U.S.A. - an overview, *U.S. Geol. Sur. Bull., 1815*, 51 pp., 1987.

Sims, P.K., and Z.E. Peterman, Early Proterozoic Central Plains orogen: A major buried structure in the north-central United States, *Geology, 14*, 488-491, 1986.

Smith, T.J., J.S. Steinhart, and L.T. Aldrich, Crustal structure under Lake Superior, in *The Earth Beneath the Continents*, edited by J. S. Steinhart, and T.J. Smith, *Am. Geophys. Un. Mono., 10*, 181-197, 1966.

Spencer, C., A.G. Green, J.H. Luetgert, More seismic evidence on the location of Grenville basement beneath the Appalachians of Quebec-Maine, *Geophys. J. R. Astr. Soc., 89*, 177-182, 1987.

van Breemen, O., and A. Davidson, Northeast extension of Proterozoic terranes of midcontinental North America, *Geol. Soc. Am. Bull., 100*, 630-638, 1988.

van Breemen, O., and S. Hanmer, Zircon morphology and U-Pb geochronology in active shear zones: Studies on syntectonic intrusions along the northwestern boundary of the Central Metasedimentary belt, Grenville Province, Ontario, *Geol. Surv. Can., Pap., 86-1B*, 775-784, 1986.

van Breemen, O., A. Davidson, W.D. Loveridge, and R.W. Sullivan, U-Pb zircon geochronology of Grenville tectonites, granulites and igneous precursors, Parry Sound, Ontario, in *The Grenville Province*, edited by J.M. Moore, A. Davidson, and A.J. Baer, *Geol. Ass. Can. Spec. Pap., 31*, 191-207, 1986.

Van Schmus, W.R., Early and Middle Proterozoic history of the Great Lakes area, North America, *Phil. Trans. R. Soc. London. Ser. A, 280*, 605-628, 1976.

Van Schmus, W.R., and M.E. Bickford, Proterozoic chronology and evolution of the mid-continent region, North America, in *Precambrian Plate Tectonics*, edited by A. Kroner, Amsterdam, Elsevier, 261-296, 1981.

Van Schmus, W.R., M.E. Bickford, and I. Zietz, Early and Middle Proterozoic provinces in the central United States, in *Proterozoic Lithospheric Evolution*, edited by A. Kroner, *Am. Geophys. Un. Geodyn. Ser., 17*, 43-68, 1987.

Van Schmus, W.R., and W.J. Hinze, The Midcontinent Rift System, *Ann. Rev. Earth and Planet. Sci., 13*, 345-383, 1985.

Van Schmus, W.R., K.D. Card, and K.L. Harrower, Geology and ages of buried Precambrian rocks, Manitoulin Island, Ontario, *Can. J. Earth Sci., 12*, 1175-1189, 1975.

Weiblen, P.W., Keweenawan intrusive igneous rocks, in *Geology and Tectonics of the Lake Superior Basin*, edited by R.J. Wold, and W.J. Hinze, *Geol. Soc. Am. Mem., 156*, 57-82, 1982.

Wold, R.J. and W.J. Hinze, editors, *Geology and Tectonics of the Lake Superior Basin*, *Geol. Soc. Am. Mem., 156*, 280 pp. 1982.

Wynne-Edwards, H.R., The Grenville Province, in *Variations In Tectonic Styles in Canada*, edited by R.A. Price, and R.J.W. Douglas, *Geol. Ass. Can. Spec. Pap., 17*, 263-334, 1972.

Zhu, T. and L.D. Brown, Consortium for continental reflection profiling Michigan surveys, *J. Geophys. Res., 91*, 11477-11495, 1986.

Zolnai, A.I., R.A. Price, and H. Helmstaedt, Regional cross-section of the Southern Province adjacent to Lake Huron, Ontario: Implications for tectonic significance of the Murray fault zone, *Can. J. Earth Sci., 21*, 447-456, 1984.

CRUSTAL EXTENSION IN THE MIDCONTINENT RIFT SYSTEM—RESULTS FROM GLIMPCE DEEP SEISMIC REFLECTION PROFILES OVER LAKES SUPERIOR AND MICHIGAN

J. C. Behrendt[1], A. G. Green[2], M. W. Lee[1], D. R. Hutchinson[3], W. F. Cannon[4], B. Milkereit[2], W. F. Agena[1], and C. Spencer[2]

Abstract. As part of the Great Lakes International Multidisciplinary Program on Crustal Evolution (GLIMPCE) the United States Geological Survey and the Geological Survey of Canada have collected about 700 km of 120-channel deep seismic reflection data across the Midcontinent (Keweenawan, 1.1 Ga) Rift system in Lakes Superior and Michigan. Results published by Behrendt et al. [1988] showed that volcanic and interbedded sediments of the rift extend to depths as great as about 32 km (10.5 s reflection time) beneath parts of Lake Superior, suggesting that this area may overlie the greatest thickness of intracratonic rift deposits on Earth. Times to Moho reflections vary from 11.5 s to 17 s (about 37-55 km depth) along the rift in Lake Superior and northern Lake Michigan. Additional GLIMPCE seismic reflection sections crossing the Midcontinent Rift system are presented here. We propose an interpretation to account for the substantial magmatic additions to the extended crust demonstrated by our data.

Introduction

Lake Superior (Figure 1) lies at the northern end of the Midcontinent Rift system of Keweenawan age (1.1 Ga), an arcuate crustal-scale structure that extends 2000 km from Kansas to south-central Michigan [Wold and Hinze, 1982; Van Schmus and Hinze, 1985]. In September 1986, as part of the Great Lakes International Multidisciplinary Program on Crustal Evolution (GLIMPCE) the United States Geological Survey (USGS) and the Geological Survey of Canada (GSC) acquired (by contract) six 26- to 30-fold seismic reflection profiles across the Midcontinent Rift system in Lakes Superior and Michigan (Figures 1 and 2). During the seismic reflection survey 120-channel data were recorded to 20 s at a 4 ms sample rate using a 3200 m streamer. The approximately 70-airgun tuned array had a capacity of 120 liters (or 7800 in^3).

We processed the data at the USGS processing center in Denver using a relatively standard sequence that included deconvolution before and after stack and dip filtering in the shot domain to suppress scattered noise from shallow sources and dip filtering in the common depth point domain to suppress multiple reflections. F-K migration was done at the GSC in Ottawa. Record sections were compiled and processed at 24- and 30-fold coverage (for the 62.5 and 50 m spaced shots, respectively), and subsequently every four adjacent traces were summed producing psuedo 96 and 120 fold stacks. For sections shown in this paper we plotted every other summed trace. Behrendt et al. [1988] reported preliminary results of the deep crustal data over the Midcontinent Rift system and presented an unmigrated section of line F. Our objective here is to present additional record sections for other lines in Lake Superior and northern Lake Michigan and to propose a model for rift magmatism.

Rift Basin Reflections

Figures 3 and 4, show the unmigrated seismic reflection sections for Lake Superior lines C and A, respectively. Figures 5 and 6 present examples of migrated data for the thickest part of rift fill for line F in eastern Lake Superior and line H in northern Lake Michigan, respectively. Seismic velocities determined from earlier coincident or nearby seismic refraction studies (Figure 2) have provided limited depth control of the interpretation of our various seismic reflection lines. Green et al. [this volume] show additional examples of the GLIMPCE profiles.

[1]U.S. Geological Survey, Denver Federal Center, Denver, CO
[2]Geological Survey of Canada, 1 Observatory Cresc., Ottawa, Ont. Canada
[3]U.S. Geological Survey, Woods Hole, MA
[4]U.S. Geological Survey National Center, Reston, Virginia

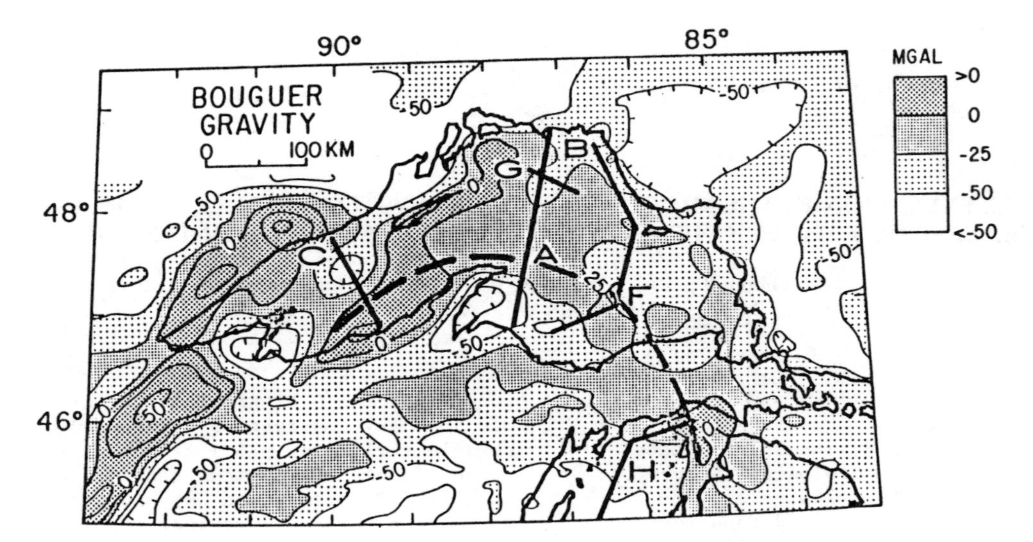

Fig. 1. Bouguer gravity anomaly map of the Lake Superior region [from Hinze et al. 1982]. Contour interval = 25 mgal. Letters indicate locations of seismic reflection profiles A, B, C, F, and G in Lake Superior (650 km) and parts of H in Lake Michigan. Brackets indicate locations for parts of migrated record sections for lines F and H shown in Figures 5 and 6. Dashed line connects the thickest rift section interpreted on lines C, A, F, and H; it is the approximate rift axis based on reflection data.

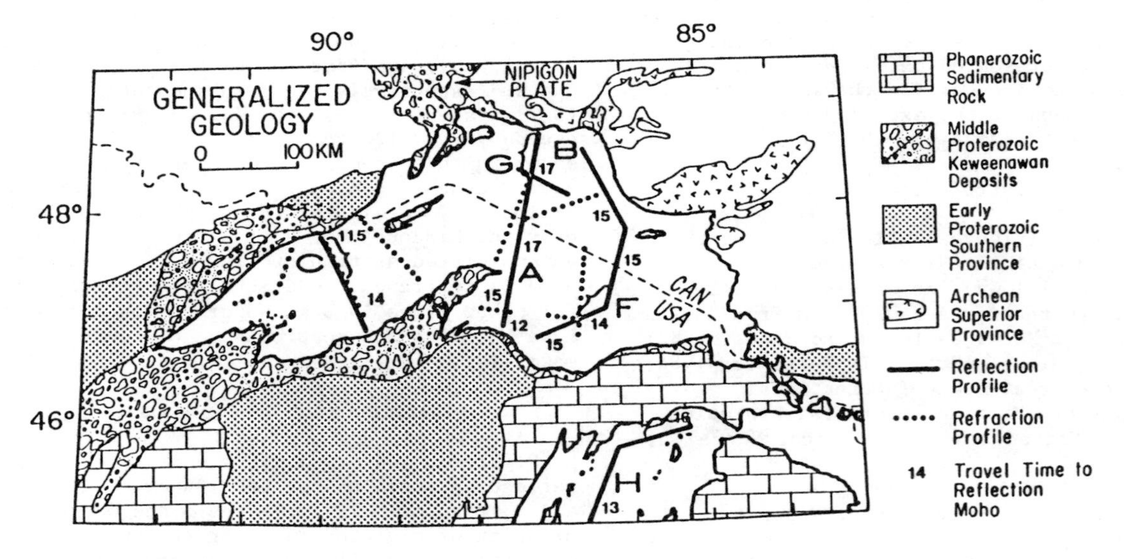

Fig. 2. Generalized geologic map showing outcrops of Keweenawan rocks. Locations of seismic reflection profiles as for Figure 1. Dotted lines = refraction profiles from earlier surveys described by Luetgert and Meyer [1982] and Luetgert [1988]. Reflection profile C is coincident with one of refraction profiles of Luetgert and Meyer (1982). Travel times to reflection Moho, are in seconds. Brackets on lines F and H indicate locations for parts of unmigrated record sections shown in Figures 7 and 8 showing Moho reflections. Brackets on lines A and C indicate locations for Moho reflections shown in Behrendt et al. [1988].

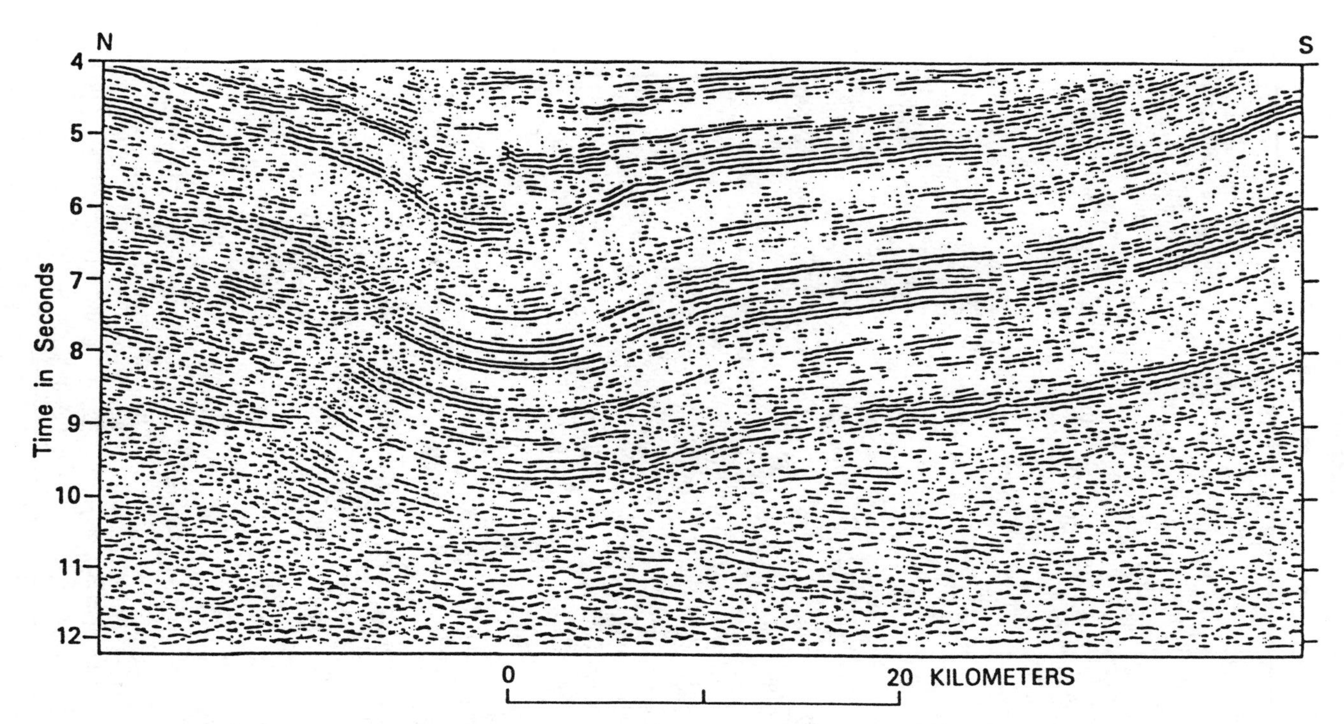

Fig. 5. Part of migrated record section for line F between shotpoints 1150-2000. Location indicated in Figure 1.

Remarkably strong and continuous reflections demonstrate that volcanic and interbedded sedimentary rocks of the rift extend to depths as great as about 32 km (e.g., 10.5 s reflection time, Figure 5) beneath parts of Lake Superior, which suggests that this area may overlie the greatest thickness of intracratonic rift deposits on Earth [Behrendt et al., 1988]. Reflections from the lava pile on lines C (Figure 3) and A (Figure 4) project almost into outcrop along the shores of the lake (Figure 2). Green [1983] mapped individual flows on land over distances up to 90 km, and we can trace reflections from inferred flows continuously for more than 70 km.

Lower Crust and Moho Reflections

Results from previous seismic experiments indicate that the depth to Moho varies considerably along the northern part of the Midcontinent Rift system. Refraction studies conducted in Lake Superior during the early 1960's [Berry and West, 1966; Smith et al., 1966; O'Brien, 1968; Halls, 1982], and combined interpretations of higher resolution seismic refraction data and regional gravity data [Luetgert and Meyer, 1982] suggest crustal thicknesses of up to 45-48 km beneath the western part of the lake and 50-56 km beneath its central and eastern regions, which compares to typical values of 35-45 km for the surrounding area. The program of GLIMPCE seismic reflection profiling in the Great Lakes was designed (i.e., powerful airgun source and 20-s recording times) to evaluate these refraction interpretations of an unusually deep Moho beneath Lake Superior.

Figures 3 and 4 show the deep reflections on lines C and A, respectively, interpreted as Moho and evidence of underplating by Behrendt et al. [1988]. Figures 7 and 8 show expanded examples of some of the deeper reflections observed on lines F and H, respectively. Although the pattern of deep reflections changes considerably along the axis of the Midcontinent Rift system, deep parts of the GLIMPCE data possess some important common features. Conspicuous reflections and diffractions occur throughout the 20 s unmigrated sections, demonstrating uniformly good signal penetration. Reflections are more prominent in the crust, with stronger concentrations of energy returned from the crust-mantle transition. Except for regions sampled by the central and northern parts of lines A (Figure 4) and G, the crust-mantle transition is usually distinguished by one or more continuous zones of reflections and diffractions (e.g., Figures 3 and 7). The 8.1 km/s mantle velocity, reported from the seismic

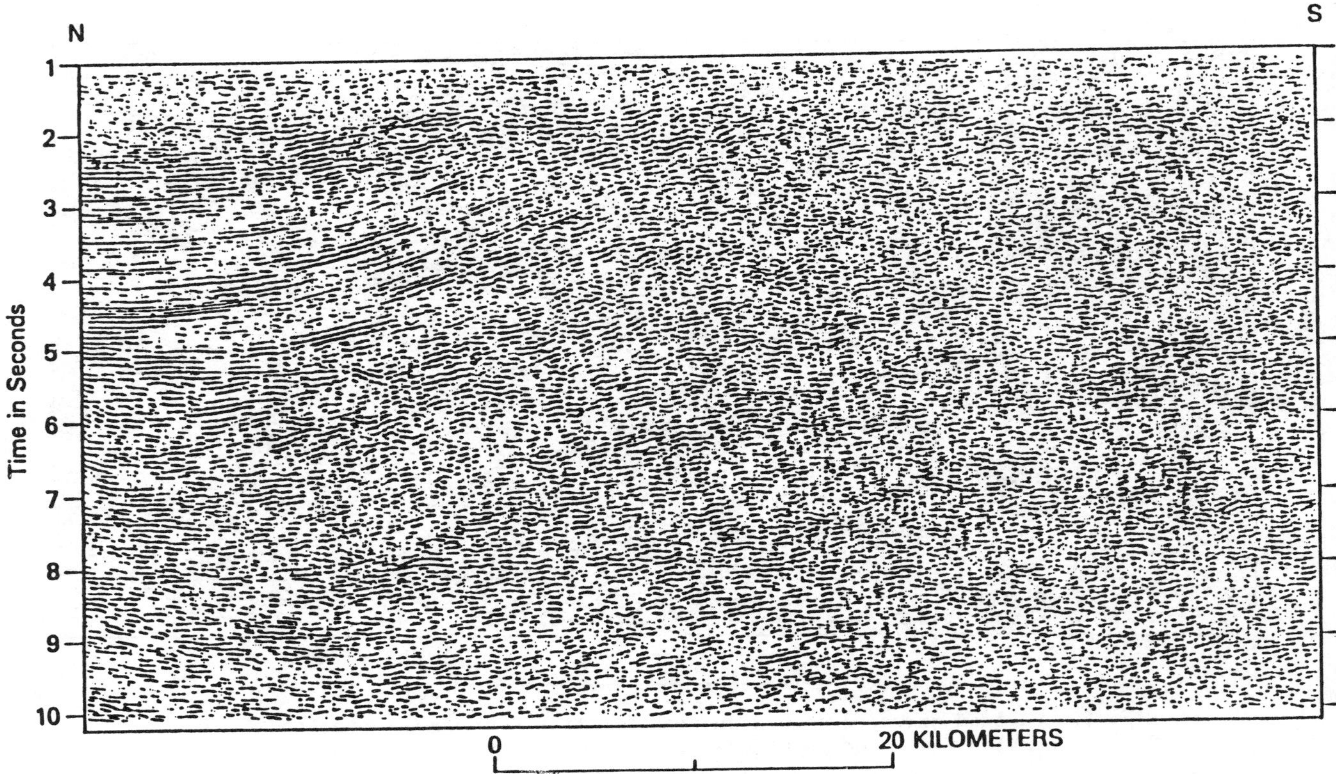

Fig. 6. Part of migrated record section for line H between shotpoints 3625-4640. Location indicated in Figure 1. Acoustically transparent section earlier than 2 s at left edge probably indicates lower velocity post-volcanic sedimentary rock.

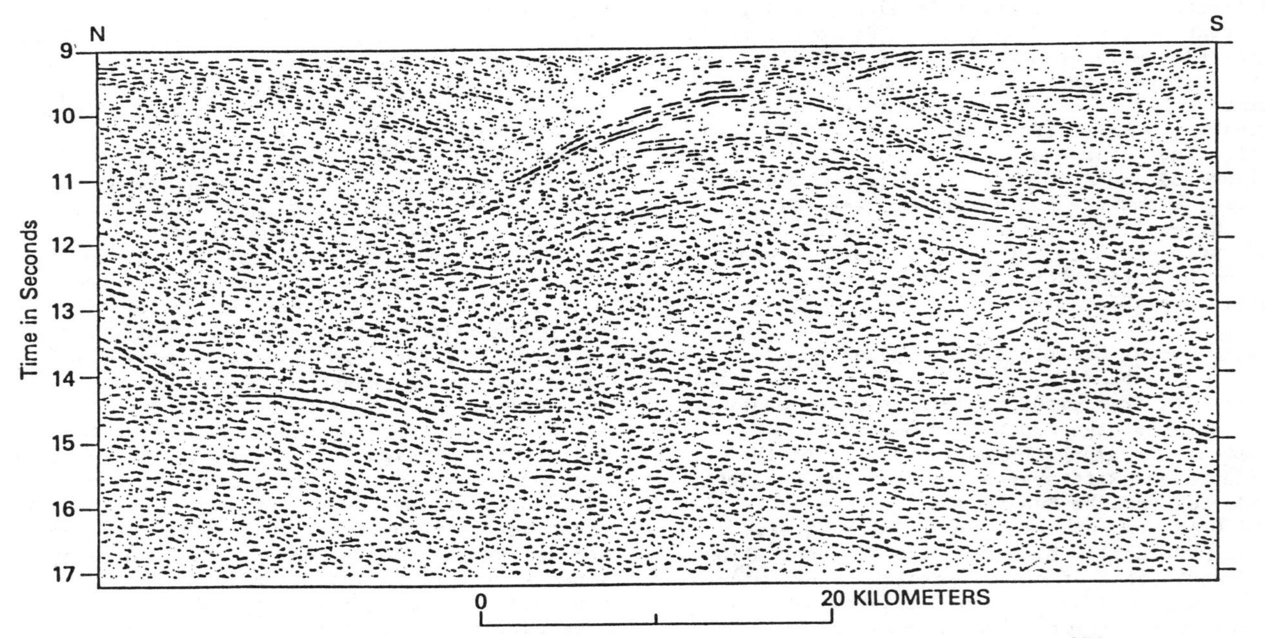

Fig. 7. Part of unmigrated record section for line F, between shotpoints 950-1750, showing deep crustal and crust-mantle reflections. See Figure 2 for location of line.

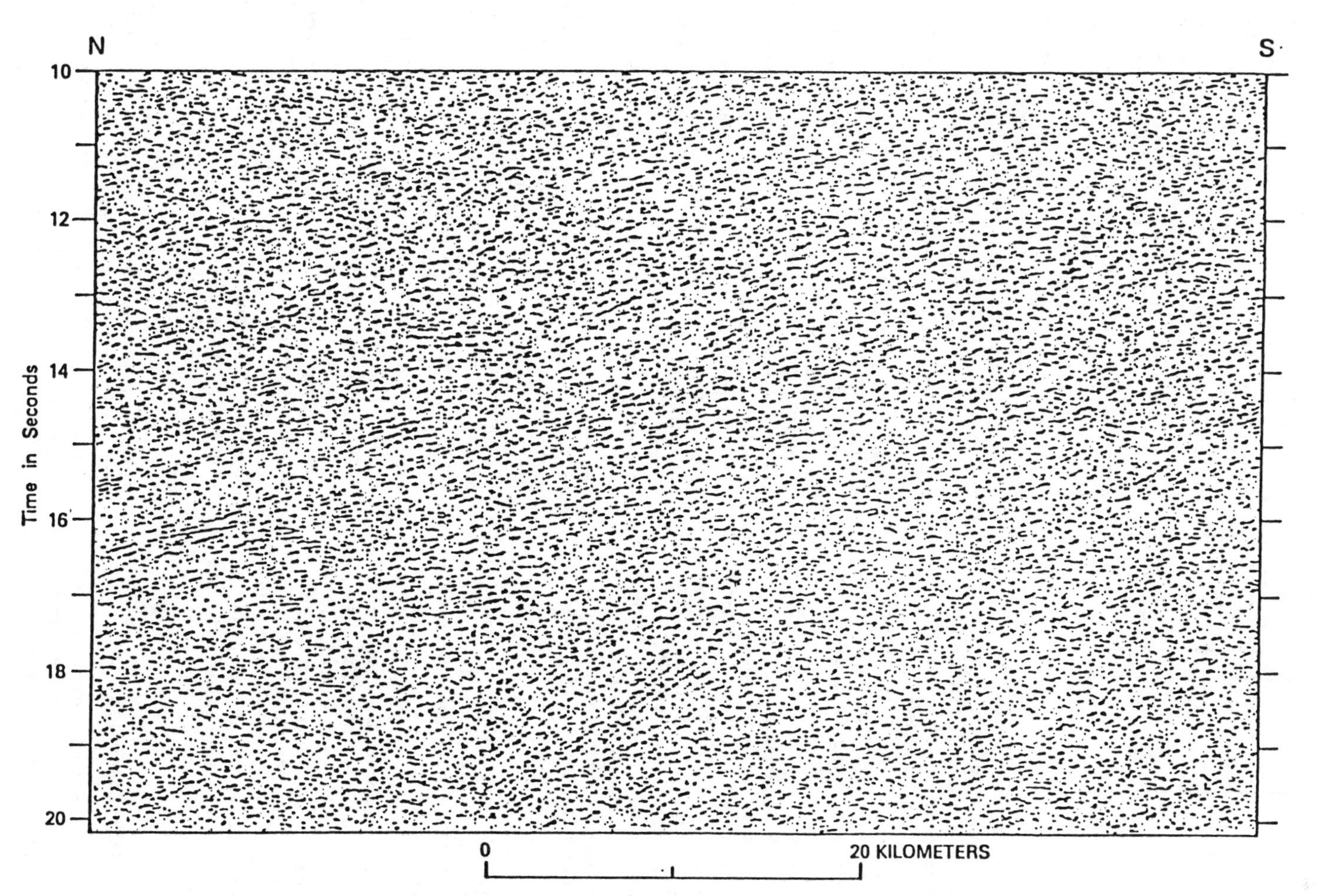

Fig. 8. Part of unmigrated record section for line H (coincident in location to Figure 6) showing deep crustal and crust-mantle reflections. Curvature on prominent reflection at about 16 s probably partly due to a low- velocity "push down" caused by lower velocity post-volcanic sedimentary rock.

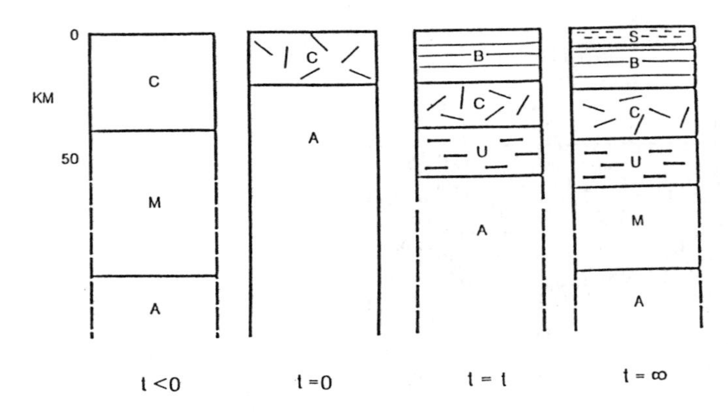

Fig. 9. Diagram showing a one-dimensional interpretation of development of the Midcontinental Rift system through time. C: Archean/Early Proterozoic crust; M: "cold" lithospheric mantle; A: asthenosphere; B: basalt flows and small amounts of interbedded volcanoclastic sedimentary rock; U: underplated material; S: post-volcanic, synrift sedimentary rock.
Time $t < 0$: prior to rifting; $t = 0$, start of rifting, thinned (stretched) Archean/Early Proterozoic crust associated with upwarped asthenosphere causing partial melting; $t = t$: end of rifting and magmatic activity; $t = \infty$ (present time) situation after cooling of rift and subsidence accompanied by post-volcanic deposition of sedimentary rock. Moho at base of underplated material (U). Depth scale approximate for upper part only; elements of model are isostatically balanced using plausible densities.

refraction experiments (Figure 2) cited above, occurs at the base of the bands of reflectors interpreted from the GLIMPCE profiles shown here and in Behrendt et al. [1988].

Reflection time to Moho varies along strike (Figure 2) from 11.5-14 s (about 37-46 km depth) in the west (Figure 3), to 17 s (about 55 km) in the center (Figure 4), and 13-15 s (about 41-49 km) in the eastern end of Lake Superior (Figure 7) and 16 s (about 52 km) in northern Lake Michigan (Figure 8). The pre-rift crust is 25-30 km thinner beneath some parts of the central rift than beneath its flanks, providing evidence for crustal extension by factors of about 3-4 [Behrendt et al., 1988].

Model for Rift Magmatism

We have considered the tectonic evolution of the Midcontinent Rift system using a conceptual approach (Figure 9) modified from work by Lachenbruch et al. [1985]. The four isostatically compensated columns (balanced using plausible densities) indicate idealized stages for the center of the rift through time. At time $t < 0$, Archean or Early Proterozoic crust (possibly similar in thickness to that presently surrounding the Lake Superior area in Figure 2) was presumably underlain by cold lithospheric mantle and hot asthenosphere.

At time $t = 0$, rifting had thinned the lithosphere (by faulting or ductile stretching) resulting in major upwarping of asthenosphere. In the extreme case the crust may have been rifted through completely. The resulting decrease in lithostatic pressure would have led to partial melting within the anomalously shallow asthenosphere.

At time $t=t$, (time of completion of active rifting) part of the magma had passed through the crust and had been extruded in the form of the pile of basaltic flows which we image in our reflection data. Magma also intruded the lower crust and uppermost mantle producing some of the deep reflections that we suggest as evidence of underplating and associate with Moho. We would expect the intruded material to have seismic velocities in the 7.0 - 7.3 km/s range [e.g., Furlong and Fountain, 1986].

Obviously the steps indicated by $t=0$ and $t=t$ are quite complex and poorly understood. Although movement on growth faults seen in our data (Figures 3 and 4, and Behrendt et al. [1988]) proceeded concurrently with the extrusion of lava flows indicating a continuous process with the surface always near sea level, we suggest substantial lithospheric extension preceded basalt extrusion. Evidence for this is found in the continuity of the deepest reflectors which demonstrate that the lowest parts of the section were little disrupted by later extension (e.g., Figure 5). Continued lithospheric thinning probably took place during basalt deposition along listric normal faults bounding the grabens seen in Figures 3 and 4.

After time $t=t$ rifting and magmatism ceased. Continued cooling and subsidence resulted in the establishment of a lithospheric mantle and the deposition of post-volcanic, synrift sedimentary rock, as illustrated for time $t = \infty$ (the present). Except for some relatively minor compression [Wold and Hinze, 1982] the area of the Midcontinent Rift system near Lake Superior has remained essentially undisturbed since cessation of the rift activity.

The Midcontinent Rift system differs from many other rifts in having a total crustal thickness equal to, or greater than surrounding (presumably unextended) regions. However, there are some striking similarities between the character of the ancient rift system in the midcontinent of North America and seaward dipping reflectors reported from several deep rift basins observed along various present continental margins as discussed by Behrendt et al. [1988].

Acknowledgments. Researchers participated in the GLIMPCE 1986 experiment from the University of Wisconsin, Madison, University of Wisconsin, Oshkosh, University of Southern Illinois, University of Northern Illinois, University of Western Ontario, University of Saskatchewan, the Geological Survey of Canada and the U.S. Geological Survey. The Geological Survey of Canada in association with LITHOPROBE and the U.S. Geological Survey jointly funded this cooperative study. Geophysical Services Inc. collected the reflection data which we processed in house. Geological Survey of Canada Contribution No. 46587.

References

Behrendt, J.C., A.F. Green, W.F. Cannon, D.R. Hutchinson, M.W. Lee, B. Milkereit, W.F. Agena, and C. Spencer, Crustal structure of the Mid-continent rift system - Results from GLIMPCE deep seismic reflection profiles, Geology, v. 16, p. 81-85, 1988.

Berry M.J., and F.G. West, A time term interpretation of the first arrival data of the 1963 Lake Superior experiment in, Steinhart, J.S., and Smith, T.J., eds., The Earth Beneath the Continents: American Geophysical Union, Geophysical Monograph 10, p. 166-180, 1966.

Furlong, K.P., and D.M. Fountain, Continental crustal underplating: thermal considerations and seismic-petrologic consequences: Journal Geophysical Research, v. 91, p. 8285-8294, 1986.

Green, A.G., W.F. Cannon, B. Milkereit, D.R. Hutchinson, A. Davidson, J.C. Behrendt, C. Spencer, M.W. Lee, P. Morel-a-l´ Huissier, and W.F. Agena, "GLIMPCE" of the deep crust beneath the Great Lakes, this volume.

Green, J.C., Geologic and geochemical evidence for the nature and development of the middle Proterozoic (Keweenawan) midcontinent rift of North America: Tectonophysics, v. 94, p. 413-437, 1983.

Halls, H.C., Crustal thickness in the Lake Superior region: in Wold, R.J., and Hinze, W.J., eds., Geology and Tectonics of the Lake Superior Basin: Geological Society of America Memoir 156, p. 239-243, 1982.

Hinze, W.J., R.J. Wold, and C.J. O´Hara, Gravity and magnetic anomaly studies of Lake Superior: in Wold, R.J., and Hinze, W.J., eds., Geology and Tectonics of the Lake Superior Basin: Geological Society of America Memoir 156, p. 203-211, 1982.

Lachenbruch, A.H., J.H. Sass, and S.P. Galanis, Jr., Heat flow in southernmost California and origin of the Salton Trough, Journal of Geophysical Research, v. 90, B8, p. 6709-6736, 1985.

Luetgert, J.H., and R.P. Meyer, Structure of the western basin of Lake Superior from cross structure refraction profiles: in Wold, R.J., and Hinze, W.J., eds., Geology and Tectonics of the Lake Superior Basin: Geological Society of America Memoir 156, p. 245-256,1982.

Luetgert, J.H., The earth´s crust beneath Lake Superior: an interpretation of crust structure seismic refraction profiles: Ph.D. Thesis, University of Wisconsin, 86 p., 1982.

O´Brien, P.N.S., Lake Superior crustal structure--An interpretation of the 1963 experiment, Journal of Geophysical Research, v. 73, p. 2669-2689, 1968.

Smith, T.J., J.S. Steinhart, and L.T. Aldrich, Crustal structure under Lake Superior in, Steinhart, J.S., and Smith, T.J., eds., The Earth Beneath the Continents: American Geophysical Union, Geophysical Monograph 10, p. 181-197, 1966.

Van Schmus, W.R., and W.J. Hinze, The midcontinent rift system: Annual Review Earth Planet Science, v. 13, p. 345-383, 1985.

Wold, R.J., and W.J. Hinze, eds., Geology and Tectonics of the Lake Superior Basin: Geological Society of America Memoir 156, 280 p, 1982.

Section II

THE LOWER CRUST FROM RESULTS OF REFRACTION/WIDE-ANGLE REFLECTION EXPERIMENTS

THE GLIMPCE SEISMIC EXPERIMENT: ONSHORE REFRACTION AND WIDE-ANGLE REFLECTION OBSERVATIONS FROM A FAN LINE OVER THE LAKE SUPERIOR MIDCONTINENT RIFT SYSTEM

Duryodhan Epili and Robert F. Mereu

Department of Geophysics, University of Western Ontario
London, Ontario, Canada N6A 5B7

Abstract. The 1986 GLIMPCE experiment (Great Lakes International Multidisciplinary Program for Crustal Evolution) was a combined on-ship seismic reflection and onshore seismic refraction experiment designed to determine the structure of the crust beneath the Great Lakes. The main tectonic targets of interest were the Midcontinent Rift System, the Grenville Front, the Penokean and Huronian Fold Belts and the Michipicoten Greenstone Belt. The source of the seismic energy came from a large air gun array fired at closely spaced intervals (50-350 m) over several long lines (150-350 km) crossing the lakes. Major participants of this experiment were the Geological Survey of Canada, the United States Geological Survey and a number of universities and research institutes on both sides of the border. The University of Western Ontario (UWO) collected data at five separate land stations using portable seismic refraction instruments.

In this paper we present the results of a fan profile which was recorded from a UWO station on Michipicoten Island for the N-S line A which crossed the axis of the Lake Superior Synclinal Basin. The azimuth and distance ranges for this profile were 237 to 321 degrees and 120 to 170 km respectively. Detailed observations of the record sections show that P* is not a simple arrival but forms a rather complex pattern of irregular multiple arrivals. The wide-angle PmP reflection signals from the Moho are strong and well observed only for the shots fired near the ends of the line. The signals from the middle of the profile arrive relatively late and form very weak complex wave trains. These results indicate that the Moho in that area is probably greatly disrupted and gives added support to the rift theory for the structure under the lake. The observations also support the results of earlier crustal studies of Lake Superior which showed that the crust under the eastern part of the lake was exceedingly thick.

Introduction

During the summer of 1986, The Great Lakes International Multidisciplinary Program for Crustal Evolution (GLIMPCE) conducted a major on-ship onshore seismic experiment in order to resolve some of the problems associated with the Earth's crust under the Precambrian Canadian Shield. The major tectonic features of interest were the Grenville Front and the Midcontinent Rift System, the Penokean and Huronian Fold Belts and the Michipicoten Greenstone Belt. Major participants of the experiment were the Geological Survey of Canada, the United States Geological Survey and a number of universities and research institutes on both sides of the border. A marine seismic crew, Geophoto Service Ltd., collected in-line near vertical seismic reflection data using a hydrophone streamer and a wide tuned air gun array, whose total capacity was 127 L consisting of 60 active air guns. The array of guns was fired at closely spaced intervals of approximately 50 to 350 m to generate the seismic energy for this survey. The location of all the lines of this experiment is shown in Figure 1. Lines C, A, F and B in Lake Superior and line H in Lake Michigan were over the Midcontinent Rift System. Line J crossed the Grenville Front and line I was on the west side of Lake Huron. The seismic refraction energy which travelled laterally over distances of up to 250 km was recorded simultaneously with a set of onshore land stations located at various sites at the end of the lines. The University of Western Ontario participated in this experiment by operating five of the onshore stations (Figure 1) using portable refraction instruments. Station 1, on Michipicoten Island recorded data for two approximately "in-line" profiles, line B, line F and one fan profile line A, in Lake Superior. Station 2 on the Bruce Peninsula, station 3, 4 and 5 at Parry Sound recorded data for a long

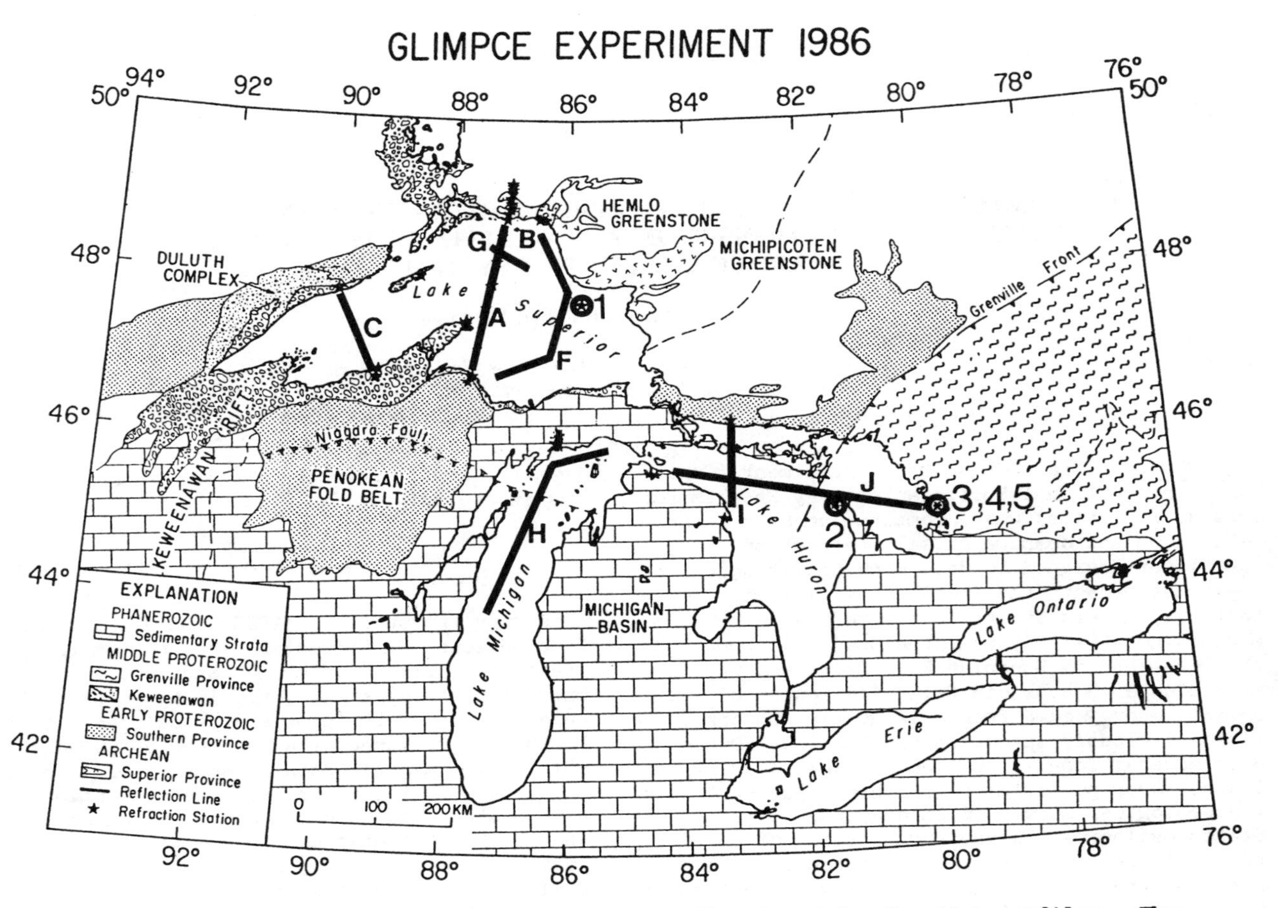

Fig. 1. Map showing the location of GLIMPCE reflection and refraction profiles. The University of Western Ontario collected data for line A, line B, line F by operating a refraction station at Michipicoten Island (station 1), for line J by operating a station at Bruce Peninsula (station 2) and 3 stations (station 3, 4, 5) at Parry Sound.

line across the Grenville Front which ran from the east side of Georgian Bay to the west side of Lake Huron. The station on Michipicoten island consisted of one vertical seismometer and recorder set up on a bedrock outcrop, a few hundred meters from the shoreline. This system was backed up by two other recorders to take care of possible technical problems in isolated area. A total of over 23,000 seismic records were obtained in this experiment during the 200 hour recording period. The results of an analysis of the data collected at Michipicoten Island for the north-south fan profile (line A) across the axis of the Midcontinent Rift in Lake Superior is presented in this paper. An interpretation of the in-line near vertical reflection data for this line is also presented by Behrendt et al and Green et al both in this volume.

Geology of the Area

The Canadian Shield is a large region of Precambrian rocks that occupies nearly half of Canada. The subdivision of the Shield into a number of structural Provinces and Subprovinces is based mainly on overall differences in internal structural trends, styles of folding, isotopic dating or orogenies (Stockwell 1964). The Lake Superior Basin belongs to the Southern Province. The Keweenawan sequence of the Lake Superior region is remarkable for the very great thickness and vast amount of basic volcanics it contains. Structurally it forms a large syncline occupying most of Lake Superior. The geology of this area was brought to the attention of geoscientists a century ago because of economic mineral deposits within the Keweenawan volcanic and sedimentary rocks.

It is broadly accepted that Lake Superior lies along one of the most interesting and complex parts of the Midcontinent Rift System (Kumarapeli and Saull 1966, Ocola and Meyer 1973, Halls 1978, Luetgert and Meyer 1982, Weiblen 1982). A triple junction in central Lake Superior with limbs that are approximately 120 degrees apart was postulated by Burke and Dewey (1973), Franklin et

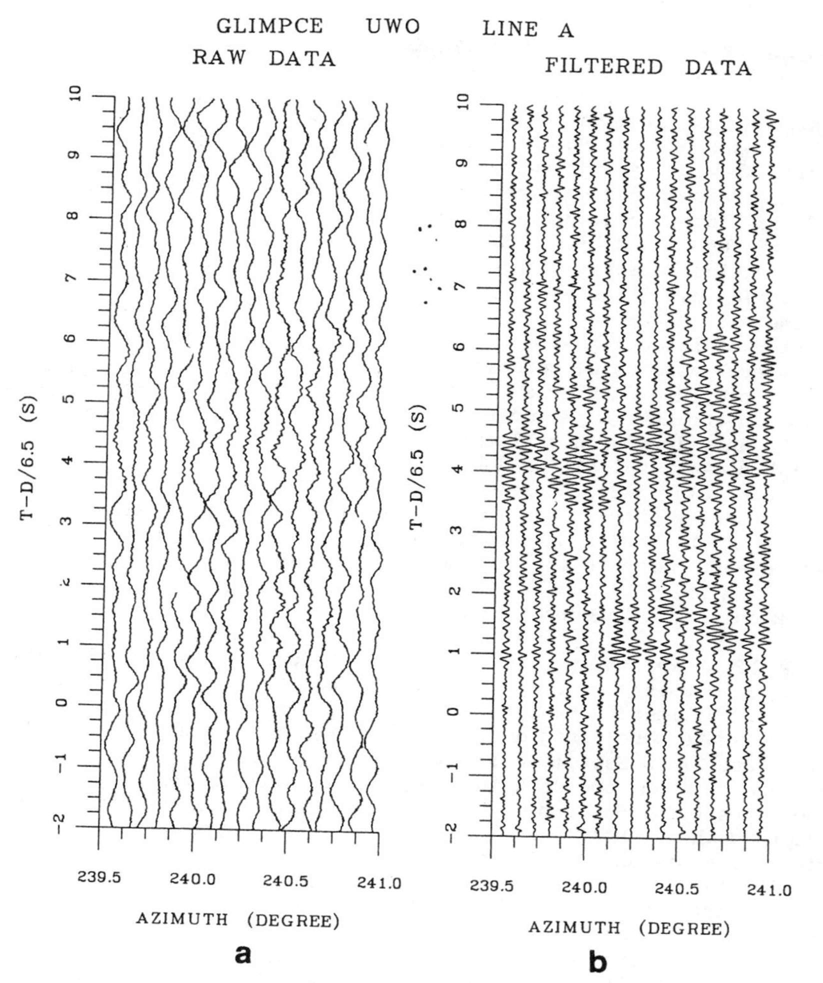

Fig. 2. An example of data enhancement by band pass filtering. a) Portion of field data from south end of line A recorded from station 1 (distance ranging from 163.79 to 160.07 km). b) Fourth order symmetric Butterworth band pass (4-16 Hz) filtered data for the same portion. Note that the seismic signal which is visible as small wiggles around 1.5 sec and 4.0 sec in reduced time scale overriding the high amplitude low frequency water wave noise, is easily recovered by the filter.

al (1980). Burke and Dewey (1973) proposed that the triple junction occurs beneath Lake Superior with the south western and the south eastern extensions of Lake Superior as two arms and the Kapuskasing structural high as the third failed arm whereas Franklin et al (1980) and Chandler et al (1982) proposed that the failed arm is represented by the middle proterozoic rocks of the Sibley Basin Nipigon Plate. Watson (1980) suggested that the Kapuskasing Subprovince developed primarily by uplift and sinistral transcurrent movement prior to 2690 Ma ago and therefore did not originate as part of the rift structure. Klasner et al (1982) also recognized the Kapuskasing structural high as a major lineament and zone of weakness in the crust. Wold and Hinze (1982), in a compilation of scientific literature pointed out some of the major unresolved problems in this region, such as the extent of Keweenawan volcanic rocks, age of formation, structure and origin of the Lake Superior Basin

Previous Geophysical Work

Two large seismic experiments the Lake Superior experiment (Steinhart 1964, Steinhart and Smith 1966, Berry and West 1966a, 1966b, Smith et al 1966a, 1966b, O'Brien 1968, Halls 1982) and Project Early Rise (Iyer et al 1969, Mereu and Hunter 1969) were performed about 25 years ago in this region. The interpretation of the results from the 1963 Lake Superior experiment is important here as it's main profile was east-west and ran over much of the axis of the Lake Superior Synclinal Basin. Nearly one hundred dynamite charges were used for the energy sources in that experiment. The main results of that experiment showed that the crustal seismic velocities were higher than expected and that the crust at the central part of Lake Superior under the axis of the rift was surprisingly very thick (greater than 50 km). Seismic and gravimetric investigations of the Midcontinent gravity high (Cohen and Meyer 1966, Ocola and Meyer 1973) reveal that middle crustal material with unexpected high seismic velocities are coincident with the axis of the rift system as defined by the gravity. A number of reconnaissance geophysical studies such as crustal seismic, gravity and magnetic provided further evidence that the crust beneath the Lake Superior Basin is highly disturbed and anomalously thick locally, with relatively high seismic velocities and densities (Hinze and Wold 1982). Hinze et al (1982) studied the changing characteristics of the Midcontinent anomaly by analysing gravity and magnetic profiles perpendicular to the rift axis.

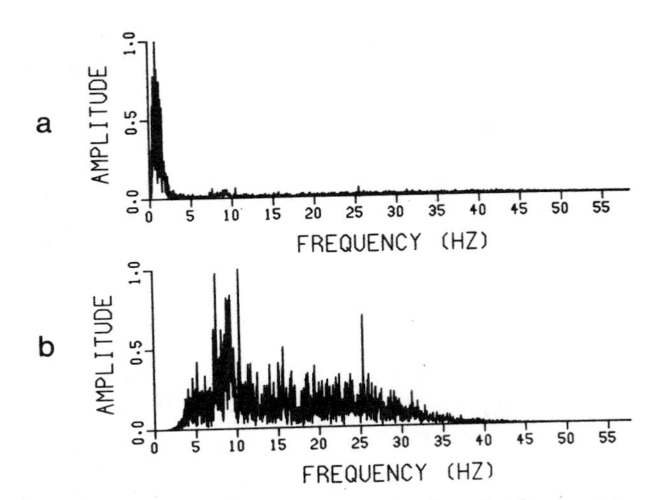

Fig. 3. Spectra of a typical seismic trace (shot no 135) for line A. a) Field data b) Filtered by Butterworth band pass (4-30 Hz) filter.

Experimental Results and Interpretation

Line A of the GLIMPCE experiment took place along a 214 km long line which ran perpendicular to the axis of the Midcontinent Rift System. A total of 640 air gun shots were fired during a 22 hour period from August 31 to September 1, 1986 to provide a coverage of about one shot in every 333 meters. The University of Western Ontario station which recorded these shots was placed on Michipicoten Island at a distance of 120 to 170 km from the line. The fan profile which was obtained covered an azimuth range from 237 to 321 degrees. The recording instruments were the same frequency modulation type refraction units which had been used in earlier Canadian long range COCRUST refraction experiments (see Mereu et al 1986) except that they were upgraded to record a higher frequency (up to 35 Hz) range. The seismometers were vertical Mark product LC 4 with a 1 Hz natural frequency. Since the refraction system used cassette tapes, an operator was employed to change the tapes every 45 minutes during the whole recording run. These tapes were later digitized in the laboratory using a sample rate of 125 samples/sec.

The station on Michipicoten Island was located on bedrock approximately 100 m from the shore line. Because of the proximity of the lake, the raw data was contaminated with low frequency water wave noise in the 1 to 3 Hz range. This noise was easy to eliminate with a simple zero phase band pass Butterworth filter of order 4 as the seismic energy was confined in the 4-16 Hz range. Figure 2 shows that the signal which appears as extremely small barely discernible wiggles around the reduced travel time of 1.5 sec and 4.0 sec (Figure 2a) could be recovered by applying the filter (Figure 2b). An example of the spectrum of a typical trace (shot 135) both before and after filtering is shown in Figures 3a and 3b respectively. After a series of numerical processing tests the complete data set for line A was band pass filtered, Nth root (N=2) stacked of 3 adjacent traces (Muirhead 1968), trace normalized and plotted in a conventional manner against azimuth with a reducing velocity of 6.5 km/s (Figure 4a). In order to highlight the large amplitudes, only amplitude values above 50% of the peak value are plotted and shaded in this figure. An alternate procedure for displaying the seismic traces is also presented in Figure 4b using true amplitudes. A normal move out (NMO) correction was computed using a crustal thickness of 55 km and average velocity of 6.6 km/s. Each trace was then shifted by an amount of time equal to the normal moveout correction and plotted with the time axis increasing downwards as is done in conventional seismic reflection industrial sections. This type of data representation removes the hyperbolic distortions of the PmP branch for two way travel time of 16.6 sec and thus enables one to visualize the nature of the Moho and its topography more clearly. Here also the data is Nth root stacked as mentioned above. Figure 5 shows a travel time diagram for a simple one layered crust. It is clear from this diagram that P* and PmP should be observed as prominent events for the distance range of 120 to 170 km which is the distance range of our fan line. The salient feature of this data set is that a double pulse appears on the P* signal around azimuths 248 to 280 degrees. This is an indication of a localized intermediate transition boundary. It's disappearance at the north extreme half and the south end of the line shows that the intermediate layer is truncated. This feature may be attributed to the synclinal

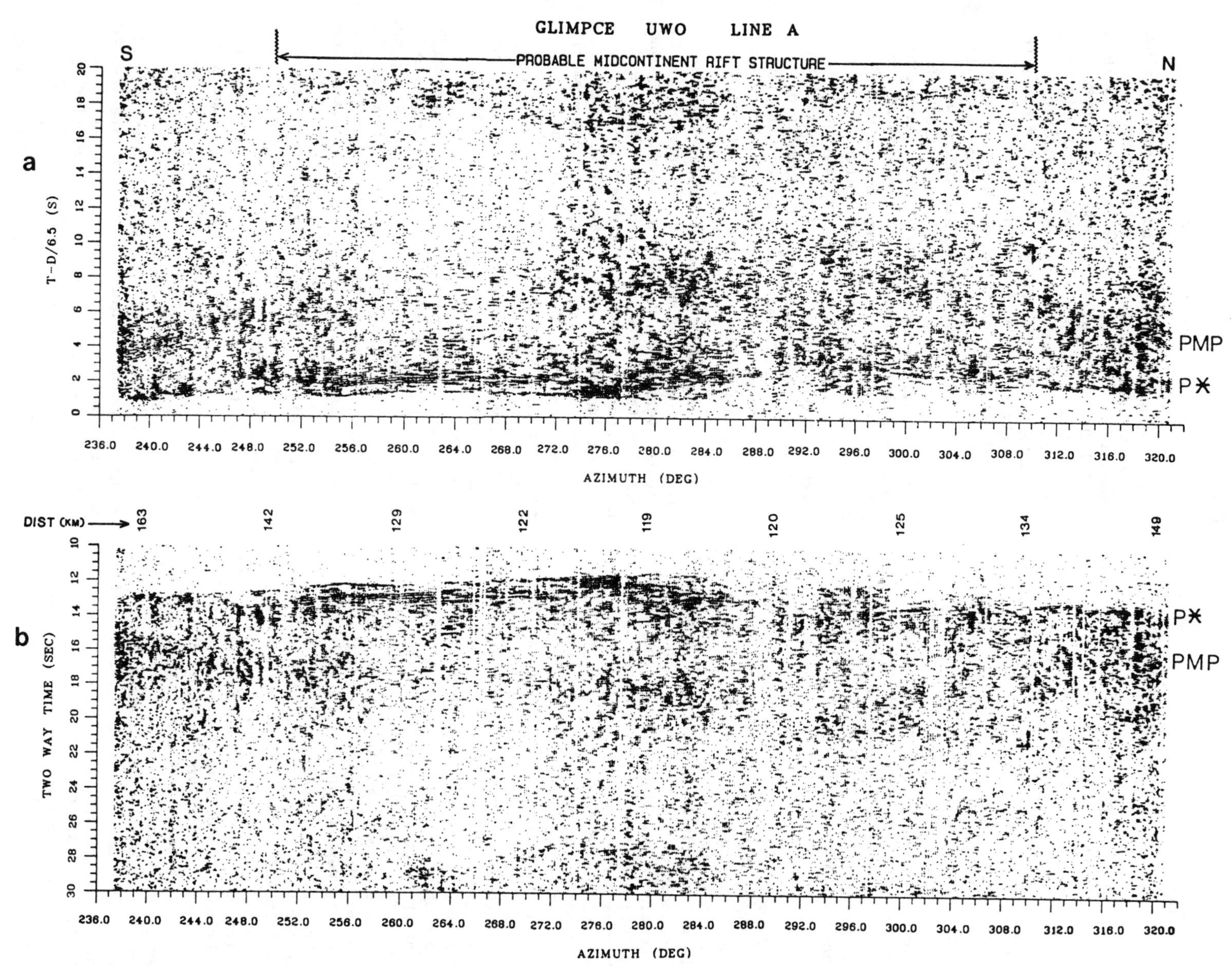

Fig. 4. Record section for line A recorded at Michipicoten Island (station 1). a) The field data was band pass (4-16 Hz) filtered and trace normalized. b) The filtered data for line A was corrected for normal move out (NMO) and true amplitude of the traces are represented. In both of the above figures, the data was Nth root (N=2) stacked (3 adjacent traces) and the amplitude values higher than 50% of the peak value are shaded and plotted to highlight the large amplitudes. Note the large amplitude PmP at south and north ends of the profile and complex behaviour in between.

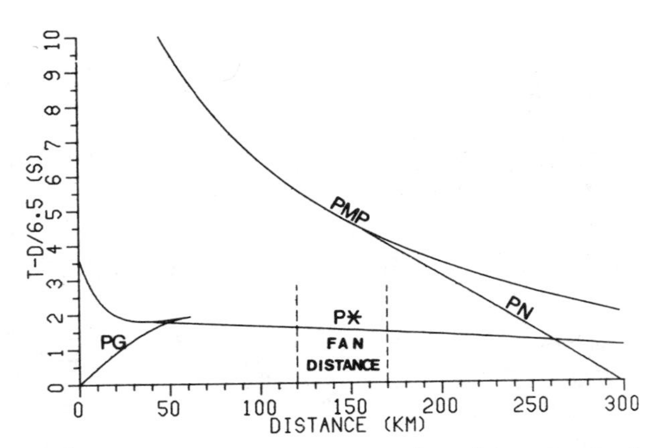

Fig. 5. Reduced travel time plot for a typical crust representing Pg, P*, PmP and Pn events. Note the presence of P* and PmP for a source to receiver distance of 120 to 170 km which is the distance range for fan line A.

feature at the centre of Lake Superior, thus the intermediate layer might have been tapered at the flanking edges. The complex character of the P* signals should come as no surprise when one examines the multiple layered structure found in the upper 7 seconds of the near vertical reflection section along line A by Behrendt et al (in this volume) and Green et al (in this volume). Precise correlation of our fan data reflections with the in-line data is not possible without detailed knowledge of the 3 D structure under the lake. The complex behaviour of the PmP wide-angle reflection is also worth noting. It was well recorded at the south end of the profile for azimuths between 237 to 250 degrees and on the north end between 310 to 321 degrees. A gradual dipping feature at both ends of the profile is probably due to the reflections from the flanks of the depression caused by the Midcontinent Rift. In between the azimuths 250 to 310 degrees the PmP phase energy does not show as a coherent event. However there is an indication of an increase in PmP signal strength at the centre of the line.

The main arrivals of the data illustrated in the time sections of Figures 4a and 4b were used to derive a single 2 D depth model using standard simple ray tracing inversion techniques. The 2 D model was derived from different models located at positions a, b, c and d on Figure 6. These models were assumed to be 1 dimensional. Since it is not possible to determine velocities from a fan profile, the velocities which were incorporated in the model were obtained from the in-line results of the 1963 Lake Superior experiment which had a long east-west line running from a point just north of Michipicoten Island to the extreme western side of the lake. The final model, Figure 6, represents our estimate of the structure on a N-S line half way between Michipicoten Island and line A (see Figure 7).

Conclusions

1. The model given in Figure 6 shows a rather complex two layered upper crust with some evidence for at least a three layered crust in the 248-280 degree azimuth range. This upper crustal complexity probably owes its origin to the thick (10-20 km) low velocity volcanic sedimentary sequence of rocks which lie along the axis of the rift. Further evidence and a more detailed structure of this sequence were given by Behrendt et al (in this volume) Green et al (in this volume) and Behrendt et al (1988b) who analysed the in-line near vertical reflection data from lines A and F. Our P* data shows that the structure perpendicular to the axis of the rift is not symmetric. It does however agree qualitatively to the asymmetric structure observed in the near vertical reflection experiment along line A. The line F reflection experiment tended to show a more symmetric structure. Complete agreement between the line A and line F structure should not be expected as our data sample the rift area in a different location as shown in Figure 7.

2. The anomalous behaviour of PmP which was clearly recorded for shots at the ends of the fan line but very weak, delayed and complex over the center of the line suggests that the Moho is badly disrupted along the axis of the rift. Possible injections of the lava flows from the upper mantle has lead to a rather diffuse poorly defined deep moho. The evidence for the anomalously thick crust (55-60 km) under the axis of the rift as opposed to 45-50 km at both ends of the line confirms previous findings from the 1963 experiment (Berry and West 1966a, Smith et al 1966a).

The depression in the upper layers, the thick crust and the weak disrupted Moho under the axis of the Lake Superior Basin all give added support to the rift hypothesis for the deep structure in the region. The weak PmP zone, shown in Figure 7, is approximately 60-70 km wide at the place where the Moho is sampled. The possibility that the weak PmP signals may be the result of a large proportion of the energy returning from the intermediate layer was considered. This hypothesis has to be rejected as the angles of incidence which should theoretically cause a large wide- angle reflected PmP are such that there are little transmission losses through intermediate layer boundaries. This probably relates to the mantle intrusions and subsequent crustal separation caused by the rifting process at approximately 1100 Ma before present which subsequently became inactive. The theories by Chase and Gilmer (1973) and Sims et al (1980) concerning the opening of the Midcontinent Rift System revealed a similar estimate for the width of the structure.

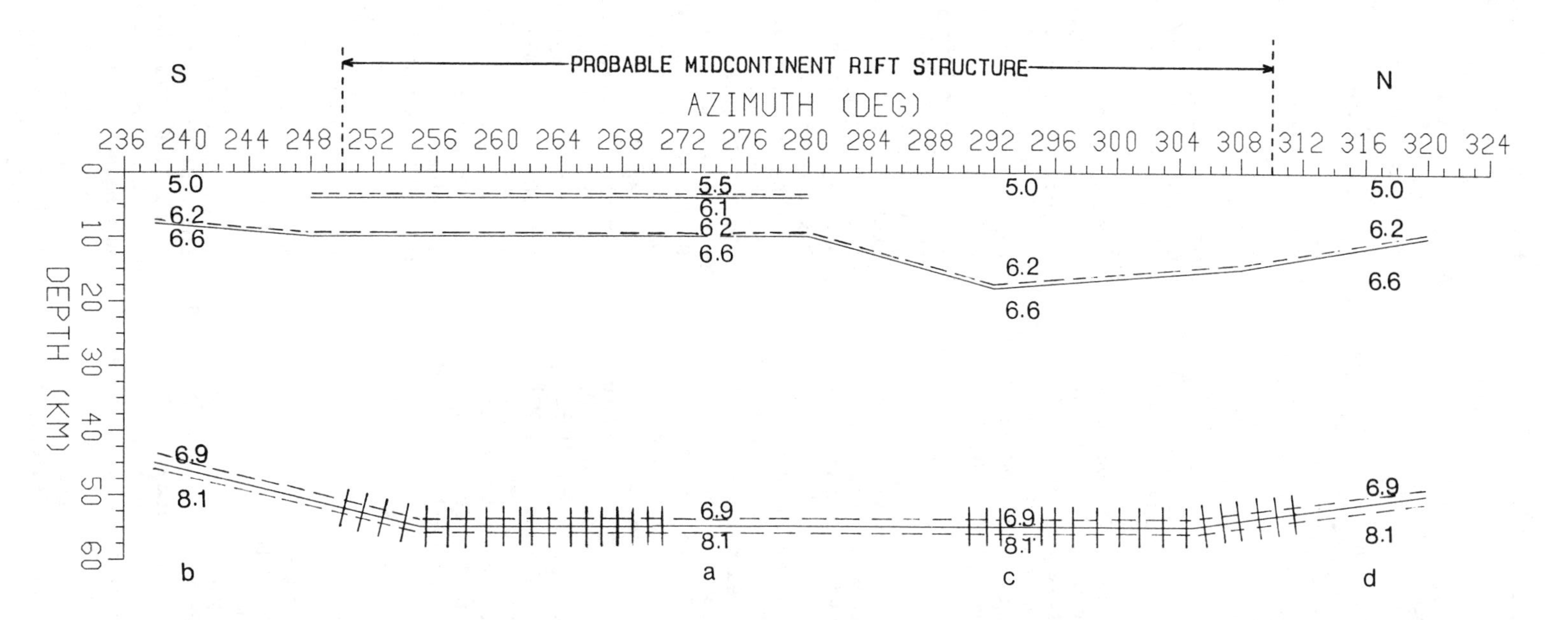

Fig. 6. Two Dimensional section as determined from one dimensional depth models (a, b, c, d). These 1 D models were determined by conventional ray-tracing inversion methods. Note the exceedingly thick crust at the center of the profile and the sloping Moho at the south and north ends of the profile. The crossed lines indicate a weak PmP event observed from the record section.

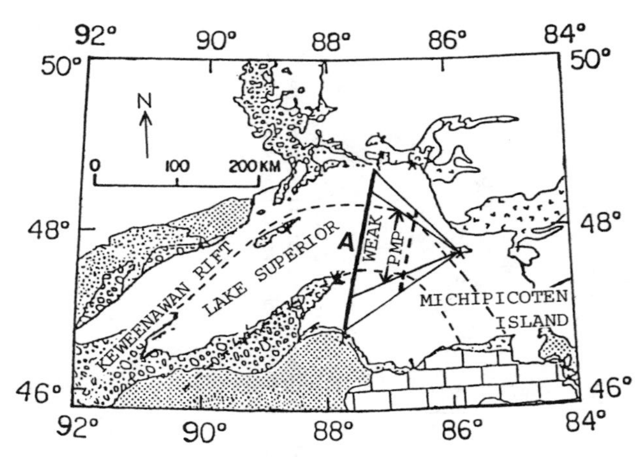

Fig. 7. Probable Midcontinent Rift zone over Lake Superior as inferred from fan line A data. The broken line indicates where we sampled the Moho. Note the weak PmP zone which is probably caused by the depression of the Moho and lava intrusions as the crust opened up during the Keweenawan rifting period.

Acknowledgments. The experiment described above was a large scale cooperative experiment involving the Geological Survey of Canada, the U. S. Geological Survey, and a number of universities and research institutes from both sides of the border. The authors are indebted to all the participants who worked together to make this experiment a success. Special thanks is extended to A. Green of the GSC who was the chief coordinator of the project, for his advice and assistance. Special thanks is given to the U. W. O technical staff J. Brunet, B. Price, B. Dunn, and C. Faust and to the U. W. O students T. P. Cox, J. Wu, T. Shortt, R. Secco, C. Spindler, P. J. Lenson and P. F. Lenson for their help with the data collection in the field and subsequent laboratory work in getting the data organized. This research was supported by the following financial sources (1) NSERC Grant A-1793 (2) EMR Grant 133/04/87 (3) DSS Contract 2322-6-0771/01 ST and (4) a Canadian Commonwealth Scholarship.

References

Behrendt, J.C., A.G. Green, M.W. Lee, D.R. Hutchinson, W.F. Cannon, B. Milkereit, W.F. Agena and C. Spencer, Crustal extension in the Midcontinent Rift System: Results from GLIMPCE deep seismic reflection profiles over lakes Superior and Michigan, *in this volume*, 1988a.

Behrendt, J.C., A.G. Green, W.F. Cannon, D.R. Hutchinson, M.W. Lee, B. Milkereit, W.F. Agena and C. Spencer, Crustal structure of the Midcontinent Rift System: Results from GLIMPCE deep seismic reflection profiles, *Geology*, *16*, 81-85, 1988b.

Berry, M.J. and G.F. West, An interpretation of the first-arrival data of the Lake Superior Experiment by the time-term method, *Bull. Seis. Soc. Am.*, *56*, 141-171, 1966a.

Berry, M.J. and G.F. West, A time term interpretation of the first arrival data of the 1963 Lake Superior Experiment, in The earth beneath the continents, edited by J.S. Steinhart and T.J. Smith, *American Geophysical Union Monograph 10*, 166-180, 1966b.

Burke, K. and J.F Dewey, Plume-generated triple junctions: Key indicators in applying plate tectonics to old rocks, *J. Geol.*, *81*, 406-433, 1973.

Chase, C.G. and T.H. Gilmer, Precambrian plate tectonics: the Midcontinent gravity high, *Earth Planet. Sci. Lett.*, *21*, 70-78, 1973.

Chandler, V.W., P.L. Bowman, W.J. Hinze and N.W. O'Hara, Long wave length gravity and magnetic anomalies of the Lake Superior region, in Geology and tectonics of the Lake Superior Basin, edited by R.J. Wold and W.J. Hinze, *Geological Society of America Memoir 156*, 223-237, 1982.

Cohen, T.J. and R.P. Meyer, The Midcontinent gravity high: Gross crustal structure, in The earth beneath the continents, edited by J.S. Steinhart and T.J. Smith, *American Geophysical Union Monograph 10*, 141-165, 1966.

Franklin, J.M., W.H. McIlwaine, K.H. Poulsen and R.K. Wanless, Stratigraphy and depositional setting of the Sibley Group, Thunder Bay District, Ontario, Canada, *Can. J. Earth Sci.*, *17*, 633-651, 1980.

Green, A.G., W.F. Cannon, B. Milkereit, D.R. Hutchinson, A. Davidson, J.C Behrendt, C. Spencer, M.W. Lee, P. Morel-A-L'Huissier and W.F. Agena, A "GLIMPCE" of the deep crust beneath the Great Lakes, *in this volume*, 1988.

Halls, H.C., The late Precambrian central North American Rift System - A survey of recent geological and geophysical investigations, in Tectonics and geophysics of continental rifts, edited by I.B. Ramberg and E.R. Neumann, *NATO Advanced Study Institute Series C*, *37*, D. Reidel, Boston, 111-123, 1978.

Halls, H.C., Crustal thickness in the Lake Superior region, in Geology and tectonics of the Lake Superior Basin, edited by R.J. Wold and W.J. Hinze, *Geological Society of America Memoir 156*, 239-243, 1982.

Hinze, W.J. and R.J. Wold, Lake Superior geology and tectonics - overview and major unresolved problems, in Geology and tectonics of the Lake Superior Basin, edited by R.J Wold and W.J. Hinze, *Geological Society of America Memoir 156*, 273-280, 1982.

Hinze, W.J., R.J. Wold and N.W. O'Hara, Gravity and magnetic anomaly studies of Lake Superior, in Geology and tectonics of the Lake Superior Basin, edited by R.J. Wold and W.J. Hinze, *Geological Society of America Memoir 156*, 203-221, 1982.

Iyer, H.M., L.C. Pakiser, D.J. Stuart and D.H. Warren, Project Early Rise: Seismic probing of the upper mantle, *J. Geophys. Res.*, *74*, 4409-4441, 1969.

Klasner, J.S., W.F. Cannon and W.R. Van Schmus, The Pre-Keweenawan tectonic history of southern Canadian Shield and its influence on formation of the Midcontinent Rift, in Geology and tectonics of the Lake Superior Basin, edited by R.J. Wold and W.J. Hinze, *Geological Society of America Memoir 156*, 27-46, 1982.

Kumarapeli, P.S. and V.A. Saull, The St. Lawrence valley system: A North American equivalent of the East African Rift valley system, *Can. J. Earth Sci.*, *3*, 639-658, 1966.

Luetgert, J.H. and R.P. Meyer, Structure of the western basin of Lake Superior from cross structure refraction profiles, in Geology and tectonics of the Lake Superior Basin, edited by R.J. Wold and W.J. Hinze, *Geological Society of America Memoir 156*, 245-255, 1982.

Mereu, R.F. and J.A. Hunter, Crustal and upper mantle structure under the Canadian Shield from Project Early Rise data, *Bull. Seis. Soc. Am.*, *59*, 147-155, 1969.

Mereu, R.F., D. Wang, O. Kuhn, D.A. Forsyth, A.G. Green, P. Morel, G.G.R. Buchbinder, D. Crossley, E. Schwarz, R. duBerger, C. Brooks and R. Clowes, The 1982 COCRUST Seismic experiment across the Ottawa-Bonnechere graben and Grenville Front in Ontario and Quebec, *Geophys. J. R. astr. Soc.*, *84*, 491-514, 1986.

Muirhead, K.J., Eliminating false alarms when detecting seismic events automatically, *Nature*, *217*, 533-534, 1968

O'Brien, P.N.S., Lake Superior crustal structure- A reinterpretation of the 1963 Seismic Experiment, *J. Geophys. Res.*, *73*, 2669-2689, 1968.

Ocola, L.C. and R.P. Meyer, The central North American Rift System, Structure of the axial zone from seismic and gravimetric data, *J. Geophy. Res.*, *78*, 5173-5194, 1973.

Sims, P.K., K.D. Card, G.B. Morey and Z.E. Peterman, The Great Lakes tectonic zone- A major crustal structure in central North America, *Geol. Soc. Am. Bull.*, *91*, 690-698, 1980.

Smith, T.J., J.S. Steinhart and L.T. Aldrich, Lake Superior crustal structure, *J. Geophys. Res.*, *71*, 1141-1172, 1966a.

Smith, T.J., J.S. Steinhart and L.T. Aldrich, Crustal structure under Lake Superior, in The earth beneath the continents, edited by J.S. Steinhart and T.J. Smith, *American Geophysical Union Monograph 10*, 181-197, 1966b.

Steinhart, J.S., Lake Superior seismic experiment, Shots and travel times, *J. Geophys. Res.*, *69*, 5335-5352, 1964.

Steinhart, J.S. and T.J. Smith, eds., The earth beneath the continents, *American Geophysical Union Monograph 10*, 663p, 1966.

Stockwell, C.H., Fourth report on structural provinces, orogenies and time classification of rocks of the Canadian Precambrian Shield, *Geol. Surv. Can. paper 64-17, part II*, 26p, 1964.

Watson, J., The origin and history of the Kapuskasing structural zone, Ontario, Canada, *Can. J. Earth Sci.*, *17*, 866-875, 1980.

Weiblen, P.W., Keweenawan intrusive igneous rocks, in Geology and tectonics of the Lake Superior Basin, edited by R.J. Wold and W.J. Hinze, *Geological Society of America Memoir 156*, 57-82, 1982.

Wold, R.J. and W.J. Hinze, eds., Geology and tectonics of the Lake Superior Basin, *Geological Society of America Memoir 156*, 280p, 1982.

THE COMPLEXITY OF THE CONTINENTAL LOWER CRUST AND MOHO FROM PmP DATA: RESULTS FROM COCRUST EXPERIMENTS

R. F. Mereu, J. Baerg and J. Wu

Department of Geophysics, University of Western Ontario,
London, Ontario, Canada, N6A 5B7

Abstract. Over the past ten years a series of long range seismic refraction and wide-angle reflection experiments was conducted by a consortium of Canadian university and government crustal seismologists (COCRUST). The main tectonic features of interest were: the Vancouver Island Subduction Zone, the Peace River Arch, the Williston Basin, the Churchill-Superior Geological Boundary, the Kapuskasing Structural Zone, the Abitibi Greenstone Belt, the Grenville Front and the Ottawa- Bonnechere Graben. A comparison of the record sections from one area to another shows significant variations in the appearance of the PmP wide-angle reflected signals. To carry out a more detailed analysis of these phases, a normal moveout corrected intensity section which flattens out the PmP reflection hyperbola and emphasizes areas of large signal complexities was plotted in addition to the conventional record section. Data from this intensity section was then applied to obtain a quantitative measurement of a complexity parameter. These measurements were used to infer or compare differences which may exist in crustal heterogeneity from one region to another.

The results show, that in most areas of the regions studied, the Moho is a poorly defined transition zone. A few exceptional profiles were, however, observed in stable tectonic regions such as that of the Peace River Arch area of Alberta and the Central Gneiss Belt of the Grenville Province just south of the Grenville Front in Quebec. Many of the seismic traces from these areas showed well defined simple PmP pulses of relatively large amplitude. As the complexity of the lower crust increased, the amplitudes of PmP not only decreased but were embedded in a scattered wave field which originated from laterally heterogeneous structures just above the Moho. The most complex lower crusts and disrupted Mohos were found under the Ottawa-Bonnechere Graben, the Kapuskasing Uplift and the Vancouver Island Subduction Zone. The results indicate that once tectonic forces disturb the Moho, the period for the Moho to re-establish itself as a sharp first order discontinuity must be very long in geological time.

Introduction

Over the past ten years, a series of long range seismic refraction and wide-angle reflection experiments was conducted by a consortium of Canadian university and Government crustal seismologists (COCRUST). The location of these experiments is shown in the map in Figure 1. The main tectonic features of interest were as follows: the Vancouver Island Subduction Zone, The Peace River Arch, the Williston Basin, The Churchill-Superior Geological Boundary, the Kapuskasing Structural Zone, the Abitibi Greenstone Belt, the Grenville Front and the Ottawa-Bonnechere Graben. The number of shots which were fired in each experiment ranged from 8 to 23. The total number of in line record sections of length 150 to 400 km was 59. Most of the recordings were made with only vertical seismic instruments. All of these data sets have been interpreted using conventional travel time and amplitude techniques to obtain crustal models. A brief summary of the results along with key geophysical and geological references is presented later in the section on the discussion of the results. In this paper, a comparison is made of all the record sections with emphasis on the nature and characteristics of the wide-angle reflected signals from the Moho (PmP) and any intracrustal discontinuities which were recorded in the super-critical distance range from 95 to 150 km. The analysis technique is based on a method of measuring coda complexities described in detail by Mereu (in press).

Method

Figure 2 shows a synthetic seismic section for a simple multilayered earth model. This section

THE LOCATION OF THE COCRUST EXPERIMENTS

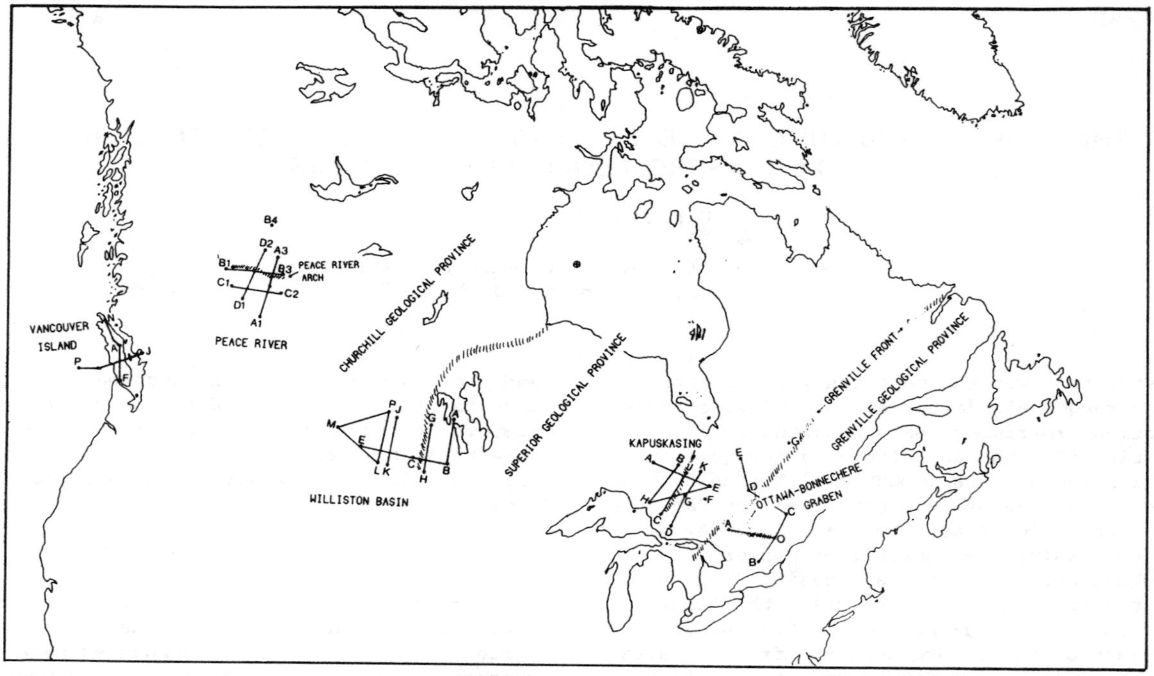

Fig. 1. Map showing the location of the COCRUST seismic refraction surveys.

was trace normalized and computed using a generalized ray method which is capable of handling laterally heterogeneous structures. This figure shows that much of the energy between the first arrivals and PmP in the distance range (50-150 km) owes its origin the effects of the intracrustal reflectors. This energy will lie along smooth hyperbolic travel-time curves as shown in Figure 2 only if the intracrustal layers are continuous and flat. In a more rigorous approach, Sandmeier and Wenzel (1986) computed synthetic seismograms using the reflectivity method of Fuchs and Muller (1971) and illustrated the signal complexities which can arise when laminated layers are inserted into the lower crust. Recent studies of the crust using near vertical reflection techniques (Brown and-Barazangi, 1986, and Mathews and Smith, 1987) show that in general the crust is exceedingly complex with numerous short reflectors occurring at various depths throughout the crust. The studies have also shown that the patterns of reflectors differ from one region to another (McGeary, 1987). When a long range refraction/wide-angle reflection experiment is carried out over a crust with numerous discontinuities, the energy reflected from these boundaries will lie along broken hyperbolic-like travel-time curves. These will show up as bursts of incoherent energy which will produce a much more complex record section than that shown in Figure 2.

Other sources of energy are also present which will appear as noise on the reflected energy. Small scale heterogeneities within the crust will create coda waves which have been studied using statistical methods (see for example Nikolaev and Tregub, 1970 and Aki and Chouet 1975). This component of the coda may arise from the backscattering effects of the heterogeneities and tend to be independent of distance. Another source of coda noise sometimes arises as signal generated noise caused by scattering from topographic features and also from the reverberation effects of low velocity sedimentary layers near the surface.

A typical crustal refraction record section may be divided into at at least four distinct distance ranges, each with a different origin to its coda complexities. These distance ranges are defined as follows:

(i) Distance range A (0 to 95 km)

Apart from reflections from near surface boundaries, most of the reflected energy is subcritical energy and relatively weak compared to other origins of scattered energy.The coda contains P waves , S waves, P to S and S to P converted phases and scattered signals from various heterogeneities.

(ii) Distance range B (95 to 150 km)

Most of the energy in this range just after the first arrivals are likely to be supercritical large amplitude reflected signals. The energy of the reflected signals in this region will in

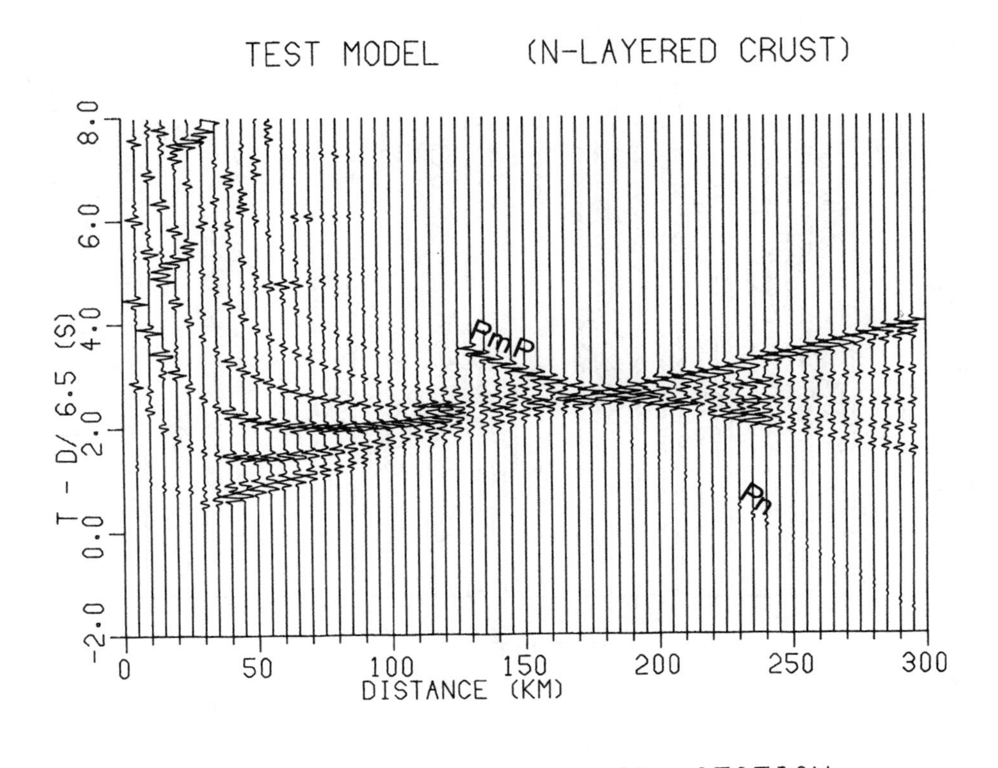

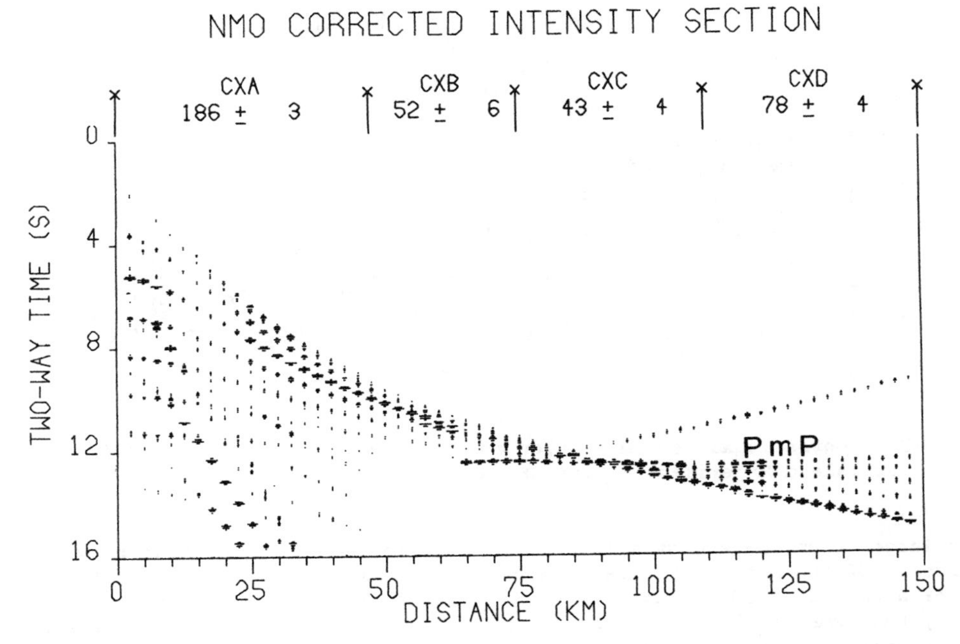

Fig. 2. Conventional synthetic record section and NMO corrected intensity section for a simple multi-layered crustal model. Velocity at the surface = 6.0 km/s. Velocity at the bottom of the crust = 6.8 km/s. Thickness of each layer = 5 km. Thickness of the crust = 40 km.

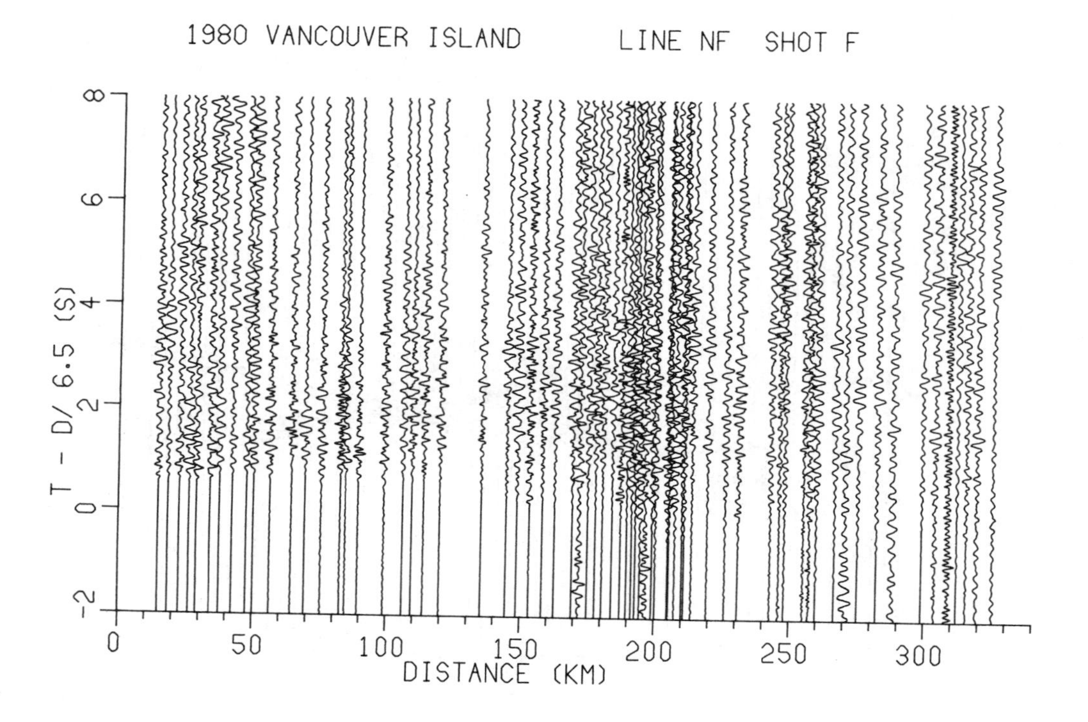

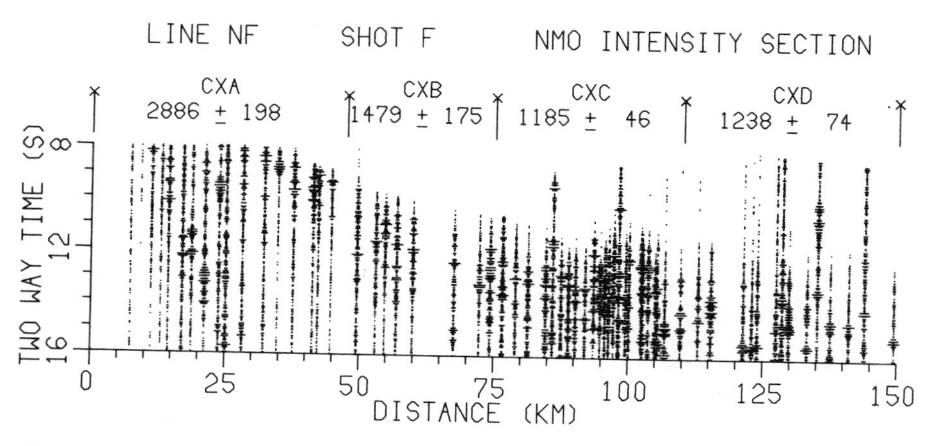

Fig. 3. Conventional record section and NMO corrected intensity section for the Vancouver Island Experiment - Line NF, Shot F.

general be much larger than the energy arising from other sources. If the Moho is a first order discontinuity, or a well defined laminated transition zone , than PmP should be clearly visible in this region. If the Moho is a weak, bumpy or a diffuse transition zone, then PmP may not be an observable signal. Intracrustal discontinuities in the crust will give rise to significant bursts of energy in the time frame following the onsets of the first arrival crustal phases and PmP. This distance range is thus considered to be the best range for lower crustal complexity studies for data from long range refraction experiments. When conventional inversion analysis is being performed on a data set, it is observations from this distance range which enables the interpreter to decide whether he should model the earth's crust with a set of layers or with a simple gradient.

(iii) Distance range C (150-220 km)

In this range the supercritical reflected waves are still present, but lie on

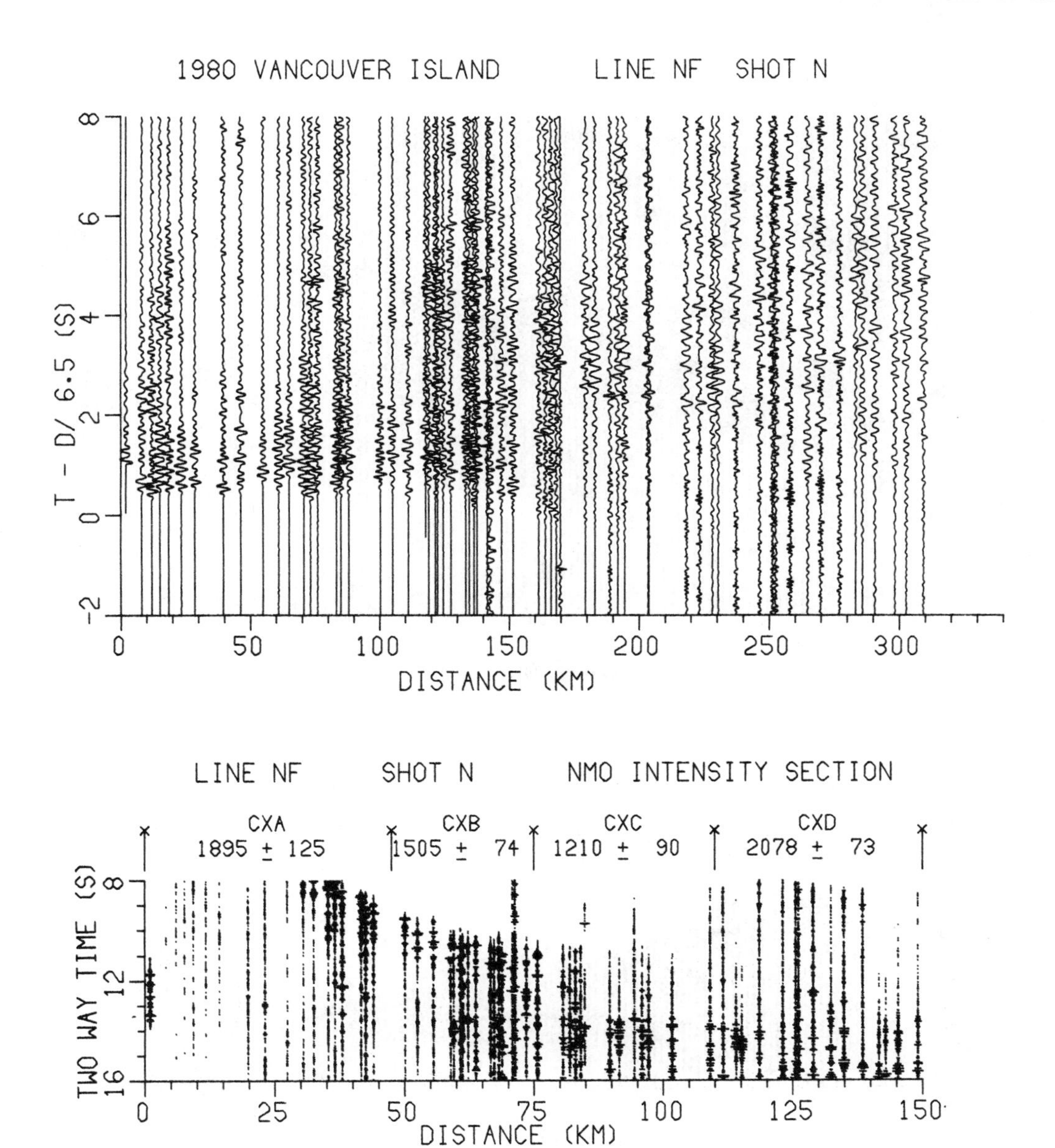

Fig. 4. Conventional record section and NMO corrected intensity section for the Vancouver Island Experiment - Line NF, Shot N.

hyperbolic-like travel-time curves which converge and intersect at a point roughly in the middle of the range. The closely spaced curves from any intracrustal discontinuities, if present, will result in a complex signature of overlapping wavelets. The interference effects of these wavelets make them very hard to interpret.

(iv) Distance range D (220-300 km)

This is the range where Pn is normally observed as a first arrival and where large amplitude reflection travel-time branches terminate in an irregular manner.

Examples of typical record sections from each of the COCRUST experiments are presented in the top portion of Figures 3 to 11. These sections show conventional trace normalized reduced travel-time plots (reducing velocity = 6.5 km/s) in which the PmP signal lies along the reflection hyperbola-like curve. In order that its characteristics may be observed and visually related to the real earth more easily, the sections are also replotted (see lower half of the Figures 2-11) using a second method in which all the traces were shifted according to a normal

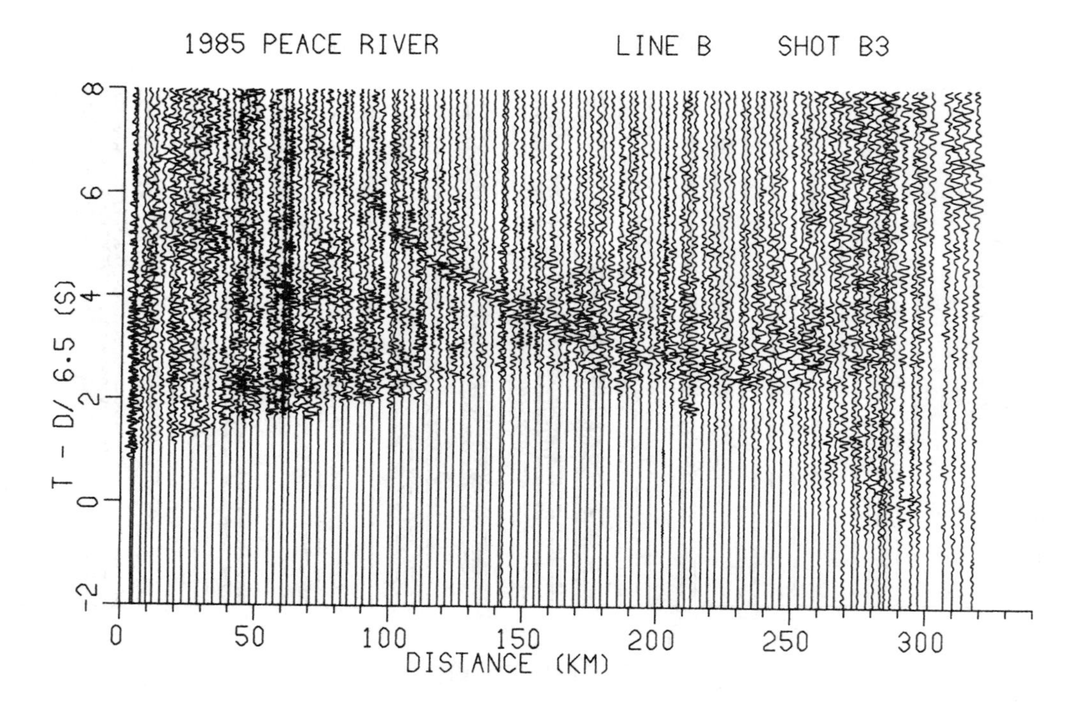

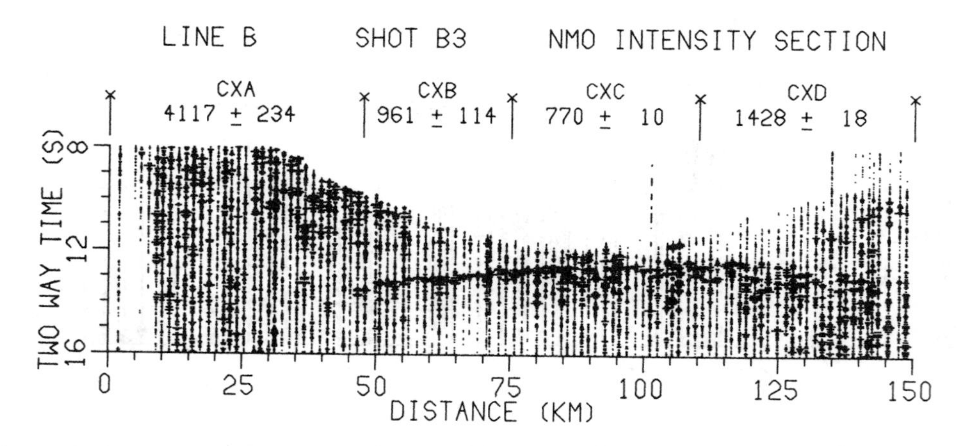

Fig. 5. Conventional record section and NMO corrected intensity section for the Peace River Arch Experiment - Line B, Shot B3.

moveout correction which was computed from known velocity models of the area under study. The traces are plotted downwards at the approximate locations where PmP would sample the Moho. (i.e. half-way between the shot and the recording stations). The time axis shown in these plots indicate the two-way travel-times to the Moho. In order that areas of the record sections with the more energetic arrivals be emphasized, the traces were normalized and then plotted using a shading method which plotted the intensity value of each peak amplitude with a short horizontal line. Only values above the threshold value of 0.1 were plotted. This method of displaying the data was adopted as it is directly related to a quantitative measurement of signal complexity which can be used to compare signal complexities of all the sections. Because of space limitations, it is not possible to reproduce the full set of record sections in this paper.

The trace signal complexity parameter (Cx) is defined as follows:

$$Cx = N \cdot E - 1 \tag{1}$$

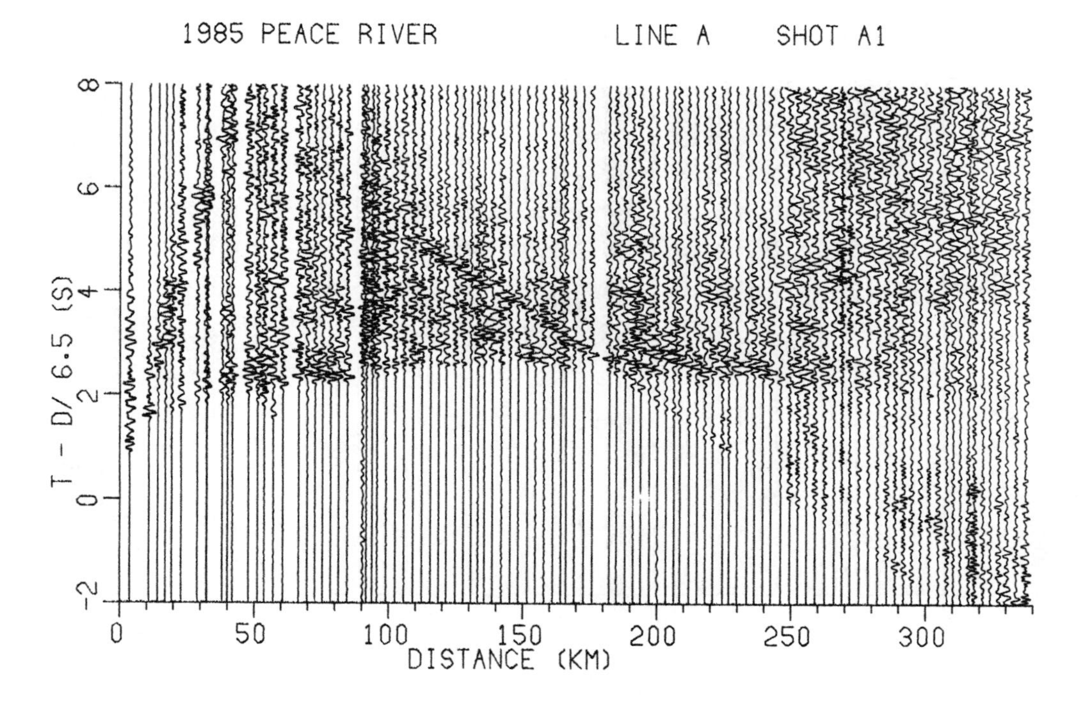

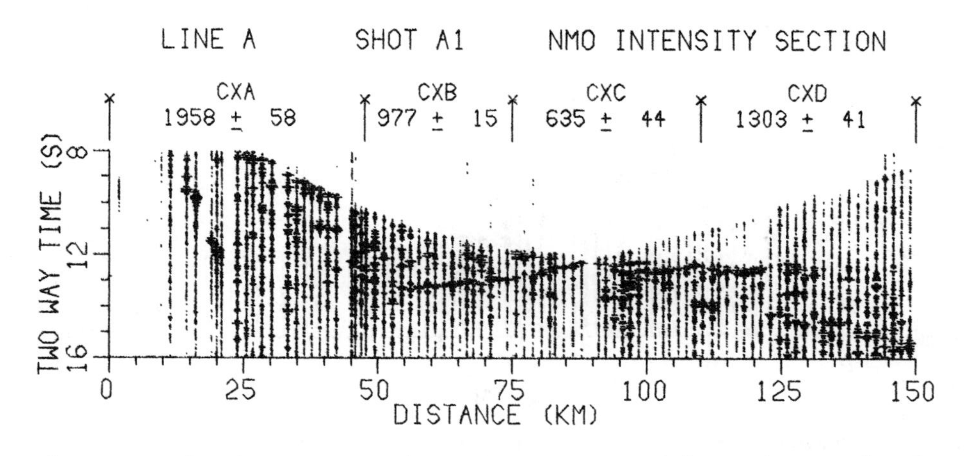

Fig. 6. Conventional record section and NMO corrected intensity section for the Peace River Arch Experiment - Line A, Shot A3.

where N = the number of peaks and troughs on the seismic trace,
E = a weight based on the trace energy.

Since all the traces were first normalized prior to plotting, a trace with a single "delta" function arrival would be the simplest trace possible. This definition of complexity gives it a minimum complexity value of 0. The complexity parameter increases with the number of events (ie peaks and troughs) on the trace. This number is weighted with the total trace energy. It is felt that if this parameter is computed for the same distance range and the same time spans from one record section to another, then its average value within a distance range can be used as a quantitative measurement for comparing relative signal complexities from one record section to another. One could also define the complexity parameter for non-normalized traces. Such a parameter would tend to be more sensitive to local site amplification effects and hence was not used here.

The average complexities (CxA,CxB,CxC,CxD) for the NMO intensity sections of Figures 2 to 11

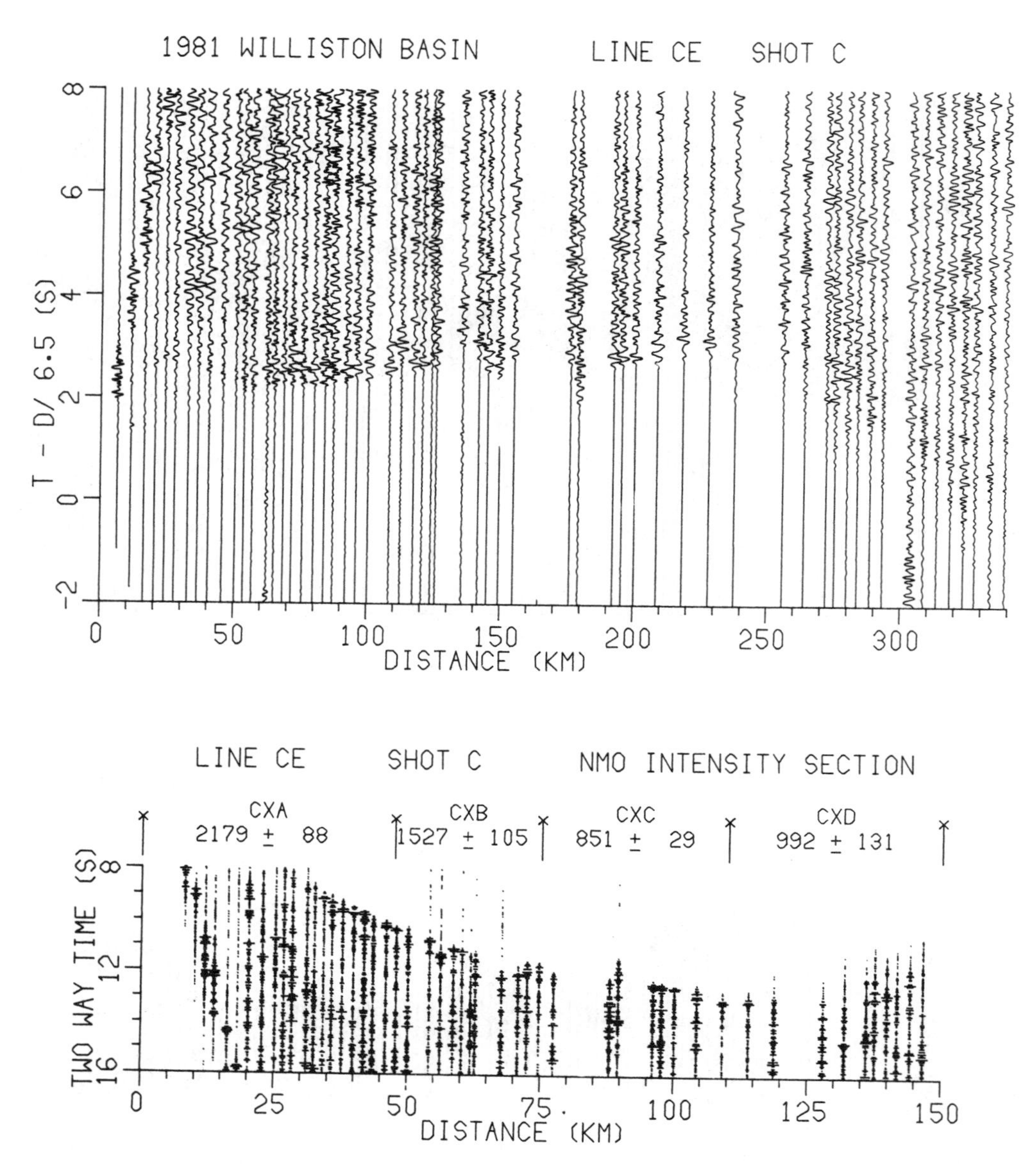

Fig. 7. Conventional record section and NMO corrected intensity section for the Williston Basin Experiment - Line CE, Shot C.

are plotted just above the traces. Since these traces were plotted at the half-way points, the distance ranges mentioned above are all halved in the lower NMO plots. The error estimates for each average are standard errors of the mean.

As was mentioned before, the most meaningful average complexity which could be related to lower crustal heterogeneities would be CxB which is computed for the distance range B. This is the range where the wide-angle reflection signals from the intracrustal boundaries dominate the coda. A comparison of this complexity parameter for the complete data set is given in Table 1 and the graph of Figure 12. This figure shows that there are consistent differences in complexity from one region to another even though there may be smaller differences from section to section within a region. Qualitative descriptions of the nature of the PmP signal is also presented in Table 1. Sharp Mohos occur in areas where PmP is well defined as a simple travel-time branch back to 95 to 110 km. If the crustal complexity parameter is less than 1500, the number of intracrustal boundaries is probably much less than average. A value greater than 1500 implies that the crust is much more heterogeneous than

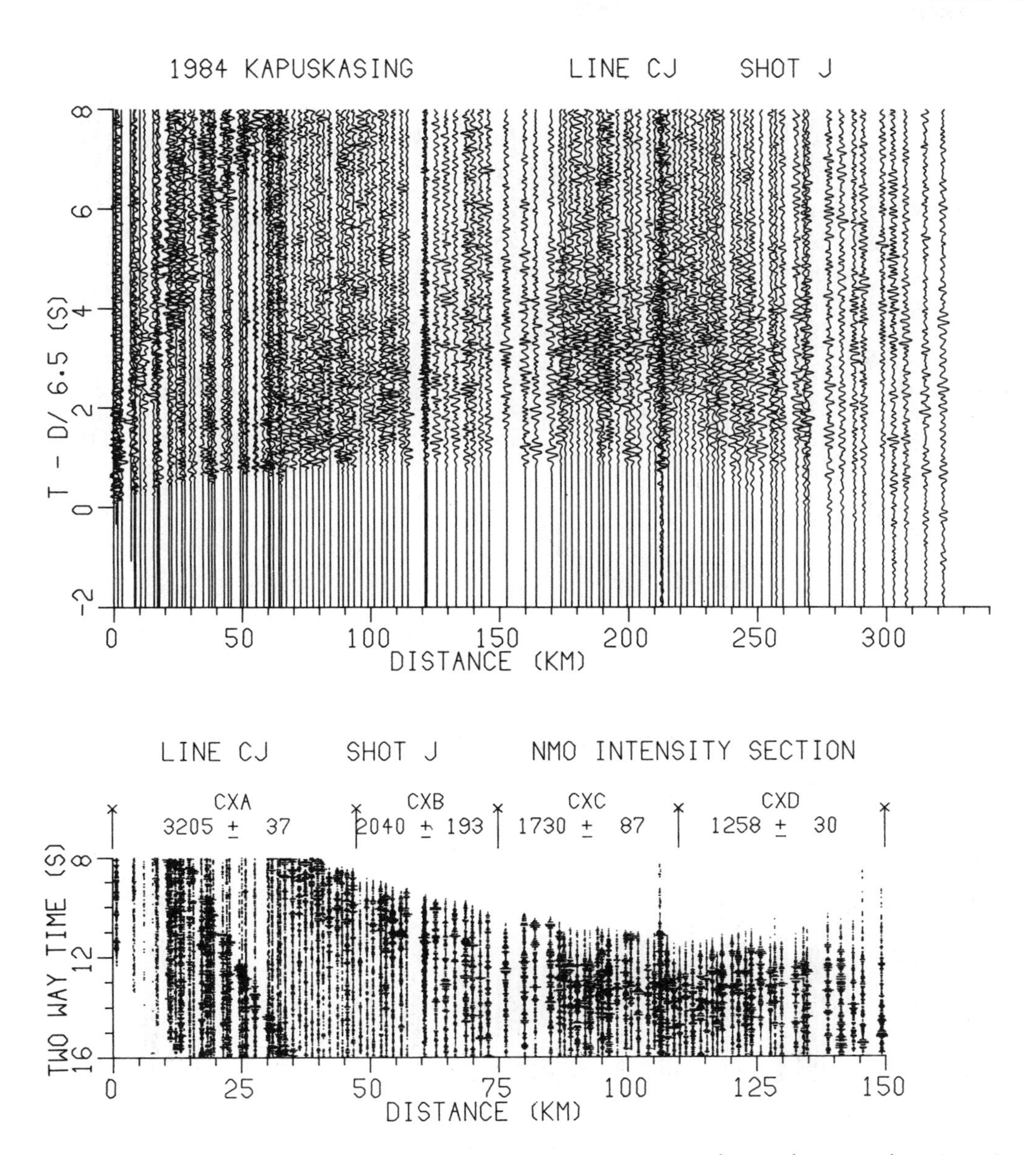

Fig. 8. Conventional record section and NMO corrected intensity section for the Kapuskasing Experiment - Line CJ, Shot J.

average. A few profiles had low complexities and poor PmP signals. These occurred in the cases where Pg or P* was a strong signal with little coda.

Discussion

(i) The 1980 Vancouver Island Experiment.

This experiment was carried out over an area where the Juan de Fuca Plate is being subducted under the North American Plate. Two main seismic lines were shot : NF along the length of Vancouver Island and PJ running across the Island. These refraction lines were also complemented with a series of on-shore and off-shore near vertical deep reflection "Lithoprobe" lines in 1984. Detailed interpretations of the data from these studies were given by Ellis et al.,1983, Clowes et al, 1987., and Green at al, 1988. The main results showed that very complex structures occur in a continental margin which lies over a subducting plate. Vancouver Island itself was found to have a continental crust with a poorly defined Moho at a depth of 40 to 50 km. None of the record sections showed any meaningful PmP signal. Two

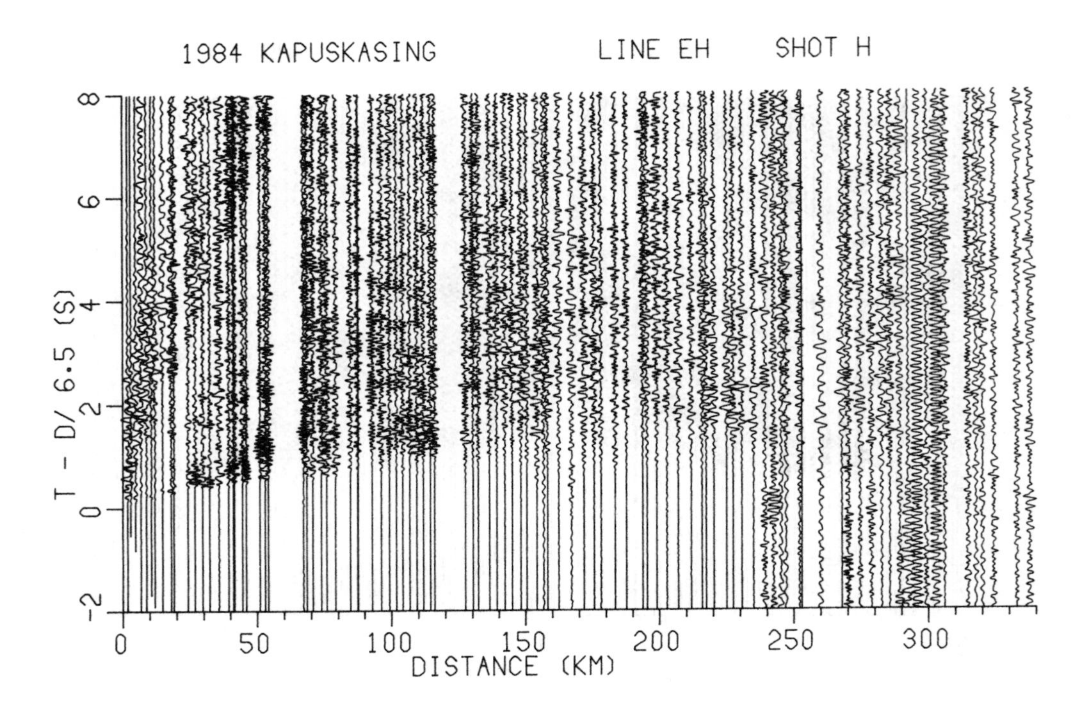

1984 KAPUSKASING EXPERIMENT

LINE EH SHOT H NMO INTENSITY SECTION

CXA 4597 ± 246
CXB 3088 ± 153
CXC 1918 ± 171
CXD 3475 ± 194

TWO WAY TIME (S)
8
12
16

0 25 50 75 100 125 150
DISTANCE (KM)

Fig. 9. Conventional record section and NMO corrected intensity section for the Kapuskasing Experiment - Line EH, Shot H.

examples are presented in Figures 3 and 4. The crustal complexity parameter was much greater than average for almost all the profiles.

(ii) The 1985 Peace River Arch Experiment.

This experiment was carried out in central Alberta over a region where a thick layer of sedimentary rocks overlies the basement rocks of the Churchill Geological Province of the Canadian Shield. The Peace River Arch is an east-west feature which appears to be mainly a structure within the sedimentary layers (de Mille,1958). Four major in line seismic lines (A,B,C,and D) were shot parallel and perpendicular to the arch. Detailed interpretations of these data sets were given by Zelt et al (1987). The main results show a rather simple crust with basement velocities increasing from 6 km/s near the surface to values of 7 km/s just above the Moho. The most significant feature of the observations was the very clear well defined PmP waves which were reflected from a very sharp Moho at a depth of 38-45 km. Figures 5 and 6 show 2 examples:

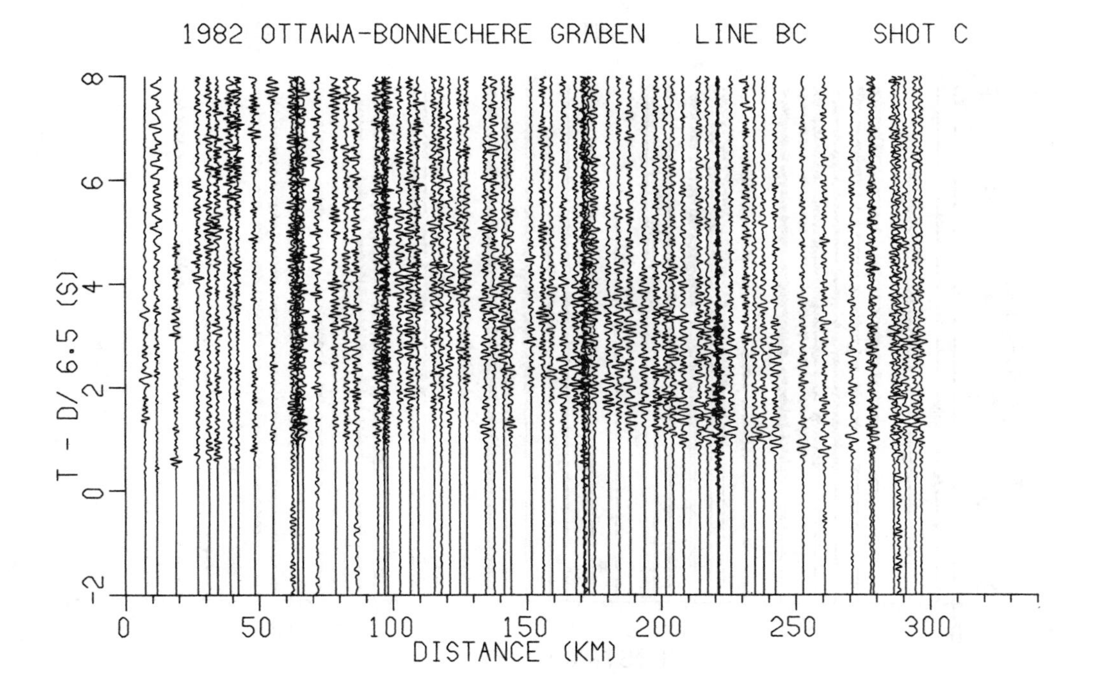

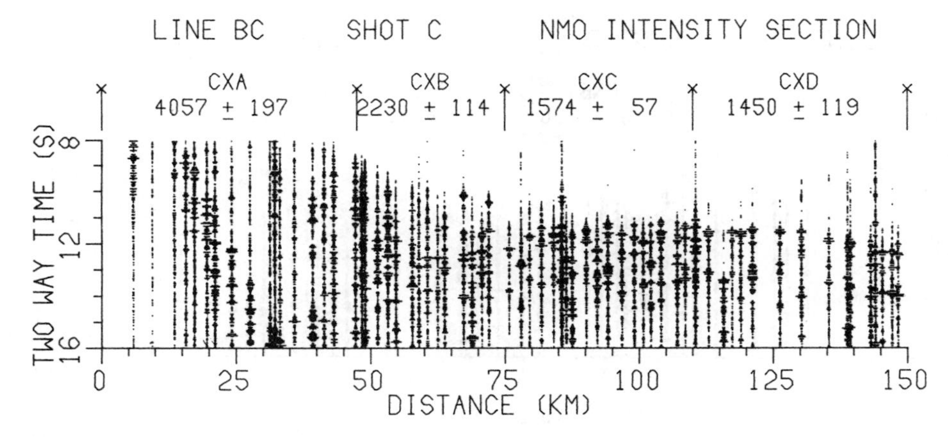

Fig. 10. Conventional record section and NMO corrected intensity section for the Ottawa Bonnechere Graben Experiment - Line BC, Shot C.

Line B along the Arch and Line A perpendicular to the Arch. The complexity parameter was much less than average for almost all the profiles. The slopes on the NMO corrected PmP travel time curves indicate that the Moho is a sloping boundary in this region.

(iii) The 1977,1979,and 1981 experiments across southern Saskatchewan and Manitoba.

These experiments were conducted over the northern part of the Williston Basin and over a region which marks the transition zone between rocks of the Churchill Geological Province from those of the older Superior Province to the east. The Williston Basin is one of the major basins in North America and has received considerable attention in both Canada and the United States because of its hydrocarbon resources. See for example Porter et al(1982), and Fowler and Nesbit (1985), Green et al (1985). The 1981 seismic lines were arranged mainly in the form of a triangle (PLM) as shown in Figure 1. Both in-line and fan profiles were obtained from shots fired at the corners of a triangle. Three north-south lines (AB,GH,and JK) and one east-west line (BCE) were shot to the east of the triangle in the 1977-1979 experiments. Detailed

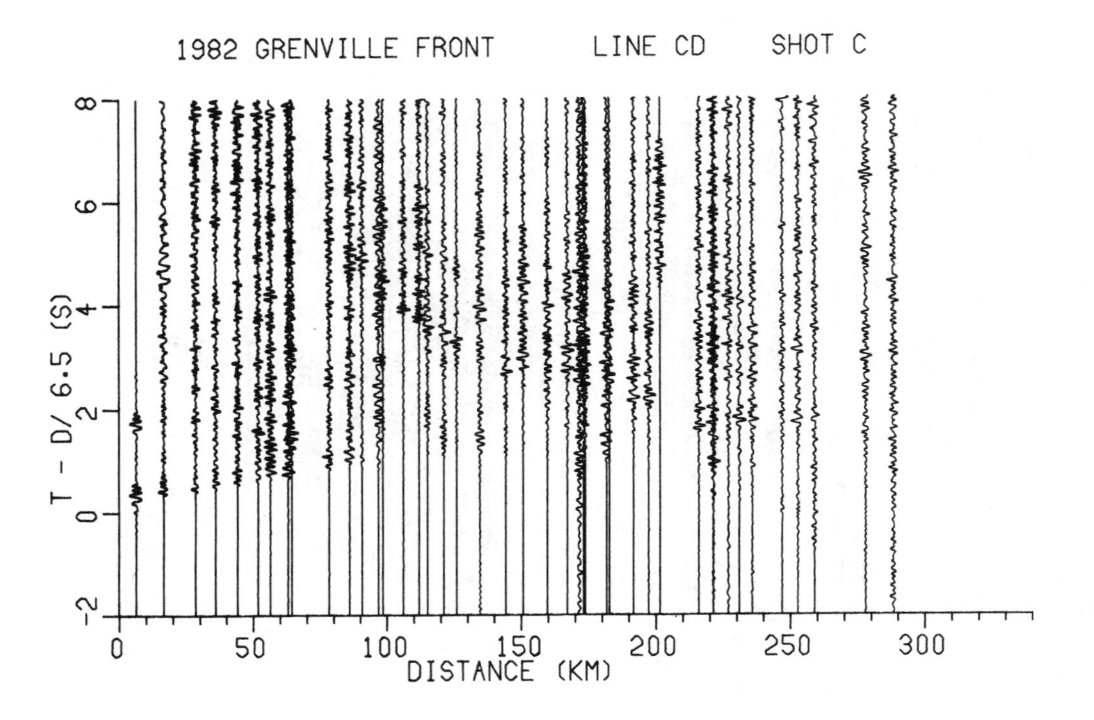

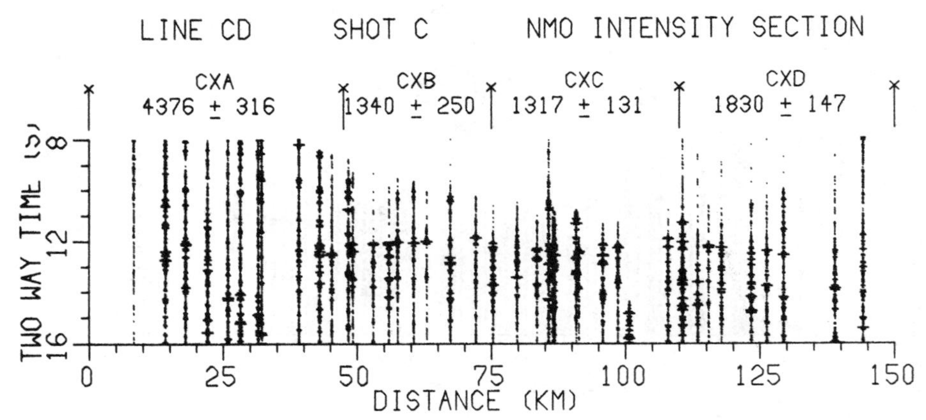

Fig. 11. Conventional record section and NMO corrected intensity section for the Ottawa Bonnechere Graben Experiment - Line CD, Shot C.

interpretations of these data sets were given by Green et al (1980), Delandro and Moon (1982), Hajnal et al. (1983), Kanasewich and Chiu (1985), Baerg (1985), Morel et al. (1987), and Kanasewich et al. (1987). The main results showed a thickening sedimentary layer towards the center of the Basin, a rather simple crust with velocities increasing from near 6 km/s in the basement rocks to values near 7 km/s above the Moho. Some of the interpretations indicated the presence of weak intracrustal discontinuities. Crustal thicknesses increased from 40 km in the Superior Province to very large values in the 45-50 km range over the Williston Basin. PmP was erratic being clear on some sections such as JK, but in general it was obscured implying that the Moho under the basin is not a very sharp boundary. The record section presented in Figure 7 is an example showing the lack of a clear PmP signal. This figure should be compared to the previous two figures from the Peace River Arch area. This may support theories such as that presented by Fowler et al that the thick complex crust under the Basin was caused by the slow subsidence of crustal material into the mantle with a slow transformation of gabbro into

TABLE 1. Comparison of PmP and Crustal Complexities From COCRUST Record Sections

EXPERIMENT	LINE	SHOT	NATURE of PmP	PmP CUSP distance (km)	CRUSTAL COMPLEXITY (CxB (distance 95- 150 km
Vancouver Island	NF	N	Extremely weak	?	1506 +/- 74
	NF	F	obscured	?	1480 +/- 175
	NA	A	obscured	?	1770 +/- 240
	AF	A	obscured	?	1840 +/- 117
	PJ	J	obscured	?	2077 +/- 219
	PJ	P	obscured	?	---
Peace River	A	A1	clear	108	978 +/- 15
	A	A2-S	very clear	100	846 +/- 172
	A	A2-N	obscured	105	1322 +/- 22
	A	A3	poor	130	1350 +/- 81
	B	B1	obscured	120	1285 +/- 51
	B	B2-W	very clear	95	785 +/- 48
	B	B2-E	very clear	95	949 +/- 25
	B	B3	very clear	95	962 +/- 114
	C	C1	poor	120	1368 +/- 79
	C	C2	very clear	95	823 +/- 69
	D	D1	very clear	95	1004 +/- 91
	D	D2	fair	95	1035 +/- 106
Williston Basin	PL	L	obscured	?	---
	PL	P	obscured	?	1283 +/- 81
	PM	P	obscured	?	1744 +/- 34
	PM	M	obscured	?	1793 +/- 523
	LM	M	clear	130	1123 +/- 45
	LM	L	fair	130	1879 +/- 66
	CE	E	very weak	150	1799 +/- 129
	CE	C	obscured	?	1528 +/- 105
Superior-Churchill Boundary	AB	A	obscured	?	---
	AB	B	obscured	?	---
	BC	B	obscured	?	---
	BC	C	obscured	?	---
	GH	G	weak	140	2181 +/- 110
	GH	H	fair	130	1395 +/- 286
	JK	J	clear	110	1454 +/- 157
	JK	K	clear	120	1141 +/- 118
Kapuskasing Structure	DK	D	clear	125	1780 +/- 71
	GD	G	obscured	?	2748 +/- 340
	GK	G	obscured	?	2334 +/- 340
	DK	K	obscured	?	1755 +/- 229
	CJ	C	obscured	?	2128 +/- 157
	CJ	J	obscured	?	2041 +/- 193
	BH	B	poor	120 ?	1440 +/- 70
	BH	H	obscured	?	2625 +/- 224
	EH	E	fair	140	2169 +/- 329
	GE	G	obscured	?	2118 +/- 96
	GH	G	obscured	?	2131 +/- 260
	EH	H	obscured	?	3089 +/- 153
	EA	E	fair	130	1846 +/- 532
	CE	C	fair	105	1578 +/- 56
	CA	C	obscured	?	1790 +/- 369
	EA	A	very clear	100	983 +/- 60
Ottawa-Bonnechere Graben	OA	A	obscured	?	3179 +/- 784
	OA	O	poor	120 ?	2746 +/-1232
	OB	O	poor	120 ?	1521 +/- 252
	OC	O	poor	120 ?	2335 +/- 422
	BC	B	fair	110 ?	2250 +/- 337
	BC	C	obscured	170 ?	2231 +/- 114
Grenville Front and Abitibi Greenstone Belt	CD	C	very clear	95	1341 +/- 250
	CD	D	very clear	95	1321 +/- 45
	DE	D	obscured	?	2110 +/- 342
	DE	E	poor	100 ?	---

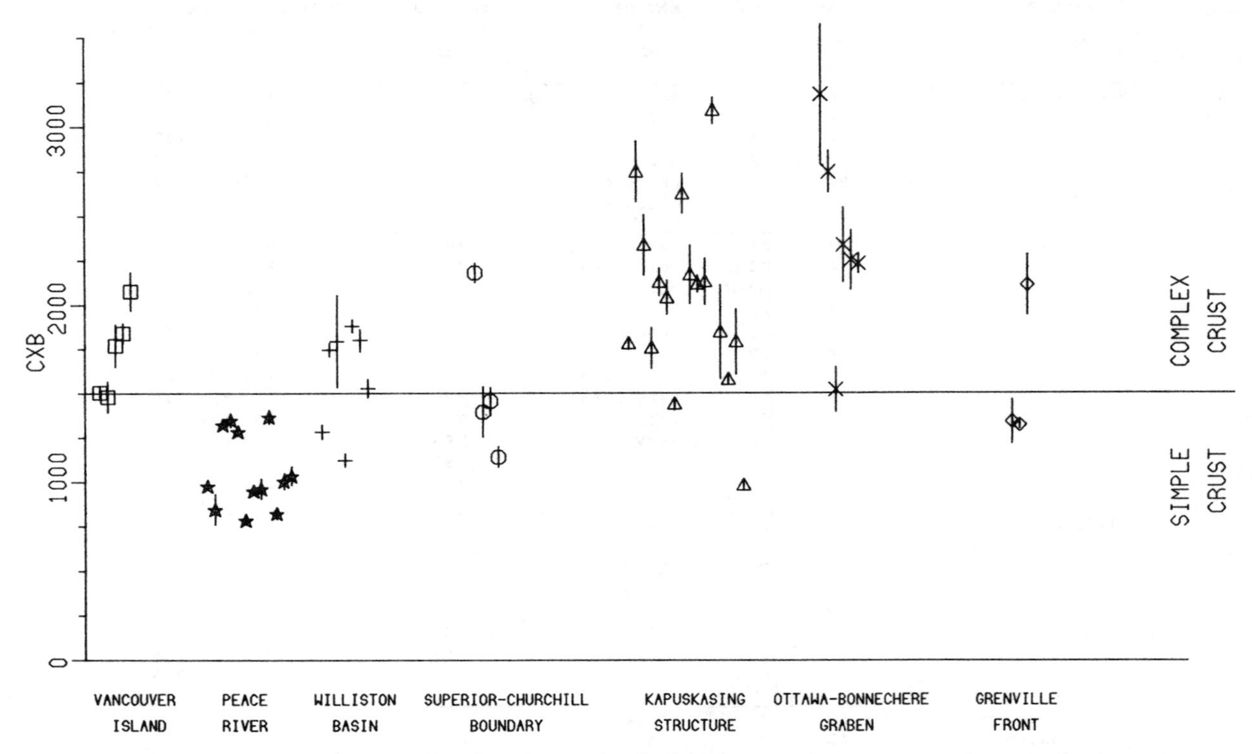

Fig. 12. Comparison of regional complexity measurements. CxB = the complexity parameter for the distance range B.

eclogite. It would appear that the process would have the effect of thickening the Moho transition zone to such an extent that the PmP amplitudes would be weakened and lost in the coda from the intracrustal boundaries.

(iv) The 1984 Kapuskasing Experiment.

This experiment was carried out in Northern Ontario over an area of the Superior Geological Province marked by a gravity high. Geological studies of the region show that the Kapuskasing Structure is a zone of high grade metamorphic rocks which may be an oblique crossection of the upper two thirds of the crust (Percival and Card ,1983). The location of the five seismic refraction lines, as shown in Figure 1, formed a pattern which sampled the crust both parallel and perpendicular to the structure. Detailed interpretations of these data sets were given by Northey and West (1985), West (1988), Wu and Mereu (1988), and Bolland and Ellis (1988). The main results showed that near surface velocities in the Shield vary from 5.7 km/s to 6.4 km/s. The largest values were found to lie along the axis of the Kapuskasing Structure giving added support to the theory that lower crustal material may have been thrust up to the surface. The crustal models showed that the Moho transition zone was not sharp and occurred at a depth of 38 to 48 km. The complexity studies shows that the crust along the axis of the structure is exeedingly heterogeneous. See Figures 8 and 9 for profiles both parallel and perpendicular to the axis. The tectonic forces which uplifted the crust probably created numerous highly reflective intracrustal shear zones which created a very complex pattern of coda waves following the Pg and P* onsets.

(v) The 1982 Ottawa-Bonnechere Graben - Grenville Front Experiment

This experiment was carried out over a rather complex region of the Canadian Shield in eastern Ontario and western Quebec. The Grenville Front is one of the major tectonic features of the Shield. It is approximately 1900 km in length extending from the northern shore of Lake Huron across Ontario and Quebec to Labrador. Over most of its length it marks the orogenic boundary between the Superior Structural Province (age 2.5 Gyr) and the much younger Grenville Province (1 Gyr). A recent review of the status of our knowledge of the Grenville Province was given by Moore et al (1986). The Ottawa-Bonnechere Graben runs parallel to the Ottawa river in the Grenville Province and has been associated with the St. Lawrence Rift System (Kumarapelli and Saull, 1966, Burke and Dewey, 1973). The seismic lines AO and BC (see Figure 1) ran parallel and perpendicular to the Ottawa-Bonnechere Graben. Lines CD and DE made up a profile which started in the Central Gneiss Belt of the Grenville Province, ran across the Grenville Front and ended in the Abitibi Greenstone Belt of the Superior Province. The main results of a seismic interpretation of this data set, (Mereu et al 1986), showed a rather complex crust with laterally varying near surface velocities in the 5.9 to 6.4 km/s range. There was little evidence for any continuous intracrustal discontinuity and the Moho itself was very poorly defined along the axis of the Ottawa- Bonnechere Graben. The complexity measurements both parallel and perpendicular to the graben ,(see Figure 10), suggests that the crust in this region is extremely heterogeneous. In contrast the Moho was found to be quite sharp in the central Gneiss area which lies between the Graben and the Grenville Front (See Figure 11). A significant thickening of the crust was found to occur along this Front indicating that the Front is a deep seated tectonic feature. An interpretation of the Abitibi Greenstone profiles by Parker (1985) showed evidence for a low velocity mid-crustal layer in the region north of the Front.

In other studies of the Shield in North America presented elsewhere in this volume, Behrendt et al., Green et al., and Smithson all found that PmP was not well observed on their near vertical record sections indicating that the Moho under the Archean Shield in Minnesota and under the Great Lakes is in general a poor reflector. Epili and Mereu, elsewhere in this volume observed a very complex pattern of poorly defined wide-angle PmP signals on a fan line across the Keweenawan Rift system under Lake Superior.

It is interesting to note that the record sections with the clearest PmP signals occur in the Peace River Arch area of Alberta which has several km of sedimentary cover on the basement rocks and that the most complex PmP signals were observed over portions of the Canadian Shield which had no sedimentary cover. From this, we may conclude that in this series of experiments, at least, reverberations within the near-surface low velocity sedimentary layers do not create enough noise to obscure any conclusions about the reflected energy from deeper portions of the crust.

Conclusions

This study has shown that the sharpest Moho was found under the Peace River Arch area of Alberta and under the Central Gneiss area just south of the Grenville Front in Quebec. Many of the seismic traces show well defined simple PmP pulses of large amplitude which are observed back as as far as 100 km from the source. The complexity parameter values for these sections lie mainly in the 800-1400 value range. The most complex lower earth structures (CxB= 1500-3000) occur in areas where major tectonic activities have occurred in the past. These are the Ottawa-Bonnechere Graben, The Kapuskasing Uplift and the Vancouver Island Subduction Zone. The Moho transition zone under the Williston Basin lies somewhere in between the extremes of the Peace River Arch area and the Vancouver Island area. No major tectonic disturbance has occurred in this area, however the slow subsidence of the region over geological time may have originated by an exchange of crust and upper mantle material at the lower crust which thickened the transition zone.

In a review of the crustal structure under old and stable tectonic provinces, Meissner (1986) concluded that, in general, the intracrustal boundaries were smooth and the Moho was a poor reflector. Our findings tend to agree with this conclusion on the Moho except for the structure under the Peace River Arch area of Alberta and a few other isolated areas of the Shield.

Acknowledgements. The authors would like to thank their colleagues in the COCRUST consortium for their efforts in making the COCRUST series of experiments a success. Special thanks is given to two unknown referees for their helpful advice in their review of this paper. This research was supported by the NSERC research grant A1793.

References

Aki, K. and Chouet, B., Origin of coda waves: source, attenuation, and scattering effects, J. Geophys. Res., 80, 3322-3342, 1975.

Baerg, J., Analysis of "COCRUST" seismic refraction experiments in Saskatchewan, M.Sc. Thesis, University of Western Ontario, 1985.

Behrendt, J. ,C. , A. ,G. ,Green, M. ,W. Lee, D. R. Hutchinson, W. F. Cannon, B. Milkereit, W. ,F. Agena, C. Spencer, Crustal extension in the Mid-continent Rift System Results form GLIMPCE deep seismic reflection profiles over Lakes Superior and Michigan, Elsewhere in this volume, 1988.

Bolland, A. and R. M. Ellis, Velocity structure of the Kapuskasing Zone from seismic refraction studies, in Project Lithoprobe Kapuskasing Structural Zone Transect, edited by G. F. West, Conf. Proc. Lithoprobe Workshop, Feb., Toronto, 1988.

Brown, L. and M. Barazangi, Editors: Reflection Seismology : A Global Perspective, Geodynamic Series, A.G.U. Vol 13, 311 pp, 1987.

Brown, L. and M. Barazangi, Editors: Reflection Seismology : The Continental Crust, Geodynamic Series, A.G.U. Vol 14, 319 pp, 1987.

Burke, K. and J. F. Dewey, Plume-generated triple junctions: Key indicators in applying plate tectonics to old rocks, J. Geol., 81, 406-433, 1973.

Clowes, R. M., M. T. Brandon, A. G. Green, C. J. Yorath, Sutherland Brown, E. R. Kanasewich, and C. Spencer, Lithoprobe-southern Vancouver Island: Cenozoic subduction complex imaged by deep seismic reflections, Canadian J. Earth Sciences, 24, 31-51, 1987.

Delandro, W. and W. Moon, Seismic structure of the Superior- Churchill Precambrian boundary zone, J. Geophys. Res., 87, 6884-6888, 1982.

de Mille, G., Pre-Mississipian history of the Peace River Arch. J. Alta. Soc. Petr. Geol., 6, 61-68, 1958.

Ellis, R. M., G. D. Spence, R. M. Clowes, D. A. Waldron, I. F. Jones, A. G. Green, D. A. Forsyth, J. A. Mair, M. J. Berry, R. F. Mereu, E. R. Kanasewich, G. L. Cumming, Z. Hajnal, R. D. Hyndman, G. A. McMechan and B. D. Loncarevic. The Vancouver Island Seismic Project: A COCRUST on shore-off shore study of a convergent margin, Can. J. Earth Sci., 20, 719-741, 1983.

Epili, D. ,and R. F. Mereu, The GLIMPCE seismic experiment: On shore refraction and wide-angle reflection observations from a fan line over the Lake Superior Midcontinent Rift System, Elsewhere in this volume, 1988.

Fowler, C. M. R. and E. G. Nisbet, The subsidence of the Williston Basin, Can. J. Earth, Sci., 22, 408-415, 1985.

Fuchs, K, and Muller, G., Computation of synthetic seismograms using the Reflectivity Method and comparison with observations, Geophys. J.R. astr. Soc., 417-423, 1971.

Green, A. G., O. G. Stephenson, G. D. Mann, E. R. Kanasewich, G. L. Cumming, Z. Hajnal, J. A. Mair, and G. F. West, Cooperative seismic surveys across the Superior-Churchill Boundary zone in southern Canada, Can. J. Earth Sci., 17, 617-637, 1980.

Green, A. G., An Evolutionary model of the Western Churchill Province and western margin of the Superior Province in Canada and the North-central United States, Tectonophysics, 116, 281-322, 1985.

Green, A. G., R. M. Clowes, and R. M. Ellis, Crustal studies across Vancouver Island and adjacent offshore margin, Conf, Proc. Controlled Source Seismology Workshop on the interpretation of crustal refraction and reflection surveys across Vancouver Island and adjacent continental margin, Whistler, BC. 1987, Geological Survey of Canada Open File Report, in press.

Green, A. G. , B. Milkereit, C. Spencer, P. Morel a l'Hussier, A. Davidson, D. Teskey, J. Behrendt, W. Cannon, D. Hutchinson, A GLIMPCE of the crust under the Great Lakes Elsewhere in this volume, 1988.

Hajnal, Z., C. M. R. Fowler, R. F. Mereu, E. R. Kanasewich, G. L. Cumming, A. G. Green, and A. J. Mair, An initial analysis of the earth's crust under the Williston Basin: 1979 COCRUST experiment, J. Geophys. Res., 89, 9381-9400, 1984.

Kanasewich, E. R. and S. K. L. Chiu, Least squares inversion of spatial refraction data, Bull. Seismol. Soc. Am., 75, 865-880, 1985.

Kanasewich, E. R., Z. Hajnal, A. G. Green, G. L. Cumming, R. F. Mereu, R. M. Clowes, P. Morel-a-l'Hussier, S. Chiu, A. M. Congram, C. G. Macrides, and M. Shahriar, Seismic studies of the crust under the Williston Basin, Can. J. Earth Sci., 24, 2160-2171, 1987.

Mathews, D. and C. Smith, Editors: Special Issue on Deep Seismic Reflection Profiling of the Continental Lithosphere, Geophys. J. R. astr., Soc., 89,1987.

McGeary, S., Nontypical BIRPS on the margin of the North Sea: The Shet Survey, Geophys. J. R. astr. Soc., 89, 231-238, 1987.

Mereu, R. F., D. Wang, O. Kuhn, D. A. Forsyth, A. G. Green, P. Morel, G. R. Buchbinder, D. Crossley, E. Schwarz, R. duBerger, C. Brooks, and R. Clowes, The 1982 COCRUST seismic experiment across the Ottawa-Bonnechere graben and Grenville front in Ontario and Quebec, Geophys. J. R. astr. Soc., 84, 491-514, 1986.

Mereu, R. F., The complexity of the crust from refraction / wide-angle reflection data, PAGEOPH, in press.

Meissner, R., The Continental crust, a Geophysical Approach, Academic Press, Inc., Toronto, 426 pp, 1986.

Moore, J. M., A. Davidson, and A. J. Baer, Editors: The Grenville Province, Geological Association of Canada Paper 31, 358 pp, 1986.

Morel a l'Hussier, P., A. G. Green, and C. J. Pike, Crustal refraction surveys across the Trans-Hudson Orogen/Williston Basin of South Central Canada, J. Geophys. Res.,92, 6403-6420, 1987.

Nikolaev, A. V. and Tregub, F. S., A statistical model of the earth's crust: Method and results, Tectonophysics, 10, 573-578, 1970.

Northey, D. and G. F. West, The Kapuskasing structural zone seismic refraction experiment-1984, Canadian Geological Survey Open File Report, 1985.

Parker, C. L., Crustal structure of the Abitibi Greenstone belt determined from refraction seismology, M.Sc. thesis. McGill University.

Percival J. A. and K. D. Card, Archean crust as revealed in the Kapuskaing structural uplift, Superior Province, Canada, Geology, 11, 323-326, 1983.

Sandmeier, K. J. , and F. Wenzel, Seismograms for a complex crustal model, Geophys. Res. Lett. , 13, 22-25, 1986.

West, G. F., Project Lithoprobe Kapuskasing Structural Zone Transect, Conf. Proc. Lithoprobe Workshop, Toronto, Feb., 1988.

Wu, J. and R. F. Mereu, Crustal Models of the Kapuskasing Structural Zone: results from the 1984 seismic refraction experiment, in Project Lithoprobe Kapuskasing Structural Zone Transect, edited by G. F. West, Conf. Proc. Lithoprobe Workshop, Feb., Toronto, 1988.

Zelt, C. A., R. M. Ellis, J. Brown, Z. Hajnal, and R. Stephenson, Crustal structure of the Peace River Arch region from a refraction survey, Abstract S5-P16, IUGG XIX General Assembly, Vancouver, Canada, 1987.

A PETROLOGICAL MODEL OF THE LAMINATED LOWER CRUST IN SOUTHWEST GERMANY BASED ON WIDE-ANGLE P- AND S-WAVE SEISMIC DATA

W. Steven Holbrook

Geophysics Department, Stanford University
Stanford, CA 94305

Introduction

On seismic reflection sections in many areas, the continental lower crust consists of subhorizontal reflections, often in such abundance that the lower crust and Moho appear layered or laminated [e.g., Hale and Thompson, 1982; Meissner and Wever, 1986; Mooney and Brocher, 1987]. The cause of this lamination remains a puzzle, but proposed explanations include compositional layering [Meissner, 1973; Hale and Thompson, 1982], lenses of partial melt [Meissner, 1967; Hale and Thompson, 1982], ductile shear zones [Jones and Nur, 1982, 1984; Smithson et al., 1986], and the presence of fluid-filled cracks [Matthews and Cheadle, 1986; Hall, 1986; Klemperer, 1987]. The lamination is often visible on wide-angle ("refraction") seismic data as highly reflective precursors to P_mP, the wide-angle Moho reflection [e.g., Sandmeier and Wenzel, 1986; Gajewski and Prodehl, 1987].

So far, most array seismic studies of the laminated lower crust have been based solely on compressional (P) waves. Important new information on the reflectivity of the lower crust can be gained, however, by combining shear (S) and compressional wave observations. In this paper I present a brief summary of a recent study of wide-angle P- and S-wave data from Southwest Germany [Holbrook et al., 1988]. The coexistence of high-quality, wide-angle P- and S-wave data with deep seismic reflection data in Southwest Germany creates a unique opportunity to model the composition of a classic laminated lower crust. In this summary I will focus on the implications of the shear-wave data on the cause of the lower-crustal reflectivity.

Wide-angle P- and S-wave Data: Contrasting Pictures of the Lower Crust

The most striking -- and unexpected -- observation in the wide-angle data is the lack of S-wave reflections from the lower crust. The Variscan lower crust of Southwest Germany is known from seismic reflection studies to be laminated [Bartelsen et al., 1982; Bortfeld et al., 1985; Meissner and Wever, 1986; Lueschen et al., 1987]. This v_p-reflectivity is prominent even on wide-angle data, producing a broad zone of high-amplitude P-wave reflections observed particularly well in the Black Forest [Sandmeier and Wenzel, 1986; Gajewski and Prodehl, 1987]. On the horizontal component, however, no corresponding S-wave reflectivity is present, even on shots which show strong P-wave reflections (Fig. 1). The lack of S-wave reflectivity cannot be attributed to high shear-wave attenuation in the lower crust, or to any difference in the wavelength contents of the P- and S-wave data [Holbrook et al., 1988]. The differing signature of P- and S-waves therefore indicates a major contrast in the detailed v_p and v_s structures of the lower crust; specifically, the P-velocities must have strong high/low alternations, while the S-velocities remain fairly constant. Any explanation of the lower crustal lamination, therefore, must account for this.

Reflectivity Modeling of the Laminated Lower Crust

One hypothesis for the cause of lower-crustal reflectivity -- compositional layering -- is capable of explaining this contrasting v_p and v_s structure. In order to demonstrate this, I constructed a compositionally layered seismic model of the lower crust and compared its predicted synthetic seismograms to the observed field data. The model consists of twelve 1-km-thick layers, each of which was assigned the v_p and v_s of an actual rock specimen measured at high pressure in the laboratory. The guiding principle was to seek out a vertical succession of rocks with strongly alternating P-velocities but gradually increasing S-velocities. The resulting velocity structure has a high/low alternation of Poisson's ratio, in which high-σ layers correspond to high-v_p layers and vice-versa (Fig. 2, Table 1). The average v_p, v_s, and σ in the model agree with the lower-crustal models of Gajewski and Prodehl [1987] and Holbrook et al. [1988].

The resulting radial-component synthetic record section, calculated by the reflectivity method [Fuchs and Mueller, 1971], is quite similar to the observed data (Fig. 3). The P-waves show a strong P_mP and lower-crustal response, while the S-waves show a clear S_mS but only a very weak lower-crustal response. Furthermore, the synthetic seismograms show a clear P_mS phase and relatively

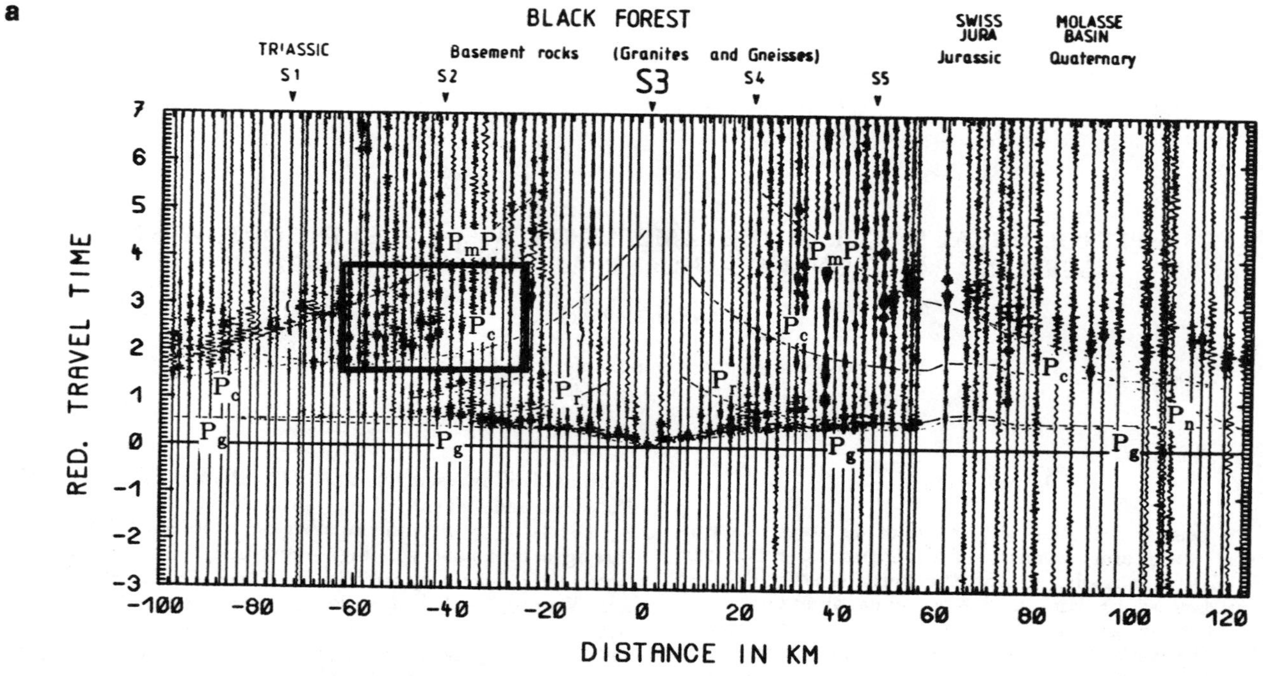

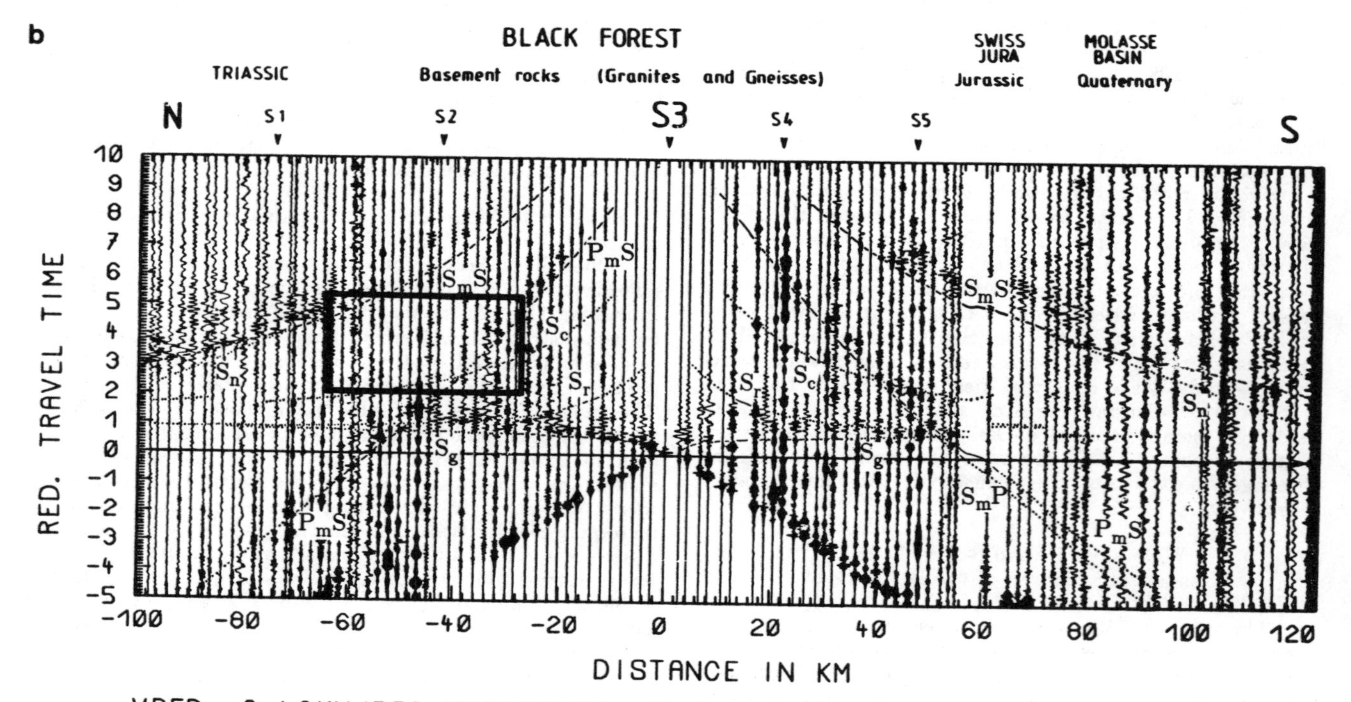

Fig. 1. (a) P-wave seismic refraction data for shot S-S3, with predicted traveltimes from the model of Gajewski and Prodehl [1987]. Traveltime curves are dashed where ray theory predicts strong amplitudes, dotted where weak amplitudes are predicted. Box encloses zone of strong P-wave reflectivity from lower crust. (b) S-wave seismic refraction data for shot S-S3, with predicted traveltimes from the model Holbrook et al. [1988]. Reduction velocity is 3.46 km/s, and time axis is compressed by $\sqrt{3}$ relative to the P-wave time axis. Note lack of lower-crustal S-wave reflectivity (box).

TABLE 1. Detailed Petrological Model of the Black Forest Lower Crust

Layer	h	v_p	v_s	σ	ρ	Ref	Mineralogy
1	0.9	6.27	3.45	0.28	2.71	h	Granite (qtz-biot-fspar)
2	0.8	5.83	3.46	0.23	2.63	a	Granite (fspar-qtz)
3	1.0	6.47	3.46	0.30	2.73	b	Gabbro-anorthosite Granulite
4	0.9	6.00	3.49	0.24	2.74	a	Diorite
5	1.0	6.78	3.55	0.31	2.78	f	Gabbro-anorthosite Granulite
6	0.9	6.24	3.59	0.25	3.03	c	Gabbroic Granulite (plg-cpx-biot)
7	1.0	6.89	3.67	0.30	3.08	c	Hornblende Granulite (hbl-plg-cpx)
8	1.0	6.36	3.66	0.25	2.93	g	Quartz Diorite
9	1.1	7.23	3.77	0.31	3.08	a	Garnet gabbroic pyriclasite(plg-px-gnt)
10	1.0	6.55	3.74	0.26	2.95	d	Al-rich Gneiss (qtz-gnt-sil)
11	1.1	7.31	3.80	0.31	2.99	e	Gabbro (plg-cpx-opx)
12	1.0	6.80	3.78	0.28	3.07	f	Gabbroic Granulite (plg-cpx-opx)
Tot	11.7	6.60	3.63	0.28	2.90		

a. Brooks, 1985
b. Christensen and Fountain, 1975
c. Chroston and Evans, 1983
d. Fountain, 1976; and Burke, 1987
e. Kroenke et al., 1976
f. Manghnani et al., 1974
g. Simmons, 1964
h. Simmons and Brace, 1965

Twelve-layer compositional layering model of the Black Forest lower crust used to calculate reflectivity synthetic of Fig. 3b. Explanation: h= thickness of layer(km), v_p= P-velocity (km/s), v_s= S-velocity (km/s), σ= Poisson's ratio, ρ= density (*GMCC*), Ref= source of velocity data.

little converted energy from within the lower crust, in agreement with the observations.

The rocks in the model -- granite, diorite, quartz-diorite, anorthosite, hornblende granulite, garnet-quartz gneiss, gabbro and gabbroic granulite -- are reasonable constituents of the lower crust [Table 1]. In general, the more quartz-rich rocks comprise the low-v_p, low-σ layers, while more mafic compositions are found in the high-v_p, high-σ laminae. There is an overall downward compositional gradation from granitic rocks through anorthosites, diorites, and hydrous mineral-bearing granulites to gabbroic and garnet granulites. This model bears a striking resemblance, in both layer thickness and compositional diversity, to the crustal cross-section of the Kapuskasing Uplift, Ontario, which contains mid-crustal granites underlain by interlayered tonalite, diorite, anorthosite, paragneiss (biot-plag-qtz), and gabbroic gneiss [Percival, 1986]. Naturally, the model of Fig. 2 is a highly simplified, non-unique picture of the lower crust in Southwest Germany. The important point, however, is that a laminated lower-crustal model consisting of realistic rock types is capable of explaining the first-order contrast in the P-wave and S-wave reflectivity of the lower crust.

The lack of S-wave laminations is an important observation, as it requires low Poisson's ratios in the low-v_p laminae of the lower crust (Fig. 2). Models of lower-crustal lamination based on partial melt and fluid-filled cracks are thus inapplicable to Southwest Germany, since they would produce high Poisson's ratios in the low-v_p laminae [Spetzler and Anderson, 1968; O'Connell and Budiansky, 1974; Christensen, 1984]. Ductile shear zones cannot be ruled out as a possible model, since low-v_p mylonites often have high quartz contents and therefore low σ [Jones, 1983]; however, since the relatively high v_p (6.7-6.8 km/s) of the lower crust in Southwest Germany requires a large proportion of mafic rocks, interspersed low-v_p mylonites are really just a special case of compositional layering. I

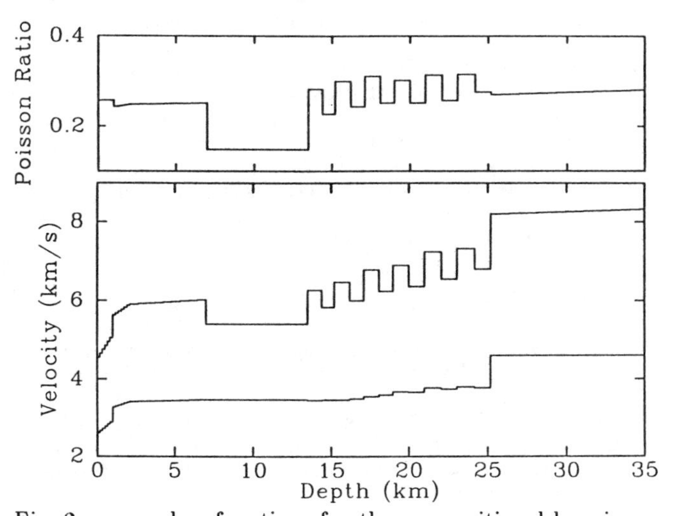

Fig. 2. v_p- and v_s-functions for the compositional layering model, used to generate the synthetic seismograms of Fig. 3(b). The strong v_p lamination and lack of v_s lamination results in a strong vertical alternation of Poisson's ratio in the lower crust. The lower crustal laminae are constructed from the lab-measured velocities of actual rocks, as compiled in Table 1.

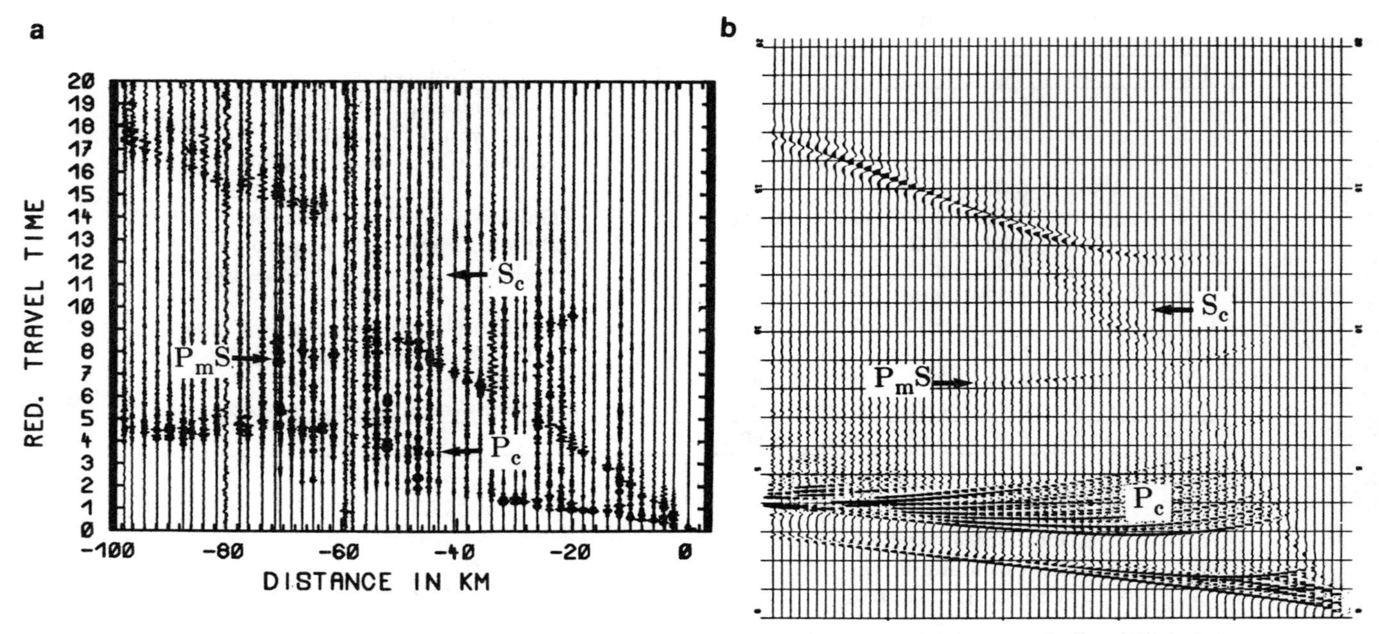

Fig. 3. (a) Transverse-component data from shot S-S3, plotted at v_{red}=7.0 km/s to display entire P- and S-wavefield. (b) Radial-component synthetic seismograms predicted by the model of Fig. 2 and Table 1. The model successfully predicts strong P_mP, S_mS, P_mS, and P-wave reverberations from the lower crust (phase P_c), as well as weak S-wave lower-crustal reverberations (phase S_c).

therefore propose a model of mafic sills in a lower crust of originally intermediate composition.

Conclusions

The striking contrast between the P- and S-wave signatures of the lower crust in Southwest Germany is a surprising observation with important bearing on the cause of the lower-crustal reflectivity. The absence of prominent S-wave reflections from the lower crust despite the presence of strong P-wave reflections implies a model of the lower-crustal lamination based on compositional layering rather than the presence of fluid-filled cracks. The extent to which these conclusions are local rather than global will become clear only as new wide-angle data from other provinces are collected and analyzed.

Acknowledgments. The seismic data used for this study were collected under the auspices of the German Research Society's research program "Stress and Stress Release in the Lithosphere" at the University of Karlsruhe. This study would not have been possible without the close cooperation of Dirk Gajewski, Claus Prodehl, and Anton Krammer. I thank Prof. Jon Claerbout for the use of the Stanford Exploration Project's Convex computer for the reflectivity modeling.

References

Bartelsen, H., E. Lueschen, Th. Krey, R. Meissner, H. Schmoll, and Ch. Walter, The combined reflection-refraction investigation of the Urach geothermal anomaly, in The Urach Geothermal Project, edited by R. Haenel, pp. 247-262, Schweizerbart, Stuttgart, 1982.

Bortfeld, R.K., and DEKORP Research Group, First results and preliminary interpretation of deep reflection seismic recordings along profile DEKORP 2-South, J. Geophys., 57, 137-163, 1985.

Brooks, S.G., Seismic velocities from crustal sections in northern Scandinavia, Ph.D. thesis, University of East Anglia, 1985.

Burke, M., Compressional wave velocities in rocks from the Ivrea-Verbano and Strona-Ceneri zones, Southern Alps, Northern Italy: Implications for models of crustal structure, M.S. thesis, University of Wyoming, 78 pp., 1987.

Christensen, N.I., Pore pressure and oceanic crustal seismic structure, Geophys. J. R. Astron. Soc., 79, 411-423, 1984.

Christensen, N.I., and D.M. Fountain, Constitution of the lower continental crust based on experimental studies of seismic velocities in granulite, Geol. Soc. Am. Bull., 86, 227-236, 1975.

Chroston, P.N., and C.J. Evans, Seismic velocities of granulites from the Seiland Petrographic Province, N. Norway: Implications for Scandanavian lower continental crust, J. Geophys., 52, 14-21, 1983.

Fountain, D.M., The Ivrea-Verbano and Strona-Ceneri Zones, northern Italy: a cross-section of the continental crust -- new evidence from seismic velocities of rock samples, Tectonophysics, 33, 145-165, 1976.

Fuchs, K., and G. Mueller, Computation of synthetic

seismograms with the reflectivity method and comparison with observations, Geophys. J. R. Astron. Soc., 23, 417-423, 1971.

Gajewski, D., W.S. Holbrook, and C. Prodehl, A three-dimensional crustal model of Southwest Germany derived from seismic refraction data, Tectonophysics, 142, 49-70, 1987.

Gajewski, D., and C. Prodehl, Seismic refraction investigation of the Black Forest, Tectonophysics, 142, 27-48, 1987.

Hale, L.D., and G.A. Thompson, The seismic reflection character of the continental Mohorovicic discontinuity, J. Geophys. Res., 87, 4625-4635, 1982.

Hall, J., Nature of the lower continental crust: evidence from BIRPS work on the Caledonides, in Reflection Seismology: A Global Perspective, Geodynamics Ser. Vol. 14, edited by M. Barazangi and L. Brown, pp. 223-231, AGU, Washington, D.C., 1986.

Holbrook, W.S., D. Gajewski, A. Krammer, and C. Prodehl, An interpretation of wide-angle shear- and compressional-wave data in Southwest Germany: Poisson's ratio and petrological implications, J. Geophys. Res., in press, 1988.

Jones, T.D., Wave propagation in porous rock and models for crustal structure, Ph.D. thesis, Stanford Univ., 223 pp., 1983.

Jones, T.D., and A. Nur, Seismic velocity and anisotropy in mylonites and the reflectivity of deep crustal faults, Geology, 10, 260-263, 1982.

Jones, T.D., and A. Nur, The nature of seismic reflections from deep crustal fault zones, J. Geophys. Res., 89, 3153-3171, 1984.

Klemperer, S.L., A relation between continental heat flow and the seismic reflectivity of the lower crust, J. Geoph., 61, 1-11, 1987.

Kroenke, I.W., M.H. Manghnani, C.S. Rai, P. Fryer, and R. Ramananantoandro, Elastic properties of selected ophiolites rocks from Papua, New Guinea: Nature and composition of oceanic lower crust and upper mantle, in The Geophysics of the Pacific Ocean Basin and its Margins, edited by G.H. Sutton, M.H. Manghnani and R. Moberly, AGU Monog. 19, AGU, Washington, D.C., 1976.

Lueschen, E., F. Wenzel, K.-J. Sandmeier, D. Menges, Th. Ruehl, M. Stiller, W. Janoth, F. Keller, W. Soellner, R. Thomas, A. Krohe, R. Stenger, K. Fuchs, H. Wilhelm, and G. Eisbacher, Near-vertical and wide-angle seismic surveys in the Black Forest, J. Geophys., 62, 1-30, 1987.

Manghnani, M.H., R. Ramananantoandro, and S.P. Clark, Jr., Compressional and shear wave velocities in granulite facies rocks and eclogites to 10 kbar, J. Geophys. Res., 79, 5427-5446, 1974.

Matthews, D.H., and M.J. Cheadle, Deep reflections from the Caledonides and Variscides west of Britain and comparison with the Himalayas, in Reflection Seismology: A Global Perspective, Geodynamics Ser. Vol. 13, edited by M. Barazangi and L. Brown, pp. 5-20, AGU, Washington, D.C., 1986.

Meissner, R., Exploring deep interfaces by seismic wide angel measurements, Geoph. Prosp., 15, 598-617, 1967.

Meissner, R., The 'Moho' as a transition zone, Geoph. Surv., 1, 195-216, 1973.

Meissner, R., and T. Wever, Nature and development of the crust according to deep reflection data from the German Variscides, in Reflection Seismology: The Continental Crust, Geodynamics Ser. Vol. 13, edited by M. Barazangi and L. Brown, pp. 31-42, AGU, Washington, D.C., 1986.

Mooney, W.D., and T.M. Brocher, Coincident seismic reflection/refraction studies of the continental lithosphere: a global review, Rev. of Geoph., 25, 723-742, 1987.

O'Connell, R.J., and B. Budiansky, Seismic velocities in dry and saturated cracked solids, J. Geophys. Res., 79, 5412-5426, 1974.

Percival, J.A., A possible exposed Conrad discontinuity in the Kapuskasing Uplift, Ontario, in Reflection Seismology: A Global Perspective, Geodynamics Ser. Vol. 14, edited by M. Barazangi and L. Brown, pp. 135-141, AGU, Washington, D.C., 1986.

Sandmeier, K.-J., and F. Wenzel, Synthetic seismograms for a complex crustal model, Geophys. Res. Lett., 13, 22-25, 1986.

Simmons, G., Velocity of shear waves in rocks to 10 kilobars, J. Geophys. Res 69, 1123-1130, 1964.

Simmons, G., and W.F. Brace, Comparison of static and dynamic measurements of compressibility of rocks, J. Geophys. Res., 70, 5649-5656, 1965.

Smithson, S.B., R.A. Johnson, and C.A. Hurich, Crustal reflections and crustal structure, in Reflection Seismology: A Global Perspective, Geodynamics Ser. Vol. 14, edited by M. Barazangi and L. Brown, pp. 21-32, AGU, Washington, D.C., 1986.

Spetzler, H., and D.L. Anderson, The effect of temperature and partial melting on velocity and attenuation in a simple binary system, J. Geophys. Res., 73, 6051-6060, 1968.

DSS STUDIES OVER DECCAN TRAPS ALONG THE THUADARA-SENDHWA-SINDAD PROFILE, ACROSS NARMADA-SON LINEAMENT, INDIA

K. L. Kaila, I. B. P. Rao, P. Koteswara Rao, N. Madhava Rao
V. G. Krishna and A. R. Sridhar

National Geophysical Research Institute, Hyderabad 500007, India

Abstract. Deep Seismic Sounding (DSS) studies were carried out during 1984-85 along a 260 km long Thuadara-Sendhwa-Sindad profile across the Narmada-Son lineament in central India. Seismic refraction and wide angle reflection data have been recorded in the analog form from a dense coverage of shot points with geophone interval of 200 m. The crustal depth section along this profile was obtained in two parts: shallow and deeper sections. The shallow depth section down to the crystalline basement has been derived from first arrival travel times data of refracted waves from 25 shot points, while the deeper crustal section has been obtained from wide angle reflections data. 1-D velocity models for P waves, derived from several shot points data by forward modeling of travel time observations of refracted and reflected phases, were used to obtain an internally consistent 2-D velocity model along the profile. A low velocity layer probably of Mesozoic sediments (velocity 3.2-4.0 km/sec) with a maximum thickness of about 1.9 km below a thin layer (900 m) of the Deccan Traps (velocity 4.8-4.9 km/sec), has been inferred south of Narmada between Sendhwa and Savkheda from the shallow depth section successively refined by ray tracing. The Deccan Traps layer, with its thickness varying from 500 to 900 m, directly overlies the crystalline basement (velocity 6.0-6.2 km/sec) in the region north of Narmada. The upper crust above the crystalline basement reveals a block structure and the Narmada-Son lineament is reflected as a fault feature, to the south of which we infer a graben consisting of the Deccan Traps and the sub-Trappean sediments above the basement.

The deeper crustal section, initially obtained by migration of wide angle reflection data, has been iteratively refined by 2-D ray tracing and modeling of travel time observations corresponding to various boundaries. The crustal section obtained in the region, reveals a large number of reflectors at various depths down to the Moho boundary indicating a uniform crustal reflectivity, rather than a model consisting of a transparent upper crust and a relatively more reflective lower crust. The Moho depth varies from 38 to 43 km along this profile. The Moho shows a gentle upwarp, with a relief of about 2 km, in the region between Narmada and Tapti where the upper crustal graben above the crystalline basement has been delineated. The P velocity in the lower crust is found to be 6.8-6.9 km/sec. Although a high velocity 'rift pillow' above the Moho has not been observed, the inferred structural features may be indicative of continental rifting in this region.

Introduction

DSS investigations were carried out by the National Geophysical Research Institute (NGRI), Hyderabad along the 260 km long Thuadara-Sendhwa-Sindad profile across the Narmada-Son lineament during 1984-85 field season. The profile covers mostly the region of the exposed Deccan Traps and is aligned in the NNW-SSE direction (Figure 1). The other geological formations exposed along this profile include, Bagh and Lameta beds (Mesozoic), Aravalli system of Sausar, Sakoli series and equivalent rocks and granitic gneisses (Archaean) in the northern part and alluvial deposits in the southern part. This region is almost flat with an average elevation of about 400 m above mean sea level.

The present study along this profile has been undertaken as a part of continuing crustal seismic investigations by the NGRI in the Narmada-Son lineament region. The objectives of the present study include determination of the thickness of the Deccan Traps cover and the underlying sediments if present, mapping the basement configuration, deriving crustal velo-

NATIONAL GEOPHYSICAL RESEARCH INSTITUTE, HYDERABAD

THUADARA-BARWANI-SENDHWA-ERANDOL-SINDAD

DSS PROFILE

ACROSS NARMADA LINEAMENT

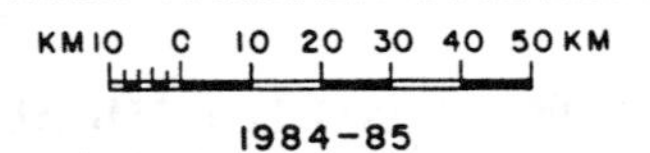

1984-85

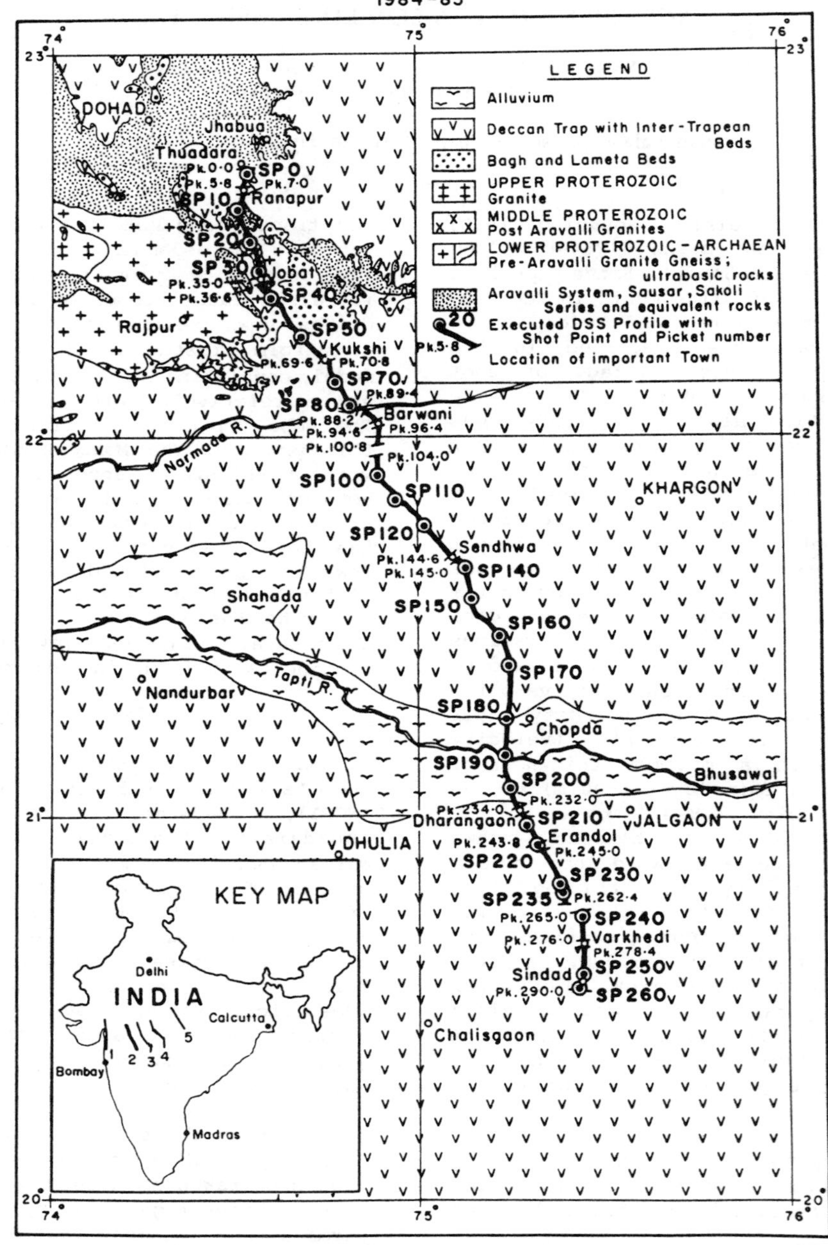

Fig. 1. Location of the Thuadara-Sindad DSS profile (profile 2) with geological information. Other DSS profiles in this region: Mehmadabad-Billimora (south Cambay), profile 1 (Kaila et al. 1981a), Ujjain-Mahan, profile 3 (Kaila et al. 1985), Multai-Pulgaon, profile 4 (Kaila and Koteswara Rao, 1986), Hirapur-Mandla, profile 5 (Kaila, Murty and Mall).

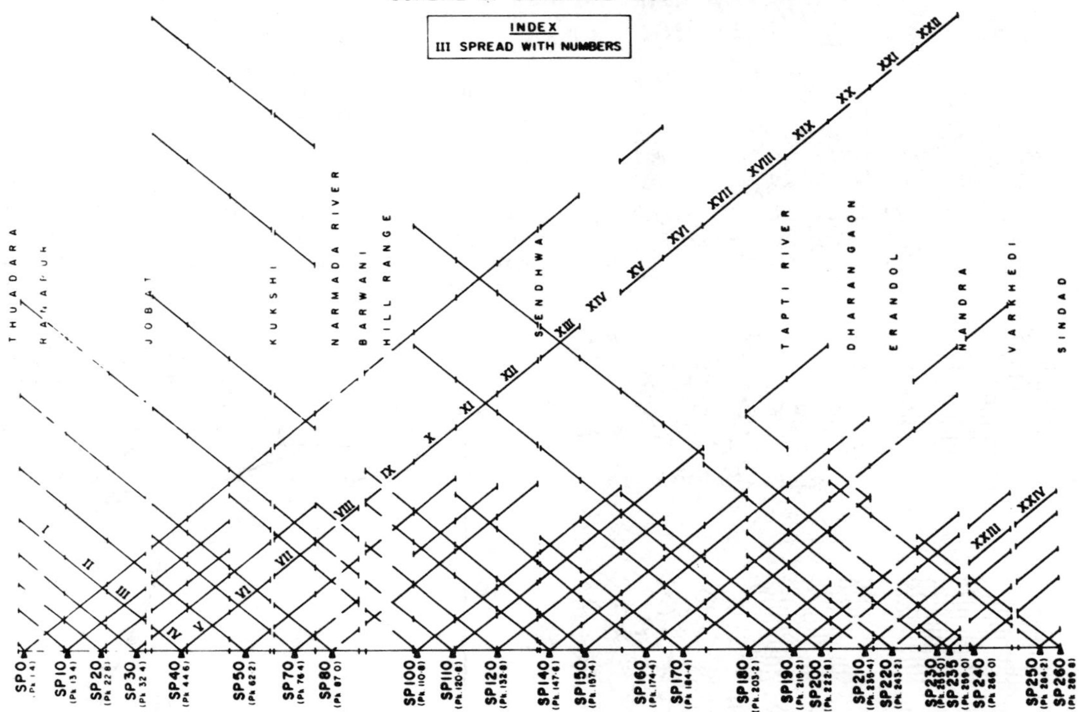

Fig. 2. Scheme of DSS coverage from all the shot points along the Thuadara-Sindad profile.

city structure in the region and understanding the deep structure and the reflectivity nature of the upper and the lower crust.

Geological and Tectonic Features of the Area

The Narmada valley is a unique feature extending about 1300 km from Broach on the western coast to Jabalpur in central India. Earlier geological studies indicate that the region was warped, uplifted and rift faulted along the Narmada valley and completely buried beneath the Deccan Trap lavas. The Narmada rift is regarded as a major crustal feature of ancient origin, reflecting sub-crustal structure and influencing the deposition and folding of the Vindhyans [Auden, 1949]. It is considered by Ahmed [1964] that the Narmada-Son line as a 'welt' (or a swell), howsoever narrow it may appear in certain parts of its length. West [1962] stated that the Narmada-Son line appears to be a zone of weakness from early times with the areas to the north and south moving up and down relative to each other along this line. According to Agarwal and Gaur [1972], an E-W rift, 70 km wide, is extending along the Narmada-Tapti system.

Origin of the Narmada valley is considered to be separated for two distinctly different periods: after the commencement of the eruption of the Deccan Trap, and the period prior to it. From various geological studies it is also clear that this lineament is situated close to the zones of fractures belonging to early Precambrian, Cretaceous and post-Deccan Trap period. The present valley dates back predominantly to the pre-Deccan Trap period, though the movement began significantly before that and still continues to a certain extent. The pre-Trappean topography is visible in places where the basalts have been completely eroded. Evidence of sedimentation along this zone is clear in the Jurassic period.

DSS and Other Geophysical Studies in the Area

The results of palaeomagnetic studies [Pal and Sreenivas, 1976] suggest that the Narmada flows along a tectonically disturbed zone. Tandon and Chowdhury [1970] showed that the epicenters of a number of shocks are aligned

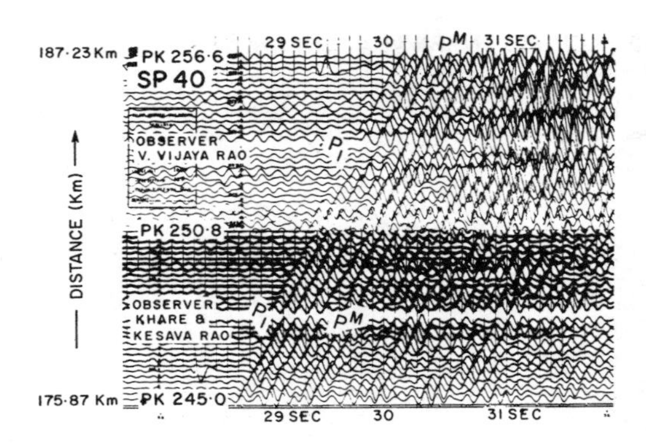

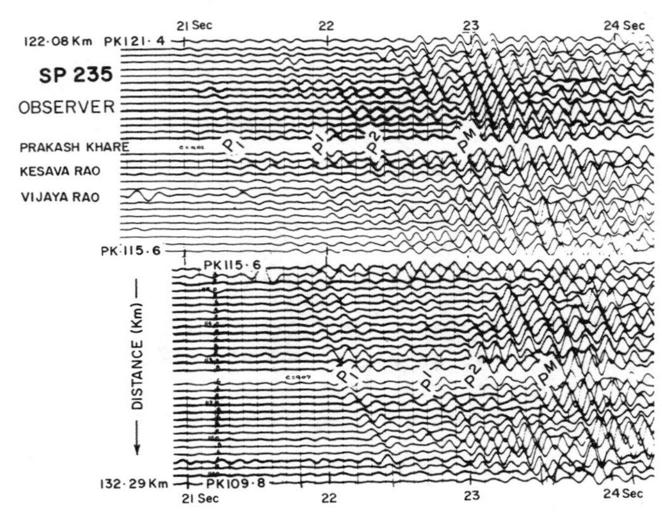

Fig. 3. Sample DSS records from SP 40 and SP 235. Refracted first arrivals (P_1), intracrustal reflections (P^1 and P^2), and wide angle reflections from the Moho discontinuity (P^M) are shown.

approximately parallel to the Narmada-Son lineament. According to Kaila et al. [1972], the Narmada-Son lineament is a zone of moderate seismicity with a maximum expected intensity of VII on the MM scale. Krishna Brahmam [1975] stated that the axis of a gravity low lies north of the fault system associated with the Narmada valley.

Crustal seismic investigations across the Narmada-Son lineament were initiated during 1976 by the NGRI along various DSS profiles as shown in Figure 1. DSS studies in the south Cambay basin (profile 1, Figure 1) revealed a major graben between Jambusar and Broach which is flanked by two deep faults, the southern one being a part of the Narmada fault system. The shallow depth section along the Ujjain-Mahan line (profile 3, Figure 1) revealed the presence of a large Mesozoic basin, with a maximum thickness of about 1.7 km of the sediments below a relatively thin cover of the Deccan Traps. The deeper crustal section in this area also revealed possible presence of deep faults related to the Narmada-Son system. Based on DSS data along the Multai-Pulgaon line (profile 4, Figure 1) about 400 m thick section of Gondwana sediments, considered to be the NW-SE extension of the Godavari graben, has been inferred beneath a very thin cover of the Deccan Traps. DSS results on the Hirapur-Mandla line (profile 5, Figure 1) indicate that the Narmada-Son lineament is a narrow horst feature bounded by faults affecting the crystalline basement in the region.

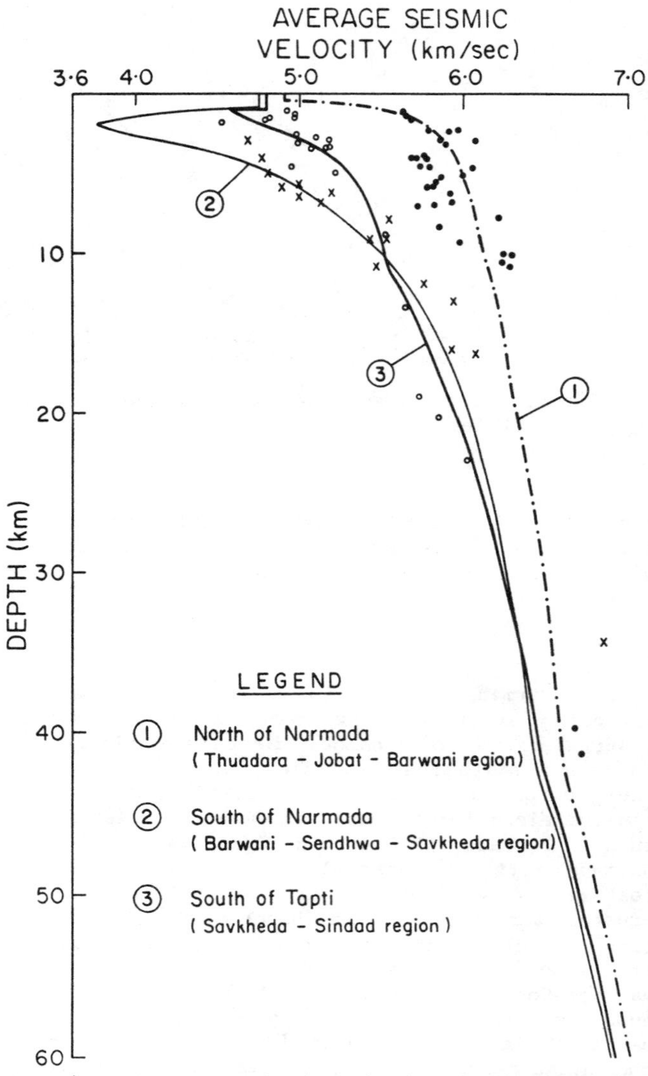

Fig. 4. Average seismic velocity-depth plots for the regions north of Narmada (.), south of Narmada (x) and south of Tapti (∘) obtained from reversed reflection data.

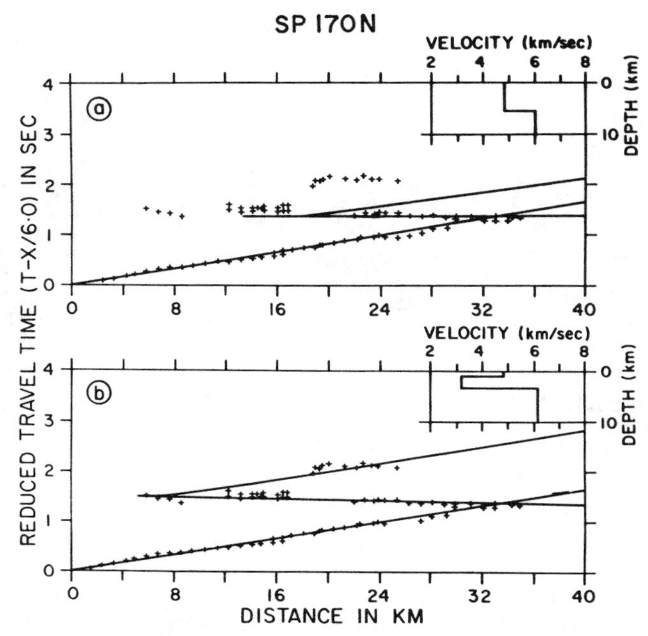

Fig. 5. P wave velocity models: (a) without LVL, (b) with LVL compared with the travel time observations from SP 170N.

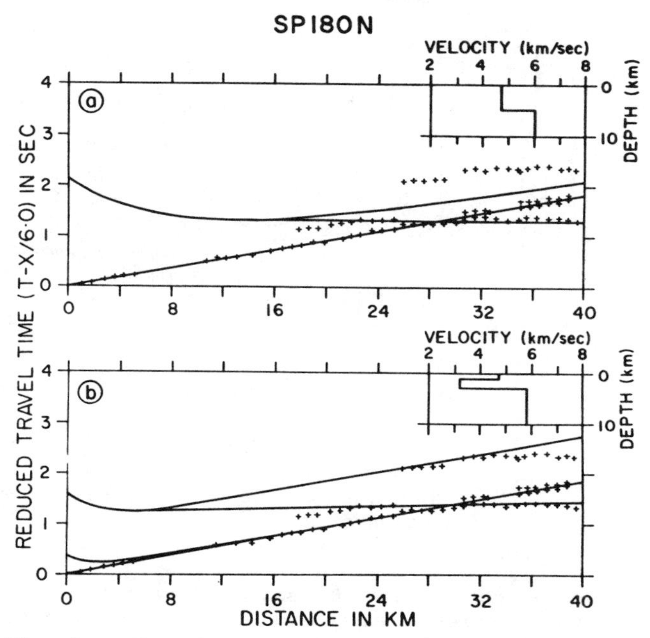

Fig. 6. P wave velocity models: (a) without LVL, (b) with LVL compared with the travel time observations from SP 180N.

Data Acquisition

The seismic refraction and wide angle reflection data with a total coverage of 2,950 line kilometers was recorded in the analog form using the POISK 48 channel seismic recording system and the explosives as the source from 25 shot points with a geophone spacing of 200 m by continuous profiling technqiue. The data redundency especially for the upper crustal layers, obtained along the present DSS profile from various shot points is shown schematically in Figure 2. As the data was recorded on magnetic tapes, necessary playback records were obtained with the desired gain and filter settings (with 15 Hz high cut). The records have been analysed, identifying prominent phases of reflected and refracted waves as shown on specimen records given in Figure 3. The first arrivals data have been picked from the original monitor records and travel time plots have been obtained for data recorded from all the shot points for further analysis and determination of velocity-depth functions.

Crustal Velocity Functions

The velocity function at shallow depth ranges was initially obtained by an analysis of the refraction travel time data. The reversed reflection data available from a large number of shot point pairs, have been analysed using the Kaila and Krishna [1979] method and the average velocity depth plot as shown in Figure 4 was obtained. As can be seen from this Figure there are at least three clusters of velocity depth points pertaining to three different regions, SP 0 to SP 80 (north of Narmada), SP 80 to SP 190 (south of Narmada)

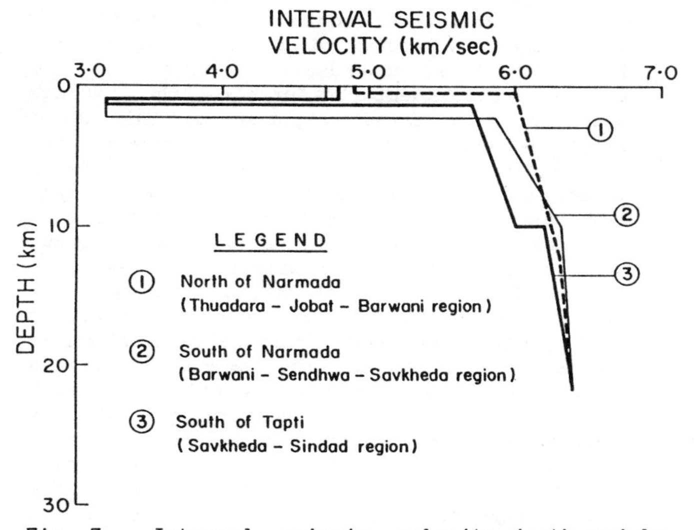

Fig. 7. Interval seismic velocity-depth models for the regions north of Narmada, south of Narmada and south of Tapti, obtained from 1-D modeling.

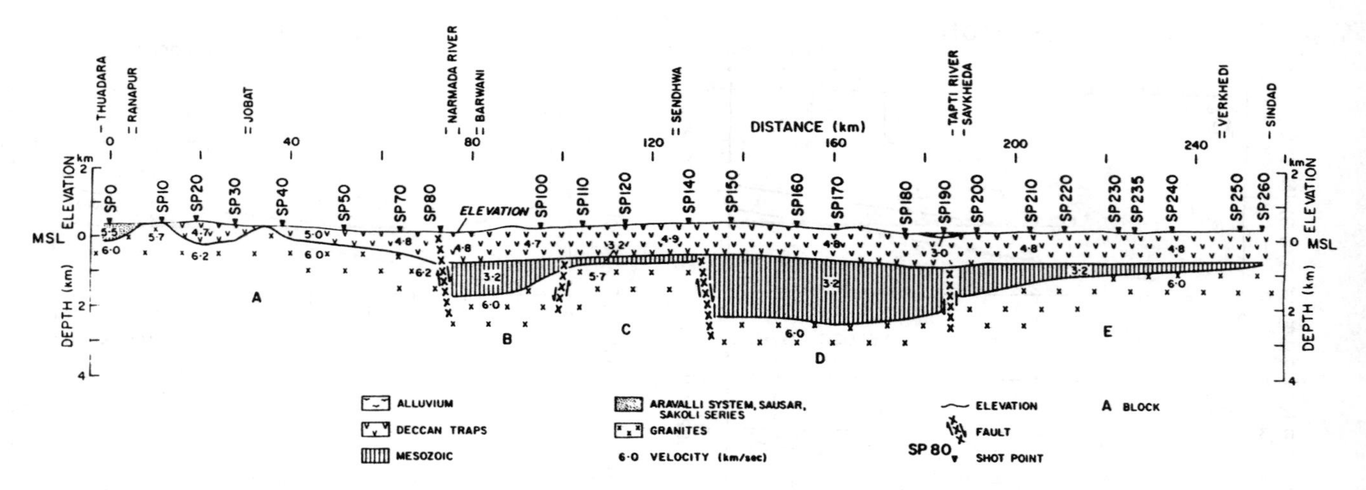

Fig. 8. Shallow depth section along the Thuadara-Sindad profile obtained from the refracation data.

and SP 200 to SP 260 (south of Tapti). It is thus inferred that there are strong lateral velocity variations at shallow depths in the upper crust, which have been resolved as will be explained later. The velocity models for the lower crust have been obtained from long distance data recorded from shot points SP 40, SP 190 and SP 235. The 1-D velocity models derived from various shot points data have been assembled together to produce an initial 2-D velocity model which has been successively refined by 2-D ray tracing.

Velocity Models for the Upper Crust

In the regions south of Narmada (SP 80 to SP 190) and further south (SP 200 to SP 260) the average velocity values are consistently lower than those towards north (SP 0 to SP 80) at comparable depths. Further, the average velocity depth curve computed for this depth range by using refraction data (direct wave in the Deccan Traps with a velocity of 4.8-4.9 km/sec and the refracted wave from the crystalline basement with a velocity of 6.0-6.2 km/sec) is found to reveal consistently higher average velocities than the observations shown in Figure 4. Based on these observations and considering the presence of Mesozoic formations in the nearby exposures where velocity shooting yielded velocities in the range of 3.2-4.0 km/sec, we have considered models with such a layer of low velocities underlying the Deccan Traps cover. Thus combinations of both refraction and reciprocal reflection data have been used to infer the possible thickness of the Deccan Traps cover and the underlying sediments applying the Kaila et al. [1981] indirect seismic method. By the above analysis we have inferred the Deccan Traps thickness of 900 m and the low velocity (3.2-4.0 km/sec) layer with a maximum thickness of 1.9 km in the region south of Narmada between SP 80 to SP 190. The travel time models in this region for SP 170 and SP 180 with and without the presence of the low velocity layer (LVL) below the Traps have been shown in Figures 5 and 6 respectively. It can be seen from these Figures that the models with the LVL reveal more satisfactory fits especially for the travel time data corresponding to the retrograde branch of reflected waves from the bottom of the LVL. A similar analysis for the data in the region south of Tapti (SP 200 to SP 260) however indicated a relatively small thickness (about 300 m) of the LVL beneath the 900 m thick Deccan Traps cover.

The 1-D interval velocity models for the upper crust in the three regions along the profile are shown in Figure 7. Assuming the thickness of 900 m of the Traps as a reasonable average from SP 80 towards south, the intercept times of the basement refracted wave recorded from various shot points have been used to infer the variable thickness of the LVL along the profile and the initial model down to the basement has thus been obtained as shown in Figure 8. The basement configuration towards north of Narmada, shown in this figure, has been obtained using the general wave front method of Rockwell [1967].

Velocity Models for the Lower Crust

To study the velocity models for the lower crust in this region, we have used the long distance recordings from shot points SP 40, SP 190 and SP 235. The upper crustal velocity models are however used as given in Figure 7 while modeling the lower crustal velocity functions. The program LAUFZEIT has been

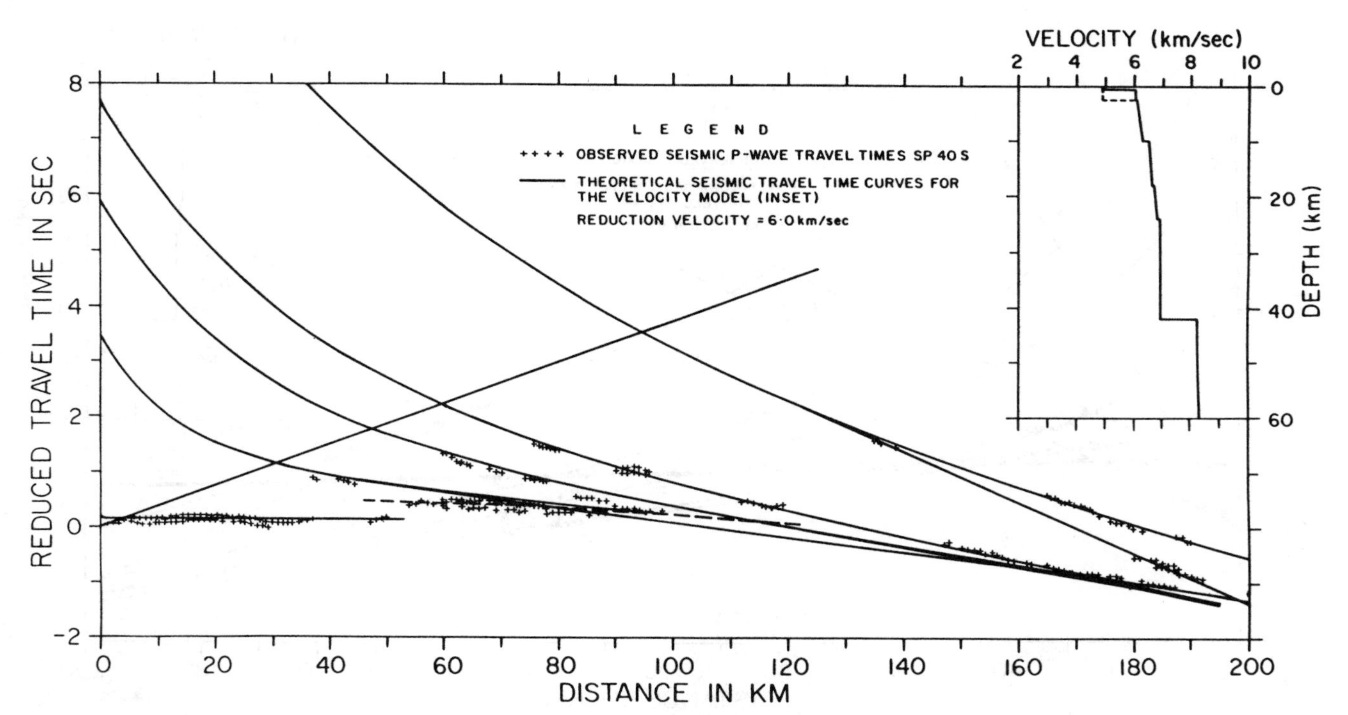

Fig. 9. P wave velocity model derived from travel times data of SP 40S by 1-D modeling. __ __ __ __ represents the model with 2.5 km Traps thickness.

used to successively refine the velocity models till a satisfactory match of the computed travel time curves with the corresponding observations is obtained. The 1-D crustal velocity models, based on travel times modeling, thus inferred using data from SP 40, SP 190 and SP 235 are shown in Figures 9, 10 and 11 respectively. In these models we have used a P_n velocity of 8.2 km/sec based on its observation in this region along other DSS profiles (profile 1, Figure 1). The velocity models shown in Figures 9 to 11 reveal some differentiation in the upper as well as the lower crustal depths as is also indicated by various prograde and retrograde travel time branches matching well with the observations.

The 1-D crustal velocity models obtained above have been used to construct the deeper crustal section (Figure 12) by migrating the wide angle reflection data applying the method of Kaila et al. [1982]. In this method travel time segments of wide angle reflections from single sided data are however used to obtain a migrated depth segment. The crustal depth section thus constructed has been further refined by 2-D ray tracing technique.

Two Dimensional Travel Times Modeling

The arrival times of different phases recorded along this profile have also been interpreted using two-dimensional ray trace modeling technique. The software package SEIS 81, developed by Cerveny and Psencik [1981] based on zero order approximation of ray theory, is employed to compute travel times through a vertically and laterally heterogeneous velocity model. For this purpose we have used the initial models of the shallow as well as the deeper crustal sections obtained above. These models have been iteratively refined until acceptable match of the computed and the observed travel times was obtained. At each iteration, the velocities or velocity gradients and structural features were perturbed in such a way to reduce the difference between computed and observed arrival times to within 0.01 to 0.02 sec. Although this precision could not be achieved for all the data points for each shot point, 90 percent of the travel time observations were fitted with this accuracy. This precision is quite satisfactory for the present data set. The upper crustal models, ray diagram, and the corresponding travel time plots for some selected shot points SP 40, SP 100, SP 170 and SP 190 are shown in Figures 13 to 16. The 2-D modeling of travel times of different phases recorded along the present DSS profile from various shot points consisted of refracted waves (diving waves) through the Deccan Traps and granitic basement and the reflected waves from top and bottom of the

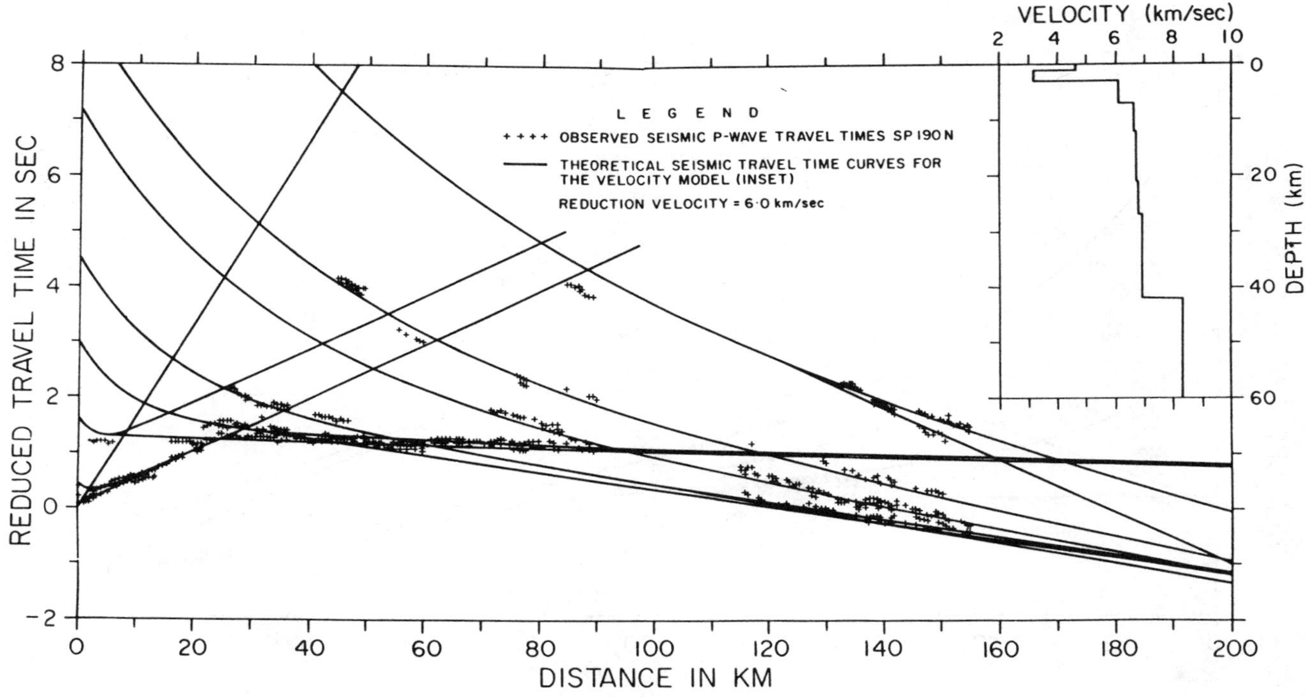

Fig. 10. P wave velocity model derived from travel times data of SP 190N by 1-D modeling.

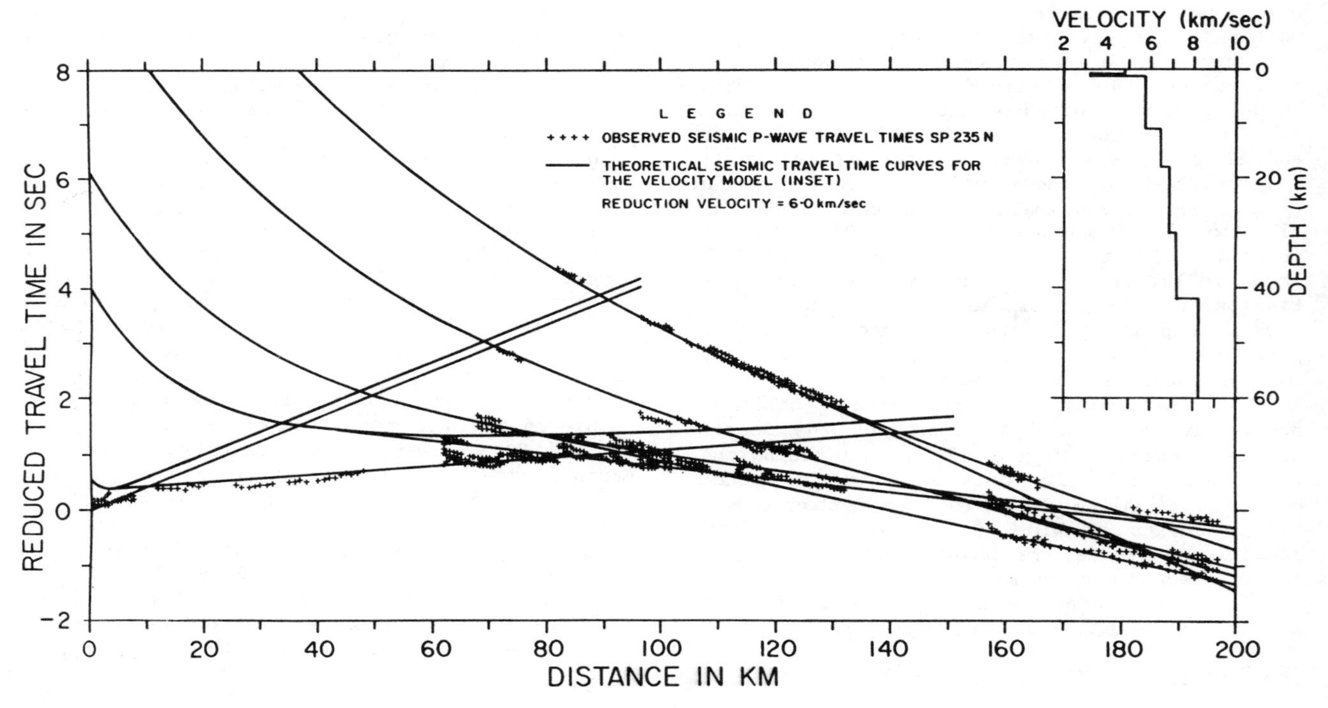

Fig. 11. P wave velocity model derived from travel times data of SP 235N by 1-D modeling.

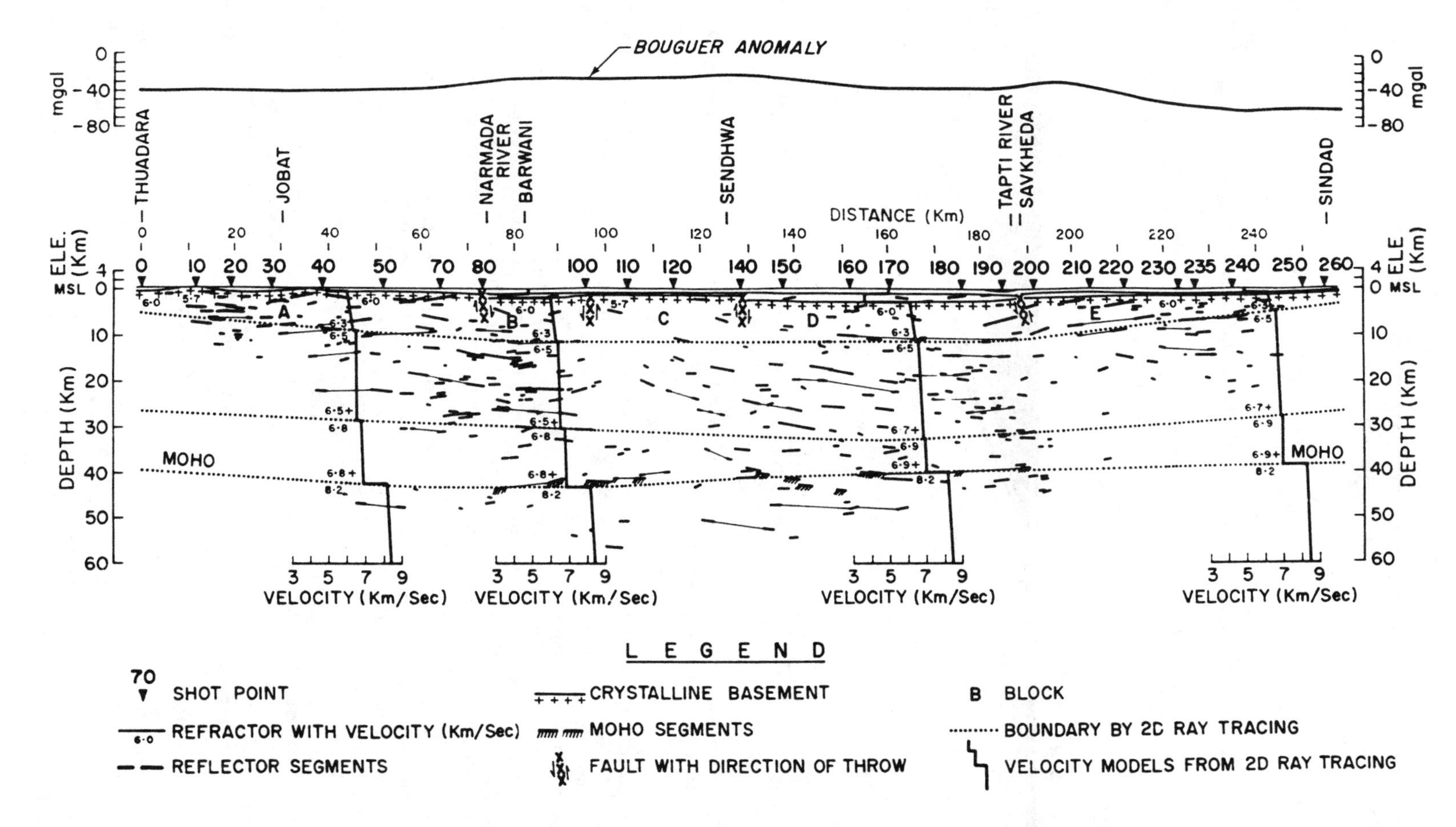

Fig. 12. Crustal cross section along the Thuadara-Sindad profile obtained from migration of wide angle reflection data............ represent the boundaries inferred by matching the reflection data using 2-D ray tracing. The velocity models shown at four locations are obtained from the final 2-D velocity structure.

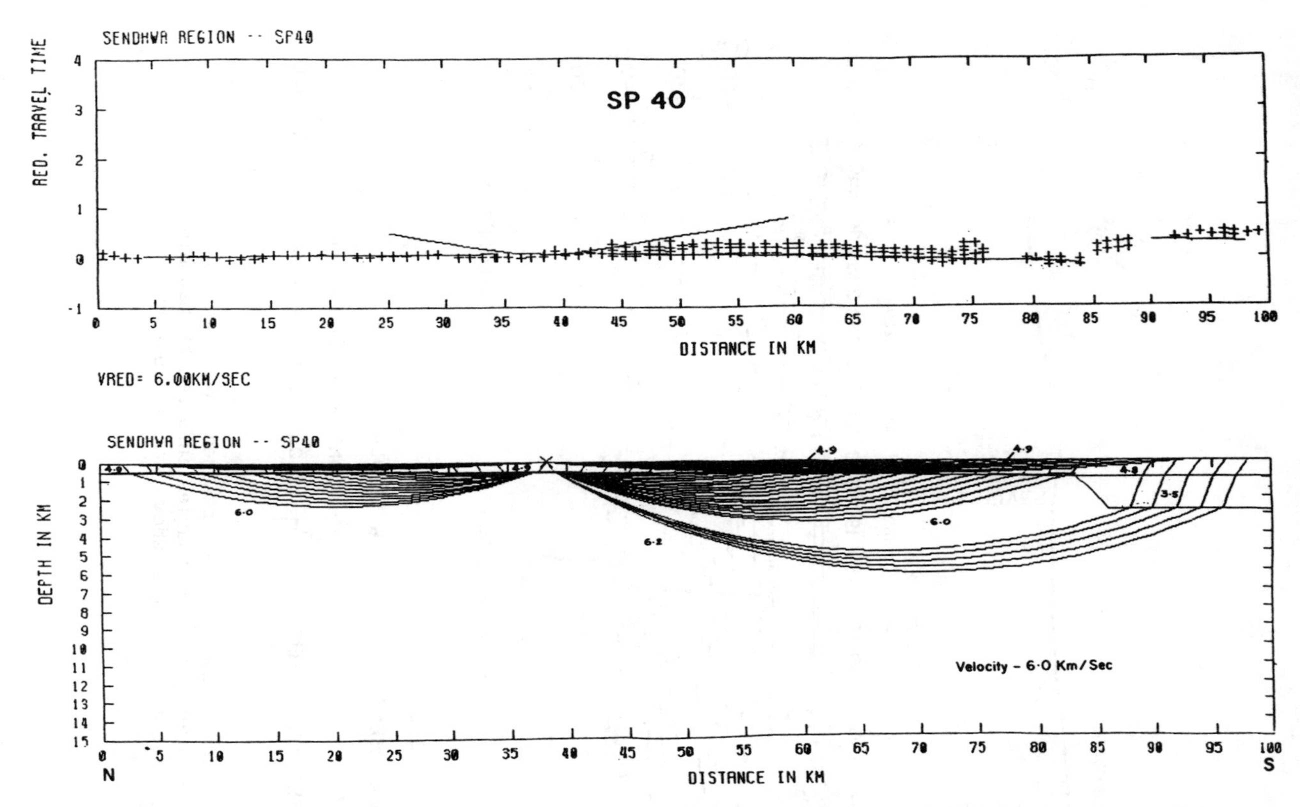

Fig. 13. Ray diagram and theoretical travel time curves with reduction velocity of 6.0 km/sec with observed travel times data (+) for SP 40.

low velocity layer (LVL) in the case of shot points SP 170 and SP 190. Modeling has been done by matching the observed data for a particular shot point and subsequently validating the model by matching the data from various other shot points sampling the same subsurface region. The travel times modeling of reversed refraction and reflection data from various shot point pairs thus yielded a reliable upper crustal section.

The final two dimensional velocity model down to the crystalline basement, thus obtained by ray tracing, consists of a 0.5 to 0.9 km thick Deccan Trap layer of velocity 4.8-4.9 km/sec and the granitic basement of velocity 6.0-6.2 km/sec. An LVL of 1.9 km maximum thickness with velocity varying from 3.2 to 4.0 km/sec underlies the Traps between SP 80 and SP 200. The modeling of the refracted arrivals from the basement in this region, consistent with the average velocities from the reversed reflection data (Figure 4) has necessiated the introduction of this LVL of 1.9 km thickness in this part of the profile. The analysis of the seismograms of a micro-spread on the outcrops of the Bagh beds (Mesozoic) near SP 40 yielded a velocity of 3.2 km/sec. Seismic refraction studies over exposed Dhrangadhra sandstones (Mesozoic) in the northern part of Saurashtra had given a velocity of 4.0 km/sec in those sediments. In the light of the above findings the velocity of the LVL in the present study is varied laterally from 3.2 to 4.0 km/sec in the 1.9 km thick Mesozoic sediments underlying the Deccan Traps. The inference of the LVL is not possible from refraction data alone because no refracted rays bottom within the LVL. In the present study, the modeling of reflections from top and bottom of the LVL with a geometrical shadow zone in between them, is a clear manifestation of the existence of the LVL as evidenced by the data sets of SP 170 and SP 190. However, conclusive evidence for this LVL may be obtained by dynamic modeling of reflection data and the results of this study will be published in a forthcoming paper.

The data from shot point SP 40 in the north and shot points SP 190 and SP 235 in the south, were primarily used to model the deeper boundaries by 2-D ray tracing. The reflected waves from the Moho boundary (P_M) are found in the wide angle range beyond 110-120 km for these shot points as well recorded events

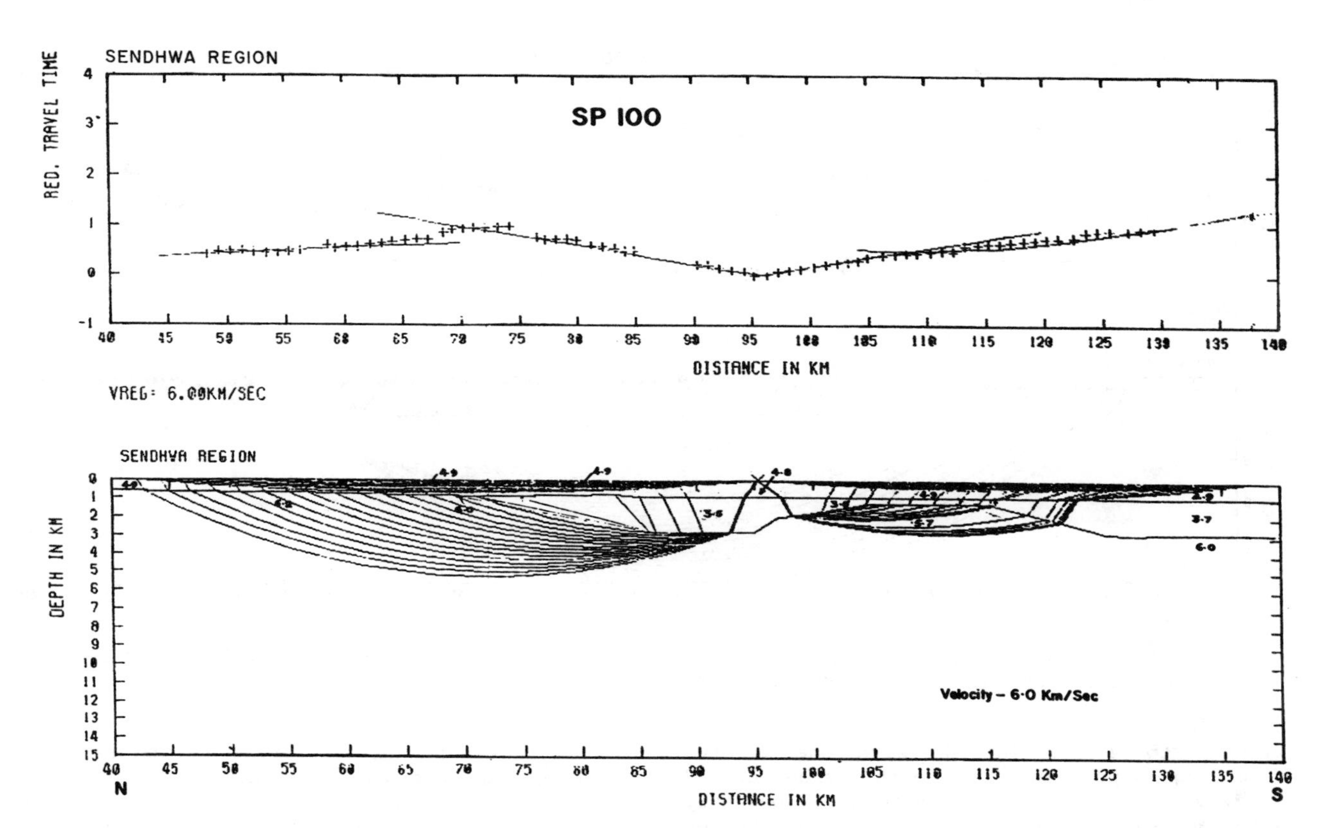

Fig. 14. Ray diagram and theoretical travel time curves with reduction velocity of 6.0 km/sec for SP 100.

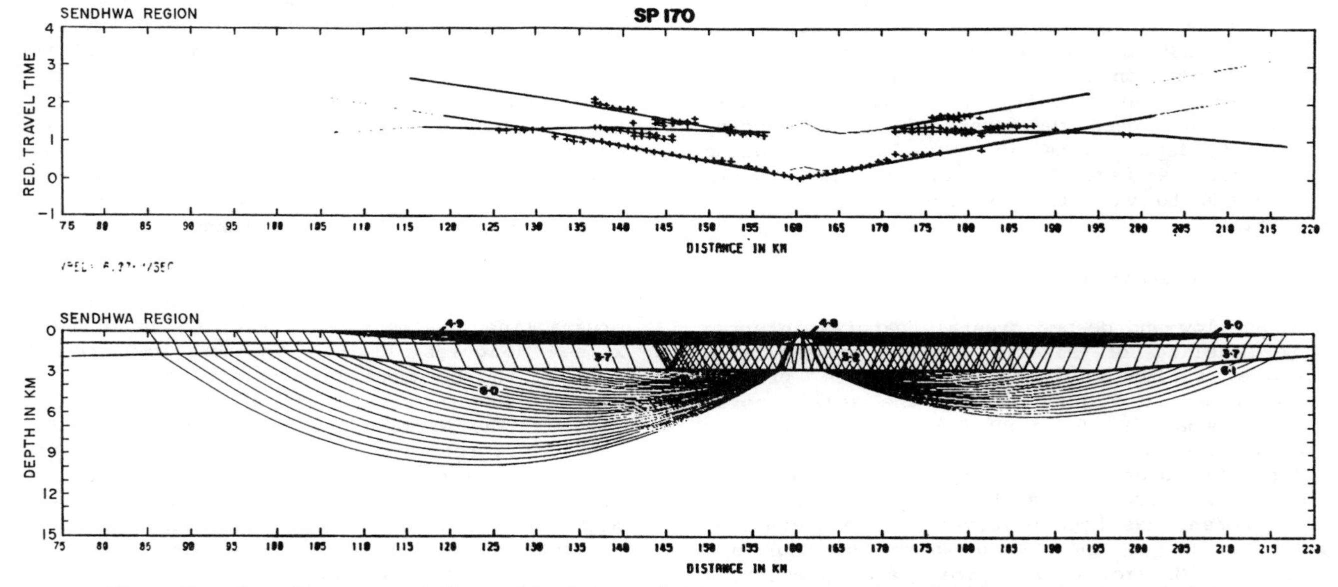

Fig. 15. Ray diagram and theoretical travel time curves with reduction velocity of 6.0 km/sec for SP 170. The travel times of refracted waves through Traps with velocity 4.8-5.0 km/sec and the basement with velocity 6.0-6.1 km/sec and the reflections from top and bottom of the low velocity layer (LVL) of velocity 3.2-4.0 km/sec are modeled.

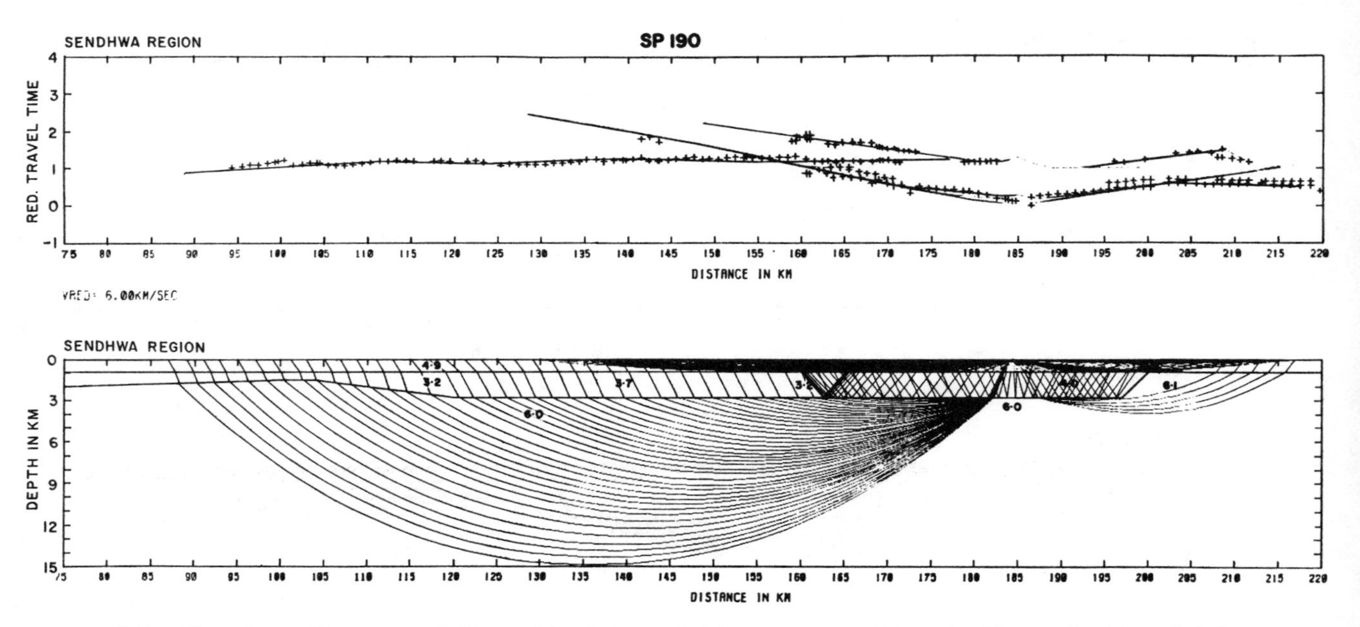

Fig. 16. Ray diagram and theoretical travel time curves with reduction velocity of 6.0 km/sec for SP 190 with reflections from top and bottom of LVL.

with 3 to 4 oscillations with one or two recording gaps. In the depth range of 27 to 32 km we recognize an intermediate boundary with a velocity contrast of 0.2-0.3 km/sec based on modeling of the observed travel times data [Figure 12]. A very satisfactory travel times fit to the observed P_M phase as well as the intermediate boundary reflection has been obtained for both the shot points SP 40 and SP 235 by two dimensional travel times modeling as can be seen from Figures 17 and 18. The final 2-D velocity model given in these Figures has been represented by corresponding velocity-depth plots at four locations along the profile (see Figure 12). The Moho depth is found to vary from 38 to 43 km along this profile.

Interpretation and Discussion of Results

Shallow and deeper crustal features inferred by modeling of seismic refraction and wide angle reflection data along this profile are given in Figure 12. In the region north of Narmada (SP 10 to SP 80) the Deccan Traps with varying thickness of 500 m to 900 m directly overlie the crystalline basement. However, near SP 0 a direct wave velocity of 5.5 km/sec has been observed in the Aravallis. Between SP 40 and SP 50 where the Bagh and Lameta (Mesozoic) exposures are present, a direct wave velocity of 3.2 km/sec has been observed only from the microspread with a geophone interval of 25 m. However, this velocity could not be found on the spreads with 200 m geophone interval in the same area indicating that the thickness of these formation is very small (within about 100 m). In the region between Narmada and Tapti, average velocities down to 10 km depth obtained from reversed reflection data are consistently smaller than those towards its north. This difference could be explained by successfully modeling an LVL (velocity 3.2-4.0 km/sec) below the Deccan Traps (velocity 4.8-4.9 km/sec). We interpret this LVL as a layer of Mesozoic sediments in view of their exposures in the nearby region (between SP 40 and SP 50). Our modeling results indicate a maximum thickness of this LVL of about 1.9 km between Sendhwa and Savkheda. A layer of Mesozoic sediments with a maximum thickness of 1.7 km beneath the Traps has been inferred from earlier DSS studies also in this region (profiles 3 and 4, Figure 1). The velocity in the crystalline basement is found to be 6.0-6.2 km/sec (Figures 13-16). The shallow section down to the basement reveals a block structure due to basement faults as shown in Figures 8 and 12. The Narmada-Son lineament across this profile is reflected as one such fault feature affecting the crystalline basement. The tectonic activity associated with this feature might be responsible for the formation of a graben structure above the crystalline basement, consisting of the low velocity formations beneath the Deccan Traps, as delineated between the Narmada and Tapti.

We recognize two intra-crustal boundaries between the basement and the Moho boundary

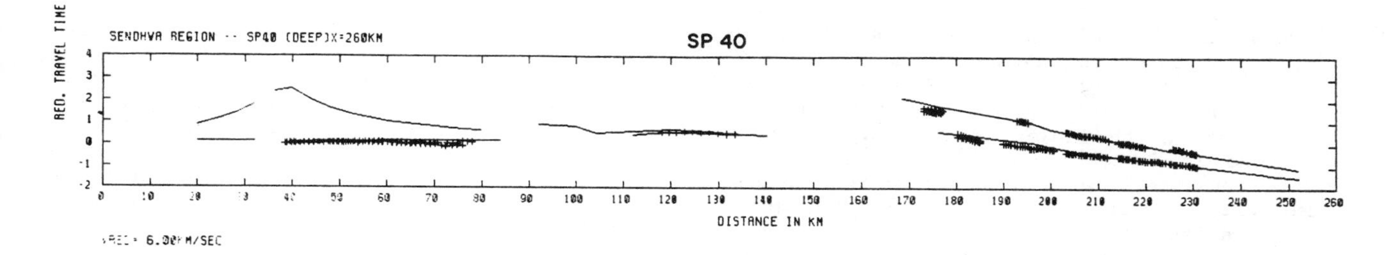

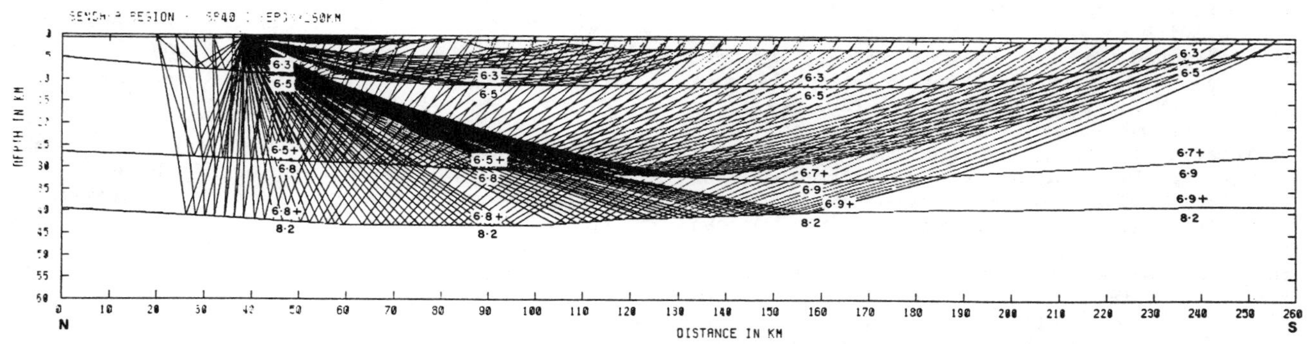

Fig. 17. Ray diagram and the 2-D velocity model along the Thuadara-Sindad profile across the Narmada-Son lineament. Computed travel times (——) and the observed travel times (++++) for SP 40 with reduction velocity of 6.0 km/sec for direct wave through Traps, basement refraction, reflection from an intermediate layer and Moho boundary are shown.

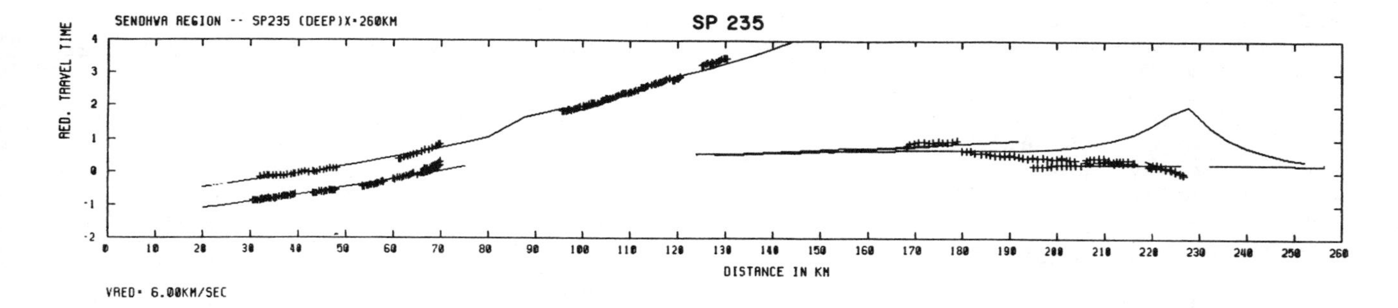

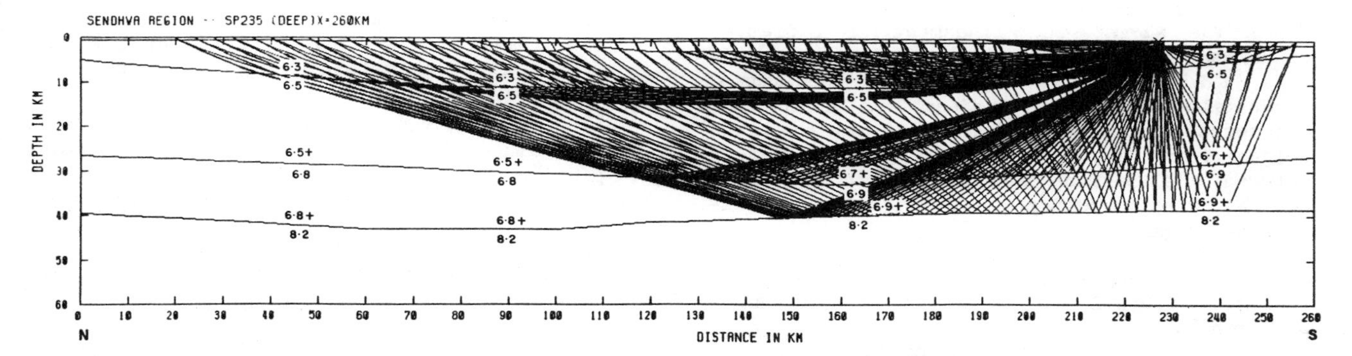

Fig. 18. Ray diagram and 2-D velocity model along the Thuadara-Sindad profile across the Narmada-Son lineament. Computed travel times (——) and the observed travel times (++++) for SP 235 with reduction velocity of 6.0 km/sec for direct wave through Traps, basement refraction, reflection from an intermediate layer and Moho boundary are shown.

with minor lateral variations as shown in Figure 12 from the deeper crustal section, obtained from migration of the wide angle reflection data and by 2-D ray tracing. The P velocity in the lower crustal layer, above the Moho is found to be 6.8-6.9 km/sec. Rock types consistent with this velocity include amphibolite, gabbro or granulite. The Moho depth varies from 38 to 43 km along this profile. Specifically in the region between Narmada and Tapti, the Moho shows a gentle upwarp with a relief of about 2 km. This is reflected in the Bouguer anomaly profile as well, as a broad gravity high between SP 50 and SP 210. However, a high velocity 'rift pillow' above the Moho has not been delineated in this region. The gravity low betwen SP 150 and SP 200 superimposed over the broad high may be related to the low velocity layer inferred beneath the Deccan Traps in this upper crustal graben above the crystalline basement. The above findings, viz. presence of an upper crustal graben above the crystalline basement, a rather thick accumulation of sub-Trappean sediments (probably Mesozoics) and a broad upwarp of the Moho seem to be the primary indicators that the crust has been rifted in this region. Other geologic and geophyhsical (gravity) observations also support this idea. The large number of reflectors shown at various depths down to the Moho boundary seem to indicate a uniform crustal reflectivity rather than a model consisting of a transparent upper crust and a relatively more reflective lower crust. It is however difficult to elucidate the nature of these reflections from the present data sets in the wide angle range. We believe that future deep seismic reflection profiling in this region may provide some clues in this regard.

Summary

The following inferences may be drawn from the present DSS study:

1) The Narmada-Son lineament is reflected as a basement fault in this region. A prominent LVL (probably of Mesozoic sediments) with velocity 3.2-4.0 km/sec and a maximum thickness of 1.9 km underlies a 0.9 km thick Deccan Traps layer (velocity 4.8-4.9 km/sec) between the Narmada and Tapti, in an upper crustal graben delineated above the crystalline basement.

2) The Deccan Traps, 500-900 m thick, directly overlie the crystalline basement (velocity 6.0-6.2 km/sec) in the region north of Narmada.

3) The velocity models obtained by 2-D ray tracing reveal some differentiation in the upper as well as the lower crustal depths. The P velocity above the Moho in the lower crust is found to be 6.8-6.9 km/sec.

4) The Moho depth varies from 38 to 43 km along this profile. A gentle upwarp of the Moho of about 2 km relief occurs between Narmada and Tapti, where an upper crustal graben above the crystalline basement has been delineated. Although a high velocity 'rift pillow' above the Moho has not been observed, the inferred structural features may be indicative of continental rifting in this region.

Acknowledgements. We are thankful to M/s Oil and Natural Gas Commission, Dehradun, for sponsoring the DSS studies along Thuadara-Sendhwa-Sindad profile with full financial support. We are thankful to the Director, NGRI for his continuous support and encouragement during the progress of these studies. Thanks are due to all the colleagues of the DSS field party without whose hard work and enthusiasm, it would not have been possible to complete the field work in time. We also thank Mr.P.J. Vijayanandam, Mr.M.Shankaraiah and Mr.B.P.S. Rana for preparing the drawings.

References

Agarwal, P.N., and V.K. Gaur, Study of crustal deformation in India, Tectonophysics, 15, 287-296, 1972.

Ahmed, F., The line of Narmada-Son valleys, Curr. Sci., 33, 362, 1964.

Auden, J.B., Geological discussion of the Satpura hypothesis, Pr. Nat. Inst. Sci. Ind., 15, 315-340, 1949.

Cerveny, V., and I. Psencik, 2D seismic ray package, Charles University, Prague, 1981.

Kaila, K.L., V.K. Gaur, and Hari Narain, Quantitative seismicity maps of India, Bull. Seism. Soc. Am., 62, 1119-1132, 1972.

Kaila, K.L., and V.G. Krishna, A new computerized method for finding effective velocity from reversed reflection travel time data, Geophysics, 44, 1064-1076, 1979.

Kaila, K.L., H.C. Tewari and V.G. Krishna, An indirect seismic method for determining the thickness of a low velocity layer underlying a high velocity layer, Geophysics, 46, 1003-1008, 1981.

Kaila, K.L., V.G. Krishna, and D.M. Mall, Crustal structure along Mehmadabad-Billimora profile in the Cambay basin, India, from deep seismic soundings, Tectonophysics, 76, 99-130, 1981a.

Kaila, K.L., K. Roy Chowdhury, and V.G. Krishna, An analytical method for crustal wide angle reflections, Studia Geophys. et Geod., 26, 254-271, 1982.

Kaila, K.L., P.R. Reddy, M.M. Dixit and P. Koteswara Rao, Crustal structure across Narmada-Son lineament, Central India from deep seismic soundings, J. Geol. Soc. Ind., 26, 465-480, 1985.

Kaila, K.L., P.R.K. Murty, and D.M. Mall,

Evolution of Vindhyan Basin vis-a-vis Narmada-Son lineament, central India, from Deep Seismic Soundings, in press, Tectonophysics.

Kaila, K.L., and P. Koteswara Rao, Crustal structure along Khajuriakalan-Rahatgaon-Betul-Multai-Pulgaon across Narmada lineament from deep seismic soundings, in Deep Seismic Soundings and Crustal Tectonics, Edited by K.L. Kaila and H.C. Tewari, AEG Publication, 43-59, 1986.

Kaila, K.L. Tectonic framework of Narmada-Son lineament - A continental rift system in central India from deep seismic soundings, Reflection Seismology: A Global Perspective, Geodynamics series, 13, 133-150, 1986.

Krishna Brahmam, N., Geophysical studies on the Deccan Traps (India), Geophys. Res. Bull., 13, 89-105, 1975.

Pal, P.C., and G. Srinivas, Narmada-Son lineament, Geophys. Res. Bull., 14, 1-7, 1976.

Rockwell, D.W., A general wavefront method in seismic refraction prospecting, Soc. . Expl. Geophys. Tulsa, 363-415, 1967.

Tandon, A.N., and H.M. Chaudhury, Seismometric study of the Koyna earthquake of December 11, 1967, Ind. J. Power River Valley Dev., Symp. on Koyna Earthquake, 1970.

West, W.D., The line of Narmada-Son valley, Curr. Sci., 31, 143-144, 1962.

SYNTHETIC SEISMOGRAM MODELING OF CRUSTAL SEISMIC RECORD SECTIONS FROM THE KOYNA DSS PROFILES IN THE WESTERN INDIA

V. G. Krishna, K. L. Kaila and P. R. Reddy

National Geophysical Research Institute, Hyderabad 500007, India

Abstract. Two deep seismic sounding profiles, each about 200 km long, were recorded in the east-west direction in the Koyna region of the Western India by the National Geophysical Research Institute acquiring wide angle crustal seismic data in the analog form. The analog DSS records, acquired by continuous profiling, have been digitized and assembled into trace normalized record sections which are displayed in the form of reduced travel time with reduction velocity of 6 km/sec. Travel times and relative amplitudes modeling, with the aid of synthetic seismograms, of the record sections obtained in the eastern and western directions on the northern profile as well as a record section in the eastern direction on the southern profile reveal consistent models of the crustal velocity structure with minor lateral variations in the Koyna region. The prominent features of the crustal velocity models inferred in this region include low velocity layers (LVL) in the upper crust (6.0 to 11.5 km depth) as well as in the lower crust (26.0 to 28.0 km depth) and at least 2 km thick transitional Moho at 35.5 to 37.5 km depth. However, the Moho is 1-2 km deeper along the southern profile. The upper crustal LVL, with its top at 6-7 km depth in these models, is consistent with the observed seismic activity concentration at 4-5 km depth and an appreciable reduction of seismic activity at greater depths in the Koyna region. The inferred crustal velocity models are also consistent with the essentially aseismic nature of the lower crust in this region. The two low velocity layers, in the upper and the lower crust in this region, may suggest rheological stratification of this part of the lithosphere with levels of increased ductility at those depths. The reflected phase PMP, from the Moho boundary, is relatively strong only beyond 90-100 km recording distance while it is almost suppressed in the subcritical distances on the record sections considered in the present study. Further the PS converted waves from the Moho are also not observable at the appropriate times in these record sections thus substantiating the transitional nature of the Moho, which is at least 2 km thick, in the Koyna region. The Pn phase, well recorded as the first arrival beyond 165 km recording distance, reveals an uppermost mantle P velocity of 8.25 km/sec in this region.

Introduction

The continental crust forms the most important zone of study for understanding the relationship between the processes in the mantle and geophysical phenomena observed at the earth's surface. In-situ information on the properties and processes of the lower crust and the nature of the crust-mantle boundary can essentially be obtained by indirect methods, although ultradeep drilling may eventually provide answers to some key questions such as continental volcanism and intraplate seismicity. Seismic investigations by refraction and reflection profiling using controlled sources provide the most valuable information on the properties of the continental crust and the subcrustal lithosphere. Since 1972, the technique of deep seismic sounding (DSS) has successfully been employed in India by the National Geophysical Research Institute (NGRI) acquiring wide angle crustal seismic data in the analog form over more than 4000 line km by continuous profiling along 17 profiles in various geological settings in the peninsular shield and in the Himalaya. Recently NGRI has also acquired the DFS V digitial recording systems and currently DSS profiling in India is accomplished by digital data acquisition and processing.

Two DSS profiles, each about 200 km long, were recorded during 1975-77 in the east-west direction in the Koyna region (Figure 1) of the Western India by the NGRI acquiring wide angle crustal seismic data in the analog form. This region is also well known for the seismic activity in this part of the subcontinent. Kaila et al [1981a,b] constructed the crustal depth sections from the DSS data along the two profiles

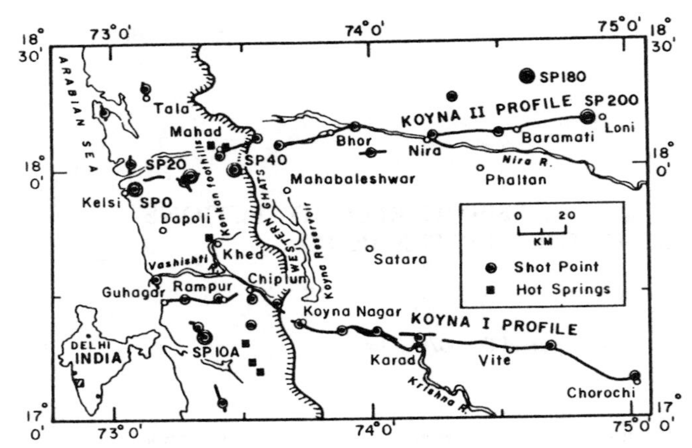

Fig. 1. Location map showing the KOYNA I and KOYNA II DSS profiles with shot points.

which reveal essentially horizontal layering in the crust. Krishna and Kaila [1986], referred here as paper I, initiated digitization of analog DSS records and reinterpretation with the aid of synthetic seismograms. In paper I the analog DSS records for SP 0 on the KOYNA II profile, after digitization and amplitude normalization, were assembled into record sections and displayed in the form of reduced travel time. These record sections were used to derive a kinematically and dynamically well constrained velocity model for the region by modeling of travel times and relative amplitudes data and successively refining the model. In this paper we present the crustal seismic record sections assembled after digitization and amplitude normalization, for SPs 0, 20, 40, 180, 200 on the KOYNA II profile and for SP 10A on the KOYNA I profile, and the synthetic seismogram modeling of these sections producing internally consistent crustal velocity models in this region. The final velocity models thus derived reveal the fine structure and nature of the seismic boundaries, leading to considerable improvements over the earlier velocity models based on inversion of travel times data alone [Kaila et al, 1981a,b].

Digitization and Presentation of the Record Sections

Typically, the DSS data acquisition has been made by continuous profiling using two seismic units of Russian make, type POISK-I-48-KMPV-OV, with a frequency range of 5 to 30 Hz. The dynamic recording range and reproducing with high cut at 30 Hz filter is about 40 db. A high cut filter with cut off frequency of 22 Hz has been used for recording in the Koyna region [Kaila et al, 1979]. The low cut filter was kept out always. Since there is provision for field recording on magnetic tapes, the field tapes have subsequently been used for reproduction of seismic records with the desired gain and filter settings. Each spread 11.6 km long with geophone spacing of 200 m utilizes a shot with varying charge size (50 to 800 kg) of chemical explosives from the shot point, depending upon the shot point to recording spread distance and the quality of the shot point being used. It may be mentioned here that the shot points considered here on both the profiles have been exceptionally good in terms of energy conversion/transmission and a charge size of only about 800 kg was adequate for recording spread distances of 180-200 km. The analog DSS data available in the form of playback records (with a high cut frequency of 15 Hz) on photographic paper, each covering 5.8 km distance, have been used for digitization. The analog magnetic tapes, however could not be used for this purpose due to compatibility problems. Although the original data was recorded with 200 m geophone spacing, we have used every alternate trace about 400 m apart for digitization with a sampling interval of 4 msec. As the entire record section for a particular shot point has to be assembled from seismograms of a large number of DSS records obtained on different spreads from shots of varying charge size, recording gain and other instrumental settings, we assembled and plotted the digitized data by trace normalization with a reduction velocity of 6 km/sec. In this normalization process, the maximum amplitude in each trace is scaled to a fixed value. In this type of presentation, the relative amplitudes of different phases in the record section can be used for comparison with synthetics for modeling. In our opinion, the models are better constrained when the observed amplitude ratios of different phases in the record sections are matched with those obtained in the synthetics. Therefore we do not consider it as a serious disadvantage that true amplitude variation with distance of individual phases cannot be extracted from this type of DSS records. The data from SP 20 and SP 40 on the KOYNA II profile towards east, which are available in the distance ranges from 0 to 70 km and from 80 to 150 km respectively, are combined together to form one continuous record section (SP 20-40) from 0 to 150 km distance. Similarly the data from SP 200 and SP 180 on this profile towards west, which are available in the distance ranges from 0 to 70 km and from 75 to 170 km respectively, are also combined together to form one continuous record section (SP 200-180) from 0 to 170 km distance. Although we have plotted the record sections for various shot points with trace interval of 400 m, 800 m, 1.6 km and 3.2 km, we present here only those with about 1.6 km trace interval to illustrate the possible correlations of different phases identified. Nevertheless we have used the record sections

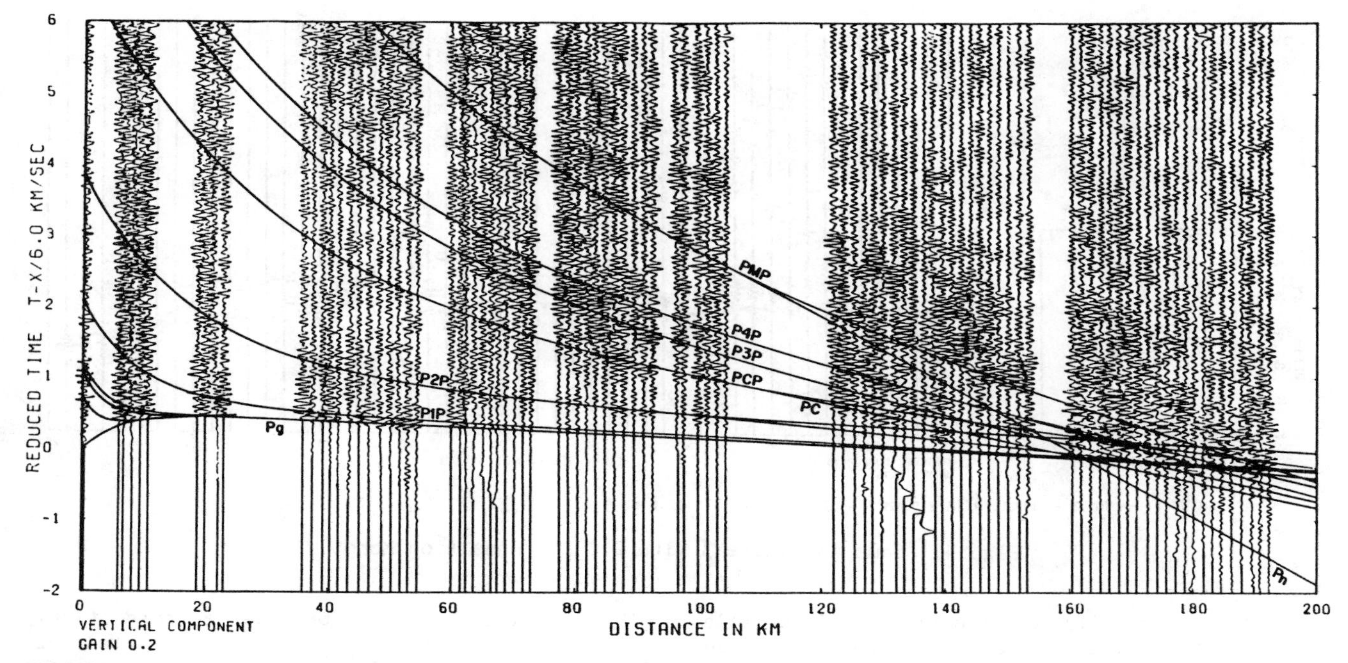

Fig. 2. Observed record section for SP 0 on the KOYNA II profile assembled from the digitized data of DSS records (trace interval ~ 1.6 km). Theoretical travel time curves, corresponding to the final velocity-depth model, are also shown for various phases correlated in the record section.

plotted with 400 m trace interval for phase correlations and the identification of various prograde and retrograde branches was rather unambiguous due to larger data density.

Analysis and Interpretation of the Record Sections

The record sections for SP 0 (Figure 2), SP 20W (Figure 8), SP 20-40 (Figure 9), SP

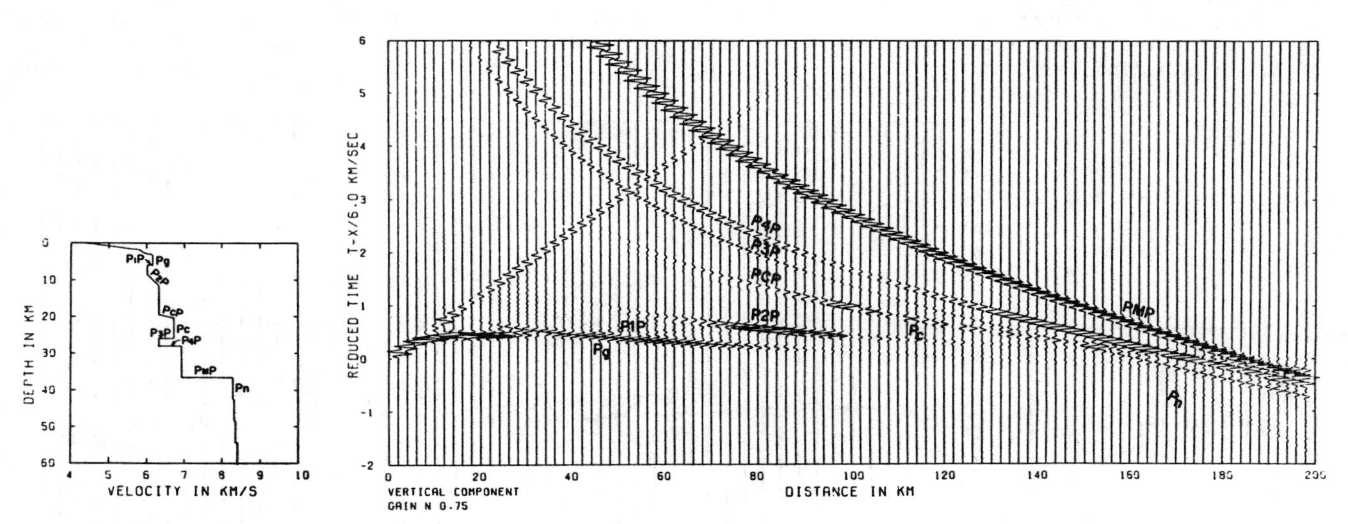

Fig. 3. Normalized record section of ray synthetic seismograms computed by using the velocity-depth model, with a first order Moho, shown for SP 0 on the KOYNA II profile. Various phases corresponding to the correlations in the observed record section are also labelled for comparison.

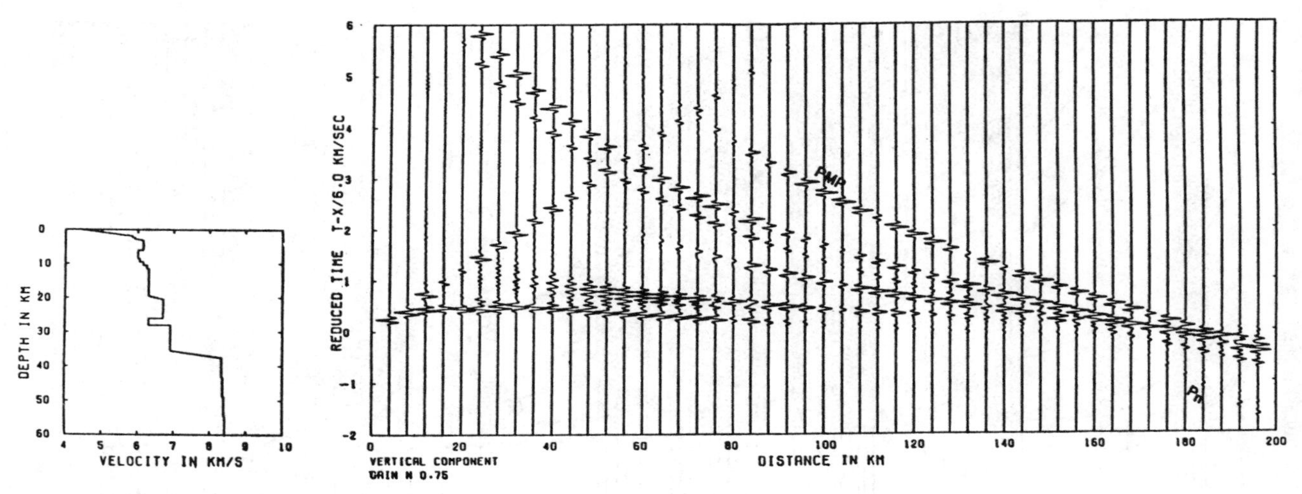

Fig. 4. Same as Figure 3 with 2.0 km transition Moho.

200-180 (Figure 11) on the KOYNA II profile and SP 10A (Figure 13) on the KOYNA I profile are shown in various Figures with the correlations made by us for all the prominent phases observed. It can be seen from these Figures that the DSS data presented in the form of record sections with relatively small trace interval offer the great advantage of reliable phase correlation especially for later arrivals. The correlations shown in Figure 2 for SP 0 record section are essentially the same as those interpreted by us in paper I. We use here the following nomenclature for various crustal phases interpreted:

Pg - refracted wave from the uppermost part of the crystalline basement,

P1P and P2P - reflected waves respectively from the top and bottom of the upper crustal LVL,

PcP - reflected wave from the midcrustal boundary,

Pc - refracted wave through the midcrustal boundary,

P3P and P4P - reflected waves respectively from the top and bottom of the lower crustal LVL,

PMP - reflected wave from the Moho boundary,

Pn - refracted wave from the uppermost mantle.

The travel time curves for various phases correlated, as shown in Figures 2, 8, 9, 11 and 13 correspond to the final velocity models inferred from these record sections. We have computed for each record section a large number of travel time models and every successful velocity model giving an acceptable travel

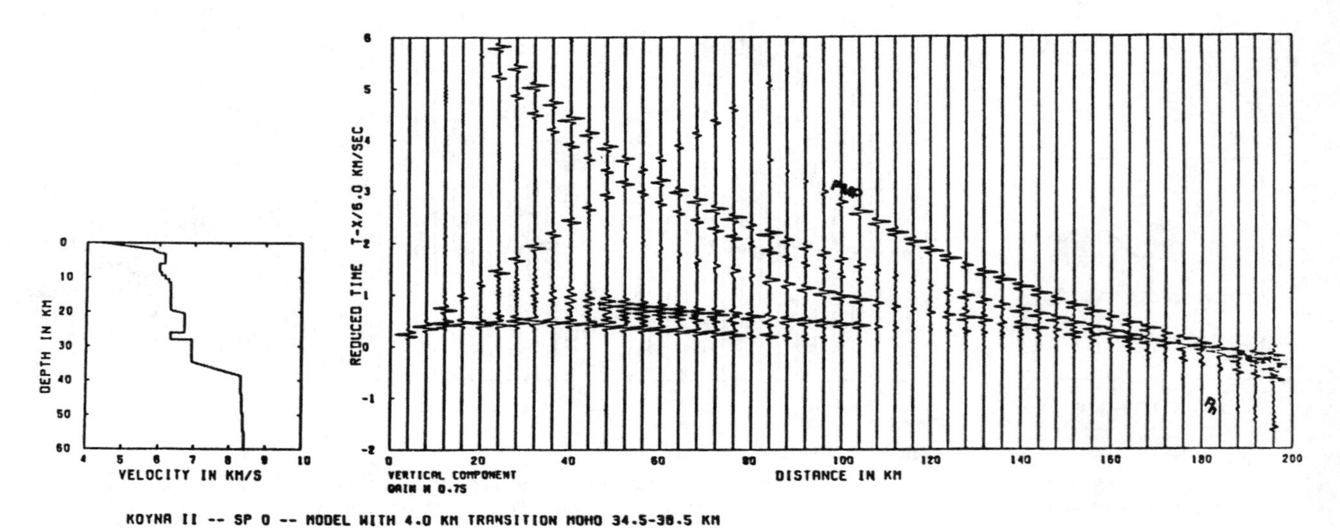

Fig. 5. Same as Figure 3 with 4.0 km transition Moho.

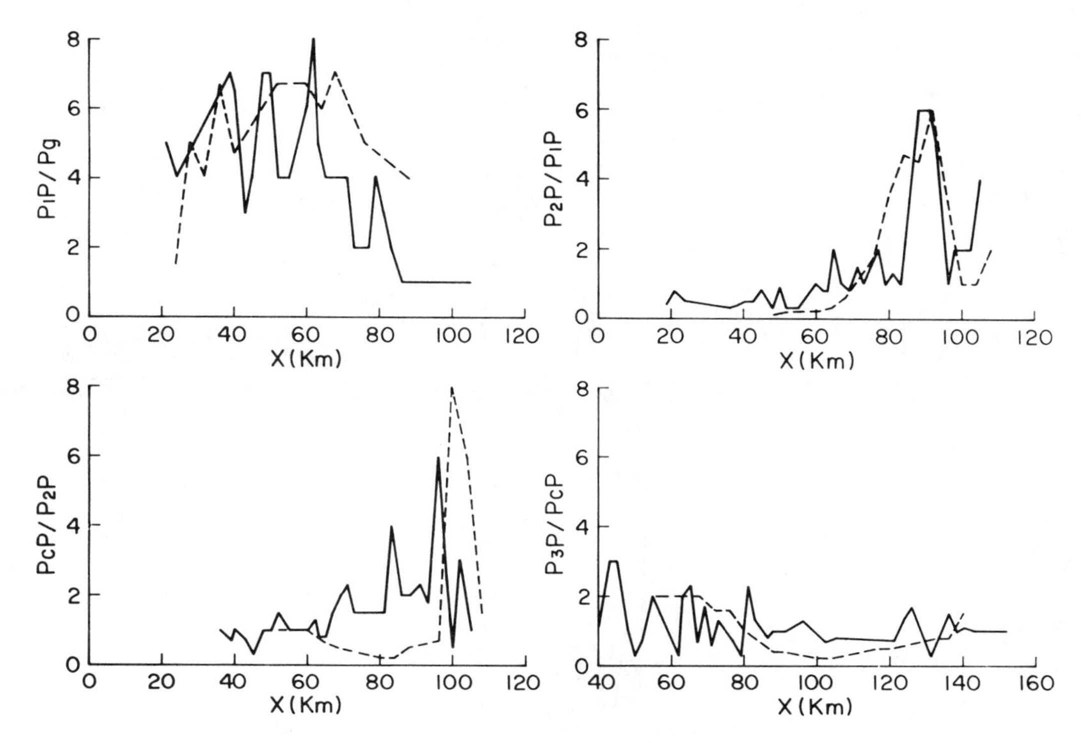

Fig. 6. Variation of the amplitude ratios with distance (X) for various combinations to the phases from the observed (_________) and the synthetic (__ __ __ __) record sections for SP 0.

times fit was tested for reproducing the relative amplitudes match by computing ray synthetic seismograms [Cerveny et al, 1977]. Thus the final velocity models are derived by successively refining both the travel times and relative amplitudes match with the observed record sections.

As can be seen from various record sections (Figures 2, 8, 9, 11 and 13), the first arrivals out to 20-25 km distance reveal continuous variation of apparent velocity of P waves from 4.4 to 6.0 km/sec due to the Deccan Traps cover. The first arrivals in the distance range from 20-25 km to 160-170 km reveal an apparent velocity of 6.0-6.2 km/sec and appear to be relatively very weak beyond 80 km distance. We have however used the original monitor records to fit the first arrivals data. We identify these first arrivals, out to 160-170 km distance, as the Pg phase (basement refraction) resulting from waves penetrating into the crystalline basement. Beyond 160-170 km distance the first arrivals are more prominent with a sharp increase of the apparent velocity to 8.25 km/sec as can be seen from the record sections for SP 0 and SP 10A on the two profiles. We identify these first arrivals, beyond 160-170 km distance, as the Pn phase resulting from waves penetrating into the uppermost mantle.

Upper Crustal Model

In the later arrivals a prominent wave group designated as P1P, shortly after the first arrivals, can be seen (with 180° phase shift) in the distance range from about 20 to 65 km in various record sections (Figures 2, 8, 9, 11 and 13). Another wave group, designated as P2P, can also be seen in the later arrivals in the distance range from about 60 to 110 km in these record sections. The prominent P1P phase indicates reflection from a velocity discontinuity in the upper crust. However, the extensive travel times and amplitudes modeling results indicate that it is almost impossible to fit any viable model with a positive velocity jump at the discontinuity corresponding to the P1P phase. Because, a very small increase of velocity at a supposed discontinuity although produces a travel times fit with the P1P phase, the corresponding synthetic amplitudes are comparatively much weaker than the observed P1P amplitudes. Further the first arrivals, computed for such a model with even a very small velocity jump, are also found to be much earlier, revealing higher apparent velocity, than the observations. Therefore, we considered the presence of a low velocity layer (LVL) in the upper crust which may explain both P1P

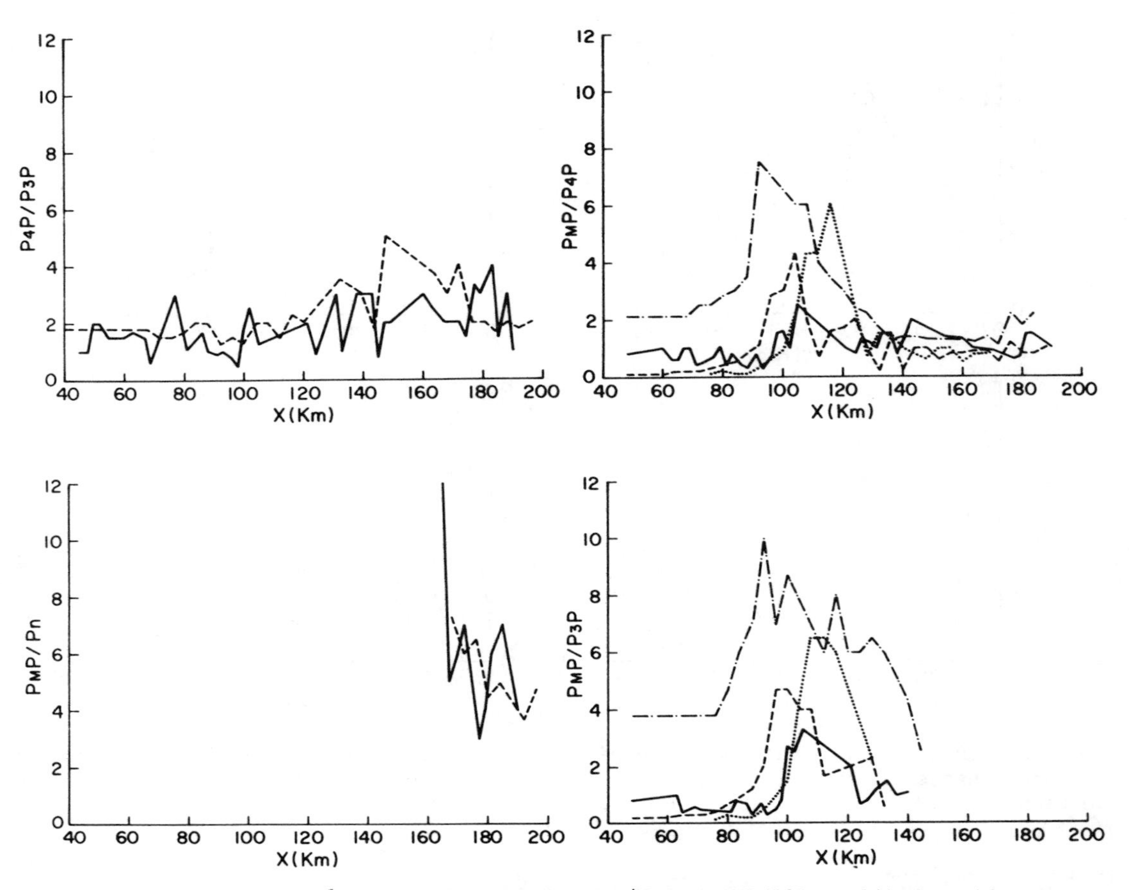

Fig. 7. Same as Figure 6. In the case of PMP/P4P and PMP/P3P amplitude ratios, __ __ __ __ corresponds to the model with 2.0 km transition Moho, ... corresponds to the model with 4.0 km transition Moho and __.__.__ corresponds to the model with a first order Moho.

and P2P phases. We could successfully model both P1P and P2P as the reflected phases respectively from the top and bottom of an upper crustal LVL. The velocity reduction of 0.2 km/sec occurs at 6.0 km depth where the top of this LVL produces a reflection corresponding to the P1P phase. The top of this upper crustal LVL which is inferred to be a sharp boundary is consistent with the observations of relatively strong amplitude P1P phase with 180° phase shift [Smith et al, 1975]. The relative amplitudes pattern of the P2P phase however requires the bottom of this upper crustal LVL to be transitional. A first order boundary at the bottom of the LVL would produce relatively strong subcritical reflections which are not evident on the observed record sections. The final models thus derived from various record sections with the aid of synthetic seismograms, indicate the presence of an upper crustal LVL with its top as a sharp boundary at 6.0 km depth, with 0.2 km/sec velocity reduction, and its bottom as a 3 km thick transitional boundary from 8.5 to 11.5 km, velocity increasing to 6.3 km/sec at 11.5 km depth. However this upper crustal LVL is found to be 1 km deeper on the KOYNA I profile, with its top at 7.0 km and the 3 km thick transitional bottom from 9.5 to 12.5 km depth. The velocity models derived from these record sections also consistently reveal a strong velocity gradient from ground surface to about 3 km depth with minor lateral variations due to the varying thickness of the Deccan Traps along the two profiles [Kaila et al, 1979]. P velocity at SP 0 (towards east) increases from 4.4 km/sec from ground surface to 5.8 km/sec at 1.8 km depth and 6.15 km/sec at 3.2 km depth, this model being also consistent at SP 20W (towards west) as shown in Figure 8. The velocity increases from 4.4 km/sec to 5.8 km/sec at 1.0 km depth at SP 20-40 (towards east) and to 5.95 km/sec at 1.0 km depth at

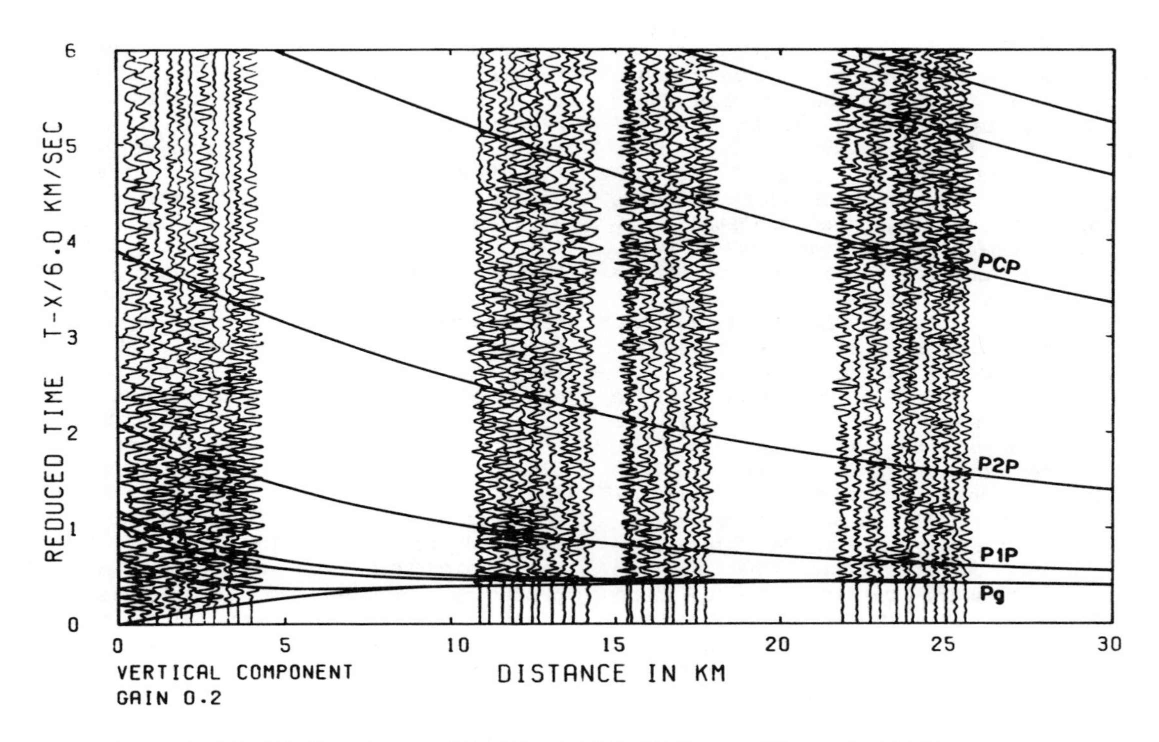

Fig. 8. Same as Figure 2 for SP 20W with trace interval ~ 400 m.

SP 200-180 (towards west). However at SP 10A (towards east) the velocity increases from 4.4 km/sec to 6.2 km/sec at 2.8 km depth.

Lower Crustal Model

We have correlated another set of wave groups in the later arrivals, designated as PcP and Pc, in the distance range from about 70 to 150 km as shown in Figures 2, 9, 11 and 13. The phase Pc appears to be tangential to the PcP phase at 100-110 km distance and indicates an apparent velocity of 6.7 km/sec. Both travel times and synthetic seismogram computations have been made to fit the PcP and Pc phases. The observed amplitudes pattern of the reflected phase PcP requires a 1 km thick transitional boundary from 19.5 to 20.5 km depth separating the upper crust from the lower crust. The computed wide angle reflections from this transitional boundary, centered at 20 km depth (at 21 km depth on the KOYNA I profile), match well with the correlations for the PcP phase. The velocity increases to 6.7 km/sec below this boundary. There are also another set of wave groups, correlated and designated as P3P and P4P at distances beyond 70-80 km with a relatively low measure of lateral continuity on these record sections. These two phases are however found to correspond to reflections from the top at 26 km and bottom at 28 km (at 29 km on the KOYNA I profile) of a lower crustal LVL by travel times and synthetic seismogram computations. The velocity in this lower crustal LVL is inferred as 6.3 km/sec. Below this LVL, the velocity increases to 6.9 km/sec. Models considered without this LVL did not give acceptable fits to the P3P and P4P phases. Attempts to fit these two phases by positive velocity jumps, however small they may be, increased the average crustal velocity beyond the acceptable range while matching the PMP phase, besides shifting the PMP critical point to larger distances than that observed at ~100 km distance. On the other hand, the wide angle reflection computed from the bottom of this lower crustal LVL matches well with the relatively strong P4P phase beyond 140 km distance. Thus we prefer to interpret P3P and P4P phases as reflections from the top and bottom of the lower crustal LVL.

The Moho Boundary

The most prominent wave group correlated in the later arrivals on all the record sections (Figures 2, 9, 11 and 13) is the phase designated as PMP, which is identified as the wide angle reflection from the Moho boundary by modeling. This phase is almost continuously observed with relatively large amplitudes at distances greater than about 100 km on all the record sections on the KOYNA II profile whereas it is relatively strong beyond about 80 km distance on the KOYNA I profile as can be seen from the record sections for SP 0, SP 20-40, SP 200-180 and SP 10A. However at smaller distances in the subcritical range, the amplitudes of the PMP phase appear to be relatively weak. By travel times modeling and considering a first order Moho at 36.5 km depth, we obtained a very satisfactory travel times fit to the observed PMP times. However the synthetic seismogram section computed by using this model with a first order Moho at 36.5 km depth reveals relatively very strong PMP arrivals in the subcritical distances also as can be seen from the synthetic record section in Figure 3. This is certainly not consistent with the amplitude observations of the PMP phase. Therefore we tried to suppress the subcritical PMP amplitudes in the synthetics by considering the Moho as a transitional boundary rather than a sharp discontinuity. There is also another good reason for us to consider the Moho as a transition zone in this region, due to the absence of otherwise observable and possibly strong PS converted waves in the record sections at about 5 sec after the PMP arrival times where one would expect such waves. While a first order discontinuity could also explain the observed travel times of the PMP, but at the same time would produce such conversions with considerable amplitudes whereas a transition zone suppresses them. Therefore a transition zone at the Moho is the most likely explanation of the observed absence of reflected PS waves from the Moho, which also appears to be a general feature of crustal record sections [Fuchs, 1975]. Thus the range of possible models that satisfy the data certainly excludes the possibility of a sharp discontinuity at the crust-mantle boundary in the Koyna region and indicates the Moho as a transition zone. We have therefore considered a transitional boundary at the Moho and progressively increased its thickness in order to obtain an acceptable fit to the observations of the PMP phase while modeling. We considered various models with 1.0, 2.0, 3.0 and 4.0 km thick transitional Moho boundary and in each case after a satisfactory travel times fit has been obtained, we computed the synthetic seismogram section and compared with the observed record section for SP 0 on the KOYNA II profile. Figures 4 and 5 show these synthetic seismogram sections for SP 0 velocity models respectively with 2.0 and 4.0 km thick transitional Moho boundary. It was found in paper I that Moho transition thickness of 1.0 km though suppresses the subcritical PMP amplitudes considerably, yet its relative amplitudes are very prominent from 84 km onwards, whereas in the observed record section for SP 0 the PMP amplitudes are prominent only from about 100 km onwards. However further increase of the transition thickness to 2.0, 3.0 and 4.0 km shifts the prominent relative amplitudes of the PMP phase respectively to 96, 100 and 104 km. Therefore we infer that the Moho in this region is a transition zone of at least 2.0 km thickness (from 35.5 to 37.5 km depth). Similarly we have also computed the synthetic seismogram sections with 2.0 km thick transitional Moho corresponding to the velocity models for SP 20-40 and SP 200-180 record sections and shown in Figures 10 and 12 respectively. It can be seen from these Figures that these models for SP 20-40 and SP 200-180 also match quite well with the observed record sections. Figure 14 shows the synthetic seismogram section computed for 1.0 km thick transitional Moho corresponding to the velocity model for SP 10A record section, on the KOYNA I profile. Comparison of the synthetic sections with the observed record section for SP 10A revealed that Moho transition thickness of 1.0 km, producing prominent PMP amplitudes from 80 km onwards, may be more acceptable than the model with Moho transition thickness of 2.0 km on this profile. We thus infer a 1.0 km thick transitional Moho (from 37.0 to 38.0 km depth) in the region of the KOYNA I profile and it is also quite evident that the Moho is slightly deeper in this region than towards north in the region of the KOYNA II profile.

The Pn velocity of 8.25 km/sec observed on the SP 0 and SP 10A record sections has been used in all the models discussed above. However we further tried some models by varying the velocity gradient in the upper mantle to produce a comparable PMP/Pn amplitudes ratio in the synthetics as that in these observed record sections. By this modeling procedure we inferred a P velocity gradient of 0.004 sec^{-1} in the uppermost mantle to a depth of about 60 km, producing a satisfactory fit to the observed PMP/Pn amplitudes ratio. It can be seen from all the record sections presented here that there is reasonably good agreement between the synthetics and the observations from various shot points considered in the present study. To demonstrate this point more convincingly we have presented in Figures 6 and 7, plots showing variation of the amplitude ratios with distance for various combinations of the phases both from the observed as well as the synthetic record sections for the SP 0 data set. It can be found from these plots that in spite of some random uncertainties, they reveal acceptable match of model outputs with the observations. Thus the models derived in the present study are internally consistent with various observations. However the frequency dependent effects may have some influence on the derived models (e.g. Cerveny's effect) which could be better resolved by application of the exact methods (e.g. reflectivity method)

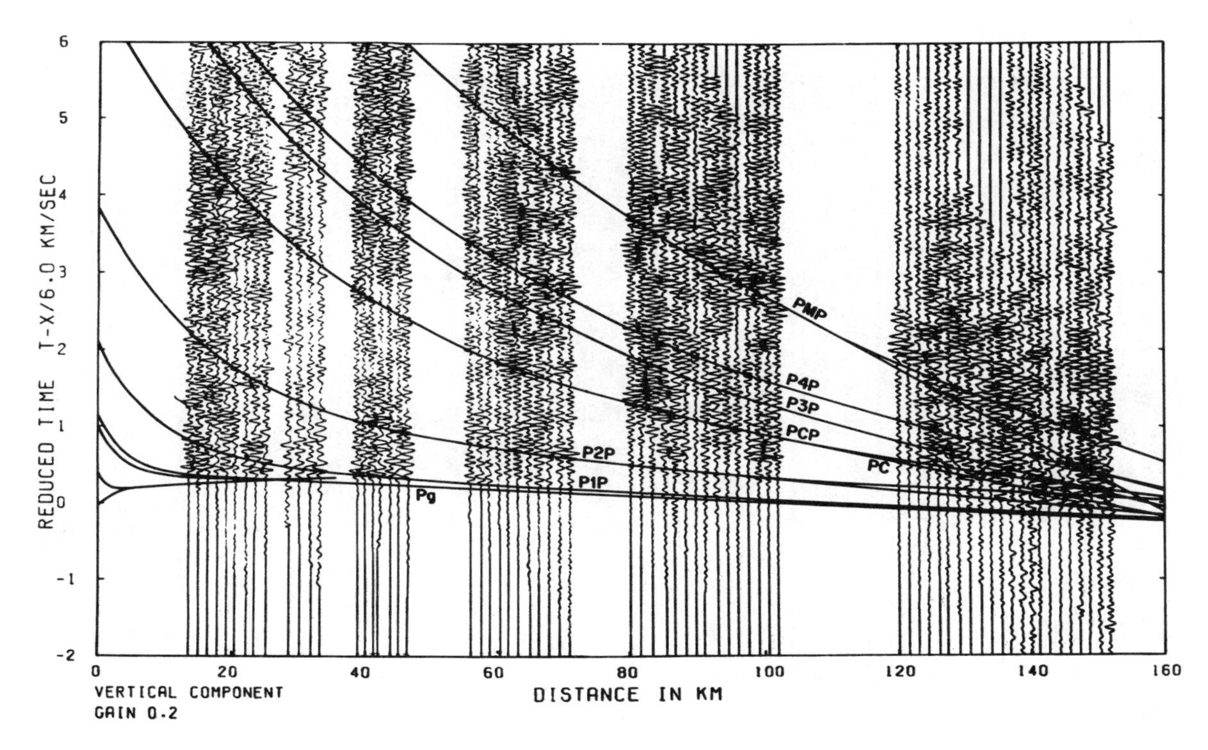

Fig. 9. Same as Figure 2 for SP 20-40 with trace interval ~ 1.6 km.

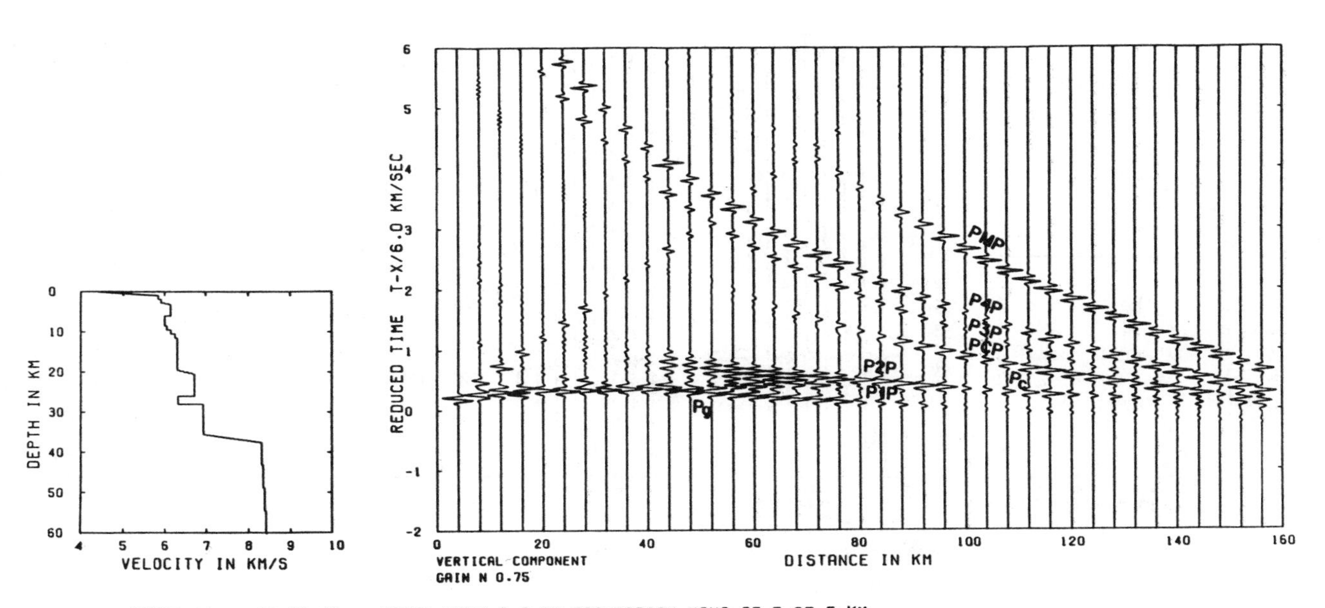

Fig. 10. Same as Figure 3 for SP 20-40 with 2.0 km transition Moho.

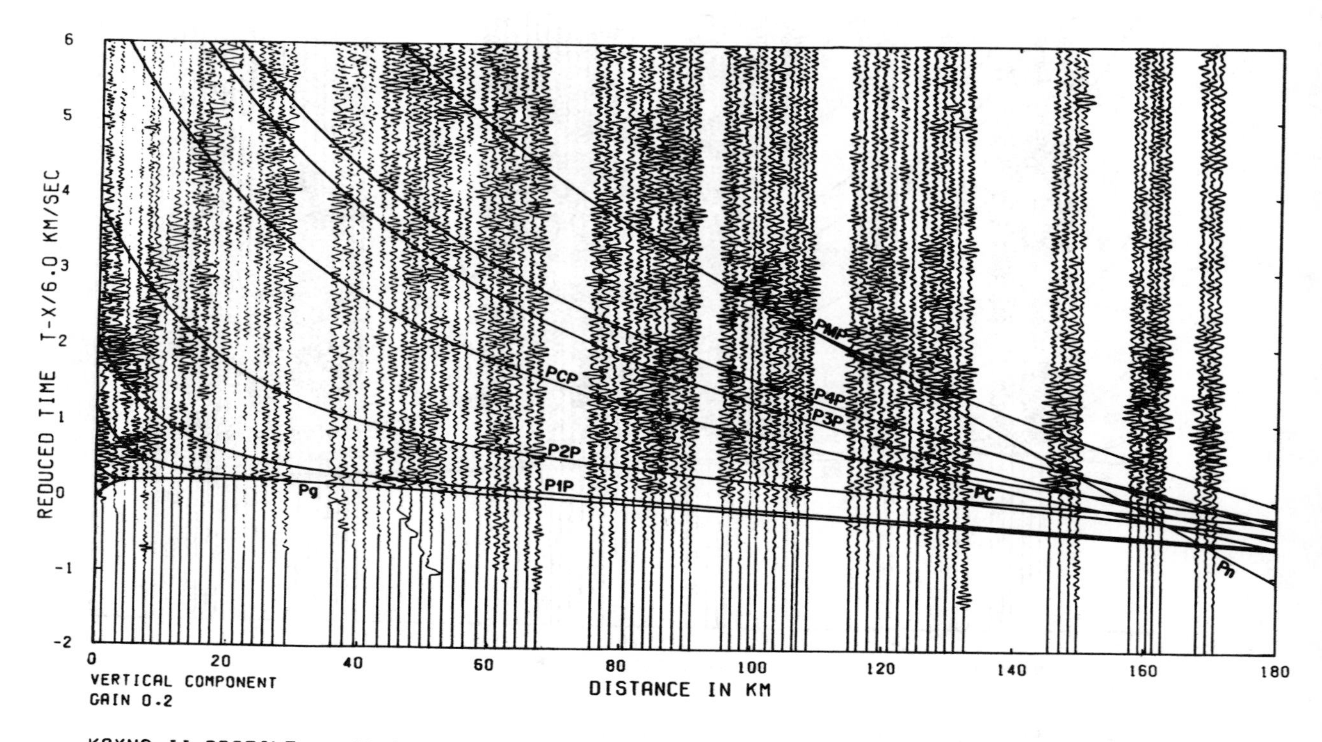

Fig. 11. Same as Figure 2 for SP 200-180 with trace interval ~ 1.6 km.

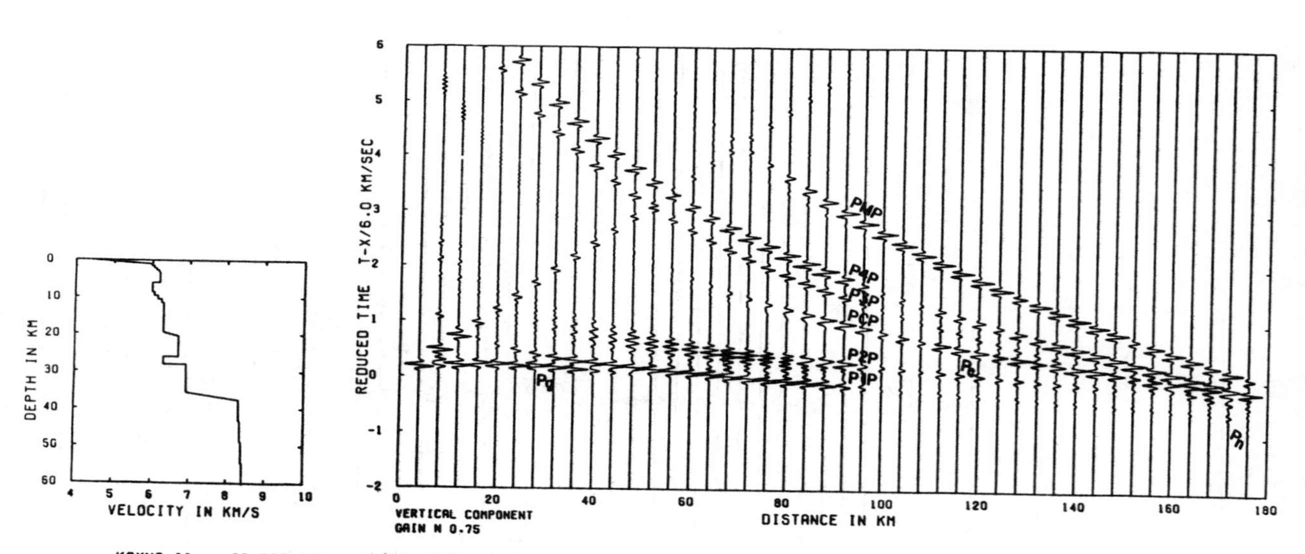

Fig. 12. Same as Figure 3 for SP 200-180 with 2.0 km transition Moho.

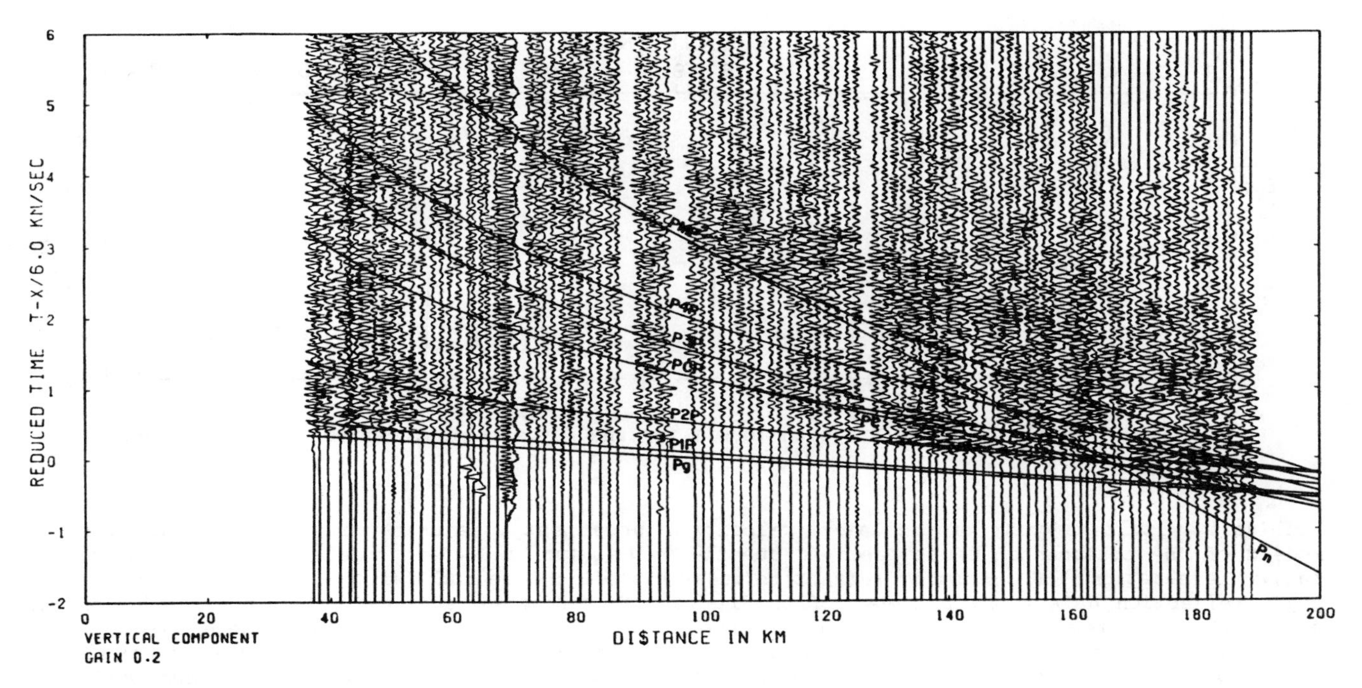

Fig. 13. Same as Figure 2 for SP 10A on the KOYNA I profile with trace interval ~ 1.6 km.

rather than the ray theoretical methods as applied in the present study.

Discussion

The final models of the crustal P velocity functions in the Koyna region as derived by modeling of the DSS record sections for SP 10A, SP 0, SP 20-40 and SP 200-180 are shown in Figure 15 for comparison. These velocity models reveal almost similar structure with minor lateral variations in this region. The prominent features of these velocity models are quite

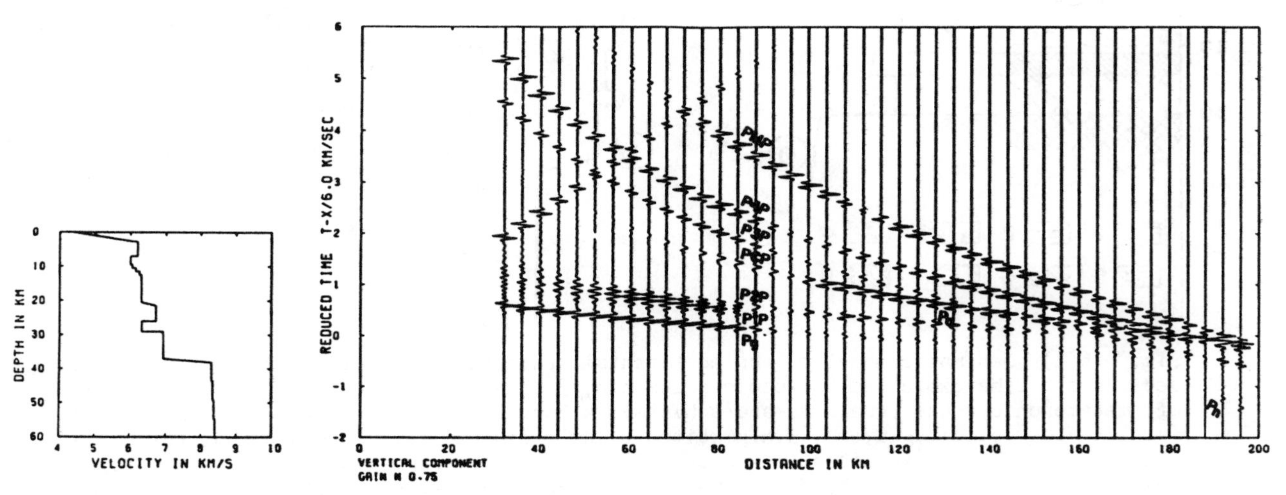

Fig. 14. Same as Figure 3 for SP 10A on the KOYNA I profile with 1.0 km transition Moho.

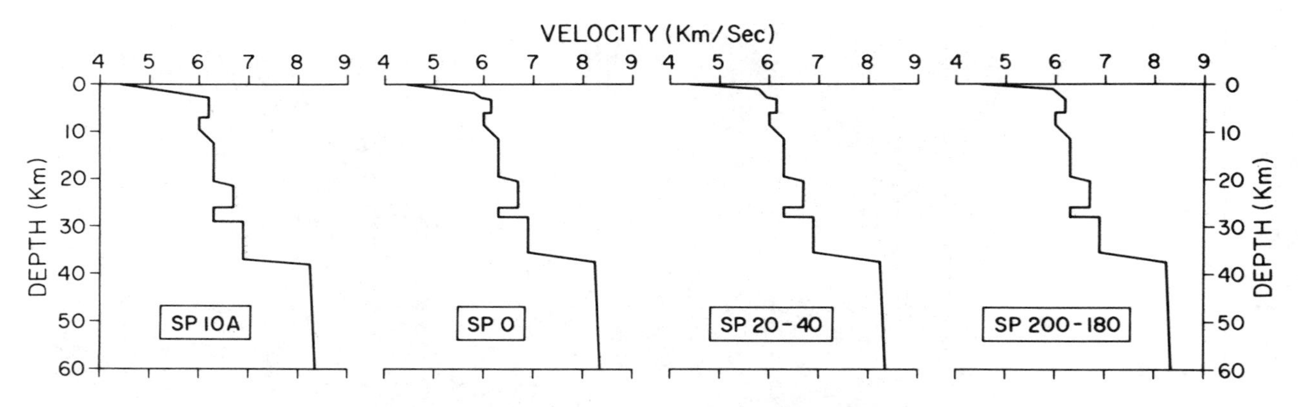

Fig. 15. Crustal P velocity-depth models for the Koyna region, derived in the present study.

consistent with a majority of the latest velocity models of the continental crust based on quantitative evaluation of seismic refraction/wide angle reflection data along a number of profiles over different regions of the earth. Mueller [1977] advanced a new model of the continental crust and the schematic model presented by him is shown in Figure 16(a) along with the model derived by us for the Koyna region from SP 0 record section shown in Figure 16(b). The similarity of these models is very striking and the prominent features are quite consistent in the two velocity models. A large number of studies dealing with detailed interpretation of refracted and reflected phases in seismic refraction surveys elsewhere, have also reported evidence for an LVL in the upper crust in the depth range of 5 to 15 km, which had originally been proposed by Mueller and Landisman [1966]. Mueller [1970] and Landisman et al [1971] pointed out that a semicontinuous laccolithic zone of granitic intrusions could well be the most

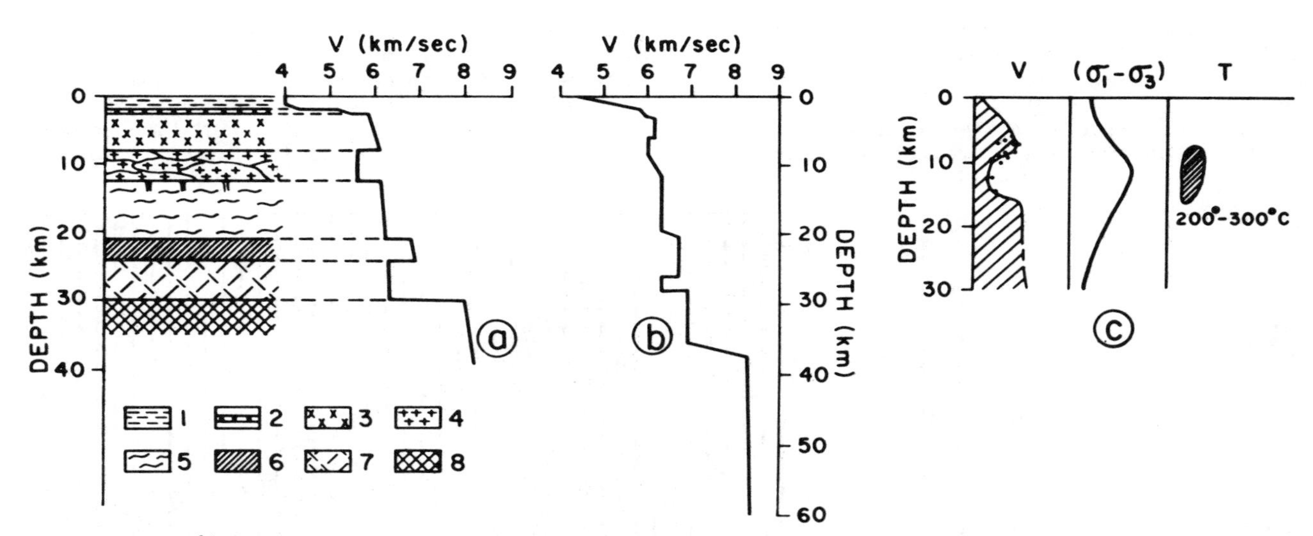

Fig. 16(a). Schematic model of the continental crust (after Mueller, 1977): 1.Cenozoic sediments, 2.Mesozoic and Paleozoic sediments, 3.Upper Crystalline basement consisting of metamorphic rocks (gneisses and schists), 4.Laccolithic zone of granitic intrusions (sialic low velocity layer), 5.Migmatites (middle crustal layer), 6.Amphibolites (high velocity tooth), 7.Granulites (lower crustal layer), 8.Ultramafics (uppermost mantle). (b): Crustal P velocity-depth model for the Koyna region, derived in the present study, revealing similar features. (c): Crustal features in the Vogtland area of the southern German Democratic Republic (after Bankwitz et al, 1985). V - P velocity model (after Knothe, 1972), dots - depth distribution of earthquakes in the area, ($\sigma 1$ - $\sigma 3$) - depth distribution of strength, T - depth distribution of temperature.

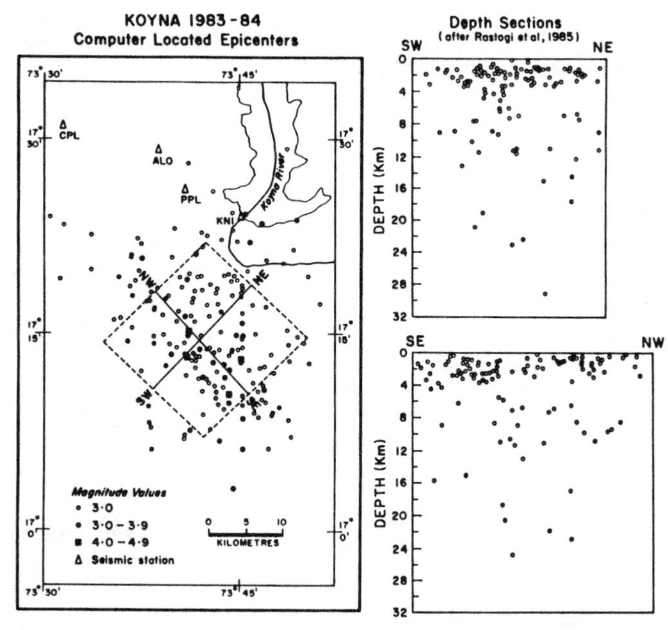

Fig. 17. Spatial and depth distribution of earthquakes in the Koyna region (after Rastogi et al, 1985).

probable explanation for the sialic LVL in the upper crust. Smith et al [1975] proposed that a plausible explanation of the sialic LVL is the effect of a high temperature gradient on the velocities of upper crustal rocks. Mueller [1977] however maintained that this upper crustal LVL could be associated with a zone of lower densities and a higher attenuation within the granitic intrusions into the surrounding basement rocks. He also pointed out that it is the water content and the pore pressure in these granites which play the major role in causing the upper crustal LVL rather than the temperature effect. In this context it may be mentioned here that hot springs occur in the Konkan foothills area of the Koyna region, as can be seen from Figure 1, which are indicative of anomalously high temperatures in the upper crust in this region.

Smith et al [1975] pointed out that there seems to exist a correlation of the sialic LVL with a decrease in the frequency of occurrence of earthquakes in the corresponding depth range which is also a tectonically 'soft' layer with decreased rigidity and high Poisson's ratio [Braile et al, 1974]. Mueller et al [1973] as well as Mueller and Rybach [1974] have also shown that in the Rhine graben area and its neighbourhood the majority of hypocenters are concentrated close to the top of the sialic LVL. More recently, Bormann et al [1985] and Bankwitz et al [1985] reported that in the Vogtland earthquake region of the southern German Democratic Republic, earthquakes occur in the depth range between 3 to 16 km with a maximum frequency around 7 to 9 km corresponding to the uppermost part of the LVL in this region. Similarly, recent analysis of the Koyna earthquakes of 1983-84 by Rastogi et al [1985] revealed that the depths are shallow, mostly within 4 km, and that there are no hypocenters deeper than 25 km in this region as shown in Figure 17. In the Koyna region, Langston [1976, 1981] as well as Langston and Franco-Spera [1985] have also found very shallow source depths, well constrained by wave form inversion studies, for the main shock of December 10, 1967 (4.5 ± 1.5 km), for its foreshock of September 13, 1967 (5 km) and also for the aftershock of December 12, 1967 (3.5-4.0 km). All these studies in the Koyna region thus suggest that the major seismic activity in this region is confined to only within 4-5 km depth. The crustal velocity models derived in the present study reveal an upper crustal LVL from 6.0 to 11.5 km depth and also a lower crustal LVL from 26 to 28 km depth consistent with the above findings on the possible relation between the seismic activity and the crustal velocity models.

According to Chen and Molnar [1983], the distribution of focal depths discussed above mainly reflects the dependencies of strength of geological materials on temperature and mineralogy. Bankwitz et al [1985] also concluded that the crustal structure has an important influence on the earthquakes occurrence. In Figure 16(c) are also shown schematically the depth distribution of earthquakes, (σ_1 - σ_3) and T depicting depth distribution of strength and temperature respectively. The earthquakes seem to be occurring in the upper rigid layer of the crust with relatively more strength and temperatures within 200°-300°C. At larger temperatures, the strength rapidly decreases and the earthquakes seem to be very scanty or absent. These observations are also consistent with the proposed upper crustal LVL in the Koyna region where the major seismic activity is confined to only within 4-5 km depth and there is absence of appreciable activity at greater depths. The velocity models in the lower crust with an LVL and constant velocity layers, derived in the present study also seem to be consistent with the essentially aseismic nature of the lower crust in the Koyna region. According to Chen and Molnar [1983], the aseismic lower crust corresponds to a zone of low strength where ductile deformation predominates. The temperature at the source region is likely to be an important factor determining whether deformation occurs seismically or not. Chen and Molnar [1983] concluded that the limiting temperatures are about 250°-450°C for the crustal materials. At higher temperatures, the strength decreases rapidly with increasing temperature and becomes so low that no stress accumulation is possible for earthquakes occurrence. Therefore, it can be inferred that temperatures higher than these limiting temperatures are probably prevailing in the lower crust which is essentially aseismic in the Koyna region and those higher temperatures are also in turn responsible for the proposed LVL and the constant velocity layers in the lower crust in this region. Thus we conclude that the crustal P velocity models derived in the present study are viable and are consistent with the observed seismic activity and the material properties and their dependence on the prevailing temperatures. The two low velocity layers within the crust in the Koyna region, as inferred from the present study, may suggest rheological stratification of this part of the lithosphere and they can be identified with levels of increased ductility at those depths.

Conclusions

Travel times and relative amplitudes modeling of a series of crustal seismic record sections, with the aid of synthetic seismograms, yielded internally consistent velocity models with minor lateral variations in the Koyna region to about 60 km depth. The prominent features of these velocity models include, (i) a strong velocity gradient at shallow depths in the Deccan Traps and the upper crystalline basement, (ii) an upper crustal LVL from 6.0 to 11.5 km depth with 0.2 km/sec velocity reduction within the LVL and a broad transition zone at its bottom, (iii) a 1 km thick transitional boundary centered at 20-21 km depth separating the upper crust from the lower crust, (iv) a lower crustal LVL from 26 to 28 km depth with 0.4 km/sec velocity reduction within the LVL, (v) constant velocity layers with velocities of 6.3 km/sec from 11.5 to 19.5 km, 6.7 km/sec from 20.5 to 26 km and 6.9 km/sec from 28 to 35.5 km depth, and (vi) at least 2 km thick transitional Moho from 35.5 to 37.5 km depth on the KOYNA II and 1 km thick transitional Moho from 37.0 to 38.0 km depth on the KOYNA I DSS profiles, and a Pn velocity of 8.25 km/sec increasing with a gradient of 0.004 sec^{-1} in the upper mantle to about 60 km depth.

The upper crustal LVL inferred from the present modeling study, from 6.0 to 11.5 km depth, is consistent with the observed seismic activity concentration at 4-5 km depth and an appreciable reduction of seismic activity at greater depths in the Koyna region. The velocity structure in the lower crust, with an LVL from 26.0 to 28.0 km depth and constant velocity layers, is also consistent with the essentially aseismic nature of the lower crust in this region. These two low velocity layers within the crust may suggest rheological stratification of this part of the lithosphere with levels of increased ductility at those depths in the Koyna region.

Acknowledgements. We are grateful to Prof. V.K.Gaur, Director, National Geophysical Research Institute, Hyderabad, for his continuing support and encouragement for this research activity and kind permission to publish this paper.

References

Bankwitz, P., E. Bankwitz, and Grunthal, Parameters of the Earth's crust and causes for the occurrence of earthquakes in the southern part of the G.D.R., Gerlands Beitr. Geophysik, 94, 294-298, 1985.

Bormann, P., A. Schulze, E. Apitz, P. Bankwitz, and K-D. Klinge, Methods and results of interpretation of deep seismic investigation in the G.D.R., Gerlands Beitr. Geophysik, 94, 259-268, 1985.

Braile, L.W., R.B. Smith, G.R. Keller, R.M.Welch, and R.P. Meyer, Crustal structure across the Wasatch front from detailed seismic refraction studies, J. Geophys. Res., 79, 2669-2677, 1974.

Cerveny, V., I.A. Molotkov, and I. Psencik, Ray method in seismology, Univerzita Karlova, Praha, 214 pp., 1977.

Chen, W.-P., and P. Molnar, Focal depths of intracontinental and intraplate earthquakes and their implications for the thermal and mechanical properties of the lithosphere, J. Geophys. Res., 88, 4183-4214, 1983.

Fuchs, K., Synthetic seismograms of PS-reflections from transition zones computed with the reflectivity method, J. Geophys., 43, 445-462, 1975.

Kaila, K.L., P.R. Reddy, P.R.K. Murty, and K.M. Tripathi, Deep seismic sounding studies along KOYNA I and KOYNA II profiles, Deccan Trap covered area, Maharashtra state, NGRI Tech. Report, 64 pp., 1979.

Kaila, K.L., P.R.K. Murty, V.K. Rao, and G.E. Kharetchko, Crustal structure from deep seismic soundings along the KOYNA II (Kelsi-Loni) profile in the Deccan Trap area, India, Tectonophysics, 73, 365-384, 1981a.

Kaila, K.L., P.R. Reddy, M.M. Dixit, and M.A. Lazarenko, Deep crustal structure at Koyna, Maharashtra, indicated by deep seismic soundings, J. Geol. Soc. India, 22, 1-16, 1981b.

Knothe, D.H., Deutsche Demokratische Republik, NKGG Geodat. Geophys. Veroff. Reihe III, 27, 59-68, 1972.

Krishna, V.G., and K.L.Kaila, Digitization of analog DSS records and their reinterpretation with the aid of synthetic seismograms: Application to the Koyna data, in Deep Seismic Soundings and Crustal Tectonics, Edited by K.L. Kaila and H.C. Tewari, AEG Publication, 99-119, 1986.

Landisman, M., S. Mueller, and B.J. Mitchell, Review of evidence for velocity inversions in the continental crust, in The Structure and Physical Properties of the Earth's Crust, Geophys. Monogr. Ser., Edited by J.G. Heacock, AGU Monograph, 14, 11-34, 1971.

Langston, C.A., A body wave inversion of the Koyna, India, Earthquake of December 10, 1967, and some implications for body wave focal mechanisms, J. Geophys. Res., 81, 2517-2529, 1976.

Langston, C.A., Source inversion of seismic waveforms: The Koyna, India, Earthquakes of 13 September 1967, Bull. Seism. Soc. Am., 71, 1-24, 1981.

Langston, C.A., and M. Franco-Spera, Modeling of the Koyna, India, Aftershock of 12 December 1967, Bull. Seism. Soc. Am., 75, 651-660, 1985.

Mueller, S., Geophysical aspects of graben formation in continental rift systems, in Graben Problems, Edited by H. Illies and S. Mueller, Schweizerbart, Stuttgart, 27-37, 1970.

Mueller, S., A new model of the continental crust, in The Earth's Crust, AGU Monograph, 20, 289-317, 1977.

Mueller, S., and M. Landisman, Seismic studies of the Earth's crust in continents, I, Evidence for a low-velocity zone in the upper part of the lithosphere, Geophys. J. R. astr. Soc., 10, 525-538, 1966.

Mueller, S., E. Peterschmitt, K. Fuchs, D. Emter, and J. Ansorge, Crustal structure of the Rhinegraben area, Tectonophysics, 20, 381-391, 1973.

Mueller, S., and L. Rybach, Crustal dynamics in the central part of the Rhinegraben, in Approaches to Taphrogenesis, Edited by H. Illies and K. Fuchs, Schweizerbart, Stuttgart, 379-388, 1974.

Rastogi, B.K., C.S.P. Sarma, R.K. Chadha, and N. Kumar, Koyna earthquakes of 1983 and 1984, NGRI Technical Report, 1985.

Smith, R.B., L.W. Braile, and G.R. Keller, Upper crustal low velocity layers: Possible effect of high temperature over a mantle upwarp at the Basin Range-Colorado Plateau transition, Earth & Planet. Sci. Lett., 28, 197-204, 1975.

STUDY OF THE LOWER CRUST AND UPPER MANTLE USING OCEAN BOTTOM SEISMOGRAPHS

Yuri P. Neprochnov

Shirshov Institute of Oceanology
USSR Academy of Sciences, Moscow 117218

Abstract. The bulk of seismic data obtained by the Shirshov Institute of Oceanology during ocean bottom seismograph experiments, as well as other published determinations of seismic structure of the lower oceanic crust and upper mantle are examined. A review of these data indicates: (1) the presence of layer 3B (velocity 7.4 km/s) under the Shatsky and Hess Rises, whereas a small quantity of sites (10%) in the Pacific Ocean basin shows layer 3B; (2) the absence of layer 3A in the studied area of the Brazil basin; (3) the lateral variation of the crustal structure in the central Indian Ocean basin correlated to the intraplate deformation. Statistical analyses of results from over 650 sites show a smaller layer 3 thickness in the Indian Ocean basin (3.8 km) than in the basins of the Pacific Ocean (4.7 km). An essential age dependence, revealing a cyclic pattern, has been defined and statistically validated for the seismic parameters of the crust in the Eastern Pacific Ocean basin. The variations in the crustal structure aré in good agreement with the global tectonic cycles, particularly with variations of sea-floor spreading. The petrology of seismic layers of the crust sampled by dredging in fracture zones is also presented.

Introduction

A number of publications on the structure of the lower oceanic crust (Maynard, 1970; Sutton et al., 1971; Christensen, Salisbury, 1975; Kosminskaya, Kapustian, 1976; Spudish, Orcutt, 1980) has summarized results of the refraction experiments. In this paper the new data of the deep seismic sounding experiments obtained in the expeditions of the Shirshov Institute of Oceanology of the USSR Academy of Sciences are presented and the published seismic refraction data on the crust of the basins of the Pacific, Atlantic and Indian oceans are examined. The seismic parameters of the lower crust and the upper mantle are the main objects of the consideration.

Data Acquisition Method

An effective method of the deep seismic sounding (DSS) by ocean bottom seismographs (OBS) and big airguns providing for the automatic obtaining and processing of data has been developed and applied in expeditions of the Shirshov Institute of Oceanology (Neprochnov et al., 1978). The confidence distance for recording deep refracted and reflected waves from airguns of 30 L is usually 50 to 80 km, which is sufficient for the study of the oceanic crust and transition between the crust and the upper mantle. The airgun shot at intervals of 150-200 m along the profile, which phase correlation of deep seismic waves possible. The profiles position was selected after geophysical survey of the area of operations. As far as possible, the profiles were made along the strike of sea-floor features. Arrays of 3 to 6 OBS's were used to determine the spatial variation of the oceanic crust.

The airgun was operated by the precision on-board crystal clock. The OBS's crystal clocks were rated against the on-board clock before and after deployment which ensured synchronization of shooting and seismic data recording. OBS's magnetic records were automatically processed on ship's computers to plot synchronous record sections. An example of the record section is shown in Figure 1.

The record sections were used for wave correlation and for the apparent velocity calculations. The travel time curves of the main refracted and reflected waves were plotted for each OBS. Afterwards reversed systems of travel time curves were plotted, which were used for the reconstruction of 2-dimension crustal sections in accordance with conventional seismic inversion

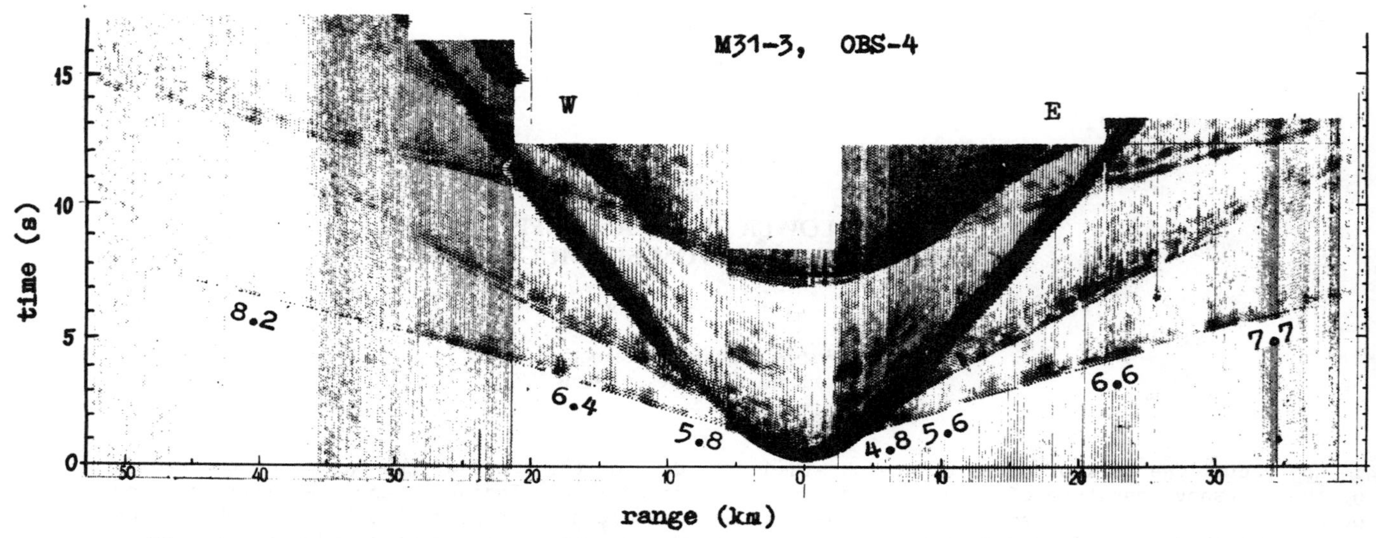

Fig. 1. Record section obtained by using OBS and big airgun in the Cape basin of the Atlantic Ocean (M31-3).

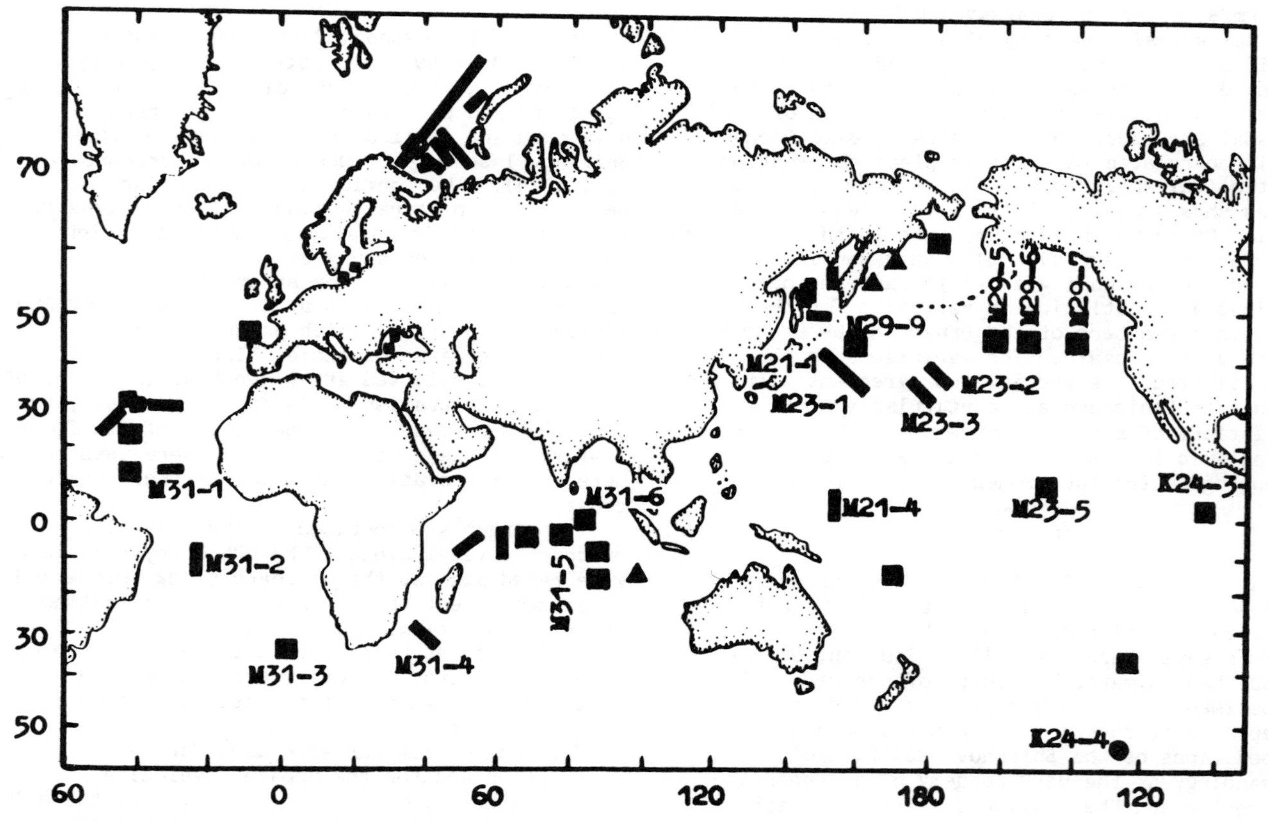

Fig. 2. World map showing locations of DSS experiments made in expeditions of the Shirsov Institute of Oceanology by using OBS's and big airguns. Single OBS (triangles), several OBS's along one profile (rectangles), system of profiles of different directions (square) and geotraverses with a number of OBS's (long solid signs) are shown. Solid circle shows the dredging area in the Eltanin F. Z. described in the last section of the paper.

TABLE 1. Field Acquisition Parameters for Deep Seismic Sounding Experiments Presented in Text and Figures 3 to 6

Experiment	Year Conducted	Geographic Location	Number of OBS's	Number of profiles	Type of profiles	References
K24-3	1977	Hess Basin, East Pacific Rise	3	three	reversed	Neprochnov et al. (1980)
M21-1	1978	North-West Basin	4	one	reversed	Neprochnov et al. (1984a)
M21-4	1978	Caroline Basin	2	one	reversed	Neprochnov et al. (1984b)
M23-1	1979	Shatsky Rise	2	one	reversed	Neprochnov et al. (1984a)
M23-2	1979	Emperor F.Z.	4	two	reversed	Neprochnov et al. (1984a)
M23-3	1979	Hess Rise	5	one	reversed	Neprochnov et al. (1984a)
M23-5	1979	North-East Basin	2	one	reversed	Neprochnov and Sedov (1984
M29-5	1982	North-East Basin	2	two	reversed	Neprochnov (1983)
M29-6	1982	North-East Basin	4	two	reversed	Neprochnov (1983)
M29-7	1982	North-East Basin	3	two	reversed	Neprochnov (1983)
M29-9	1982	North-West Basin	5	two	reversed	Neprochnov (1983)
M31-1	1983	Cape Verde Basin	5	one	reversed	Neprochnov et al. (1986)
M31-2	1983	Brazil Basin	3	one	reversed	Neprochnov et al. (1986)
M31-3	1984	Cape Basin	5	two	reversed	Neprochnov et al. (1986)
M31-4	1984	Mozambic Basin	3	one	reversed	Neprochnov et al. (1986)
M31-5	1984	Central Basin	7	three	reversed	Neprochnov et al. (1986)
M31-6	1984	Central Basin	5	two	reversed	Neprochnov et al. (1986)

technique. The amplitude studies were applied for the next stage of interpretation, in particular for determination of velocity gradients in the seismic layers.

The method of automated DSS was used with success during the past ten years to study the oceanic crust of various tectonic structures in the ocean. Figure 2 shows the location of deep seismic sounding sites made in expeditions of the Shirshov Institute of Oceanology by using OBS's and big airguns. About 380 deployments of OBS's were made and more than 120 refraction lines were shot of total length of about 12,000 kilometers. The acquisition parameters for the OBS's experiments used below in the text and Figures are presented in Table 1.

The Main Results of DSS

Some results in the form of the averaged crustal columns for the Pacific Ocean sites are shown in figure 3 from experiments made during the 24th cruise of the R/V ACADEMIK KURCHATOV (K24), the 21st and the 23rd cruises of the R/V DMITRY MENDELEEV (M21 AND M23) (Neprochnov and Sedov, 1984; Neprochnov et al., 1980; 1984a; 1984b). In addition to these, the Shatsky Rise model of Den et al. (1969)and the Central Pacific basin model of Shor et al. (1970) are shown for comparison.

The structural pattern of oceanic basins is presented in Figure 3 by crustal columns of the North-West, North-East, Central and Caroline basins. The Emperor F.Z. region may also be included into this group. The crustal structures of the above regions have much in common. The layer 3 shows the velocity of 6.7-6.8 km/s. Despite fairly detailed observations and high quality of the OBS's records, our profiles, however, did not detect a high-velocity layer 3B.

The model M23-3 for the central part of the Shatsky Rise in general fairly well correlates with the Den et al. (1969) model for the southern part of this rise.

The Hess Rise was previously studied by the Hawaii Institute of Geophysics (HIG), USA, in the cruise of the R/V KANA KEOKI using sonobuoys technique (L. K. Wippeman, HIG, unpublished data, 1975). As a result the crustal layers 2B, 3A and 3B were detected beneath the sediments. The M-discontinuity did not reach under the Rise. The model M23-2 for this region is based on the reversed profile which yields good information on the main crustal layers (except for layer 3B) and the upper mantle.

The crustal thickness in the studied regions of the Shatsky and Hess Rises is aproximately the same (16-18 km) which is 2-3 times thicker than the crust in the adjoining basins. Hussong et al. (1979) by dividing the Shatsky Rise actual crustal structure by factor 2.7 and comparing the result with the typical Pacific basin structures suggested that this rise consists merely of bloated normal oceanic crust. Our seismic studies and dredging results in the Shatsky and Hess Rises (Rudnik et al., 1981; Rudnik and Melankholina, 1984) also support the oceanic origin of these rises.

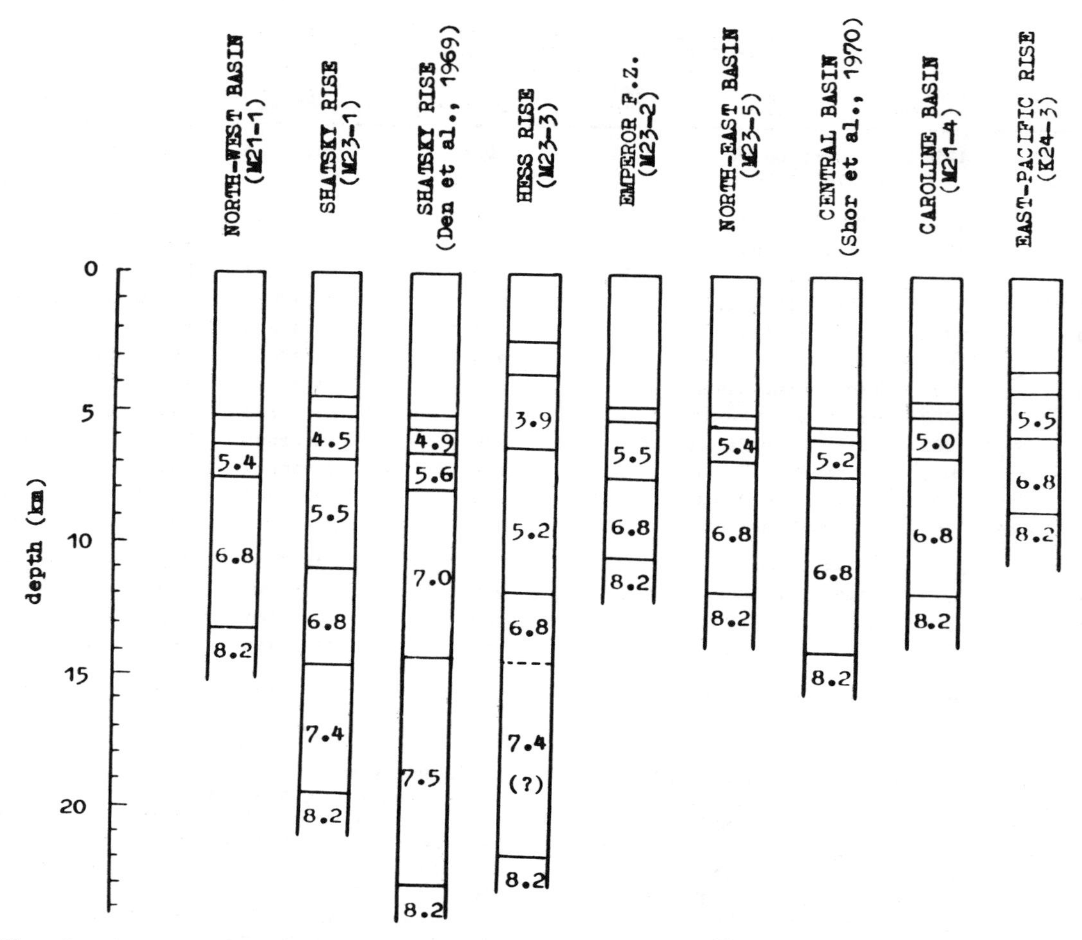

Fig. 3. Averaged crustal columns for the Pacific Ocean (Neprochnov and Sedov, 1984). See Fig. 2 for locations.

During the 29th cruise of the R/V DMITRY MENDELLEEV detailed deep seismic soundings were carried out at four sites in the North-East and North-West Pacific basins in the areas with the age of sea floor of 29,56,70 and 110 million years (figure 2, M29-7, M29-6, M29-5 and M29-9). Two perpendicular seismic profiles using system of OBS's were shot at each site (Neprochnov, 1983).

Figure 4 shows the averaged seismic models for the studied areas. Only one site (M29-6) of the four indicates the existence of high-velocity layer 3B. The analysis of reversed travel time curves and 2-dimension crustal sections inside each of the studied areas indicates that layer 2 is most variable in thicknesses and velocities. Layer 3 is more uniform: the variations of velocity at one site are small (about 0.1 km/s) which suggests its homogeneity. Anistotropy for layer 3 and the upper mantle has not been established in these experiments.

During the recent 31th cruise of the R/V DMITRY MENDELEEV detailed DSS with OBS's and big airguns were carried out in the basins of the Atlantic and Indian oceans (Figure 2, M31). The averaged crustal columns for the studied areas are presented in Figure 5.

The Cape Verde basin (Figure 2, M31-1) and Cape basin (Figure 2, M31-3) have a normal three-layer crust.

The anomalous crust without layer 3A was found in the studied area of the Brazil basin (Figure 2, M31-2). One possible explanation of this feature is a displacement of the crustal blocks by faults resulting in exposing of layer 3B under layer 2 as can be seen in Figure 6 (Kazmin et al., 1986).

The lateral variation of the crustal structure was discovered in the Central basin of the Indian Ocean (Figure 2, M31-5 and M31-6). These experiments were conducted inside of the unique intraplate deformation area (Weissel et al.,

1980; Levchenko et al., 1985). Results of a detailed continuous seismic profiling survey reveal that this area consists of extremely tectonically disturbed individual crustal blocks which alternate with less deformed portions of the sea-floor. According to DSS experiments M31-5 and M31-6 weakly deformed parts of the area with smooth sea-floor are characterized by normal oceanic crust (Figure 5, M31-5A), whereas the folded and fractured basement rise has anomalous crust and upper mantle (Figure 5, M31-5B and M31-6). These features make it possible to suggest that intraplate deformations are not confined to the crust but extend to greater depth (the layer with velocity of 7.4-7.6 km/s is supposed to be an unconsolidated upper mantle).

Comparison of DSS Results

Several authors have indicated that the seismic structure of the oceanic crust varies with age (Le Pichon, 1969; Goslin et al., 1972; Christensen and Salisbury, 1975; Houtz and Ewing, 1976), in particular the thickness of layer 3 increases with age to about 40 m.y.

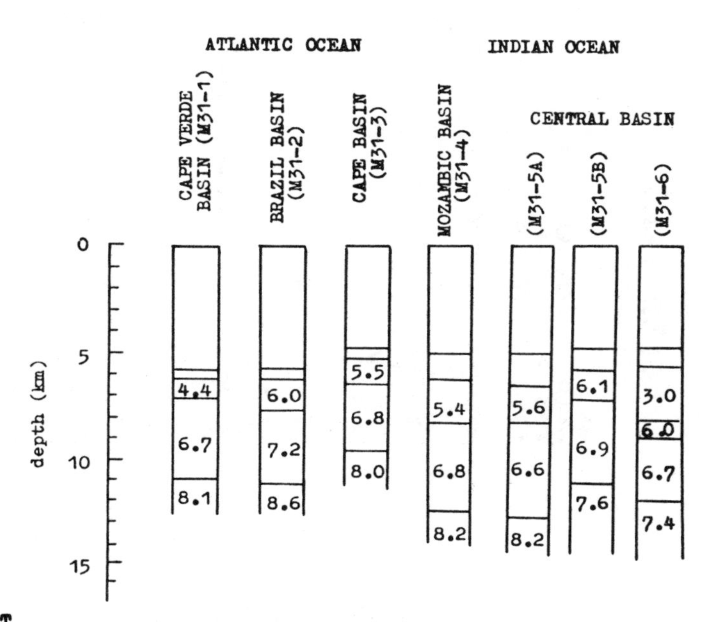

Fig. 5. Averaged crustal columns for the basins of the Atlantic and Indian Oceans obtained during the 31st cruise of the R/V DMITRY MENDELEEV (Neprochnov et al., 1986). See Fig. 2 for locations.

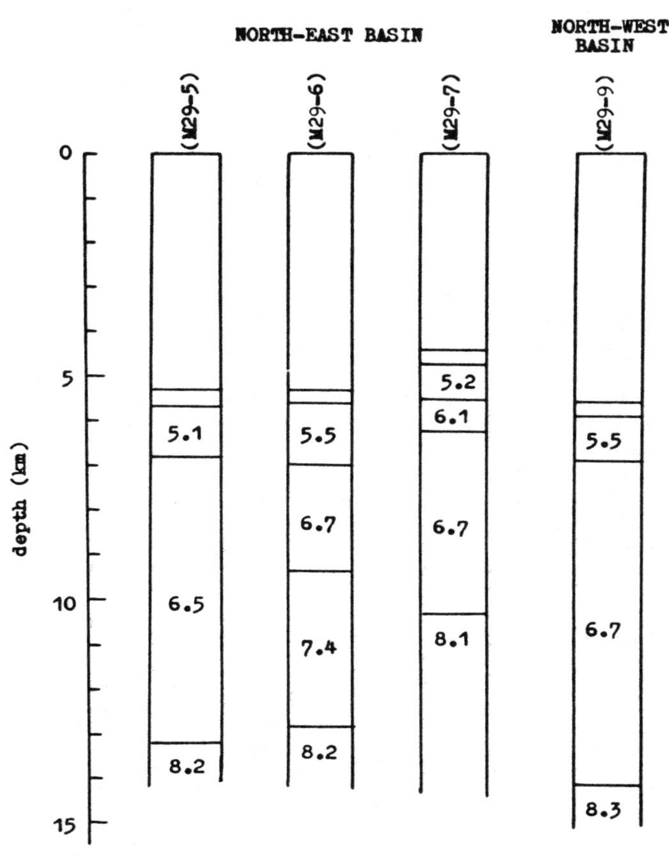

Fig. 4. Averaged crustal columns for the areas with plate age of 29 (M29-7), 56 (M29-6), 70 (M29-5) and 110 (M29-9) million years obtained in the North Pacific Ocean during the 29th cruise of the R/V DMITRY MENDELEEV (Neprochnov and Sedov, 1984). See Fig. 2 for locations.

The detailed statistical analysis of 200 published refraction sites revealed the age dependences of crustal thickness (H), mean seismic velocity in the crust (V_k) and upper mantle velocity (V_m) in the East Pacific basin and the adjacent areas (Neprochnov and Kuzmin, 1984), which have pronounced cyclic variations (Figure 7a). As it can be seen in Figure 7b these variations are in good agreement with the global tectonic cycles (Kunin and Sardonnokov, 1976). They also correlate with variations of the sea-floor spreading described by Frutos (1981).

The comparison of results of 2-dimension crustal sections for sites M29-7, M29-5, M29-9 and the age dependence curve presented in Figure 7a indicates that variations of seismic parameters of the crust and the upper mantle within each site are significantly less than the age dependence variations. The interchanging of areas with low and high values of seismic parameters is the important feature of the crustal structure in the vast part of the East Pacific ocean.

Figure 8 shows the averaged seismic models of the crust and the upper mantle of oceanic basins (Kuzmin, 1987; Shishkina and Neprochnova, 1987). 447 published sites have been used from the Pacific Ocean; 69 sites from the Atlantic Ocean,

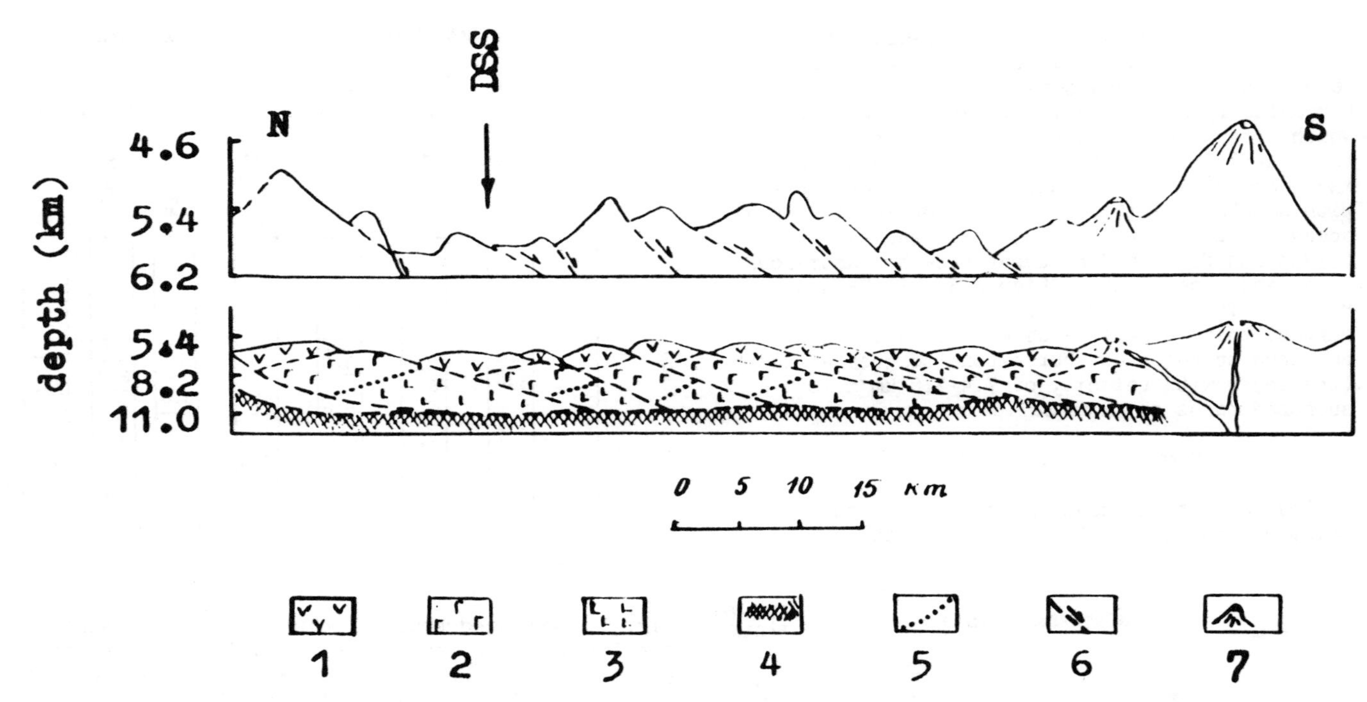

Fig. 6. Structure of the basement and crust across the Brazil basin (Fig. 2, M31-2). Upper section is shown with vertical scale exaggeration and lower one in the same horizontal and vertical scales (Kazmin et al., 1986). 1 - layer 2, 2 - layer 3A, 3 - layer 3B, 4 - upper mantle, 5 - supposed boundary between layer 3A and layer 3B, 6 - fault, 7 - volcano.

136 sites - from the Indian Ocean. The depth is indicated from the top of the basement, i.e. the surface of the second layer. Velocities and thicknesses for the main crustal layers, as well as the standard deviations (σ) of these parameters are presented in Table 2. It can be seen that the averaged seismic models of the basins of different oceans are similar both in thicknesses and in compressional velocities of the main layers. The three-layer crust characterizes all the basins: the sedimentary layer, layer 2 (seismic velocities from 4.9 to 5.4 km/s) layer 3 (seismic velocities from 6.6 to 6.9 km/s). It may only indicate some lower seismic parameters of the crust and the upper mantle of the Atlantic Ocean basins compared to the Pacific and Indian Oceans. Statistical analysis indicates also that the basins of the Indian Ocean have lower thickness of layer 3 (3.8 km) than basins of the Pacific (4.5 km) and

TABLE 2. Comparison of Crustal Seismic Parameters of the Oceanic Basins

Region	Layer 1			Layer 2				Layer 3				Upper Mantle	
	V_1 (km/s)	h_1 (km)	σh_1	V_2 (km/s)	σV_2	h_2 (km)	σh_2	V_3 (km/s)	σV_3	h_3 (km)	σh_3	V_m (km/s)	σV_m
Atlantic Ocean	(1.8)	1.0	0.54	4.9	0.29	1.3	0.62	6.6.	0.26	4.7	1.15	8.0	0.18
Indian Ocean	(2.0)	0.5	0.18	5.4	0.48	1.9	0.47	6.9	0.17	3.8	0.78	8.1	0.05
Pacific Ocean	(2.0)	0.4	0.15	5.2	0.40	1.6	1.02	6.9	0.22	4.5	1.50	8.2	0.17

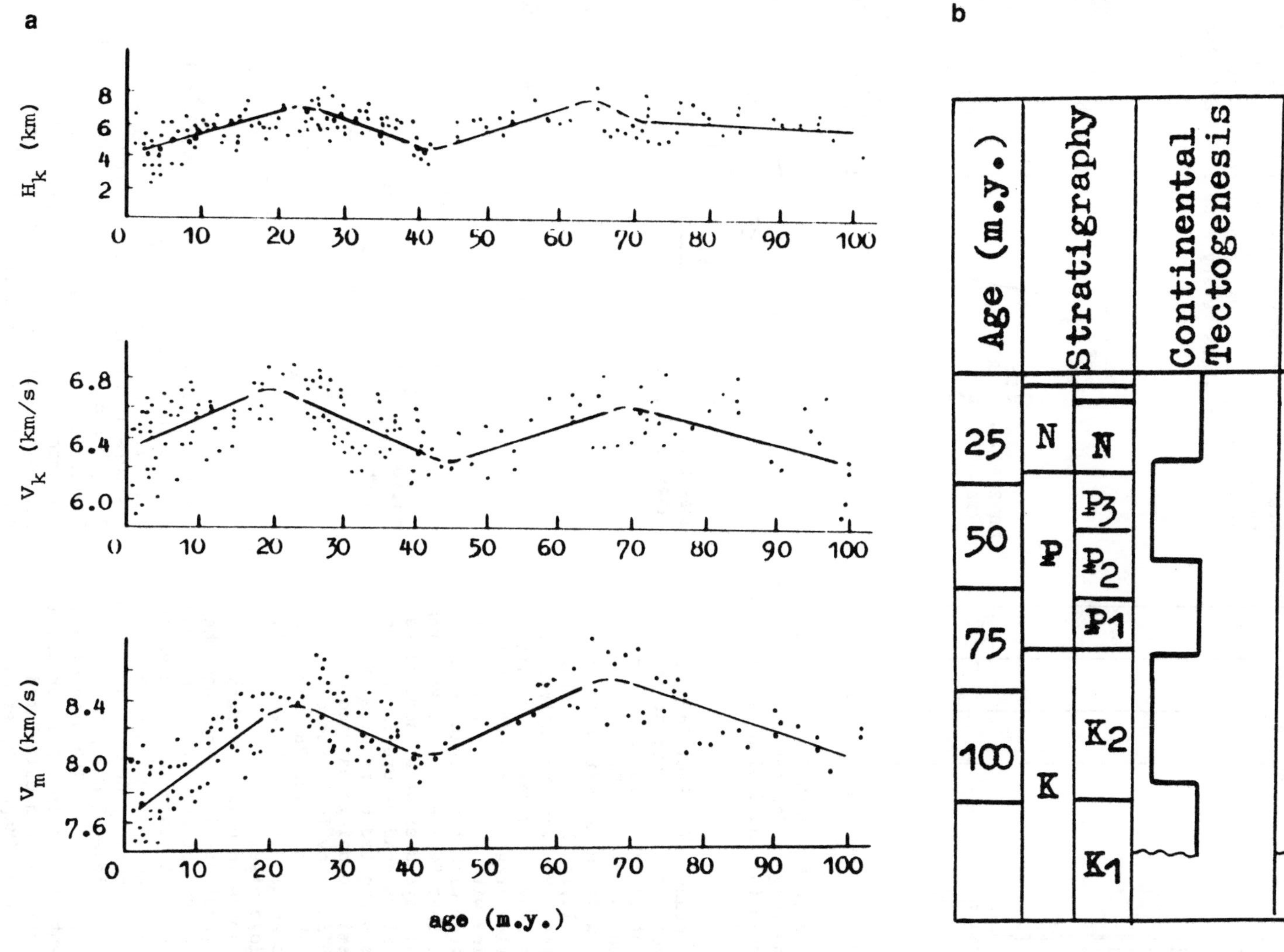

Fig. 7. (a) Age dependence for thickness of the crust (H_k), mean seismic velocities of the crust (V_k) and upper mantle seismic velocities (V_m) in the East Pacific Basin. (b) Comparison of cyclicity of the continental tectogenesis and age dependence of the crustal seismic parameters. The figure is taken from Neprochnov and Kuzmin (1984).

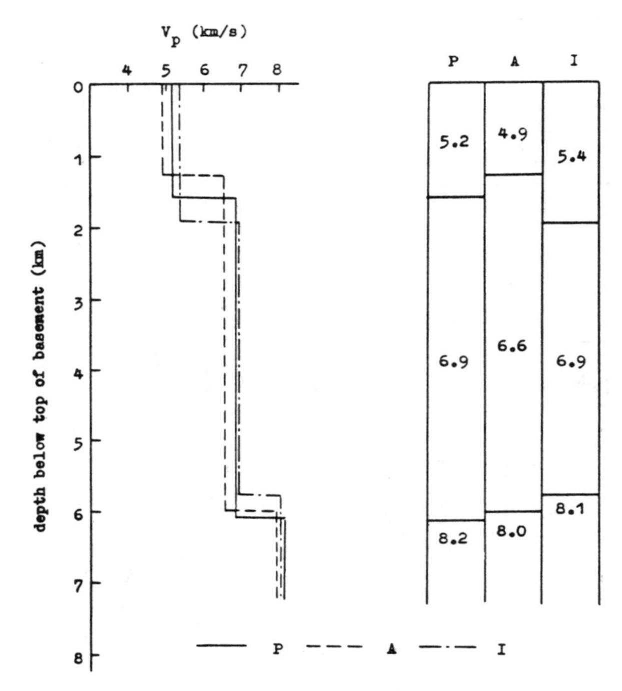

Fig. 8. Averaged velocity-depth functions (left) and crustal columns (right) for the basins of the Pacific (P), Atlantic (A) and Indian (I) Oceans.

Atlantic (4.7 km) oceans. The Pacific Ocean basins are distinguished by the largest mean square deviations of layer 3 thickness (1.5 km).

Airgun - sonobuoy studies of Sutton et al. (1971) suggest the wide-spread existence of a high-velocity basal layer (layer 3B) in the Pacific Ocean basins. However the analysis of all published data shows only small quantity of refraction sited (less than 10 percent) in the Pacific Ocean basins with layer 3B. Perhaps it is necessary to agree with the opinion of Spudich and Orcutt (1980), that "while it is likely that high-velocity basal layers exist at some sites in the oceans, their widespread nature is in doubt". This conclusion is supported by results of detailed deep seismic soundings carried out in expeditions of the Shirshov Institute of Oceanology. They show that layer 3B is more typical of oceanic areas with thickened crust, for example, offridge rises.

Now we shall examine in brief the data on vertical velocity gradients in layer 3 and the upper mantle and on transition between the crust and the mantle. The last years many studies of these parameters were made using a synthetic seismogram modeling. Spudich, Orcutt (1980) and others have shown that layer 3 has gentle velocity gradients (0-0.2 s^{-1}). The determination of velocity gradients is possible also by using comparison of experimental and calculated amplitude-distance curves (Magnitsky et al., 1970). Using this method for the refraction data from the Indian Ocean basins we found velocity gradients for layer 3 -0.03-0.06 s^{-1}, and for the upper mantle 0.02-0.03 s^{-1} (Neprochnov, 1976).

The velocity boundary between the crust and the mantle is apparently sharp or has a thin transition zone as evidenced by observations of nearcritical and postcritical reflected waves from the mantle at distance interval from 20 to 40-50 km at almost every site in oceanic basins.

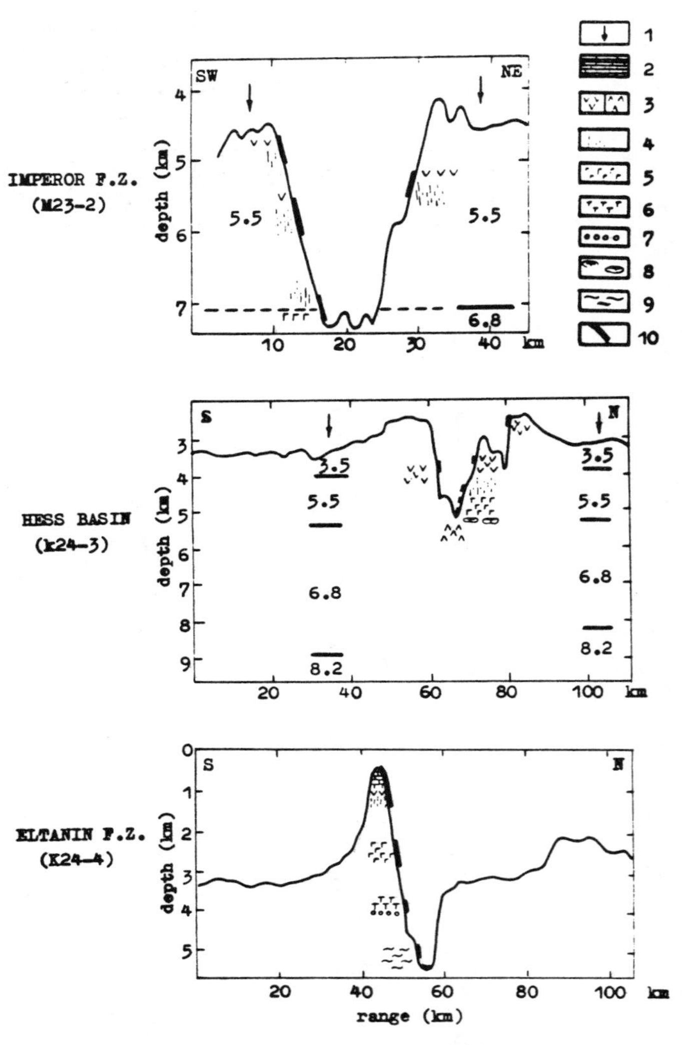

Fig. 9. Sections of the crust obtained by dredging and DSS in the areas of the Imperor F. Z. (top), Hess basin in Galapagos F. Z.(middle) and Heezen Fracture in Eltanin F. Z. (bottom). 1 - DSS profile, 2 - limestone, 3 - basalt, 4 - dolerite, 5 - gabbro, 6 - peridotite, 7 - granulite, 8 - olivinite, 9 - amphibolite, 10 - dredging interval. The Figure is taken from Kashintsev and Rudnik (1984), Neprochnov and Gorodnitsky (1984).

Composition of the Crust

The discussion is continuing about the petrological composition of the oceanic crust, especially of its lower parts. Recent information on this problem is based on the deep-sea drilling which now penetrates only the uppermost part of the oceanic crust, on the dredge samples and on the correlation with the onshore ophiolites (Christensen and Salisbury, 1975; Spudich et al., 1978; Spudich and Orcutt, 1980).

During expeditions of the Shirshov Institute of Oceanology a lot of samples of the rocks were dredged near the seismic refraction sites, which allow to make comparison between the seismic and petrographic data.

The dredging in the slopes of the Shatsky and Hess Rises indicates that layer 2 is composed of alkali basaltoids and tholeitic basalts (Rudnik et al., 1981). The authors of this work suggest that magmatism of these regions changed from tholeite at early stages of development to differentiated alkali at later stages.

The most complete sections of the crust were obtained in the areas of the Emperor F.Z., Hess basin in Galapagos F.Z. and Heezen Fracture in Eltanin F.Z., where steep slopes of 3-5 km high have been discovered (Figure 9).

On the basis of the obtained data it is concluded (Neprochnov and Kashintsev, 1978; Kashintsev and Rudnik, 1979; 1984), that layer 2 is composed of pillow basalts and flow-breccias in the upper part and of basalt-dolerite dikes in the lower part. Layer 3 is composed mainly of gabbros; the ultramafics and metamorphic rocks were also dredged from the lower parts of this layer. The upper mantle consists of ultramafics; here also metamorphic rocks(amphilites and granulites) were found. These zones may be recommended for further more detailed seismic studies and rock sampling from submersibles.

Acknowledgments. The author is grateful to the convenor of the symposium Lower Crust Properties and Processes of IUGG XIX General Assembly, Vancouver, Canada, August 9-22, 1987, Prof. R.F. Mereu for inviting to present this paper, to T.Z. Grigoryan who drafted the figures, and to V. V. Gerasimova who typed the manuscript. The author would like to thank two anonymous reviewers who greatly improved the paper.

References

Christensen, N. I., and M.H. Salisbury, Structure and constitution of the Lower oceanic crust, Rev. Geophys. and Space Phys., 13, 57-86, 1975.

Den, N., W. J. Ludwig, S. Murauchi, J. Ewing, H. Hotta, N. T. Edgar, T. Yoshii, T. Asanuma, K. Hagiwara, T. Sato and S. Ando, Seismic refraction measurements, in the Northwest Pacific Basin, J. Geophys. Res., 74, 1421-1434, 1969.

Frutos, G., Andean tectonics as a consequence of sea-floor spreading, Tectonophysics, 72, 21-32, 1981.

Goslin, J., P. Beauzart, J. Francheteau, and X. Le Pichon, Thickening of the oceanic layer in the Pacific Ocean, Mar. Geophys. Res., 1, 418-427, 1972.

Hussong, D. M., L. K. Wippeman, and L. W. Kroenke, The crustal structure of the Ontong Java and Manihiki Oceanic Plateau, J. Geophys. Res., 84, 6003-6010, 1979.

Kashintsev, G. L., and G. B. Rudnik, Earth's crust beneath the ocean, Priroda, 4, 42-47, Moscow, 1979 (in Russian).

Kashintsev, G. L., and G. B. Rudnik, Magmatic rocks associated with the fractures of the earth's crust, in Deep Fractures of the Oceanic Floor, edited by Yu. P. Neprochnov, pp. 174-197, Nauka, Moscow, 1984 (in Russian).

Kazmin, V. G., O. V. Levchenko, L. R. Merklin, Yu. P. Neprochnov, and V. V. Sedov, Some peculiarities of the oceanic crust structure as exemplified by the Brazil basin, Geotektonika, 2, 46-55, 1986 (in Russian).

Kosminskaya, I. P., and N. K. Kapustian, Generalized seismic model of the typical oceanic crust, Volcanoes and Tectonosphere, edited by H. Aoki and S. Iizuka, Tokai University Press, pp. 143-156, 1976.

Kunin, N.Y., and N. M. Sardonnikov, Global cycle pattern of tectonic movements, Bull. MOIP, geol., 3, 5-20, 1973 (in Russian).

Kuzmin, P. N., Generalized seismic models of the earth's crust and upper mantle of the Pacific and Atlantic oceans, in Problem of the Oceanic Floor Geophysics, 2, edited by Yu. P. Neprochnov, p. 22, Institute of Oceanology USSR Academy of Sciences, Moscow, 1987 (in Russian).

Le Pichon, X., Models and structure of the oceanic crust, Tectonophysics, 7, 385-401,1969.

Magnitsky, V. A., Yu. P. Neprochnov, and L. N. Rykunov, On the gradients of the seismic velocities and temperature under the M-discontinuity, Report of USSR Academy of Sciences, 195, 85-88, 1970 (in Russian).

Maynard, G. L., Crustal layer of seismic velocity 6.9 to 7.6 kilometers per second under the deep oceans, Science, 168, 120-121, 1970.

Neprochnov, Yu. P., 29th cruise of the R/V DMITRY MENDELEEV, Okeanologiya, 23, 365-368, 1983 (in Russian).

Neprochnov, Yu. P., Seismic Researches in the Ocean, Nauka, Moscow, 178 p., 1976 (in Russian).

Neprochnov, Yu. P., and A. M. Gorodnitsky, Geological-geophysical feature and genesis of the main deep fractures in the World Ocean, in Deep Fractures of the Oceanic Floor, edited by Yu. P. Neprochnov, pp. 197-211, Nauka, Moscow, 1984 (in Russian).

Neprochnov, Yu. P., and G. L. Kashintsev, On composition of the main layers of the oceanic crust. Doklady AN SSSR, 239 1222-1225, 1978 (in Russian).

Neprochnov, Yu. P., and P. N. Kuzmin, On dependence of the seismic parameters of the earth's crust and upper mantle in Pacific Ocean upon lithospheric age, Doklady AN SSSR, 275, 704-707, 1984 (in Russian).

Neprochnov, Yu. P., and V. V. Sedov, The crustal structure of major tectonic features of the North Pacific, in Geology of the world Ocean, Reports to 27th Intern. Geol. Congress, 6, part 1, pp. 40-51, Nauka, Moscow, 1984 (in Russian).

Neprochnov, Yu. P., V. V. Sedov, and B. N. Grinko, Experience of the Computer processing of OBS's records of DSS in the ocean, Okeanologiya, 8, 939-944, 1978 (in Russian).

Neprochnov, Yu. P., V. V. Sedov, G. A. Semenov, I. N. Elnikov, and V. D. Feofilaktov, Crustal structure and seismicity of the Hess Basin region, Okeanologiya,20, 485-494, 1980 (in Russian).

Neprochnov, Yu. P., V. V. Sedov, B. V. Kholopov, and I. N. Elnikov, The Earth's crust structure according to Deep Seismic sounding data, in Structure of North-West Pacific Ocean Floor (Geophysics, Magmatism, Tectonics),pp. 89-101, Nauka, Moscow, 1984, (in Russian).

Neprochnov, Yu. P., V. V. Sedov, andP. N. Kuzmin, The structure of the earth's crust near Dmitry Mendeleev seamount in the East-Caroline basin,Okeanologiya, 24, 493-497, 1984b (in Russian).

Neprochnov, Yu. P., V. V. Sedov, A. A. Pokryshkin, L. G. Akentiev, B. N. Grinko, A. A. Ostrovsky, and B. V. Kholopov, New data on the earth's crustal structure and seismicity of the basins of the Atlantic and Indian Oceans, Doklady AN SSSR, 290 , pp. 1448-1453, 1986 (in Russian).

Rudnik, G. B., E. N. Melankholina, D. I. Kudryavtsev, O. S. Lomova, V. G. Safonov, S. A. Silantyev. and O. A. Shmidt, Rocks of the second oceanic layer in the sections in the Shatsky and Hess Rises (Pacific Ocean), Okeanologiya, 11, 21-33, 1981 (in Russian).

Rudnik, G. B., and E. N. Melankholina, Rock composition of the Oceanic crust in the North-West Pacific (according to the data on drilling, dredging and study of deep-seated xenoliths), in Structure of the North-West Pacific Ocean Floor (Geophysics, Magmatism, Tectonics), edited by Yu. M. Pushcharovsky and Yu. P. Neprochnov, pp. 127-157, Nauka, Moscow, 1984 (in Russian).

Shishkina, N. A. , and A. F. Neprochnova, Comparative features of the earth's crustal structure in the basins of the Atlantic and Indian oceans, Problems of the Oceanic Floor Geophysics, 2, edited by Yu. P. Neprochnov, p. 40, Institute of Oceanology USSR Academy of Sciences, Moscow, 1987.

Shor, G. G., H. W. Menard, R. W. Raitt, Structure of the Pacific basin, in The sea, 4, part 2, edited by A. E. Maxwell, pp 3-27, Wiley-Interscience, New York, 1970.

Spudich, P., and J. Orcutt, A new look at the seismic velocity structure of the ocean crust, Rev. Geophys. and Space Phys., 18, 627-645, 1980.

Spudich, P., M. H. Salisbury, and J. A. Orcutt, Ophioliotes found in oceanic crust? Geophys. Res. Lett., 5, 341-344, 1978.

Sutton, G. H., G. L. Maynard, and D. M. Hussong, Wide spread occurence of a hing-velocity basal layer in the Pacific crust found with repetitive sources and sonobuoys, The Structure and Physical Properties of the Earth'sCrust, Geophys. Monogr. Ser., 14, edited by J. G. Heacock, pp. 193-209, Washington, F.C., 1971.

Weissel, J. K., R. N. Anderson, and C. A. Geller, Deformation of the Indo-Australian plate, Nature, 287, 284-291, 1980.

A SEISMIC REFRACTION STUDY OF THE CRUSTAL STRUCTURE OF THE SOUTH KENYA RIFT

W. Henry[1], J. Mechie[2], P. K. H. Maguire[1], J. Patel[3],
G. R. Keller[4], C. Prodehl[2], M. A. Khan[1],

Abstract. This extended abstract of a forthcoming paper [Henry et al, unpublished manuscript] describes the results from the explosion part of the Kenya Rift International Seismic Project (KRISP 85). This consisted of two lines, each with 50 3-component recording stations [KRISP Working Group,1987;Henry, 1987]. Along the N-S line, the stations were deployed along the rift axis at 3.5 km intervals for about 200 km from Lake Baringo (BAR) to Susua (SUS) (Fig 1), with a few additional stations at the southern end of the line near Magadi (MAG). On the E-W line the stations were deployed at 1 km intervals from Mt. Margaret (MAR), beneath the eastern margin of the rift, westwards for 50 km to Ntulelei (NTU) at the western edge. A few additional stations were deployed further west to Ewaso Nyiro (EWA).

Of the thirteen shots fired at the locations shown by the large crosses, two were in Lake Baringo, one in Crater Lake near Naivasha (NAI), one on the surface in shallow water (SOL), and the rest in boreholes. The lake shots were much more efficient than the borehole shots. Thus a 1.0 tonne shot at 20 m depth in Lake Baringo gave Pn arrivals at Magadi 280 km away, but a 60 m deep borehole shot of the same size from Makuyu (MAK) did not show any arrivals on the EW line whose nearest-shot station was at 68 km. Only shallow information was obtained along the EW line.

[1] Department of Geology, The University, Leicester LE1 7RH, UK

[2] Geophysikalisches Institut, Universität Karlsruhe, Hertzstrasse 16, Karlsruhe, F.R. Germany

[3] Department of Physics, The University, PO Box 30197, Nairobi, Kenya

[4] Department of Geological Sciences, The University of Texas, El Paso, Texas 79968, USA

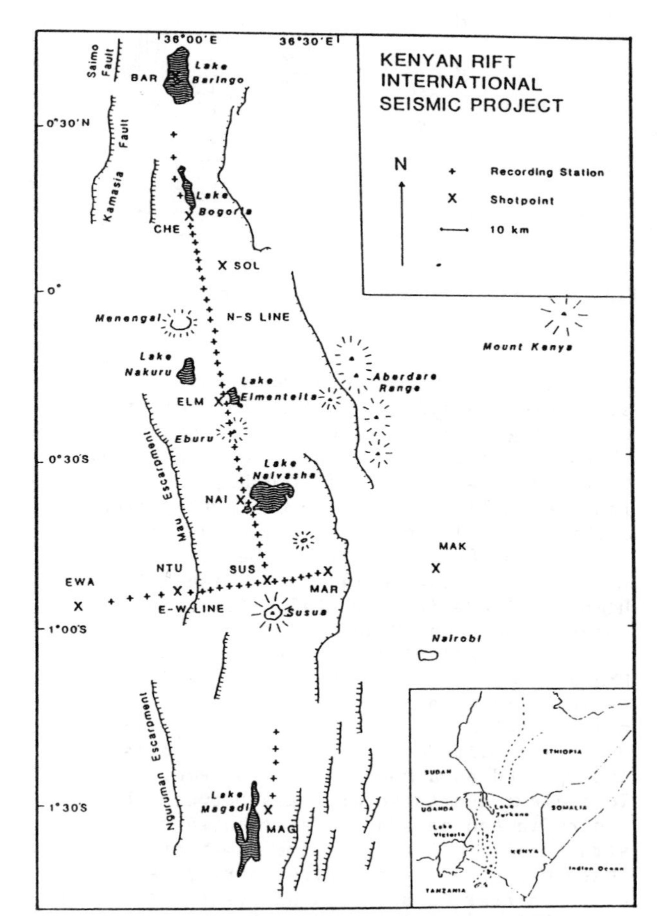

Fig. 1. Location map of the seismic refraction lines and shot points.

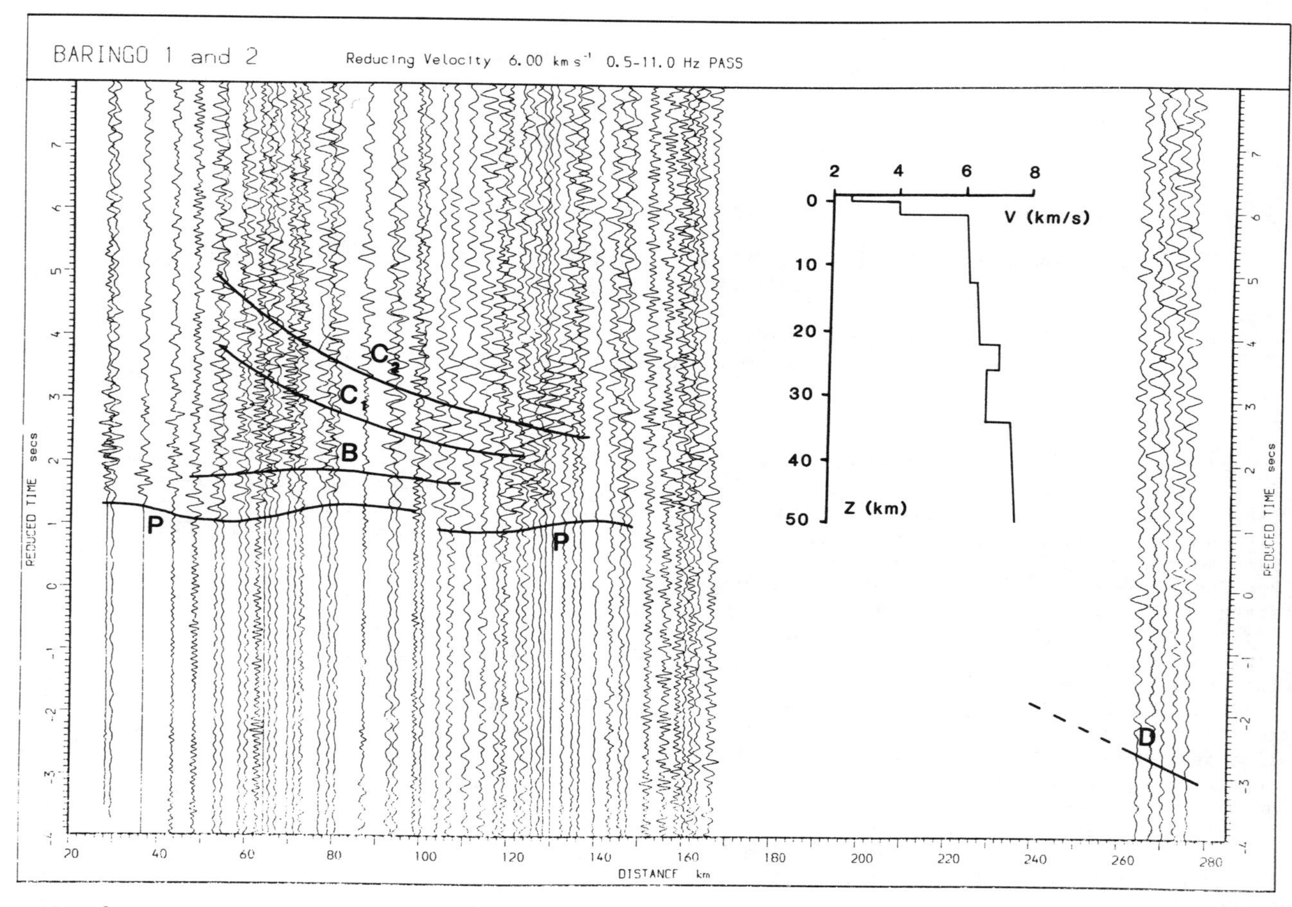

Fig. 2. Record section from a Baringo shot with phase correlations and velocity-depth function.

Most of the information on the deep structure beneath the NS line has come from the Baringo shots. In the record section shown in Fig 2, the traces have been band-pass filtered between 0.5 and 11.0 Hz, trace-normalised with respect to the largest amplitude, and plotted with a reducing velocity of 6.0 km/s. Pg can be seen as a first arrival out to 140 km. Phase B is a high amplitude intracrustal reflection observed between 60 and 100 km. It is also seen in several of the other record sections. Phase C_1 has a maximum amplitude at 90 km and is interpreted as a reflection from a mid-crustal discontinuity. Phase C_2 has a maximum amplitude at about 130 km and is interpreted as a MOHO reflection PmP. The less clear scattered arrivals between these two reflected phases are attributed to structural complexity of the lower crust. Phase D is seen as the first arrival at the stations near Magadi. Its apparent velocity is 7.5 $\pm$ 0.2 km/s and is interpreted as the upper mantle refraction Pn. The other Baringo shot, 1.5 km away gives almost identical results. The data from the Crater Lake shot (NAI) can be interpreted in terms of the same 1-D model as the Baringo shots.

The data from the NS line have been used to produce the model shown in Fig 3. Straight line fitting, time-term analysis and ray-trace-modelling of the near-surface refractions (down to Pg) from all the shots along the NS line show:

1. A basement velocity of about 6.0 km/s.
2. The depth to the crystalline basement varies between 2 and 6 km.
3. The velocity of the volcanics and sediments filling the rift varies between 2.5 and 5.1 km/s.
4. There is a high velocity region in the vicinity of the large volcano Menengai. This is inferred from the travel time advancement of both N and S travelling waves in this region.

Forward modelling using 2-dimensional ray-tracing methods have been used to model the travel times and amplitudes of the later

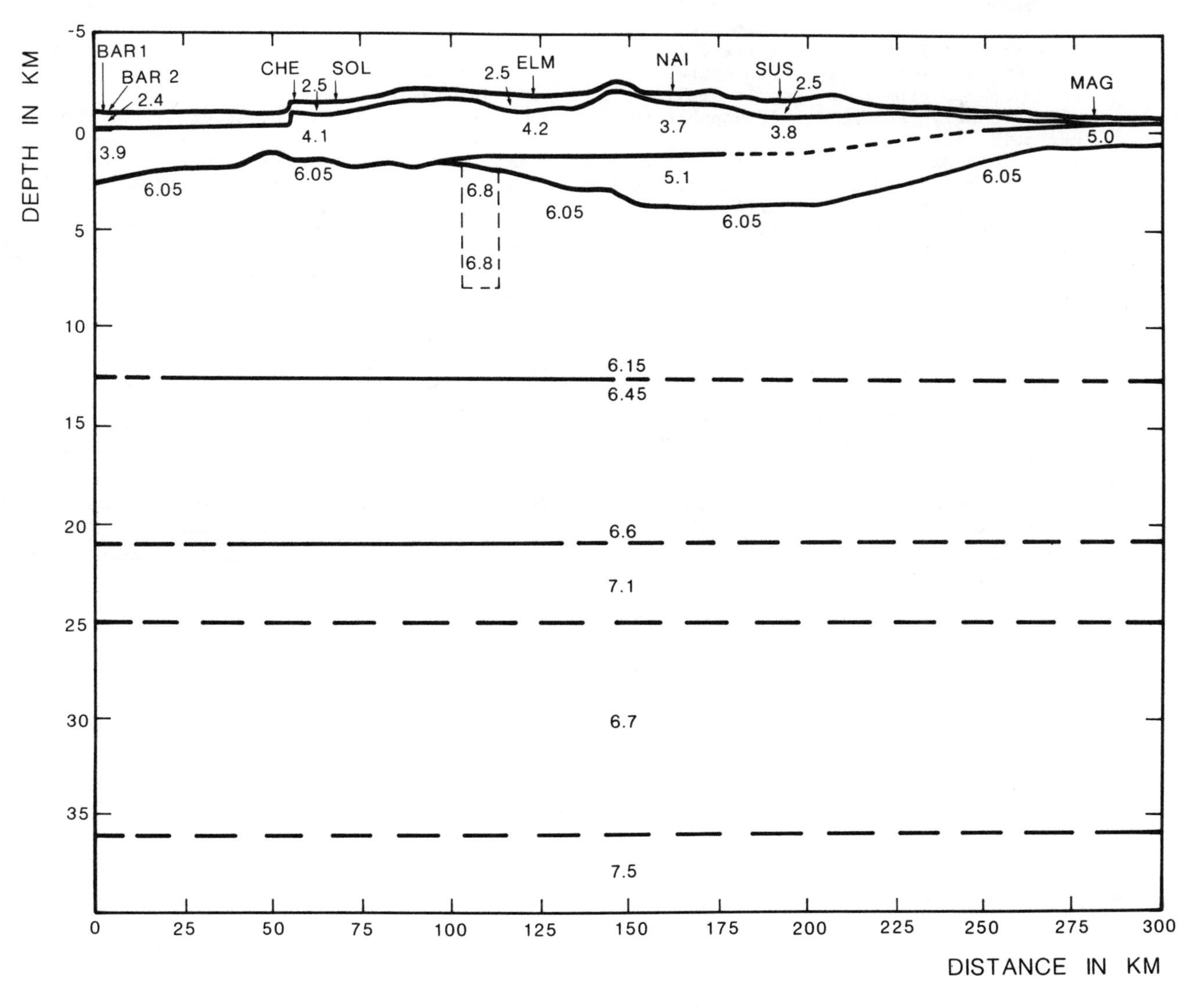

Fig. 3. 2-dimensional velocity-depth structure beneath the axis of the rift from Lake Baringo to Magadi. For mid- and lower crustal layers: —— is reversed ray coverage; --- is unreversed ray coverage. Velocities in km.s^{-1}. Shotpoints as NAI.

phases shown on the Baringo record section. Phase B is produced by a sharp increase in velocity at 12-13 km depth. Phase C_1 results from an increase to high velocity (7.1 km/s) at just over 20 km depth. The velocity variation below this depth is not well constrained by the data but a significant reduction in velocity is required to explain the high amplitude Moho reflection. The constraints provided by the travel time and amplitude behaviour of phase C_2, together with the velocity of phase D, restrict the Moho depth to be 34 $\pm$ 2 km.

The data from this pilot experiment in the south Kenya Rift therefore suggest:

1. There is considerable crystalline basement relief. There is no evidence in the region for the massive axial intrusion at shallow depth suggested previously [Searle,1970; Baker and Wohlenberg,1971;Khan and Mansfield,1971; Savage and Long, 1985] to explain gravity and teleseismic data. There is a high velocity zone of limited lateral extent in the vicinity of the volcano Menengai.
2. There is a reflector at 12-13 km depth.

This coincides with the brittle-ductile transition inferred from hypocentres determined by the Lake Bogoria small aperture array [Cooke et al, unpublished manuscript].

3. There is a high velocity layer at 22-23 km. This may be due to mafic instrusions or the cumulate residue from the siliceous volcanics observed in the rift.
4. The crustal thickness of 34 ± 2 km shows that the crustal thinning beneath the rift is not as great has been reported further north [Griffiths et al, 1971].

References

Baker,B.H., and Wohlenberg, J., Structure and evolution of the Kenya Rift Valley, Nature, 229, 538-542, 1971.

Cooke,P.A.V., Maguire,P.K.H.,Evans, J.R. and Laffoley,N.d'A., Seismicity near Lake Bogoria, Kenya Rift Valley, Unpublished manuscript, Leicester University.

Griffith,D.H., King,R.F., Khan,M.A., and Blundell,D.J., Seismic Refraction Line in the Gregory Rift. Nature Phys. Sci, 229, 69-71, 1971.

Henry,W.J., A seismic investigation of the Kenya Rift Valley, Unpublished PhD thesis, University of Leicester, 1987.

Henry,W.J., Mechie,J., Maguire,P.K.H., Patel,J., Keller, G.R., Prodehl,C., Khan,M.A., A seismic refraction study of the South Kenya Rift Valley, Unpublished manuscript, Leicester University.

Khan,M.A. and Mansfield,J., Gravity Measurements in the Gregory Rift, Nature Phys. Sci., 229, 72-75, 1971.

KRISP Working Group, Structure of the Kenya Rift from seismic refraction, Nature, 325, 239-242, 1987.

Savage,J.E.G., and Long,R.E., Lithospheric structure beneath the Kenya Dome. Geophys. J.Roy Astr.Soc., 82, 461-477, 1985

Searle,R.C., Evidence from gravity anomalies for thinning of the lithosphere beneath the Rift Valley of Kenya, Geophys. J Roy. Astr. Soc., 21, 13-31, 1970.

Section III

THE LOWER CRUST FROM RESULTS OF SEISMICITY STUDIES

SEISMICITY NEAR LAKE BOGORIA IN THE KENYA RIFT VALLEY

Philippa Cooke[1], Peter Maguire[1], Russ Evans[2] and Nicholas Laffoley[1]

Abstract. An analysis of a local earthquake data set from within the Kenya Rift Valley has provided constraints on the crustal structure and rheology of the Rift as a whole. A 15 station seismic network operated for three months near Lake Bogoria in the central trough of the Kenya Rift (Fig.1). The project was part of the Kenya Rift International Seismic Project of 1985 (KRISP 85). The principal aim of the network was to record local seismicity. The network covered a 20 x 20 km^2 area including the southern part of Lake Bogoria and had a station spacing of approximately 5 km. This extended abstract of a forthcoming paper [P.A.V.Cooke et al.,unpublished ms.] describes activity which occurred within an area of about 70 km diameter centred on the network.

The study area was 50 km north of the apex of the Kenya dome near Nakuru. In this region the trend of the Rift changes from generally NNE to NNW. The faults are predominantly of normal dip-slip type, although a minor strike-slip component may be assumed where obliquely intersecting contemporaneous faulting occurs [King,1978]. In common with much of the Rift there is intense young "grid" faulting in the central trough. This type is marked by numerous, often sinuous faults with small displacements, and throws forming miniature grabens and horsts. The older, large faults form the Rift shoulders. Bosworth et al. [1986] visualize this region to be an "accommodation zone" between two half-grabens of opposite polarity, formed by the juxta-position of two opposing detachment systems. The Lake Bogoria region is characterised by geothermal activity with hot springs and fumaroles at various sites in the locality [McCall,1967], and in general the Rift is characterised by high heat flow values [Morgan,1982].

The overall lack of major seismic events in the Kenya Rift has been taken to indicate an elevated geotherm, with stress release occurring generally aseismically or by microseismic activity alone [Searle,1970; Fairhead and Girdler,1981]. In 1928 an event of ISC unified magnitude 6.0 [Shah,1986] caused cracking at the base of the Laikipia-Marmanet escarpment, 10 km east of Lake Bogoria [Richter,1958;McCall,1967]. A previous local earthquake study conducted in the Lake Bogoria region [Hamilton et al.,1973] suggested that microseismic activity occurred to depths of 19 km. However, because many events occurred outside the network, and only P-wave arrivals were used in the location procedure, depth constraint was probably poor.

Approximately 600 local events within 30 km of the present network were large enough to be located, and the hypocentral coordinates were determined using HYPOINVERSE, the necessary velocity model being obtained from the results of the KRISP 85 seismic refraction profile [KRISP Working Group,1987;Henry,1987] which crossed the Lake Bogoria network. The construction of Wadati plots for 72 well located events provided a mean Vp/Vs ratio of 1.74 ± 0.09 km/s. Local magnitude estimates ranged from 0.0 to 3.2, the majority being less than 1.0. Epicentres with horizontal position standard errors of less than 2 km are shown in Figure 1. Although the bulk of the activity is found outside the network, a synthetic analysis of the location errors suggests that the epicentres are well enough constrained to interpret the seismicity distribution.

Most of the activity located by the 1985 Lake Bogoria network appears to be associated with the larger, older faults on the Rift flanks, rather than the younger grid faulting on the Rift floor. The seismicity parallels the predominantly north-south fault trends, and there is no evidence for any seismic trend crosscutting the Rift, which might have indicated an accommodation zone. A well located line of seismicity within the network follows the trend of small scale faulting, but does not correlate with any surface feature. A similar event group observed 15 years

[1]Department of Geology, University of Leicester, Leicester, LE1 7RH, England

[2]Global Seismology Research Group, British Geological Survey, Murchison House, West Mains Road, Edinburgh, Scotland

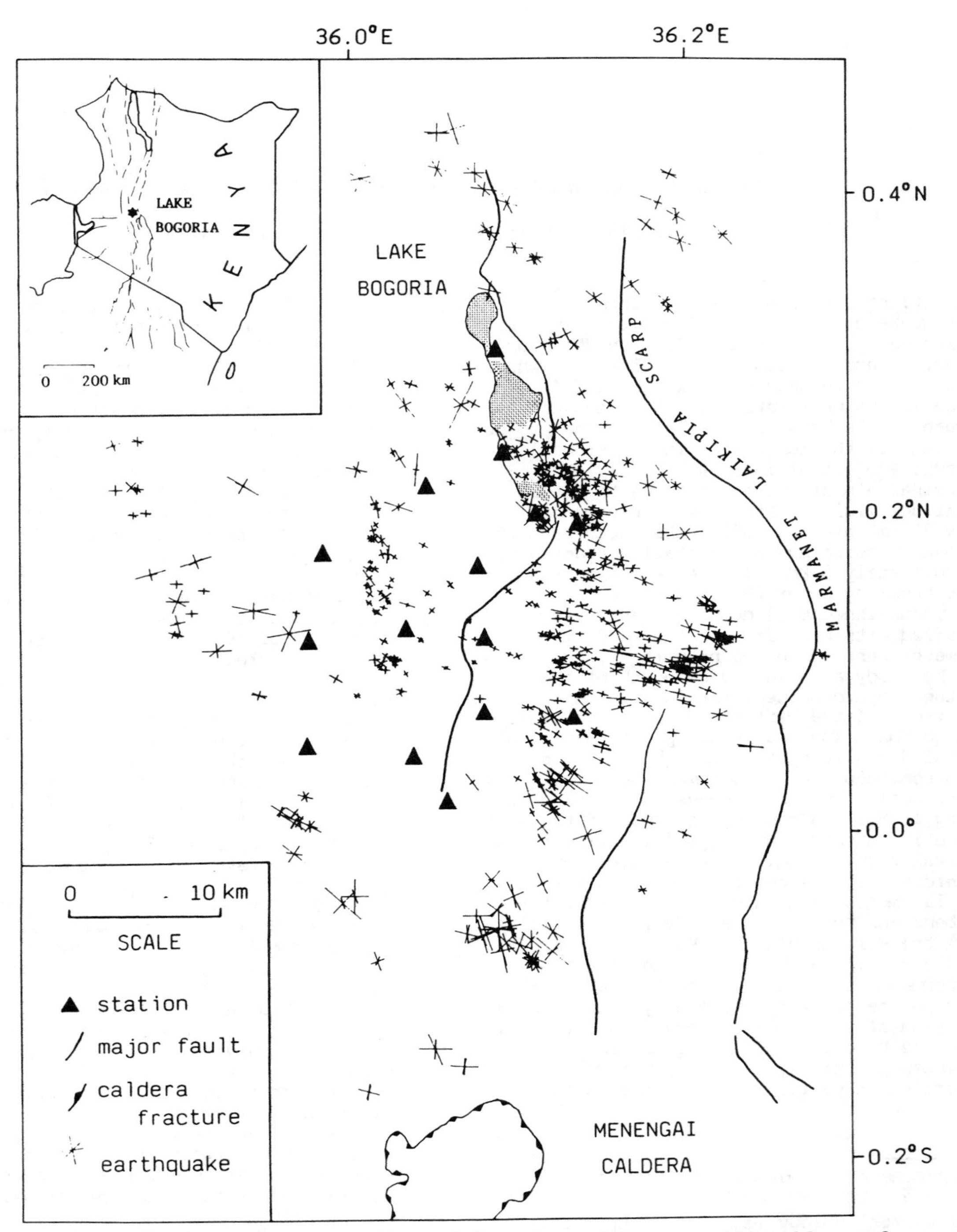

Fig. 1. Lake Bogoria network and epicentral locations with error ellipsoids. Only events with an epicentral error less than 2 km are plotted.

previously [Hamilton et al.,1973] originated approximately 5 km to the south-east of the present group, suggesting the presence of at least two sub-parallel deep faults in this region. Immediately to the south of the south-west element of the network there is a small cluster of unusually deep earthquakes with focal depths between 22 and 26 km, and

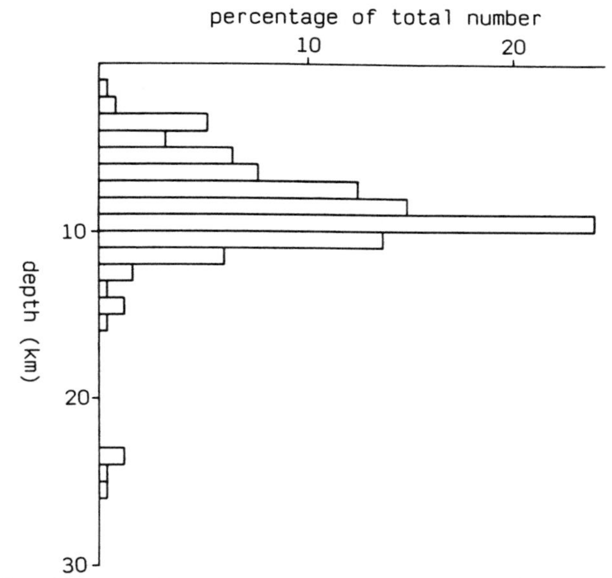

Fig. 2. Depth-frequency distribution of 250 earthquakes which have vertical and epicentral errors less than 1 km, and which occurred within 10 km of the network.

which have abnormally low frequency seismograms. Such events are common in areas of active volcanism [McNutt,1986] and may indicate movement of magma [Walter,1986].

The earthquake depth distribution in Figure 2 shows that most activity occurs above a depth of 12 km, and that no "normal" activity occurs below 16 km. The brittle-ductile transition is thus defined at 12 - 16 km. The presence of a generally aseismic lower crust within the Kenya Rift is in contrast to observations of other parts of the East African Rift System [Shudofsky et al.,1987]. A cut-off depth of 12 - 15 km is normal for earthquakes in "young" intra-continental areas [Chen and Molnar,1983] with medium to high heat flow, [Sibson,1982;Meissner and Strehlau,1982]. Such a brittle-ductile transition depth, and the peak in earthquake activity just above it are as predicted by models for quartz-based crustal material [Sibson,1982;Meissner and Strehlau,1982]. These observations, together with the recently determined crustal thickness of about 35 km [KRISP Working Group,1987] suggests that the crust beneath the Kenya Rift has normal rheology.

References

Bosworth,W., Lambiase,J. and Keisler,R., A new look at Gregory's Rift: the structural style of continental rifting, EOS, 67,577-582,1986.

Chen,W. and Molnar,P., Focal depths of intracontinental and intraplate earthquakes and their implications for the thermal and mechanical properties of the lithosphere, J. Geophys. Res., 88, 4183-4214, 1983.

Cooke,P.A.V, Maguire,P.K.H, Evans,J.R. and Laffoley,N.d'A., Seismicity near Lake Bogoria, Kenya Rift Valley. Unpublished manuscript, Leicester University.

Fairhead,J.D. and Girdler,R.W., The seismicity of Africa, Geophys. J. Roy. Astr. Soc., 24, 271-301, 1971.

Hamilton,R.M., Smith,B.E. and Knapp,F., Earthquakes in geothermal areas near Lakes Naivasha and Hannington, Kenya. UNESCO unpublished report, 1973.

Henry,W.J., A seismic investigation of the Kenya Rift Valley, Unpublished PhD thesis, University of Leicester, 1987.

King,B.C., Structural and volcanic evolution of the Gregory Rift Valley, in Geological Background to Fossil Man, ed: W.W.Bishop, pp.29-54, 1978.

KRISP Working Group, Structure of the Kenya Rift from seismic refraction, Nature, 325, 239-242, 1987.

McCall,G.J.H., Geology of the Nakuru - Thompson's Falls - Lake Hannington area. Kenya Geol. Survey Report No 78, 1967.

McNutt,S.R., Observations and analysis of B-type earthquakes, explosions and volcanic tremor at Pavlof Volcano, Alaska, Bull. seis. Soc. Am., 76(1), 153-175, 1986.

Meissner,R. and Strehlau,J., Limits of stresses in the continental crust and their relation to the depth density distribution of shallow earthquakes, Tectonics, 1, 73-90, 1982.

Morgan,P., Heat flow in Rift zones, in Continental and Oceanic Rifts, ed: G.Pálmason, AGU Geodynamics Series Volume 8, 107-122.

Richter,C.F., Elementary Seismology, W.H.Freeman, San Francisco, 768 pp., 1958.

Searle,R.C., Evidence from gravity anomalies for thinning of the lithosphere beneath the Rift Valley of Kenya, Geophys. J. Roy. Astr. Soc., 21, 13-31, 1970.

Shah,E.R., Seismicity of Kenya, Unpublished PhD thesis, University of Nairobi, 1986.

Shudofsky,G.N., Cloetingh,S., Stein,S. and Wortel,R., Unusually deep earthquakes in East Africa: constraints on the thermo-mechanical structure of a continental rift system, Geophys. Res. Lett., 14(7), 741-744, 1987.

Sibson,R.H., Fault zone models, heat flow, and the depth distribution of earthquakes in the continental crust of the United States, Bull. seis. Soc. Am., 72(1), 151-163, 1982.

Walter,S.R., Long-period earthquakes at Lassen Peak - evidence for magma movement, EOS, 67, 1264, 1986.

ANELASTIC PROPERTIES OF THE CRUST IN THE MEDITERRANEAN AREA

A. Craglietto,[1,2] G. F. Panza,[1,2] B. J. Mitchell[3] and G. Costa[1]

Abstract. The single-station multimode method for the estimation of Q_β is applied to earthquakes recorded at stations operating in the Mediterranean area. Because relatively short paths are used the procedure allows studies on a regional scale. The effect of lateral variations on Q estimates can be minimized compared to what can be achieved with single station-single mode methods. This is particularly important in tectonically active regions like the Mediterranean.

The main result of our study is the identification of four major categories of regions of anelastic properties in the region of study. The Eastern Po Valley characterized by a thick cover of sediments has a high Q, up to 1000, whereas the North-Central Adriatic Sea and the Alps have medium Q values of about 500. The Apennines, the Rhinegraben, the Tyrrhenian Sea, and the Mediterranean region between Crete and Southern Italy are characterized by low Q values, 80-100, in the upper 20-25 km of the crust. Q values increase to 250-300 in the lower crust in all of the regions except the Rhinegraben and the Tyrrhenian Sea where a Q of about 100 seems to characterize the whole lithosphere.

Introduction

The attenuation of seismic energy, being a direct measure of anelasticity, is also an important source of information regarding the composition, state and temperature of the Earth interior. Unfortunately, amplitude information, reliable for attenuation studies, cannot easily be obtained since instrumental effects, local geology, phase conversion, scattering and source characteristics tend to obscure the amplitude variation due to true energy dissipation.

In this paper the single source-single station multimode method (Chen and Mitchell, 1981; Kijko and Mitchell, 1983) will be extended to the time domain. The analysis in the time domain, making use of phase information, usually neglected when comparing amplitude spectra, allows better constraints on possible fault-plane solutions than the analysis in the frequency domain. Since the effect of attenuation on amplitudes is usually small, especially for high-Q regions, longer paths lead to more accurate measurements. Longer paths are, however, also likely to include more lateral heterogeneities. Here, applications will be shown for the study of the anelastic properties of the crust along relatively short paths. By utilizing short paths, effects of lateral variations as well as those of refraction, reflection and multipathing can be minimized. Furthermore, utilizing the result of Levshin (1985), it is possible to approximately define the period ranges where the effect of the lateral variations is more or less the same on the amplitude of each mode. This may definitely help in satisfying the crucial assumption that the mechanism which causes seismic waves to attenuate affects fundamental and higher modes in the same way.

The Single Source-Single Station Multimode Method

The method consists in generating theoretical seismograms formed by fundamental and the first 20 higher modes of surface waves and comparing them to corresponding observed signals. Since only vertical components will be considered the attention will be limited to Rayleigh waves and P-SV higher modes.

Theoretical seismograms were calculated using the formulation of Panza (1985a). The calculations require a knowledge of the source depth and fault-plane solution as well as the velocity-density model along the path of propagation. The synthetic seismograms are very sensitive to variations of focal depth and elastic structural parameters, while anelastic parameters and fault plane geometry represent a second order effect. Potentially serious is the effect of source finiteness, but a

[1]Istituto di Geodesia e Geofisica, Università di Trieste, I-34100 Trieste, Italy

[2]International School for Advanced Studies, P. O. Box 586, I-34100 Trieste, Italy

[3]Department of Earth and Atmospheric Sciences, Saint Louis University, Saint Louis, Missouri 63156, USA

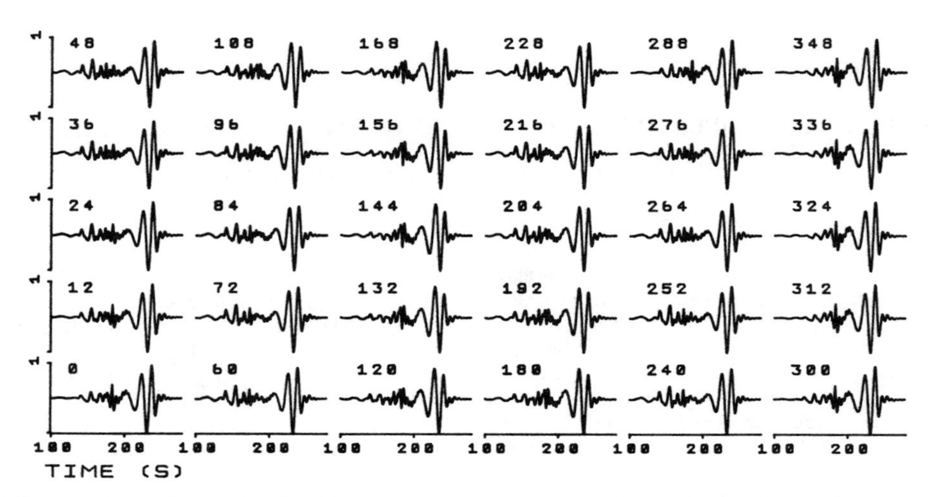

Fig. 1. Variation of synthetic time series, each normalized to unity, with changes in strike. The different angle values, in degrees, are given on the left of each seismogram. Dip, rake and focal depth are kept fixed at the value shown in the figure.

proper choice of the size of the earthquake and of the period range of investigation makes this effect negligible.

The determination of the internal friction profile, Q_β (z), for the lithosphere can therefore be made in two stages:

1) choose the elastic structural parameters and range of focal depths,

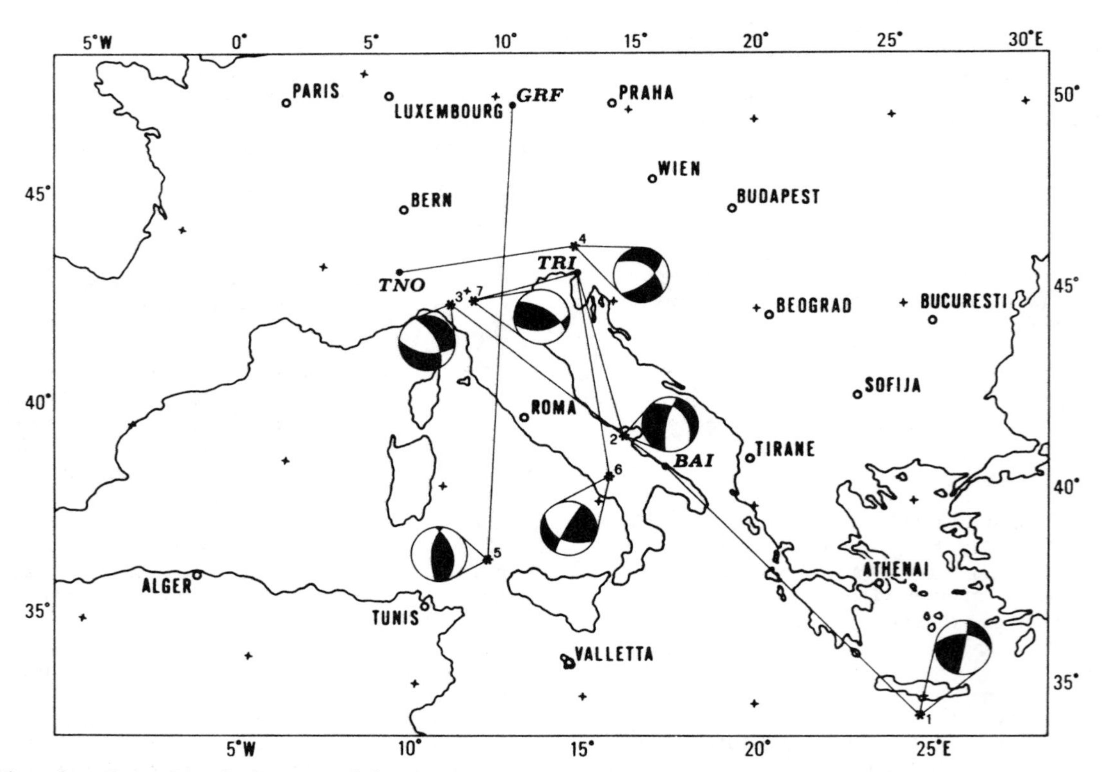

Fig. 2. Investigated area with stations, epicenters and focal mechanisms obtained from first motions in this study.

TABLE 1. Events Used

Station n		Date m/d/y	Origin time h : m : s	Lat. °N	Long. °E	Magn. M_b
BAI	1	04-07-74	14:22:47.1	34.7	24.7	4.7
TRI	2	06-19-75	10:11:13.0	41.6	15.7	4.9
BAI	3	11-16-75	13:04:24.4	44.7	9.6	4.9
TNO	4	05-07-76	00:23:50.4	46.2	13.3	4.5
GRF	5	12-08-79	04:06:30.3	38.2	11.7	5.4
TRI	6	11-24-80	00:23:59.2	40.9	15.3	4.8
TRI	7	11-09-83	16:29:51.8	44.7	10.3	4.9

2) perturb the anelastic structural parameters and the fault plane solution.

The initial values for the elastic parameters can be chosen either on the basis of already existing information or, by means of multiple-filter techniques (Levshin et al., 1972), directly from the record, which will be used for the determination of $Q_\beta(z)$. The initial value of the focal depth can be taken from the available seismological bulletins (e.g. ISC or NEIS). If the event was not studied earlier, from the bulletins it is also possible to extract some information about the focal mechanism. With these input data it is possible to construct a multimode synthetic seismogram to be compared with the experimental record.

By trial-and-error, the depth range of the source can be defined, as for instance from the period of the maximum of the fundamental mode (Panza et al., 1973, 1975a,b). The elastic structural parameters can also be determined by finding those which give satisfactory agreement between theory and observations.

Once the first-order parameters have been chosen, keeping them fixed, it is possible to see the effects of variations in anelastic structural parameters and in fault-plane geometry. Since the method does not make use of absolute amplitude values, the parameters defining the source geometry whose effect can be visible in the time series are dip and rake. The strike of the fault can be kept constant and equal to a value determined from the first P-wave arrivals; in fact, as can be seen in Figure 1, varying the strike as much as about 40° does not significantly affect the wave forms. On the basis of the resolving power of the data it is reasonable to restrict our studies to two-layered $Q_\beta(z)$ models (Cheng and Mitchell, 1981; Kijko and Mitchell, 1983), steps in dip of 20°, and steps in rake variable between 20° and 40°. For the definition of the parameters describing the source geometry, see for instance Panza (1985a). The ranges of $Q_\beta(z)$ and of the fault-plane geometry are then determined by trial-and-error. The availability of reliable fault-plane solutions, determined in other ways, may help in reducing the uncertainties.

The possible sources of error in this method may be separated into two groups: those which are associated with uncertainties in the source specifications and those which may be due to propagation effects between the source and the receiver. Errors due to source effects include uncertainties in focal depth, strike, dip and rake as well as the effects of finiteness of the fault in space and time. Errors due to propagation effects include those due to an incorrect velocity model and lateral variations of elastic and anelastic properties along the path of propagation.

The effects of uncertainties in the strike, dip and rake are relatively small while the effect of focal depth is quite severe (e.g. Panza et al., 1973, 1975a,b; Kijko and Mitchell, 1983). If standard long-period instruments are used, the effect of source finiteness is relevant only when using earthquake sources which produce significant fault movement.

The effects of changes in the velocity model can be controlled by satisfying the group velocity dispersion of the fundamental mode. In fact, in the period range of interest, this quantity has a very weak dependence on the source apparent initial phase (e.g. Panza et al., 1973). What we may expect are changes in individual phases but not in the general shape of the wave train represented by the signal envelope. In fact the effects of

TABLE 2. Focal Plane Solution

Event	plane A			plane B		
m/d/y	strike	dip	rake	strike	dip	rake
04-07-74	278°	39°	174°	14°	86°	50°
06-19-75	306°	32°	208°	185°	72°	232°
11-16-75	323°	58°	308°	86°	48°	225°
05-07-76	300°	45°	225°	45°	60°	235°
12-08-79	334°	37°	70°	178°	57°	75°
11-24-80	107°	40°	280°	203°	80°	230°
11-09-83	70°	40°	128°	297°	60°	62°

TABLE 3. Stations Co-ordinates

Station	Latitude	Longitude	Altitude
BAI	40.878 N	17.204 E	280 m
TRI	45.709 N	13.759 E	116 m
TNO	45.556 N	7.697 E	260 m
GRF	49.692 N	11.215 E	525 m

changes in the velocity model upon spectral amplitudes have been found to be easily separable from those produced by changes of $Q_\beta(z)$ model, as long as the velocity model is reasonably close to the true structure (Mitchell, 1980). In any case our experience has shown that even using different models consistent with the dispersion data, the results inferred about $Q_\beta(z)$ do not vary significantly.

The effect of lateral changes in elastic properties may cause fluctuations in amplitude which, if not considered, will lead to incorrect determination of Q. Levshin (1985) developed a method which allows a semi-quantitative estimate of the effect of lateral inhomogeneities on mode amplitudes. He analyzed quantities of the type

$$R(\omega) = (\sqrt{uI_1})_{M1}/(\sqrt{uI_1})_{M2} \quad (1)$$

where ω is the angular frequency, u is the group velocity, I_1 is the energy integral (e.g. Panza, 1985a) and M1 and M2 indicate the medium where the receiver and the source are respectively located.

In general, $R(\omega)$ is strongly frequency dependent and the measure of the effects of lateral variations is given by how different R is from unity. In other words R can be taken as a rough estimate of the frequency-dependent, percentage variation in amplitude due to lateral inhomogeneities.

The numerical examples given by Levshin (1985) clearly indicate the magnitude of the bias which can be introduced by lateral heterogeneities when single mode amplitude measurements are performed. The situation is much better when the multimode method is applied. In this case, in fact, the absolute amplitude is not relevant, and meaningful results can be obtained as long as the comparison of amplitudes is limited to the frequency ranges over which the values of $R(\omega)$, for the different modes, are close to each other. Thus, for the first few modes, a preliminary analysis of $R(\omega)$ computed for structural models representative of the considered region may help in controlling the effect of lateral inhomogeneities on amplitude measurements.

These effects of lateral changes in elastic properties, at present, are the most difficult to assess. On the basis of the existing literature and from the consideration that, in general, the anelastic behavior is a second order property of Earth materials, we do not expect that anelastic lateral variations may significantly affect the records we are processing.

Data Processing for the Determination of Q Models

In this section the complete data processing process will be described, making use of a real example. Ideally, the method should be applied to situations where the elastic properties along the path are known, as well as the location, depth and fault-plane solution of the source. For intermediate-sized events, i.e. for events for which the point-source approximation is valid in the period range of interest, in general only the source location is available with sufficient accuracy, and a set of starting values must be chosen for the other parameters.

The earthquakes considered are listed in Table 1 and their fault-plane solutions are shown in Figure 2 and Table 2. The coordinates of the stations used are listed in Table 3.

Analog records produced by the standard LP instruments were digitized with a sampling interval of .5 sec, while the digital broad-band record was decimated to a sampling interval of .25 sec. Then all records were corrected for the instrumental response and processed by means of a multiple-

TABLE 4. Focal Depths Determined by Agencies (ISC, NEIS) and by Waveform Fitting (this study)

Date	Focal depth (km)		
m/d/y	ISC	NEIS	This study
04-07-74	38 ± 7	29	13 - 15
06-19-75	18 ± 7	16	22 - 28
11-16-75	20 ± 7	19	15 - 20
05-07-76	20 ± 10	10	15 - 20
12-08-79	4 ± 4	33	20 - 30
11-24-80	26 ± 3.3	10	20 - 25
11-09-83	37 ± 2.5	37	30 - 35

filter technique called FTAN (Levshin et al., 1972). An example of the resulting plots of spectral amplitudes, contoured as a function of period and group velocity, are shown in Figure 3. The event in the example is number 6 of Table 1. In the left side of this figure the time series is plotted, after removal of the instrument transfer function. From the group velocity dispersion of the fundamental mode it is possible to infer, by inversion, the depth distribution of elastic parameters. We have chosen to invert group velocity and not phase velocity to minimize the error arising from uncertainties in the apparent initial phase of the source.

The parameter to be fixed next is the hypocentral depth. This can easily be done choosing a set of values properly spaced (few kilometers) around the depth given by bulletins. For the different focal depths, using the source mechanism obtained from P-waves polarities, synthetic seismograms are computed. From a comparison of the observed time series with the synthetic ones it is possible to determine the focal depth of the equivalent point source with a better precision then the one given by bulletins (Tsai and Aki, 1970; Panza et al., 1973, 1975a,b). In Figure 4 the experimental signal is compared with the synthetic seismograms computed for several depths.

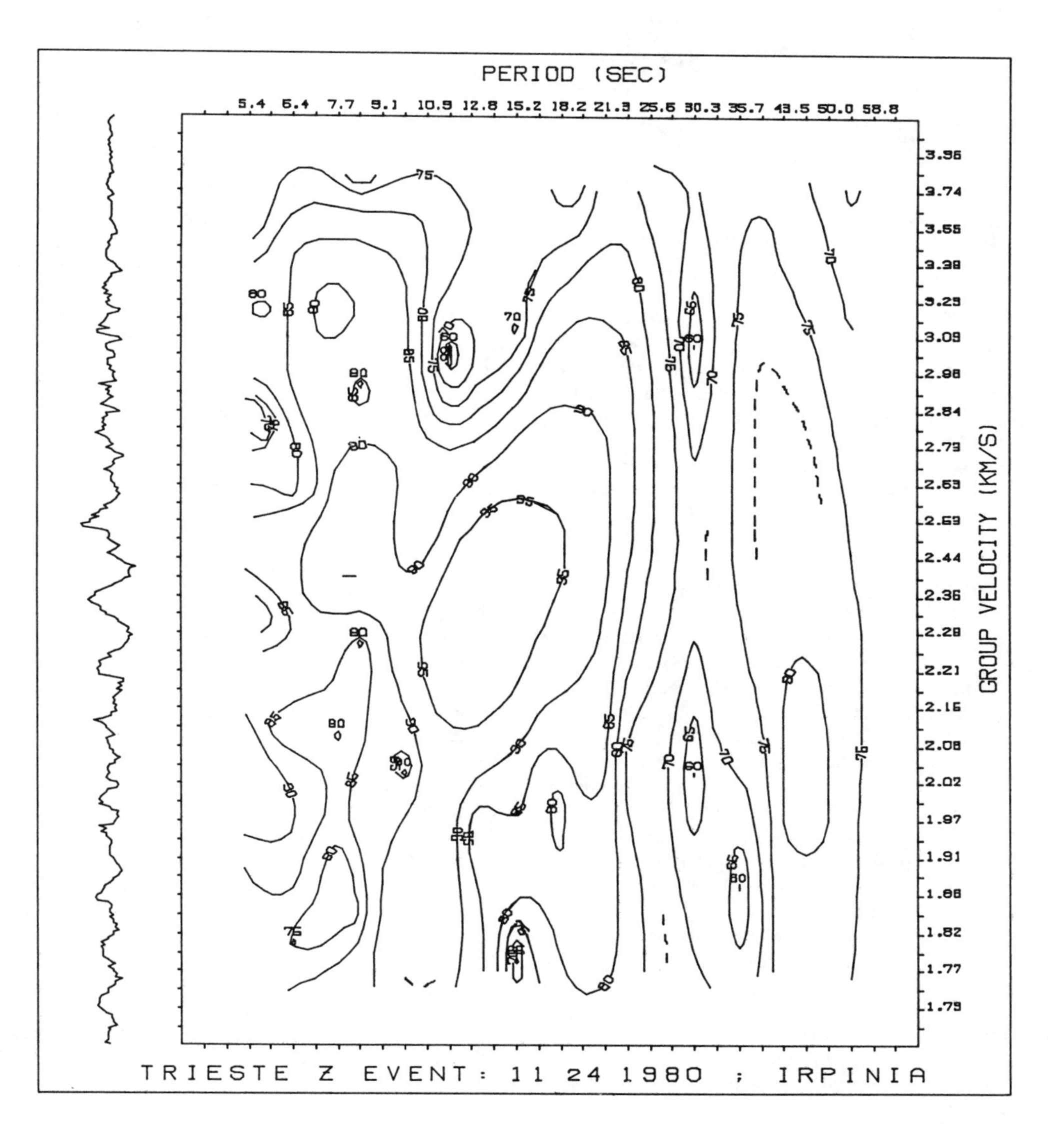

Fig. 3. Spectral amplitudes contoured as a function of period and group velocity for the event n. 6 of Table 1. Epicentral distance equal to 551 km. The time series, after removing instrumental response, is plotted on the left side of the figure.

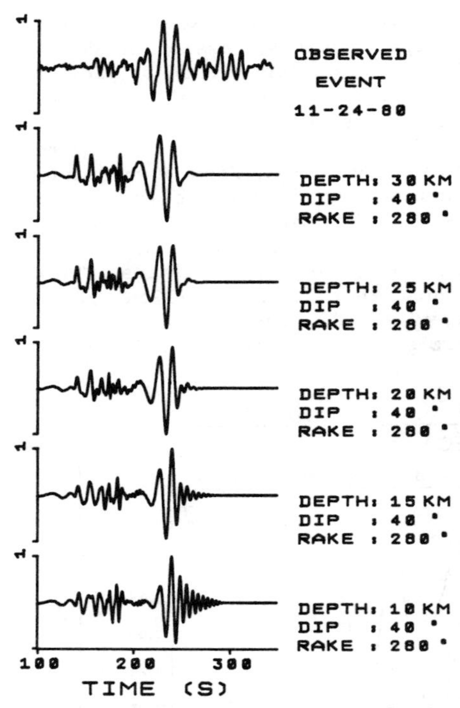

Fig. 4. Synthetic seismograms, convolved with the instrument response of the station TRI, compared with the record of the event n. 6. Computation has been performed for different values of focal depth. Epicentral distance equal to 551 km. The input structure at 1 Hz is given in Figure 10f, while the input fault-plane solution is given in Table 2.

From this comparison it can be deduced that the focal depth varies from 20 km to 25 km.

It is now possible to start the procedure for determining $Q_\beta(z)$. It consists in generating matrices of synthetic seismograms where each element corresponds to a value of rake (λ) and dip (ω). These computations are performed in a way corresponding to those used to find extremes of the focal depth range. Dip angle is variable between 0° and 90°, while rake angle is varied around a value consistent with the predetermined fault-plane solution, namely 40°. These computations are performed for two extreme values of $Q_\beta(z)$: about 100 and about 2000 (Figure 5 and Figure 6). In general the extreme values of Q can be chosen in such a way that the comparison with experimental data allows us to exclude one of the two models. In the example the case with high Q (Q=2000) can be excluded. Therefore, by iteration, the range of possible variations of $Q_\beta(z)$ is reduced. The result of this trial and error procedure defines a single-layer Q model.

Now an attempt can be made to get a finer $Q_\beta(z)$ structure. On the basis of the resolving power of the available data, the best we can do is to consider a two-value model, with the surface of separation, between the two values, placed at variable depth. In this way $Q_\beta(z)$ is found to be very low, about 100, in the upper 25 km of the crust and about 250 at greater depths (see Figure 7).

In Figure 8 the experimental trace is compared with some of the elements of the space of solutions extracted from Figure 7. The choice of these elements is based on a qualitative comparison of the time series and therefore suffers some subjectivity. At present we consider this approach the only applicable when comparing time series of high complexity like the ones obtainable summing some tens of modes. To minimize the integral of the square of the difference between the experimental and the synthetic seismograms will give highly misleading results. In fact, for crustal shocks, the contribution to the integral from the fundamental mode part dominates over everything else obscuring the contribution from higher modes. To avoid this, weighting functions must be introduced whose definition, at present, is missing a satisfactory theoretical basis.

The procedure has been applied to all the events listed in Table 1. The observed traces, compared with a set of possible solutions extracted from the final matrices, are shown in Figures 9a-f.

Modeling

The data of Figure 8 and Figure 9 are consistent also with the elastic properties given in Figures 10a-g which shows, for each path, one of the elements of the space of solutions. Keeping in mind the limited meaning of an average structure along the profiles considered, it must be stressed that, in general, the crustal thickness is in good agreement with the available DSS data (Mostaanpour, 1984). The model SADR shown in Figure 10a and determined from the analysis of event n.1 (Table 1) has a crustal thickness of about 38 km and contains a gentle low velocity layer in the lower crust. The Moho discontinuity is quite sharp. The crust of the model ADRI shown in Figure 10b and determined from the analysis of event n.2 (Table 1) does not contain a low velocity layer and, due to a thinning of the sedimentary layer, is about 36 km. The model PALP shown in Figure 10d and determined from the analysis of event n.4 (Table 1) has a crustal thickness of about 40 km, the Moho discontinuity is quite sharp, and low velocity sediments are present in the first 3 km.

A peculiarity of the model PAD shown in Figure 10g and determined from the analysis of event n.7 (Table 1) is the very low velocity sediments in the first 5 km followed by a very sharp increase of velocity, associated with carbonate rocks and

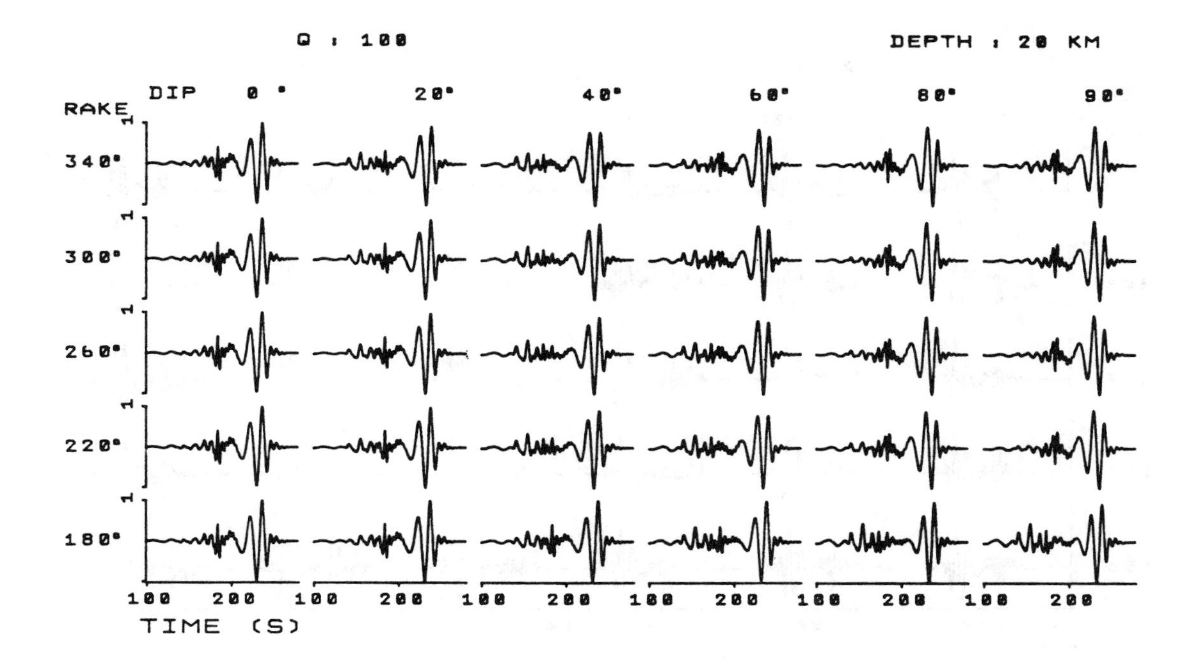

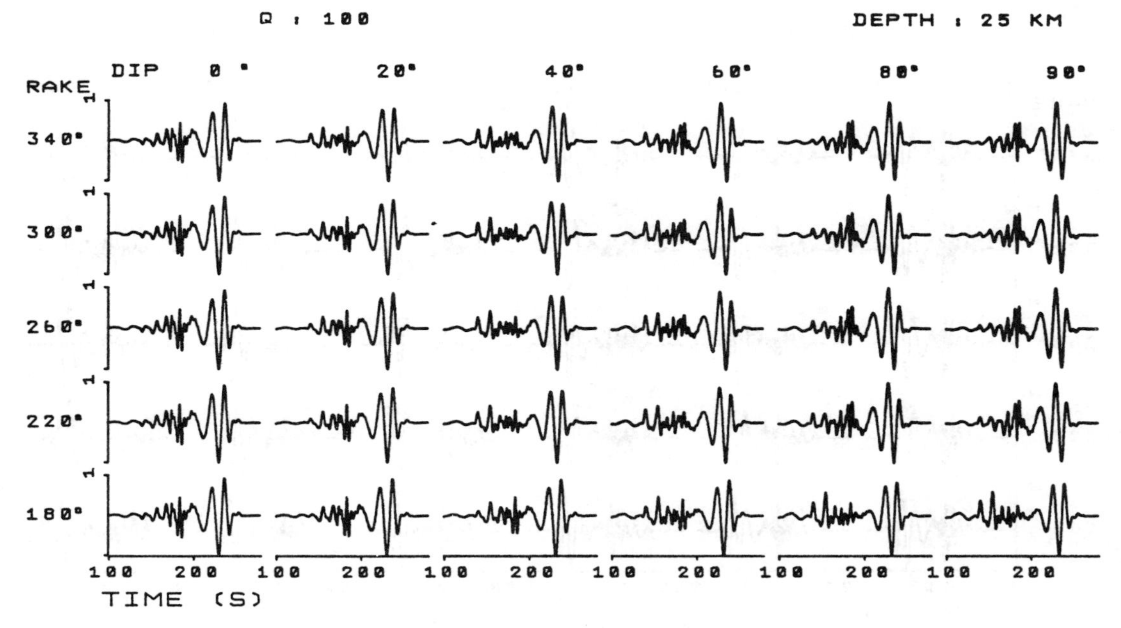

Fig. 5. Examples of matrices of synthetic seismograms, convolved with the instrument response of station TRI, for the event n. 6. Input elastic model, at 1 Hz, given in Figure 10f. The body wave dispersion relations used for P-and S-waves are given by Panza, 1985a. The upper matrix is computed for focal depth, h, of 20 km, the lower for h=25 km, $Q_{\beta}(z)=100$.

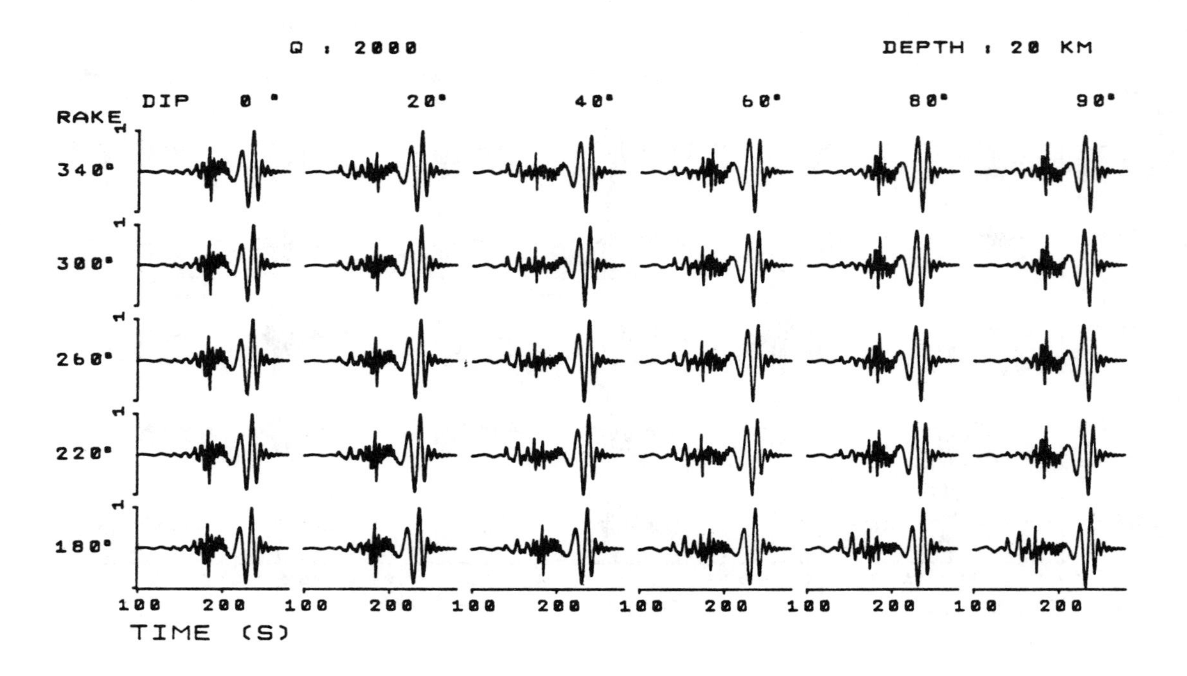

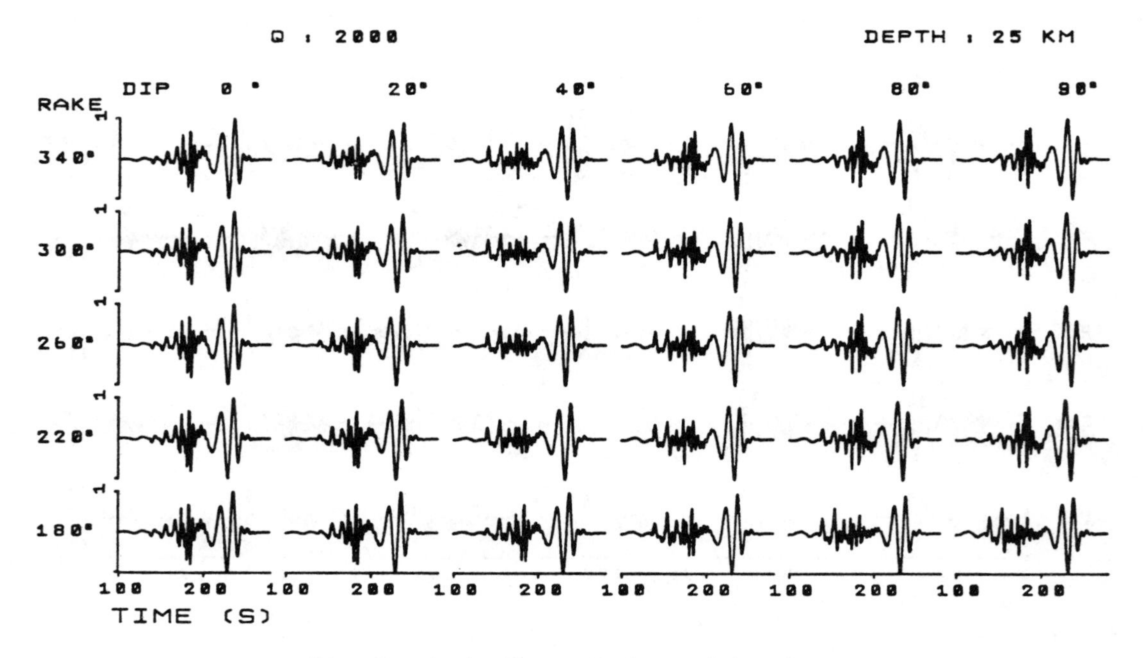

Fig. 6. As in Figure 5, but $Q_\beta(z)$=2000.

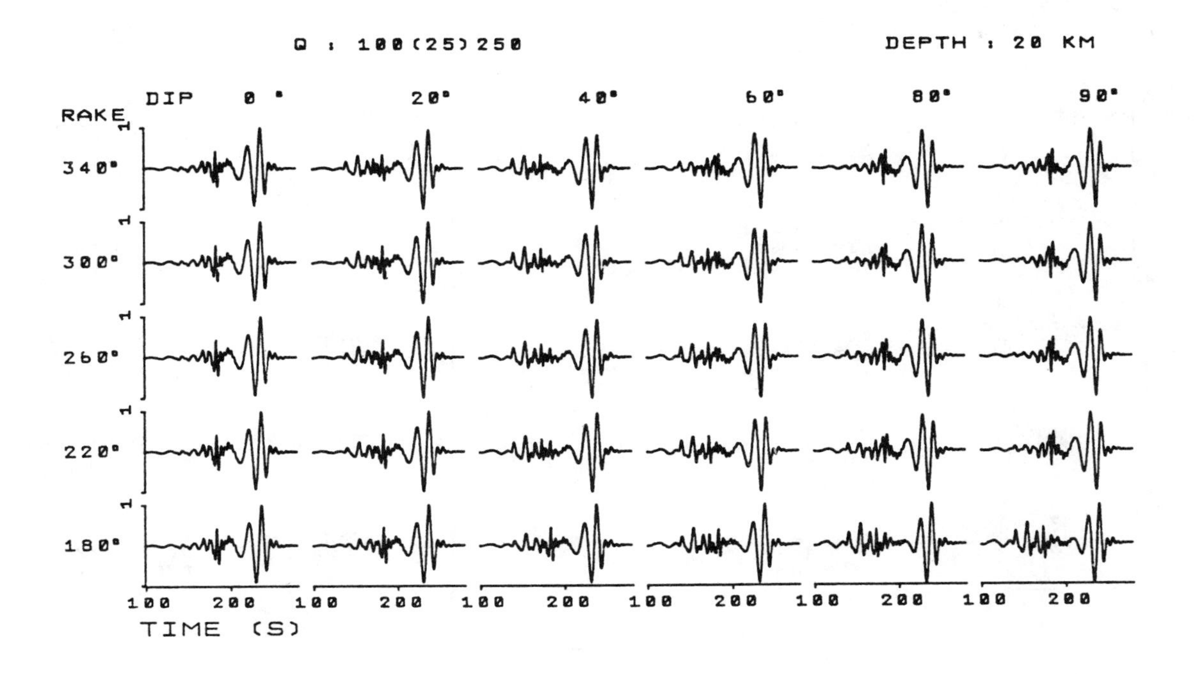

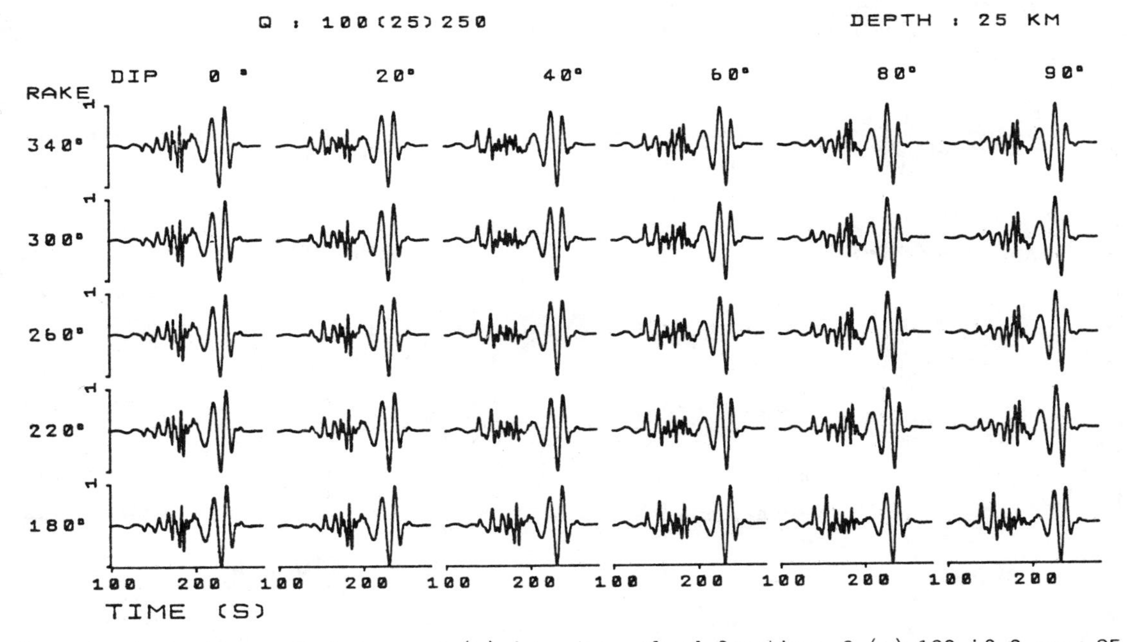

Fig. 7. As in Figure 5, but now $Q_{\beta}(z)$ is a two valued function: $Q_{\beta}(z)=100$ if $0 < z < 25$ km and $Q_{\beta}(z)=250$ if $z > 25$ km.

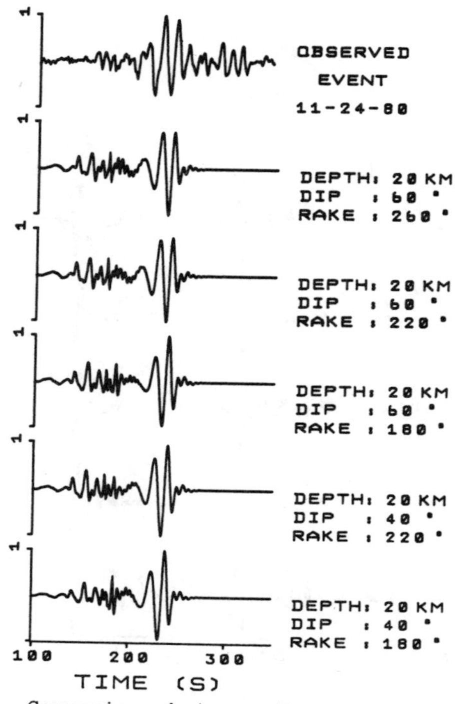

Fig. 8. Comparison between the record of the event n. 6 with some theoretical time series choosen by trial-error from Figure 7. Input elastic parameters at 1 Hz given in Figure 10f, epicentral distance 551 km.

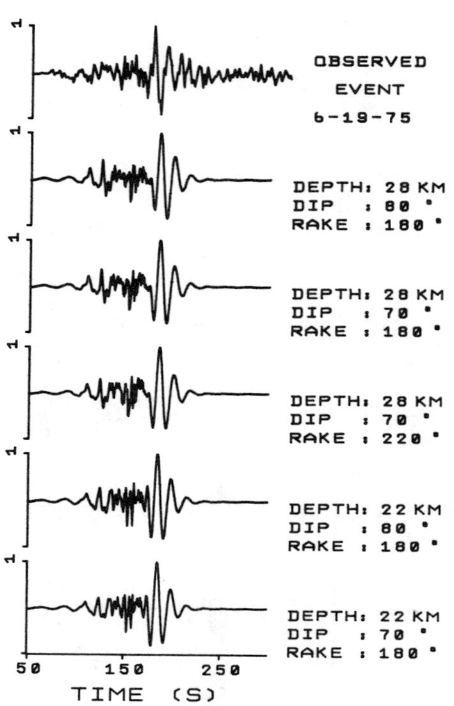

Fig. 9b. As in Figure 8, but for event n. 2. Input elastic parameters at 1 Hz given in Figure 10b, epicentral distance 478 km.

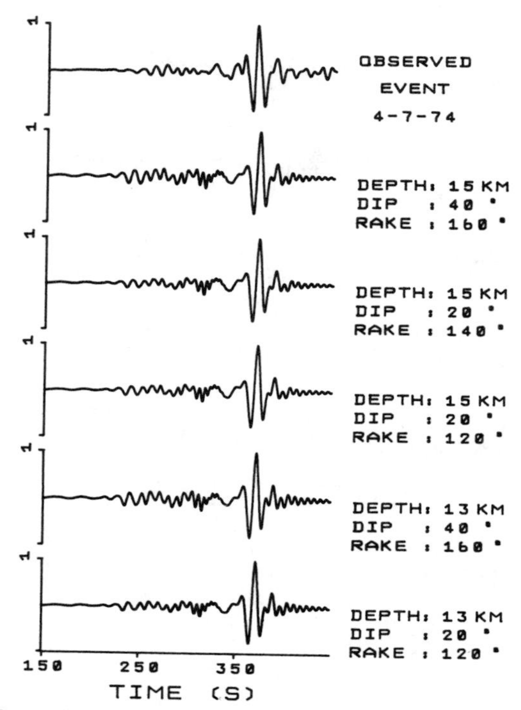

Fig. 9a. As in Figure 8, but for event n. 1. Input elastic parameters at 1 Hz given in Figure 10a, epicentral distance 922 km.

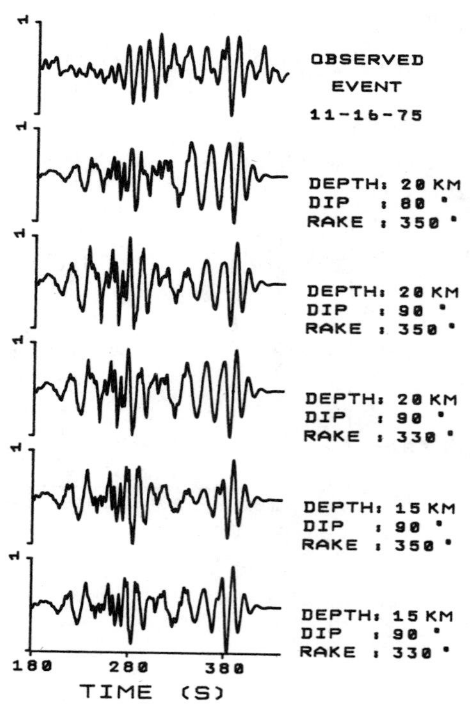

Fig. 9c. As in Figure 8, but for event n. 3. Input elastic parameters at 1 Hz given in Figure 10c, epicentral distance 752 km.

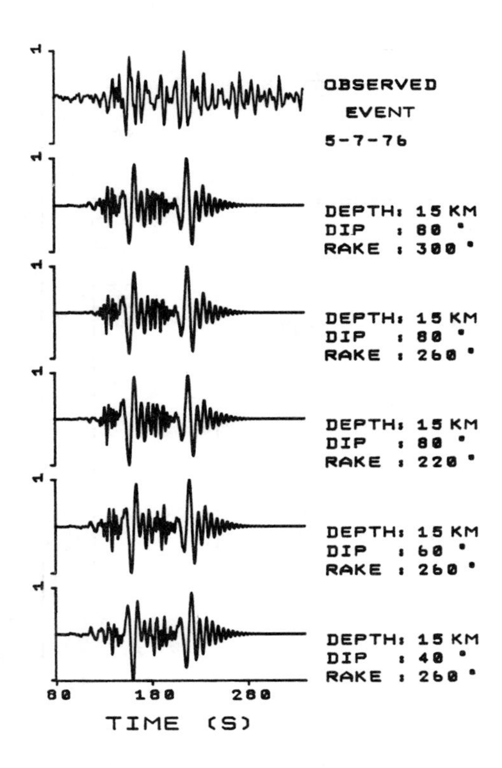

Fig. 9d. As in Figure 8, but for event n. 4. Input elastic parameters at 1 Hz given in Figure 10d, epicentral distance 453 km. The wave-train arriving at about 230 s has been interpreted as the effect of multipathing since multiple events can be excluded on the basis of existing strong motion data. The corresponding synthetic signal has been obtained considering a travel distance of 625 km, simulating, in an approximate way, a signal leaving the source with an azimuth of 100° and reaching the station after a reflection in correspondence of the Apennines.

the crystalline basement. The interpretation of the crustal thickness for this structure is ambiguous. If the velocity discontinuity placed at a depth of 26 km is interpreted as the Moho then the thickness of the crystalline and lower crust does not exceed 10-15 km. This is a small thickness for a continental crust. The anomaly can be removed if we include in the crust the relatively high velocities extending to depths of about 36 km.

The average P and S-wave velocities of model APNC shown in Figure 10c and determined from the analysis of event n.3 (Table 1) are rather low in comparison with the values obtained along the other paths. The Moho discontinuity is rather smooth. The model ADRAP shown in Figure 10f and determined from the analysis of event n.6 (Table 1) has intermediate properties between model ADRI for the lower part of lithosphere and model APNC for the low velocity sediments in the upper crust.

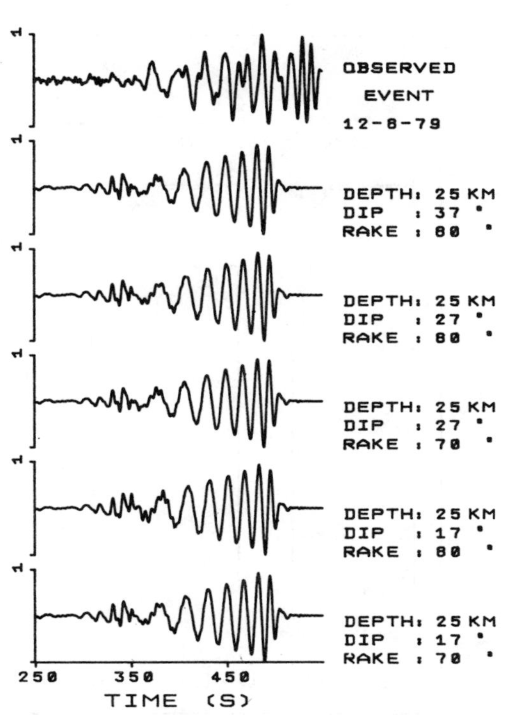

Fig. 9e. As in Figure 8, but for event n. 5. Input elastic parameters at 1 Hz given in Figure 10e, epicentral distance 1278 km.

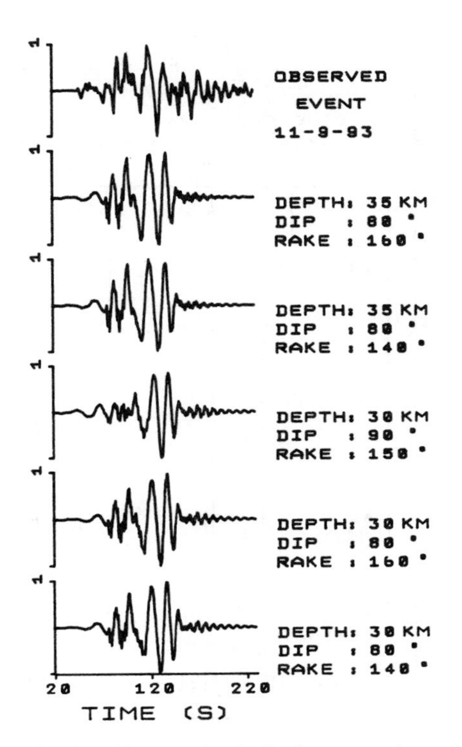

Fig. 9f. As in Figure 8, but for event n. 7. Input elastic parameters at 1 Hz given in Figure 10g, epicentral distance 294 km.

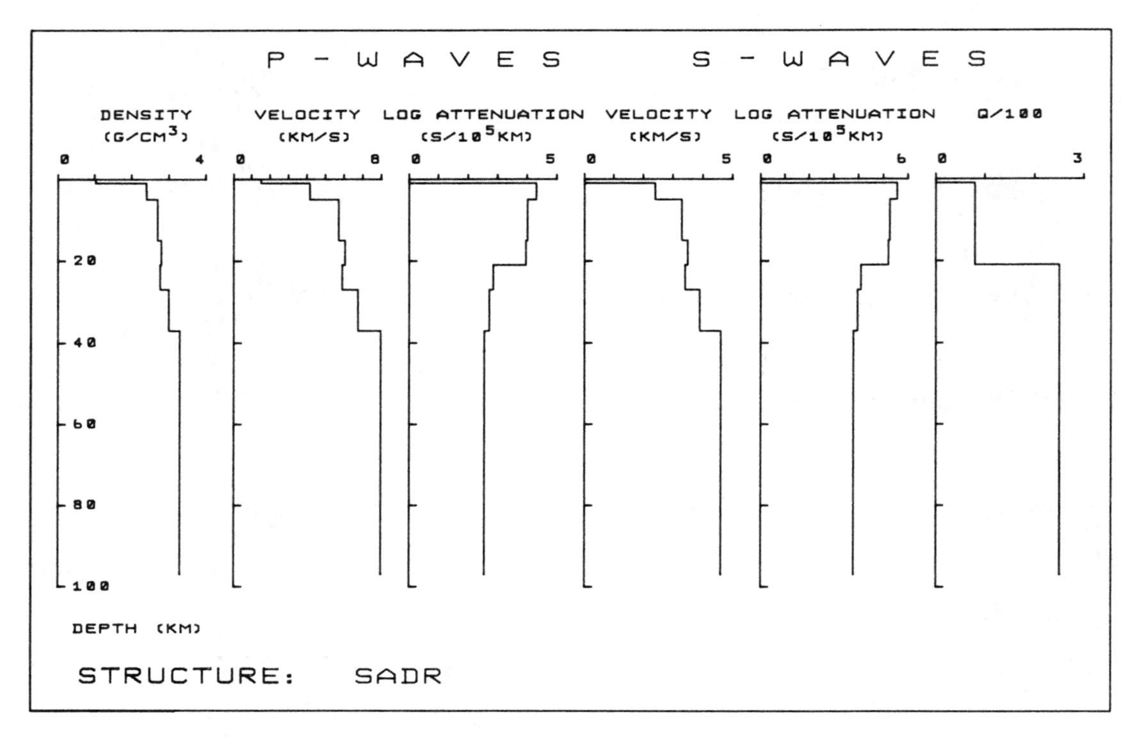

Fig. 10a. Elastic and anelastic models used to compute synthetic seismograms for the event n. 1.

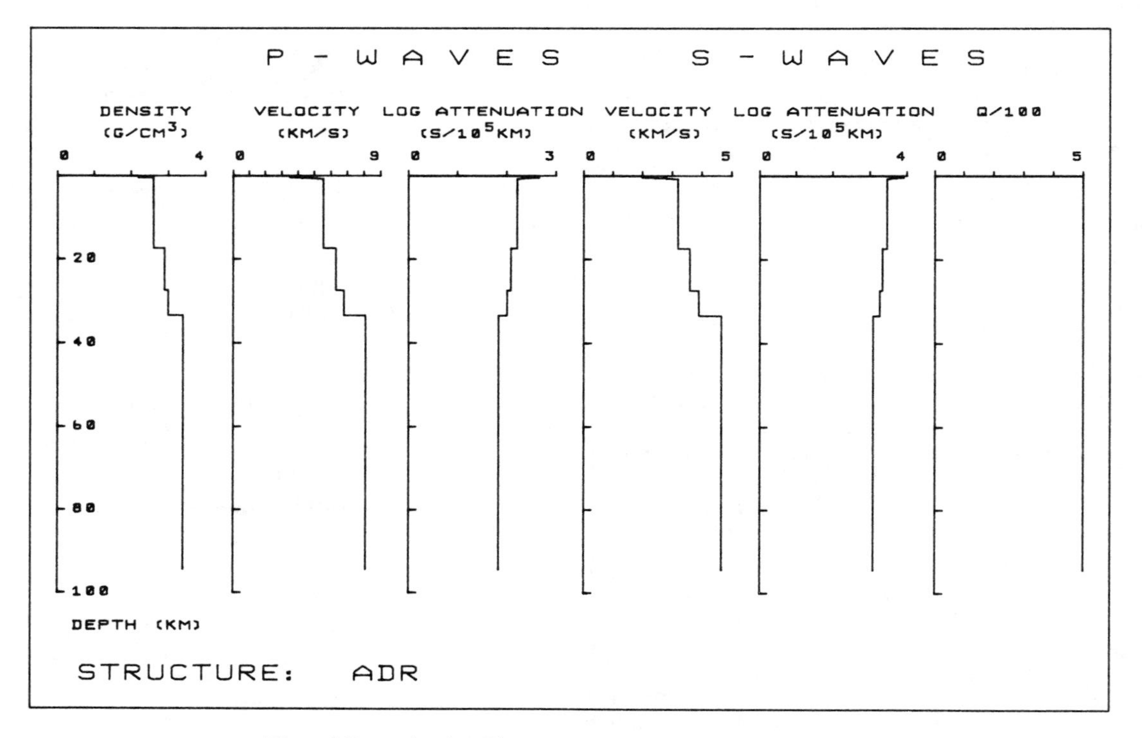

Fig. 10b. As in Figure 10a but for event n. 2.

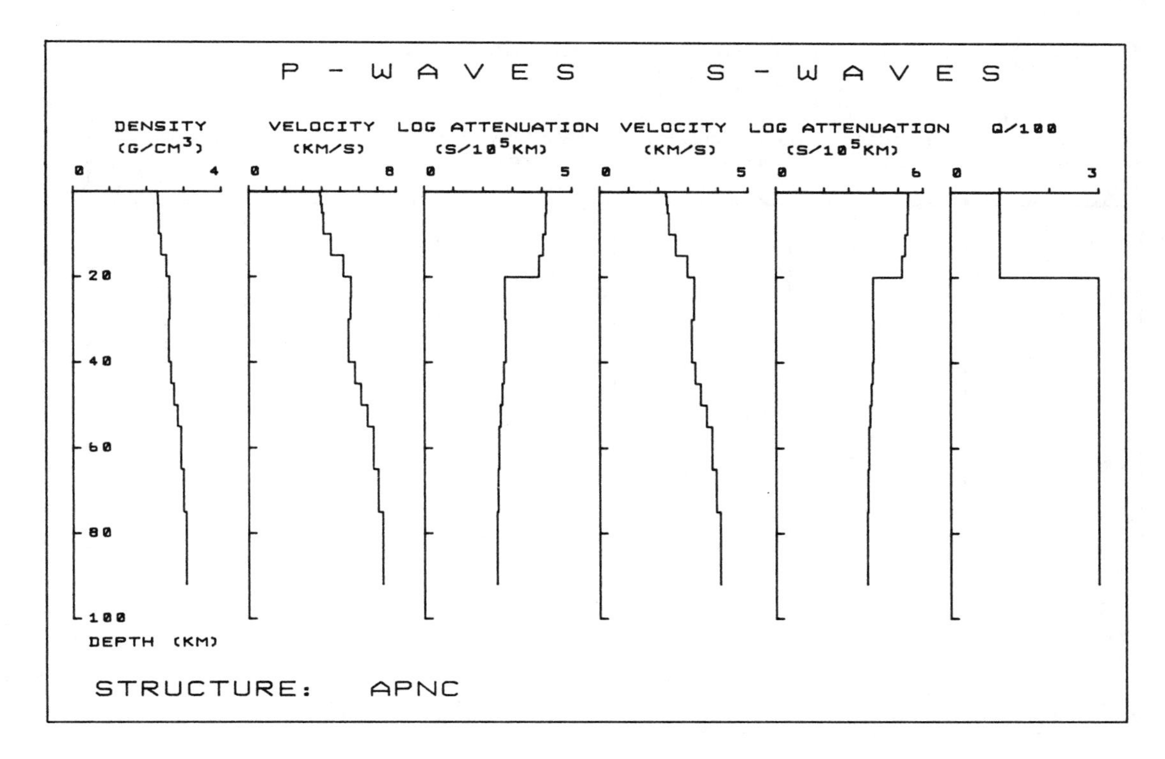

Fig. 10c. As in Figure 10a but for event n. 3.

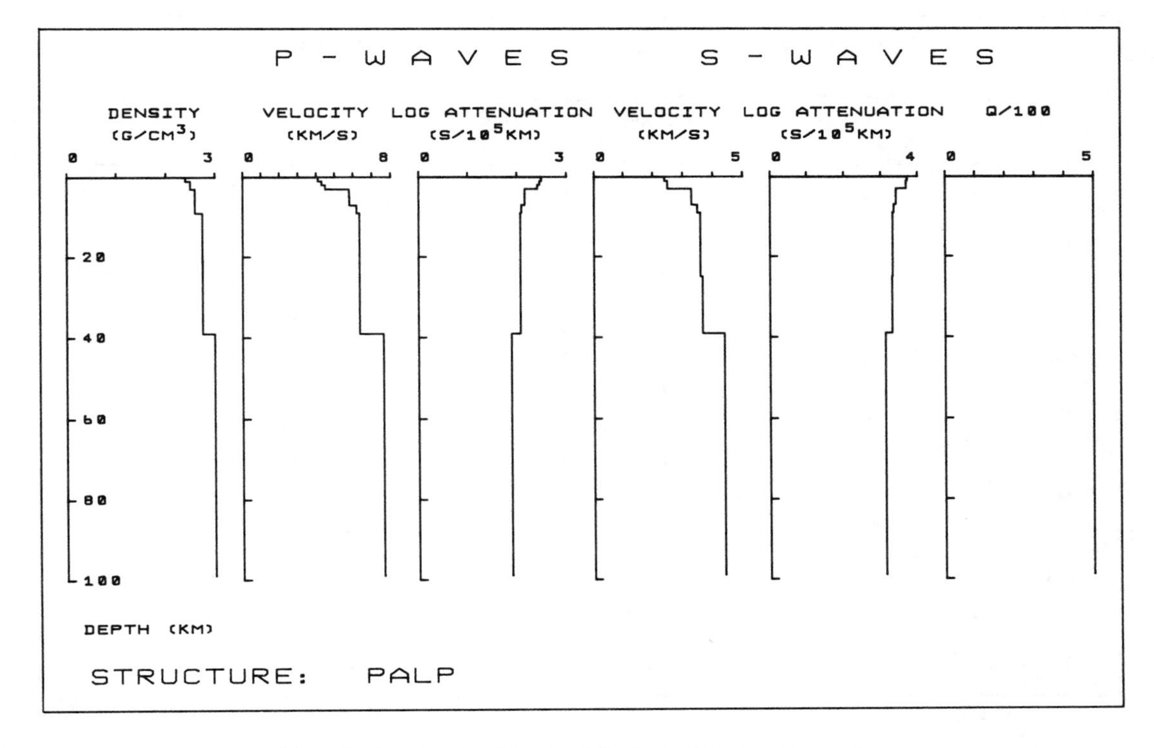

Fig. 10d. As in Figure 10a but for event n. 4.

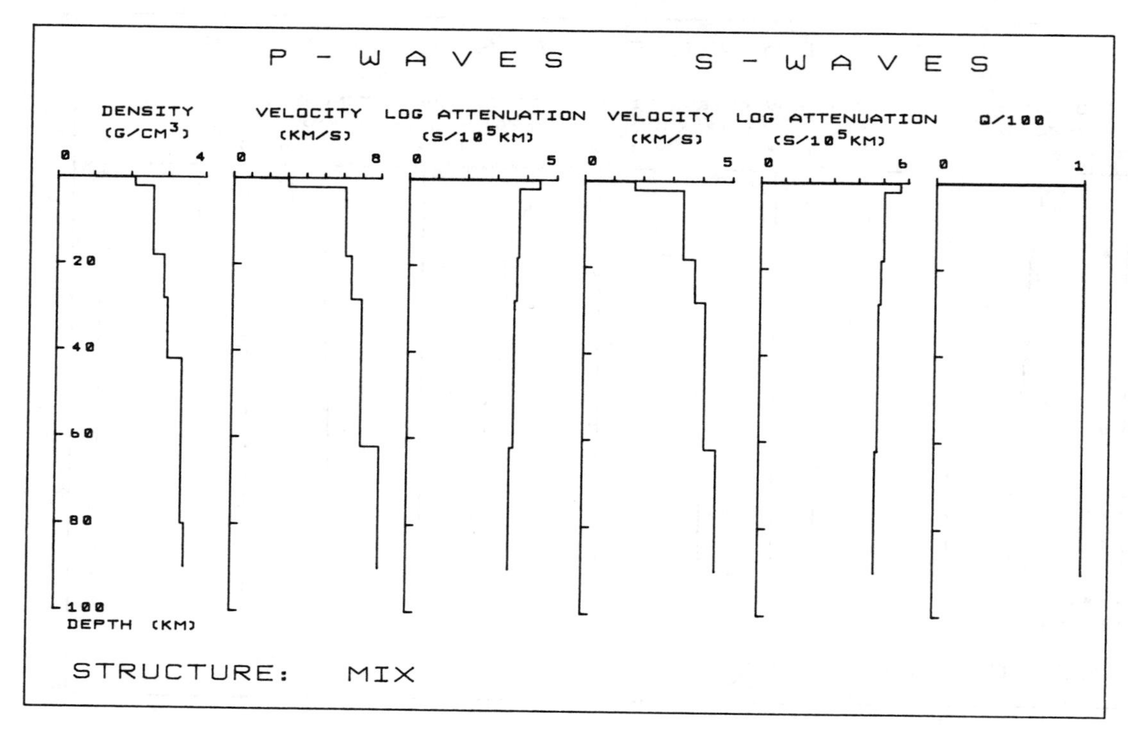

Fig. 10e. As in Figure 10a but for event n. 5.

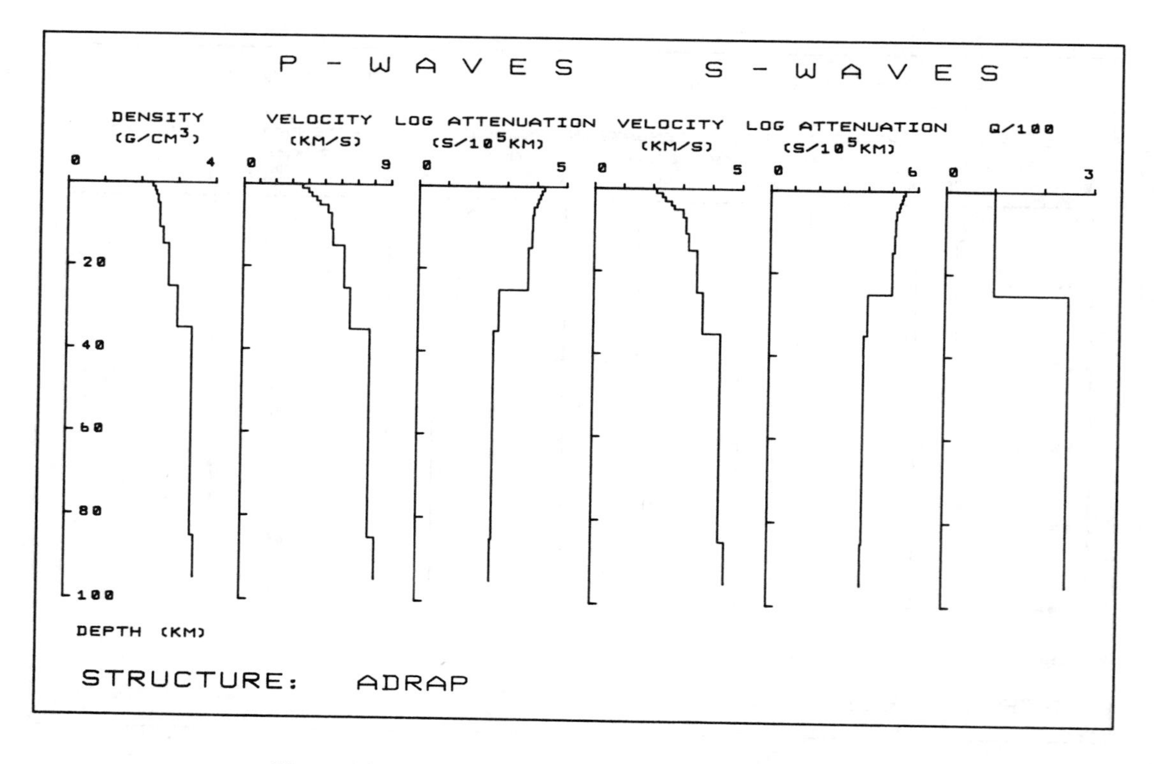

Fig. 10f. As in Figure 10a but for event n. 6.

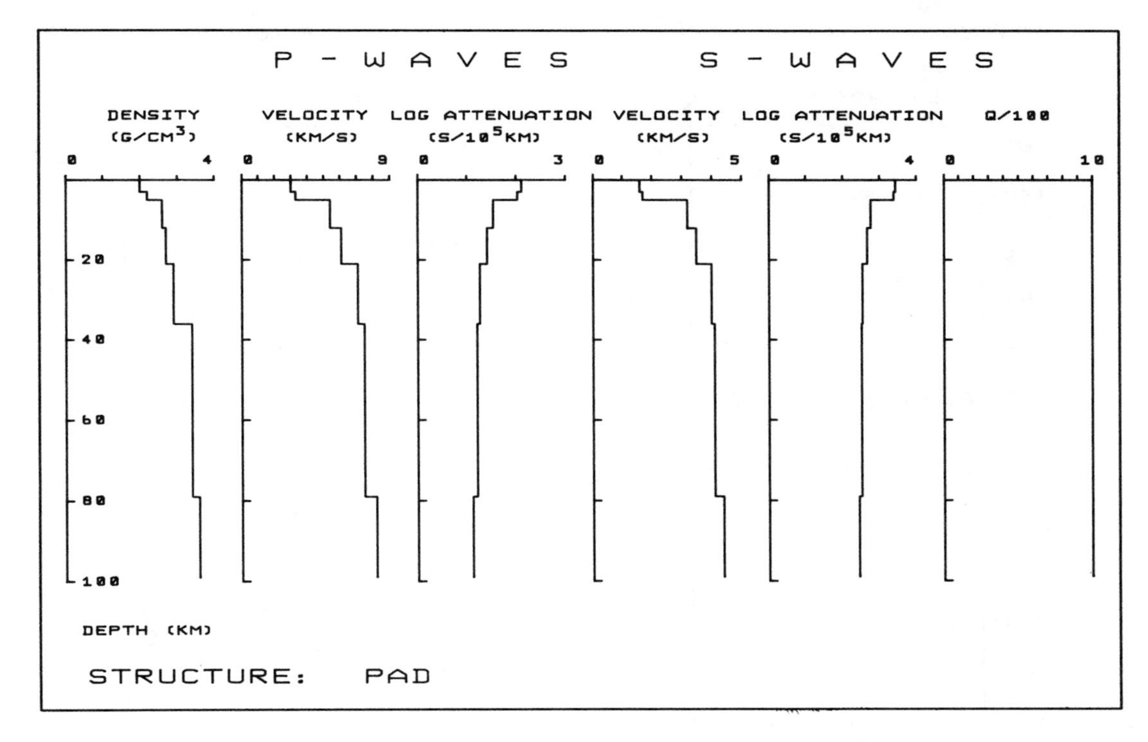

Fig. 10g. As in Figure 10a but for event n. 7.

Finally the structure MIX shown in Figure 10e and determined from the analysis of event n.5 (Table 1) is meant to simulate the average properties along a path where the crustal properties are extremely variable (Mostaanpour, 1984), while the variations in the first part of the upper mantle are relatively small (Panza, 1985b). Furthermore for all of the models of Figures 10a-g the velocity in the upper mantle at depths around 50 km are not very different.

Other important information which can be extracted from the procedure of wave form fitting proposed in this paper concerns the source depth and mechanism in the point source approximation.

In Table 4 the comparison between focal depth determinations by the agencies (ISC, NEIS) and those determined in this study is reported. It is interesting to observe that the discrepancy between the values given by the agencies and those determined in this study decreases for the more recent events, an improvement which is probably related to the improvement of the number and quality of recording instruments. This is especially true considering only ISC data. The exception to this rule evidenced by event n. 5 is only apparent since its location, being in the Southern Tyrrhenian Sea, is still surrounded by very few seismic stations.

The focal mechanism obtained from P-wave polarities and from wave form fitting are compared in Figure 11. The figure includes examples of good agreement and strong discrepancy. This is not a surprise if one takes into account the possible bias in fault-plane solutions based on P-wave arrivals and strongly suggests that for intermediate-size shocks the use of alternative procedures to retrieve fault-plane parameters may give very useful results.

Analysis of Factor R

Making use of equation (1) it is possible to assess some of the effects of lateral variations in velocity structures on attenuation measurements made using the multimode method. For this purpose we have computed the value of R for several combinations of the models of Figures 10a-g. Some of the results are shown in Figures 12a-d.

In each figure, the ratio between the value of R for the N-th higher mode, R(N), to that for the fundamental mode, R(F), is given. In Figure 12a the results referring to the coupling of the structures PAD and APNC are illustrated; from the figure it is evident that the comparison of the amplitudes of the fundamental mode with the ones of the higher modes may give meaningful results only in the frequency ranges around .07-.1 Hz, .13-.17 Hz and .19-.25 Hz. The situation is definitely more satisfactory if coupling of the structures APNC and ADRI is considered (Figure 12b). In this case, in fact, the comparison may give satisfactory results over most of the fre-

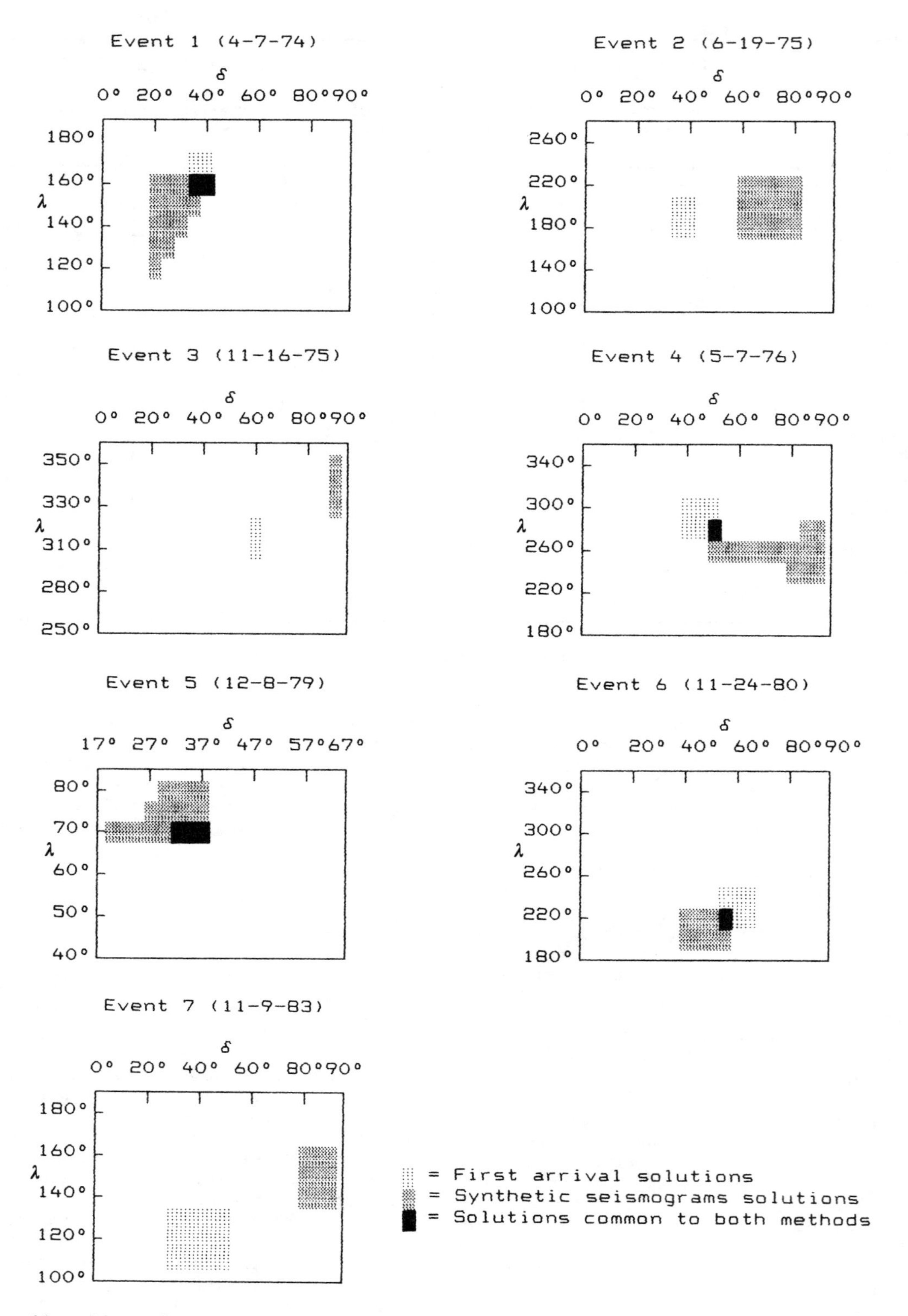

Fig. 11. Dip and rake angles as determined from P-wave polarities and from waveform fitting.

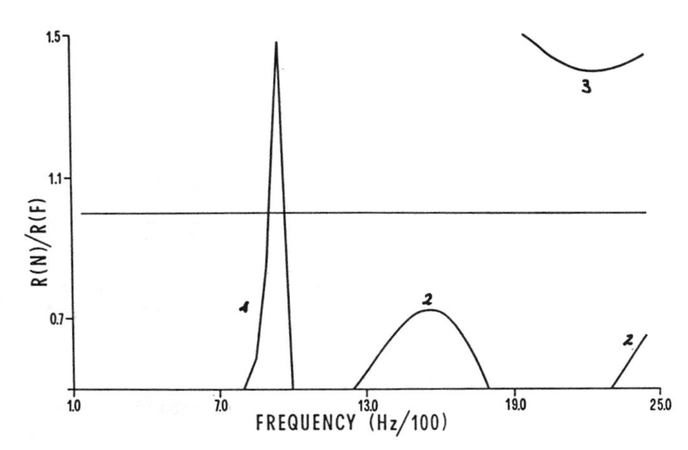

Fig. 12a. Plot of R(N)/R(F) for the coupling between tne structures PAD-APNC. F indicates the fundamental mode, N=1,2,3 indicates the first three higer modes.

quency band considered. On the other hand, the coupling between the structures PAD and ADRI (Figure 12c) would give absolutely unsatisfactory results. In conclusion let us consider Figure 12d where coupling between two quite similar structures is given, the variation being limited to the first 17 km, where the S-wave velocities differ by .1 km/s. In this case satisfactory results could be achieved over the entire frequency band considered.

The plots of Figures 12a-d indicate that very large variations in amplitude ratios can occur if surface wave paths are used which traverse two or more greatly different structures. This result emphasizes the importance of using paths which lie within a single tectonic province, or if they traverse more than one province, those provinces should not differ greatly from one another in velocity structure.

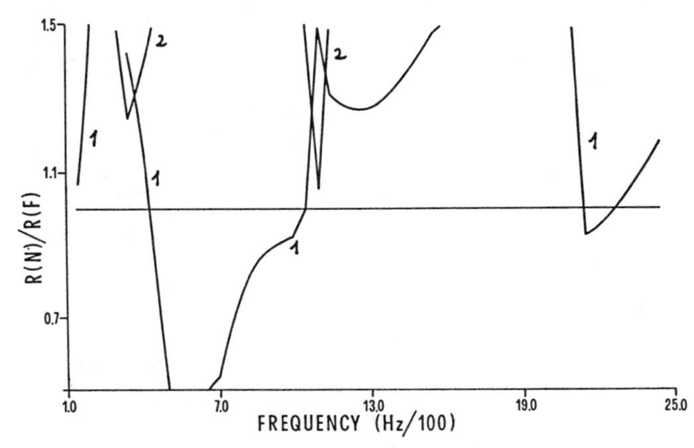

Fig. 12b. As in Figure 12a but for structures APNC-ADRI.

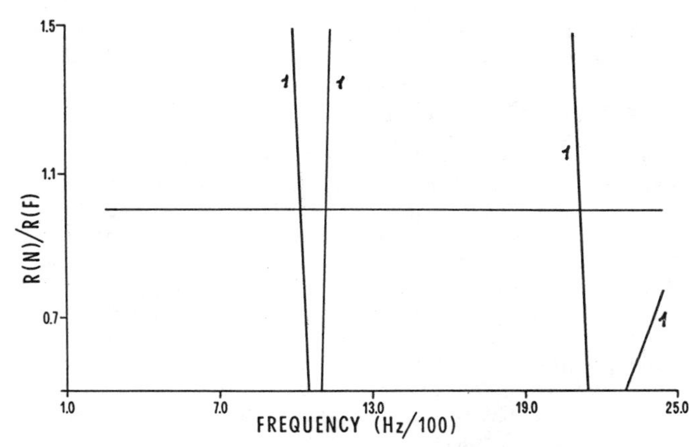

Fig. 12c. As in Figure 12a but for structures PAD-ADRI.

The extension to the time-domain of the single source-single station method for the determination of the anelastic properties of the Earth's lithosphere allows the simultaneous estimation of source parameters (depth and mechanism) and of $Q_\beta(z)$. For path lengths not exceeding 10°, using standard LP data, it is possible to resolve only the gross layering in anelastic properties of the lithosphere. However, relevant information can also be extracted about the elastic properties a-

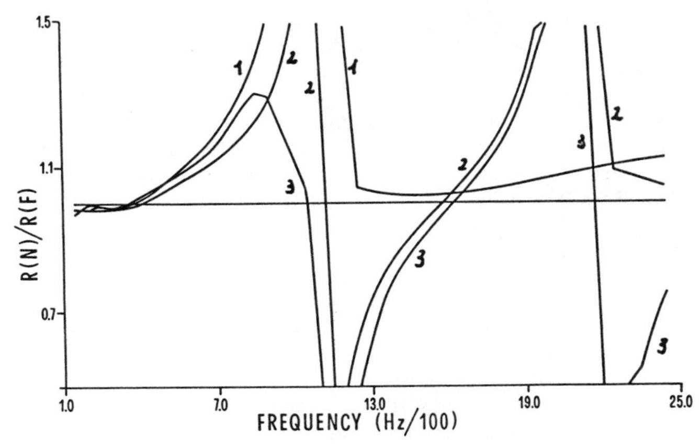

Fig. 12d. As in Figure 12a but for two variants of a possible structural model for the Adriatic Plate.

Conclusions

long the path and about the source in the point source approximation.

As far as the anelastic properties are concerned the main result of this preliminary study consists in the identification of four major categories of Q structures in the Mediterranean region.

The Eastern Po Valley, which is characterized by a thick sedimentary cover, consisting of alluvium and Mesozoic carbonates over a crystalline basement, has a high Q, up to 1000. This high Q value occurs in a region of very low (40 mW/m^2) heat flow. Here and in the following discussion all the information about the heat flow is taken from Cermak and Hurtig (1979).

The North-Central Adriatic Sea which seems to behave as a rigid block of continental type (Panza, 1984, 1985b) has a medium Q value of about 500. A similar value is obtained for the PreAlps from the analysis of event n. 7. The path analyzed runs parallel to the Alpine Chain half way between this and the axis of the Po Valley. Along the path the crustal thickness is about 40 km. This relatively large crustal thickness can be well understood taking into account the southward position of the Alpine thrust front, well inside the Po Valley (Castellarin, 1985). Although the Adriatic and the PreAlps are geologically very different regions they are both characterized by low heat flow values (less than 50 mW/m^2).

Low Q values, about 80-100, occur in the upper 20-25 km of the crust in the Apennines, which are essentially formed by a series of carbonate and flysch nappes which are very highly tectonized. At greater depths Q values increase up to 250-300. A similar distribution of anelastic properties is found in the Mediterranean region between Crete and Southern Italy. The low Q values in the uppermost crust of those areas can partly be explained in terms of scattering associated with the fracturing due to tectonization.

The Tyrrhenian Sea is a basin characterized by a thin oceanic crust and high heat flow values. In this region low Q values, about 100, characterize the entire lithosphere. A similar Q distribution is present in the Rhinegraben region characterized by a thin continental crust and high heat flow. The low Q values obtained from the analysis of event n. 5 may be affected by the lateral heterogeneities along the path (Levshin, 1985); however we consider low Q values representative of the Tyrrhenian and Rhinegraben areas since the path considered also crossed the PreAlps which are characterized by Q values as high as 500. The gross features of the anelastic properties just outlined can be better resolved considering a greater number of events in the region.

The next step, required to reach a better resolving power using the method described in this paper, is the computation of multimode synthetic seismograms for laterally heterogeneous models (e.g. Gregersen et al., 1987) and a definition of proper criteria to be used in the comparison between observed and synthetic signals.

These topics are the subjects of research presently in progress.

Acknowledgments. This research has been performed with the financial support of CNR (grants 85.00033.05 and 86.00666.05) and MPI (40% and 60% funds).

References

Castellarin, A., and G. B. Vai, Southalpine versus Po Plain Apenninic Arcs, in: "The Origin of Arcs", edited by F. C. Wezel, Amsterdam, 253-280, 1986.

Cermak, V., and E. Hurtig, Heat flow map of Europe, enclosure for: "Terrestrial Heat Flow in Europe", edited by V. Cermak and L. Rybach, Berlin, 1979.

Cheng, C. C., and B. J. Mitchell, Crustal Q structure in the United States from multi-mode surface waves, Bull. Seismol. Soc. Am., 71, 161-181, 1981.

Gregersen, S., G. F. Panza, and F. Vaccari, Developments toward computations of synthetic seismograms in laterally inhomogeneous anelastic media, Phys. Earth Planet. Int., 00, 1988.

Kijko, A., and B. J. Mitchell, Multimode Rayleigh wave attenuation and Q_β in the crust of the Barents shelf, J. Geophys. Res., 88, 3315-3328, 1983.

Levshin, A. L., Effects of lateral inhomogeneities on surface wave amplitude measurements, Ann. Geophys., 3,4, 511-518, 1985.

Levshin, A. L., V. F. Pisarenko, and G. A. Pogrebinsky, On a frequency-time analysis of oscillations, Ann. Geophys., 2, 211-218, 1972.

Mitchell, B. J., Frequency dependence of shear wave internal friction in the continental crust of Eastern North America, J. Geophys. Res., 85, 5212-5218, 1980.

Mostaanpour, M. M., Einheitliche Auswertung krustenseismischer Daten in Westeuropa: Darstellung von Krustenparametern und Laufzeitanomalien, Berliner Geowiss. Abh., B10, Berlin, 1984.

Panza, G. F., The deep structure of the Mediterranean-Alpine Region and large shallow earthquakes, Mem. Soc. Geol. It., 29, 3-11, 1984.

Panza, G. F., Synthetic seismograms: the Rayleigh waves modal summation, J. Geophys., 58, 125-145, 1985a.

Panza, G. F., Lateral variation in the lithosphere in correspondence of the southern segment of EGT, in: Second EGT Workshop "The Southern Segment", edited by D. A. Galson and St. Mueller, European Science Foundation, Strasbourg, 47-51, 1985b.

Panza, G. F., F. Schwab, and L. Knopoff, Multimode surface wave response for selected focal mechanisms I. Dip-Slip sources on a vertical fault plane, Geophys. J. R. astr. Soc., 34, 265-278, 1973.

Panza, G. F., F. Schwab, and L. Knopoff, Multimode surface wave response for selected focal mechanisms II. Dip-Slip sources, Geophys. J. R. astr. Soc., 42, 931-943, 1975a.

Panza, G. F., F. Schwab, and L. Knopoff, Multimode surface wave response for selected focal mechanisms III. Strike-Slip sources, Geophys. J. R. astr. Soc., 42, 945-955, 1975b.

Tsai, Y. B., and K. Aki, Precise focal depth determination from amplitude spectra of surface waves, J. Geophys. Res., 75, 5729-5743, 1970.

EARTHQUAKES AND TEMPERATURES IN THE LOWER CRUST BELOW THE NORTHERN ALPINE FORELAND OF SWITZERLAND

N. Deichmann and L. Rybach

Institute of Geophysics, ETH-Zürich, CH-8093 Zürich, Switzerland

Abstract. A detailed study of the recent seismicity in northern Switzerland reveals that earthquakes with magnitudes between 0.9 and 4.2 occur not only in the upper part of the crust, but, contrary to observations in most other intra-continental settings, seismicity extends down to depths of about 30 km in the lower crust. Based on an analysis of the asymptotic velocity of the direct waves, the P-wave velocity in the focal region of even the deepest earthquakes is about 6.2 km/s. The average value of Poisson's ratio for the entire focal-depth range (6 - 30 km) lies between 0.23 and 0.24. Focal mechanisms are mostly a combination of strike-slip and normal faults, with a consistent orientation of P- and T-axes, but without any systematic dependence on focal depth.

The pattern of focal-depth distributions is compared to the results of two-dimensional temperature field calculations, based on new heat-flow data and on plausible lower-crustal petrological models. The temperature modelling takes into account the temperature dependence of thermal conductivity as well as the depth dependence of radioactive heat production. Variations in mineral composition are accounted for with an experimentally established relation between seismic velocity, heat production and rock type. The calculated temperatures for the lower crustal seismogenic zone are above 450°C and are thus higher than what is generally considered compatible with brittle failure.

Introduction

The structure and composition of the lower continental crust has recently been the subject of extensive research. Most of the information that has served as input for the current models has been based on the results of refraction and reflection surveys, complemented to a small degree by geoelectric and gravimetric measurements. Despite this extensive data set, the petrology of the lower crust, which is key to an understanding of the dynamics of the earth's continents, has remained a subject of intense debate and much speculation.

One of the difficulties lies in the non-uniqueness of assigning a rock type to the compressional wave velocities derived from seismic experiments. Shear-wave velocities can help constrain the range of compositional possibilities, but, because shear waves are more difficult to generate artificially and more complicated to identify and interpret in the data, they have only been analyzed systematically in a few cases. Moreover, in most areas the velocity contrast between the middle and lower crust is much smaller than the contrast at the Moho, and thus the velocities of the lower crust can hardly ever be determined directly from first arrivals. Consequently, the lower-crustal velocities themselves are already subject to the inevitable non-uniqueness inherent in most modelling procedures.

Because the source is situated below the earth's surface, earthquake waves sample the crust in a different way than waves from surface explosions. In many cases they can provide a direct measure of seismic velocities, which can only be inferred indirectly from reflection or refraction surveys. Moreover, shear waves are generated much more efficiently from earthquakes than from explosions, so that even simple techniques can give reliable values of Poisson's ratio.

The existence or absence of earthquakes together with a quantitative estimate of the temperature at a particular depth within the crust is an indication of the rheological behaviour, and thus of the material properties, of the focal region. In most intra-continental areas, focal depths suggest a direct correlation between the lower depth limit of earthquakes and heat flow. In high heat flow areas the seismicity is typically restricted to the upper 10 to 20 km of the crust, whereas earthquakes at greater depths are associated with a cold lower crust. This suggests the general existence of one or more well-defined transitions between brittle behaviour and ductile flow within the crust, depending on temperature, chemical composition, strain rate, fluid content and tectonic regime (e.g. Sibson,

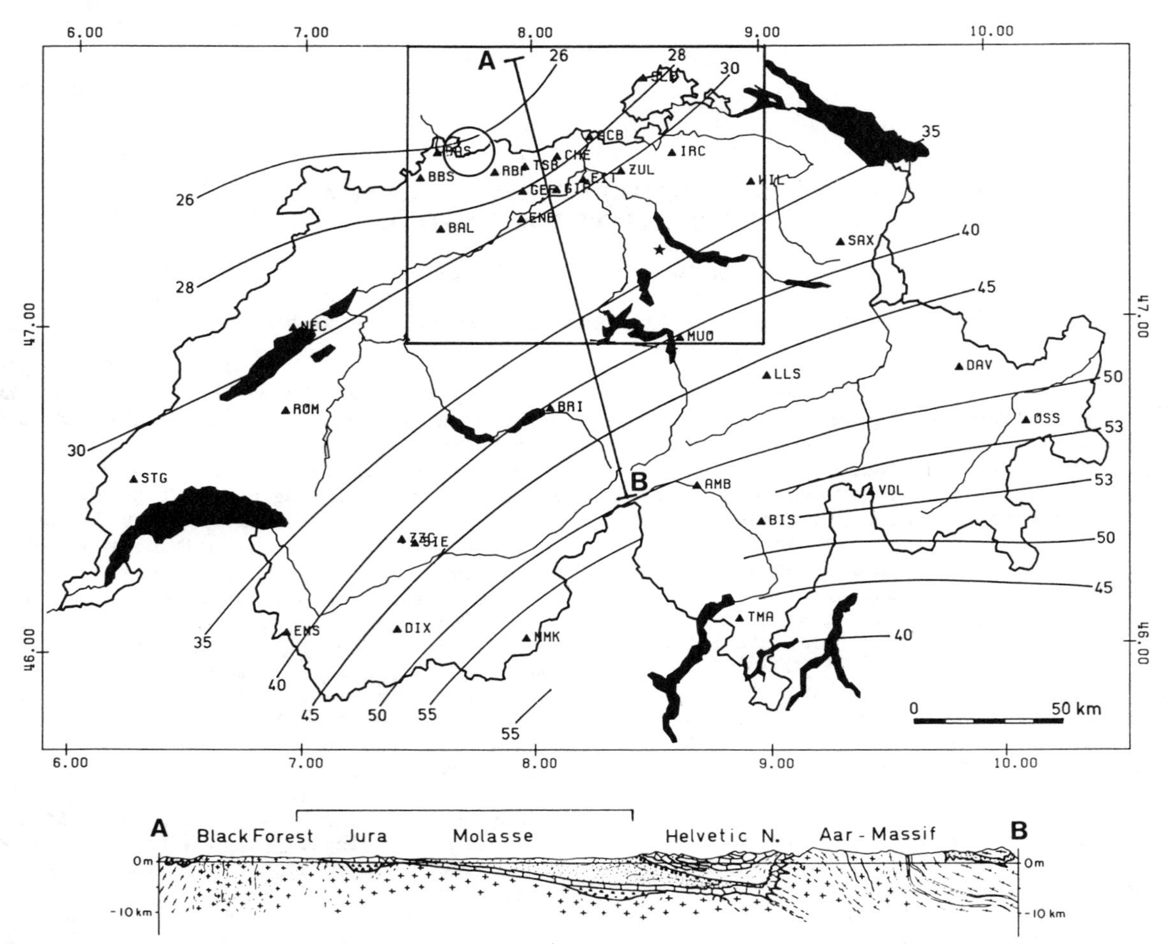

Fig. 1. Map of Switzerland with Moho-depth isolines (km) (after Mueller et al. 1980). Solid triangles indicate seismometer stations in operation since 1984 and earlier; the circle surrounds the area covered by the temporary array near Basel, in operation since 1987. The rectangle borders the region discussed in this paper. The star indicates the epicenters of the Albis events (seismograms in Figures 4 and 5). Line A-B marks the trace of the geological cross-section shown below (after Diebold and Müller, 1984). The bracket above the cross-section delimits segment C-D in Figures 2 and 3.

1982; Meissner and Strehlau, 1982; Chen and Molnar, 1983; Smith and Bruhn, 1984).

In a previous paper (Deichmann, 1987a), it was shown that earthquakes below northern Switzerland occur down to depths of 30 km, in some places even reaching the base of the crust. This paper presents additional data from ongoing seismicity studies in this area. These data support the earlier evidence and give more detailed information regarding P- and S-wave velocities in the lower crust. In addition, we present the results of two-dimensional temperature calculations, constrained by measured bore-hole heat flow values. These calculations show that, under reasonable assumptions of heat production and thermal conductivity, earthquakes in the lower crust below northern Switzerland occur in a depth range where temperatures are higher than what is generally considered compatible with brittle failure.

Tectonic Setting and Instrumentation

The region under consideration comprises part of the northern Alpine foreland, consisting mainly of the the Jura Mountains and the Molasse basin. It is bordered in the north by the Rhinegraben and by the crystalline massif of the Black Forest and in the south by the Helvetic nappes, which were thrust northward during the Alpine orogeny (Figure 1). Crustal thickness across Switzerland varies from 26 km in the north-west to over 50 km below the Alps (Mueller et al., 1980). Moho depths below northern Switzerland are actually interpolated from the results of refraction surveys situated to the north and south, but the Moho isolines shown in Figure 1 follow the general trend of the Bouguer anomalies (Klingelé and Olivier, 1980; Kahle et al., 1976), so that crustal thickness is known to within ± 2 km.

As can be seen in Figure 1, the seismicity in Switzerland is monitored by a relatively dense seismograph network, distributed throughout the country. Most of the data is transmitted continuously via FM-telemetry to a central recording site at the Swiss Seismological Service, where it is stored directly in digital form for subsequent analysis on a computer. In order to lower the detection threshold and improve the location accuracy in the northern part of the country, a temporary FM-telemetry network, consisting of an additional 9 stations with digital recording on three Kinemetrics, PDR-2, seismic event recorders, was installed at the end of 1983 (Mayer-Rosa et al., 1984). For selected events the azimuthal coverage, and hence the accuracy of locations and the reliability of fault-plane solutions, could be further improved by using recordings from stations in Southern Germany, obtained from the Universities of Karlsruhe and Stuttgart, and from a small seismograph array near Basel. This last array, consisting of 8 stations with local continuous recording on analog magnetic tape, was put into operation in the beginning of 1987 and is situated between the stations BAS and RBF (Figure 1).

Epicenter Locations

The location procedure for local earthquakes is based on the widely used computer program HYPO-71 (Lee and Lahr, 1972) with a velocity-depth model consisting of a three-layer crust over a mantle half-space. In order to account for the effect of Moho topography on the travel-time calculations, the thickness of the third crustal layer is adjusted for the Moho depth beneath each station. The travel times of the compressional and shear waves refracted in the upper mantle (Pn and Sn) are calculated assuming a horizontal Moho at an average depth between each station and the station closest to the epicenter. This constitutes only a rough approximation; therefore, the events presented here are located with a distance weighting that gives most weight to the direct P-waves (Pg) and practically excludes the Pn- and Sn-arrivals from the calculations. In general, Sg-arrivals are also used, assuming a constant ratio between P- and S-wave velocities (Vp/Vs) throughout the crust. Since readings of S-arrivals are usually less reliable than those of P-arrivals, they are given a weight of 0.5 for stations with horizontal components and 0.25 when only vertical component seismograms are available.

With this procedure, epicenters of events in northern Switzerland, recorded with good azimuthal coverage, can be determined with an accuracy of about ±1 km. This has been verified on several occasions by locating quarry blasts and refraction shots (Deichmann and Renggli, 1984).

The location results are shown in Figure 2. To ensure uniform data quality, the time period under consideration begins with the operation of the new temporary telemetry network in January of 1984. Despite the relatively short time span covered by the data, the overall distribution of the epicenters, when compared to data recorded over longer time periods, can be regarded as representative of the seismicity in the region (e.g. Mayer-Rosa et al., 1983).

Focal Depths

Given a sufficiently uniform azimuthal station distribution, epicenter locations are not overly sensitive to the velocity-depth model. Focal depths and origin times, on the other hand, can vary rather strongly as a function of the model parameters. Therefore, the routinely calculated depths of selected events, especially of the deeper ones which are of particular interest, were verified both by a model-independent procedure and by modelling travel-time differences between Pg and Pn or Pg and PMP with dipping layers and more accurate Moho depths (Deichmann, 1987a). In general, agreement between the results of the different methods lies within about 2 km.

Focal depths of all but seven events located within the dashed rectangle in Figure 2 are plotted in two vertical cross-sections, one running roughly perpendicular and the other parallel to the trend of the Moho-depth isolines (Figure 3). Of the seven events which were not included in the depth cross-sections, five were not localizable with sufficient accuracy and two may possibly have been explosions. Table 1 lists the relevant parameters of the HYPO-71 locations for the events included in Figure 3. Focal depth uncertainties were estimated on the basis of the procedures mentioned above and on the basis of the number of P- and S-readings (NP and NS), the azimuthal distribution of the stations (GAP) and the number of stations with epicentral distances less than twice the focal depth (NZ).

About 50% of the earthquakes recorded between 1984 and 1987 in the area in question occurred in clusters of similar events. These clusters comprise between 2 and 17 individual events whose seimograms at a given station are practically identical, except for a magnitude dependent amplitude difference. By applying a cross-correlation technique to these signals, relative arrival time differences between events in each cluster could be determined with an accuracy of a few milliseconds. It was thus possible to calculate hypocentral locations relative to a chosen master event in each cluster with standard errors of about 20-30 m (Deichmann 1987b, 1988). In all cases the events occurred within a few hundred meters from each other. For the clusters near Günsberg, Läufelfingen and Winterthur (see Figure 2) the results showed that the hypocenters lie exactly on one of the nodal planes determined from the fault-plane solutions

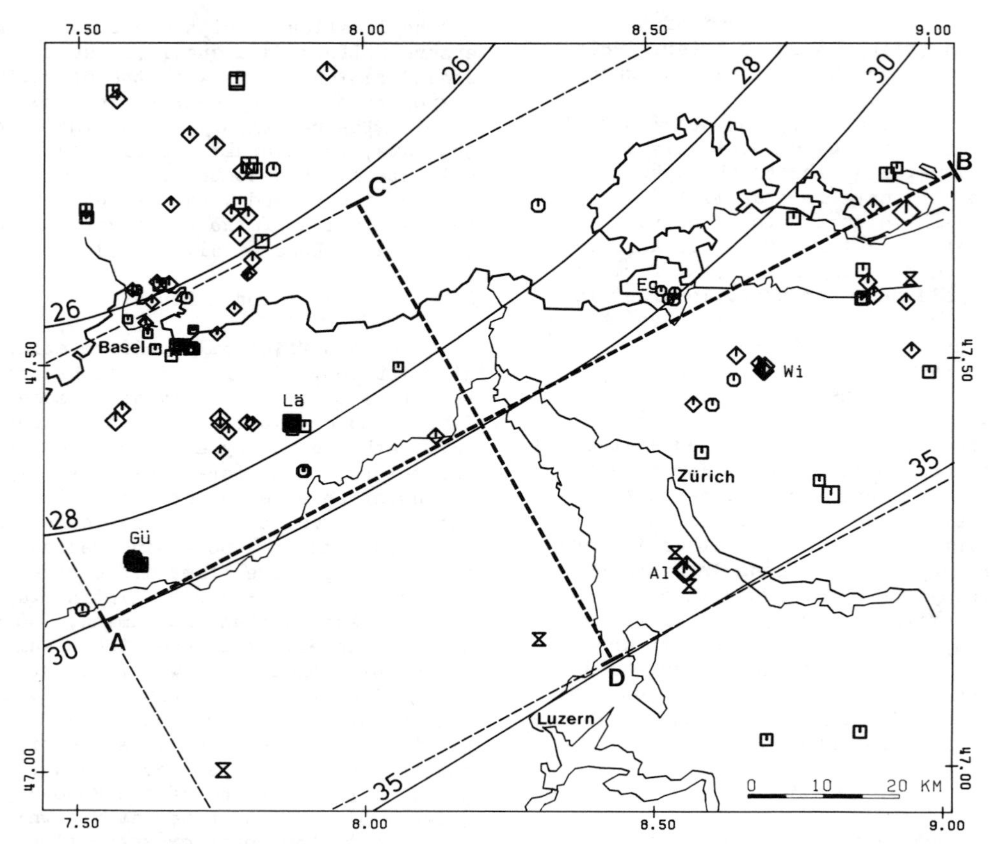

Fig. 2. Epicenter map for the period 1984 - 1987, with Moho-depth isolines (km). The outlined area includes the events projected onto the depth cross-sections A-B and C-D in Figure 3. The different epicenter symbols correspond to different focal-depth ranges (see Figure 3). Locations mentioned in the text: Al - Albis, Eg - Eglisau, Gü - Günsberg, Lä - Läufelfingen, Wi - Winterthur.

of the corresponding master event. Thus, clusters of similar earthquakes are obviously associated with repeated slip on the same fault. In table 1 only the strongest event of each cluster is listed, and the number of individual events that it comprises is indicated in parenthesis after the location name.

For the four weak events near Eglisau the focal depth could not be calculated, but was fixed at 1 km. The shallow focus of these events is deduced from the pronounced surface waves, which suggest that the sources lie in the sedimentary layer, and from the fact that one of them was clearly felt by people living in the vicinity of the calculated epicenter in spite of its small magnitude (M=1.4). The earthquakes in the southern Black Forest and Rhinegraben region (upper left corner of Figure 2) are not included in Figure 3, because they lie too far outside the station array and, consequently, their focal depth determinations are not sufficiently reliable.

The results clearly show that earthquakes in northern Switzerland are not restricted to a particular depth range within the crust, but that they occur at all depths from close to the surface all the way down to the Moho.

Compressional Wave Velocities

In order to facilitate reliable identification of the various phases, the seismograms of selected events were plotted in the form of time-reduced record sections, similar to crustal refraction profiles (Figure 4).

In general, the main P-phases visible in seismograms of crustal earthquakes recorded at epicentral distances up to about 300 km are the direct wave (Pg), the wave refracted in the upper mantle (Pn) and a reflection from the lower crust or the crust-mantle boundary (PMP). Strictly speaking, PMP denotes the reflection from the Moho, but in practice it is often difficult to determine whether the particular phase is reflected from the top or bottom of a lower-crustal layer, or whether it is actually a diving wave from within a lower-crustal gradient zone.

TABLE 1. HYPO-71 location results

DATE	TIME	LON LAT	M	RMS	NP	NS	NZ	GAP	Z	E	LOCATION
84. 1.11	9:17:35.6	702.4/245.3	1.4	.21	3	5	1	100	10	3	Wetzikon
84. 1.11	14:11:57.8	703.9/243.4	3.2	.11	21	8	1	69	11	2	Wetzikon
84. 4.10	16:50:53.2	609.5/253.3	2.6	.17	14	11	9	75	22	2	Breitenbach
84. 4.12	0:50:40.5	623.3/253.7	2.5	.16	18	12	11	68	21	2	Bubendorf
84. 4.20	4:34:04.6	627.6/252.9	1.3	.16	7	11	8	97	14	2	Hölstein
84. 7.12	8:11:55.1	698.9/281.2	2.0	.05	6	2	0	277	8	4	Diessenhofen
84. 8.26	19:30:45.7	683.3/270.9	1.4	.01	3	0	0	148	1	F	Eglisau
84. 8.31	7:25:00.4	695.0/260.3	2.4	.10	14	10	8	131	23	2	Winterthur (6)
84. 9. 5	5:16:49.3	685.0/233.4	4.0	.10	21	5	2	95	15	2	Albis
84. 9.14	22:30:29.4	684.6/232.9	2.9	.13	19	8	9	61	24	2	Albis
84. 9.17	8:48:10.1	634.4/252.6	1.7	.04	6	1	4	167	7	3	Zeglingen
84. 9.19	23:38:09.9	682.4/270.1	1.2	.11	4	2	0	126	1	F	Eglisau
85. 1. 7	9:52:31.5	665.5/223.7	2.1	.09	14	8	11	86	27	2	Hochdorf
85. 2. 3	2:15:38.9	681.5/271.2	1.0	.00	3	0	0	137	1	F	Eglisau
85. 2.24	4:57:55.4	714.5/263.1	1.6	.08	4	3	2	232	14	4	Wil
85. 2.24	21:44:14.7	651.7/251.3	1.6	.06	8	4	8	182	16	2	Schafisheim
85. 3.15	23:31:45.1	683.2/270.0	1.6	.12	4	0	0	126	1	F	Eglisau
85. 5.15	7:45:31.8	634.4/246.5	1.8	.10	5	3	0	271	1	4	Olten (2)
85. 7. 7	0:08:57.9	623.8/205.7	2.7	.05	11	3	8	108	30	2	Langnau
85. 9.19	1:57:18.4	626.8/253.2	1.5	.11	13	14	9	96	14	2	Hölstein
85.11.21	12:07:44.9	714.3/272.9	2.2	.10	12	9	4	234	25	3	Frauenfeld
86. 2.27	12:07:06.8	713.8/282.0	4.2	.08	16	8	2	136	17	2	Steckborn
86. 3.19	23:53:47.8	716.8/260.1	1.8	.16	5	5	1	184	12	3	Wil
86. 3.22	15:37:27.9	623.3/249.0	1.1	.10	7	5	5	282	16	2	Hauenstein
86. 4.10	11:52:13.9	691.4/262.4	2.1	.12	13	7	3	130	13	2	Winterthur
86. 5.30	3:44:06.8	623.3/252.8	1.8	.10	11	5	7	187	17	2	Bubendorf
86. 6. 5	23:41:46.0	708.1/274.1	2.0	.12	3	4	0	198	9	4	Frauenfeld
86. 8.31	9:56:24.2	646.7/260.7	0.9	.05	7	2	4	136	6	2	Frick
86.10. 8	3:12:07.3	683.4/235.6	2.0	.13	16	9	11	62	28	2	Albis
86.11. 1	4:01:08.8	625.1/268.7	1.2	.15	16	6	14	93	17	2	Rheinfelden
86.11. 5	5:02:51.0	711.2/287.2	2.4	.11	21	11	0	182	12	3	Radolfzell (2)
86.11.27	16:03:25.6	619.5/265.7	0.5	.06	4	5	5	231	10	4	Pratteln (2)
86.12.12	22:49:29.3	613.4/266.7	1.0	.13	11	8	7	147	13	2	Basel (2)
87. 1. 8	19:24:20.8	612.5/233.6	2.6	.15	27	16	1	71	6	2	Günsberg (13)
87. 1.13	18:36:48.8	611.8/233.8	2.3	.14	18	12	1	122	6	2	Günsberg (5)
87. 2.18	21:28:00.8	707.9/270.2	2.0	.06	4	4	2	210	11	4	Frauenfeld (3)
87. 4.11	3:14:39.9	632.5/253.0	3.4	.09	19	4	7	89	7	2	Läufelfing.(17)
87. 4.11	9:59:29.9	632.4/253.0	1.1	.09	11	5	6	121	7	2	Läufelfing. (2)
87. 4.11	13:16:13.1	632.4/252.8	1.8	.11	13	9	6	122	7	2	Läufelfing.(14)
87. 4.11	14:50:39.0	632.1/252.7	1.4	.11	13	7	6	124	7	2	Läufelfingen
87. 4.13	22:18:43.9	613.4/265.1	0.7	.11	7	6	7	233	8	3	Muttenz
87. 4.14	17: 6:17.8	632.5/252.8	1.3	.11	9	8	7	121	7	2	Läufelfingen
87. 4.15	13:25:42.5	632.6/252.5	1.3	.09	7	5	5	180	7	2	Läufelfingen
87. 4.17	17:41:45.2	632.8/252.7	1.1	.09	8	5	5	177	7	2	Läufelfingen
87. 4.26	16:06:04.1	626.8/273.5	1.0	.10	7	4	7	229	15	3	Minseln (2)
87. 5. 5	20:29:02.8	685.1/230.9	2.3	.15	15	19	13	62	29	2	Albis
87. 5. 9	17:40:07.7	617.6/263.7	1.2	.14	11	9	9	130	11	2	Pratteln
87. 5.23	7:18:16.3	622.5/265.1	1.2	.09	22	18	19	53	19	2	Rheinfelden
87. 5.27	15:11:54.9	610.1/254.9	1.6	.07	9	8	9	193	20	2	Grellingen
87. 6. 5	18:49:50.7	709.3/270.6	2.4	.08	7	4	2	192	13	4	Frauenfeld
87. 6.22	19:45:53.5	619.5/263.0	1.4	.06	7	6	7	95	10	2	Pratteln
87. 6.22	23:06:06.3	619.2/263.2	1.5	.06	7	6	7	102	11	2	Pratteln
87. 7. 6	19:53:10.6	618.8/263.1	1.3	.05	5	5	6	110	11	2	Pratteln
87. 7.20	18:21:34.0	708.5/272.3	2.1	.15	12	4	3	194	18	3	Frauenfeld
87.10.19	6:46:45.3	686.7/249.1	1.9	.16	7	6	2	185	12	4	Zürich
87.11.23	14:30:49.1	709.2/282.7	1.9	.17	12	7	3	219	20	4	Stein am Rhein
87.12.16	9:36:01.1	617.7/263.2	2.7	.11	27	18	11	36	9	2	Pratteln
87.12.16	10:50:50.2	617.5/263.2	0.6	.04	5	5	5	136	10	2	Pratteln
87.12.20	7:06:24.2	618.5/270.0	1.4	.13	19	9	3	75	3	3	Riehen
87.12.31	15:16:11.2	617.0/263.7	1.2	.06	19	11	11	134	11	2	Pratteln

LON/LAT = Swiss coordinates (km), (bottom left corner of Fig. 2 = 600/200);
M = magnitude; RMS = root-mean-square travel-time residuals (s);
NP= number of P-arrivals (weight $\geq$0.5); NS= number of S-arrivals (weight $\geq$0.25);
NZ= number of stations with epicentral distance $\leq$ twice the focal depth;
GAP= largest azimuth interval between neighbouring stations;
Z = focal depth (km); E = estimated uncertainty of Z (km), (F = depth fixed).

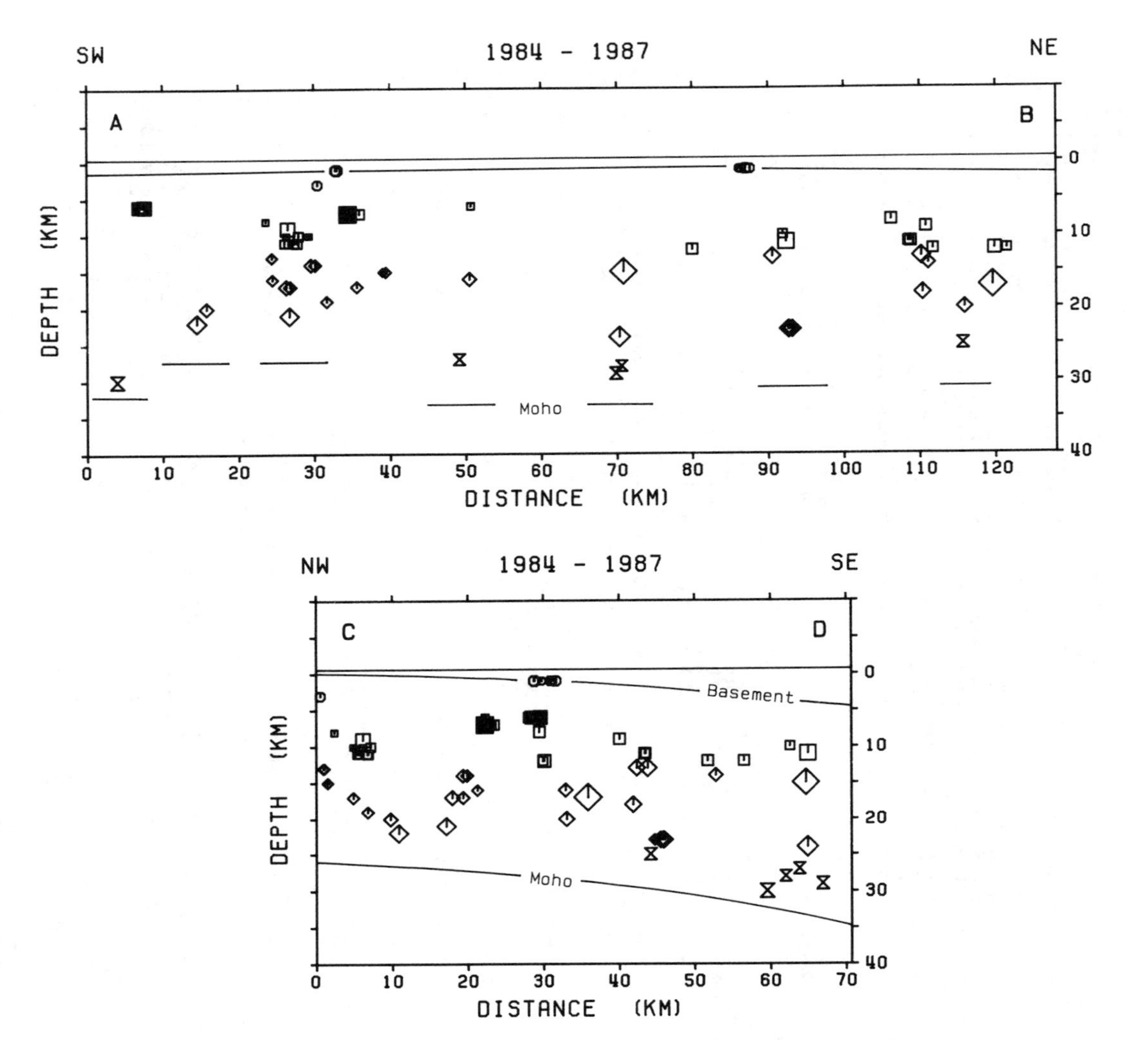

Fig. 3. Focal depths projected onto cross-sections, parallel (above) and perpendicular (below) to the trend of the Moho isolines (see Figure 2). The horizontal line segments in the top diagram mark the Moho depth below the location of the deepest events.

Since epicentral distances are generally fairly well constrained by the HYPO-71 location, the main source of error in the construction of record-sections is the inaccurate origin time, which causes a vertical offset equal for all traces. Consequently, due to errors in origin time and variations of the P-wave velocity structure (including possible undetected velocity reversals) above the source, modelling travel times of Pg alone can not improve depth determinations. However, the relative travel times of Pg and Pn (or PMP and Pg) constitute additional independent information to constrain focal depths, provided the Moho topography and Pn-velocity is sufficiently well known. In the case of deeper earthquakes, errors in the velocity between source and Moho of about 0.5 km/s are less significant than depth differences of one or two km. Even if crustal thickness is known only roughly, one can at least obtain an independent estimate of focal depths relative to the Moho.

The examples in Figure 4 are record sections of three of the four earthquakes with nearly identical epicenters, located near the Albis Pass (see Figures 1 and 2), but with focal depths varying between 15 and 29 km. The travel-time curves were calculated with a program developed by Gebrande (1976). In the model below, are shown only the ray paths for the deepest of the three events.

The seismograms are plotted as a function of epicentral distance regardless of azimuth. Depending on the source-receiver direction, the dipping Moho strongly affects the Pn arrival-times; thus two-dimensional travel-time modelling only matches those stations situated in a restricted azimuth range. In these examples the travel-time curves were calculated to match the arrival-time differences between Pn and Pg at station ROM, situated 130 km SW of the epicenters (see Figure 1). With increasing depth the Pn-phase (marked by horizontal arrows in Figure 4) arrives earlier with respect to the Pg-phase, which clearly corroborates the relative focal depths calculated with HYPO-71 for these three events. While station ROM is situated in the western part of the Molasse Basin, where crustal thickness amounts to about 32 km, the stations at distances around 100 km are located in the Alps, where the Moho dips to depths of more than 50 km. Because of the greater crustal thickness below these Alpine stations, Pn is delayed by such an amount that it is not visible at all on these records. For the same reason, the prominent second arrival at station AMB in the middle record-section, which corresponds to PMP, is late by 0.3 s relative to the calculated travel-time curve. Conversely, in the lower record section, the Pn arrival at station BAL, marked by the oblique arrow in Figure 4, is about 0.2 s earlier than calculated, because the crust is only 28 or 29 km thick beneath this station.

The Pg-arrivals, on the other hand, follow the travel-time curves for a one-dimensional model very closely; this is not merely an artifact of the least-squares fit in the location algorithm, since it is also the case for the more distant stations, which were not used for routine location. In fact, the epicenter coordinates used to plot the three record sections shown in Figure 4 were calculated using only stations situated at epicentral distances smaller than 50 km (for clarity of presentation, only a small selection of all the available seismograms are actually shown here). The asymptotic velocity of the Pg-phase is a direct measure of the P-wave velocity in the depth range of the source, or, if the source is situated in a low-velocity layer, of the maximum velocity between the source and the surface. Figure 4 illustrates that this velocity does not exceed 6.2 km/s, even for the deepest event. This result is corroborated both by record sections plotted out to greater distances than shown here, and by all other events examined in this fashion (Deichmann, 1987a). For comparison, and to allow for possible errors due to station elevation and unknown near-station geology, the dashed line in Figure 4 shows the asymptote of the Pg-phase for a velocity of 6.4 km/s in the focal region. It is clear that such errors have even less effect on the calculated velocities at greater epicentral distances.

It is also important to note that there is no additional reflected or refracted phase visible between the Pn- and Pg-arrivals in any of the seismograms. This indicates that a transition layer between crustal velocities of 6.2 km/s and upper mantle velocities above 8.0 km/s can not consist of a continuous layer with a constant velocity, but, if present at all, must be a zone with a strong velocity gradient below the depth of the deepest earthquake in a given region. As shown in the ray-trace diagram (Figure 4) such a gradient zone produces a diving wave and restricts the super-critical reflection from the Moho to a distance range where it can not be identified as a separate arrival. Consequently the velocities and the thickness of this crust-mantle transition zone, as suggested by this model, are compatible with the available data, but are by no means unique. Nevertheless the range of possible models is small and does not affect the main conclusions

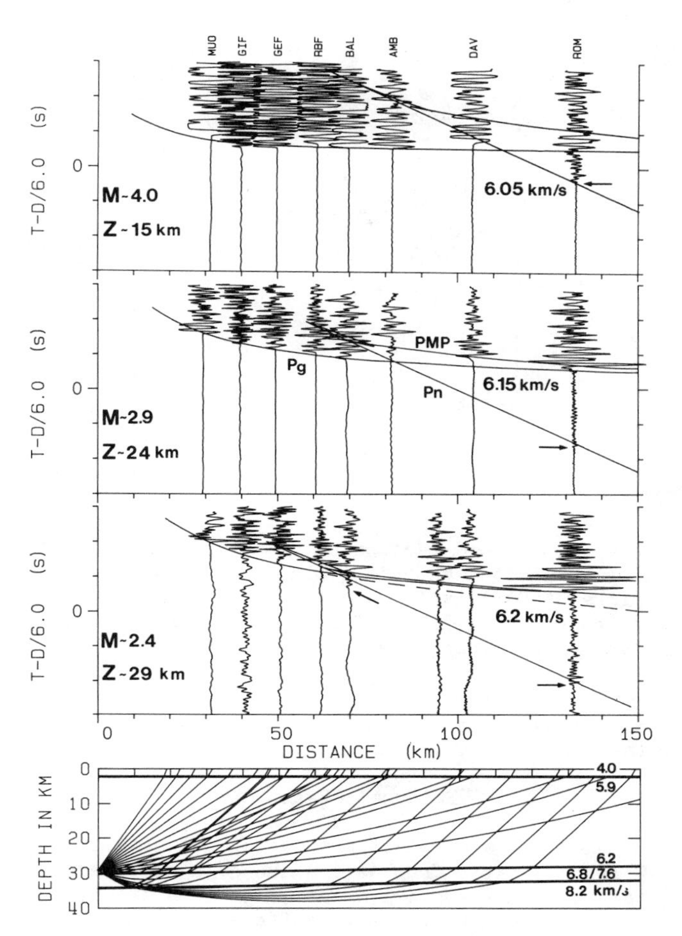

Fig. 4. Time-reduced record sections of three earthquakes with similar epicenters and different focal depths. The continuous lines mark the travel-time curves corresponding to the ray-trace model shown below, and the velocities correspond to the asymptotic velocity of the direct wave. The dashed line indicates an asymptotic velocity of 6.4 km/s. Only the rays for the deepest event are shown. For further explanations, see text.

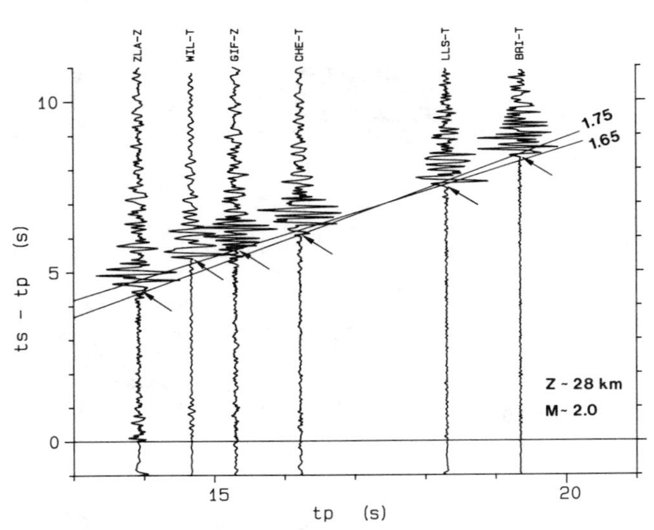

Fig. 5. Record section of an earthquake at 28 km depth, plotted as Wadati diagram. Arrows indicate picks of S-arrivals, and the lines delineate the range of possible TS/TP ratios as determined from these six seismograms alone.

regarding depths of earthquakes and velocities in the focal region.

Poisson's Ratio

One of the simplest methods of estimating Poisson's ratio from earthquake data is based on the well-known Wadati diagram. It consists of plotting the arrival-time differences between the S- and P-arrivals (ts-tp) against the P-wave arrival time (tp) for each station. Under the assumption that Poisson's ratio is constant for all layers traversed by the considered rays, the points of the Wadati diagram will lie on a straight line with a slope equal to (TS/TP - 1), where TS/TP is the ratio of the the S- and P-wave travel times (arrival times minus origin time). In this case, the velocity ratio Vp/Vs, from which Poisson's ratio can be calculated directly, is also constant and equal to TS/TP (e.g. Kisslinger and Engdahl, 1973). The advantage of this method is that Vp/Vs can be determined without prior knowledge of either the location of the hypocenter or the velocity structure. If the assumption of a constant velocity ratio does not hold, the P- and S-phases will follow slightly different ray paths between source and receiver, and, consequently, Vp/Vs is not identical to TS/TP. However, Kisslinger and Engdahl (1973) have shown that, even in this case, the slope of the Wadati diagram gives a useful approximation for the velocity ratio in the source layer. Consequently, 14 events out of the data set presented here, having a sufficient number of clear S-arrivals and reliable focal depths, were analyzed by this method.

As an example of the available data, Figure 5 shows a selection of seismograms of the northernmost of the four earthquakes located near the Albis Pass. Its focal depth is 28 km. The signals are plotted as a Wadati diagram, with the arrival times of the Pg-phase on the horizontal axis. Where available, the horizontal records were combined and rotated to give the transverse component of motion, thus enhancing the SH-phase. For reference, the lines corresponding to TS/TP ratios of 1.65 and 1.75 have been drawn through the Sg-arrivals. Even considering only these few signals recorded over a short arrival-time range, and taking into account possible mispicks as well as unknown lateral variations of Vp/Vs due to near-station effects, it is clear that the most likely fit to the data lies somewhere between these two lines. In fact, the regression line, fitted with a least-squares algorithm through all the available data points of this event, gives a TS/TP ratio of 1.69 ± 0.02 (figure 6, the third line from the right). For all 14 events care was taken to include only Pg- and Sg-arrivals, and each point was weighted according to the confidence with which the corresponding S-arrival time could be picked. In general, this weight was highest for readings that matched closely on all three components, and lowest for readings from vertical records alone. The residuals to some of the data points are greater than the estimated uncertainties, presumably due to significant differences in near-surface Vp/Vs ratios below individual stations. Such differences are not entirely unexpected, considering that some of the stations are situated over several km of poorly consolidated Molasse sediments, while others are located directly on compact Mesozoic rocks or crystalline basement.

Despite these complexities, the data set as a whole gives a rather consistent and uniform picture: TS/TP ratios for all the events lie between 1.68 and 1.73. Bearing in mind that the TS/TP ratios determined in this way are to be regarded as an approximation of Vp/Vs in the source layer, these results seem to suggest a mean Vp/Vs ratio of 1.70. Thus a Poisson's ratio between 0.23 and 0.24 is indicated for the entire focal depth range of 6 to 30 km (Figure 7). An apparent decrease in Poisson's ratio below about 27 km is suggested by the three deepest events. While this observation is not based on sufficient data to draw any further conclusions, there is clearly no evidence for any significant increase of Poisson's ratio with depth. This conclusion contrasts with results from the Black Forest and adjacent areas in southern Germany, where values around 0.26 and greater were determined for the lower crust from high-quality refraction data (Holbrook et al.,1987).

Focal Mechanisms

Fault-plane solutions were constructed for events recorded since January 1984, for which a

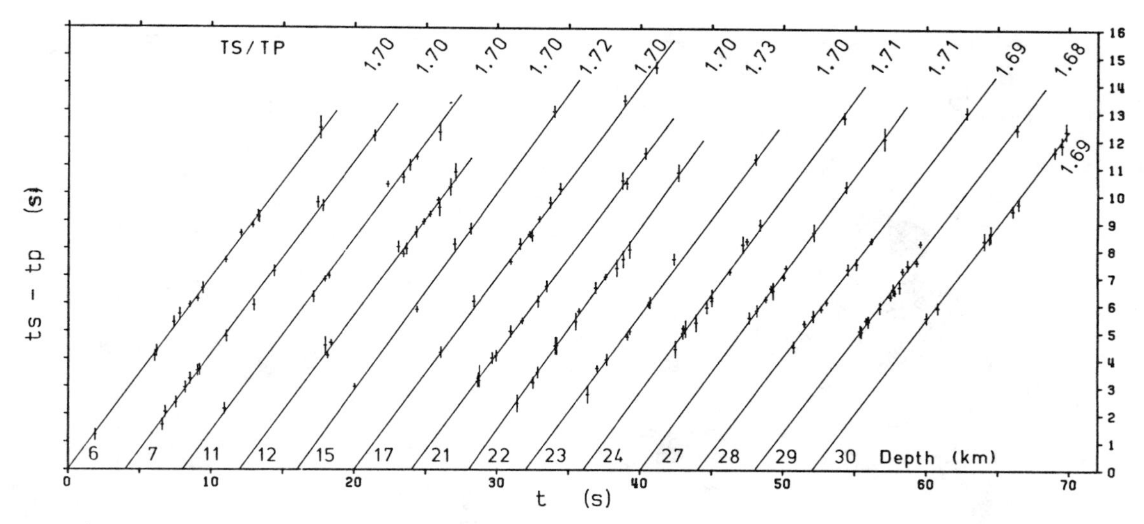

Fig. 6. Wadati diagrams of the 14 events analyzed in this paper, with the corresponding focal depth and the calculated TS/TP ratio. The numbering on the horizontal axis is given only for scale and the intercept of each line is arbitrary.

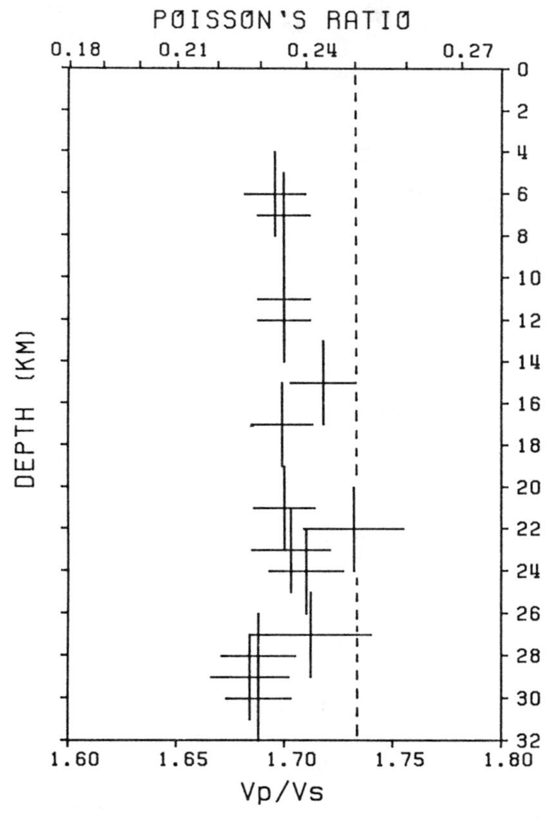

Fig. 7. Poisson's ratio and VP/VS as a function of depth. The size of each cross is proportional to the calculated standard deviation of each value; the dashed line corresponds to a Poisson's ratio of 0.25 (VP/VS=1.732).

sufficient number of reliable first motions could be determined (Figure 8 and Table 2). Strike-slip mechanisms predominate, but there are also a significant number of events with a normal faulting component. However, there is no evidence for any systematic depth variation of either focal mechanism type or orientation of the P- and T-axes. Assuming the rough 45 degree rule for the angle between fault planes and principal stress axes, all focal mechanisms are compatible with the well established regional stress field, characterized by a NNW-SSE oriented compression and an ENE-WSW oriented extension (Pavoni, 1980; 1987).

The fault-plane solutions for the events at depths of 6, 7 and 23 km correspond to the master events of the earthquake clusters near Günsberg, Läufelfingen and Winterthur mentioned before. Based on the high-precision relative hypocenter locations of the events within each cluster, it was possible to identify the nodal plane corresponding to the actual fault (Deichmann, 1987b, 1988). The six earthquakes near Winterthur ranged in magnitude between 1.2 and 2.4 and occurred over a period of 2 years, with time intervals between individual events ranging from a few hours to several months. The hypocenters were located within 150 m of each other on a plane that coincides exactly with the nearly north-south striking and eastward dipping fault plane determined from the focal mechanism solution (Deichmann, 1987b). The close match between hypocentral locations and fault plane for this 23 km deep cluster suggests that earthquake mechanisms are compatible with the traditional model of repeated brittle friction failure occurring on pre-existing faults, even at depths below 20 km.

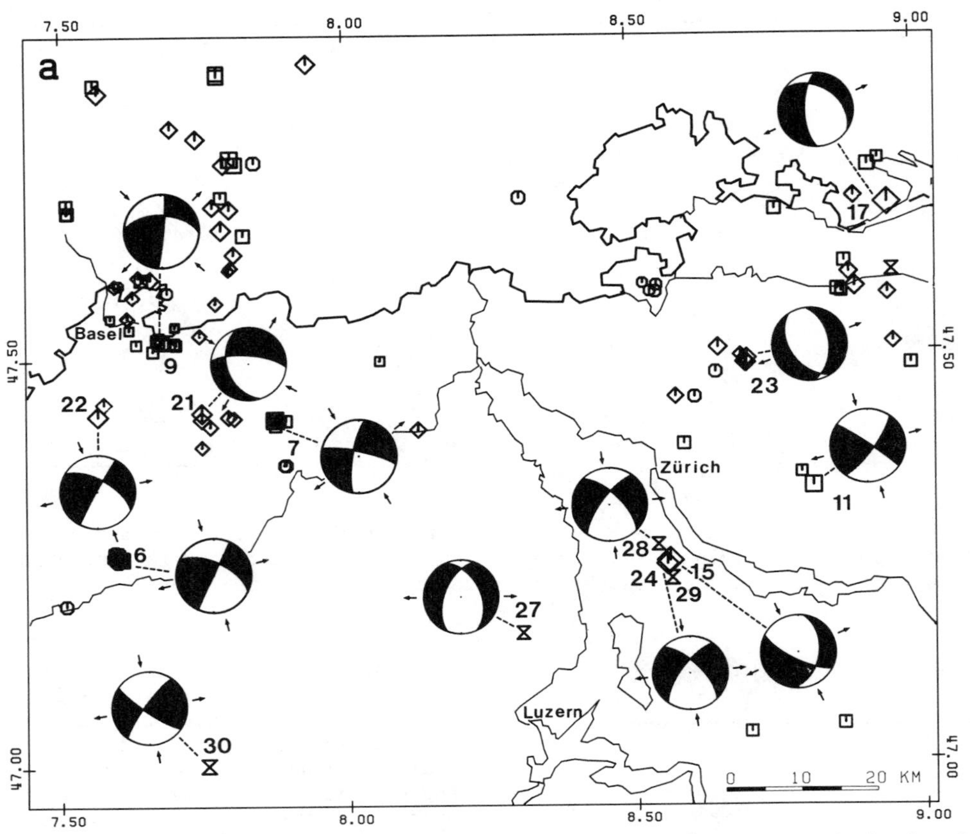

Fig. 8a. Epicenter map with fault-plane solutions. Arrows indicate horizontal directions of P-axes (inward) and T-axes (outward).

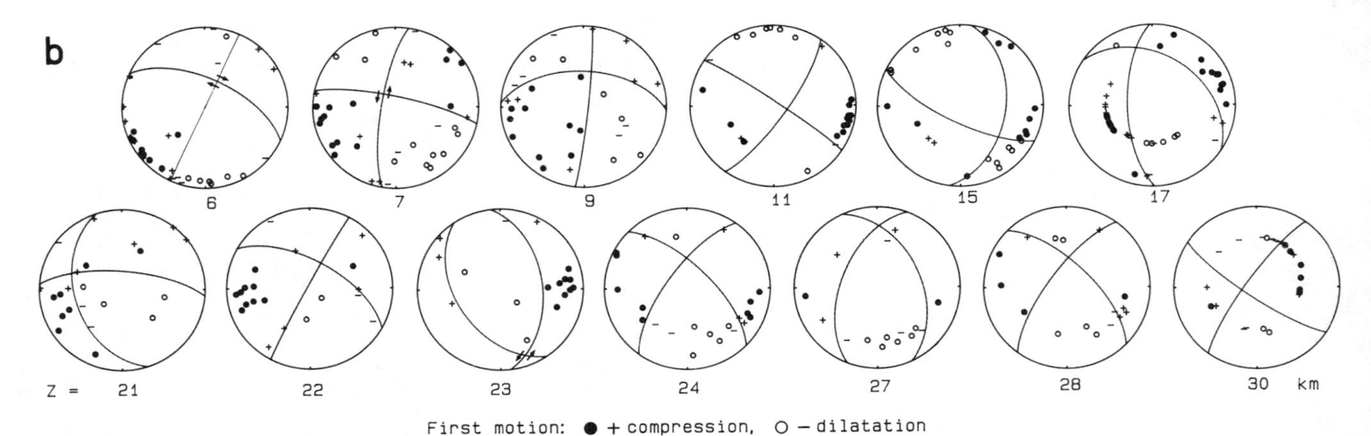

Fig. 8b. Scatter plots of the fault-plane solutions (lower hemisphere, equal area projections). Individual events are identified by their focal depth (km).

Crustal Temperatures

Northern Switzerland is characterized by a pronounced positive thermal anomaly. The thermal gradient, determined from bore hole measurements at a depth range between 750 and 1700 m, reaches a maximum value of 45°C/km (Figure 9). The corresponding surface heat flow values, along the NW-SE trending profile across the region (Figure 9), lie in the range between 75 and 145 mW/m^2 (Figure 11). Numerous hot springs are evidence for extensive hydrothermal activity. Water that

TABLE 2. Fault-plane solutions

Date	Time	Lat. Lon.	Depth	Mag.	Nodal Planes		P-axis	T-axis
87.01.08	19:24:20.8	47.255N/7.605E	6	2.6	295/61NE	25/90	156/20	254/20
87.04.11	03:14:39.9	47.428N/7.870E	7	3.4	190/76NW	282/79NE	146/18	56/02
87.12.16	09:36:01.1	47.521N/7.675E	9	2.7	273/54NE	6/86SE	134/21	236/28
84.01.11	14:11:57.8	47.335N/8.815E	11	3.2	36/76SE	304/85NE	351/06	259/13
84.09.05	05:16:49.3	47.247N/8.562E	15	4.0	8/44SE	117/72SW	345/46	236/17
86.02.27	12:07:06.8	47.680N/8.955E	17	4.2	304/38NE	179/65SW	131/58	247/15
84.04.12	00:50:40.5	47.435N/7.748E	21	2.5	162/42SW	275/71NE	143/49	32/17
84.04.10	16:50:53.2	47.432N/7.565E	22	2.6	300/62NE	208/87NW	160/21	258/17
84.08.31	07:25:00.4	47.488N/8.700E	23	2.4	146/46SW	353/48NE	336/76	70/01
84.09.14	22:30:29.4	47.243N/8.557E	24	2.9	315/67NE	216/70NW	175/31	266/02
85.01.07	09:52:31.5	47.162N/8.303E	27	2.1	336/46NE	201/54NW	170/65	270/04
86.10.08	03:12:07.3	47.267N/8.542E	28	2.0	315/66NE	217/72NW	174/30	267/04
85.07.07	00:08:57.9	47.003N/7.753E	30	2.7	124/79SW	216/79NW	350/00	80/15

Nodal Planes: strike/dip; P- and T-axis: azimuth/plunge.

is drained from higher elevations to the north and south rises to the surface at the margins of the Northern Swiss Permocarboniferous Trough. Thus part of the enhanced surface heat flow is caused by convective heat transport.

The extent of the hydrothermal effect was quantified by two-dimensional numerical calculations based on a coupled thermo-hydraulic model (Figure 10). The assumptions and constraints of these model calculations are discussed in

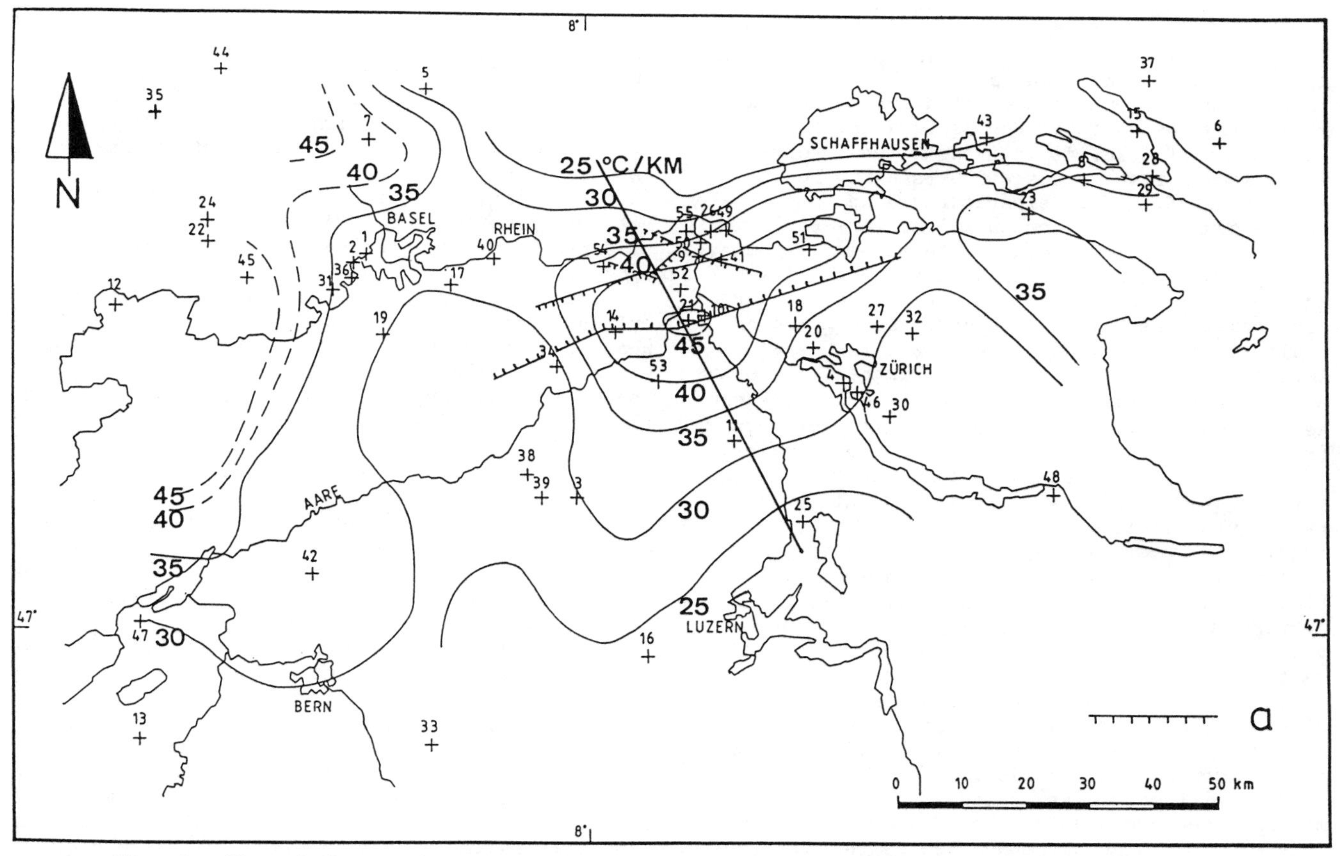

Fig. 9. Map of the geothermal gradient of northern Switzerland for the depth range between 750 and 1700 m (from Rybach et al., 1987). Crosses mark locations of bore holes and a) the border faults of the Northern Swiss Permocarboniferous Trough. Solid line delineates the trace of the NW-SE focal-depth cross-section (Figures 3 and 11).

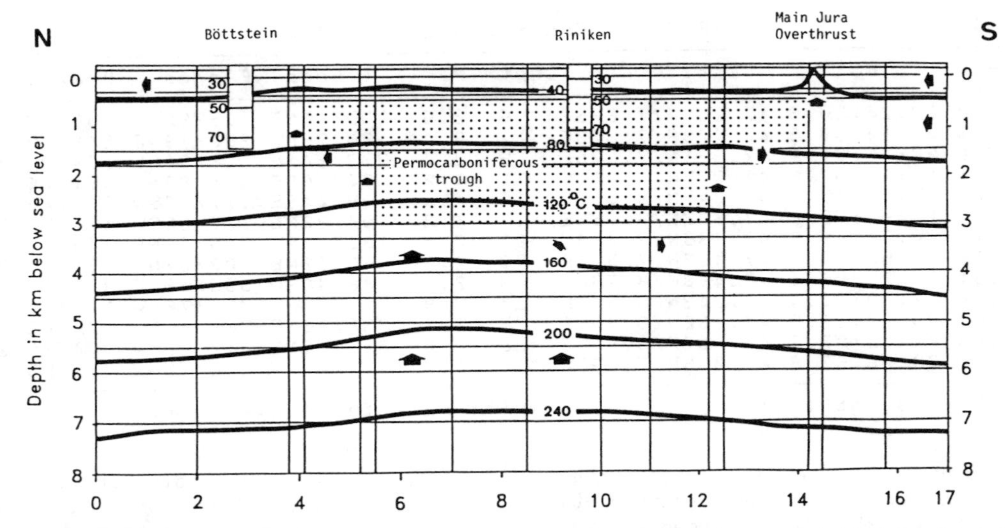

Fig. 10. Calculated temperature field in the realm of the Permocarboniferous trough (from Griesser and Rybach, 1988). Dominant directions of ground water flow are indicated schematically by arrows. Measured temperatures are shown for comparison (drillholes Böttstein and Riniken).

detail by Rybach et al. (1987) and by Griesser and Rybach (1988). The results show that the local thermal anomaly can be accounted for by ground water flow that affects the temperature field down to a depth of 8 km. At this depth the inferred temperatures reach about 250°C (see Figure 10).

In order to infer temperatures down to the base of the crust, further calculations were performed by two-dimensional, steady-state models, assuming purely conductive heat flux. These models take into account the temperature dependence of thermal conductivity as well as the depth dependence of radioactive heat production. The latter was inferred from the velocities in Figure 4, by means of an experimentally established relation between seismic velocity, heat production and rock type (Rybach and Buntebarth, 1982). For the numerical calculations, the crustal section along the investigated profile was subdivided into 9x14 rectangular elements. The upper boundary conditions were given by the laterally varying temperatures at 8 km depth from the model in Figure 10. The lower boundary conditions, corresponding to laterally varying heat flow values at 33 km depth, were based on a constant mantle heat flow of 30 mW/m^2, typical of Hercynian Europe (Čermák, 1982). Mantle heat flow values signficantly lower than 30 mW/m^2 are unrealistic considering the general tectonic setting of the area (Vitorello and Pollack, 1980).

Crustal temperatures calculated for these models reach 500°C at depths of about 20 km, and, in the depth range where the deepest earthquakes are found, temperatures are well above 600°C (Figure 11). These results are in close agreement with calculations by Čermák and Bodri (1986) along a line ending in the foreland of the Eastern Alps, and constitute a smooth continuation of the temperature field determined by Stiefel et al. (1986) for the Black Forest, just to the north. These temperatures are significantly higher than the values of 300 to 450°C, which are typical for estimates for the transition from brittle to ductile behaviour in a continental crust based on both experimental rheologies and geologic observations (e.g. Chen and Molnar, 1983; Scholz, 1988).

Since any thermal model of the crust depends critically on thermal conductivity and radioactive heat production, we have investigated effects of variation in these parameters. Specifically, we considered the range of parameters needed to obtain a cooler lower crust, while still satisfying the upper boundary conditions given by the measured surface heat flow. For this purpose we have performed one-dimensional model calculations with a fixed temperature of 450°C at a depth of 35 km. This constitutes a reasonable upper temperature limit for brittle faulting (Chen and Molnar, 1983; Scholz, 1988). Thus, increased heat production (A), increased thermal conductivity (B), or a combination of both (C) are the three options which would give lower temperatures at the base of the crust.

Option A calls for a heat production of 3.2 μW/m^3 at 18 km depth and leads to a negative mantle heat flow. Hence, this model is unrealistc. Option B yields an unreasonably high thermal conductivity of 5.0 W/m°K at 18 km depth. Assuming that this large value is due mostly to the high conductivity of quartz and allowing for the temperature dependence of thermal conductivity, this result corresponds to a quartz content of 75% (by volume); this constitutes a highly unlikely crust. Finally, option C yields a heat production of 2.0 $\mu W/m^3$ and a thermal conductivity of 3.0 W/m°K; the latter would correspond to a quartz content of 40%. However the last option would still lead to an unreasonably low mantle heat flow of only 10.5 mW/m^2.

Discussion

Definitions of the term "lower crust" can differ significantly, depending on the context in which it is used. The term can denote 1) the lower half or third of the crust, 2) a zone of enhanced reflectivity at the base of the crust, 3) the ductile part of the crust, 4) a depth range with P-wave velocities greater than about 6.5 km/s, or 5) any combination of the four. There is evidence from both refraction and reflection surveys in northern Switzerland for the existence of a series of reflective horizons below about 18 km depth (Mueller et al., 1987, Figure 3). Thus, by definitions 1) and 2) earthquakes beneath northern Switzerland extend well into the lower crust. On the other hand, for some as yet unknown reason the lower crust in this region could be abnormally thin. Then by definitions 3) and 4), the lower crust would be restricted to a less than 5 km thick transition zone with enhanced P-wave velocities, situated between the depth of the deepest earthquakes and the Moho. However, independently from this largely semantic question, there remains the problem of reconciling the existence of seismicity in a depth range where temperatures are expected to be high, and where both P-wave velocities and Poisson's ratio are low.

The first experimental extrapolations regarding the rheology of the lithosphere used data for quartz and olivine, on the assumption that the creep resistance of these two minerals control the rheological behaviour of the crust and upper mantle, respectively (Brace and Kohlstedt, 1980). This led to the first models of a single brittle-ductile transition within the crust, whose depth decreases with lower strain rates and higher temperatures. The depth of this transition was found to be in good agreement with focal depth determinations from many parts of the world: seismicity in most intraplate continental settings is limited to the upper 10 to 20 km of the crust, while the lower crust is generally aseismic, and this depth limit correlates well with heat flow (Sibson, 1982; Meissner and Strehlau, 1982; Chen and Molnar, 1983). Meanwhile it has been recognized that, as a function of varying mineralogical composition with depth, creep strength within the continental crust is not controlled solely by a quartz rheology, but that rocks of intermediate to basic composition offer greater resistance to steady-state creep than quartz-rich granitic rocks at the same temperature (Kirby, 1983; Smith and Bruhn, 1984; Strehlau and Meissner, 1987; Meissner and Kusznir, 1987; Ranalli and Murphy,1987; Carter and Tsenn, 1987).

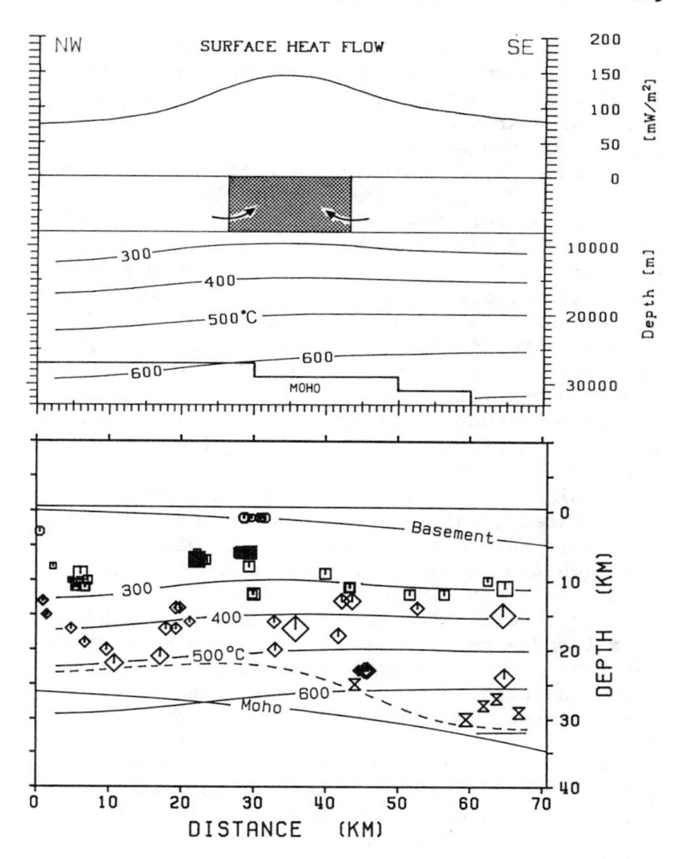

Fig. 11. Measured surface heat flow and calculated crustal temperature field along the profile in Figure 9. Top: calculated isotherms for the depth range 8-33 km; shaded area corresponds to Figure 10 and arrows indicate deep ground water circulation. Bottom: isotherms with earthquake focal depths. The dashed line indicates the lower bound of seismicity.

Thus, for a lithologically layered crust there is likely to be more than one brittle-ductile transition within the crust, and brittle failure becomes possible at the higher temperatures expected in the lower crust. However, none of these stronger rocks are compatible with the observed seismic velocities in the crust below northern Switzerland.

Under the assumption of a completely dry crust, both low P-wave velocities and low Poisson's ratios generally imply a granitic or a quartz-rich gneissic composition (Birch, 1960; Simmons, 1964; Kern, 1982). Granulite facies rocks, thought to be likely constituents of the lower continental crust, exhibit a wide range of mineral composition and seismic velocities. Some granulites of gabbroic composition (quartz-free) closely match the P-wave velocities and the values of Poisson's coefficient found in this study (S. Holbrook, personal communication). Thus, assuming a granulitic lower crust, the resistance to ductile deformation would be controlled by the rheology of plagioclase. Approximate temperatures for the onset of quartz and feldspar plasticity are 300°C and 450°C, respectively (Scholz, 1988).

In general, the combined effect of increasing pressure and temperature will result in an increase of both P-wave velocities and Poisson's ratio with depth (Spencer and Nur, 1976; Kern and Richter, 1981). However, under the influence of a sufficiently high temperature gradient, some rocks can exhibit the opposite trend. In this study Poisson's ratio does not increase with depth, suggesting that the assumption of a dry crust is indeed unrealistic.

The behaviour of seismic velocities for a given rock in the presence of water is quite complex, because it depends critically on the degree of saturation and on pore pressure. Measurements on a saturated granite at room temperature show that, with no pore pressure, both Vp and Poisson's ratio approach the values of dry rocks; in contrast, with pore pressure equal to the confining pressure, Poisson's ratio is significantly higher than for dry rocks, despite a 10% decrease of Vp (Nur and Simmons, 1969). In similar experiments at 1 kb confining pressure with variable temperatures, Vp/Vs decreases as temperature increases in the case of low pore pressure, while, for a pore pressure equal to the confining pressure, Vp/Vs first increases before it decreases (Spencer and Nur, 1976). Since these measurements in presence of water have not been extended to pressures and temperatures pertinent to the lower crust, the measured seismic velocities can not supply any further evidence for or against the presence of water at greater depths within the crust.

This lack of experimental data is an important problem, because the role of water is also critical for the assessment of the crust's rheology: in the presence of water, both the frictional resistance to failure and the resistance to ductile deformation are decreased relative to a dry rock.

One possible source of water in the lower crust could be dehydration reactions. There are several minerals, which undergo dehydration (or decarbonation) reactions at higher temperatures, thereby releasing fluids into existing or newly formed pore spaces (e.g. Fyfe et al., 1978). Pore pressure can increase so markedly as a consequence of this process, that friction on existing faults is lowered sufficiently to enable brittle failure to occur at temperatures where ductile flow would be expected. This mechanism has been suggested in conjunction with the transformation of serpentinite to olivine, as a possible explanation for deep subduction-zone earthquakes (Raleigh and Paterson, 1965). However, serpentinite weakening is unlikely to be a factor controlling lower-crustal seismicity below northern Switzerland, since any significant amount of both serpentinite and olivine is inconsistent with the low Poisson's ratios measured in this study (Christensen, 1966).

An additional important factor influencing rock deformation is strain rate. Northern Switzerland's position at the front of the still active continent-continent collision zone could imply somewhat larger strain rates than the value of $10^{-14}s^{-1}$ usually associated with young but stable continental interiors. A regional lithospheric bending of the Alpine foreland, due to the load caused by the northward thrust of the Alps, seems to be an unlikely cause for increased strain rates: such bending should manifest itself by widespread normal-fault focal mechanisms with a NW-SE oriented T-axis, exactly opposite to what is observed (Figure 8). For the Jura overthrust, Müller and Briegel (1980) calculated a strain rate of $6x10^{-13}s^{-1}$, while Neugebauer et al. (1980) derived strain rates between $10^{-14}s^{-1}$ and $10^{-12}s^{-1}$ for the lower crust by modelling the uplift of the Alps. Pfiffner and Ramsey (1982) argue that strain rates must be confined between $10^{-15}s^{-1}$ and $10^{-12}s^{-1}$, on the basis of finite strain observed in various rocks, although they do not exclude the possible existence of higher strain-rate concentrations in localized mylonite zones. Therefore, as temperatures increase with depth and thus tend to counteract the hardening effect of higher strain rates, an increase in strain rate of one or two orders of magnitude will lower the brittle-ductile transition by only a few km (Sibson, 1982; Smith and Bruhn, 1984). Consequently, higher strain rates alone are an unlikely explanation for the observed lower-crustal seismicity.

The apparent paradox represented by the occurrence of earthquakes in the lower crust which is expected to be ductile is, of course, based on the assumption that earthquakes are caused by brittle deformation. An alternative hypothesis regards earthquakes under certain conditions as manifestations of plastic instability (G. Ranalli, personal communication). However, as mentioned above in the context of the focal mechanisms, such a process of plastic instability would have to be practically indistinguishable from the traditional model of brittle friction failure.

As discussed above, our knowledge of temperatures in the lower crust is based on extrapolations of measurements performed over a very limited depth range just below the earth's surface. If we can constrain the possible range of mineral compositions more tightly and improve our knowledge of the temperature and pore-pressure dependence of thermal conductivity, such extrapolations can be carried out with more confidence. Our model calculations are based on pure heat conduction below a depth of 8 km. If significant amounts of fluids are present in the lower crust and permeability is sufficiently high, convective heat transfer might play an important role down to greater depths. Furthermore, our model calculations were performed on the assumption of steady-state conditions. As models of the geodynamic and thermal history of the Alpine orogeny are refined (e.g. Kissling et al., 1983), it may ultimately become possible to estimate the extent of temperature perturbations below the Alpine foreland, associated with the collision process that formed the Alps.

An additional question raised by the observations presented here concerns the P-wave velocities. The in situ velocities determined from the direct waves of the earthquakes suggest a gradual and continuous velocity increase from about 5.9 km/s, at a few km below the top of the crystalline basement, to 6.2 km/s, just above the crust-mantle transition zone. Whereas these results are in relatively good agreement with observations from refraction surveys in northern Switzerland and southern Germany as far as the upper crust is concerned, the velocities in the lower crust are significantly lower than what has been derived from refraction interpretations. Moreover, the enhanced reflectivity below about 18 km depth, observed in both wide-angle and normal incidence reflection data (Mueller et al., 1987), does not seem to show up in the earthquake data. This discrepancy could be due to anisotropy, which influences seismic velocities determined from earthquakes and reflection data in different ways: the direct rays from earthquakes in the lower crust, recorded at epicentral distances around 100 km, travel through the source layer at almost horizontal angles over a relatively long distance (see Figure 4), whereas rays from surface explosions sample the same layers at angles closer to vertical. It is thus important for our understanding of the nature of the lower crust to determine whether this discrepancy between the two results is due to a different methodological bias, or is actually due to petrophysical reasons.

Conclusions

Although it has not yet been possible to give a satisfactory explanation for the rather anomalous occurrence of lower-crustal earthquakes below the northern Alpine foreland of Switzerland, the additional observations regarding P-wave velocities, Poisson's ratio and temperatures have served to constrain the range of possible interpretations and to point out directions for further research. There is, in particular, urgent need for more experimental data on the rheology of feldspar dominated rocks and on the effect of pore fluids on Vp and Vs at temperatures and pressures pertinent to the lower crust. As more heat-flow data and reliable hypocenter determinations become available, a three-dimensional picture of temperatures and seismicity will provide better insight into the problem.

The results presented here should have served to demonstrate that the clear correlation between the lower depth limit of seismicity and heat flow, evident in many regions, and the simple model of a brittle upper crust over a ductile lower crust can not be generalized indiscriminately. Moreover, from a more methodological point of view, the present project illustrates that our understanding of the nature of the lower continental crust could be improved significantly by searching systematically for other areas with lower-crustal seismicity, in which to carry out detailed reflection and refraction surveys.

Acknowledgements. We gratefully acknowledge the contributions of M. Baer for the seismic data acquisition software and the routine data evaluation, M. Dietiker for development and maintenance of the seismograph arrays, and W. Eugster for running the two-dimensional temperature calculations. S. Holbrook kindly searched his computer data base for rock types that match the seismic velocities. Thanks are due to our colleagues at the Universities of Karlsruhe and Stuttgart for providing numerous seismograms and to M. Garcia for the data from the local network near Basel. We have greatly benefitted from stimulating discussions with V. Čermák, B. Della Vedova, D. Hill, G. Ranalli and S. Schmid. We also appreciate the painstaking review by T. Pavlis. Part of this work was carried out under contract by the National Cooperative for the Disposal of Nuclear Waste (NAGRA). Contribution No. 575 of the Institute of Geophysics, ETH-Zürich.

References

Birch, F., The velocity of compressional waves in rocks to 10 kilobars, part 1. J. Geophys. Res., 65, 1083-1102, 1960.

Brace, W. F., Kohlstedt, D. L., Limits on lithospheric stress imposed by laboratory experiment. J. Geophys. Res., 85, 6248-6252, 1980.

Carter, N. L., Tsenn, M. C., Flow properties of continental lithosphere. Tectonophysics, 136, 27-63, 1987.

Čermák, V., Crustal temperature and mantle heat flow in Europe. Tectonophysics, 83, 123-142, 1982.

Čermák, V., Bodri, L., Two-dimensional temperature modelling along five East-European geotraverses. J. Geodynamics, 5, 2, 133-164, 1986.

Chen, W.-P., Molnar, P., Focal depths of intra-continental and intraplate earthquakes and their implications for the thermal and mechanical properties of the lithosphere. J. Geophys. Res., 88, 4183-4214, 1983.

Christensen, N., Elasticity of ultrabasic rocks. J. Geophys. Res., 71, 5921-5931, 1966.

Deichmann, N., Focal depths of earthquakes in northern Switzerland. Ann. Geophysicae, 5B, 4, 395-402, 1987a.

Deichmann, N., Seismizität der Nordschweiz, 1983-1986. Technischer Bericht NTB 87-05, NAGRA, Baden, 95 pp., 1987b.

Deichmann, N., Rupture geometry from high-precision relative hypocenter locations (Abstract). EGS, XIII General Assembly, Bologna, Ann. Geophysicae, Special Issue, p 12, 1988.

Deichmann, N., Renggli, K., Mikrobeben-

Untersuchung Nordschweiz, Teil 2: Seismizität, Jan.1983-Sept.1984. Technischer Bericht NTB 84-12, NAGRA, Baden, 81 pp., 1984.
Diebold, P., Müller, W. H., Szenarien der geologischen Langzeit-sicherheit: Risikoanalyse für ein Endlager für hochaktive Abfälle in der Nordschweiz. Technischer Bericht NTB 84-26, NAGRA, Baden, 110 pp., 1984.
Fyfe, W. S., Price, N. J., Thompson, A. R., Fluids in the Earth's Crust. Elsevier, Amsterdam, 383 pp., 1978.
Gebrande, H., A seismic-ray tracing method for two-dimensional inhomogeneous media. In: Explosion Seismology in Central Europe, ed. Giese, P., Prodehl, K., Stein, A., Springer: Berlin, Heidelberg, New York, 162-167, 1976.
Griesser, J.-C., Rybach, L., Numerical thermohydraulic modelling of deep ground water circulation in crystalline basement: an example of calibration. In: Hydrogeological Regimes and their Subsurface Thermal Effects, ed. A. E. Beck, L. Stegena, G. Garven, AGU Monograph, in press, 1988.
Holbrook, W. S., Gajewski, D., Prodehl, K., Shear-wave velocity and Poisson's ratio structure of the upper lithosphere in southwest Germany. Gephys. Res. Let., 14, 3, 231-234, 1987.
Kahle, H.-G., Klingelé, E., Mueller, S., Egloff, R., The variation of crustal thickness across the Swiss Alps based on gravity and explosion seismic data. Pageoph, 144, 479-493, 1976.
Kern, H., Elastic-wave velocity in crustal and mantle rocks at high pressure and temperature: the role of the high-low quartz transition and of dehydration reactions. Phys. Earth and Planet. Int., 29, 12-23, 1982.
Kern, H., Richter, A., Temperature derivatives of compressional wave velocities in crustal and mantle rocks at 6 kb confining pressure. J. Geophys., 49, 47-56, 1981.
Kirby, S. H., Rheology of the lithosphere. Rev. Geophys. and Space Phys., 21, 6, 1458-1487, 1983.
Kissling, E., Mueller, St., Werner, D., Gravity anomalies, seismic structure and geothermal history of the Central Alps. Ann. Geophysicae, 1, 1, 37-46, 1983.
Kisslinger, C., Engdahl, E. R., The interpretation of the Wadati diagram with relaxed assumptions. Bull. Seis. Soc. Am., 63, 5, 1723-1736, 1973.
Klingelé, E., Olivier, R., Carte gravimetrique de la Suisse - Anomalie de Bouguer - 1/500000. Service Topographique Federal, Wabern, 1980.
Lee, W. H. K., Lahr, J. C., HYPO-71, a computer program for determining hypocenter, magnitude and first motion pattern of local earthquakes. U. S. Geol. Surv. Open File Rep., 100 pp., 1972.
Mayer-Rosa, D., Benz, H., Kradolfer, U., Renggli, K., Inventar der Erdbeben 1910-1982 und Karten der Magnitudenschwellen-Werte 1928-1982. Technischer Bericht, NTB 83-08, NAGRA, Baden, 30 pp., 1983.
Mayer-Rosa, D., Dietiker, M., Deichmann, N., Renggli, K., Brändli, J., Studer, J., Rutishauser, G., Mikrobeben-Untersuchung Nordschweiz, Teil 1: Technische Unterlagen, Stationsnetz. Technischer Bericht, NTB 84-11, NAGRA, Baden, 58 pp., 1984.
Meissner, R., Kusznir, N. J., Crustal viscosity and the reflectivity of the lower crust. Ann. Geophysicae, 5B, 4, 1987.
Meissner, R., Strehlau, J., Limits of stresses in continental crusts and their relation to the depth-frequency distribution of shallow earthquakes. Tectonics, 1, 1, 73-90, 1982.
Mueller, St., Ansorge, J., Egloff, R., Kissling, E., A crustal cross section along the Swiss Geotraverse from the Rhinegraben to the Po Plain. Eclogae geol. Helv., 73, 2, 463-483, 1980.
Mueller, St., Ansorge, J., Sierro, N., Finckh, P., Emter, D., Synoptic interpretation of seismic reflection and refraction data. Geophys. J. R. astr. Soc., 89, 345-352, 1987.
Müller, W. H., Briegel, U., Mechanical aspects of the Jura overthrust. Eclogae geol. Helv., 73, 1, 239-250, 1980.
Neugebauer, H. J., Brötz, R., Rybach, L., Recent crustal uplift and the present stress field of the Alps along the Swiss Geotraverse Basel-Chiasso. Eclogae geol. Helv., 73, 2, 489-500, 1980.
Nur, A., Simmons, G., The effect of saturation on velocity in low porosity rocks. Earth and Planet. Sci. Let., 7, 183-193, 1969.
Pavoni, N., Crustal stresses inferred from fault-plane solutions of earthquakes and neotectonic deformation in Switzerland. Rock Mechanics, Suppl. 9, 63-68, 1980.
Pavoni, N., Zur Seismotektonik der Nordschweiz. Eclogae geol. Helv., 80, 2, 461-472, 1987.
Pfiffner, O. A., Ramsey, J. G., Constraints on geological strain rates: arguments from finite strain states of naturally deformed rocks. J. Geophys. Res., 87, B1, 311-321, 1982.
Raleigh, C. B., Paterson, M. S., Experimental deformation of serpentinite and its tectonic implications. J. Gephys. Res., 70, 16, 3965-3985, 1965.
Ranalli, G., Murphy, D. C., Rheological stratification of the Lithosphere. Tectonophysics, 132, 281-295, 1987.
Rybach, L., Buntebarth, G., Relationships between the petrophysical properties, density, seismic velocity, heat generation, and mineralogical constitution. Earth Planet. Sci. Lett., 57, 367-376, 1982.
Rybach, L., Eugster, W., Griesser, J.-C., Die geothermischen Verhältnisse in der Nordschweiz. Eclogae geol. Helv., 80, 2, 521-534, 1987.
Scholz, C. H., The brittle-plastic transition and the depth of seismic faulting. Geologische Rundschau, 77, 1, 319-328, 1988.
Sibson, R. H., Fault zone models, heat flow and depth distribution of earthquakes in the continental crust of the United States. Bull. Seis. Soc Am., 72, 1, 151-163, 1982.

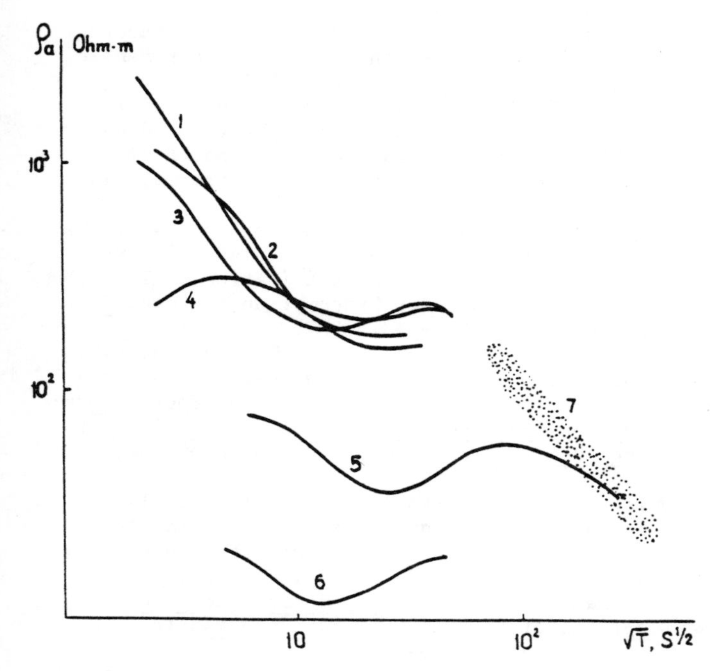

Fig. 3. Typical apparent resistivity curves. 1-Baltic shield, 2-Central Karelia, 3-Canadian Shield, 4-Eastern part of the Siberian Platform, 5-Western United States, 6-Pannonian Basin, 7-global apparent resistivity.

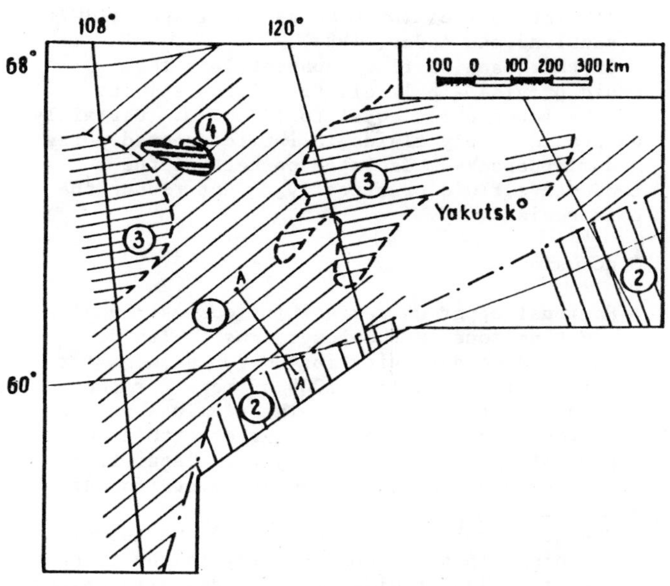

Fig. 4. The geographical distribution of two ranges of the crustal integrated conductivity at Eastern Siberia. 1 and 2-integrated conductivity equal to 400 and 700 siemens respectively, 3-areas unfavourable for deep soundings due to high value of the sediment cover conductance, 4-low resistivity anomaly of the upper part of the basement.

One can see in Figure 4 the geographical distribution of the two crustal conductance values at the eastern part of the Siberian platform. Areas of thick sediment cover and highly conductive anomalies in the upper part of the basement are also shown.

Discussion

According to geothermal models, the temperature at the Moho discontinuity is 400-500°C for stable regions. Since these temperatures are lower than solidus, the anomalous resistivity of the lower crust cannot be attributed to partial melting. The most probable explanation suggests that there is a highly conductive fluid in pores and microcracks in the lower crust (Vanyan, 1980; Shankland and Ander, 1983; Gough, 1987).

Electrical conductivity of the supercritical water solutions was investigated by Quist and Marshall (1968). Figure 5 shows an example for 0.1 mol concentration of NaCl; this dependence was constructed from the original data of Quist and Marshall (Vanyan, 1980). Conductivity behaviour is rather complicated. For instance, at small pressures conductivity decreases with temperature increase from 300 to 700°C. However, at pressures higher than 6 kbar (i.e., depths greater than 20 km) all isothermal curves are close to the conductivity value of 6 siemens/m. There is sparse information about salinity of lower crust fluids. If we assume that salinity is about 10 g/litre, one can, by Archie's Law, calculate the electrical conductivity in the

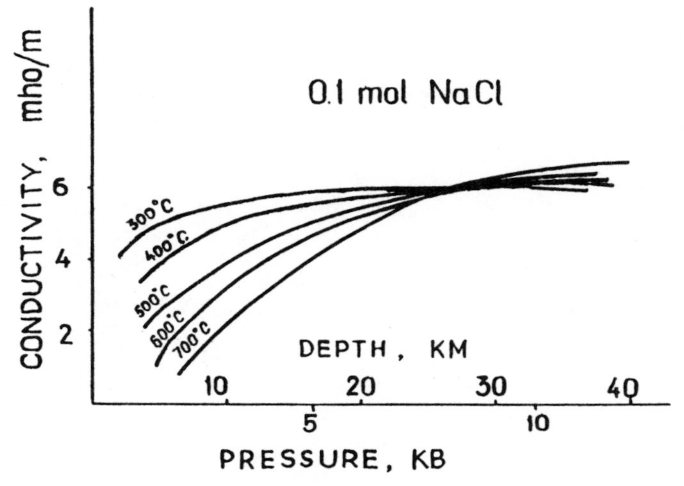

Fig. 5. Water solution conductivity versus pressure and depth. Curves are labelled by constant temperature.

depth interval of 20-40 km as 10 siemens/m, (Shankland and Ander, 1983).

In this case, a fluid content less than 1% could provide a bulk electrical resistivity of the lower crust equal to 10 ohm-m. Certainly, we consider only fluids in interconnected pores and microcracks. Isolated pockets of the conductive fluid practically do not reduce the bulk resistivity.

Conclusion

Regional apparent resistivity curves reveal a conductive zone in the lower crust. Its depth-integrated conductivity is 400-700 siemens for stable regions and 1500-2000 siemens for tectonically active regions. These values are two or three orders of magnitude greater than the depth-integrated conductivity calculated from the results of laboratory measurements for dry samples.

The most probable explanation of the anomalously high "in situ" conductivity is the presence of supercritical fluids in pores and microcracks.

References

Adam, A., D.A. Varlamov, I.V. Yegorov, A.P. Shilovski and P.P. Shilovski, Depth of crustal conducting layer and asthenosphere in the Pannonian basin determined by mangetotellurics, Phys. Earth Planet. Int., 28, 251-260, 1982.

Berdichevski, M.N., Dmitriev, V.I., Mershikova, N.A., Barashkov, I.S. and Kobzova, V.M., On the magnetotelluric sounding of the conducting zones in the earth crust and upper mantle, Fizika Zemli, No. 7, 55-68, 1982.

Gough, D.I., (ed.), Interim report on electromagnetic lithosphere-asthenosphere soundings (ELAS) to coordinating committee no. 5 of the International Lithosphere Programme, Am. Geophys.Un., 1987.

Hjelt, S.E., Deep electromagnetic studies of the Baltic Shield, J.Geophys., 55, 144-152, 1984.

Jones, A.G., On a type classification of lower crustal layers under Precambrian regions, J. Geophys., 49, 226-233, 1981.

Osipova, I.L., Berdichevski, M.N., Vanyan, L.L. and Borisova, V.P. Asthenosphere of North America from electromagnetic data, Fourth Workshop on Electromagnetic Induction in the Earth and Moon, Murnau, Federal Republic of Germany, 1987.

Quist, A.S. and Marshall, W.L., Electrical conductances of aqueous sodium chloride solutions from o to 800°C and at pressure to 4000 bars, J.Phys.Chem., 72, 684-703, 1968.

Rokityansky, I.I., Geoelectromagnetic Investigations of the Earth's Crust and Mantle, Springer-Verlag, Berlin, pp.381, 1982.

Vanyan, L.L., Progress report on ELAS-project, IAGA News No. 19, 1980.

Parkhomenko, E.I., Electrical resistivity of minerals and rocks at high temperature and pressure, Revs. Geophys.Space Phys., 20, 193-218, 1982.

Shankland, T.J. and M.E. Ander, Electrical conductivity, temperatures, and fluid in the lower crust, J.Geophys.Res., 88, 9475-9484, 1983.

THE MAGNETISATION OF THE LOWER CONTINENTAL CRUST

Albrecht G. Hahn

Niedersächsisches Landesamt für Bodenforschung, Hannover, F.R.G.

Hans A. Roeser

Bundesanstalt für Geowissenschaften und Rohstoffe, Hannover, F.R.G.

Abstract. In order to obtain a range of magnetisations for the lower crust, two approaches are adopted.

- Interpretation of aeromagnetic and satellite measurements displaying anomalies more than 100 km across gives magnetisations in the range 0 to 6 A/m.
- Among the interpretations of Magsat data, a magnetisation model for the whole crust of the Earth was developed which uses magnetisations in the range 0 to 2.5 A/m.

The magnetisation of rock samples pertaining to the probable inventory of the lower crust is then considered, in order to come to conclusions about the composition of the lower crust. In doing so, it was found that rock samples of old shields exhibit magnetisations great enough to explain almost all magnetisation values required by distinct model bodies for the interpretation of field anomalies. On the other hand, samples of the lower crustal part of elevated crustal blocks generally possess too small a magnetisation compared with this. Xenoliths have magnetisations of sufficient strength, but there are doubts about their being really representative.

Outline of the Problem and Method of Approach

The Magsat anomaly maps proved the existence of widely extended magnetic anomalies, the sources of which must lie in the crust or, eventually, in the upper mantle. These data, together with the study of long-wavelength magnetic anomalies obtained by ground, shipborne or airborne surveys, revealed a problem concerning the sources of these anomalies. They require, in many cases, a much higher magnetisation of large volumes of the Earth's crust than it was anticipated. This problem has been formulated by Hall et al. [1979] and by Williams et al. [1985]. In these papers, the results of airborne surveys and of studies of rock magnetism in parts of Manitoba and Ontario, Canada were compared.

In the present paper, we attempt to continue the discussion of the problem on the basis of more data and, in particular, a global model of the Earth's crust which was compared with Magsat anomaly maps [Hahn et al., 1984].

We take the "lower crust" to mean the part of the continental crust situated between a depth of 10 to 15 km and the Moho at about 30 to 50 km.

In order to determine the range of average magnetisation of rock units measuring at least several km in all directions, two approaches are adopted:

i) Interpretation of near surface and satellite measurements displaying anomalies more than 100 km across.

ii) Consideration of a magnetisation model of the whole Earth's crust [Hahn et al., 1984], which should correspond to the Magsat anomaly maps [Langel et al., 1982; Coles, 1985].

From i) and ii), a range of average magnetisation values for volumes with horizontal diameters of 100 km or more and thicknesses of ca. 10 to 30 km consisting of lower crustal rocks will follow. Magnetisation values of rock samples which pertain, or might pertain, to the constituents of the lower crust will then be considered. It will be discussed whether or not the rock magnetisation values harmonise with the results of i) and ii). From that, implications about the composition of the lower crust can possibly be formulated.

Ranges of Magnetisation

Interpretation of Magnetic Anomalies in Particular Regions

Due to the great depth of the lower crust, magnetic anomalies measured at, or a few km above, the Earth's surface can only yield average

TABLE 1. Magnetisation of the Lower Crust for Some Areas

No	Area	Magnetisation values A/m	Type of survey	Reference
1	Ukrainian Shield	0.5 - 2.3	airborne	Hahn and Bosum [1986]
2	Manitoba, Ontario	5	airborne	Hall [1974]
3	Northwest Germany	2	airborne	Hahn et al. [1976]
4	Fennoscandia	2	ground	Elming and Thorne [1976]
5	United States	3.5	satellite	Schnetzler [1985]
6	Utah	3	airborne	Shuey et al. [1973]
7	Kentucky	4	airborne	Mayhew et al. [1982]
8	Sweden	3 - 4	airborne	Riddihough [1972]
9	British Columbia	3 - 5	airborne	Coles and Currie [1977]
10	British Columbia	5	airborne	Coles [1976]
11	Central African Republic	3.5	satellite, airborne	Regan and Marsh [1982]
12	Broken Ridge	6	satellite	Johnson [1985]
13	Alpha Ridge	3	satellite, airborne	Taylor [1983]
14	Lord Howe Rise	5	satellite	Frey [1985]
		5	shipborne	Roeser (unpubl. report, 1985)
15	Ross Sea	2.6	airborne	Bosum (unpubl. report, 1987)
16	Southeastern United States	3 - 4	satellite	Ruder and Alexander [1986]

magnetisations for rock units within the lower crust measuring at least 10 km in all dimensions. In contrast to this, from Magsat anomaly maps one cannot reasonably derive model bodies with horizontal dimensions smaller than 200 km. There is no chance of resolving contributions of the upper and the lower crust from satellite data in the vertical direction.

Mayhew et al. [1985] compiled a list of 16 magnetisation values of extensive parts of the lower crust or of large bodies situated at least partly within the lower crust. We make use of this list with a few modifications and addition of examples (Table 1).

In place of the value given for the Ukrainian Shield ((1) in the list of Mayhew et al. [1985]), Hahn and Bosum's [1986] interpretation is used, which only uses the two-dimensional part of the anomalies with respect to the lower crust. The figure determined by Shuey et al. [1973] for Utah (6) is a susceptibility value; this, multiplied by the field strength for that area, yields M_i = 3 A/m (induced magnetisation). Similarly, the figures of Regan and Marsh [1982] (11) correspond to M_i = 3.5 A/m. Among the model bodies computed by Riddihough [1972] for the large anomaly in Central Sweden (8), only those with a maximum magnetisation of 4 A/m would reach the depth of the lower crust. Johnson's [1985] models for the Broken Ridge (12) vary from 6 to 42 A/m. Only the value 6 A/m, corresponding to a mean magnetisation of the whole crust, pertains to Table 1.

In order to confine ourselves to continental crust, we eliminated volcanic arcs and mid-Pacific seamounts. Furthermore, we have omitted a value by Caner [1969]. Caner's paper was aimed at the investigation of tectonic processes. It was not his intention to find out whether the magnetic bodies responsible for his anomalies would lie in the lower or the upper crust and to determine their magnetisation. In fact, his data is insufficient for the determination of reliable values.

In general, Table 1 suffers from the problem that the magnetisation of the lower crust is investigated and published mainly for places where it gives rise to conspicuously high-amplitude, long-wavelength magnetic anomalies. Thus, Table 1 may contain predominantly exceptional bodies. Only for Table entries 1 to 5 are we sure that they cover areas not specially selected. Thus, we can state that the bulk of the rocks of the lower crust are magnetised in the range 0 to 3 A/m, but, in some cases, values as high as 6 A/m are attained by bodies at least 10 km across within the lower crust.

Model of the Earth's Crust as a Whole Compared with Magsat Data

A model of the Earth's crust as a whole was established in the following way [Hahn et al., 1984]: The Earth's crust was subdivided into blocks of 2° x 2°. Each block was assigned to one of 16 crustal types, e.g. shield, sedimentary basin, folded mountain chain etc. These types were characterised (mainly on the basis of refraction seismics) as sequences of two or three layers (cover, upper crust, lower crust). A susceptibility value was estimated for each layer of each crustal type.

The resultant magnetisation represents the induced magnetisation (susceptibility multiplied by the magnetic field at each place) plus the component of remanent magnetisation parallel to the Earth's field. In a field, B = 50,000 nT, the range of magnetisation of this model is M = 0 to 2.0 A/m.

For comparing the magnetic "mass" of the model with that of the actual Earth's crust, its magnetic field is represented by a series of spherical harmonics like the Magsat field data. In the spectrum of the Magsat data, the contribution from crustal sources dominates from degree and order 14 onwards [Langel and Estes, 1982]. In this spectral interval, both spectra would fit together if the magnetisations of model layers were multiplied by a factor of 1.25 giving the magnetisation range

$$M = 0 \text{ to } 2.5 \text{ A/m} \qquad (1)$$

at a middle latitude where the total field amounts to 50,000 nT.

Since the 2° x 2° blocks are, in most cases, larger than the bodies causing the magnetic anomalies referred to in the previous section, it is plausible that the range of magnetisation required for the crustal model is smaller than for these single bodies. In reviewing the Magnetic Anomaly Map of North America [1987], for instance, one gets the impression that bodies with magnetisations exceeding 2.5 A/m are, only in rare cases, extended over areas of more than the half of the 2° x 2° blocks.

Since the paper of Hahn et al. [1984] many new facts about the magnetisation of the lower crust were gained. It seems now that the mean magnetisation of the lower crust is higher than that of the upper crust. Although the magnetisation values of the upper part of the crustal types in Fig. 2b of Hahn et al. [1984] were chosen very carefully, there is enough room for reducing them in favour of a higher magnetisation of the lower crust.

Thus, the magnetisation values of the crustal type layers, as shown in Fig. 2b (l.c.), could be changed for several types by generally reducing the values assigned to the upper crust and enhancing those of the lower crust. This would entail that if, in type B, for example, the magnetisation of the lower crust were chosen to 2.0 A/m and that of the upper crust to 1.3 A/m (or after adjustment of the model spectrum to the Magsat spectrum to 2.5 A/m for the lower and 1.6 A/m for the upper crust), there would not be a change in the anomaly map and in the spectrum. Also in the types A, K and G one should shift magnetic "mass" from the upper to the lower crust, however, a magnetisation of 2.0 A/m (or adjusted to Magsat 2.5 A/m) could here only be assigned to parts of the lower crust without a change of the spectrum. This latter remark can be taken as a hint for a future refinement of the model where in some types the lower crust should be subdivided in two layers.

Summarising one can state that the magnetisation of the lower crust averaged over 2° x 2° blocks should vary in the interval 0 to 2.5 A/m.

Estimations from Rock Magnetics

Samples of the lower crust are available in the form of xenoliths, in elevated blocks ("chips") of lower crust or in deeply eroded areas of ancient shields. However, in most cases they left their original environment of high temperature and high pressure long ago, and that implies that their magnetic properties may differ from those down in the lower crust.

Frost and Shive [1986] investigated the question whether the magnetic minerals observed in now accessible rocks from the lower crust are phases which developed during the time since their elevation out of lower crustal conditions. They concluded that this is not the case.

Wasilewski and Mayhew [1982] concluded from the investigation of xenoliths, that metabasic rocks in the granulite facies from several tectonic settings have the magnetisation necessary for modelling the long-wavelength magnetic anomalies. However, as they state, investigations of xenoliths have severe disadvantages [Kay and Kay, 1981]: the conveying kimberlite and lava may not sample all the wall rocks, some of the samples may be destroyed more easily than others, and the depth from which the xenoliths originate is often difficult to determine.

Studies of elevated crustal blocks provide a valuable independent source of information. One of their disadvantages, however, is that the rock complexes elevated from the lower crust may not really be representative. Those properties which make them apt for elevation may also be responsible for their lower magnetisation [see also Kay and Kay, 1986, p. 148]. Indeed, these blocks do have only partly the required mean magnetisations:

- Wasilewski and Fountain [1982] observe that in the Ivrea Zone mafic-ultramafic granulite facies rocks and amphibolites from the granulite-amphibolite facies transition zone have natural remanent magnetisations of more than 1 A/m whereas all other lithologies show much weaker magnetisations. Wasilewski and Fountain [1982] consider the mafic-ultramafic granulite facies rocks as more significant than the amphibolites for regional studies. Hahn et al. [1984] model the Ivrea magnetic anomaly with an inclined body 15 km high and 10 km thick and a magnetisation of 1 A/m. The top of the body corresponds almost perfectly with the outcrop of the granulite facies basic rocks.
- A rock series in S-Calabria, Italy which,

according to the rocks outcropping there, also indicates the presence of an elevated crustal block [Kern and Schenk, 1985] does not show any anomaly in the geomagnetic survey of Italy which could be assigned to the steeply dipping lower crustal layer.

From the study of areas where rocks crop out which have previously formed the lower crust, petrologists suppose that the most abundant rocks in the lower crust are (a) granites, granitoids and gneisses, and, in the lowermost 3 to 10 km, (b) basic to intermediate metamorphic and igneous rocks.

An estimate of the ranges of magnetisation was carried out on the basis of a collection of 77,000 rock samples from the Ukrainian Shield [Krutikhovskaya et al., 1982] from which susceptibility values had been determined. From these samples 45,300 were selected as possible lower crustal rocks. Their susceptibility values were according to the rock types given by the authors distributed among the rock groups (a) and (b) above.

We obtained the ranges of magnetisations shown in Table 2.

Williams et al. [1985] investigated 56 samples from the Pikwitonei granulite domain and the adjacent Cross Lake subprovince in Manitoba. Eighty percent of the section are silicic plutonic rocks with a mean induced magnetisation of 0.4 A/m whereas the mafic rocks are much less magnetised. This shows that mafic rocks are not always more magnetic than silicic ones.

Schlinger [1985] investigated 4,435 samples from Lofoten and Vesteralen, Norway, which gave a mean induced magnetisation of 1.4 A/m. The granulite facies gneisses have 1.9 A/m, mafic and ultramafic rocks 2.7 A/m.

Susceptibility values from an investigation of ca 50,000 rock samples from Scandinavia including samples from Lofoten/Vesteralen were presented by Henkel during the XIX IUGG Assembly 1987 in the form of a histogram. These values show a tendency of being a little smaller than the Ukrainian Shield data.

Summarising all these data, we obtain the following ranges of magnetisation induced in a field of 50,000 nT:

$$M_i(a) = 0 \text{ to } 0.7 \text{ A/m}$$
$$M_i(b) = 0 \text{ to } 3.1 \text{ A/m} \quad (2)$$

Hall et al. [1979] suspect that some rocks may have much higher magnetisation at the high temperatures of the lower crust and that in addition different mineralogies may play an important role.

In contrast to this, the above mentioned results by Frost and Shive [1986] indicate that the magnetisations observed under surface conditions can be transferred to lower crustal conditions.

An increase of the susceptibility by the Hopkinson effect as proposed by Dunlop [1974] is presumably not important because high-temperature susceptibility measurements by Williams et al. [1985] and by Schlinger [1985] do not show this effect for samples whose composition is similar to lower crustal rocks.

A viscous remanent magnetisation might, however, under lower crust conditions play an important role. Williams et al. [1985] discuss the problem that for the Superior Province of Manitoba/Canada the magnetisations observed in samples are too small in comparison with the long-wavelength magnetic anomalies. They tentatively propose a high-temperature viscous remanent magnetisation in the lower crust. Treloar et al. [1986] have investigated the magnetisation of now exposed rocks from the lower crust at the temperatures of the lower crust. They found that viscous remanence is the only important remanence and that it is about half the induced magnetisation at the same temperature. Coles and Currie [1977] state from experiments at 300 °C that the viscous component could add about 30 % of the

TABLE 2. Magnetisation Ranges for Rock Types of the Lower Crust

Rock type	Range of induced magnetisation (B = 50 000 nT) in A/m	Number of samples
Group (a) - felsic rocks		
granites	0.0 - 0.7	18,200
Group (b) - mafic rocks		
basic gneiss	0.2 - 2.6	13,200
amphibolites	0.4 - 1.3	4,500
basic magmatic rocks	0.7 - 3.1	3,600
ultrabasic rocks	0.8 - 2.5	5,800
Total (b)	0.2 - 3.1	27,100

induced magnetisation. As with all conclusions on the in-situ magnetisation of the lower crust, it is quite risky to generalise this proportion, but presently there is no other information available. Increasing the induced magnetisations of (2) by 50 %, we obtain the following total magnetisations:

$$M\ (a) = 0 \text{ to } 1.0\ A/m$$
$$M\ (b) = 0 \text{ to } 4.6\ A/m. \qquad (3)$$

Implications for the Composition of the Lower Crust

We will now use the material presented in the foregoing discussion to attempt a guess at the relative abundance of rock types (a) and (b) in the lower crust. Attempts of this kind are not new, (see Williams et al. [1985]); however, we are using a wider range of information.

Schnetzler [1985] determined for the United States the magnetisation 2 to 6 A/m for the lower crust. In his paper he has considered a minimum positive magnetisation due to a certain susceptibility, which should be present in all elementary blocks (150 km x 150 km) of the area under investigation. From a comparison of the thickness of the lower crust and the dipole strength of the elementary blocks he obtained 1.5 A/m for this magnetisation. If we assume that the magnetisation of the lower crust is not independent of its thickness, this value may be lowered, let us say, to 0.5 A/m. With this basic magnetisation there were several areas where the rocks of the lower crust could possibly consist of rock type (a) only. And the ratio 1:1 between types (a) and (b) which is compatible with the range M = 0 to 2.8 A/m would be possible in about 2/3 of the area investigated by Schnetzler [1985, Fig. 1].

The anomalies displayed in the European region of the Magsat anomaly map of Coles et al. [1982] show shapes and amplitudes which are not essentially different from those of the United States. Therefore, these relations should in general also hold for Europe.

In contrast to this result obtained from Magsat data, the interpretations of airborne surveys in Europe (areas (1), (3) and (4) of Table 1) are compatible with the ratio rock type (a):rock type (b) = 2:1 in the lower crust in general:

$$M = (2\ M(a) + 1\ M(b))/3 = 0 \text{ to } 2.2\ A/m, \qquad (4)$$

a relation which is expected by petrologists for Europe. There is obviously a discrepancy in the estimate of the ratio rock type (a):rock type (b) in the lower crust:

- 1:1 for 2/3 of the area from Magsat anomaly interpretation

and

- 2:1 for the whole area from airborne survey interpretation and from the comparison of Magsat data with a global model of the Earth's crust with the help of spectra of spherical harmonics.

The cause of this discrepancy might be sought in the characteristics of the modelling procedures. The anomaly transformation as used by Mayhew et al. [1980], for example, seems to have a tendency to yield greater values of magnetisation than the other approach.

But even so we can state the following: in wide areas, the lower crust shows a magnetisation (averaged over volumes of more than 100 km across) in the range M = 0 to 3 A/m which is between the range of felsic and mafic lower crustal rocks. This makes probable that the lower crust is composed by rocks of felsic and of mafic type in spatially varying proportions.

There are also large bodies in the lower crust composed essentially of mafic rocks which may have magnetisations of M = 4.6 A/m.

Table 1 shows several bodies which might fall into this group. But for some of these strongly magnetised bodies, the magnetisations reach 5 A/m for surveys at or near the ground, and 6 A/m for satellite surveys which are clearly above the upper limit of our magnetisation range of M = 0 to 4.6 A/m for rock type (b). However, in our magnetisation values as well as in the calculated model parameters there is a certain range of variability, thus we do not consider this a serious discrepancy. In any case, the composition of coherent masses several 100 km across, 10 km or more thick with a magnetisation of 6 A/m is not easily understood.

Conclusions

1. The magnetisation values required in the lower crust for modelling single long wavelength anomalies from near surface surveys and anomalies obtained from Magsat data are in the range M = 0 to 3 A/m, in some cases values up to M = 6 A/m are reached.
2. Rocks pertaining to the lower crust can be sampled
   - as xenoliths,
   - from the respective part of elevated lower crustal sections which are in general steeply dipping,
   - from deeply eroded areas of old shields.

   Up to now only old shield samples have yielded magnetisations which are sufficiently strong for covering an interval of M = 0 to 4.6 A/m. The upper end of the interval is the maximum induced magnetisation multiplied by 1.5 in order to account for a probable viscous magnetisation at temperatures of the lower crust. Crustal sections seem to have only weakly magnetised lower crustal parts: M = 0 to 1 A/m. According to Wasilewski and Mayhew [1982] lower crustal xenoliths show susceptibilities in the range $(0.045 \text{ to } 5.56)\cdot 10^{-3}$ (cgs), corresponding to induced magnetisations of M = 0.02 to 2.8 A/m, and natural remanent

magnetisations in the range M_{rem} = 0.1 to 88.7 A/m.
Some of these magnetisations might be acquired during the uplift of the xenoliths. Thus their relevance on conclusions on the magnetisation in the lower crust is limited.

3. Statements no. 1 and 2, above, are compatible with the concept that the lower crust is composed of mafic as well as felsic rocks with similar amounts of contribution but varying spatially.
In this situation it is not very likely that there is a considerable additional amount of magnetic rocks arranged in global shells so as to show no magnetic field anomaly.

Acknowledgements. We are grateful to Peter N. Shive and an anonymous reviewer for many important hints.

References

Caner, B., Long aeromagnetic profiles and crustal structure in western Canada, Earth Planet. Sci. Lett., 7, 3-11, 1969.

Coles, R. L., A flexible interactive magnetic anomaly interpretation technique using multiple rectangular prisms, Geoexploration, 14, 125-131, 1976.

Coles, R. L., Magsat scalar magnetic anomalies at northern high latitudes, J. Geophys. Res., 90, 2576-2582, 1985.

Coles, R. L., and R. G. Currie, Magnetic anomalies and rock magnetizations in the southern coast mountains, British Columbia: Possible relation to subduction, Can. J. Earth Sci., 14, 1753-1770, 1977.

Coles, R.L., G.V. Haines, G. Jansen van Beek, A. Nandi and J.K. Walker, Magnetic anomaly maps from 40°N to 83°N derived from Magsat satellite data, Geophys. Res. Lett., 9, 281-284, 1982.

Committee for the Magnetic Anomaly Map of North America, Magnetic Anomaly Map of North America: Boulder, Colo; Geological Society of America, 4 sheets, scale 1:5,000,000, 1987.

Dunlop, D.H., Thermal enhancement of magnetic susceptibility, J. of Geophysics, 40, 439-451, 1974.

Elming, S.-A., and A. Thorne, The blue road geotraverse: A magnetic ground survey and and the interpretation of magnetic anomalies, Geol. Foeren. Stockholm Forh., 98, 264-270, 1976.

Frey, H., Magsat and POGO magnetic anomalies over the Lord Howe Rise: Evidence against a simple continental crustal structure, J. Geophys. Res., 90, 2631-2639, 1985.

Frost, R.B. and Shive, P.N., Magnetic mineralogy of the lower continental crust, J. Geophys. Res., 91, 6513-6521, 1986.

Hahn, A., E. G. Kind, and D. C. Mishra, Depth estimation of magnetic sources by means of Fourier amplitude spectra, Geophys. Prospect., 24, 287-308, 1976.

Hahn, A., H. Ahrendt, J. Meyer, and J. H. Hufen, A model of magnetic sources within the Earth's crust compatible with the field measured by the satellite Magsat, Geol. Jb.,A 75, 125-156, 1984.

Hahn, A., and W. Bosum, Geoexploration Monographs, Series 1 - No 10: Geomagnetics - Selected Examples and Case Histories, 166 pp., 1986.

Hall, D. H., Long-wavelength aeromagnetic anomalies and deep crustal magnetization in Manitoba and northwestern Ontario, Canada, J. Geophys., 40, 403-430, 1974.

Hall, D.H., Coles, R.L. and Hall, J.M., The distribution of surface magnetization in the English River and Kenora subprovinces of the Archean shield in Manitoba and Ontario, Can. J. Earth Sci., 16, 1764-1777, 1979.

Harrison, C.G.A, Carle, H.M. and Hayling, K.L., Interpretation of setellite elevation magnetic measurements, J. Geophys. Res., 91, 3633-3650, 1986.

Johnson, B. D., Viscous remanent magnetization model for the Broken Ridge satellite magnetic anomaly, J. Geophys. Res., 90, 2640-2646, 1985.

Kay, R.W. and Kay, S.M., The nature of the lower continental crust: Inferences from geophysics, surface geology, and crustal xenoliths, Rev. Geophys. Space Physics, 19, 271-297, 1981.

Kay, R.W. and S.M. Kay, Petrology and geochemistry of the lower continental crust: an overview, in The Nature of the Lower Continental Crust, edited by J.B. Dawson, D.A. Carswell, J. Hall and K.H. Wedepohl, 147-159, Geological Society Special Publication No. 24, Blackwell, Oxford, 1984.

Kern, H. and V. Schenk, Elastic wave velocities in rocks from a lower crustal section in southern Calabria (Italy), Phys. Earth Planet. Int., 40, 147-160, 1985.

Krutikhovskaya, Z. A., I. K. Pashkevich, and I. M. Silina, Magnitnaya Model i Struktura Zemnoi Kory Ukrainskovo Shtshita (A Magnetic Model and the Structure of the Earth's Crust of the Ukrainian Shield), 216 pp., 1982.

Langel, R. A., J. D. Phillips, and R. J. Horner, Initial scalar anomaly map from Magsat, Geophys. Res. Lett., 9, 269-272, 1982.

Langel, R. A. an R. H. Estes, A geomagnetic field spectrum, Geophys. Res. Lett., 9, 250-253, 1982.

Mayhew, M. A., H. H. Thomas, and P. J. Wasilewski,Satellite and surface geophysical expression of anomalous crustal structure in Kentucky and Tennessee, Earth Planet. Sci. Lett., 58, 395-405, 1982.

Mayhew, M. A., B. D. Johnson and P. J. Wasilewski, A review of problems and progress in studies of satellite magnetic anomalies, J. Geophys. Res.,90, 2511-2522, 1985.

Mayhew, M.A., B.D. Johnson and R.A. Langel, An equivalent source model of the satellite-altitude magnetic anomaly field over Australia, Earth. Planet. Sci. Lett., 51, 189-198, 1980.

Regan, R. D. and B. D. Marsh, The Bangui magnetic anomaly: Its geological origin, J. Geophys. Res., 87, 1107-1120, 1982.
Riddihough, R. P., Regional magnetic anomalies and geology in Fennoscandia: A discussion, Can. J. Earth Sci., 9, 219-232, 1972.
Ruder, M.E. and Alexander, S.S., Magsat equivalent source anomalies over the southeastern United States: implications for crustal magnetization, Earth Planet. Sci. Lett., 78, 33-43, 1986.
Schlinger, C.M., Magnetization of lower crust and interpretation of regional magnetic anomalies: example from Lofoten and Vesteralen, Norway, J. Geophys. Res., 90, 11484-11504, 1985.
Schnetzler, C. C., An estimation of continental crust magnetization and susceptibility from Magsat data for the conterminous United States, J. Geophys. Res., 90, 2617-2620, 1985.
Shuey, R. T., D. D. Schellinger, E. H. Johnson, and L. B. Alley, Aeromagnetics and the transition between Colorado Plateau and the Basin Range province, Geology, 1, 107-110, 1973.
Taylor, P. T., Magnetic data over the Arctic from aircraft and satellites, Cold Reg. Sci. Technol., 7, 35-40, 1983.
Treloar, N.A., Shive, P.N. and Fountain, D.M., Viscous remanence acquisition in deep crustal rocks, EOS, 67, 266, 1986.
Wasilewski, P. and Fountain, D.M., The Ivrea Zone as a model for the distribution of magnetization in the lower crust, Geophys. Res. Lett., 9, 333-336, 1982.
Wasilewski, P. and Mayhew, M.A., Crustal xenolith magnetic properties and long wavelength anomaly source requirements, Geophys. Res. Lett., 9, 329-332, 1982.
Williams, M.C., Shive, P.N., Fountain, D.M. and Frost, B.R., Magnetic properties of exposed deep crustal rocks from the Superior Province of Manitoba, Earth Planet. Sci. Lett., 76, 176-184, 1985.

GENERALIZED INVERSION OF SCALAR MAGNETIC ANOMALIES: MAGNETIZATION OF THE CRUST OFF THE EAST COAST OF CANADA

J. Arkani-Hamed

Department of Geological Sciences, Brock University, St. Catharines, Ontario, Canada, L2S 3A1*

J. Verhoef

Atlantic Geoscience Centre, Geological Survey of Canada, Bedford Institute of Oceanography, Dartmouth, Nova Scotia, Canada, B2Y 4A2

Abstract. A generalized inversion technique is presented in this paper which transforms scalar magnetic anomalies into crustal magnetization while taking into account variations in the directions of the geomagnetic field and the magnetization of the crust over the area considered. It also accounts for the effects of the topography of the upper and the lower surfaces of the crustal magnetic layer on the magnetic anomalies. The crustal magnetic susceptibility models show better correlation with geological features than do the observed magnetic anomalies. The ocean-continent boundary does not have a pronounced magnetic signature, rather the gradient in the anomalies corresponds to the deepening of the basement. The basement topography in the oceanic regions has only a minor influence on the magnetic anomalies. This suggests that the magnetic anomalies are largely due to the lateral variations in the magnetization of the crust.

Introduction

Magnetic anomalies are commonly displaced with respect to their causative magnetic sources, because the directions of the geomagnetic field and the magnetization of the crust are generally not vertical. This emphasizes the fact that direct correlation of magnetic anomalies with geological features can be misleading. Therefore, it is necessary to convert magnetic anomalies into lateral variations in the magnetization of the crust, which delineate the basic physical properties of the rocks. This is usually accomplished through an inversion process. The standard inversion procedures [Bhattacharyya 1980; Letros et al., 1983; amongst others] are based on the assumption that the directions of the geomagnetic field and the magnetization of the crust are constant over the studied region. This assumption is plausible over small areas, but it is certainly not reasonable over large areas where the directions of the geomagnetic field and the magnetization of the crust may change appreciably. Regional magnetic anomalies presently derived from satellite magnetic data [Benkova et al., 1973; Regan et al., 1975; Cain et al., 1984; Arkani-Hamed and Strangway, 1985; Mayhew and Galliher, 1982; Ridgeway and Hinze, 1986] or from the combination of many aeromagnetic and/or marine magnetic data [Hinze and Zietz, 1985; Dods et al., 1985; also the North American magnetic anomaly map recently published jointly by the United States Geological Survey and the Geological Survey of Canada, Committee, 1987] cover large areas. Arkani-Hamed and Strangway [1986a] developed an inversion technique which inverts regional scale magnetic anomalies into the lateral variations in the magnetization of the crust while taking into account the variations in the directions and intensities of both the geomagnetic field and the magnetization of the crust. The technique is suitable for satellite data inversion but may not be useful for the inversion of aeromagnetic and marine magnetic data, mainly because it does not account for the effects of the topography of the crustal magnetic layer. Topography has, indeed, little effect on magnetic data gathered at satellite elevations, but it may have a significant effect on aeromagnetic and marine magnetic data due to the proximity of magnetic sources to the data-collecting plane. The inversion technique developed by Parker [1973] takes into account the topography of the magnetic layer boundaries, but it assumes that the directions of the geomagnetic field and the magnetization of the crust are constant over the region.

This paper presents a generalized inversion technique which inverts regional magnetic anomalies to lateral variations in the magnetization of the crust while considering variations in the directions of the geomagnetic field and the magnetization of the crust, and accounting for the undulations of the upper and the lower boundaries of the crustal magnetic layer. The present technique combines the method developed by Arkani-Hamed and Strangway [1986a] and that developed by Parker [1973]. It formulates the inversion process on the basis of a spectral frequency domain algorithm, which requires less computer time owing to the use of the fast Fourier transform method [Cooley and Tukey, 1965]. Furthermore, the spectral frequency domain algorithm delineates the main sources of the instabilities which arise when low-latitude magnetic anomalies and high-frequency components of magnetic anomalies are inverted into the magnetization of the crust. Also the present method does not

* now at: Department of GeologicalSciences, McGill University, Montreal, Quebec, Canada, H3A 2A7

develop the instability inherent in the space domain inversion techniques as discussed by Mayhew [1979].

The first section of this paper develops the mathematical formulas for the generalized inversion process. In the second section these formulas are applied to the marine magnetic anomalies of Canada's eastern offshore area, in order to provide lateral variations in the magnetic susceptibility of the crust in this region. The interpretations of the magnetization pattern of the crust are presented in the third section.

Theory

Let $\mathbf{M}$ denote the magnetization at a point $\mathbf{r_o}$ in a magnetic layer which is bounded by a lower surface S_1 and an upper surface S_2. S_1 and S_2 are piecewise continuous surfaces of variable depth. The magnetic potential, A, of the layer at a point $\mathbf{r}$ is defined by

$$A = -\int\int\int \mathbf{M} \bullet \nabla \left[\frac{1}{|\mathbf{r}-\mathbf{r_o}|} \right] dv_o, \tag{1}$$

where the integration is carried over the entire volume (v_o) of the layer. Applying the Fourier transformation to this equation yields [Arkani-Hamed and Strangway, 1986a]

$$A_{uv} = -2\pi \frac{e^{-Kz}}{K} \mathbf{G} \bullet \mathbf{M}_{uv}, \tag{2}$$

where u and v are wave numbers along the x and the y axes of the coordinate system respectively. The x, y and z axes are eastward, northward, and upward, respectively. In equation (2), z denotes the elevation of the observation plane, and K is

$$K = (u^2 + v^2)^{1/2}, \tag{3}$$

and $\mathbf{G}$ is given by

$$\mathbf{G} = (iu, iv, -K). \tag{4}$$

A_{uv} and $\mathbf{M_{uv}}$ are the Fourier transforms of A and $\mathbf{M}$ defined by

$$\xi_{uv} = \int\int_{-\infty}^{+\infty} \xi(x, y)\, e^{-i(ux+vy)} dxdy, \tag{5}$$

where ξ stands either for A or for $\mathbf{M}$. The vector $\mathbf{M}$ is given by

$$\mathbf{M} = \int_{S_1}^{S_2} \mathbf{M} e^{Kz_o} dz_o, \tag{6}$$

which shows that the magnetic potential depends on the integral of the magnetization over the thickness of the layer. Equation (6) illustrates the basic non-uniqueness encountered in determining the magnetization of the layer. Any distribution of magnetization which makes the integral in the right hand side of the equation vanish, called annihilator [Parker and Huestis, 1974; Harrison et al., 1986], can be added to a given magnetization of the layer without affecting the magnetic potential. A classic example of an annihilator is a flat layer with a uniform magnetization which produces no magnetic anomaly, and thus it can be added to any magnetization model without affecting the observed magnetic anomalies. This implies that magnetic anomalies are basically produced by the lateral variations in the magnetization of the magnetic layer. The magnetic anomaly analysis alone can yield the magnetization which is averaged over the thickness of the layer. It is, therefore, sufficient to calculate the mean value of the magnetization, $\mathbf{m}$, at a given x and y position and reduce equation (6) into

$$\mathbf{M} = \frac{\mathbf{m}}{K}\left[e^{KS_2} - e^{KS_1} \right]. \tag{7}$$

The flat-earth assumption made in the derivation of equation (2) does not introduce significant error to the results [Vint et al., 1970].

The magnetic anomaly at the point $\mathbf{r}$ is defined by

$$T = -\hat{B} \bullet \nabla A \tag{8}$$

where $\hat{B}$ is the unit vector along the geomagnetic field at that point.

Over large areas the directions of $\mathbf{m}$ and $\hat{B}$, and the depth to S_1 and S_2 change appreciably and the terms in the right hand sides of equations (2) and (8) couple different spectra together. We adopted a linear perturbation procedure to solve these equations, and let

$$\begin{aligned} \mathbf{m} &= m\hat{M}, \\ \hat{M} &= \mathbf{M_o} + \delta\mathbf{M}, \\ S_2 &= S_2^{\,0} + \delta S_2, \\ S_1 &= S_1^{\,0} + \delta S_1, \end{aligned} \tag{9}$$

and

$$\hat{\mathbf{B}} = \mathbf{B_o} + \delta\mathbf{B},$$

where m and $\hat{M}$ are the intensity and the unit vector along $\mathbf{m}$, respectively. $\mathbf{M_o}$, $\mathbf{B_o}$, $S_2{}^0$ and $S_1{}^0$ are the mean values of $\hat{M}$, $\hat{\mathbf{B}}$, S_2 and S_1 and $\delta\mathbf{M}$, $\delta\mathbf{B}$, δS_2 and δS_1 denote the deviations of $\hat{M}$, $\hat{\mathbf{B}}$, S_2 and S_1 from their mean values, respectively. Assimilating equations (9) into equations (2) and (8) and making use of the identity

$$\nabla A_{uv} = \mathbf{G} A_{uv}, \tag{10}$$

yields

$$A_{uv} = -\frac{1}{(\mathbf{B_o} \bullet \mathbf{G})}\left[T_{uv} + (\delta\mathbf{B} \bullet \nabla A)_{uv} \right] \tag{11}$$

and

$$m_{uv} = \frac{-K}{\left(e^{KS_2^0} - e^{KS_1^0}\right)(\mathbf{G} \bullet \mathbf{M_o})}\left[\frac{K}{2\pi} e^{KZ} A_{uv} + \mathbf{G} \bullet \zeta_{\mathbf{uv}} \right], \tag{12}$$

where

$$\zeta = m\left(\frac{e^{KS_2^0} - e^{KS_1^0}}{K}\right)\delta\mathbf{M} + \mathbf{m}\left[e^{KS_2^0}\sum_{n=1}^{\infty}\frac{K^{n-1}}{n!}\delta S_2^n - e^{KS_1^0}\sum_{n=1}^{\infty}\frac{K^{n-1}}{n!}\delta S_1^n\right]. \tag{13}$$

The Fourier transform of the magnetic potential can be obtained from that of the magnetic anomalies by solving equations (10) and (11) iteratively. Likewise, the Fourier transform of the effective magnetization of the magnetic layer can be calculated from that of the magnetic potential by solving equation (12) iteratively. Assuming that the magnetization is of an induced origin, the lateral variations in the effective magnetic susceptibility may be determined by

$$\sigma = \frac{m}{B}, \tag{14}$$

where B is the intensity of the geomagnetic field inside the magnetic layer.

The spectral frequency domain algorithm suggests a natural coordinate system and provides a better understanding of the instability of the inversion process. Equation (12) develops instability at high wave numbers owing to the exponential term exp (Kz). This is similar to the instability inherent in the standard inversion method which is unavoidable though it may be suppressed by filtering [Pilkington and Crossley 1986]. The special instability arising in the inversion of low latitude magnetic anomalies develops when $(\mathbf{G} \bullet \mathbf{B_o})$ or $(\mathbf{G} \bullet \mathbf{M_o})$ or both vanish. These instabilities are unavoidable and also develop in the standard inversion procedure. The first singularity, $(\mathbf{G} \bullet \mathbf{B_o}) = 0$, suggests that

$$B_z = 0 \tag{15}$$

and

$$uB_x + vB_y = 0, \tag{16}$$

where B_x, B_y and B_z are the x, y and z components of $\mathbf{B_o}$, respectively. The instability develops only at the geomagnetic equator and for wavenumbers along the line given by equation (16). The instability is not confined to high-frequency components, rather it may develop at any frequency satisfying equation (16). Equation (16) suggests a natural coordinate system to be employed in order to simplify the analysis and to suppress the instability. The natural coordinate system is the one with the x axis toward east and the y axis toward north. In this coordinate system B_x vanishes and equation (16) is reduced to $v = 0$ implying that all the components with constant amplitudes in the geomagnetic north-south direction develop instability regardless of their east-west variations [see Arkani-Hamed and Strangway 1986a, for simple examples]. These non-crustal components appear in the original magnetic anomalies due to errors made while taking measurements or during data reduction processes. A classic example is the magnetic data acquired by the magnetometer satellite (MAGSAT) which was a polar orbiter, having north-south trending passes over the equatorial zone. The flight line levelling noise produces magnetic anomalies which are elongated in the north-south direction and are strongly enhanced in the inversion process [Arkani-Hamed and Strangway, 1985]. The erroneous spectra can be filtered out by a directional filter which removes all the non-crustal harmonics with constant amplitude in the north-south direction.

The second instability $(\mathbf{G} \bullet \mathbf{M_o}) = 0$, arises when

$$M_z = 0 \tag{17}$$

and

$$uM_x + vM_y = 0 \tag{18}$$

where M_x, M_y and M_z are the x, y and z components of $\mathbf{M_o}$, respectively. These equations are similar to equations (15) and (16) and thus have similar implications, that instability develops when the crust is magnetized horizontally with a uniform intensity along the direction of the magnetization. This situation arises, for example, when oceanic crust is formed at the oceanic ridge axis located at the geomagnetic equator and oriented along the geomagnetic north-south direction.

Magnetization of the Crust off Canada's East Coast

Systematic geophysical mapping of the continental shelf and ocean basins off Canada's east coast (Figure 1 displays the area and the locations referred to in the text) over a period of about 25 years has resulted in a large set of gravity, magnetic and bathymetric data. Recently, the magnetic data have been reduced to the DGRF reference field (IAGA Working Group, 1986) calculated for the period of observations [Verhoef and Macnab, 1988]. Plate 1 shows the magnetic anomalies on an equal area grid with an interval of 5.6 km. The plate covers an area of about 1400 x 2800 km, over which the geomagnetic field changes appreciably both in intensity and in direction. This is illustrated in Figure 2, which shows the direction cosines and the intensity of the geomagnetic field over the area based on the 1975 DGRF model [IAGA Working Group, 1986]. This model is adopted because the major part of the magnetic surveys was carried out between 1970 and 1980. The east and north components of the field are small and highly variable compared to the vertical component because of the proximity of the area to the geomagnetic north pole. This variability of the field emphasizes the fact that a generalized inversion procedure is needed in order to determine the lateral variations in the magnetization of the crust.

In the inversion process we assumed that the magnetization of the crust is along the present geomagnetic field direction, due to the lack of information about the magnetization of the crust, especially in deeper regions. Marine magnetic anomalies are generally dominated by the effects of the remanent magnetization of the oceanic crust, which usually is not orientated along the present geomagnetic field direction, and thus this assumption may not hold. However, in the region considered in this paper, the above assumption may not introduce a significant error on account of the proximity of the geomagnetic pole. The location of the pole has not changed by more than 20 degrees in the last 80 Ma [Harrison and Lindh, 1982] during which time Greenland separated from the North American continent and the Labrador Sea was created. Moreover, the vertical component of the geomagnetic field over

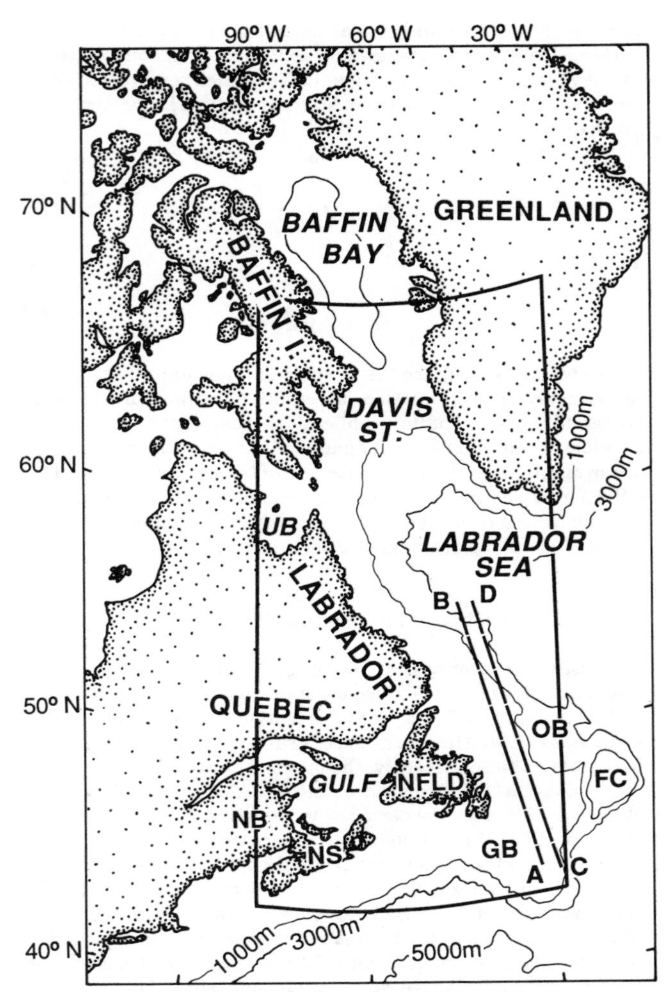

Fig. 1. Location diagram showing the study area (box) and geographical names referred to in the text. Also shown are the 1, 3 and 5 km bathymetry contours. The dashed lines AB and CD are the profiles shown in Figure 7. Abbreviations stand for: NB, New Brunswick; NS, Nova Scotia; GB, Grand Banks; FC, Flemish Cap; NFLD, Newfoundland; GULF, Gulf of St. Lawrence; OB, Orphan Basin; UB, Ungava Bay.

the region was always the most dominent component because the pole has moved mainly in an almost east-west direction. Therefore, the remanent magnetization of the oceanic crust in the study area is approximately in the direction of the present geomagnetic field for normally magnetized crust and in the opposite direction to the field for reversely magnetized crust. This is also the case for continental shelves in the Labrador Sea, because magnetic anomalies over these areas are likely caused by Cretaceous volcanism [Umpleby, 1979] associated with the initial rifting of the Sea. Consequently, reversely magnetized regions appear as low magnetic areas and normally magnetized regions manifest themselves as high magnetic areas.

Figure 2D shows that the intensity of the geomagnetic field changes by about 17% over the studied region. This implies that the magnetic anomaly created by a given magnetic body has a larger amplitude if the body is located in the northern part of the area than if it is in the southern part. Therefore, the effects of variations in intensity of the geomagnetic field have to be accounted for when determining magnetic properties of the crust.

Other parameters involved in the inversion process are depth and shape of the upper and lower boundaries of the magnetic layer. We approximated the upper boundary of the magnetic layer by the oceanic basement, i.e. by the upper surface of the oceanic basaltic layer 2A (Figure 3). The depth to the basement in the Orphan Basin and in the Labrador Sea was obtained from contoured seismic reflection travel times published by Grant et al. [1986] and Srivastava [1986], respectively. The two-way seismic reflection travel times were converted to depth using a seismic velocity of 1500 m/s for the water and the seismic velocity functions given by Tucholke [1986] for ocean floor sediments. In other parts of the area, such as on the Grand Banks, the Scotian shelf, and in the Gulf of St. Lawrence, older depth to basement maps existed [Wade et al., 1977]. New compilations of depth to the basement, not yet released, show significant changes in some locations which makes the validity of the older data doubtful. For the present, we have used the bathymetry data in these areas as a first approximation to the depth of basement. The discrepancy at the contact of the bathymetry and basement data sets was smoothed by tapering the two sets within a transition zone of about 50 km width. Using bathymetry as an approximation to the depth to basement would lead to an underestimate of the crustal magnetization. This should be considered in the interpretation of the resulting magnetization pattern of the crust. We will come back to this point later when we discuss the inversion results.

The depth to the basement shallows towards the central Labrador Sea (Figure 3). However, a trough about 1 to 2 km deep runs roughly along a northwest-southeast trend which corresponds to the extinct sea-floor spreading axis. Also, deep basins are located off the Labrador coast and off Newfoundland with depths of more than 10 km, but not along the Greenland margin. Sediments in the central Labrador Sea are about 2 km thick, as can be observed at about 62° N latitude, where the discontinuity between the basement and bathymetry clearly shows.

The depth to the lower boundary of the magnetic layer is not certain. It is a matter of controversy whether the Moho discontinuity denotes the lower boundary of the magnetic layer, or whether the boundary is located in the the uppermost mantle [Haggerty, 1979; Wasilewski et al., 1979; Haggerty and Toft, 1985; Arkani-Hamed and Strangway, 1986b]. Sea-floor spreading in the Labrador Sea probably ceased about 35 Ma ago [Srivastava and Tapscott, 1986] and the upper part of the upper mantle has since cooled below the Curie temperature of magnetic minerals [Arkani-Hamed and Strangway, 1987]. It is therefore possible that the lower boundary of the magnetic layer beneath the oceanic crust is deeper than Moho, and if so is probably at the same depth as the lower boundary of the magnetic layer beneath the continental crust. Therefore, three different models are considered in this paper. The first model adopts the Moho as the lower boundary of the magnetic layer, whereas the second model assumes that no lower boundary is present. In the third model, the lower boundary is taken at a constant depth inside the upper mantle.

In the first model, the depth to the Moho beneath oceanic

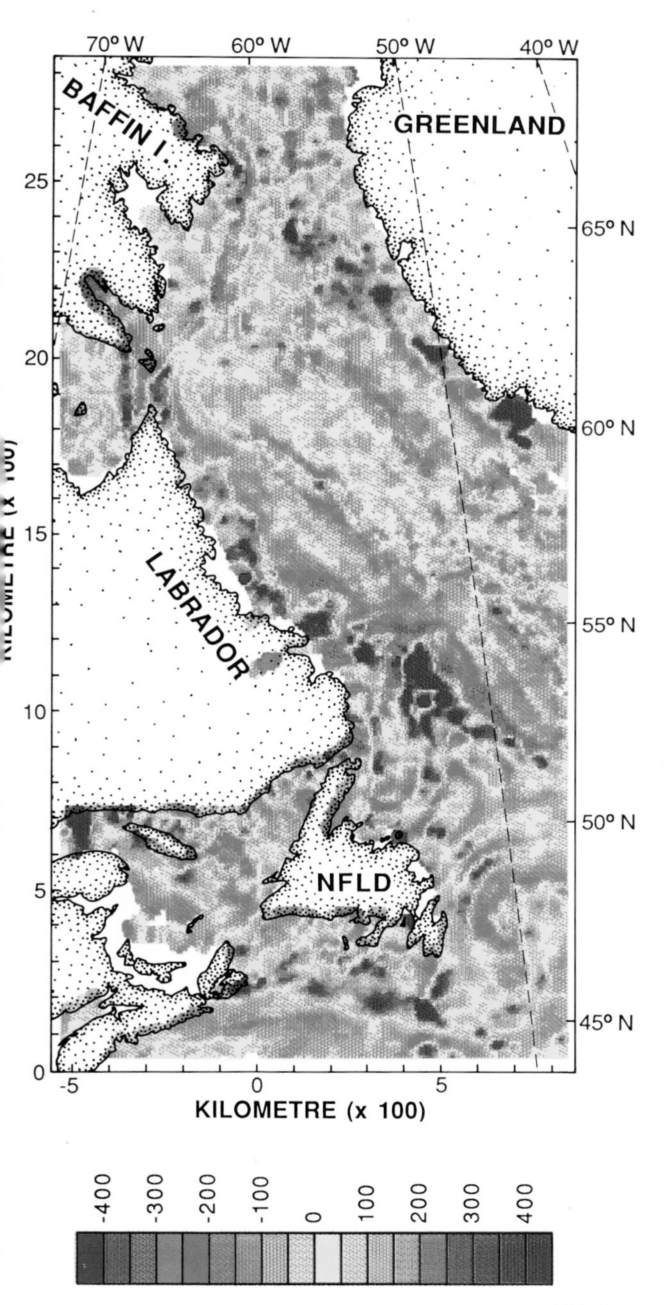

Plate 1. Observed magnetic anomalies (in nT) gridded onto an equidistant grid of 5.6 km. The lower and left axes give the distance to the reference origin in km, while the right and top axes show the geographical coordinates.

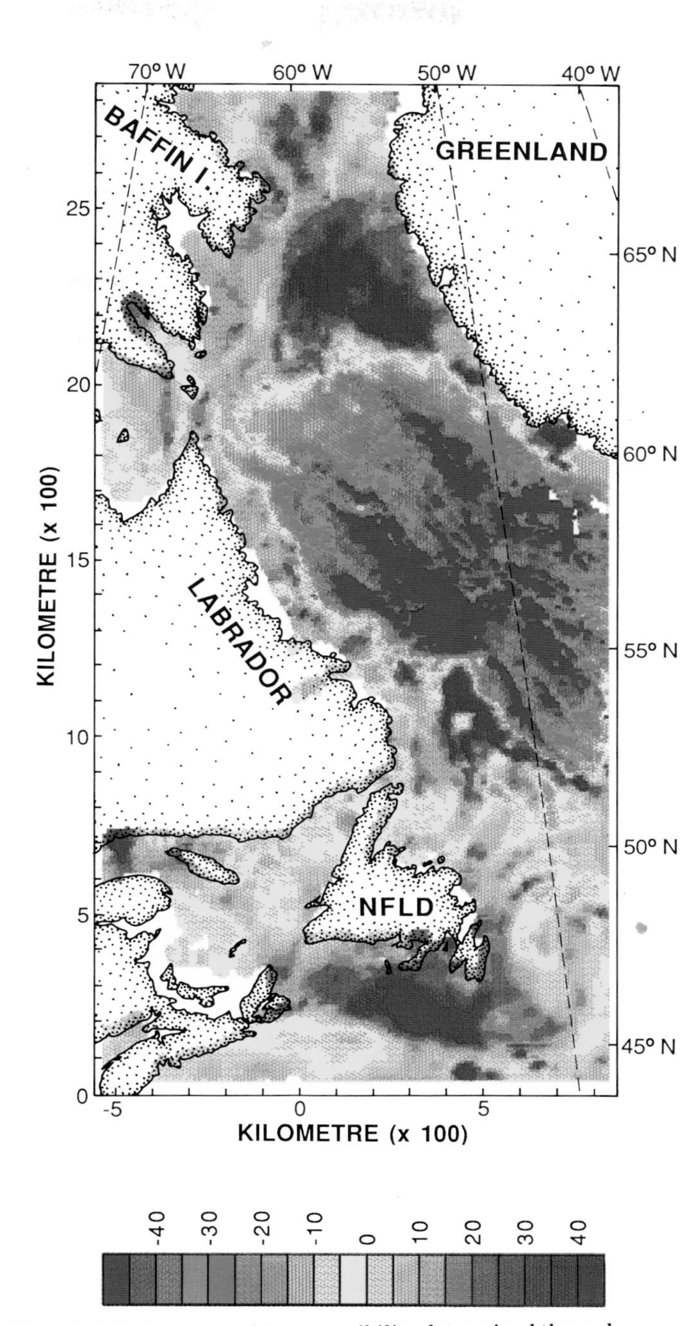

Plate 2. Effective magnetic susceptibility determined through a generalized inversion technique, assuming that the magnetic sources are located between the basement (shown in Figure 3) and the Moho (shown in Figure 4). Units are in 10^{-3} SI.

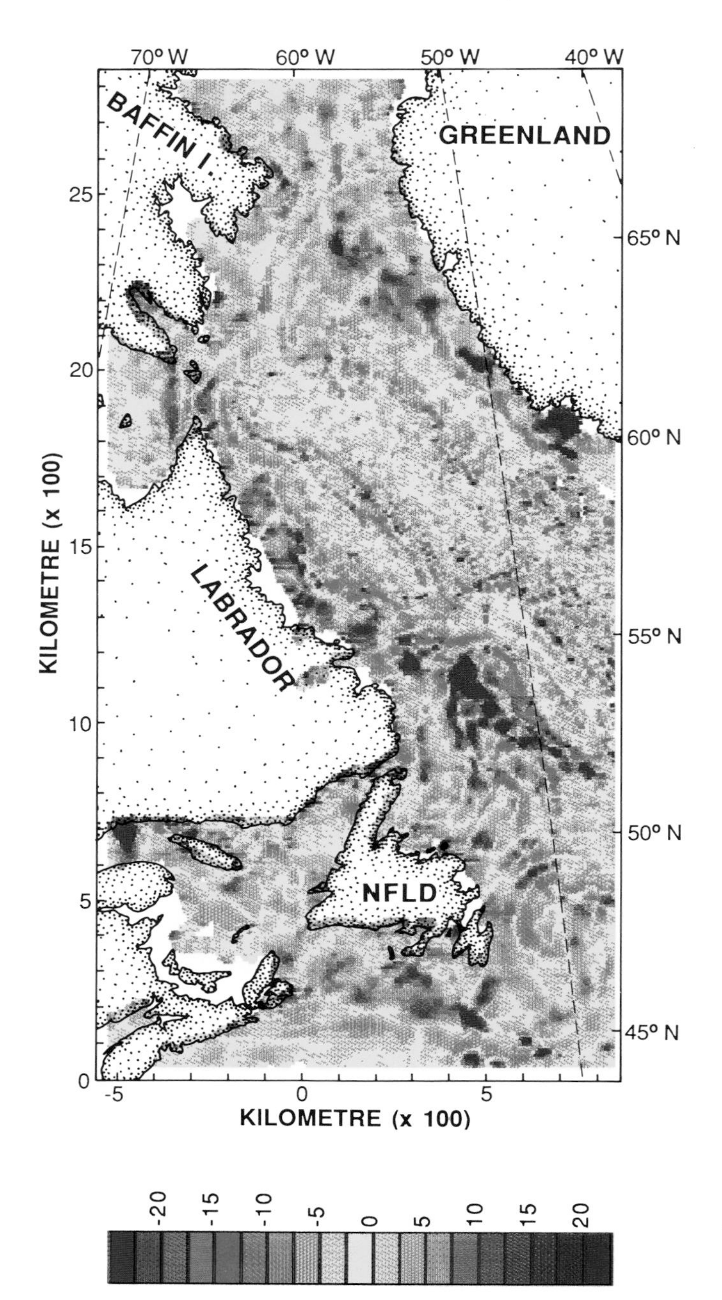

Plate 3. Effective magnetic susceptibility, assuming that the magnetic sources are located between the basement and infinity, i.e. with no lower boundary. Units are in 10^{-3} SI.

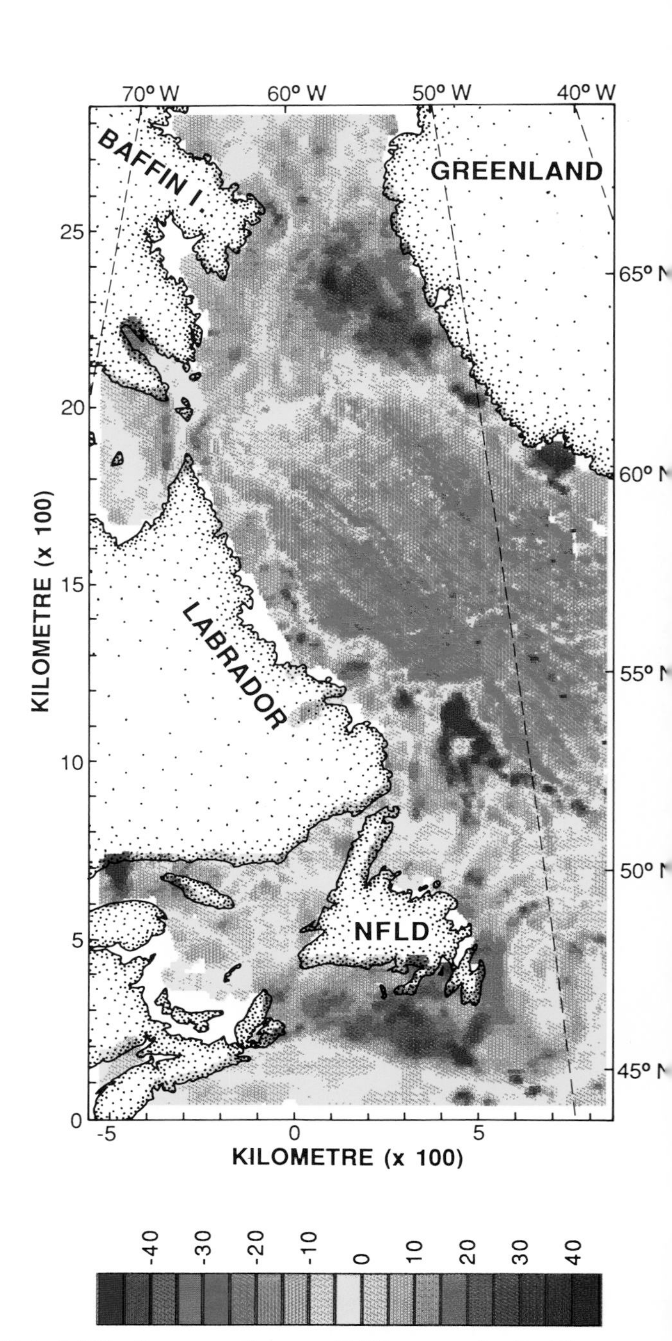

Plate 4. Effective magnetic susceptibility, assuming that the magnetic sources are located between the basement and a lower boundary at a constant depth of 35 km. Units are in 10^{-3} SI.

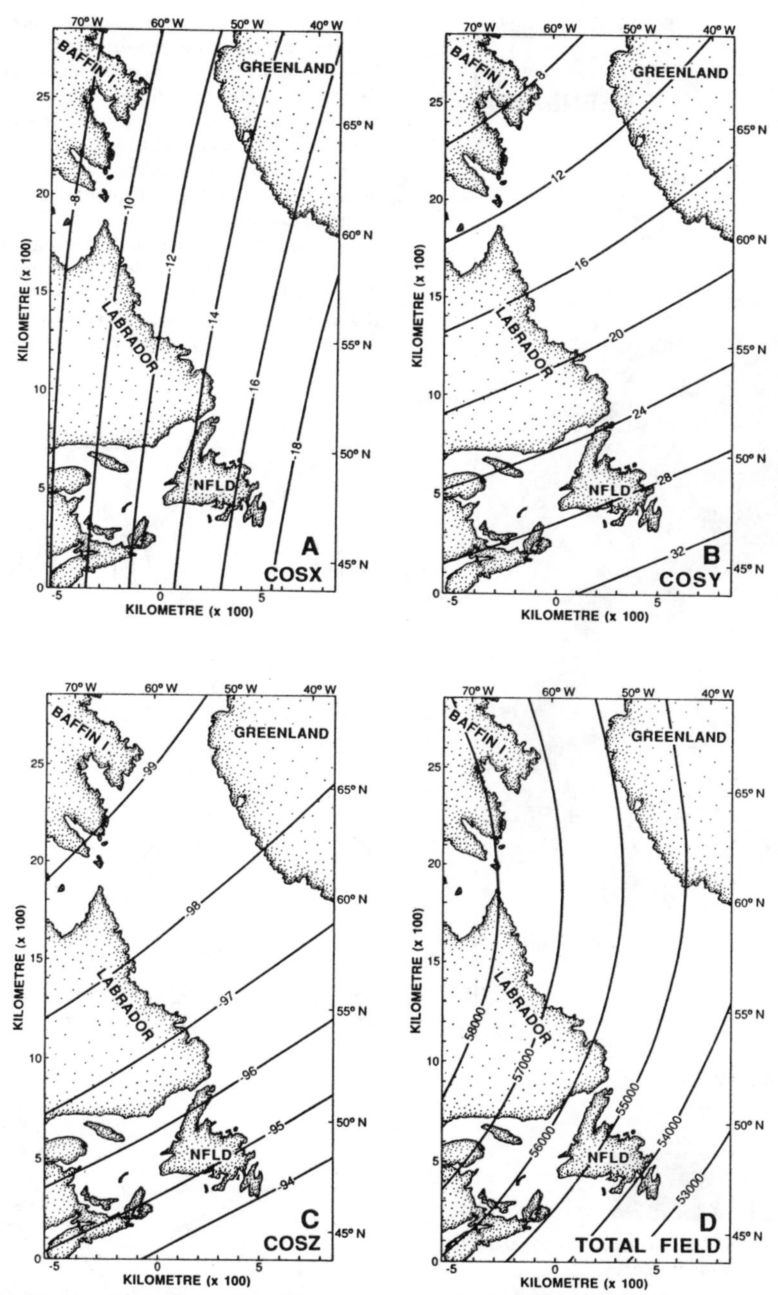

Fig. 2. Direction cosines and total intensity of the geomagnetic field based on the DGRF field of 1975. A, B and C show the cosine directors along the east, north and vertical directions, respectively (values are multiplied by 100). D depicts the total field intensity in nT.

areas was determined from the thickness of the crust [Shih et al., in press] and the depth to the basement. In the central Labrador Sea, where no data were available, a crustal thickness of 10 km is adopted, whereas the depth to the Moho is set at 35 km beneath continents, which represents a reasonable average below continental areas. Figure 4 displays the depth to the Moho thus obtained. The crust-mantle boundary deepens to more than 20 km across Davis Strait (65° to 67° N) and shallows again further north in Baffin Bay. In the southeastern part, around 50° N, the depth to the Moho increases because of the presence of the Flemish Cap continental fragment.

For numerical calculations the area under consideration was

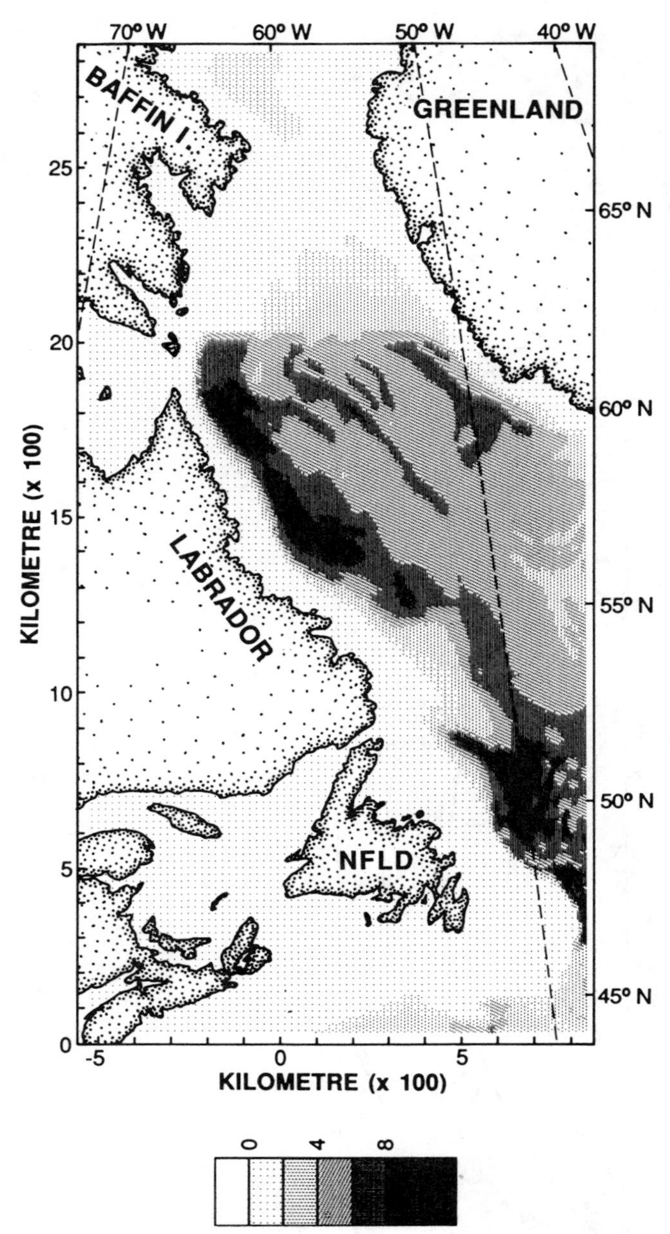

Fig. 3. Depth to basement (in km). At locations where no depth to basement was available we used bathymetry.

subdivided into 256 × 512 square grid cells of 5.6 km sides. The Nyquist wavelength thus retained is 11.2 km. Over continental regions depth to the basement is set to zero km. In the case of magnetic anomalies the grid values were obtained from the observations along the ship tracks by using a gridding algorithm that applies a first-order Butterworth filter to account for the varying spacing of the observations [Verhoef et al., 1986]. We used a cut-off wavelength of 35 km for the filter. Also, the mean value of the magnetic anomalies was removed and the resulting anomalies were then extrapolated beyond the boundaries of the original data using a two-dimensional extrapolation technique. The extrapolated values were tapered to zero by a two-dimensional Hanning function over the first 10 neighboring and initially blank cells. The magnetic anomalies over the rest of the blank cells were set to zero. The resulting magnetic anomaly map was also tapered to zero within 10 cells from the frame boundaries using a one-dimensional Hanning function, in order to suppress the edge effects in the Fourier transformation of the anomalies. The Fourier transformations were carried out by a fast Fourier transformation technique developed by Cooley and Tukey [1965]. Equation (12) was solved by iteration. At each iteration the perturbation terms in the right-hand side were calculated in the space domain and then were transformed to the

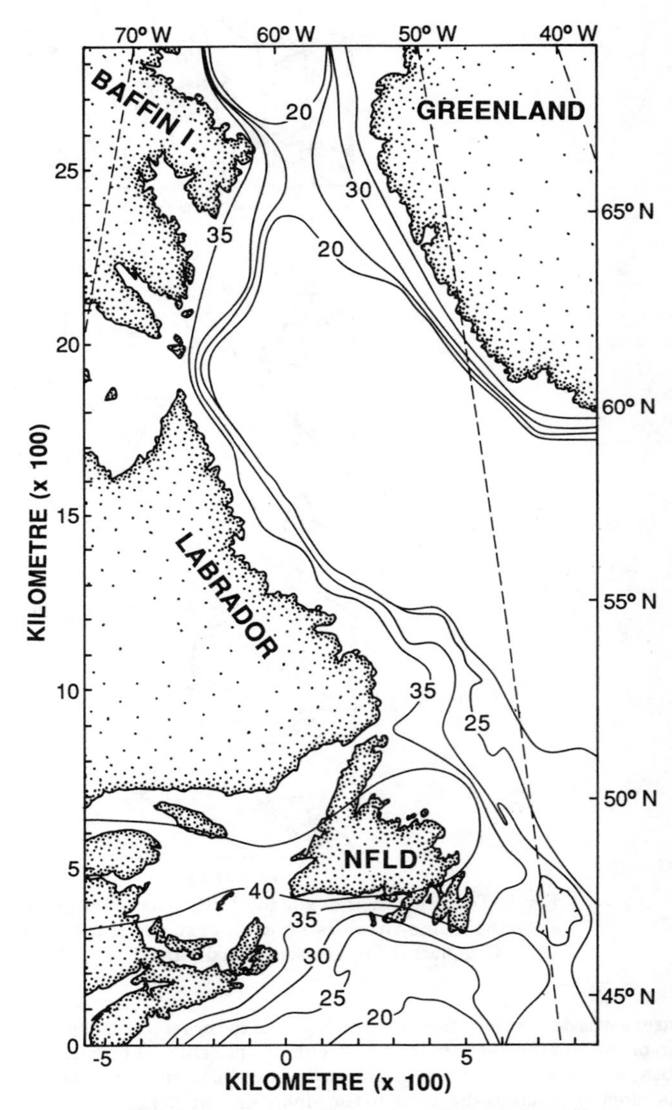

Fig. 4. Depth to Moho (in km). Data were obtained from the crustal thickness map given by Shih et al. [in press] and the depth to basement data they used.

Simmons, G., Velocity of shear waves in rocks to 10 Kilobars, 1. J. Geophys. Res., 69, 6, 1123-1130, 1964.

Smith, R. B., Bruhn, R. L., Intraplate extensional tectonics of the eastern Basin-Range: inferences on structural style from seismic reflection data, regional tectonics, and thermal-mechanical models of brittle-ductile deformation. J. Geophys. Res., 89, B7, 5733-5762, 1984.

Spencer, J. W., Nur, A. M., The effects of pressure, temperature, and pore water on velocities in Westerly granite. J. Geophys. Res., 81, 5, 899-904, 1976.

Stiefel, A., Dornstädter, J., Reifenstahl, F., Bried, M., Jäger, K., Geothermische KTB-Voruntersuchungen im Mittleren Schwarzwald. In: Sitzungsberichte der 16. Sitzung der FKPE-Arbeitsgruppe "Ermittlung der Temperaturverteilung im Erdinneren", ed. G. Buntebarth, TU Clausthal, 88-107, 1986.

Strehlau, J., Meissner, R., Estimation of crustal viscosities and shear stresses from an extrapolation of experimental steady state flow data. In: Composition, Structure and Dynamics of the Lithosphere-Asthenosphere- System, ed. Fuchs, K., Froidevaux, G., Geodynamics Series, 16, AGU, 69-87, 1987.

Vitorello, I., Pollack, H. N., On the variation of continental heat flow with age and the thermal evolution of continents. J. Geophys. Res., 85, 983-995, 1980.

HEAT FLOW, ELECTRICAL CONDUCTIVITY AND SEISMICITY IN THE BLACK FOREST CRUST, SW GERMANY

H. Wilhelm[1], A. Berktold[2], K.-P. Bonjer[1], K. Jäger[1], A. Stiefel[1], and K.-M. Strack[3]

Abstract. Several geophysical site surveys have been performed in the Black Forest, southwest Germany, for the German Continental Deep Drilling Program (KTB) during 1984 to 1986. Besides seismic reflection and refraction measurements and 1D and 2D gravity and magnetic field surveys investigations of surface heat flow and of the crustal electrical conductivity structure were carried out. Also, the seismicity record and results from repeated levellings since 1922 in the Rhinegraben region were inspected. This paper concentrates on the results of heat flow and seismicity observations and of magnetotelluric, magnetovariational and electromagnetic investigations. The surface heat flow shows a regional decline from about 90 mW/m^2 in the North to about 70 mW/m^2 in the central Black Forest on a horizontal scale of 30 km. The electrical conductivity is anomalously high at midcrustal level, and in the envisaged drilling area near Haslach a well-conducting layer was detected at 7 to 9 km depth by long offset transient electromagnetic (LOTEM) recordings. A correlation to a pronounced midcrustal seismic low velocity layer is presumed. The seismicity shows a regional variation in frequency and hypocenter depth distribution. A common interpretation of the results of the different kinds of observations is facilitated if the crustal properties are significantly influenced by the presence of fluids.

Introduction

The Black Forest (BF) in SW-Germany is part of the Rhinegraben system as shown in Figure 1. It has been a focus of geophysical and interdisciplinary investigations during 1984 to 1986 because it had been proposed as one of two possible drilling sites in the German Continental Deep Drilling Program KTB. In order to prepare a sound basis for the decision between the two sites (the other site being situated in the Oberpfalz near the Czechoslovakian border), a broad spectrum of geophysical, geochemical, geological, petrological, and mineralogical investigations were performed in both areas. Geophysical data were obtained by reflection and refraction seismic surveys, by

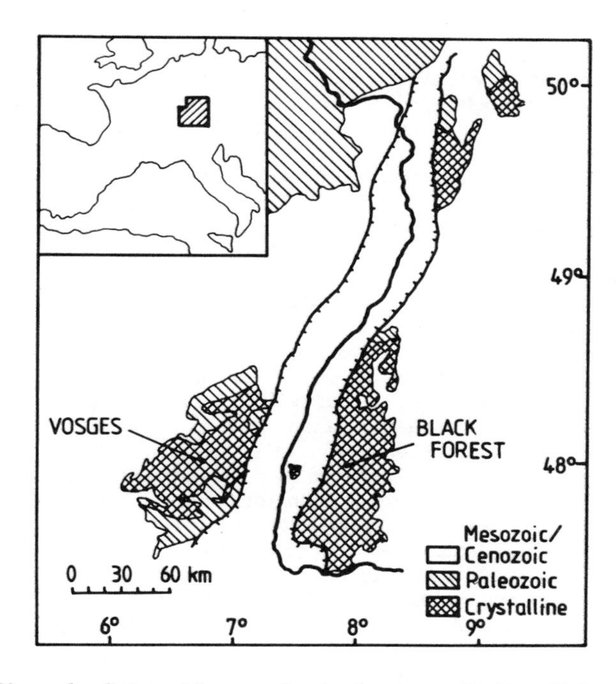

Fig. 1. Schematic geological map of the Rhinegraben rift region. Cross-hatching indicates exposed crystalline basement rock. Western and eastern flanks of the graben are Vosges and the investigation area Black Forest. Northern latitude and eastern longitude give at the frame. In the centre the Rhine and the graben faults are indicated by solid lines.

[1]Geophysikalisches Institut, Universität Karlsruhe, Hertzstr. 16, D-7500 Karlsruhe 21, F.R. Germany

[2]Institut für Allgemeine und Angewandte Geophysik, Universität München, Theresienstr. 41/IV, D-8000 München 2, F.R. Germany

[3]Institut für Geophysik und Meteorologie, Universität Köln, Albertus-Magnus-Platz, D-5000 Köln 1, F.R. Germany

electromagnetic, magnetovariational and magnetotelluric soundings and by geothermal investigations. Also, a surface gravimetric and magnetic survey was performed. In the Black Forest special attention was paid to a continuous record of seismicity in the Rhinegraben region for more than 15 years and to repeated surveys of vertical crustal movements since 1922 in SW-Germany.

The results of the seismic surveys are discussed in Fuchs et al. (1987), Gajewski and Prodehl (1987) and Gajewski et al. (1987). A detailed description of the near-vertical and wide-angle seismic investigations is given in Lüschen et al. (1987). Here, the results of the investigations of heat flow and electrical conductivity and of the seismicity will be reported.

In the past, geophysical investigations in SW-Germany were mainly concentrated on the Rhinegraben proper in order to reveal its structure beneath the surface (Illies and Mueller, 1970; Illies and Fuchs, 1974; Edel et al., 1975; Prodehl et al., 1976; Kahle and Werner, 1980; Werner and Kahle, 1980). With the advent of the KTB program it became possible to obtain a detailed view of the crystalline Variscan basement which crops out in the Black Forest.

The main results of the geophysical KTB surveys in the Black Forest may be summarized as follows: There is

- a well defined structural difference between upper and lower crust (Lüschen et al., 1987),
- a pronounced low velocity layer in the lower part of the upper crust (Gajewski and Prodehl, 1987),
- a midcrustal layer of high electrical conductivity (Schmucker, 1987; Tezkan, 1988),
- a systematic regional variation of heat flow with no apparent correlation to Bouguer gravity anomaly and depth of Conrad discontinuity (Stiefel and Jäger, 1987),
- a systematic variation in frequency and depth of earthquakes from N to S and from the graben proper to the shoulders (Bonjer et al., 1984).

Although these results established attractive geophysical research aims for a superdeep drill hole, the decision was taken against the Black Forest at a conference at Seeheim/Odenwald in September 1986 (Emmermann and Behr, 1987). This decision was strongly determined by arguments concerning the risk of drilling into a possible high temperature region. For present day drilling and logging techniques high temperatures present a serious bound so that the envisaged drilling aim of 12 to 14 km could probably not be reached. Nevertheless, the Black Forest continues to be a research area where important contributions to the "investigation of the physical and chemical conditions and processes in the deeper continental crust to achieve a better understanding of the structure, composition, and evolution of intracontinental crustal regions" (research aim of the KBT program) can be obtained.

Geothermal Investigations

The Black Forest and the Vosges (see Fig. 1) are part of the European Hercynian fold belt consisting of two external zones, the Rhenohercynian and the Saxothuringian which at the surface are mainly composed of Paleozoic sedimentary rocks, and the internal Moldanubian zone where larger parts of the crystalline basement are exposed at the surface in the Bohemian Massif, the Black Forest, the Vosges and the Massif Central (Behr et al., 1984). The crystalline basement of the BF consists of Prevariscan high grade gneisses and migmatites in the central part. In the north, east and south of the BF these gneisses were intruded by Hercynian granitoids, compare Fig. 2. In the west a graben system formed in the early Tertiary due to the stress regime in the foreland of the Alpine collision zone. Subsidence in the Rhinegraben started 45-40 My ago and uplift of the graben shoulders by more the 2000 m caused exposure of the Hercynian basement. As the seismicity of the upper Rhinegraben is rather low compared to other continental graben systems the graben formation process is considered to be almost finished. However when this process was active during the Tertiary and Quaternary it probably influenced the crustal structure in that region deeply (Fuchs et al., 1987).

The geologic and tectonic evolution of the Black Forest crust is of fundamental importance for its structural and compositional properties which are intimately related to the crustal temperature field and heat flux. Also its hydrological properties which can substantially influence the mechanism of heat transfer are essentially based on the tectonic style which has evolved from the Variscan orogeny and the Rhinegraben rift process.

The present day thermal surface heat flux may still contain signatures of the past. Hydrothermal alterations and thrust tectonics may have changed substantially the thermal and seismic properties of the rocks at depth. Granitisation and granulitisation may have caused enhanced or reduced concentrations of radiogenic elements respectively and a redistribution mainly of quartz with significant consequences for the thermal conductivity of the corresponding rocks.

Prior to the geothermal KTB investigations almost nothing was known about surface heat flow in the BF. There was only one value, i.e. 68.8 mW/m^2, observed in the sediments of the Feldsee (Hänel, 1983). Temperature extrapolations to 10 km depth were therefore largely influenced by surface heat flow determinations in the adjacent Rhinegraben and Swabian Jura.

Temperature Measurements

The BF crust at the surface is mainly composed of different types of gneisses and granites which have been investigated with respect to their thermal properties during the KTB survey by probing surface rocks and core samples of 6 drill holes which were drilled to a depth between 180 m to 300 m into the crystalline basement of the central BF gneiss complex and into the Triberg granite (Jenkner et al., KTB-Report 1986, Geol. Landesamt Baden-Württemberg, 1987). These holes were completely cored for geological, mineralogical, petrological and geophysical laboratory investigations. Holes No. 1 (Hechtsberg) and No. 2 (Schönmatt) were prepared in the vicinity of the proposed deep drilling site near Haslach. Hole No. 3 (Moosengrund) was positioned to yield compositional and structural information about the properties of granites in the surroundings of Haslach and hole No. 4 (Kunklerwald) was chosen to explore the properties of a special granulite gneiss facies which is supposed to represent the main rock type of the lower crust. With hole No. 5 (Ettersbach) anatectic gneisses enriched in biotite were explored and hole No. 6 (Geschahse) was drilled into heterogeneous rock comprising granites, gneisses and syenites. By this selection of drill sites samples from the main rock types of the central BF gneiss complex could be collected for geothermal laboratory investigations. All wells except No. 3 are approximately situated on a continuous NS geothermal profile. In Figure 2 these holes can be identified by their corresponding encircled heat flow values listed in Table 1.

In these holes high resolution temperatures (HRT) at 10 cm intervals and bottom hole temperatures (BHT) were measured during pauses in drilling at weekends (Stiefel, KTB-Report 1985, Geophysik. Inst. Karlsruhe, 1986). Thus fairly undisturbed temperature profiles could be obtained even for depth regions where, by subsequent drilling, the temperature logs later got severely disturbed when circulating water infiltrated the hole (Burkhardt et al., 1988).

From other already existing holes temperature profiles could also be obtained, mainly from mineral and thermal water wells. In these commercial wells temperature was measured if possible, otherwise the data had to be accepted as published, for example from Zoth (1985) for the Kirchzarten well and from Kiderlen (1977, 1981) for thermal wells in the northern BF. The quality of these data is, however, not comparable to that of the data obtained from the KTB-holes.

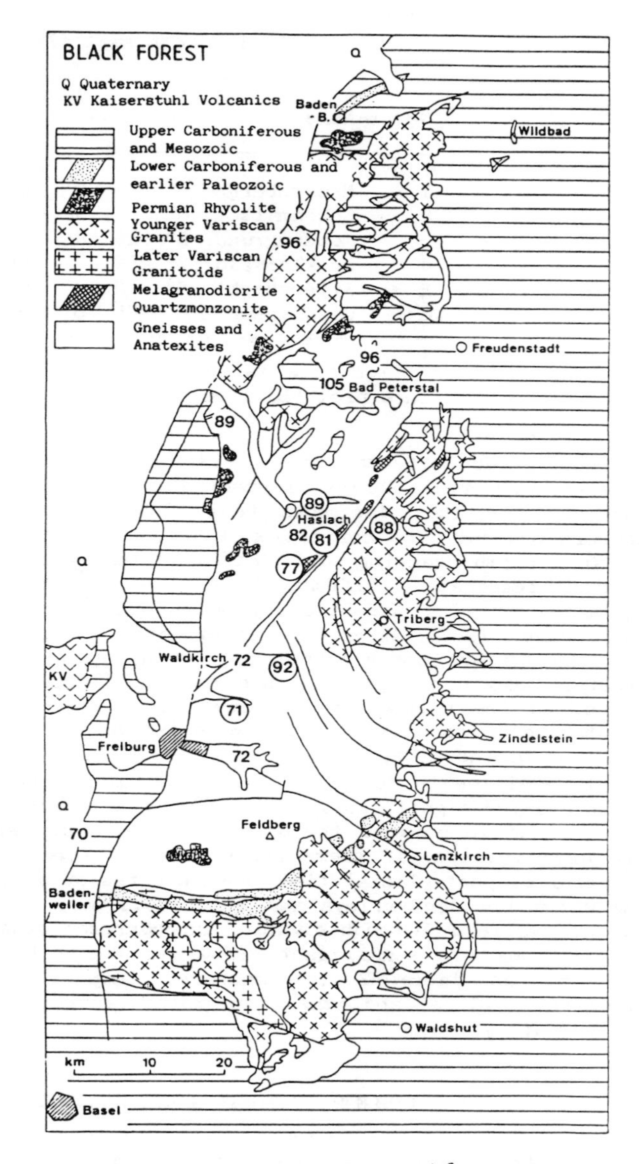

Fig. 2. Surface heat flow in mW/m^2 in the Black Forest. Encircled numbers are topography corrected values determined in the scientific drillholes of Table 1.

Corrections

Topographic corrections had to be applied to the temperature gradients obtained in the KTB drill holes. The analytical 1D-Bullard method (Bullard, 1940) and a numerical 2D-finite difference correction were used separately for verification of the results. A comparison shows that the topographic effects are mostly overcorrected by the Bullard method because of the assumed radial symmetry.

Climatic corrections were estimated from temperature time series measured at Gengenbach, Strasbourg, Freiburg and Basel. The effect of the variation of the mean annual temperature between 1755 and 1985 has been calculated to be smaller than 0.5 K/km (corresponding to about +2% of the observed temperature gradient) below 250 m.

TABLE 1. Geothermal KTB-Drillholes in the Black Forest

No.	Name	Longitude	Latitude	Altitude (m)	Depth (m)	HFD [1] (mW/m²)
1	Hechtsberg	8° 8.04'	48° 16.90'	243	300	89
2	Schönmatt	8° 7.95'	48° 14.91'	310	272	81
3	Moosengrund	8° 16.69'	48° 14.05'	650	265	88
4	Kunklerwald	7° 59.27'	48° 1.64'	630	300	71
5	Ettersbach	8° 3.09'	48° 4.69'	567	182	92
6	Geschahse	8° 5.76'	48° 12.55'	547	190	77

[1] Heat flow values computed from thermal resistivity versus temperature plot (Bullard, 1939) with 2D-topography corrected temperatures.

Paleoclimatic effects were determined from a temperature time series derived from different sources for the time span from 1700 to 50 My before present. The correction is strongest above 1500 m depth reaching -7.8 K/km near the surface for a temperature diffusivity of $1.2\ 10^{-6}\ m^2/s$ (Stiefel et al., KTB-Report 1986, Geophys. Inst. Karlsruhe, 1987). However, the expected increase of the temperature gradient with depth near the surface did not appear in the borehole temperature, so the paleoclimatic corrections were only tentatively calculated but were actually not applied to the data.

In addition to topographic effects erosion may have changed significantly the subsurface temperature field in some of the borehole sites situated in deep valleys. If these valleys were formed in glacial or post-glacial time erosion rates of 0. 5 mm/y for broader valleys and 1.5 to 4.5 mm/y for narrow valleys are conceivable and corresponding corrections are significant for holes No. 1 and No. 5. In Table 2 the measured temperature gradient values and the topography corrected values are given.

Thermal Conductivity

From surface rocks and from cores of the scientific holes water-saturated samples were taken to measure thermal conductivity at atmospheric pressure and room temperature using a needle probe (Sattel, Diploma thesis, Geophys. Inst. Karlsruhe, 1979) and a QTM probe, which was kindly supplied by Dr. Rybach, Zürich University (ETH). No statistically significant deviations between the results of both kinds of thermal conductivity measurements could be detected. Measurements were performed parallel and normal to the texture of the rocks. Anisotropy is significant in the holes Nos. 1, 4 and 5 with a corresponding anisotropy factor between 1.13 and 1.29 derived from conductivity measurements parallel and perpendicular to the texture. The corresponding values are given in Table 2.

In addition an in-situ measurement of thermal conductivity was performed in hole No. 1 applying a temperature relaxation method TAV (Dornstädter and Sattel, 1985; Dornstädter, Diploma thesis, Geophys. Inst. Karlsruhe, 1987). The mean value of

TABLE 2. Thermal Gradients and Conductivites and Surface Heat Flow HDF

Hole No.	Geothermal Gradient[1] (K/km)			Depth Interval (m)	Thermal Conductivity (W/(mK))		n[7]	HFD (mW/m²)		
	a)[2]	b)[3]	σ[4]		c)[5]	d)[6]		a)[8]	b)[9]	σ[10]
1	36.0	33.7	1.8	175-300	2.85±0.30	2.25±0.28	53	95.7	89.4	2.4
2	34.7	30.0	0.4	41-272	3.00±0.29	2.69±0.19	34	97.3	80.6	1.4
3	32.0	30.1	1.5	110-265	2.96±0.22		52	94.2	88.4	2.6
4	26.0	24.4	1.5	50-300	3.01±0.38	2.63±0.39	32	73.2	70.5	2.4
5	45.6	35.3	6.1	120-182	3.29±0.29	2.64±0.45	9	113.9	91.5	9.1
6	27.0	26.9	1.9	120-190	2.84±0.27		11	77.5	76.8	3.2

[1] Temperatures determined at 10 cm intervals in the indicated depth range. [2] Temperature gradient obtained from measurements. [3] Topography corrected values. [4] Standard deviation of thermal gradient. [5] Thermal conductivity parallel to texture with standard deviation. [6] Thermal conductivity perpendicular to texture with standard deviation. [7] Number of conductivity measurements. [8] Surface heat flow determined from Bullard plot. [9] Surface heat flow corrected for topography. [10] Standard deviation of surface heat flow.

thermal conductivity in hole No. 1 determined by TAV measurements was 2.57±0.28 W/m/K which is within the standard error of the mean thermal conductivity 2.76±0.25 W/m/K perpendicular to the borehole axis derived from the foregoing mentioned laboratory measurements.

Surface Heat Flow

The surface heat flow values determined at the holes No. 1-6 by evaluating the corresponding Bullard plots (Bullard, 1939) and the topography corrected values are displayed in Table 2. The overall distribution of surface heat flow in the northern and central BF is shown in Figure 2. In the northern BF down to the Kinzig valley passing by Haslach the heat flow is at a level of 90 mW/m^2 partly exceeding this value whereas south of the Kinzig valley the heat flow decreases to about 70 mW/m^2 across a distance of approximately 30 km. This strong variation turns out to be a severe constraint to geothermal models. With exception of hole No. 5 where topographic and erosion effects are difficult to be exactly determined the mean errors are smaller than ±3 mW/m^2.

Usually a linear relationship is assumed to be valid between surface heat flow and surface heat production rate for sites with a common basement and with the same tectonic and thermal history (Birch et al., 1968; Roy et al., 1968) although this is questioned for larger regions (Lachenbruch and Sass, 1977; Nielsen, 1987; Furlong and Chapman, 1987; Drury, 1987). This relation simply means that there is a common basal heat flow q* from the mantle and the lower crust given by the intersection with the ordinate and that the vertically integrated heat production reduced by its surface value A_o is the same for the whole region under consideration. In Figure 3 the surface heat flow corrected for topography is plotted against surface heat production at the five scientific holes indicated by black points. The radiogenic heat production has been measured at surface rocks and core samples. A description of the results is given in Burkhardt et al. (1988).

Because of the large scatter no consistent linear relationship is obtained. A better fit can be arranged if a correction for erosion is applied to the data. The resulting (open) points are correlated (correlation coefficient r = 0,934 with 4 degrees of freedom) yielding a linear relationship with a basal heat flow as high as 61 mW/m^2 and a characteristic depth of only 4 km. If conduction and radiogenic heat production would determine the surface heat flow alone this result would mean that the contribution of radiogenic heat production is rather small and that the basal heat flux is as high as for the Basin and Range province for which the original linear relationship is displayed in the upper part of Figure 3 (Roy et al., 1968; Lachenbruch et al., 1985). Indeed, according to Lachenbruch and Sass (1977)

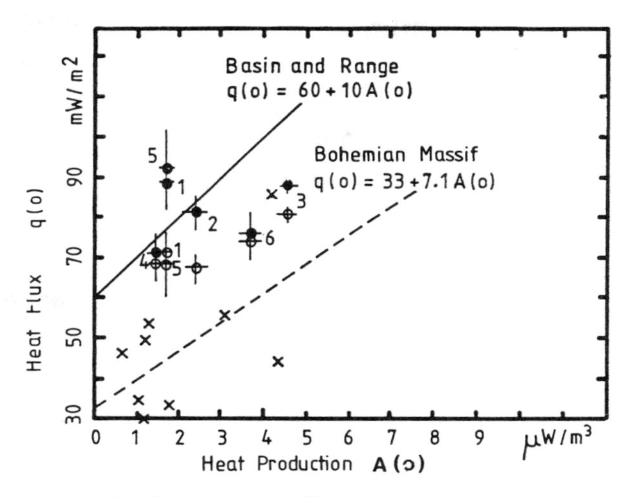

Fig. 3. Surface heat flow and heat production measured in the scientific holes Nos. 1-6 (Table 1) respectively on corresponding cores. Black circles: heat flow corrected for topographic effects; open circles: correction for erosion effects applied in addition; crosses: values for Bohemian Massif from Cermak (1980). Linear relation for Basin and Range from Roy et al. (1968).

no linear relation exists at all for the Basin and Range province, but the available data mostly plot above this line, so that the reduced heat flow which actually must exist is presumed not be smaller than the corresponding value from the linear relationship of Roy et al. (1968). The upper Rhinegraben is much less tectonically active than the Basin and Range province so the reduced heat flow in the BF should also be less. From this consideration it is inferred that in the BF crust there are additional heat sources or that heat must be advected, perhaps at midcrustal level, to supply the observed surface heat flux. Normal heat flow values in the BF region would be expected to correspond to the relation obtained by Cermak (1975, 1982) for the Bohemian Massif (Figure 3), perhaps with a somewhat higher basal heat flow between 40 and 50 mW/m^2 in order to account for the present rift situation.

Geothermal Models

There are two aspects constraining geothermal models for the central BF crust, i.e. the lateral variation of 20 mW/m^2 on a horizontal distance of 30 km and a high basal heat flow which presumably is partly a result of heat advection. There is no possibility to model the regional surface heat flow variation by lateral variations in the mantle heat flux. The cause of this change has to be sought at a shallower depth. Possible explanations are variations in the heat production rate or convective heat transport at high or midcrustal level.

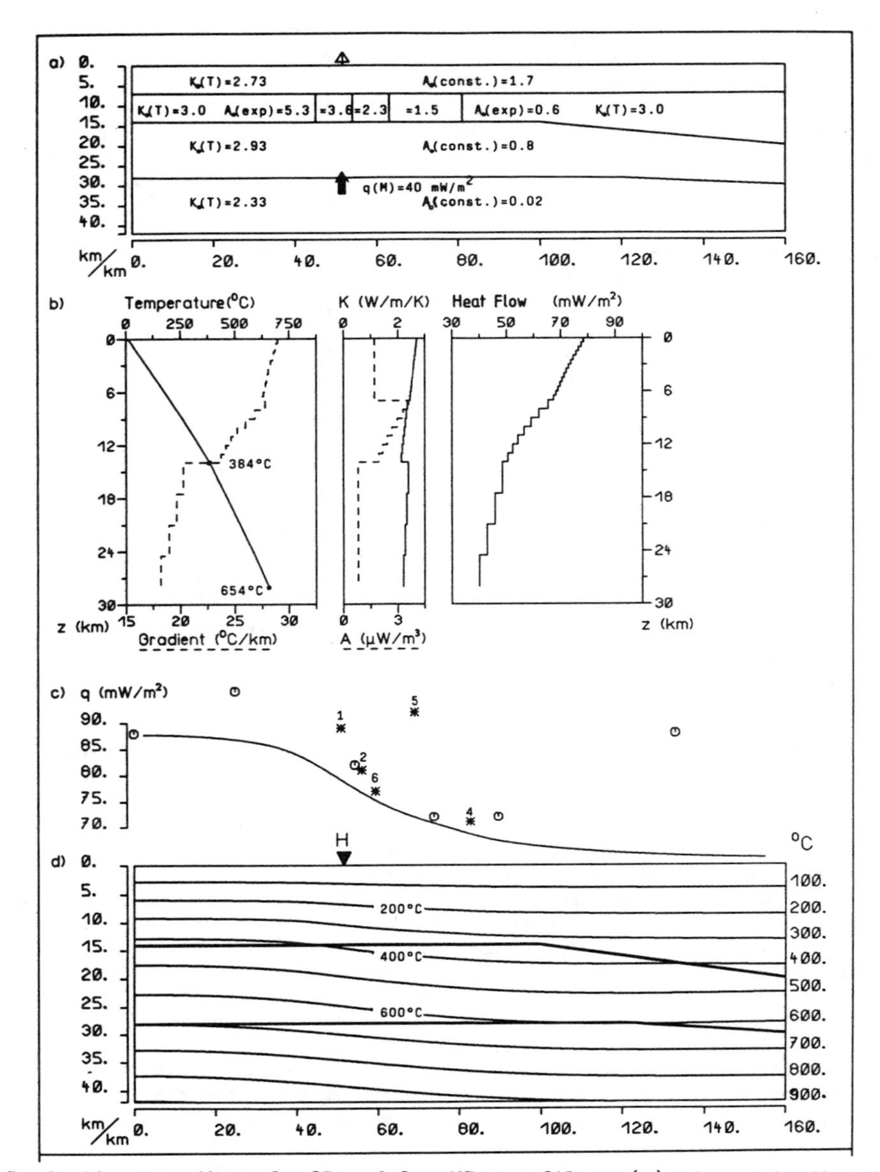

Fig. 4. Conductive geothermal 2D-model, NS profile. (a) Assumed distribution of geothermal parameters. K thermal conductivity (W/m/K), temperature dependent; A heat production ($\mu W/m^3$) constant or exponentially varying, basal heat flow 40 mW/m^2. (b) Depth distribution of temperature, temperature gradient, thermal conductivity K, heat production A and heat flow at assumed deep drilling site, indicated by H in (d). (c) Surface heat flow; values measured in scientific holes (stars) and in commercial holes (circles), calculated heat flow (solid line). (d) Temperature isolines and structural crustal model.

Two-dimensional simulations were performed with the finite element (FE) code ADINAT (Bathe, 1978). Structural information was drawn from reflection and refraction seismic observations (Lüschen et al., 1987; Gajewski and Prodehl, 1987). Thermal conductivity and heat production were assigned to the blocks according to different assumptions about their lateral and vertical distribution. Surface temperature (10°C), basal heat flow (50 mW/m^2) in 26 km depth and insulating vertical boundaries at the sides were assigned as boundary conditions. Figure 4 shows an example where at midcrustal level in the low velocity zone a strong variation in the heat production rate was assumed. This distribution of heat production is a purely fictitious assumption as it would imply the presence of a large granitic batholith in the low velocity zone for which, until now, there are no plausible model interpretations of the gravity observations (Plaumann et al., 1986). On the other hand Mueller (1977) has postulated a granitic midcrustal layer causing the frequently observed reduction of P-wave velocity at the corresponding depth. On top (a) of Figure 4 the assumed conductivity and heat production distribution is shown. In the centre (b) the calculated vertical profile of temperature and temperature gradient and in addition of thermal conductivity, heat production and of the resulting heat flux are displayed for the proposed deep drilling site Haslach. Below in (c), the calculated surface heat flux together with the observed values only corrected for topography and in (d), the thermal field in the crust is shown. The trend of the observed surface values has been accurately modelled assuming rather high values of heat production in the crust but even higher values would have to be assumed in the northern (left-hand) part of the model to produce a much better fit to the observations. Although such a model is not completely unreasonable, there is some doubt concerning the positional stability of such a largely extended granitic batholith within the crust.

Another possibility to produce the observed variation in the surface heat flux is by convection. In order to get quantitative results two-dimensional FE-calculations for forced convection were performed with different assumptions about the permeability distribution and the forcing hydraulic potential $H = p/(\rho g)-z$ with p pressure, ρ density, g gravity, and z depth. Figure 5 shows the results for an example where the hydraulic potential was assumed to vary linearly corresponding to the regional surface topography which is lowest in the central BF. Surface temperature and mantle heat flux were prescribed as upper and lower boundary conditions, the vertical sides of the model were assumed to be impermeable and isolating. To the upper crust the permeability of granitic rocks (10^{-19} m^2) was assigned, the lower crust is practically impermeable (10^{-27} m^2). The corresponding hydraulic conductivities are 10^{-12} m^2 for the upper crust and 10^{-20} m^2 for the lower crust. The resulting velocity field (a) shows recharge and downward flow induced by the high topography in the N and S and discharge and upward flow in the central BF.

The resulting surface heat flow (c) is in correspondence with the observed values in the central BF, if erosion corrections are applied to the flux at the holes No.1 and No. 5. The large discrepancies in the N and S can be eliminated by accounting for the observed heat production rate. Obviously the calculated flow field is based on simplifying assumptions, however, the model shows that the observed regional variation of surface heat flux can be produced by fluid flow in the crust. Also it has to be stated, that the surface heat flow critically depends on the chosen values of the permeability. This has been verified in one-dimensional simulations, following a procedure described in Mansure and Reiter (1979).

Discussion

The geothermal KTB-investigations in the BF have shown that in the northern and central part the surface and the basal heat flux are greater than in the Bohemian Massif and that there is a considerable lateral variation in the surface heat flow which can be modelled either by an extended midcrustal granitic layer, for which, however, there are no indications from gravity measurements, or by forced convective heat transport. Both conductive and convective model calculations have shown that there are lateral components up to 10% of the temperature gradient and of the heat flux.

Corrections for topography and in some places for erosion have to be applied to the observed surface temperature gradients in order to get reliable starting values for temperature and heat flow extrapolations. For the Haslach area temperature extrapolations from conductive modelling with a temperature dependent thermal conductivity yield a value of 400°C at 14 km depth with estimated bounds of $\pm$80°C. This spread mainly reflects the uncertainties in the assumed conductivity profile and in the surface heat flux.

From purely conductive models it appears that variations in temperature down to a depth of 14 km are mainly determined by variations in thermal conductivity and surface heat flux, whereas variations in the heat production of the upper crust do not affect the temperature as much. Deviations of $\pm$20% in the assumed thermal conductivity distribution are conceivable as a consequence of differences in type and composition of rocks, which may have changed e.g. by hydrothermal mineralisation, granitisation and mylonitisation. At 14 km depth this could approximately result in a temperature variation of $\pm$80°C, and at 10 km depth of $\pm$60°C. This variation was estimated from a simple one-dimensional model, assuming $k = 3$ W/(mK), $\sigma(k)/k = 18\%$, $q_o = 86$ mW/m², $\sigma(q_o)/q_o = 8\%$, $A = 3$ µW/m³ for thermal conductivity k, surface heat flow q_o and heat production A using

$$\sigma(T) = q_o z/k \left[\left((q_o)/q_o \right)^2 + \left(1+(Az)^2/(2q_o)^2\right)\left(\sigma(k)/k\right)^2 \right]^{1/2}$$

More complicated models would not yield essentially different results for the estimated uncertainty of the extrapolated temperature. For models taking convection into account no serious estimation of temperatures and of corresponding error bounds can be given.

Thermal conductivity and heat production have been determined within statistical error bounds from rock samples (Burkhardt et al., 1988). These

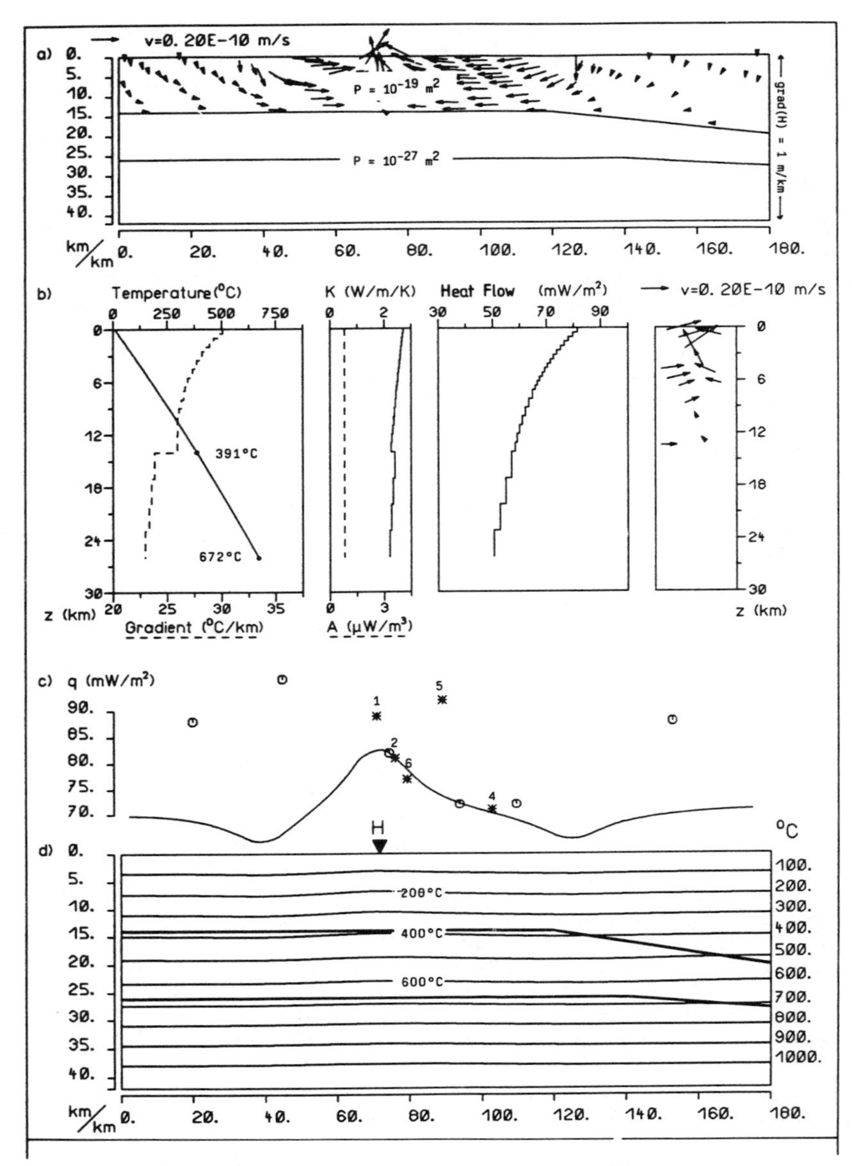

Fig. 5. Convective geothermal model, convection forced by linearly varying hydraulic potential according to a corresponding topography. (a) Resulting velocity field for a crustal model with permeabilities $10^{-19}m^2$ and $10^{-27}m^2$ for upper crust and space beneath respectively; no heat production assumed, thermal conductivity assumed constant. (b) Depth distribution of temperature, temperature gradient, thermal conductivity K, heat production A, heat flow and velocity field at assumed deep drilling site, indicated by H in (d). (c) Surface heat flow, measured values in scientific holes (stars) and in commercial holes (circles), calculated heat flow (solid line). (d) Temperature isolines and structural crustal model.

errors do not tell us anything about the statistics of these variables at depth. Therefore, no attempt has been made to apply statistical methods in modelling the error of the measured surface heat flow in the BF. Many more temperature gradient, conductivity and heat production measurements would be necessary at the surface and at depth to obtain the observational base for a statistical description of the thermal variables at depth in a complicated model.

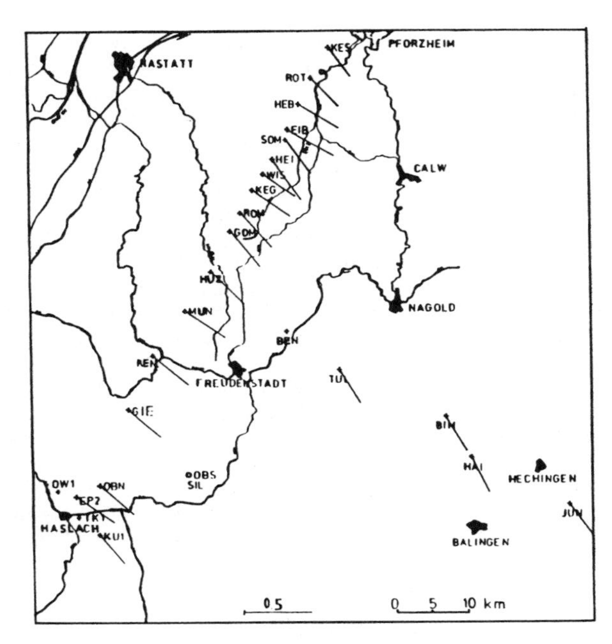

Fig. 6. Direction of horizontal magnetic component with maximum coherence to the vertical component at 500 s period for stations in the northern and central Black Forest (more precisely: real part of transfer function).

Electrical Conductivity

Independent results concerning a possible fluid content in the crystalline basement of the BF can be inferred from magnetotelluric, magnetovariational, and electromagnetic depth soundings performed as part of the KTB survey (Berktold et al., DFG-Report 574/10-1, Inst. Geophysik München, 1986; LOTEM Working Group, BMFT-Report RG 86028, Inst. Geophysik Köln, 1986). Near the envisaged KTB drilling area in the central BF vertical electric soundings were carried out at 16 sites, active audiomagnetotellurics (controlled source frequency domain electromagnetic measurements) at 12 sites, LOTEM long offset transient electromagnetic measurements (controlled source time domain electromagnetic measurements) at 57 sites and passive audiomagnetotellurics at 35 sites. Magnetotelluric and magnetovariational measurements were undertaken at 36 sites all over the BF.

From the geological point of view it was expected that the well-conducting sediments of the Rhinegraben (RG) would produce a polarization of the time-varying electrical and magnetic fields which is parallel to the RG direction N25°E and perpendicular to it. The electrical field component parallel to the strike of the RG (TE-mode) should be coherent to the magnetic field component perpendicular to the strike and vice-versa for the other two corresponding field components (TM-mode). Also a telluric anisotropy decreasing with increasing distance from the RG and depending on period was inferred. This simple view was not confirmed in the survey.

Rather surprisingly it was observed that the direction of maximum coherence between the corresponding magnetic and electrical field components was NW-SE and perpendicular to it (Tezkan, 1988). The anisotropy was found to be not decreasing with distance from the RG, the conductivity being highest in NE-SW direction, the resistivity being strongest in NW-SE direction. In Figure 6 the direction of the horizontal magnetic component with optimum coherence to the vertical one is shown for stations in the northern and central BF at periods of about 500 s from magnetovariational

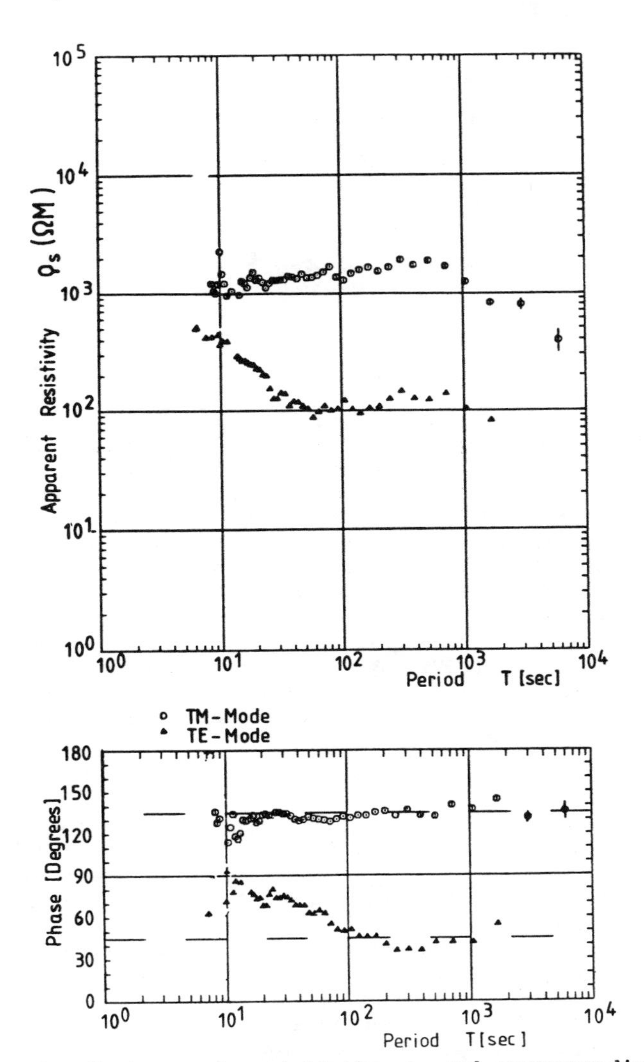

Fig. 7. Apparent resistivity ρ_s and corresponding phase as function of period at station KU1 near Haslach, compare Fig. 6. TE-mode designated by triangles, TM-mode by circles. Note static shift between TE-mode (triangles) and TM-mode (circles).

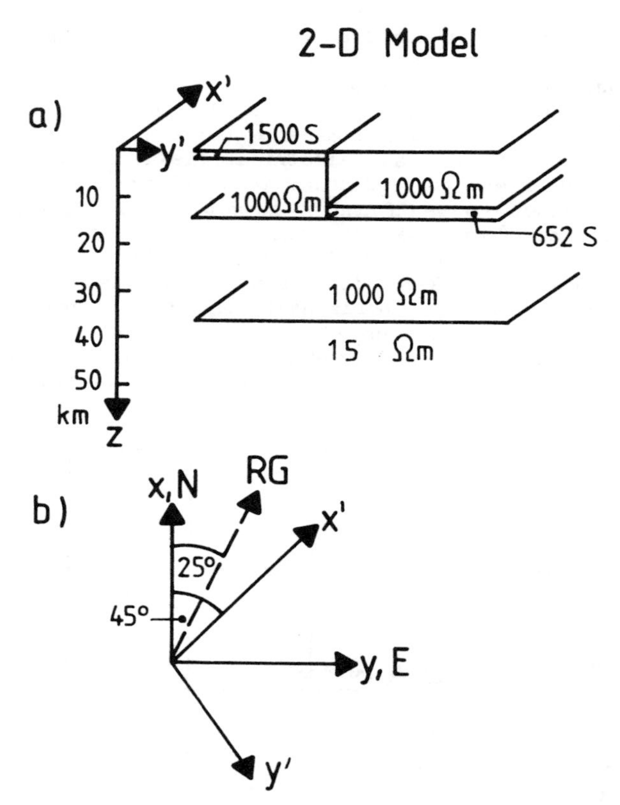

Fig. 8. 2-D resistivity model (Tezkan, 1988), left side Rhinegraben, right side Black Forest. (a) Resistivity of crystalline basement 1000 Ω m, integrated conductivity (conductance) of sedimentary fill of Rhinegraben 1500 S (1/Ω). At midcrustal level in Black Forest (probably between 12 and 18 km depth) low resistivity layer with conductance 650 S, well-conducting half space (resistivity 15 Ω m) below. (b) Horizontal coordinates x versus N, y versus E, x' strike direction of 2D-model of Fig. 8a, y' direction of maximum coherence derived from magnetotelluric and magnetovariational measurements; RG Rhinegraben direction N25°E.

measurements. This direction is also NW-SE in the mean and deviates by 20 degrees from the expected value perpendicular to the RG strike direction (N25°E). Similar results are obtained within a broad spectral range.

The apparent resistivity derived from the ratio of the most coherent components of the induced electrical and magnetic fields is a complex function of period. As an example its amplitude and phase are displayed in Figure 7 for a station near Haslach for the TM-mode (white circles) and the TE-mode (black triangles) in the preferred x',y'-coordinate system (compare Figure 8b) for periods from 1 to 10^4 s. Obviously there is a static shift between the apparent resistivity in the two preferred directions, the resistivity being smaller in NE-SW direction (TE-mode) than in NW-SE direction (TM-mode). The phase of the TE-mode is about 80° at 10 s period and decreases with increasing period to nearly 45°. In correspondence to this the apparent resistivity of the TE-mode decreases with increasing period. Both results are found at all sites in the BF. They can only be modelled by introducing an anomalous zone of high electrical conductivity between 12 and 18 km depth in the BF crust (Tezkan, 1986, 1988), as displayed in Figure 8. In the 2D-model the sedimentary fill of the RG was assigned an integrated conductivity of 1500 S ($S = \Omega^{-1}$) according to borehole measurements and former magnetotelluric results. The crystalline basement on both sides of the model was given a resistivity of 1000 Ω m with a well-conducting midcrustal layer having an integrated conductivity of 650 S on the BF side. To fit the results for longer periods a 15 Ω m halfspace was assumed beneath 40

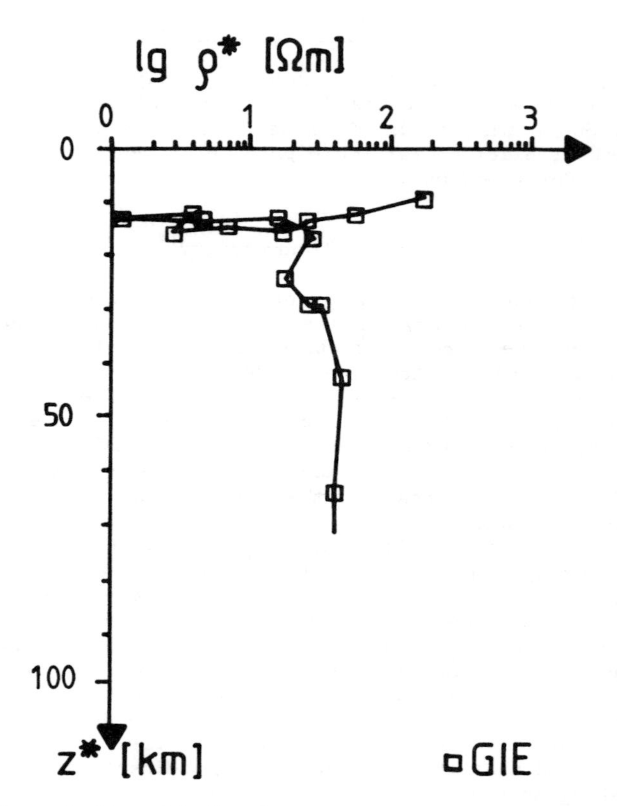

Fig. 9. 1D- ρ*(z*) distribution derived from TE-mode data for station GIE between Haslach and Freudenstadt. Approximate depth z* of the centre of the induced current system and corresponding apparent resistivity ρ* as a function of increasing period T (T parameter) denoted by squares. Note strong concentration of squares at 12-15 km current centre depth z*.

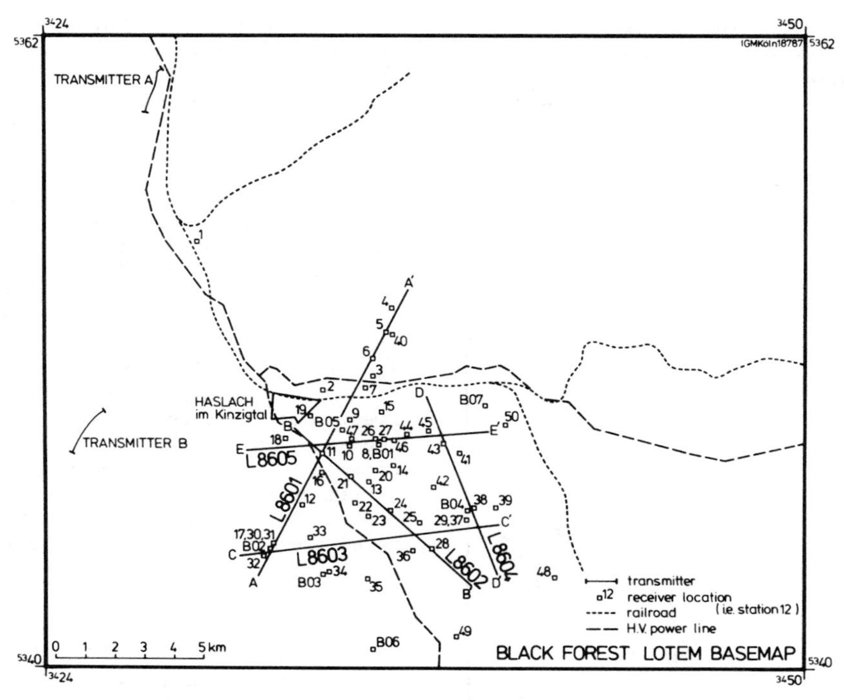

Fig. 10. Map of LOTEM recording stations (1-50, B01-B07) and transmitter stations (A and B) in the Haslach area.

km depth. Model calculations were performed by Tezkan (1988) for a profile which was assumed to be in the direction of maximum coherence x'. The anisotropy can be explained sufficiently well by a 2D-model consisting of a sequence of isolated box shaped well-conducting structures embedded in higher resistivity material striking in Variscan direction and rising to a depth of 7 km (Schmucker and Tezkan, 1987; Schmucker, 1987). The intention is to create a horizontal pseudo-anisotropy by this conductivity structure. The discrepancy of 20 degrees between the RG direction and the Variscan strike direction NW-SE is still unexplained. It is suggested that this discrepancy may result from a superposition of the RG influence and a regional effect caused by an elongated inductive structure in the Moldanubian striking along the Moldanubian /Saxothuringian thrust zone from the BF to the Bohemian Massif.

The most interesting magnetotelluric result concerning the scientific aims of a deep hole in the BF is the midcrustal high conductivity layer. A typical example showing the effect of this layer is displayed in Figure 9 for a station in the surrounding of Haslach. In this $\rho^*(z^*)$ plot, the position of the centre of the electric current system is given by z^* whereas ρ^* is an approximate expression for the apparent resistivity. Both expressions are determined from a complex transfer function for a 1D-conductivity model (Schmucker and Weidelt, Electromagnetic Induction in the Earth, Inst. Geophysik Göttingen, 1975). The resistivity ρ^* is displayed as a function of increasing period. The resulting points are concentrated at a depth z^* of about 12-15 km. Therefore, at this depth the electrical conductivity must be anomalously high.

From the magnetotelluric investigations it was impossible to obtain precise enough information about the position of the anomalous midcrustal conductivity layer to find out whether it would be accessible in a deep hole or not. In order to get independent information an attempt was made to perform a long offset transient electromagnetic (LOTEM) survey in the envisaged drilling area around Haslach, see Figure 10. In the LOTEM method a controlled source is used to generate strong induction currents in the subsurface (Strack, 1985) and the response to these downward and outward diffusing currents is recorded. At approximately 60 sites electromagnetic transients, which were artificially excited at two transmitter stations, were recorded and stacked in order to obtain a high signal-to-noise ratio which was extremely difficult in that region because of strong industrial electromagnetic noise. The interpretation of the results has been done for different profiles in the station network using one-dimensional models and an inversion algorithm (LOTEM Working Group, BMFT-Report RG 86028, Inst. Geophysik Köln, 1986) based on singular value decomposition (Jupp and Vozoff, 1975; Vozoff and

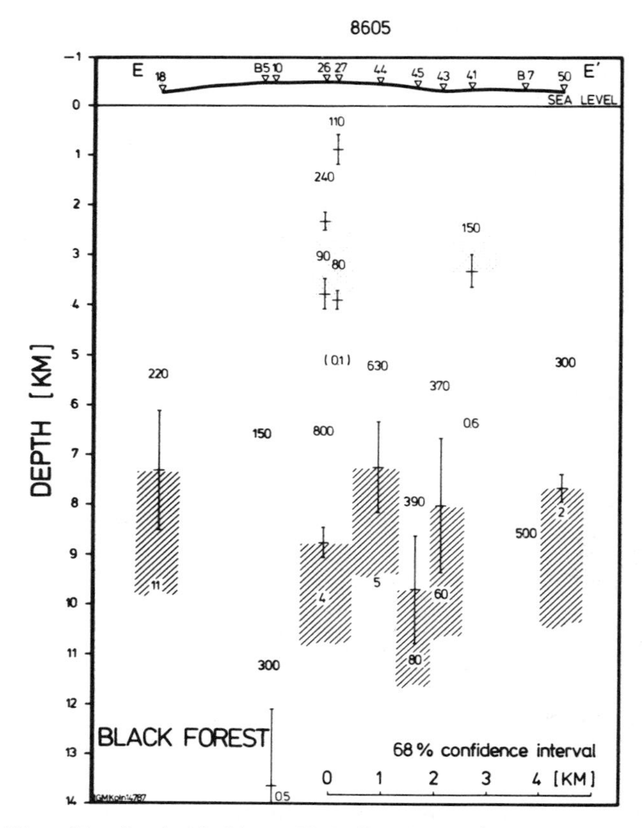

Fig. 11. Resistivity distribution (numbers are values in Ωm) on profile EE' of Fig. 10, derived from 1D-inversion for each recording station. Error bars indicate 68% confidence interval. Horizontal scale for profile right at the bottom. Hatching draws attention to low resistivity values.

Jupp, 1975; Raiche et al., 1985). By this electromagnetic survey a highly conductive layer was detected at a depth varying between 7-9 km. Figure 11 shows the resistivities in Ωm and the layer depths with 68% confidence intervals for the stations on the EW-profile EE' of Figure 10. The depth variation appears to be mainly statistical because of the noisy environment. The LOTEM survey did not reveal the anisotropy appearing in the magnetotelluric measurements. Both transmitters, A and B in Figure 10, gave approximately the same results at the two stations 27 and 47 situated close to each other. A slight dip to NW is indicated for profile BB', according to Figure 12. This direction coincides with the magnetotelluric main resistivity direction.

On the other hand a pronounced low velocity layer has been determined between 7 and 14 km depth (Gajewski and Prodehl, 1987), and consequently the question arose whether there is a common cause for both, the enhanced electrical conductivity and the reduced seismic velocity. Until now, this is an unsolved problem.

As for periods greater 10 s the conductivity-depth distribution in the uppermost 12 km cannot be resolved by magnetotellurics, the well-conducting zone detected by the LOTEM soundings did not appear in the magnetotelluric measurements. Strack et al. (submitted to Geophysics) conclude that the electromagnetically determined anomalous layer must at least be 500 m thick in order to be detectable by LOTEM soundings. It would be interesting to learn whether the LOTEM layer is a local structure or a regional one like the midcrustal conductivity layer.

Layers of high electrical conductivity have frequently been observed in the crust. A possible tectonic explanation could be that the difference between the smallest and the largest principal stress and hence the shear stress increases with depth. Eventually the rigidity is reduced by the development of microcracks which create a perme-

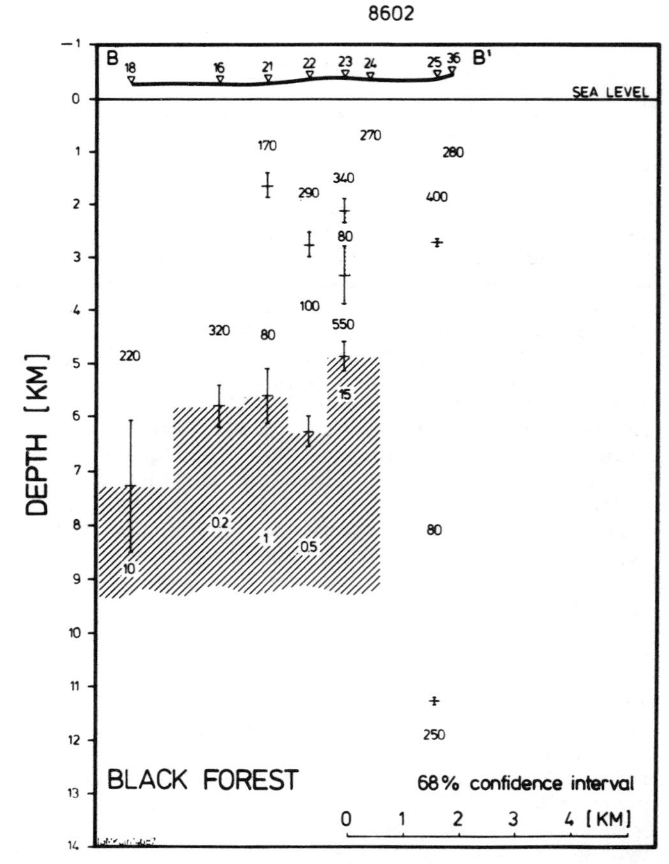

Fig. 12. Resistivity distribution on profile BB' of Fig. 10, derived from 1D-inversion for each recording station. Note dip of conducting bottom layer to NW. For further explanation, see Fig. 11.

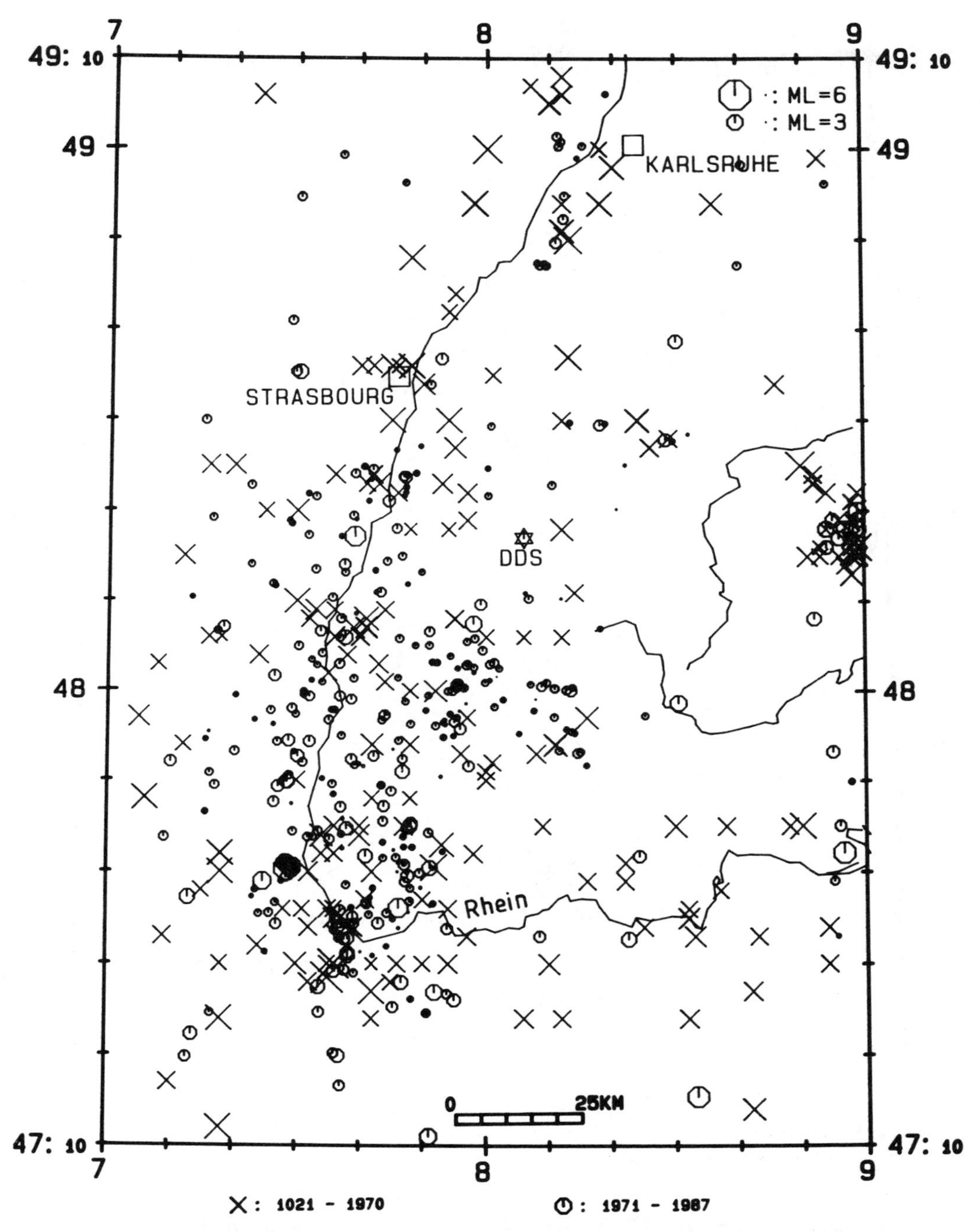

Fig. 13. Seismicity pattern in the southern Rhinegraben region. Data for the period 1021 to 1970 AD, mainly based on macroseismic events, are taken from Leydecker (1986). Instrumentally determined epicentres for 1971 to 1987 and local magnitudes ML from Bonjer et al. (1984) and Bonjer and Apopei (1986, 1987, cited in the text). Star (DDS) marks proposed site of ultradeep drillhole in the central Black Forest. Northern latitude and eastern longitude given at the frame.

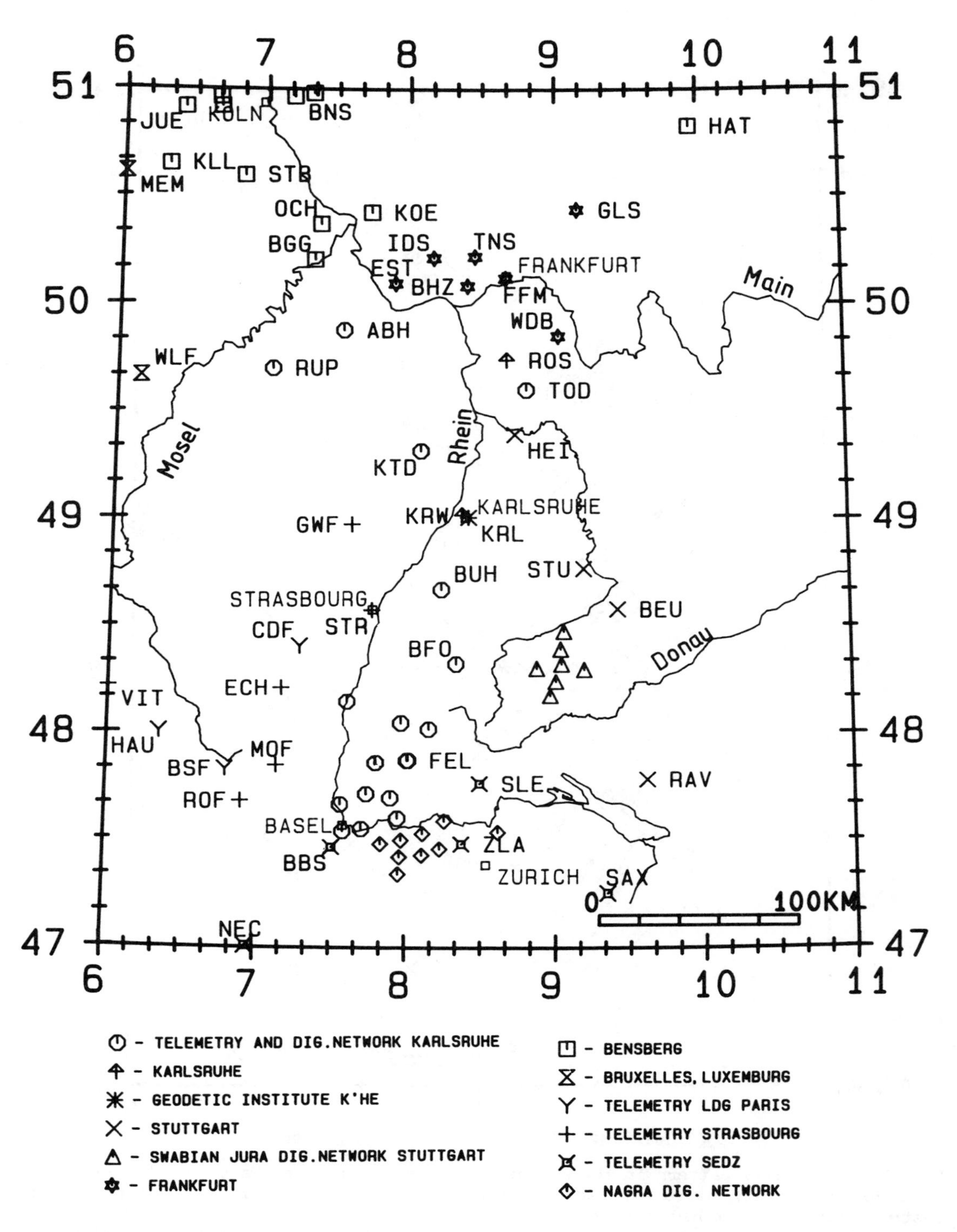

Fig. 14. Actual state of the seismic network in the Rhinegraben area. Participating institutions are named at bottom. Northern latitude and eastern longitude given at the frame.

able layer allowing fluids to move horizontally, thus enhancing the electrical conductivity. If the material is only partly saturated with fluids a reduction of the P-velocity would follow, resulting in a low velocity layer.

In the BF the layers of unusually high electrical conductivity may also be a relic of overthrusted, graphite-containing metasediments, which recently have been found among surface rocks, or as indicated above, they may be caused by ionic conduction in salt fluids residing in a sufficiently permeable region of the crust which might, at least partly, overlap with the low velocity zone. Either of these suggestions or both may be valid, however, if fluids were present in a permeable layer of the crust, this would be in agreement with a corresponding explanation of the geothermal results.

Almost nothing is known about lateral variations of the electrical conductivity in the BF crust. The magnetotelluric studies only indicate smooth lateral variations mainly in the central part of the BF. On the other hand structural differences are apparent in the seismic records (Lüschen et al., 1987) and a change of thermal properties in the heat flow measurements. Variations are especially strong in the seismicity distribution with depth and lateral distance.

Seismicity

The seismicity pattern of the upper RG region is shown in Figure 13 (Bonjer et al., 1984; Bonjer and Apopei, Report 1984-1986, Special Research Program 108, pp. 99-113, University Karlsruhe, 1986). The historical earthquakes, indicated by crosses, are compiled and reviewed by Leydecker (1986). Earthquakes from 1971 until 1987 (Bonjer and Apopei, BMFT-Report 1500666/1, Geophys. Inst. Karlsruhe, 1987) indicated by hexagons, are recorded by the RG seismic network which is supported by different institutions in Germany, France and Switzerland (Figure 14).

Bonjer et al. (1984) have projected the focal depths within five RG regions to corresponding profiles of RG earthquakes from 1971-1980, in order to reveal the maximum crustal hypocenter depth distribution (Figure 15a-d). In the central BF (Fig. 15a) and in the graben proper (Fig. 15b) earthquakes mainly occur at shallow midcrustal depths. Deep earthquakes (h > 15 km) are mainly restricted to the northern and southern BF. Also in the central and northern BF the seismicity is comparatively low whereas in the southern BF the seismic activity is stronger (Fuchs et al., 1987). Clearly the BF profile in Figure 15a shows a striking variation in frequency and maximum

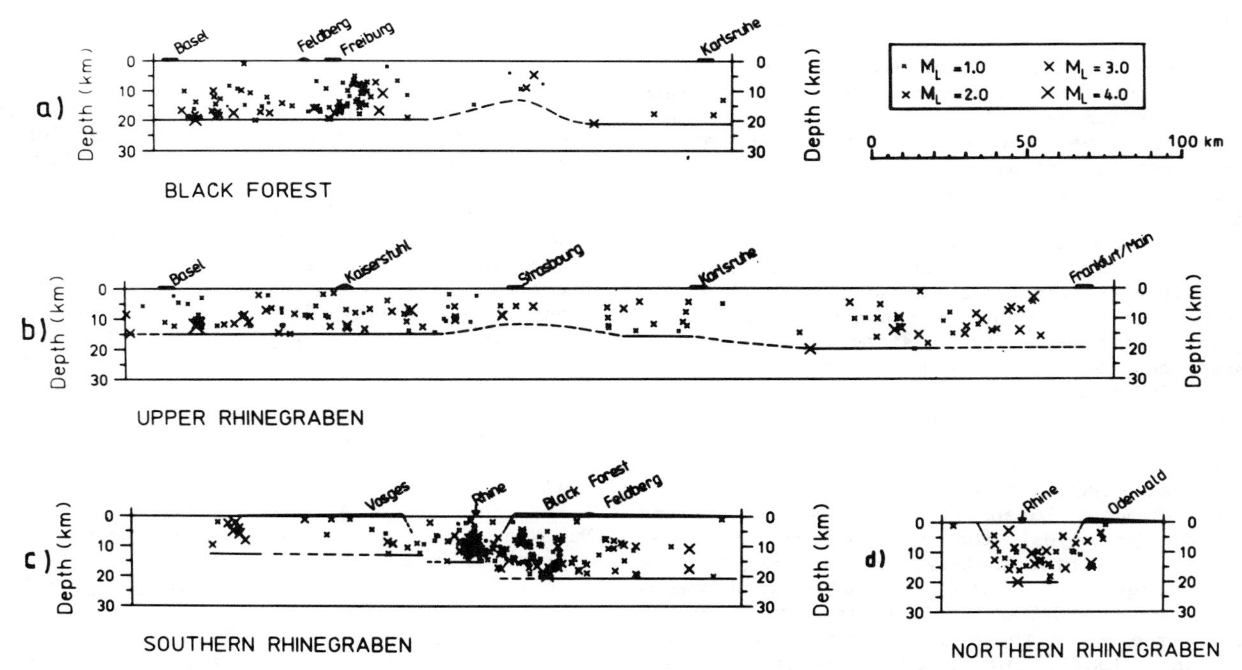

Fig. 15. Depth distribution of foci (1971-1980), hypocentres projected to different profiles. Boundaries of greatest focal depth are marked by solid or broken lines. (a) Black Forest profile, parallel to graben axis. (b) Graben proper profile. (c) Profile perpendicular to strike of southern Rhinegraben, hypocentres between Strasbourg and Basel projected. (d) Profile perpendicular to strike of northern Rhinegraben, hypocentres between Strasbourg and Frankfurt projected.

hypocentre depth of earthquakes. Because of the low seismicity in the central and northern BF, the maximum hypocentre depth is, admittedly, not very well defined there. However, a rising of the maximum focal depth in the central BF cannot be denied.

To explain this rise one might infer that the tectonic stress is reduced at midcrustal level or that the stress release is mainly non-seismic in the central BF. In both cases an explanation would be offered by an enhancement of fluid activity in the crust. If fluids were present at that depth stress release by solution, precipitation and recrystallization could result in a reduced seismic activity. This suggestion would also be in accordance with the observed enhanced electrical conductivity and heat flux and with the reduced P-wave velocity at midcrustal depth.

Conclusions

The results of the geophysical KTB-survey and of the seismicity investigations described in the preceeding sections have produced the impression that the physical properties of BF crust are modified by presently acting fluid processes. An apparently reduced seismic stress regime in the central BF, a midcrustal layer of anomalously high electrical conductivity and reduced seismic velocity and a comparatively high basal heat flow with strong lateral variations in the surface heat flow are indicative for this view.

However, there is no definite conclusion to be drawn in this respect because there is no unique explanation for each of these phenomena. It cannot be excluded that the enhanced heat flow arises from granitic intrusions or concentrations of radiogenic elements in the upper crust, that the enhanced electrical conductivity is caused by a layer of overthrusted well-conducting metasediments and that the reduced seismicity is a consequence of enhanced aseismic stress release or enhanced rock strength or a consequence of a lower degree of heterogeneity of strength or stress in comparison to the southern BF and to the RG proper. The crustal low velocity layer may also be explained by an uplift process described in Fuchs et al. (1987). All these facts derived from different kinds of geophysical investigations may accidentally happen to come together.

An ultradeep borehole in the central BF was expected to reveal the real causes of these observations. Unfortunately, drilling and measuring techniques are strongly limited at temperatures above 250-300°C. This was an argument against the BF in the decision process (Emmermann and Behr, 1987). Future developments in these techniques are expected from the KTB hole in the Oberpfalz. Then, hopefully, there will be a chance to surmount the technical difficulties expected for a corresponding hole in the BF. Until then, additional geophysical observations, interpretations and modelling results should render further knowledge of the structural and material crustal properties in this region, which is especially attractive because these investigations can also be performed in the Vosges on the other side of the RG. There is an asymmetry of both heat flow values and maximum hypocenter depth between the Black Forest and the Vosges, and it is an intriguing question whether this asymmetry is correlated to corresponding differences in the thermal, electrical and seismic structure of the crust.

Acknowledgements. The work described in this paper was funded by the Ministry of Research and Technology (BMFT) of the Federal Republic of Germany and by the German Research Society (DFG) within the framework of the Continental Deep Drilling Program (KTB). We are grateful to L. Rybach, D.S. Chapman, U. Schmucker, B. Tezkan, K. Fuchs, E. Lüschen, I. Apopei, K. Schädel, A. Hahn, H. Burkhardt and C. Clauser for fruitful discussions. Numerous scientists were involved in the field measurements. Their personal engagement is gratefully acknowledged. We also thank for the support by federal, provincial, and local authorities whose cooperation was an indispensable aid for the field work and for the preparation of the boreholes. Finally we thank Gaby Bartman for typing the manuscript and two known referees and one unknown referee for their really helpful and critical comments.

References

Bathe, K.J., A finite element program for automatic dynamic incremental nonlinear analysis of temperatures, M.I.T., Report 82448-5, 1978.

Behr, H.J., Engel, W., Franke, W., Giese, P., and K. Weber, The Variscan belt in Central Europe: main structures, geodynamic implications, open questions. Tectonophysics, 109, 15-40, 1984.

Birch, F., Roy, R.F., and E.R. Decker, Heat flow and thermal history in New York and New England. Studies of Appalachian Geology: Northern and Martime, pp. 437-451, Wiley, New York, 1968.

Bonjer, K.-P., Gelbke, C., Gilg, B., Rouland, D., Mayer-Rosa, D., and B. Massinon, Seismicity and dynamics of the Upper Rhinegraben. J. Geophys., 55, 1-12, 1984.

Bullard, E.C., Heat flow in South Africa, Proc. Roy. Soc. London A, 173, 474-502, 1939.

Bullard, E.C., The disturbance of the temperature gradient in the earth's crust by inequalities of height. Monthly Not. Roy. Astr. Soc., Geophys. Suppl., 4, 300-362, 1940.

Burkhardt, H., Haack, U., Hahn, A., Honarmand, H., Jäger, K., Stiefel, A., Wägerle, P., and H. Wilhelm, Geothermal investigations in the KTB areas Oberpfalz and Schwarzwald, in: Observation of the Continental Crust through Drilling, Vol. IV, eds. H.-J. Behr, R. Emmermann, J. Wohlenberg, Springer, in press, 1988.

Cermak,V., Combined heat flow and heat generation

measurements in the Bohemian Massif, Geothermics, 4, 19-26, 1975.
Cermak, V., Crustal temperature and mantle heat flow in Europe. Tectonophysics, 83, 123-142, 1982.
Dornstädter, J., and G. Sattel, Thermal measurements in the Rhenish Borehole Konzen, Hohes Venn (West Germany), N.Jb. Geol. Paläont. Abh., 171, 117-130, 1985.
Drury, M., Heat flow provinces reconsidered. Physics Earth Planet. Inter., 49, 78-96, 1987.
Edel, J.B., Fuchs, K., Gelbke, C., and C. Prodehl, Deep structure of the southern Rhinegraben area from seismic refraction investigations. J. Geophys., 41, 333-356, 1975.
Emmermann, R., and H.-J. Behr, Location for super-deep borehole confirmed. Tectonophysics, 139, 339-340, 1987.
Fuchs, K., Bonjer, K.-P., Gajewski, D., Lüschen, E., Prodehl, C., Sandmeier, K.-J., Wenzel, F., and H. Wilhelm, Crustal evolution of the Rhinegraben area. I. Exploring the lower crust in the Rhinegraben rift by unified geophysical experiments. Tectonophysics, 141, 261-275, 1987.
Furlong, K.P. and D.S. Chapman, Crustal heterogeneities and the thermal structure of the continental crust, Geophys. Res. Let., 14, 314-317, 1987.
Gajewski, D., and C. Prodehl, Crustal evolution of the Rhinegraben area II. Seismic refraction investigation of the Black Forest, Tectonophysics, 142, 27-48, 1987.
Gajewski, D., Holbrook, W.S., and C. Prodehl, Three-dimensional crustal model of southwest Germany, derived from seismic refraction data. Tectonophysics, 142, 49-70, 1987.
Illies, J.H., and K. Fuchs (editors), Approaches to Taphrogenesis, Schweizerbart Stuttgart, 460 pp., 1974.
Illies, J.H., and S. Mueller (editors), Graben Problems. Schweizerbart Stuttgart, 316 pp., 1970.
Hänel, R., Geothermal investigations in the Rhenish Massif, pp. 228-246 in Plateau Uplift, eds. Fuchs et al., Springer, Berlin, 1983.
Jupp, D.L., and K. Vozoff, Stable iterative methods for the inversion of geophysical data, Geophys. J. Roy. Astr. Soc., 42, 957-976, 1975.
Kahle, H.-G., and D. Werner, A geophysical study of the Rhinegraben. Part II: Gravity anomalies and geothermal implications. Geophys. J.R. Astron. Soc., 62, 631-647, 1980.
Kiderlen, H., Die Thermalquellen von Wildbad (Schwarzwald), ihre Mechanik und Genese, Jb. Geol. Landesamt Baden-Württemberg, 19, 165-217, 1977.
Kiderlen, H., Die thermalen Mineralquellen von Bad Liebenzell (Schwarzwald), Jb. Geol. Landesamt Baden-Württemberg, 22, 7-34, 1981.
Lachenbruch, A.H. and J.H. Sass, Heat flow in the United States, in: The Earth's Crust, ed. J.G. Heacock, Am. Geophys. Union Monograph, 20, 626-675, Washington, D.C., pp. 754, 1977.
Lachenbruch, A.H., Sass, J.H., and S.P. Galanis, Jr., Heat flow in southernmost California and the origin of the Salton Trough, J. Geophys. Res., 90, 6709-6736, 1985.
Leydecker, G., Erdbebenkatalog für die Bundesrepublik Deutschland mit Randgebieten für die Jahre 1000-1981. Geol. Jb., E36, 1986.
Lüschen, E., Wenzel, F., Sandmeier, K.-J., Menges, D., Rühl, T., Stiller, M., Janoth, W., Keller, F., Söllner, W., Thomas, R., Krohe, A., Stenger, R., Fuchs, K., Wilhelm, H., and G. Eisbacher, Near-vertical and wide-angle seismic surveys in the Black Forest, SW Germany, J. Geophys., 62, 1-30, 1987.
Mansure, A.J., and M. Reiter, A vertical groundwater movement correction for heat flow, J. Geophys. Res., 84, 3490-3496, 1979.
Mueller, S., Evolution of the earth's crust, in: The Earth's Crust, ed. J.G. Heacock, Am. Geophys. Union Monograph, 20, 11-28, Washington D.C., pp.754, 1977.
Nielsen, S.B., Steady state heat flow in a random medium and the linear heat flow-heat production relationship, Geophys. Res. Let., 14, 318-321, 1987.
Plaumann, S., Groschopf, R., and K. Schädel, Kompilation einer Schwerekarte und einer geologischen Karte für den mittleren und nördlichen Schwarzwald mit einer Interpretation gravimetrischer Detailvermessungen, Geol. Jb., E33, 15-30, 1986.
Prodehl, C., Ansorge, J., Edel, J.B., Emter, D., Fuchs, K., Mueller, St., and E. Peterschmitt, Explosion seismology research in the central and southern Rhinegraben - a case history. In: P. Giese, C. Prodehl, and A. Stein (editors). Explosion Seismology in Central Europe - Data and Results. Springer, Berlin, 313-328, 1976.
Raiche, A.P., Jupp, D.L.B., Rutter, H., and K. Vozoff. The joint use of coincident loop transient electromagnetic and Schlumberger sounding to resolve layered structures. Geophysics, 50, 1618-1627, 1985.
Roy, R.F., Blackwell, D.D., and F. Birch, Heat generation of plutonic rocks and continental heat flow provinces, Earth Planet. Sci. Letters, 5, 1-12, 1968.
Schmucker, U., Directional dependence of telluric variations in western and southwestern Germany. XIX General Assembly International Union of Geodesy and Geophysics, Abstracts V.1, 71, 1987.
Schmucker, U., and B. Tezkan, Zur Deutung regional einheitlicher Richtungsabhängigkeiten tellurischer Variationen. 47th Annual Meeting Deutsche Geophysikalische Gesellschaft, Abstracts, 195, 1987.
Stiefel, A., and K. Jäger, Black Forest: new geothermal aspects for SW Germany. XII Gen. Ass. Europ. Geophys. Soc., Strasbourg, Terra Cognita, 7, 479, 1987.
Strack, K.-M., Das transient-elektromagnetische Tiefensondierungsverfahren angewandt auf die Kohlenwasserstoff- und Geothermieexploration.

Mitteilungen Inst. Geophysics, University Köln, No. 42, 1985.

Tezkan, B., Erdmagnetische und magnetotellurische Untersuchungen auf den hochohmigen Kristallinstrukturen des Hochschwarzwaldes und des Bayerischen Waldes bei Passau. Ph.D.-Thesis, University Göttingen, 1986.

Tezkan, B., Electromagnetic sounding experiments in the Schwarzwald central gneiss massif. J. Geophys., 62, 109-118, 1988.

Werner, D., and H.-G. Kahle, A geophysical study of the Rhinegraben. Part I: Kinematics and geothermics. Geophys. J. Astro. Soc., 62, 617-629, 1980.

Vozoff, K., and D.L.B. Jupp, Joint inversion of geophysical data. Geophys. J. Roy. Astr. Soc., 42, 977-991, 1975.

Zoth, G., Untersuchungen zum Temperaturfeld in Raum Kirchzarten, Geol. Jb., E28, 175-190, 1985.

Section IV

THE LOWER CRUST FROM RESULTS OF OTHER GEOPHYSICAL STUDIES

ON THE VERTICAL DISTRIBUTION OF RADIOGENIC HEAT PRODUCTION IN THE CONTINENTAL CRUST AND THE ESTIMATED MOHO HEAT FLOW

Vladimír Čermák[1] and Louise Bodri[2]

Abstract. Experimental evidence suggests a certain relationship between seismic velocity and radiogenic heat production. The results of explosion seismology can thus be used to estimate the distribution of crustal heat sources. Assuming the general exponential depth dependence of the radioactivity, $A(z) = A_o \exp(-z/D)$, in which however, the logarithmic decrement D need not necessarily be constant in the whole crust, we studied its possible vertical behaviour for a number of $v_p(z)$ profiles in various tectonic units in Central and Eastern Europe. The D-parameter seems to decrease with depth, which contradicts the expected increase of the D-value with depth based on the geochemically established succession $D_U \leq D_{Th} < D_K$ corresponding to the composite U, Th and K contributions. There must be a considerable difference in the radioactive structure of the upper and lower crust and in the upper part up to a depth of 10-15 km mainly U might have undergone a certain redistribution due to deep groundwater circulation. Below this depth the radioelements maintain their primordial composition. The data obtained also suggest certain dependence of the A(z) on the geological history and the estimated mantle heat flow shows considerable variability, ranging from 15-25 $mW.m^{-2}$ under the Precambrian shields to more than 40 $mW.m^{-2}$ in the young tectonic regions.

This work is a condensed summary of several papers discussing the application of the $v_p(z) \longrightarrow A(z)$ conversion [Rybach and Čermák, 1987] , and its attempted interpretation [Čermák and Rybach, 1988], 2-D crustal temperature modelling [Čermák and Bodri, 1986, 1987] and Moho heat flow estimates in Central and Eastern Europe [Čermák, 1988] .

Introduction

A number of simple functional relationships, such as e.g. step, linear, or exponential, have been proposed to describe the depth distribution of the radiogenic heat production in the continental crust. However, crustal radioactivity may be considerably affected by igneous, metamorphic and tectonic activity, uplift and erosion, subsurface fluids, etc. It is becoming increasingly clear that no such simple distribution is applicable to the whole earth's crust without the local crustal features being taken into account. Nevertheless, a certain general and reasonably simple relationship is needed in order to be able to compare the individual tectonic provinces, to rate them and to assess the long-term geothermal consequences of the crustal evolution. The experimental relationship between seismic velocity (v_p) and heat production (A), first proposed by Rybach [1973] , and functionally formulated thereafter [Rybach and Buntebarth, 1984] , provided an independent method to evaluate crustal heat sources. Compared with other geophysical information, our knowledge of the seismic velocity distribution, $v_p(z)$, and the corresponding seismo-tectonic model of the middle and lower crust is probably the most reliable. Thus in addition to the possibility of constructing the relevant geothermal model of the specific crustal type, and of determining the deep temperatures or heat flow contribution from depth, the $v_p \rightarrow A$ conversion gives also a unique chance to check the validity of various functional relationships.

The proposed $v_p \rightarrow A$ conversion [Rybach and Buntebarth, 1984] is based on numerous combined laboratory measurements of both parameters on a set of rocks varying in composition from granite to ultrabasites. It was shown that the $A(v_p)$ relationship is part of a general systematics of the interdependence of petrophysical properties [Rybach and Buntebarth, 1982] . As the geological age is likely to be another parameter, which affects the observed relationship, two formulae were proposed for Precambrian and Phanerozoic rocks : $\ln A = 12.6 - 2.17 v_p$ (Precambrian) and/or $\ln A = 13.7 - 2.17 v_p$ (Phanerozoic), respectively, with A given in $\mu W.m^{-3}$, v_p in km/s. Laboratory measurements of v_p were performed at 100 MPa pressure and at room temperature; before applying the above relationships to in-situ conditions, the value of the corresponding "correction" is to be evaluated [Rybach and Buntebarth, 1984] . This means to consider the pressure (dv_p/dP) and temperature (dv_p/dT) dependences. This procedure is described in detail in Čermák [1988] .

The conversion technique was applied to ample material obtained for five long-run continental

---

[1]Geophysical Institute, Czechosl.Acad.Sci., 141-31 Praha-Sporilov, Czechoslovakia

[2]Geophysical Research Group of the Hung.Acad.Sci., Geophysical Dept., Eötvös University, 1083 Budapest, Hungary

geotraverses, the result of the joint deep seismic sounding programme in several countries in Central and Eastern Europe [Sollogub et al., 1980] (Fig.1). Using the published seismo-tectonic data the corresponding block structure models were derived, for which two-dimensional crustal temperature distributions were calculated [Čermák and Bodri, 1986,1987]. In addition, the detailed 1-D seismic velocity-versus-depth profiles were revealed at 49 sites along the above geotraverses, which define the crustal structure of various tectonic units ranging in age from Precambrian to Alpine. As a result of the conversion, a number of corresponding A(z)-profiles were obtained, which were then analyzed and compared with a series of simple models. The single, double and triple step-models, linear model, and several exponential models were successively examined and the statistical criteria applied to finding the best fit between the observed and theoretical values [Čermák and Bodri, in preparation] . It was found that the exponential function corresponded relatively best to the studied A(z)-profiles, indicating however a certain systematic depth deviation, which might be explained by a depth depending logarithmic decrement (D). In the following work the general exponential form of the heat source distribution was therefore assumed and its finer structure was studied with the help of the converted $A(v_p)$ data.

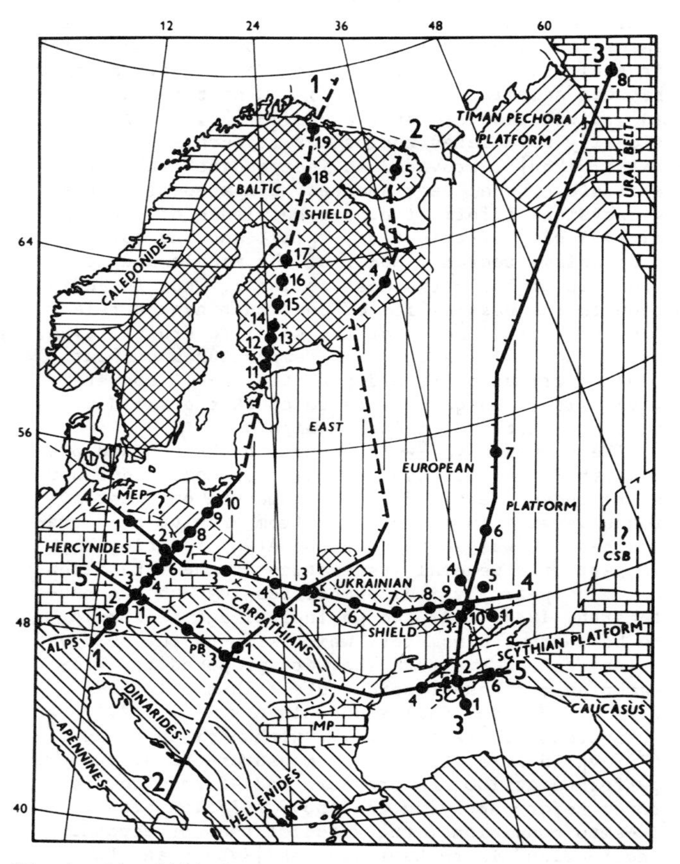

Fig.1. Simplified tectonic setting of Central and Eastern Europe together with the position of five geotraverses and the locations of sites for which the crustal seismic velocity profiles were converted into heat production profiles.

Conversion Technique

To demonstrate the procedure used we show in Fig.2 the application of the conversion $v_p \rightarrow A$ for an arbitrary $v_p(z)$ profile which summarizes all specific cases which may occur in practice.

- If the site investigated is located in a sedimentary basin and the thickness of the sedimentary cover exceeds 1 km, this fact was taken into account. The radiogenic heat production was assumed to be constant with depth and equal 1.0 and/or 1.2 $\mu W.m^{-3}$ for Phanerozoic or Precambrian sediments, respectively, according to mean values given by Haack [1982] .
- The uppermost part of the crust is most heterogeneous and deep groundwater circulation in this zone might have lead to a considerable redistribution of radioelements. Futhermore, a highly variable pressure derivative of seismic velocity at pressures of up to a few hundred MPa makes the $A(v_p)$ conversion problematic. However, there is an independent way of evaluating the near-surface heat production. The surface heat flow (Q_o) - heat generation (A_o) relationship : $Q_o = q_o + DA_o$ [Roy et al., 1968, Lachenbruch, 1968] can be combined with another empirical relationship between reduced heat flow (q_o) and the mean surface heat flow ($\overline{Q}$) : $q_o = 0.6\ \overline{Q}$ [Pollack and Chapman, 1977] . If we formally assume $Q_o = \overline{Q}$, and also $A_o = \overline{A}_o$ in the area to be investigated [Čermák, 1988], then $A_o = 0.4\ Q_o/D$. As a typical D-value, the value of 10 km was taken, which equals the mean of all reported heat flow provinces (see e.g. Morgan and Sass [1984]). At the depth of about 10 km the crustal structure is more homogeneous , and also most microfractures which allow water penetration from the surface are closed [Costain, 1978] . Therefore, for the "upper-ten-kilometres" layer we supposed the standard exponential distribution of heat production $A(z) = A_o\exp(-z/D)$ with a fixed value D = 10 km and with A_o proportional to the observed heat flow as shown above.
- Below 10 km, the reported seismic velocity was converted into heat production by using the formulae of Rybach and Buntebarth [1984] , after the effects of pressure and temperature had been accounted for. The general exponential character of A(z) was assumed and D-value calculated for each layer; the obtained D-values varied within an interval from a few kilometres to about 50 km [Čermák and Rybach, 1988] .
- Certain problems may exist for the $v_p \rightarrow A$ conversion within the inversion zones (low velocity layers) observed in some $v_p(z)$ profiles. Low velocity layers in the crust may depend on local physical conditions (e.g.temperature produced alpha/beta quartz transition, partial melting, dehydration reaction,etc.)

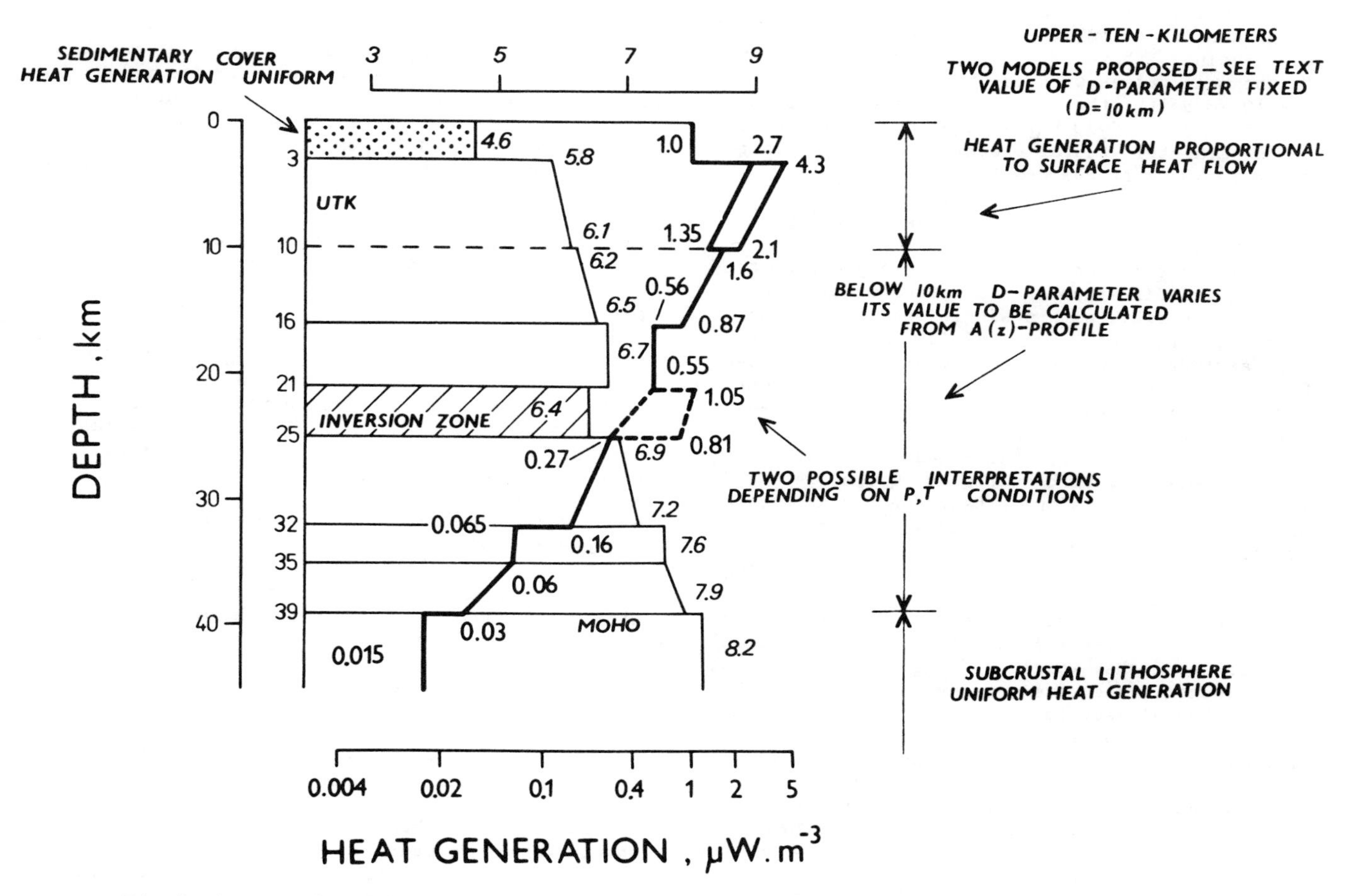

Fig.2. An example of how the seismic velocity versus depth profile is converted to heat production versus depth profile (see text),

[Fielitz, 1976, Kern, 1982] or may reflect the compositional changes produced by e.g. underthrusting. The nature of the existing low velocity layers may thus limit the applicability of the $A(v_p)$ conversion and local conditions should be assessed in each specific case. The simple criterion was applied by [Čermák 1988] , which follows Kern's [1982] empirical knowledge acquired from laboratory measurements of elastic parameters of rocks, particularly the requirement that pressure be increased by a minimum of 100 MPa per 100°C to prevent thermal cracking. If the pressure (in MPa) was lower or approximately equal to that required by the temperature (in °C) at the top of the inversion zone, we have interpreted this zone as produced by the local physical conditions, and the heat production value was calculated by interpolation from these in the neighbouring layers. If the pressure was substantially greater than that required by the temperature, the inversion zone was interpreted as a layer of different composition with a lower velocity and thus also of higher heat production. Even though this is a purely speculative approach, it separated well the Precambrian profiles containing a low velocity layer from the Phanerozoic ones.

D-parameter Variation with Depth

The logarithmic decrement D itself has the dimension of length and for any given A(z) distribution it can be formally calculated from two heat production values A_1 and A_2 at depths z_1 and z_2, respectively : $D = (z_2 - z_1)/(\ln A_1 - \ln A_2)$.

As mentioned above the conversion technique was applied for a number of 1-D seismic velocity versus depth profiles reported along the geotraverses [Sollogub et al., 1980] , the sites of which are shown in Fig.1. For the resulting heat production versus depth profiles, the D-value was calculated for each 1 km thick layer below the depth of 10 km, these data were then grouped into 5 km intervals (i.e. 10-15, 15-20, etc.) for the whole crustal thickness. In spite of a considerable scatter of the D-values, there is a certain systematic change in the mean values with depth. Immediately below the 10 km depth, the characteristic value of the D-parameter is 20-30 km, whereas deeper in the crust, the D-parameter gradually decreases with depth to about 10-15 km at the crustal base.

The investigated geotraverses cross various tectonic provinces of Central and Eastern Europe, ranging from Precambrian shield areas to young Alpine-Carpathian units. The seismic profiles traverse quite different crustal structures, which may not be compatible with a single set of parameters. Therefore, the set of 47 profiles (two profiles were excluded as probably not characteristic, see Čermák and Rybach [1988]) was divided into four sub-groups : Precambrian profiles (n=27) and Phanerozoic profiles in areas of low heat flow, Q less than 62 $mW.m^{-2}$ (n=11), medium heat flow, 62 to 74 $mW.m^{-2}$ (n=5), and high heat flow, Q over 74 $mW.m^{-2}$ (n=4). This subdivision is based on the existence of specific crustal features, such as age and mean crustal thickness. The decreasing tendency of the D-parameter with depth well characterizes all four sub-groups and thus supports the above general statement. In addition, areas of higher heat flow (higher crustal temperature) show lower D-values at all comparable depths, which may suggest an enrichment of radioactive elements towards the surface. Rybach [1976] has already suggested that, for an exponential distribution of crustal radioactivity, the D-parameter is inversely proportional to heat flow.

The depth dependence of the D-parameter was further evaluated by using different arrangement of the seismic velocity data set [Čermák and Rybach, 1988] . Contrary to the above procedure, we now have calculated first the characteristic (mean) $v_p(z)$ profiles for each above tectonic/heat flow subset (v_p were averaged over 5 km depth intervals) and four resulting profiles were then converted directly into A(z) distributions (Fig.3).Irrespective of uncertainties in the D-values thus calculated (up to 40 %), the decreasing tendency of the D(z) clearly prevails.

It is necessary to stress that the relevance of the obtained results on D-values should not be overestimated. All parameters involved in the conversion are subject to uncertainties and the error of the resulting D-value is the product of the noise propagation along the chain : v_p(in situ) $\rightarrow$ v_p(corrected for P,T) $\rightarrow$ A(z) $\rightarrow$ D(z). The error of a calculated D-value is at least $\pm$ 20-30 %. In the presence of such uncertainties, differentiating between the three Phanerozoic crustal types may be difficult, but the decreasing tendency of D(z) seems to be in all cases beyond doubt. If one accepts a general exponential distribution of radiogenic heat production within the continental crust, such result is surprising and must be explained (see further).

Moho Heat Flow

The value of heat flow at the crust-upper mantle boundary, the so-called Moho heat flow (Q_M), determines the energy budget of the upper mantle deep processes. In order to evaluate its value, it is necessary to subtract the crustal contribution due to radioactivity from the surface heat flow.

The area studied is dominated by the Precambrian European craton with a generally good data coverage, the sites belonging to Phanerozoic realm are less numerous and are rather scattered. The lowest surface heat flows are observed in both Baltic and Ukrainian shields (40 $mW.m^{-2}$), and a slightly higher heat flow (46 $mW.m^{-2}$) characterize the remaining part of the craton, the East European platform. The individual calculated Moho heat flows [Čermák,1988] vary from about 17 to 30 $mW.m^{-2}$, again with the minimum values in both shields; the mean value is 20 $\pm$ 1 (s.e.) $mW.m^{-2}$ (n=11) in the Baltic shield, and 23 $\pm$ 1 (s.e.) $mW.m^{-2}$ (n=7) in the Ukrainian shield. The slopes of the shields have slightly higher Moho heat flows which approach the mean Moho heat flow of the platform (26 $\pm$ 2 $mW.m^{-2}$, n=9). While there is practically no difference in the surface heat flow observed in both shields, the lower crust of the Ukrainian shield seems to be less radioactive than the crust of the Baltic shield and consequently the mean Moho heat flow of the Baltic shield is lower by 3 $mW.m^{-2}$ than that of the Ukrainian shield.
It is difficult to decide whether this difference is structurally important, but the larger and more stable Baltic shield forms the oldest part of the original huge craton (~ 1750 Ma), while the smaller Ukrainian shield in its southern sector may have experienced certain rejuvenation in the Late Paleozoic when graben structures, now dividing both shields, were formed.

There are no significant regional differences in the contribution of the lower crust between the platform and the shields. The calculated Moho heat flow values within the platform vary from 20 to 30 $mW.m^{-2}$, with the higher values corresponding to depressed structures, such as Dnieper-Donetz aulacogen and to Pachelmskiy aulacogen, and to border zones of the platform.

All Phanerozoic units surrounding the craton generally show higher Moho heat flows ranging in a broad interval of less than 20 to more than 50 $mW.m^{-2}$, mean from 20 values is 31 $\pm$ 4 (s.e.)$mW.m^{-2}$. For these data a considerable scatter is typical. Usually only one or a few A(z)-profiles are available for each specific tectonic unit, thus any detailed statistical conclusions are out of the question. The outer Eastern Carpathians profiles gave low Moho heat flow of only 17 $mW.m^{-2}$, which was not explained correspondingly, but all other Phanerozoic profiles clearly confirmed the increasing outflow of heat from the upper mantle in younger terrains. The Variscan (Hercynian) structures in Central Europe represents a complex system, which includes the Bohemian Massif, Paleozoic platform and the

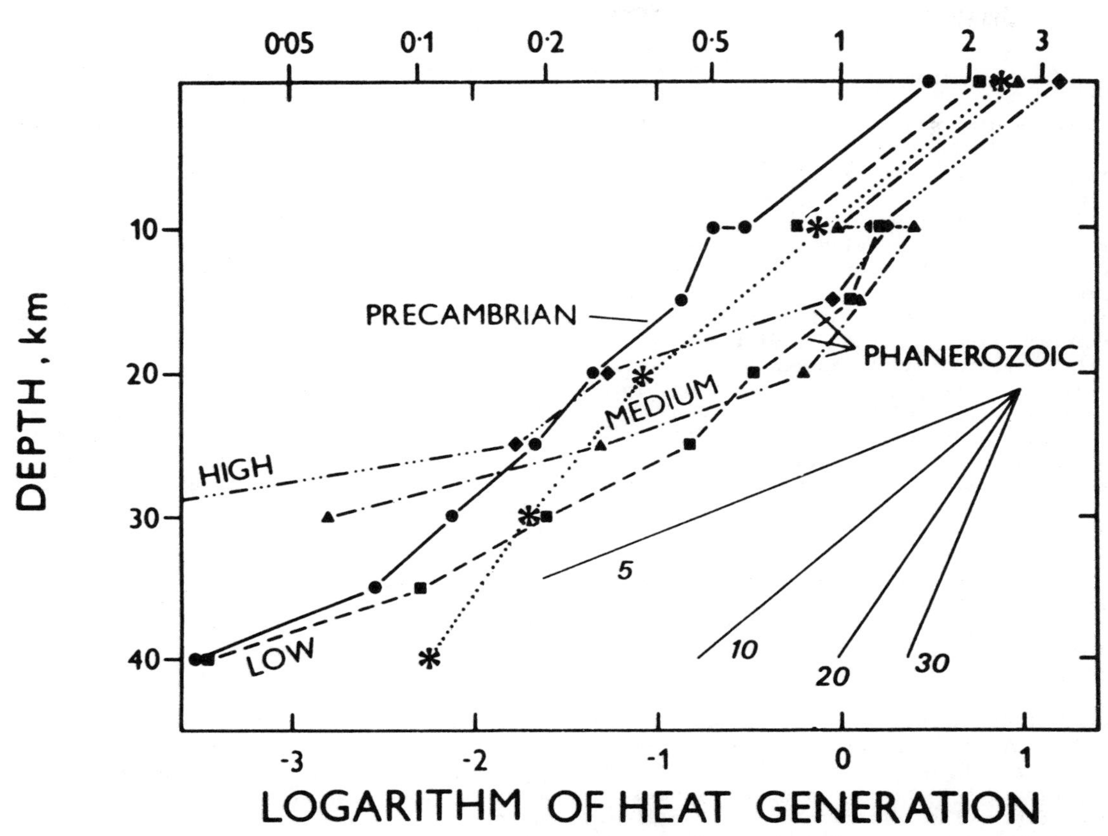

Fig.3. The variation of heat production with depth in four crustal types: Precambrian and Phanerozoic with low, medium and high heat flow. The lines in the lower right corner correspond to D-values (=logarithmic decrement,in km) of exponential depth dependence.Asterisks correspond to extrapolated near surface heat production $A(z)=A_0\exp(-z/D)$, where D=10 km at surface and increases with depth to correspond to the composite contribution of U, Th and K, with respective D-values: D_U=6, D_{Th}=10, and D_K= 30 km ; the near surface concentrations of radioelements are C_U : C_{Th} : C_K = 4.0 ppm : 18 ppm : 3.5 % [Čermák and Rybach, 1988] .

East Labe Massif. Local anomalies may be detected here as being producted by underground water movement, salt tectonics, etc. The mean surface heat flow amounts to 61.4 ± 8.4 $mW.m^{-2}$, with lower crust contributing about 18 $mW.m^{-2}$; the calculated Moho heat flow of 29 ± 4 (s.e.) $mW.m^{-2}$ (n=9) corresponds well with an increased value expected of its tectonic age and history [Čermák, 1988] . An increased outflow of heat from the upper mantle in the area framing the ancient craton was indicated by a single value calculated for the Vorkuta foredeep in the northern part of the Ural, Q_M = 24 $mW.m^{-2}$.

Alpine molasse is characterized by a Moho heat flow of 39 $mW.m^{-2}$, the Inner Western Carpathians by Q_M = 37 $mW.m^{-2}$. The Pannonian Basin wedged between two branches of the Alpine mountain range has an anomalous structure of a thin crust and a very high surface heat flow, the calculated Moho heat flow being more than 50 $mW.m^{-2}$ (see also Horváth et al., [1979] .

All the above-mentioned standard-error values correspond to the scatter of the individual data belonging to specific tectonic units. Due to the uncertainties of the individual parameters of the $v_p \rightarrow A$ conversion (about 10-15 %), to error bounds of the reported seismic velocities (minimum ± 0.1 km/s in the Precambrian strata, and up to probably maximum ± 0.5 km/s in some Alpine regions) as well as the uncertainty in the depth of each seismic layer (± 0.1 - 0.5 km) and in the reported surface heat flow (5-10 %) ; the actual uncertainty of the calculated Moho heat flow would be much larger.The systematic increase of its value from older to younger tectonic units, however, seems to be existing.

Discussion and Conclusions

The radiogenic heat production A(z) is a contribution of three radioactive elements uranium, thorium and potassium. Their depth distribution within the crust is not necessarily governed by the same law. If each element has its own depth dependence, then the relative contributions change with depth. This would cause a certain depth deviation of the D-parameter from a constant value determined from surface or near-surface observations.

Jaupart et al. [1981] derived individual depth scales of all three radiogenic elements by analizing combined heat flow - heat generation relationship, and they revealed a general trend, $D_{Th} \leq D_U < D_K$. They also concluded that certain redistribution of the radioactive elements had occured by magmatic or metamorphic fluid circulation, to which thorium is especially sensitive, and by alteration due to meteoric water down to several km depth, which affects particularly uranium distribution. Haack [1982] reported the degree of relative enrichment of the upper crustal rocks by uranium and thorium compared with potassium. Similar conclusions can be drawn from data publised by Nicolaysen et al. [1981] or Swanberg [1972] , which support $D_U \leq D_{Th} < D_K$.

Regardless of the actual depth scales of the individual radioelements, the value of the D-parameter, which characterizes the exponential distribution of total heat production in the whole crust, should increase with depth. The concentration of the element with smaller D decreases more quickly and its relative contribution thus becomes less important with depth. Finally, towards the crustal base, the element with the highest D-value takes over and controls the heat production. This is, however, in clear disagreement with the obtained tendency found from the converted heat generation-versus-depth profiles discussed here. The only explanation for this contradiction is that there must be another distribution of the radioelements in the lower crust and that it is not possible to extrapolate near-surface observations to greater depths. Here, the lower crust is defined as the bulk of the crust below the upper part of the thickness of approximately D km.

The original distribution of the radiogenic elements, a result of differentiation processes of protocrustal material, might have due to considerably changed later processes, with probably the exception of that of potassium. Besides recrystalization, which accompanies magmatic and metamorphic processes and which may affect the primary distrubution of uranium and thorium, it is mainly the circulation of meteoric water that causes alteration and that contributes to the upward concentration of uranium. Costain [1978] first suggested that the zone of microcrack porosity played certain role in uranium redistribution by the penetration of underground waters to a depth of at least 7 km. The importance of interactions between crustal fluids and the rock in the uppermost crust was also demonstrated by detailed studies of the Idaho batholith [Gosnold and Swanberg, 1980] .

The following model was proposed [Čermák and Rybach, 1988] . The "original" distribution of heat sources was preserved in the lower crust and can hardly be identified from surface observations. If approximated by a general exponential curve, the D-parameter decreases with depth, i.e., deeper in the crust the heat production decreases relatively more rapidly with depth than in the intermediate parts of the crust. The decrease with depth is more pronounced within younger or high heat flow provinces than in old shields and platforms. Near the surface, at depths comparable with the D-parameter obtained from the combined heat flow-heat generation observations, i.e. in the upper 7-14 km, the subsequent processes (magmatism, metamorphism, hydrogeological alteration, etc.) caused a certain redistribution of radiogenic elements. Groundwater circulation plays a decisive role in producing upward migration and enrichment of uranium [Jaupart et al., 1981; Etheridge et al., 1983] . Although there is hardly any distinct transition zone separating the uppermost crust (defined here as the region of meteoric water penetration) from the underlying unaffected crustal layers, it is not possible to preclude the existence of some inversion zone at depths close to about 10 km. This zone would correspond to the inflection in the heat production-depth curve at the lowest depth of interaction between plutons and groundwater [Gosnold and Swanberg, 1980] . The heat production within the upper crust can be evaluated from combined heat flow - heat generation studies where the D-parameter corresponds to the slope of the linear relationship. The uppermost crustal rocks are enriched in radioelements while the rocks at the base of this layer are depleted with respect to the original distribution.

The conversion of seismic velocity into heat production provides a unique possibility of assessing the distribution of heat sources within the crust and may thus significantly help improve our knowledge on the outflow of heat from the upper mantle. Notwithstanding a certain uncertainty in the reliability and thus limits of validity of the proposed conversion technique, and the scatter of the present data, it was possible to obtain some estimate of the Moho heat flow in various tectonic units on a large scale. It has been shown that the Moho heat flow is generally low and stable over large regions of the Precambrian crust and that its value increases towards younger units [Čermák, 1988] .

As opposed to the earlier ideas of relative constancy of the mantle heat flow over all continental areas [Clark and Ringwood, 1964] , it is now generally accepted that there are some regional variations in its distribution and that the younger and tectonically more active areas are characterized by increased outflows [Pollack and Chapman, 1977] . Lesser agreement, however, was reached in the specification of the characteristic value of the Moho heat flow in the stable continental crust, which is the key parameter for any model of the continental lithosphere.

The equating of reduced and mantle heat flows leads to a relatively high value of the mantle heat flow of 25-28 $mW.m^{-2}$ in the areas of continental shields. Similar values were put forward e.g. by Vitorello and Pollack [1980], who gave the value of 27 $mW.m^{-2}$ as a background heat flow, or by Sclater et al. [1981] who proposed 25-29 $mW.m^{-2}$ for the mantle heat flow below a continental craton. Contrary to the above data, some other authors arrived at considerably lower Moho heat flows, e.g. Smithson and Decker [1974] offered value of 12-21 $mW.m^{-2}$, Nicolaysen et al. [1981] restricted the Moho heat flow to lie between 12 and 17 $mW.m^{-2}$.

The former group of models is above all based on correlating surface data on heat flow, heat production and the mean heat flow within the corresponding province, and extrapolating these results downwards. The vertical distribution of heat sources is described by a simple exponential function and thus relates to a uniform crustal evolution by the differentiation of the primordial magma. The latter models employ stratified crustal structures and compute crustal heat production using the characteristic rock properties according to the presumed petrological composition. While there are no substantial differences in surface conditions and both groups of models have similar values of A_o, D, or in the heat production of the lower crust, values of parameters for the middle parts of the crust may depart significantly as was mentioned by Nicolaysen et al. [1981].

Our data gave Moho heat flow values of 20-23 $mW.m^{-2}$ for the shield and about 26 $mW.m^{-2}$ for the platform, which is somewhere between the two groups of models. As the lowest part of the crust has negligible contribution to the heat flow, our data support the idea of a certain hump in heat sources distribution [Nicolaysen et al., 1981] , which we shall locate just beneath the "upper-ten-kilometres" layer (uppermost crust), corresponding to the upper zone of exponential decline of Nicolaysen et al. [1981] . This hump may be more pronounced in young terrains, but cannot be excluded in the craton either. The crust immediately below the near-surface layer may have higher heat production than the base of this layer, the inversion being mainly the result of redistribution of uranium by migrating groundwaters [Čermák and Rybach, 1988] . This hypothesis seems to be supported by the experimental results from some $v_p(z)$-profiles; however, the problem, especially the detailed D-parameter variation with depth, needs further study and experimental confirmation from other parts of the world.

Acknowledgments. The manuscript and the original papers referenced here were read by several colleagues who offered valuable comments. We are particularly indebted to Prof.D.S.Chapman, Prof.H.N. Pollack and Dr.T.Lewis for their thorough criticism, and to Prof.L.Rybach who participated in preparation of some original works. Also an anonymous reviewer provided useful comments that significantly improved the manuscript.

References

Čermak, V., Crustal heat production and mantle heat flow in Central and Eastern Europe, Tectonophysics, in press, 1988.

Čermák, V., and Bodri, L.,Two-dimensional temperature modelling along five East-European geotraverses, J.of Geodynamics, 5, 133-163, 1986.

Čermak, V., and Bodri, L., Közep- és Kelet-Európa geotermikus modellje, Magyar Geofizika, 28,153-186, 1987.

Čermák, V., and Rybach, L., Vertical distribution of heat production in the continental crust, Tectonophysics, in press, 1988.

Clark, S.P., and Ringwood, A.E., Density distribution and constitution of the mantle, Rev.Geophys. Space Phys.,2 , 35-88, 1964.

Costain, J.K., A new model for the linear relationship between heat flow and heat generation, EOS, Trans.Am.Geophys.Union, 59, 329, 1978.

Etheridge, M.A., Wall, V.J., and Vernon, R.H., The role of the fluid phase during regional metamorphism and deformation, J.metamorphic Geol., 1, 205-226, 1983.

Fielitz, K., Compressional and shear wave velocities as a function of temperature in rocks at high pressure. In: P.Giese, C.Prodehl and A.Stein (Eds.), Explosion Seismology in Central Europe, Springer Verlag, Berlin, Heidelberg and New York, 40-44, 1976.

Gosnold, W.D., and Swanberg, C.A., A new model for the distribution of crustal heat sources, EOS, Trans.Am.Gephys.Union, 61, 387, 1980.

Haack, U., Radioactive isotopes in rocks. In: G.Angenheister (Ed.), Landolt-Börnstein New Series, Vol.Vlb, Physical Properties of Rocks, Springer Verlag, Berlin, Heidelberg and New York, 433-560, 1982.

Jaupart, C., Sclater, J.G., and Simmons, G., Heat flow studies: constrains on the distribution of uranium, thorium and potassium in the continental crust, Earth Planet.Sci.Lett., 52, 328-344, 1981.

Kern, H., Elastic wave velocities and constants of elasticity of rocks at elevated pressures and temperatures. In: G.Angenheister (Ed.), Landolt-Börnstein New Series, Vol.Vlb, Physical Properties of Rocks, Springer Verlag, Berlin, Heidelberg and New York, 99-140, 1982.

Lachenbruch, A.H., Preliminary geothermal model of the Sierra Nevada, J.Geophys.Res, 73, 6977-6989, 1968.

Morgan, P., and Sass, J.P., Thermal regime of the continental lithosphere, J.of Geodynamics, 1, 143-166, 1984.

Nicolaysen, L.O., Hart, R.J., and Gale, N.H.,The Vredefort radioelement profile extended to subcrustal strata at Carletonville, with implications to continental heat flow, J.Geophys.Res., 86, 10653-10661, 1981.

Pollack, H.N., and Chapman, D.S., Mantle heat flow, Earth Planet.Sci.Lett., 34, 174-184, 1977.

Roy, R.F., Blackwell, D.D., and Birch, F., Heat generation of plutonic rocks and continental heat flow provinces, Earth Planet.Sci.Lett., 5, 1-12, 1968.

Rybach, L., Wärmeproduktionsbestimmungen an Gesteinen der Schweizer Alpen, Beitr.Geol.Schweiz.Geotechn.Ser., 51, 43 pp., 1973.

Rybach, L., Radioactive heat production: A physical properties determined by the chemistry of rocks. In: R.G.J.Strens (Ed.), The Physics and Chemistry of Rocks. Wiley and Sons, London, 309-318, 1976.

Rybach, L., and Buntebarth, G., Relationship between the petrophysical properties, density, seismic velocity, heat generation and mineralogical constitution, Earth Planet.Sci.Lett., 57, 367-376, 1982.

Rybach, L., and Buntebarth, G., The variation of heat generation, density and seismic velocity with rock type in the continental lithosphere, Tectonophysics, 103, 335-344, 1984.

Rybach, L., and Cermák, V., The depth dependence of heat production in the continental lithosphere, derived from seismic velocities, Gephys.Res.Lett., 14, 311-313, 1987.

Sclater, J.G., Parsons, B., and Jaupart, C., Oceans and continents: similarities and differences in the machanismus of heat loss, J.Geophys.Res.,86, 11535-11552, 1981.

Smithson, S.B., and Decker, E.R., A continental crustal model and its geothermal applications, Earth Planet Sci.Lett., 22, 215-225, 1974.

Sollogub, V.B., Guterch, A., and Prosen, D.,(Eds.), Struktura zemnoy kori i verhkney mantiyi po dannim geofizicheskikh issledovaniyi. Naukova Dumka, Kiev, 208 pp.(in Russian), 1980.

Swanberg, C.A., Vertical distribution of heat generation in the Idaho batholith, J.Geophys.Res., 77, 2508-2513, 1972.

Vitorello, I., and Pollack, H.N., On the variation of continental heat flow with age and the thermal evolution of the continents, J.Geophys.Res., 85, 983-995, 1980.

FLUIDS IN THE LOWER CRUST INFERRED FROM ELECTROMAGNETIC DATA

Leonid Vanyan and Andrej Shilovski

Shirshov Institute of Oceanology,
Moscow 117218, USSR

Abstract. Electromagnetic data suggest an anomalously low resistivity for the lower crust. The most plausible explanation is super-critical fluid in pores and microcracks.

Introduction

Electrical resistivity is probably the most sensitive physical property of rocks to fluids. Even 0.1 volume percent of salt water may reduce the resistivity of crystalline rock by several orders of magnitude. Therefore, information which is obtained by magnetotelluric soundings in many instances can be considered as "in situ" indication of fluids in the lower crust.

Crustal conducting layers were discussed by Jones (1981). Recently, the new results obtained in the frame of the International ELAS programme (Electrical conductivity of the lithosphere and asthenosphere) were presented by Gough (1987).

The purpose of the paper is to consider anomalously low electrical resistivity of the lower crust both in stable and in tectonically active regions and to discuss the role of super-critical fluids in the microcracks of rocks.

Experimental Data

Laboratory measurements suggest that the larger portion of dry rocks constituting the Earth's crust have electrical resistivity as high as 10^4 ohm-m. Electronic-type conductors such as graphite and sulphides are an exception. For instance, the specific resistivity of graphite-bearing schists may be as low as 0.1 ohm-m (Parkhomenko, 1982). This result of laboratory measurements is supported by field measurements. Long but relatively narrow bands of low resistivity are revealed by magnetotelluric (MT) and magnetovariational (MV) methods in many regions of the upper part of the crystalline basement. The results of MV-arrays in Finland (Hjelt, 1984) are shown in Figure 1.

The crustal conductive anomalies of a second type have quite different features. These broad anomalies are hundreds of kilometers in size and are revealed mainly by MT-soundings. Since the electrical field measurements are influenced by near-surface inhomogeneities, the first problem is to separate the deep geoelectrical information. In many cases this problem can be solved by smoothing of the irregular anomalies caused by the near-surface inhomogeneities. One can see in Figure 2 sharp changes of the apparent resistivity caused by surficial anomalies. Typical apparent resistivity curves are shown in Figure 3. Each curve was obtained by statistical averaging of several tens of single soundings. Two groups of curves are presented. The first group contains apparent resistivity data from several stable areas (Eastern Siberian Platform, Baltic, and Canadian Shields). All apparent resistivity curves of the first group are of the same character. The apparent resistivity decreases to values about 150-200 ohm-m at 4-7 min periods. At longer periods the apparent resistivity slightly increases and forms a maximum; this increasing apparent resistivity suggests the lower boundary of the conductive zone.

The second group includes soundings from Western United States (Osipova et al, 1982) and from the Pannonian Basin (Adam et al, 1982). A sharp minimum of apparent resistivity again indicates a conductive zone. In this group, which corresponds to tectonically active regions, the apparent resistivity in the minimum is as low as 30-50 ohm-m.

At long periods the apparent resistivity values both for the first and for the second groups coincide well with the global apparent resistivity curve which was obtained by spherical analysis

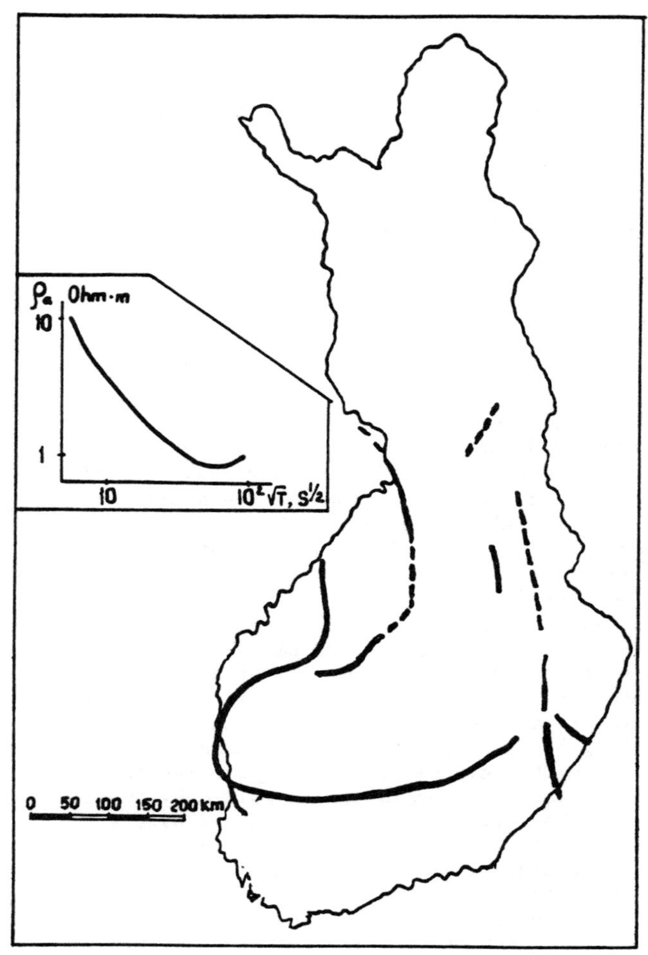

Fig. 1. Low resistivity anomalies of the upper part of the basement in Finland. Insert shows a typical apparent resistivity curve.

of data from geomagnetic observatories (Rokityansky, 1982). An important feature of geomagnetic variations is a rather weak dependence on small-scale surficial inhomogeneities. Agreement between regional MT-curves and global apparent resistivity suggests that the background MT-response obtained by smoothing procedures is really free of the influence of the irregular near-surface anomalies.

The global apparent resistivity at periods longer than 1 hour corresponds to depths greater than 300 km. Note that the resolution of MT-sounding decreases dramatically with increasing depth. Let us assume that the temperature inhomogeneities at depths of 300 km change the conductivity by a factor of 2-4 times. If the horizontal dimensions of such a zone are of an order of hundreds of km, the anomaly in apparent resistivity is estimated to be several percent or at best tens of percent (Berdichevsky, Dmitriev, 1984).

Thus, the apparent resistivity for both stable and active regions is approaching the same global curve.

Interpretation

After filtering the geological noise, which is due to irregular near-surface inhomogeneities, one can consider the deep MT-sounding. Due to nonuniqueness of the MT-inversion a large family of resistivity models can be constructed. Each model of the family satisfies the experimental MT-responses.

If we had enough a priori information, we could choose the most probable model of the family. Unfortunately, our knowledge about the nature of the deep crustal conductivity is limited, which makes it difficult to choose an adequate model. Another way is to use parameters of the family of models, which are estimated as more or less valid. The first important parameter is the depth-integrated electrical conductivity of the Earth's crust, while the second parameter is the depth of "mass-centre" of the conductive zone or the depth of the resistivity minimum. Estimations show the conductance value of the crust is about 400-700 siemens for the stable regions and 1500-2000 siemens for the active ones.

The depth of the resistivity minimum is 35-40 km and 25-30 km, respectively. It means that the conductive zone belongs to the lower crust. The largest specific resistivity value of the conductive layer which satisfies the experimental data is 50-100 ohm-m for the stable regions and 10-20 ohm-m for the active ones. These values of the specific resistivity are by 2-3 orders of magnitude smaller than the resistivity of crustal rocks under temperatures typical of the lower crust (Parkhomenko, 1982).

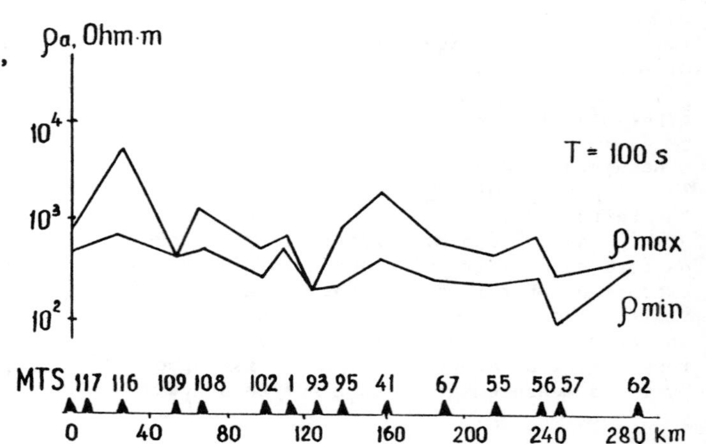

Fig. 2. An example of sharp changes of the apparent resistivities along the profile AA in Eastern Siberia (see Fig. 4).

spectral frequency domain. The number of terms retained in the binomial series appearing in the right hand side of the equation was determined from the maximum amplitude of S_1 and S_2. In a given series the last term retained was equal to or less than 5 percent of the term with a maximum value. Convergence of the iteration was very fast and after four iterations the results were almost identical. The fast convergence may be due in part to the fact that the Nyquist wavelength of the magnetic anomalies was comparable to the depth to the basement, and in part to the smoothing of the magnetic anomalies by the Butterworth filter.

Plate 2 shows the lateral variations in the effective magnetic susceptibility (magnetic susceptibility averaged over the thickness of the magnetic layer) of the first model determined through the generalized inversion technique. The long wavelength components of the magnetic susceptibility are more enhanced compared to those of the magnetic anomalies. This is especially true over the oceanic areas where the layer is thin. The enhancement can be explained in terms of the exponential factors in equations (2) and (7), by putting them into the following form:

$$e^{-Kz}\left(e^{KS_2}-e^{KS_1}\right)=e^{-Kh}\left(1-e^{-KH}\right), \tag{19}$$

where h and H are the depth to the upper boundary and the thickness of the magnetic layer, respectively. The first exponential term in the right hand side of this equation causes the attenuation of the magnetic anomalies with increasing h, in particular the stronger attenuation of the short wavelength components relative to the long wavelength ones. This term is responsible for the enhancement of the short wavelength components through an inversion process. However, the Nyquist wavelength, 11.2 km, of the magnetic anomalies seen in Plate 1 is greater than the depth to the basement everywhere except at a few locations, and the enhancement of the short wavelength components is less than a factor of 2.

The second term in the right hand side of equation (19) shows the attenuation of the magnetic anomalies with decreasing thickness of the magnetic layer. The thinner the magnetic layer, the more pronounced the attenuation. Also, the long wavelength components of the anomalies attenuate more rapidly than the short wavelength ones. To appreciate this point consider the magnetic anomalies produced by a vertically magnetized layer with flat boundaries as a simple example. The anomalies arise from the distribution of magnetic poles placed along the lower and the upper boundaries of the layer [Baranov, 1975]. The magnetic fields of the boundaries tend to cancel each other when the wavelengths are appreciably longer than the thickness of the layer. A classic example is the DC term of the fields which are canceled completely regardless of the intensity of the magnetization of the layer. This is not, however, the case for the short wavelength components of the magnetic anomalies, because the components due to the lower boundary are attenuated more than those due to the upper boundary, when detected at sea-level. Consequently, the second term in the right hand side of equation (19) results in the enhancement of the long wavelength components of the magnetization upon inversion. As the thickness of the magnetic layer increases, the second term in the right hand side of equation (19) approaches unity and the effects of the lower boundary diminish. This phenomenon was examined by calculating the magnetization of a magnetic layer with a flat lower boundary. The lower boundary was set at depths of 15, 20, 30, and 50 km. As the depth to the lower boundary increased the long wavelength components of the magnetization decreased. This is illustrated by the second model where the thickness of the magnetic layer approaches infinity; it becomes a bottomless magnetic layer. Plate 3 shows the lateral variations in the magnetic susceptibility of this model. The long wavelength components are not enhanced and the anomalies are dominated by the short wavelength features. This model is similar to the inversion models which interpret magnetic anomalies in terms of magnetic monopoles distributed at the upper surface of the magnetic layer [Urquhart and Strangway, 1985]. Plate 3 is equivalent to a high-pass filtered version of Plate 2, suggesting that the monopole models represent the high-frequency components of the crustal magnetization.

Plate 4 shows the effective magnetic susceptibility of the third model, where the flat lower boundary of the magnetic layer is located at a depth of 35 km, which is approximately the depth to the Moho beneath continental shelves [Keen et al., in press]. The long wavelength components of the magnetization of the layer are not enhanced as much as those of the first model, illustrated in Plate 2. The third model is preferred because the long wavelength marine magnetic anomalies from other studies [Harrison, 1976; Harrison and Carle, 1981] and the satellite magnetic anomalies derived from the magnetometer satellite (MAGSAT) data suggest that the oceanic upper mantle is magnetic [Arkani-Hamed and Strangway, 1986b].

Plate 4 shows a distinct difference in magnetic susceptibility beneath the oceanic part of the Labrador Sea and the adjacent continental shelves. The former is an overall low magnetic region which is in good agreement with the long wavelength magnetic susceptibility of the region derived from MAGSAT data [Arkani-Hamed and Strangway, 1985; Hahn and Bosum, 1986]. The latter is characterized by both high and low magnetic susceptibilities. However, it is implausible to generalize these ocean-continent magnetic differences to other regions, because the MAGSAT data do not show a consistent difference between the magnetic properties of the continental and oceanic areas elsewhere. The special case of the Labrador Sea will be discussed in the next section.

There is a general degradation of the magnetic susceptibilities towards the north in Plate 4 compared to the magnetic anomalies depicted in Plate 1. For example, the amplitudes of the north-south striking positive magnetic anomalies in Ungava Bay, between 60° and 62° N, are comparable to those of the anomalies south of Newfoundland, whereas the magnetic susceptibilities of the former are less than those of the latter. This arises from the fact that a given magnetic body produces a greater magnetic anomaly in the north, where the inducing magnetic field is stronger than in the south.

As mentioned before, at some places the upper surface of the magnetic layer was approximated by bathymetry. The effect of this approximation is illustrated in Figure 5, which shows the bathymetry, depth to basement and the resulting magnetic susceptibilities for the third model mentioned above and for a fourth model along profile AB (see Figure 1 for a location). The fourth model is identical to the third one except for the upper surface of the magnetic layer where bathymetry is used instead of basement topography. The results demonstrate that taking into account the deep basin with a basement depth of about 8 km, significantly influences the amplitude of the obtained susceptibilities. However, the susceptibilities show similar patterns in both cases, implying that the basement topography alone cannot account for the anomalies.

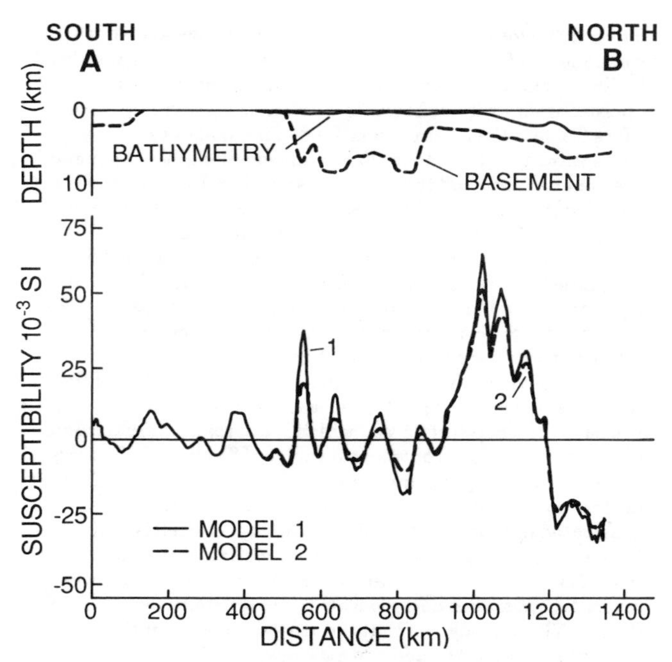

Fig. 5. Synthetic profile AB calculated from the grid values; see Figure 1 for location. The profile shows the bathymetry and depth to basement, together with the resulting susceptibilities obtained using the following model: 1) upper boundary of the magnetic layer is the depth to basement and the lower boundary is a flat layer at 35 km depth; 2) upper boundary is the bathymetry and the lower boundary is the same as for 1.

Geophysical Interpretation

This section discusses the general features of the magnetic anomalies (Plate 1) and the crustal magnetic susceptibility models (Plates 2, 3 and 4) in relation to the geology of the area. The entire region is divided into three parts: the submerged continental area, the oceanic area, and the rift basins and ocean-continent boundary.

Submerged Continental Area

Figure 6 shows the approximate location of the ocean-continent boundary (OCB) obtained from gravity, seismic and magnetic data [Srivastava, 1978, 1983]. Almost the entire southern part of the area is continental, where a large west-east trending magnetic anomaly, the Collector anomaly [Haworth, 1975] delineates the offshore extension of the Cobequid-Chedabucto transcurrent fault which runs across Nova Scotia and eastern New Brunswick, marking the boundary between the Avalon and the Meguma terranes of the Paleozoic Appalachian orogen [Haworth, 1975; Williams and Hatcher, 1983]. The southern boundary of this anomaly is better defined on the magnetic susceptibility maps, where it is shifted northwards by about 30 km with respect to the observed magnetic anomaly. Several strong but localized northeast trending anomalies are located north of the Collector anomaly; these can be continued onshore east of the 56° W meridian, where they correlate with Precambrian volcanic rocks of the Avalon terrane [Haworth and Lefort, 1979]. South of the Collector anomaly, the magnetic anomalies are subdued except for a few small and isolated anomalies probably associated with volcanics.

In the Gulf of St. Lawrence, a magnetic lineation runs from the Gaspe Peninsula to Magdalen Island. It is shifted

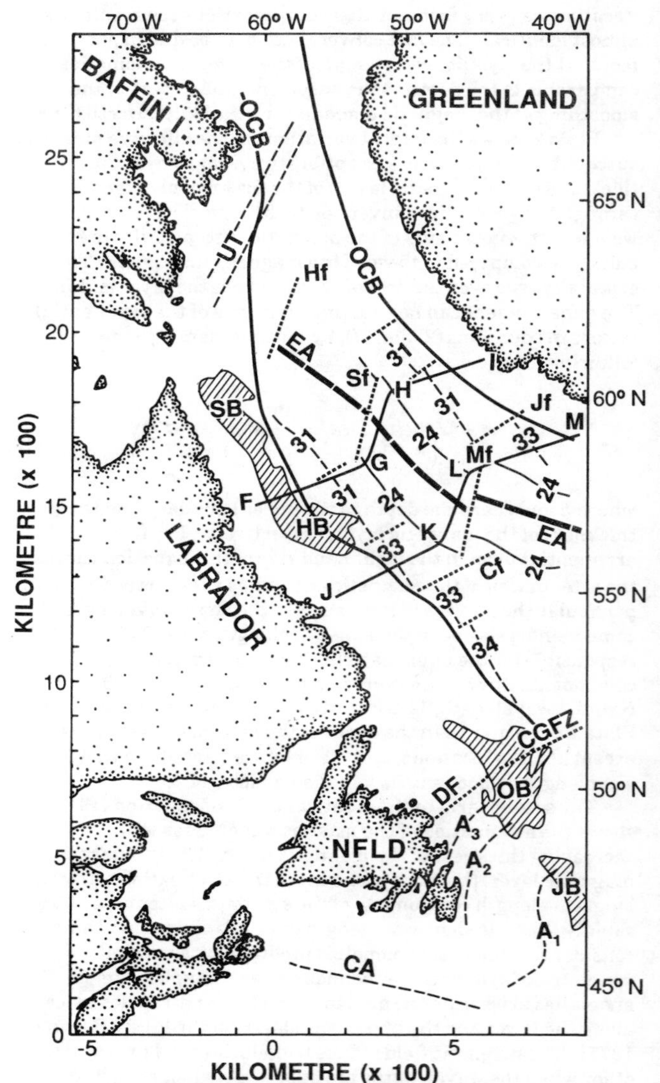

Fig. 6. Interpretation map. The major basins in the area (shaded) are outlined by their 8 km depth to basement contour; OB: Orphan Basin, JB: Jeanne d'Arc Basin, HB: Hopedale Basin, SB: Saglek Basin. EA: Extinct spreading axis. Fracture zones are denoted by dotted lines; CGFZ: Charlie Gibbs fracture zone, Cf: Cartwright fracture zone, Mf: Minna fracture zone, Jf: Julianehaab fracture zone, Sf: Snorri fracture zone, Hf: Hudson fracture zone. Sea-floor spreading magnetic anomalies are shown as solid thin lines where they are certain and by dashed lines where they are uncertain. Fracture zones and identified sea-floor spreading anomalies after Srivastava and Tapscott [1986]. DF: Dover fault. CA: Collector anomaly. A1. A2. and A3: Avalon basement trends. Synthetic profiles FGHI and JKLM are shown in Figure 8. UT: Ungava transform zone. OCB: approximate location of the ocean-continent boundary [after Srivastava, 1978 and 1983].

northwards by about 25 km in the magnetic susceptibility maps with respect to the observed anomaly, where it coincides with the St. Lawrence promontory, which is a dextral offset of the Appalachian orogen from New Brunswick and Quebec to Newfoundland [Stockmal et al., 1987]. The inversion of the anomalies into the magnetic susceptibility hardly refined the lineation because we used bathymetry as the depth to basement in this area, and hence did not take into account the sedimentary basins.

The arcuate magnetic trends seen to the east of Newfoundland fade away towards the northeast. These eastward veering anomalies are separated from northward veering Gander zone magnetic trends by the Dover fault [Haworth and Lefort, 1979]. Figure 7 shows two parallel profiles, AB and CD (see Figure 1), across this area. Profile AB crosses the western edge of the deep Orphan Basin (see OB in Figure 6), while profile CD crosses almost through the center of the basin where the depth to the basement is more than 10 km. The corresponding anomalies in profile CD are smaller than those of profile AB. The eastward decrease in the amplitude of the anomalies cannot completely be accounted for by the overall deepening of the basement. Figure 7 shows a decrease in the magnetic susceptibility of the crust under the deeper part of the basin. The positive anomalies A2 and A3 correspond to basement lows on profile AB, whereas they show no obvious correlation with the almost smooth basement along profile CD. This suggests that details of the basement topography in the Orphan Basin have some influence on the magnetic anomalies,

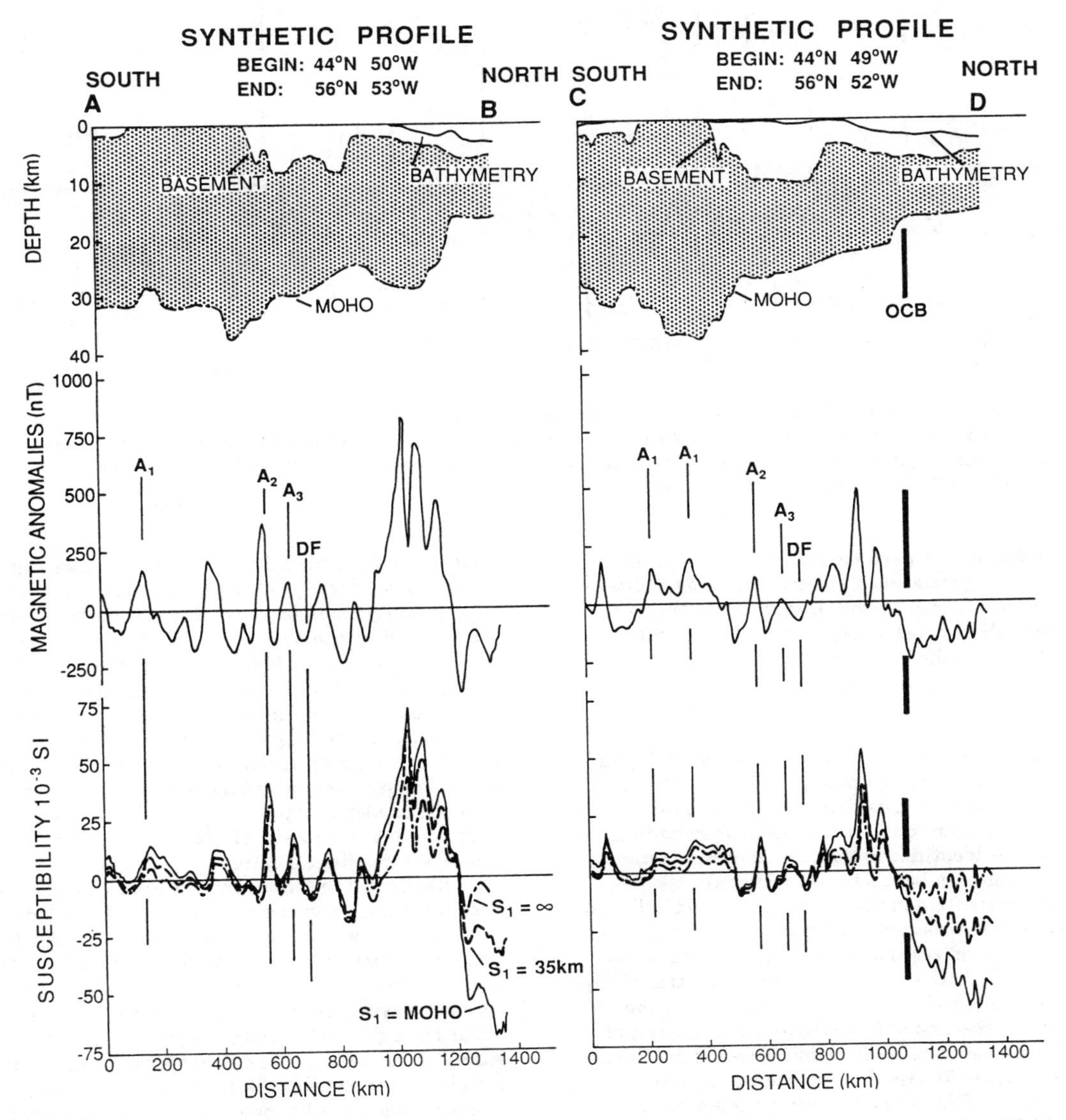

Fig. 7. Synthetic profiles AB and CD calculated from the grid values; for location see Figure 1. A1, A2 and A3: Avalon basement trends. DF: Dover fault, OCB: approximate location of the ocean-continent boundary [after Srivastava, 1978].

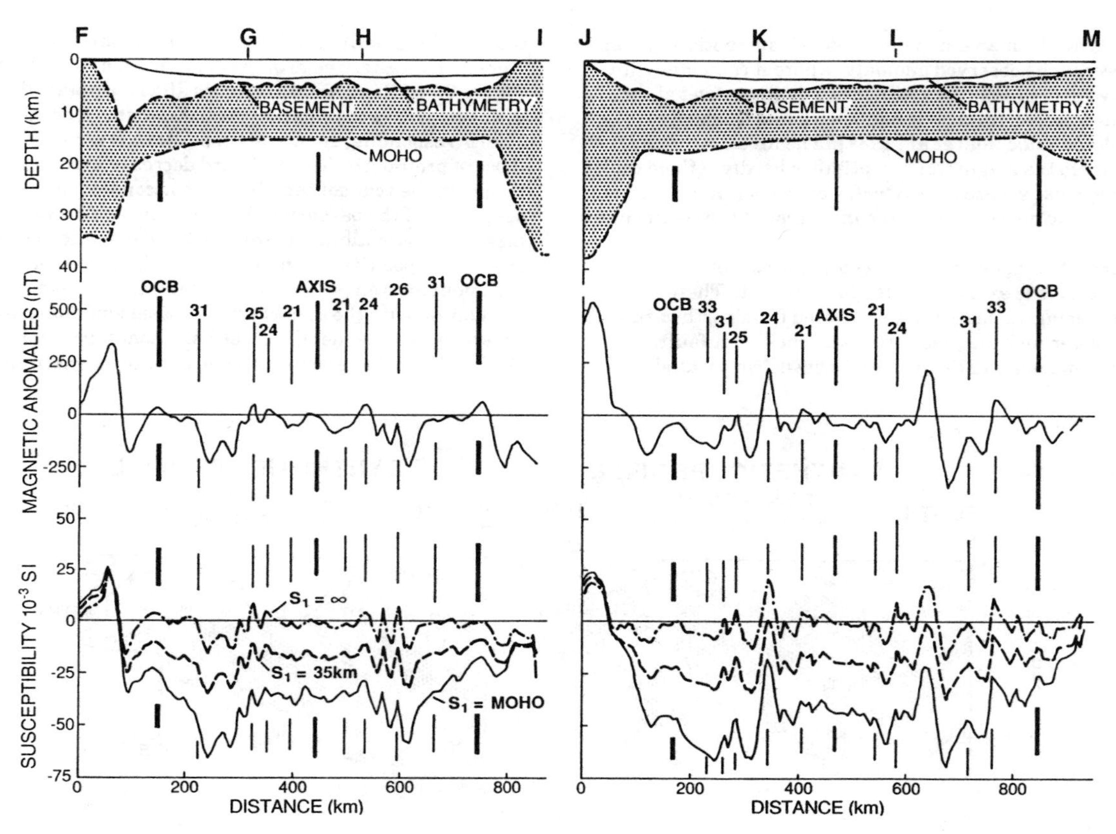

Fig. 8. Synthetic profiles across the Labrador Sea; see Figure 6 for location. The profiles follow synthetic flow lines obtained from the rotation poles given by Srivastava and Tapscott [1986]. Identified spreading anomalies are shown, together with the ancient spreading axis and the approximate location of the ocean-continent boundary (OCB) [after Srivastava, 1978].

and that the anomalies are partly associated with lateral variations in the magnetization of the crust. However, due to the large sediment thickness in the Orphan Basin, the real basement is difficult to detect, and often the deepest reflector is used as an indicator for the basement.

Oceanic Areas

Rifting of the Labrador Sea started in Cretaceous, because the oldest identified sea-floor spreading magnetic anomaly is anomaly 34 just north of the Charlie Gibbs fracture zone [Srivastava, 1978]. Figure 6 shows the sea-floor spreading magnetic anomalies identified in the Labrador Sea. Active spreading in the southern Labrador Sea occurred simultaneously with rifting in the northern part and in Davis Strait. The spreading direction changed at about the time of anomaly 25, which is reflected in the fracture zone patterns depicted in Figure 6. At the time of anomaly 24 another change in the direction of the sea-floor spreading resulted in oblique spreading in Baffin Bay and in Davis Strait [Srivastava and Tapscott, 1986], probably in the direction shown in the Ungava transform zone (Figure 6). Sea-floor spreading stopped prior to anomaly 13 time, and Greenland started to move with the North American plate. The northwest-southeast trending magnetic anomalies of the Labrador Sea (Plate 1) suggest that they are mainly of sea-floor spreading origin. The magnetic susceptibility maps, however, do not show the details of the sea-floor spreading anomalies, due to the enhancement of the long wavelength components of the lateral variations in the magnetic susceptibility of the crustal layer as discussed before, and due to the gridding of the original data. The grid cell of 5.6 × 5.6 km adopted in this study spans an area which is formed during more than one core field polarity period, especially in the locations with a low sea-floor spreading rate. Therefore, the magnetic anomaly at each grid point was calculated by averaging magnetic data of opposite directions, which would suppress the sea-floor spreading signatures.

The effect of the basement topography on the identified sea-floor spreading magnetic anomalies is illustrated in two profiles (Figure 6) along flow lines obtained using the rotation poles proposed by Srivastava and Tapscott [1986]. Profile FGHI (see Figure 8) shows a clear basement expression of the extinct spreading axis. The magnetic anomalies are almost symmetric about this axis. This feature is even clearer in the magnetic susceptibility profiles. Sea-floor spreading anomaly 21 is hardly seen in the magnetic profiles, whereas it is quite apparent in the magnetic susceptibility profiles. The southern profile (profile JKLM in Figure 6) shows no clear evidence of the spreading axis. Along these profiles, the similarity between the magnetic anomalies and the crustal magnetic susceptibility emphasizes the fact that the anomalies are largely due to the lateral variations in the magnetization of the crust, rather than to the topography of the basement. This suggests that the basement topography is probably produced by the vertical displacement of the oceanic crust due to graben formation near the ridge axis, and not by variations in the thickness of the crust.

The observed sea-floor spreading magnetic anomalies, especially anomalies 31 and 33, show skewnesses similar to those seen elsewhere in the Atlantic and Pacific Oceans [Cande, 1976; Cande and Kristoffersen, 1977]. These large skewnesses cannot be due to the effect of the geographic locations of the anomalies because the magnetic susceptibility profiles also exhibit similar skewnesses. The assumption made in the inversion process that the magnetization is along the present geomagnetic field direction may not introduce this amount of skewness because of the proximity of the region to the geomagnetic pole. Analysis of the long wavelength marine magnetic data from both the Atlantic and Pacific Oceans [Harrison, 1976; Harrison and Carle, 1981], and the magnetic anomalies of the Earth derived from the MAGSAT data [Arkani-Hamed and Strangway, 1985] suggest that the oceanic lower crust and the upper part of the oceanic upper mantle are magnetic. Also, theoretical studies show that possible remanent magnetization of the oceanic lower crust and the upper part of the oceanic upper mantle induce skewnesses similar to those seen in Figures 7 and 8 [Arkani-Hamed, 1988]. Therefore, we suggest that the skewnesses seen in Figure 8 arise from remanent magnetization of the oceanic lower crust and the upper part of the oceanic upper mantle.

The magnetic anomaly map and the crustal magnetic susceptibility models show that the oceanic region is an area of overall low magnetization. The magnetic anomalies in the oceanic region are dominated by sea-floor spreading anomalies arising from the remanent magnetization of the oceanic crust and probably the oceanic upper mantle. The core field has changed its polarity many times during the formation of the Labrador Sea, and the reversed polarities were usually longer than the normal ones in this period [Cox, 1980]. This may explain the overall negative amplitudes of the long wavelength components of the magnetic anomalies and the magnetization models of the crust. The sea-floor spreading rate was quite variable; it was about 10 mm/year between anomaly 33 and anomaly 25, about 7 mm/year between anomaly 25 and anomaly 21, and less than 2 mm/year between anomaly 21 and anomaly 13 when the spreading ceased [Srivastava, 1986]. The thermal evolution models of the oceanic lithosphere [McKenzie, 1969; Sclater et al., 1981; Phipps Morgan et al., 1987; among many others] show that the oceanic lower crust and upper mantle cool more slowly than the oceanic upper crust, and thus they become magnetized more gradually; during this time the core field may change its polarity several times [Arkani-Hamed, 1988]. This implies that the magnetization of the oceanic lithosphere beneath a given location may have opposite directions at different depths. The magnetization pattern of the older lithosphere, i.e., between anomaly 33 and anomaly 21, is most likely skewed because of the appreciable spreading rate. This suggests that the oppositely magnetized sublayers at depth may not completely overlie, whereas the younger lithosphere, i.e., younger than anomaly 21, is probably characterized by oppositely magnetized and almost overlying sublayers, on account of the very slow spreading rate. As already mentioned in the previous section, magnetic anomaly analysis alone provides information about the effective magnetization, i.e., the magnetization integrated over the entire thickness of the magnetic layer. Therefore, the effective magnetization of the younger lithosphere becomes weaker, because of the canceling effect of the oppositely magnetized sublayers, than that of the older lithosphere. This may explain the subdued magnetic anomalies in the central part of the Labrador Sea, compared to the pronounced anomalies over the older parts of the Sea.

The continental shelves are characterized by strong but localized magnetic anomalies largely associated with intrusive volcanic rocks, which have been stationary with respect to their surroundings since their formation. The intrusives have cooled slowly and have acquired their magnetization gradually, especially in the deeper part of the crust. This implies that different parts of the intrusives may have been magnetized in opposite directions because of the changes in the polarity of the core field, and the integral effect of the remanent magnetization on the magnetic anomalies is probably small [Celetti and Arkani-Hamed, 1988]. Based on these assumptions it is, therefore suggested that the anomalies are largely due to the induced magnetization of the intrusives. The induced magnetization of the oceanic lithosphere may not produce a significant magnetic anomaly because it is relatively uniform.

Rift Basins and Ocean Continent Boundary (OCB)

The margins in the study area are formed by rifting of the North American plate during which the continental crust stretched, thinned, and finally ruptured. The rifting stage in the Labrador Sea was accompanied by volcanism [Srivastava, 1978]. The stretching and thinning resulted in the development of rift basins such as the Orphan, Hopedale, and Saglek basins (Figure 6). The post-rift phase is characterized by subsidence and sedimentation, in response to thermal contraction due to cooling of the lithosphere. This has resulted in sediment thicknesses of as much as 10 km in the rift basins [Keen et al., in press].

The location of final breakup, i.e., the ocean-continent boundary (OCB), has been the subject of several studies [e.g. Srivastava, 1978; Keen and de Voogd, 1988; Srivastava et al., in press]. Due to the diversity of tectonic styles involved, the definition of the OCB cannot be based upon any single criterion. The mode of rupture changed from region to region and the OCB varied from a sharp boundary to a gradual and diffuse one, with a transition zone of up to 50 km in width [Srivastava, 1978]. In a recent investigation just to the east of our study area [Srivastava et al., in press], it was shown that the OCB separates two zones of different magnetic characteristics over the northeast Newfoundland margin: long wavelength and small amplitude magnetic anomalies on the landward side, and quasi-lineated and high-amplitude anomalies on the seaward side. This result is similar to that observed across the Goban Spur margin off the British Isles [Scrutton, 1985], which is a conjugate of the northeast Newfoundland margin.

The magnetic anomalies along the Labrador and Greenland margins (Plate 1) have high amplitudes and decrease abruptly at a distance of about 100 km from the shore. The sharp gradient along the Labrador margin is located in the middle of the continental shelf and does not correlate with the OCB, as depicted in Figure 6, which is about 100 km further seaward. However, there is a good correlation between this sharp gradient and the fairly abrupt deepening of the basement across the hinge zone of the Hopedale Basin (Figure 8). Further south, between

52° and 55° N, the OCB lies seaward of a series of large positive magnetic anomalies, where drilling in wells has sampled early Cretaceous rocks [Umpleby, 1979]. The ocean-continent transition zone there is probably narrow and occurs beneath the continental shelf [Fenwick et al., 1968]. Along the Greenland margin, the sharp gradient in the magnetic anomalies also correlates with the abrupt deepening of the basement, at least south of 62 N where depth to basement information is available. Again the OCB is located further seaward relative to the sharp magnetic gradient, although the shift here is less than that along the Labrador margin. These features are clearly visible along the profiles in Figure 8, where the OCB is shifted seaward with respect to the sharp decrease in crustal thickness [Shih et al., in press].

The OCB has a consistent, though less pronounced, magnetic anomaly. It is generally marked by a local increase in the magnetic anomalies on both the Greenland and the Labrador margins (see Figure 8). This is similar to the observations along the northeastern Newfoundland margin [Srivastava et al., in press], where the OCB is marked by a large-amplitude magnetic anomaly.

The general correlation between the sharp gradient in the magnetic anomalies and the deepening of the basement along the margins of the Labrador Sea, could indicate that this gradient is largely due to the deepening of the magnetic sources. Then one would expect the general inversion of the anomalies to give susceptibilities with hardly any gradient at these locations. However, Plates 2, 3 and 4, together with the profiles in Figure 8, show that there also is a susceptibility change at these locations, a situation similar to the one observed at Orphan Basin (see Figure 7). This result indicates that the decrease in the magnetic anomalies over the deep sedimentary basins arises from both the deepening of the magnetic sources and the decrease in the effective magnetization of the crust.

Conclusions

The generalized inversion technique presented in this paper determines the lateral variations in the magnetization of the crust using observed magnetic anomalies, while taking into account variations in the directions of the geomagnetic field and the magnetization of the crust, and the topography of the upper and lower boundaries of the magnetic layer. The technique is based on the spectral frequency domain algorithm which delineates the sources of instability encountered in the inversion. It also suggests a natural coordinate system to be employed in order to suppress the instability.

The generalized inversion technique was applied to marine magnetic anomalies off the east coast of Canada. Three different models were considered for the magnetic crustal layer. The upper boundary of the layer coincides with the basement topography, or with sea-floor where the basement data is not available, in all three models. The lower boundary of the layer differs among the models: in the first model, it coincides with the Moho discontinuity; in the second, it is at an infinite depth; and in the third, the boundary is flat at a depth of 35 km. Lateral variations in the magnetic susceptibility of the first and the third models are dominated by long wavelength components because the layer has a small thickness compared to these wavelengths. The second model is dominated by high-frequency components because the lower boundary has no effect on the magnetic anomalies.

The general characteristics of the magnetic anomalies are reflected in the lateral variations of the magnetic susceptibility of the crustal layer in the oceanic regions. This indicates that the topography of the upper surface of the magnetic layer has minor influence on the anomalies and that the topography is probably produced by vertical displacement of the oceanic crust, as opposed to the variations of the crustal thickness by, for example, erosion. Also, the magnetic anomaly and the crustal magnetic susceptibility profiles show skewnesses. This evidence suggests that the anomalies are mainly due to the lateral variations in the magnetization of the oceanic crust, with probable contributions from the upper part of the oceanic upper mantle.

The pronounced magnetic gradient parallel to the shore line in the Labrador Sea does not correlate with the ocean-continent boundary. Rather, it delineates the deepening of the continental shelf due to the formation of the rift basins. The decrease in the magnetic anomalies over the rift basins may not be completely accounted for by the overall deepening of the basement; a decrease in the effective magnetic susceptibility of the crust is also required.

Acknowledgments. This research was partly supported by the National Science and Engineering Research Council of Canada (NSERC) under operating grant # A2037 and partly by the Frontier Geoscience Program of the Department of Energy, Mines and Resources. Uwe Brand from Brock University and Charlotte Keen, Ron Macnab and John Woodside from the Atlantic Geoscience Centre kindly read the first draft of the manuscript and made many helpful suggestions. The drafting and illustration section of the Bedford Institute of Oceanography prepared some of the figures.

References

Arkani-Hamed, J., Remanent magnetization of the oceanic upper mantle, Geophys. Res. Lett., 15, 48-51, 1988.

Arkani-Hamed, J. and D.W. Strangway, Intermediate scale magnetic anomalies of the earth, Geophysics, 50, 2817-2830, 1985.

Arkani-Hamed, J. and D.W. Strangway, Magnetic susceptibility anomalies of lithosphere beneath eastern Europe and the middle East, Geophysics, 51, 1711-1724, 1986a.

Arkani-Hamed, J. and D.W. Strangway, Effective magnetic susceptibility of the oceanic upper mantle derived from MAGSAT data, Geophys. Res. Lett., 13, 999-1002, 1986b.

Arkani-Hamed, J. and D.W. Strangway, An interpretation of magnetic signatures of subduction zones detected by MAGSAT, Tectonophysics, 133, 45-55, 1987.

Baranov, W., Potential Fields and Their Transformations in Applied Geophysics Geoexplor. Monogr., Ser. 1, No. 6, Gebrueder Borntraeger, Berlin, 1975.

Benkova, N.P., Sh. Sh. Dolginov, and T. N. Simonenko, Residual geomagnetic field from the satellite COSMOS 49, J. Geophys. Res., 78, 798-803, 1973.

Bhattacharyya, B.K., A generalized multibody model for inversion of magnetic anomalies, Geophysics, 45, 255-270, 1980.

Cain, J.C., D.R. Schmitz and L. Muth, Small scale features observed by MAGSAT, J. Geophys. Res., 89, 1070-1076, 1984.

Cande, S.C., Anomalous behavior of the paleomagnetic field inferred from the skewness of anomalies 33 and 34, Earth Planet. Sci. Lett., 40, 275-286, 1976.

Cande, S.C. and Y. Kristoffersen, Late Cretaceous magnetic anomalies in the North Atlantic, Earth Planet. Sci. Lett., 35, 215-224, 1977.

Celetti, G. and J. Arkani-Hamed, Remanent magnetization of intrusive bodies (abstract), EOS, Trans. Am. Geophys. Un., 69, 335, 1988.

Committee for the Magnetic Anomaly Map of North America, Magnetic Map of North America, Boulder, CO, Geol. Soc. of America, 5 sheets, scale: 1:5,000,000, 1987.

Cooley, J.W. and J.W. Tukey, An algorithm for the machine calculation of complex Fourier series, Math. Comp., 19, 297-301, 1965

Cox, A.V., Magneto-stratigraphic time scale, in: W.R. Harland, A.V. Cox, P.G. Llewellyn, C.A.G. Pickton, A.C. Smith and R. Walters (eds.), A Geologic Time Scale, Cambridge University Press, 74-75, 1980.

Dods, S.D., D.J. Teskey and P.J. Hood, the new series of 1:1000,000-scale magnetic-anomaly maps of the Geological Survey of Canada: Compilation techniques and interpretation, in: The Utility of Regional Gravity and Magnetic Anomaly Maps, W.J. Hinze (ed.), Society of Exploration Geophysicists, 69-87, 1985.

Fenwick, D.K.B., M.J. Keen, C.E. Keen and A. Lambert, Geophysical studies of the continental margin northeast of Newfoundland, Can. J. Earth Sci., 5, 483-500, 1968.

Grant, A.C., K.D. McAlpine and J.A. Wade, The Continental Margin of eastern Canada: Geological framework and petroleum potential, in: Future Petroleum Provinces of the World, M.T. Halbouty (ed.), Amer. Assoc. Petrol. Geolog. Memoir 40, 177-205, 1986.

Haggerty, S.E., The aeromagnetic mineralogy of igneous rocks, Can. J. Eart. Sci., 16, 1281-1293, 1979.

Haggerty, S.E. and Toft, P.B., Native iron in the continental lower crust: Petrological and geophysical implications, Science, 229, 647-649, 1985.

Hahn, A. and W. Bosum, Geomagnetism - Selected examples and case histories, Geoexplor. Monogr. Ser. 1, No. 10, Gebrueder Borntraeger, Berlin, 1986.

Harrison, C.G.A., Magnetization of the oceanic crust, Geophys. J.R. astr. Soc., 47, 257-283, 1976.

Harrison, C.G.A. and T. Lindh, A polar wandering curve for North America during the Mesozoic and Cenozoic, J. Geophys. Res., 87, 1903-1920, 1982.

Harrison, C.G.A. and H.M. Carle, Intermediate wavelength magnetic anomalies over ocean basins, J. Geophys. Res., 86, 11585-11599, 1981.

Harrison, C.G.A., H.M. Carle and K.L. Hayling, Interpretation of satellite elevation magnetic anomalies, J. Geophys. Res., 91, 3633-3650, 1986.

Haworth, R.T., The development of Atlantic Canada as a result of continental collision - evidence from offshore gravity and magnetic data, in: Canada's continental margins and offshore petroleum exploration, C.J. Yorath, E.R. Parker and D.J. Glass (eds.), Canadian Society of Petroleum Geologists, Memoir 4, 59-77, 1975.

Haworth, R.T. and J.P. Lefort, Geophysical evidence for the extent of the Avalon zone in Atlantic Canada, Can. J. Earth Sci., 16, 552-567, 1979.

Hinze, W.J. and I. Zietz, The composite magnetic anomaly map fo the conterminous United States, in: The Utility of Regional Gravity and Magnetic Anomaly Maps, W.J. Hinze (ed.), Society of Exploration Geophysicists, 1-24, 1985.

IAGA Division 1, Working Group 1, International Geomagnetic Reference Field revision 1985, EOS, Trans. Am. Geophys. Un., 67, 523-524, 1986.

Keen, C.E. and B. de Voogd, The continent-ocean boundary at the rifted margin off eastern Canada: new results from deep seismic reflection studies, Tectonics, 7, 107-124, 1988.

Keen, C.E., B.D. Loncarovic, I. Reid, J. Woodside, R.T. Haworth, and H. Williams, Tectonic and geophysical overview, in: The Geology of North America: Geology of the Continental margin Off Eastern Canada, M.J. Keen and G.L. Williams (eds.), Dec. N. Am. Geol. (DNAG) Ser., in press.

Letros, S., D.W. Strangway, A.M. Tasillo-Hirt, J.W. Geissman and L.S. Jensen, Aeromagnetic interpretation of the Kirkland Lake-Larder Lake portion of the Abitibi greenstone belt, Ontario, Can. J. Earth Sci., 20, 548-560, 1983.

Mayhew, M.A., Inversion of satellite magnetic anomaly data, J. Geophys. Res., 45, 119-128, 1979.

Mayhew, M.A. and S.C. Galliher, An equivalent layer magnetization model for the United States derived from MAGSAT data, Geophys. Res. Lett., 9, 311-313, 1982.

McKenzie, D.P., Speculations on the consequences and causes of plate motions, Geophys. J. R. astr. Soc., 18, 1-32, 1969.

Parker, R.L., The rapid calculation of potential anomalies, Geophys. J. R. astr. Soc., 31, 447-455, 1973.

Parker, R.L. and S.P. Huestis, The inversion of magnetic anomalies in the presence of topography, J. Geophys. Res., 79, 1587-1593, 1974.

Phipps Morgan, J., E.M. Parmentier and J. Lin, Mechanism for the origin of mid-ocean ridge axial topography: implications for the thermal and mechanical structure of accreting plate boundaries, J. Geophys. Res., 92, 12823-12836, 1987.

Pilkington, M. and D.J. Crossley, Determination of crustal interface topography from potential fields, Geophysics, 51, 1277-1284, 1986.

Regan, R.D., J.C. Cain, and W.M. Davis, A global magnetic anomaly map, J. Geophys. Res., 80, 794-802, 1975.

Ridgeway, J.R. and W.J. Hinze, MAGSAT scalar anomaly map of South America, Geophysics, 51, 1472-1479, 1986.

Sclater, J.G., B. Parsons and C. Jaupart, Oceans and continents: Similarities and differences in mechanisms of heat loss, J. Geophys. Res., 86, 11535-11552, 1981.

Scrutton, R.A., Modelling of magnetic and gravity anomalies at Goban Spur, Northeast Atlantic, in: P.C. de Graciansky, C.W. Poag, et al., Initial Reports DSDP, 80, Washington (U.S. Government Printing Office), 1141-1151, 1985.

Shih, K.G., W. Kay, J.M. Woodside and H.R. Jackson, Crustal thickness map, appendix 1, in: The Geology of North America: Geology of the Continental Margin off eastern Canada, M.J. Keen and G.L. Williams (eds.), Dec. N. Am. Geol. (DNAG) Ser., in press.

Srivastava, S.P., Evolution of the Labrador Sea and its bearing on the early evolution of the North Atlantic, Geophys. J. R. astr. Sco., 52, 313-357, 1978.

Srivastava, S.P., Davis Strait: Structures, Origin and Evolution, in: Structure and Development of the Greenland-Scotland Ridge, M.H.P. Bott, S.H. Saxov, M. Talwani and J. Tiede (eds.), Plenum. Publ. Corp., 159-187, 1983.

Srivastava, S.P., Geophysical maps and geological sections of the Labrador Sea, Geol. Surv. Can. Paper 85-16, 11 p, 1986.

Srivastava, S.P. and C.R. Tapscott, Plate Kinematics of the North Atlantic, in: The Geology of North America: The Western Atlantic Region, P.R. Vogt and B.E. Tucholke (eds.), Dec. N. Am. Geol. (DNAG) Ser., vol. M, 379-404, 1986.

Srivastava, S.P., J. Verhoef and R. Macnab, Results from a detailed aeromagnetic survey across the northeast

Newfoundland margin, part I: spreading anomalies and the relationship between magnetic anomalies and the ocean continent boundary, Marine and Petroleum Geology, in press.

Stockmal, G.S., S.P. Colman-Sadd, C.E. Keen, S.J. O'Brien and G. Quinlan, Collision along an irregular margin: a regional plate tectonic interpretation of the Canadian Appalachians, Can. J. Earth Sci., 24, 1098-1107, 1987.

Tucholke, B.E., Structure of the basement and the distribution of sediments in the western North Atlantic, in: The Geology of North America: The western North Atlantic Region, P.R. Vogt and B.E. Tucholke (eds.), Dec. N. Am. Geol. (DNAG) Ser., vol. M, 331-341, 1986.

Umpleby, D.C., Geology of the Labrador Shelf, Geol. Surv. Canada, Pap. 79-13, 34 p., 1979.

Urquhart, W.E.S. and D.W. Strangway, Interpretation of part of an aeromagnetic survey in the Matagami area of Quebec, in: The Utility of Regional Gravity and Magnetic Anomaly Maps, W.J. Hinze (ed.), Society of Exploration Geophysicists, 426-438, 1985.

Verhoef, J., B.J. Collette, P.R. Miles, R.C. Seale, J.-C. Sibuet and C.A. Williams, Magnetic anomalies in the northeast Atlantic ocean (35-50° N), Mar. Geophys. Res., 8, 1-25, 1986.

Verhoef, J. and R. Macnab, Magnetic Data over the Continental Margin of Eastern Canada: Preparation of a data base and construction of a 1:5,000,000 magnetic anomaly map, Geol. Surv. Canada, Open File 1504, 1988.

Vint, B.D., V.I. Pochtarev and R. Sh. Rakhmatulin, Method of computing the geomagnetic field upward in near-earth space, Geomagn. and Aeron., 10, 90-98, 1970.

Wade, J.A., A.C. Grant, B.V. Sanford and M.S. Barss, Basement structures, eastern Canada and adjacent areas, Geol. Surv. Canada, Map 1400A, 1977.

Wasilewski, P.J., H.H. Thomas and M.A. Mayhew, The Moho as a magnetic boundary, Geophys. Res. Lett., 6, 541-544, 1979.

Williams, H. and R.D. Hatcher Jr., Appalachian suspect terranes, in: Contributions to the tectonics and geophysics of mountain chains, R.D. Hatcher Jr., H. Williams and I. Zietz (eds.), Geol. Soc. Am., Memoir 159, 33-53, 1983.

INVERSION OF MAGNETIC AND GRAVITY DATA IN THE INDIAN REGION

B. P. Singh, Mita Rajaram and N. Basavaiah

Indian Institute of Geomagnetism
Colaba, Bombay 400 005 India

Abstract. The study is based on the premise that long-wavelength magnetic field anomalies can characterize the nature of the lower crust. Magnetisation obtained from inversion of the MAGSAT Z-anomaly clearly demarcates the five Archean-Proterozoic foldbelts that form the basement of the Indian platform. The difference in character of these five foldbelts appears to extend into the lower crust. Of the five, the three oldest ones are areas of positive magnetisation contrasts, and the remaining two are areas of negative magnetisation. The earlier suggestion of a lowering of the Curie isotherm beneath the Dharwar craton seems correct. That the Peninsular shield, the Ganga basin and the Himalayas are three different geotectonic blocks is clearly reflected in the magnetisation distribution. Flexuring of the crust under the Ganga basin due to root formation of the Himalayas is supported by the study. Areas of negative magnetisation over the south Indian region seem to be associated with a thin crust, high heat flow or seismotectonic activity. In general, gravity anomaly and heat flow data agree characteristically well with the magnetisation distribution.

Introduction

The evolution of the crust of the Indian subcontinent has resulted from processes that span the period Archean to Neogene, i.e. ranging in age from the present to 3.8 Ga. Geologically the landmass is divided into three major units: the Peninsular shield, the Indo-Gangetic plain and the Himalayan mountain system. The Peninsular shield is considered to consist of three cratons: the Dharwar, the Aravalli and the Singhbhum. The Narmada-Son and Godavari grabens demarcate these three protocontinents (Figure 1). (Singhbhum forms a part of the Satpura folding lying to the east of the Mahanandi graben). Eremenko and Negi (1968) suggest that the base of the Indian platform is constituted by five Archean-Proterozoic folded regions: the Dharwar, Aravalli, Eastern Ghat, Satpura and Delhi foldings (Figure 1).

Anomalies in gravity and magnetic field over the subcontinent at satellite height are characteristically different over the three major geological provinces (Qureshy and Midha, 1986). Such a division is not so clear in the ground data because of the presence of both short-and long-wavelength components. Features of short-wavelength arise from near-surface structures like faults, sedimentary basins, intrusive bodies, etc.; whereas long-wavelength features arise from changes in composition, thermal state or thickness of the crust and mantle. Space-borne measurements are an effective means of studying the long-wavelength part independently since they contain very little contribution from short-wavelength components. The latter get totally attenuated at satellite heights. The lower crust being the major source of long-wavelength magnetic anomalies (Wasilewski and Mayhew, 1982), space-borne measurements of the magnetic field are of relevance to studies related to properties and processes of the lower crust. However, the magnetic anomaly alone does not uniquely determine the crustal structure, but in conjunction with the gravity anomaly it will provide a better estimate. One must remember that sources of the gravity anomaly can extend down into the mantle as against the magnetic sources that are limited to the depth of the Curie isotherm, i.e. the Moho, in general. Even then, their correlation or non-correlation has a diagnostic value. Knowledge of the long-wavelength part from satellite data has also made the ground data more useful. With analytic continuation this part can be removed from the ground data thereby accentuating the short-wavelength features and making the estimates of shallow features more realistic.

The present study deals with the analysis of data collected with MAGSAT (Magnetic Field Satellite) launched by NASA, U.S.A. Efforts have been made to use this data for studies related to the composition and thermal state of the lower

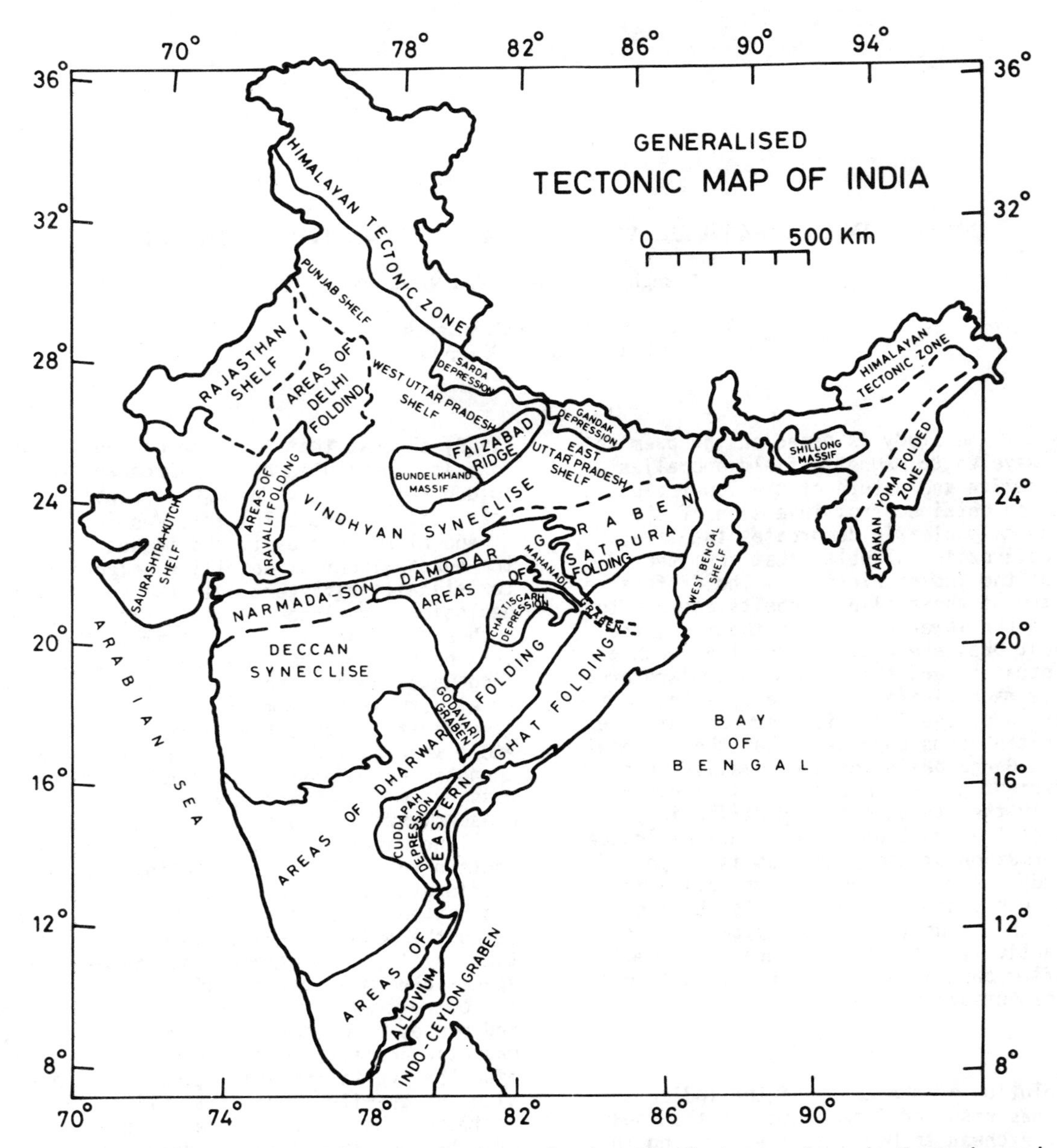

Fig. 1. Tectonic map of India redrawn to emphasize the features given in Eremenko and Negi (1968).

crust in India. Gravity anomalies at ground level and satellite heights along with heat flow data (wherever available) have also been examined to supplement the findings.

The Method

The mantle will not contribute to magnetic anomalies, if we assume that the Curie isotherm is reached near the base of the lower crust. The Moho and Curie isotherm may not be coincident everywhere but such exceptions are not many. Sources in the upper crust mainly produce short-wavelength anomalies. Unless their spatial structures are such that they coalesce into an anomaly of longer wavelength, these do not significantly interfere with satellite measurements. From the informations available on the character of the ground anomaly in Qureshy and Midha (1986), one can rule out such a possibility in India. Thus, we can work with the assumption of Wasilewski and Mayhew (1982) that MAGSAT anomalies are from sources in the lower crust. The zoning of crustal blocks directly with the anomaly map is difficult here, because the angle of inclination (I) changes from zero over the southern tip of the peninsula to about 45° over the Himalayas. One could circumvent this limitation by calculating the magnetisation of the crust. This was done through the equivalent source technique of Mayhew et al. (1980) wherein an array of dipole sources is placed on the Earth's surface such that their moments collectively give rise to a field which makes a least-squares best fit to the observed anomaly. We assume that the magnetisation arises from induction only, i.e. the dipoles have been assumed to be magnetised in the direction of the main field. The moment calculated in this manner being proportional to the product of thickness of the magnetised crust and the effective susceptibility of the constitutent rocks, becomes a useful parameter to study the nature of the crust.

The method for calculating the dipole moment is described in Bapat et al. (1987) and is based on the ridge-regression approach. Moments were calculated for 2° x 2° blocks using anomalies averaged over 1° x 1° blocks. The studied area extends from 6°S to 38°N and 60°E to 100°E. We give here in Figure 2 the moments derived from the inversion of anomalies in the vertical component (Z). This distribution is slightly different from that of Bapat et al. (1987) because in the latter work the whole area was split into six overlapping blocks of 20° x 20° size, whereas in the present case the whole of 46° x 40° was inverted as one block. We consider the results given here to be more reliable since the edge effects are expected to be much less in our inversion procedure.

Magnetisation alone cannot reveal much on the nature of the lower crust, as the quantity is a function of both the thickness and susceptibility. However, if one could find the thickness of the crust independently, by other means the susceptibility could be inferred. Some estimates of the crustal thickness have been made by Qureshy (1970) from Bouguer gravity anomaly maps. The presence of a long-wavelength component (Qureshy and Midha, 1986) necessitates its removal from ground data to arrive at an estimate of crustal thickness. This component was derived from the gravity anomaly map for a height of 400 km prepared by Prof. R.H. Rapp of Ohio State University, U.S.A. Using the equivalent point source technique, a mass distribution at a depth of 600 km was selected such that it reproduces the gravity anomaly at the height of 400 km again in the sense of least squares. The choice of the depth in the vicinity of 600 km is not so critical, but we selected this value on the consideration that lateral heterogeneities in mass distribution are confined largely up to the base of the upper mantle, which is at about 700 km depth. The expression for the field produced by

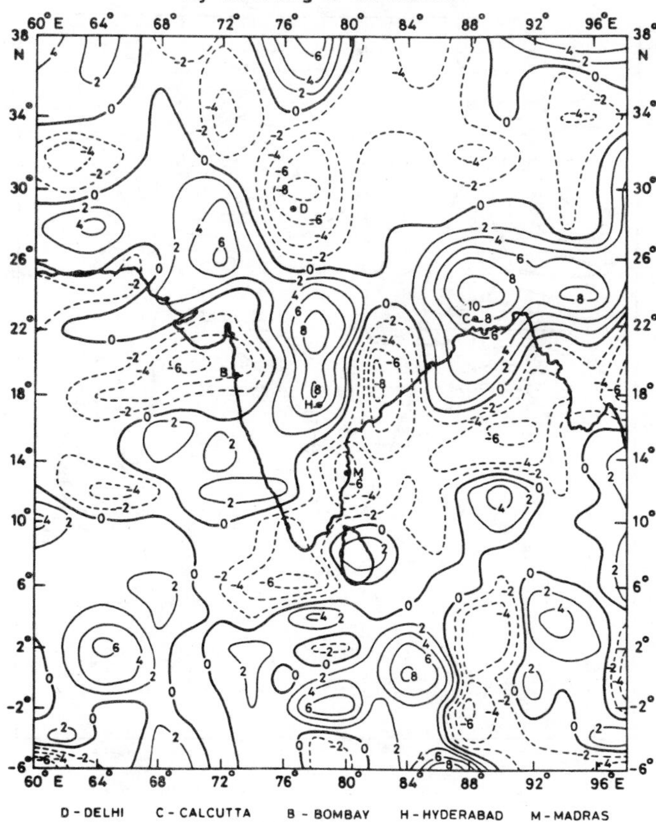

Fig. 2. Magnetic moment (x 10^{14} Am^2) distribution of 2° x 2° blocks for India and its contiguous regions calculated by inverting anomalies in the vertical component of the magnetic field (Z), observed at MAGSAT height.

such an equivalent mass distribution is given in von Frese et al. (1981). Sources were distributed at 2° x 2° intervals and their masses were calculated so as to reproduce the gravity anomalies averaged for the 1° x 1° blocks at 400 km above the surface of the Earth. In this case, too, we find the normal least-squares solution to be unstable. Like the magnetic case (Bapat et al., 1987) stable solutions were obtained through the method of ridge-regression. When the damping factor (k) is 0.2 or above the values of masses stabilise and do not change with respect to k. Hence we accepted the results with k = 0.2 as a reasonable estimate of the mass distribution. The gravity field at the ground level estimated with this mass distribution is given in Figure 3. A well-pronounced minimum over the Indian ocean geoid low and a maximum over the Himalayas are the two characteristic features of the field. We termed these the long-wavelength component and removed them

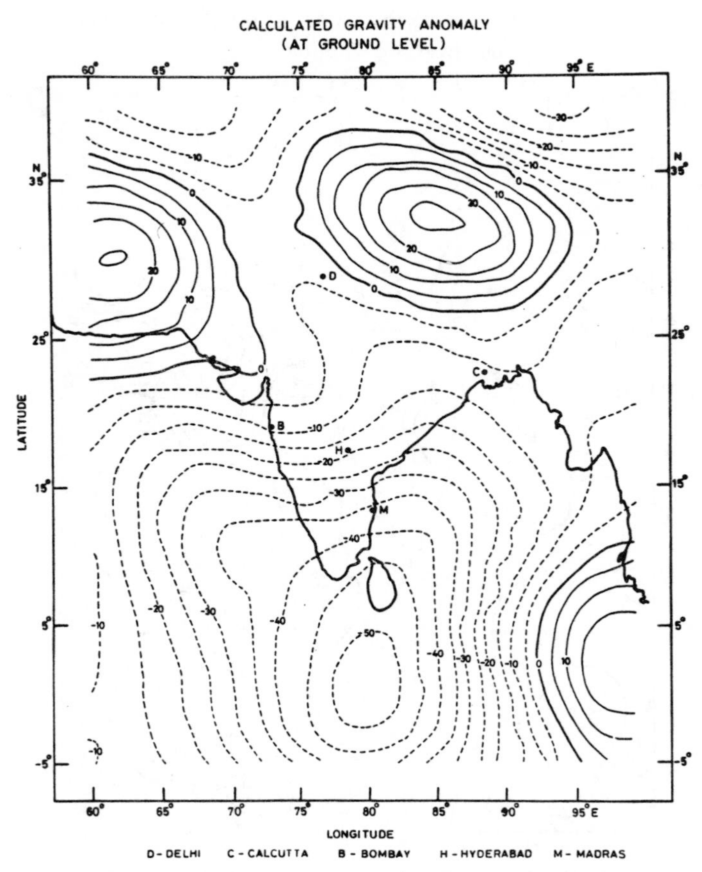

Fig. 3. Gravity field anomaly (in mGal) at the surface of the Earth recalculated from the equivalent mass distribution obtained by inverting gravity data at a height of 400 km.

from the Bouguer anomaly map given in Qureshy and Midha (1986). The resulting residuals are plotted in Figure 4 and have been named the modified Bouguer anomaly map after removal of the long-wavelength component. This map is expected to reflect the crustal features more realistically.

Results and Discussion

The geotectonic structure of the sub-continent can be classified into three cratons (Dharwar, Aravalli and Singhbhum; see Figure 1) and two mobile belts, i.e. Eastern Ghats and the areas of Delhi folding. The three cratons appear in Figure 2 as areas of positive magnetisation and the two mobile belts as areas of negative magnetisation. Theoretically the induced magnetisation should be positive everywhere; however, this does not happen because the calculated values are essentially relative contrasts between different blocks. A negative magnetisation in this sense signifies one or more of: thin crust, crustal materials of low susceptibility or higher crustal temperature. In this context the results are in general agreement with the heat flow results given in Gupta (1982). The three cratons have low heat flow values and positive magnetisation, whilst the mobile belts have high heat flow and low (negative) magnetisation values.

The results correlate equally well with known variations in crustal thickness. Gupta (1982) mentions that deep seismic sounding (DSS) and earthquake studies indicate that the Moho beneath the Peninsula is relatively flat with a depth of 37 ± 3.5 km. Qureshy (1970) from the study of gravity anomalies and Narain (1973) from the study of body wave travel times, surface wave dispersion and gravity data report that the crust is 35-40 km thick in the Peninsular shield, 30-35 km in the Indo-Gangetic plains and 60-80 km in the Himalayan-Tibetan plateau. Conforming to the characteristics defined earlier, in the Indo-Gangetic plains, where the crust is thinnest, the magnetisation is negative. Both to the north and south of it, the magnetisation becomes higher (positive) as the crust thickens. However, an examination of the magnitude of the maxima and

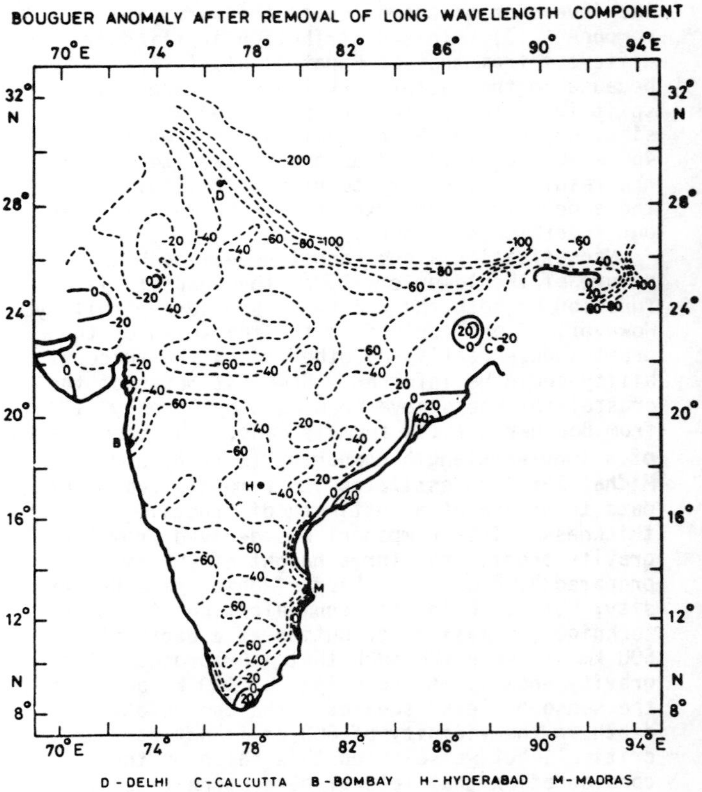

BOUGUER ANOMALY AFTER REMOVAL OF LONG WAVELENGTH COMPONENT

D - DELHI C - CALCUTTA B - BOMBAY H - HYDERABAD M - MADRAS

Fig. 4. Bouguer anomaly map of India obtained after removal of the long-wavelength component in Fig. 3.

minima of the magnetic moments (Figure 2) over the Peninsular shield reveals that such a large variation in their magnitude cannot be accounted for by the small (~10%) change of crustal thickness. Either the composition or the thermal state or both must vary significantly in the lower crust to account for the magnetisation characteristics. The thermal state seems to be a plausible cause, because Gupta (1982) has noticed a variation in heat flow from 26 mWm^{-2} to 107 mWm^{-2}. Compositional changes are equally likely, because the evolution of the Indian platform has taken place through varied tectonic activities. Vertical movements, reactivation of ancient faults and redistribution of masses must have caused a transfer of material between crust and mantle leading to lateral gradients in the composition of the crust.

Amongst the three cratons, the magnetisation should have been maximum in the Dharwar because of its largest crustal thickness and lowest heat flow. On the other hand, the maximum overlies the Singhbhum. Reduced magnetisation intensity over the Dharwar could be associated with the composition of its lower crust. Negi et al. (1987) mention that the presence of titanoferrous magnetite intrusions reported by Jafri et al. (1983) have caused a lowering of the Curie temperature to as low a value as 350°C. Such a lowering of the Curie isotherm will substantially reduce the thickness of the magnetised crust and consequently the total magnetisation of the crust in Dharwar. In the Aravalli, the comparatively reduced magnetisation is due to the crust being thin there (Qureshy 1970, Rajaram and Singh, 1986). A thin crust in this area is also evident from the Bouguer anomaly map (Figure 4). The maximum of magnetisation in Singhbhum is postulated to arise from a gabbroic intrusive body underneath. Such a hypothesis is further supported by the fact that the Bouguer anomalies over Singhbhum and Aravalli are similar (Verma and Subrahmanyam, 1984). Verma et al. (1986) have postulated the presence of a gabbroic intrusive body in the lower crust of the Aravalli to explain the observed gravity anomalies.

Negative magnetisation (Figure 2) over the areas in between the three cratons are related to high heat flow, thin crust, sedimentary deposits or down flexuring of the crust. A comparison of Figure 2 and the seismo-tectonic map of the South Indian region (Figure 5) published by Narain and Subrahmanyam (1986) brings out a one-to-one correspondence between areas of negative magnetisation and seismically active zones. For example, the Eastern Ghats, the Kerala-Tamil Nadu-Karnataka granulite terrain, the Panvel flexure and the Cambay basin are all seismically active and negatively magnetised. Evidence for a thin crust in the Eastern Ghats is seen in the Bouguer anomaly map (Figure 4). A high heat flow (55 mWm^{-2}) at Karadikuttam (Gupta et al., 1987) has been

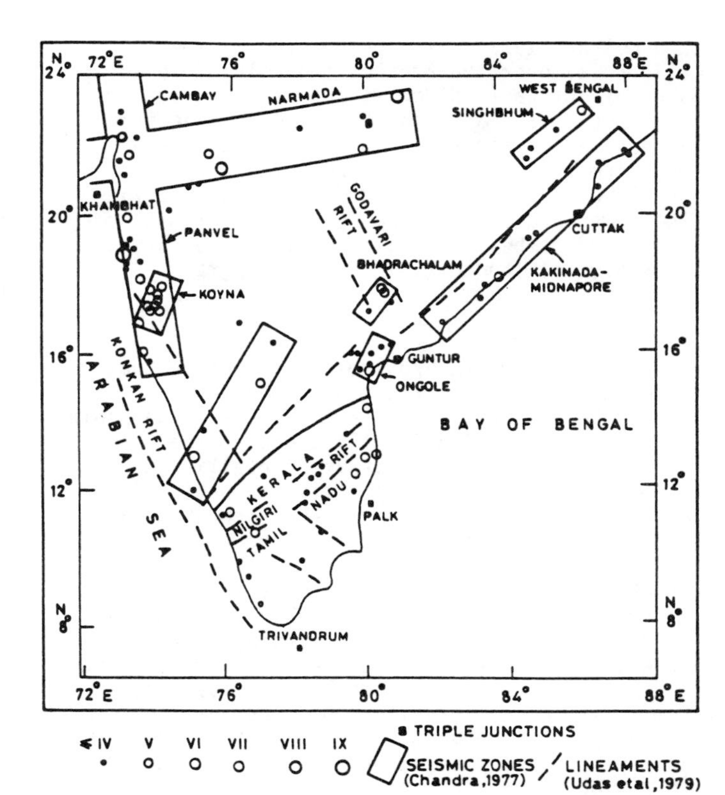

Fig. 5. Seismotectonic map of South India with epicentres, their intensities and seismic zones taken from Chandra (1977); lineaments (after Udas et al., 1979) and rift zones are also shown. The map is taken from Narain and Subrahmanyam (1986).

observed in the Kerala-Tamil Nadu-Karnataka granulite terrain. The Khambhat hot spot, a gravity high north of Bombay (denoted by B in the figure) and over the Cambay basin are evidence of a thin crust in this region. Unusually high temperatures have been encountered in the Cambay oil field (Krishnaswamy, 1981). The areas of negative magnetisation over the South Indian region thus seem to be associated with a thin crust, high heat flow and/or seismotectonic activity.

Negative magnetisation over the Eastern Ghats continues into the Gondwana basins that separate the Dharwar and Singhbhum cratons. Measured heat flow in the area ranges from 49 to 107 mWm^{-2} with a mean value of 75 mWm^{-2} (Gupta 1982). The tectonic activity of the region (Godavari rift in Figure 5) is also high. Both high heat flow and tectonic activity, correspond well with the negative magnetisation of the region. Negative magnetisation over the Ganga basin is significant in the sense that it supports thick sedimentary fills and a crustal downwarp resulting from the root formation underneath the Himalayas (Warsi and Molnar, 1977). The pronounced minimum (Figure 2) over the area of the Delhi folding is also of interest, since it appears as a low in the Bouguer anomaly map (Figure 4). A negative magnetisation and a low in the gravity field reaffirm the presence of a conductive zone in the lower crust (Arora and Mahashabade, 1987). The lower crust under the Delhi foldbelt is concluded to be in a state of high temperature, possibly related to rising mantle magma.

Conclusions

The five Archean-Proterozoic foldbelts that form the basement of the Indian platform (Eremenko and Negi, 1968) are clearly discernible on the magnetisation map (Figure 2). The three oldest amongst these (Dharwar, Aravalli and Singhbhum) are areas of positive magnetisation, whereas the two younger ones (Eastern Ghats and areas of the Delhi folding) are negatively magnetised. The Dharwar, though being the thickest and oldest crust, is not the strongest magnetised region. The presence of a titanoferrous magnetite intrusion is considered to have lowered its Curie temperature to as low as 350°C and hence decreased the total magnetisation. Similarity of the gravity anomalies over the Aravalli and Singhbhum has been postulated earlier to indicate the presence of a large-scale intrusion of gabbros in the lower crust of Singhbhum. This is confirmed by a relatively high intensity of magnetisation seen in these areas.

Negative magnetisation over the Eastern Ghats, the Tamil Nadu-Karnataka-Kerala (TTK) granulite terrain, the Panvel flexure and the Cambay basin are due to a thin crust and seismotectonic activity. Furthermore, the Cambay basin and the TTK are also areas of high heat flow. A similar situation also exists under the Gondwana basins. The broad negative magnetisation over the Indo-Gangetic plain supports sedimentation on top and downwarp of the crust. The temperature of the lower crust under the Delhi folding is postulated to be high. The Himalayas, being a tectonically active zone, still have a positive magnetisation. This is so because the region has a large crustal thickness arising from the subduction of the Indian crust under the Eurasian crust.

The present work has highlighted the utility of space-borne measurements of the magnetic field in the study of the nature and properties of the lower crust. This approach becomes more useful when undertaken jointly with available gravity and heat flow data. We conclude, by repeating Qureshy and Midha (1986), that the Indian shield despite its antiquity is not as stable and rigid as is usually presumed.

Acknowledgements. We wish to thank the National Aeronautics and Space Administration (NASA), Goddard Space Flight Centre (GSFC), for supplying us the MAGSAT data, and Prof. R.H. Rapp of Ohio State University, U.S.A., for providing gravity anomaly maps. To M.N. Qureshy we are extremely grateful for many useful discussions and help in completing the work. We thank our colleagues B.R. Arora and Nandini Nagarajan for critically reading the manuscript and our Director, R.G. Rastogi, for his continued interest in this work. Comments of the two referees have greatly improved both the content and presentation of the paper and we sincerely thank them for their suggestions.

References

Arora, B.R. and Mahashabde, M.V. : A transverse conductive structure in the northwest Himalaya Phys. Earth Planet. Inter., 45, 119-127, 1987.

Bapat, V.J., Singh, B.P. and Rajaram, Mita, : Application of ridge-regression in inversion of low-latitude magnetic anomalies derived from space measurements. Earth Planet Sci. Lett., 84, 277-284, 1987.

Chandra, U. : Earthquakes of Peninsular India - a seismotectonic study. Bull. Seis. Soc. Amer., 67, 328-332, 1977.

Eremenko, A.N. and Negi, B.S. (Chief Editors), : Tectonic Map of India. Oil and Natural Gas Commission, Dehradun, India, 1968.

Gupta, M.L. : Heat flow in the Indian peninsula - its geological and geophysical implications. Tectonophysics, 83, 71-90, 1982.

Gupta, M.L., Sharma, S.R., Sunder, A. and Singh, S.B. : Geothermal studies in the Hyderabad granitic region and the crustal thermal structure of the Indian shield. Tectonophysics, 140, 257-264, 1987.

Jafri, S.H., Khan, N., Ahmed, S.N. and Saxena, R.: Geology and Geochemistry of Nuggihalli Schist belt, Dharwar craton, Karnataka, India, in Precambrian of South India, eds., Naqvi, S.M. and Rogers, J.J.W., Geol. Soc. India, Memoir, 4, 110-120, 1983.

Krishnaswamy, V.S. : The Deccan volcanic episode, related tectonics and geothermal manifestations, in Deccan Volcanism and Related Basalt Provinces in Other Parts of the World, eds., K.V. Subha Rao and R.N. Sukeshwala, pp.1-7, 1981.

Mayhew, M.A., Johnson, B.D. and Langel, R.A. : An equivalent source model of the satellite altitude magnetic anomaly field over Australia. Earth Planet. Sci. Letts., 51, 189-198, 1980.

Narain, H. : Crustal structure of Indian subcontinent. Tectonophysics, 20, 249-260, 1973.

Narain, H. and Subrahmanyam, C. : Precambrian tectonics of the South Indian Shield inferred from geophysical data. Jour. Geology, 94, 187-198, 1986.

Negi, J.G., Agarwal, P.K. and Pandey, O.P. : Large variation of Curie depth and lithospheric thickness beneath the Indian sub-continent and a case for magnetothermometry. Geophys. J.R. astr. Soc., 88, 763-775, 1987.

Qureshy, M.N. : Relation of gravity to elevation,

geology and tectonics in India. Proc. Second Upper Mantle Symposium, Hyderabad, India, 1-23, 1970.

Qureshy, M.N. and Midha, R.K. : Deep crustal signatures in India and contiguous regions from satellite and geophysical data, in Reflection Seismology: The continental crust, M. Barazangi and L. Brown, eds., Am. Geophys. Union, Geodyn. Ser., 14, 77-94, 1986.

Rajaram, Mita and Singh, B.P. : Spherical earth modelling of the scalar magnetic anomaly over the Indian region. Geophys. Res. Lett., 13, 961-964, 1986.

Udas, G.R., Narayana, B.L., Das, D.R. and Sharma, C.V. : Carbonatites of India in relation to structural setting. Geol. Surv. India, Misc. Publ., 34, pt. 2, 77-94, 1979.

Verma, R.K., Mitra, S. and Mukhopadhyay, M. : An analysis of gravity field over Aravallis and the surrounding region. Geophys. Res. Bull. 24, 1-12, 1986.

Verma, R.K. and Subrahmanyam, C. : Gravity anomalies and the Indian lithosphere: review and analysis of existing gravity data. Tectonophysics, 105, 141-161, 1984.

von Frese, R.R.B., Hinze, W.J. and Braile, L.W. : Spherical Earth gravity and magnetic anomaly analysis by equivalent point source inversion. Earth Planet. Sci. Lett., 53, 69-83, 1981.

Warsi, W.E.K. and Molnar, P. : Plate tectonics and gravity anomalies in India and the Himalaya. Collog. Int. C.N.R.S. Ecologie et Geologie de l' Himalaya, C.N.R.S., Paris, p. 269, 1977.

Wasilewski, P. and Mayhew, M.A. : Crustal xenolith magnetic properties and long wave-length anomaly source requirements. Geophys. Res. Lett., 9, 329-332, 1982.

GRAVITY FIELD, DEEP SEISMIC SOUNDING AND NATURE OF CONTINENTAL CRUST UNDERNEATH NW HIMALAYAS

R. K. Verma and K. A. V. L. Prasad

Indian School of Mines, Dhanbad, India

Abstract. A large number of earth scientists believe that the Himalayas have evolved as a result of collision of the Indian and Eurasian plates, sometime during Cretaceous to Eocene times [Gansser, 1964, 1977; Dewey and Bird, 1973; Powell and Conaghan, 1973]. As a result of the collision, an estimated crustal shortening of the order of at least 300 Km took place apparently through movements along major Himalayan thrusts, such as the Main Central Thrust (MCT), the Main Boundary Thrust (MBT) and several others. These movements have resulted in thickening of the crust to 65-75 Km as shown by Gupta and Narain [1967], Kaila et al. [1982, 1984]. A vital question concerning the Himalayas is how the crustal thickening has taken place.

An analysis of the gravity field in the Northwestern Himalayas and the Kohistan region has been carried out in order to throw some light on this problem. The results of this study have peen presented in detail by Verma and Prasad [1987]. A brief summary is given herewith. Figure 1 shows a simplified geological map of the NW Himalayas and the Kohistan regions. The figure shows the location of major geological formations such as the Upper and Lower Tertiary sediments (the siwaliks) constituting the Lower Himalaya and foothills, the Mesozoic and Paleozoic formations located in the Higher Himalaya, the Panjal trap forming the Pir Panjal range, the tourmaline granite in Ladakh and the Kohistan sequence (KS), mostly consisting of 10-15 Km thick complex of hypersthene gabbro, amphibolites, diorite, hornblende gneisses etc. located between the Main Mantle Thrust (MMT) and the Main Karakoram Thrust (MKT) in NW Pakistan, west of Nanga Parbat [Brookfield and Reynold, 1981]. Alluvium covers the Panjab plains and the Kashmir Valley. Location of the Main Boundary Thrust (MBT) and Main Central Thrust (MCT) is also shown in the figure. Location of two gravity profiles passing from Gujranwala in the south towards Nanga Parbat-Haramosh massif (Profile AA') and Gujranwala to Ghizar in Kohistan region (Profile CC') is also shown.

Figure 2 shows the Bouguer anomaly of Himalaya-Hindukush region prepared using data from Gulatee [1950], Farah et al. [1977], Ebblin et al. [1983] and Chugh [1978, SOI unpublished report]. It is apparent from the figure that the Bouguer anomaly is nearly -50 mGal near the edge of the Indian shield and reaches values of the order of -400 to -450 mGal near Nanga Parbat and Haramosh massif. To the west of the Nanga Parbat a major gravity high of about +100 mGal amplitude is located over the Kohistan region. A Deep Seismic Sounding (DSS) profile shot in Nanga Parbat region under the Indo-Soviet Geodynamics Project [Beloussov et al., 1980, Kaila et al., 1982, 1984] has provided vital information concerning the P wave velocity vs. depth relationship and the configuration of the Moho in the region. Location of the DSS profile is shown in Fig. 2.

Using the velocity information available along different parts of the profile, the density of the lower crust has been determined from the relationship given by Ludwig, Nafe and Drake [1970]. A schematic geological section along profile AA', the Moho configuration obtained from the DSS [Kaila et al., 1984] and density of the lower crust inferred from DSS data is shown in figure 3.

The Bouguer anomalies observed along the profile have been interpreted satisfactorily as shown in Fig. 4 using the Moho configuration and the density contrast between the crust and the mantle as shown in the Figure 3. It may be observed that the Moho increases from a depth of about 30 Km near Gujranwala to about 40 Km near the Murree thrust, increasing further to 60-65 Km over the Higher Himalayas, north of the Kashmir Valley. The Moho has an upward dip near Nanga Parbat but is downfaulted towards the Haramosh massif. The gravity model shows that the roots of the High Himalayas are made up of highly basic to ultrabasic rocks. The results do not support continental underthrusting as a possible mechanism of thickening of the crust, underneath the Higher Himalayas. In case crustal shortening has

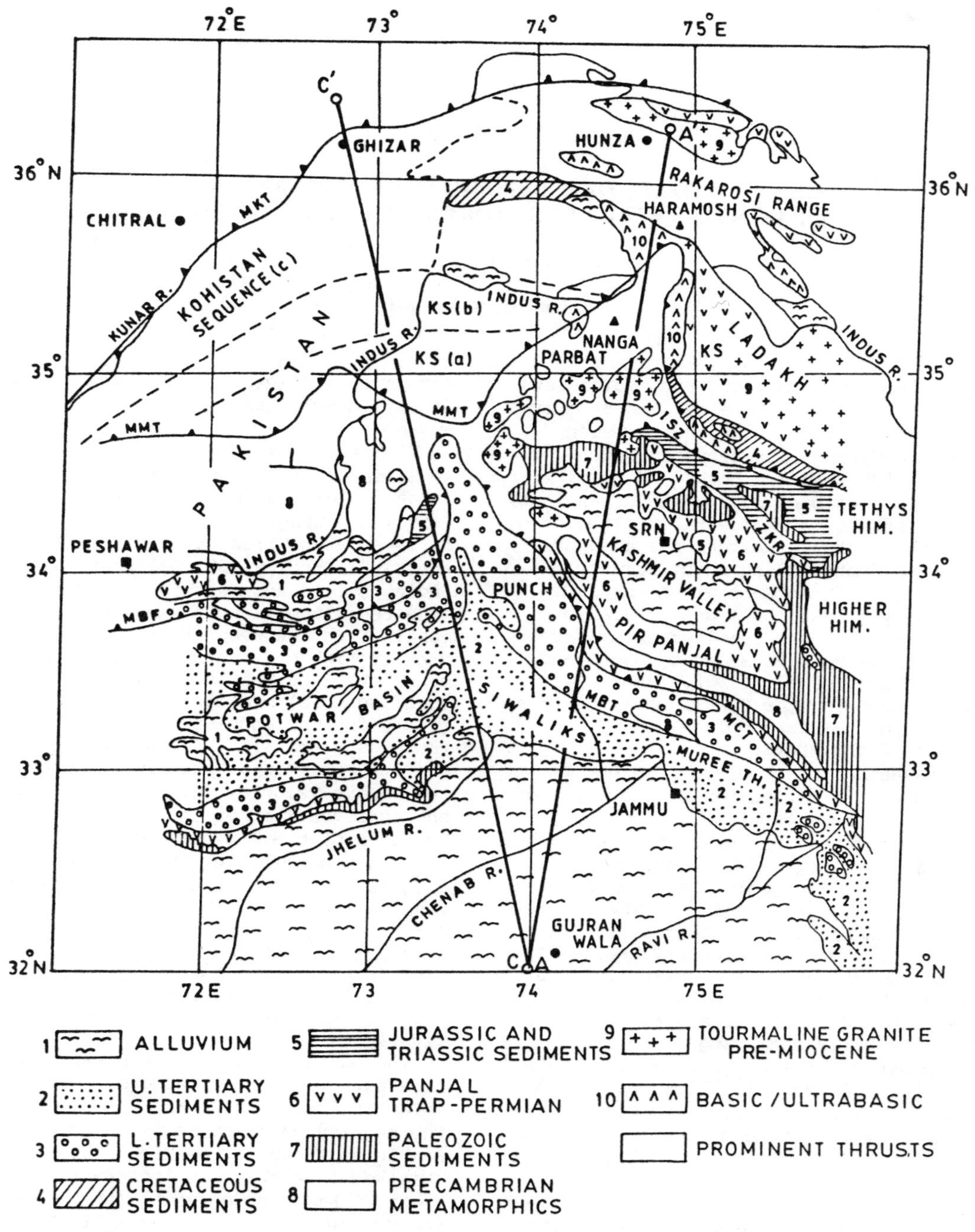

Fig. 1. Simplified geological map of NW Himalayas and Kohistan region, after Gansser (1964). Profiles AA' extending from Gujranwala to Haramosh massif and CC' from Gujranwala to Ghizar are also shown in the figure.
Abbreviations: ISZ - Indus Suture Zone, KS - Kohistan Sequence, MBT - Main Boundary Thrust, MCT - Main Central Thrust, MKT - Main Karakorum Thrust, MMT - Main Mantle Thrust, SRN - Srinagar, ZKR - Zanskar Range.

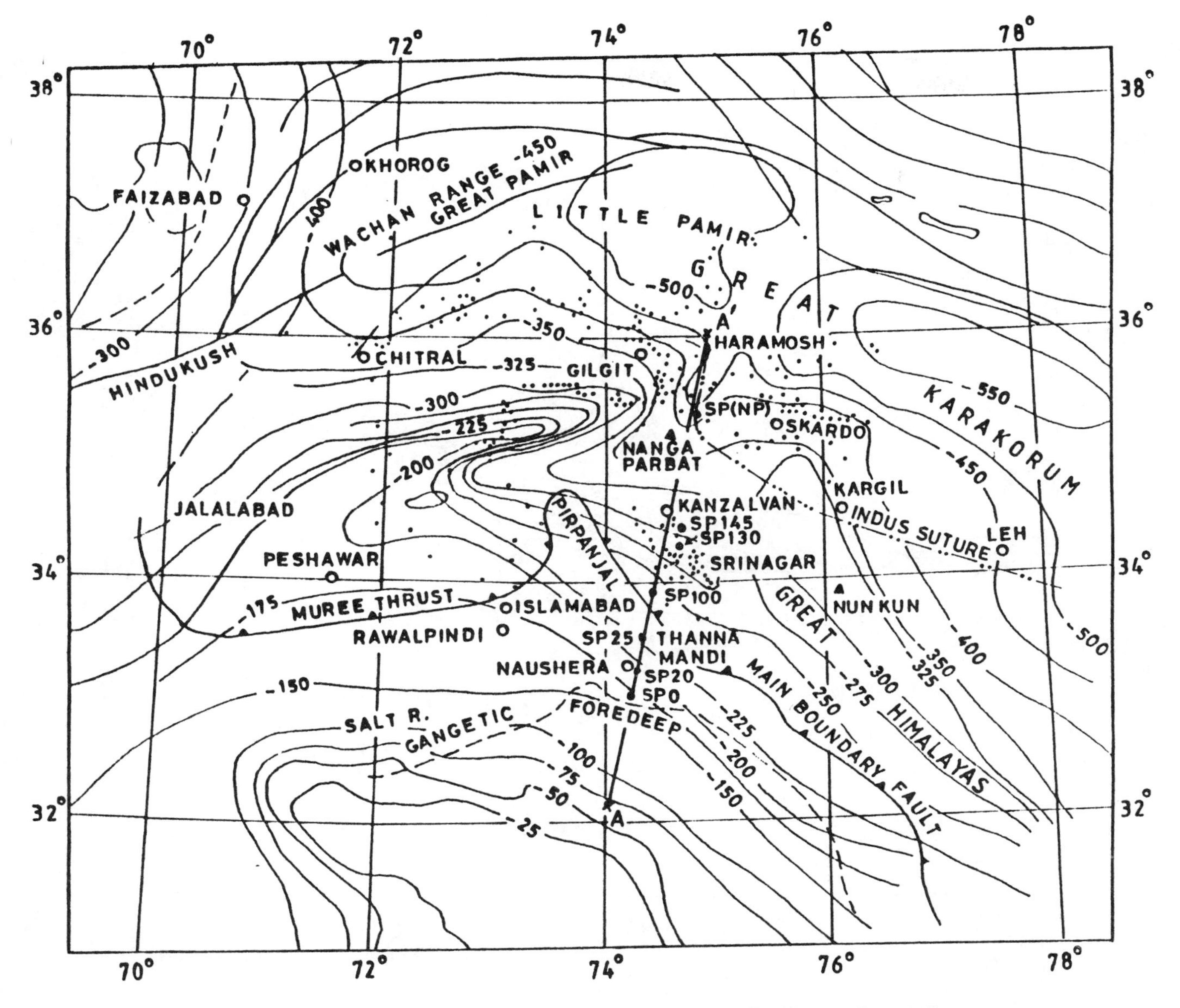

Fig. 2. Bouguer anomaly map of NW Himalaya-Hindukush region prepared from various sources including Gulatee (1950), Marussi (1964), Farah et al. (1977), Ebblin et al. (1983) and Chugh (1978). Only the stations in the Higher Himalaya given by Ebblin et al. (1983) have been plotted in the figure. Reduction density is 2.67 g/cm^3 above sea level.

taken place, the composition of the crust has been considerably changed with the addition of basic materials from the upper mantle.

Interpretation of the gravity profile CC' in terms of a two-dimensional model is shown in Figure 5. No DSS information is available along this profile. The Moho assumed for the model is based on average elevation vs. Moho relationship obtained for the Nanga Parbat profile discussed above. In order to explain a major gravity high

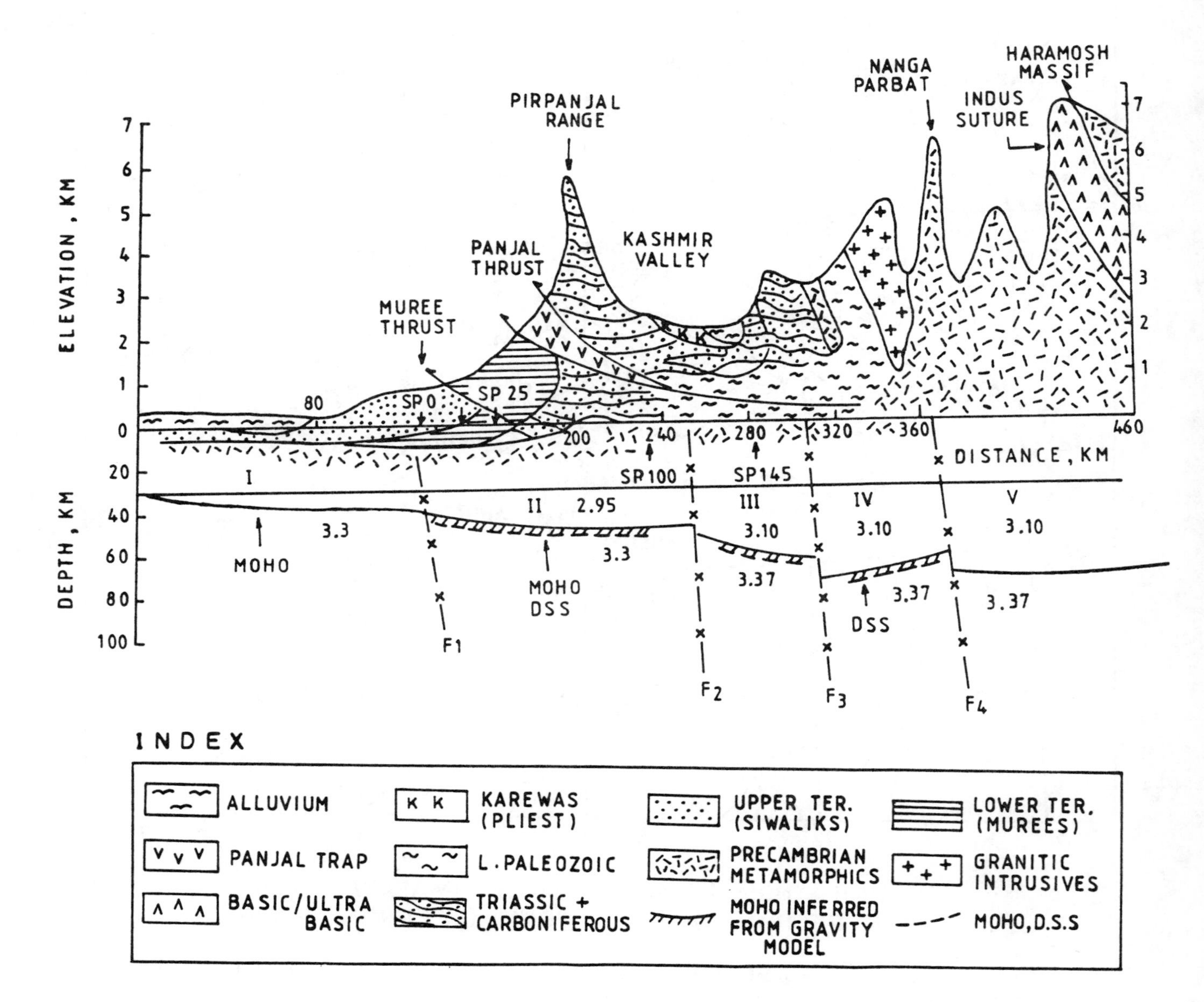

Fig. 3. Geological section along profile AA' (after Wadia 1975). Moho assumed for gravity interpretation and that obtained from DSS is also shown in the figure. See text for explanation.

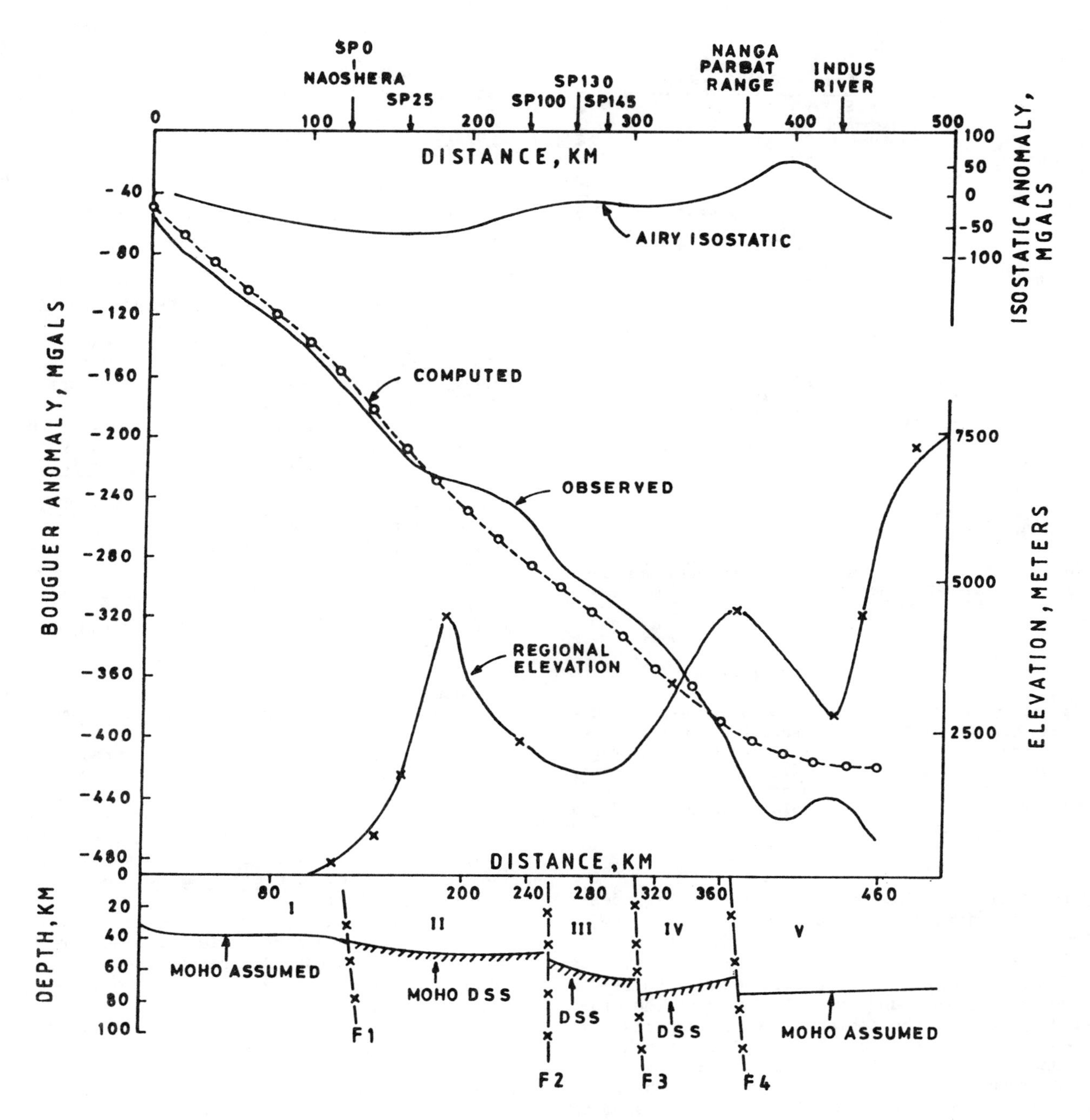

Fig. 4. Observed and computed gravity values for model shown in Fig. 3 along profile AA', using density contrasts between crust and the mantle for different blocks of the Himalayas as shown in the figure. The Airy isostatic anomaly (T = 30 Km) and the regional elevation are also shown in the figure.

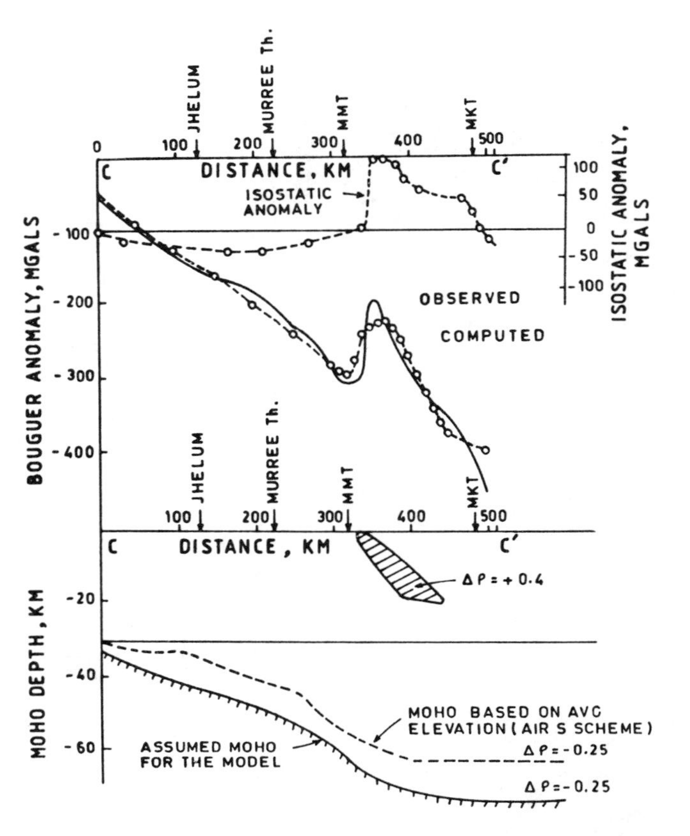

Fig. 5. Model for root formation underneath the Kohistan region (based on the general relationship between Moho and average elevation for the Nanga Parbat profile) is shown in the figure. A high density mass in the upper part of the crust, close to the MMT, is needed in order to explain the nature of gravity field observed.

present in the Kohistan region, a very high density intrusive, (perhaps a thrust slice of the lower crust or the upper mantle) close to the MMT is invoked as shown in Fig. 5. The intrusive has a density contrast of about 0.4 g/cm^3 with respect to its surroundings. Analysis of the gravity field over the Kohistan region supports the idea of collision of the Indian plate with an island arc, which existed between India and Eurasia, resulting in the emplacement of high density lower crust/upper mantle rocks up to a depth of about 20 Km along the MMT.

It appears from Figures 4 and 5 that the process of continental collision of India with Eurasia has been very different to the east of the Nanga Parbat than to its west.

References

Beloussov, V.V., Belyaevsky, N.A., Borisov, A.A., Volvovsky, B.S., Volvovsky, I.S., Rosvoy, D.P, Tal-Virsky, B.B, Khamrabaev, I., Kh., Kaila, K.L., Narain, H., Marussi, A., and Finetti, J., 1980. Structure of the lithosphere along the Deep Seismic Sounding profile. Tien Shan, Pamir, Karakorum, Himalaya, Tectonophysics, 70, 193-221.

Brookfield, M.E. and Reynold, P.H., 1981. Late Cretaceous emplacement of the Indus Suture zone ophiolite melange and an Eocene-Oligocene magmatic arc on the northern edge of the Indian place. Earth and Planet Sci. Lett. 55, 157-162.

Dewey, J.F. and Bird, J.M., 1970. Mountain belts and new global tectonics, J. Geophy. Res. 75, 2625-2647.

Ebblin, C., Marussi, A., Poretti, G., Rahim, S.M., and Richard, P., 1983. Gravity measurements in the Karakorum, Bulletino di Geofisica, Teor. Appl., 25, 303-315.

Farah, A., Mirza, M.A., Ahmed, M.A., Butt, M.H., 1977. Gravity field of the buried shield in the Panjab plains, Pakistan, Geol. Soc. Am. Bull., 88, 1147-1155.

Gansser, A., 1964. Geology of the Himalayas, Interscience Publishers, N.Y.

Gansser, A., 1977. The great suture zone between Himalaya and Tibet; A preliminary note. In : Himalaya, Science de la Terre. CNRS, 268, 181-192.

Gulatee, B.L., 1950. Gravity data in India. Tech. Paper, 10, Survey of India.

Gupta, H.K. and Hari Narain, 1967. Crustal structure in Himalayan and Tibet plateau region from surface wave dispension, Bull. Seism. Soc. Am., 57, 235-248.

Kaila, K.L., Roy Choundhury, K., Krishna, V.G., Dixit, M.M. and Hari Narain, 1982. Crustal structure of Kashmir Himalaya and inferences about the asthenosphere layer from DSS studies along the international profile, Qarrakol-Zorkol-Nanga-Parbat-Srinagar-Pamir Himalaya Monograph, Bulletino di Geofisica Teorica et Applicata, 25, 221-234.

Kaila, K.L., Tripathi, K.M. and Dixit, M.M, 1984. Crustal structure along Wular lake-Gulmarg-Naoshera profile across Pir Panjal Range of the Himalayas from deep seismic soundings, Jour. Geol. Soc. India, 25, 706-719.

Ludwig, W.J., Nafe, J.E. and Drake, C.L., 1970. Seismic refraction. The Sea, Vol. 4, Pt. I, pp. 53-81.

Powell, C.M. and Congahan, P., 1973. Plate tectonics and the Himalayas, Earth and Planet Sci. Lett., 20, 1-22.

Verma, R.K. and Prasad, K.A.V.L., 1987. Analysis of gravity field in Northwestern Himalayas and Kohistan Region using Deep Seismic Sounding data. Geophy. J. Roy. Astr. Soc., 91, 869-889.

Section V

THE LOWER CRUST FROM STUDIES OF GEOLOGICAL PROCESSES

GROWTH AND MODIFICATION OF LOWER CONTINENTAL CRUST IN EXTENDED TERRAINS: THE ROLE OF EXTENSION AND MAGMATIC UNDERPLATING

David M. Fountain

Program for Crustal Studies, Department of Geology and Geophysics
University of Wyoming, Laramie, Wyoming 82071

Abstract. The parameters of extension rate, amount of lithospheric extension (ß) and the temperature of the underlying asthenosphere exert a strong control on the evolution of the continental lower crust in extended terrains. The extreme case is represented by regions characterized by high extension rates, large ß and high asthenospheric temperatures. In this case, upwelling of asthenosphere causes accretion of mafic magmas to the base of the crust (magmatic underplating) thereby raising temperatures to values high enough to drive granulite facies metamorphism, anatexis, migmatization and to also increase the ductility of the lower crust during extension. The mafic magmas crystallize in the pyroxene granulite and/or garnet granulite facies. The resultant rocks are characterized by high compressional wave velocities similar to those observed in the lower crust in continental rift zones and passive continental margins. The Ivrea-Verbano zone in northern Italy, an exposed granulite facies terrain, may represent an example of this type of lower crust. There will be little growth or modification of the continental lower crust for low extension rates, small ß or low asthenosphere temperatures. Extended lower crust will exhibit a high degree of diversity reflecting the relative importance of these parameters.

Introduction

A predominant view of crustal evolution holds that the lower levels of continental crust form and evolve within the deeper regions of magmatic arcs either along continental margins or oceanic island arcs [e.g., Hamilton, 1981; Kay and Kay, 1987; Bohlen, 1987]. Others [Fountain and Salisbury; 1981; Dewey, 1986], however, suggest that lower crust evolves in differing tectonic environments. In fact, the densest seismic reflection and refraction coverage of the continental crust is within Phanerozoic extensional terrains such as the Basin and Range in the U.S. [e.g., Allmendinger et al., 1987] and the offshore regions of Great Britain [e.g., Cheadle et al., 1987]. Because of this coverage, seismologists delineated several interesting seismic characteristics of extensional terrains. Of particular note are the pronounced seismic reflectivity of the lower crust [e.g., Allmendinger et al., 1987], strong Moho reflections [e.g., Klemperer et al., 1986; McGeary, 1987] and the comparatively high P-wave velocities in the lowermost crust [e.g., Mooney et al., 1983; Pakiser, 1985]. These features, although not unique to extensional terrains, appear to be prevalent enough in these regions to require some unified explanation for their genesis.

Geophysical observations such as these have stimulated Earth scientists to more carefully consider the effects extensional processes have on the overall evolution of the lower crust. There is growing evidence that the lower crust of extensional terrains grows by magmatic additions [e.g., Lister et al., 1986; Okaya and Thompson, 1986; Gans, 1987] and that high grade metamorphism occurs in these regions [Dixon et al., 1981; Wickham and Oxburgh, 1985; Sandiford and Powell, 1986; Chapman, 1986; Frost and Frost, 1987]. Various contrasting theories concerning lower crustal deformational style in extension have also been advanced [Smith, 1978; Eaton, 1979; Wernicke, 1981; Hamilton, 1982; Miller et al., 1983; Kligfield et al., 1984; Smith and Bruhn, 1984].

In this paper, I review some of the current ideas about lower crustal evolution in extensional terrains with the intent of developing an integrated view of the nature of extended lower crust. This paper explores various thermal-mechanical models of extension and discusses the consequences of these models in terms of magmatic underplating, crustal metamorphism, magma genesis, lower crustal deformation and seismic properties. Finally, I discuss a possible exposed example of a deep crustal terrain that illustrates these processes. Although this paper is intended as a synthesis of diverse views of lower crustal evolution in extended terrains, it emphasizes the view that, in some regions, extensional lower crust grows by magmatic additions, experiences high grade metamorphism and is weak in regions characterized by high extension rates, large amounts of extension or high potential temperatures of the asthenosphere. Lower crust developed in regions characterized by low extension amounts or rates with little or no magmatic underplating are likely to exhibit different geophysical and geological characteristics. High lower crustal diversity can, therefore, be expected even in regions sharing a generally similar tectonic history.

Thermal-Mechanical Models of Extension

An important aspect in the evolution of the lower continental crust in extensional regimes is the temperature field as temperature is a first-order control of metamorphic reactions, anatexis and steady state creep in the lower crust. Estimation of thermal gradients in the lower crust depends upon the appropriate choice of thermophysical parameters (e.g., thermal conductivity, heat production) and, importantly, an assumption about the heat transfer mechanisms that operate in the crust [e.g., Chapman, 1986]. This section reviews the various types of heat transfer mechanisms postulated to operate in extensional terrains and presents examples of resultant geothermal gradients.

One-Dimensional Conductive Heat Transfer

Simple and commonly used thermal gradients for extensional terrains are calculated for a steady state conductive regime characterized by a high surface heat flow [Pollack and Chapman, 1977; Chapman, 1986] with no particular assumption about the cause of high heat flux from the mantle. These calculations employ the equation for one-dimensional heat flow in a homogeneous and isotropic medium with a prescribed distribution of heat production and thermal conductivity (see Chapman [1986] for a review). A partition in the continental heat sources is generally assumed; Chapman [1986] ascribes 40% of continental heat sources to the upper crust and the balance to deeper sources. Resultant geotherms are commonly presented for a given surface heat flux and values in the neighborhood of 80 to 90 mW/m^2 (Figure 1) are often assumed to be representative of rift or extensional terrains.

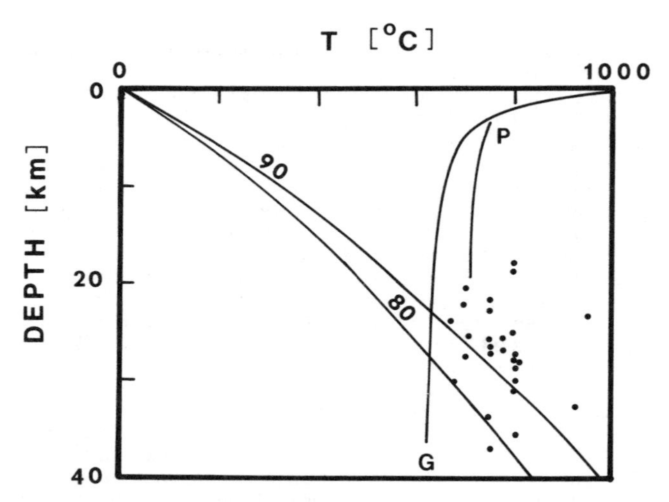

Fig. 1. One-dimensional, conductive steady state geothermal gradients corresponding to surface heat flow values of 80 and 90 mW/m^2 [Pollack and Chapman, 1977; Chapman, 1986]. Also shown on the diagram are P-T data for granulite facies rocks [Bohlen, 1987], solidus curves for pelitic rocks (P) and H_2O-saturated granite (G) from Grant [1985] and Wyllie [1977], respectively.

Use of these one-dimensional, steady state geotherms for extensional terrains, however, poses a few problems. Chapman [1986] points out that this set of gradients predicts wholesale melting of the lower continental crust for anhydrous and hydrous conditions if extrapolated to depths greater than 30 km. More importantly, heat transfer in extensional regimes is likely to occur, at least partially, by mechanisms other than one-dimensional heat conduction, such as advection.

Extension without Magmatism

There have been numerous studies of the thermal-mechanical response of the lithosphere to thinning [e.g., McKenzie, 1978; Royden and Keen, 1980; Sclater et al., 1980; Chenet et al., 1982; Beaumont et al., 1982; Steckler and Watts, 1982] in order to understand the thermal and subsidence history of continental basins and passive continental margins. Many of these models generally assume that the lithosphere is uniformly stretched by a certain amount (Figure 2a) or that lithosphere

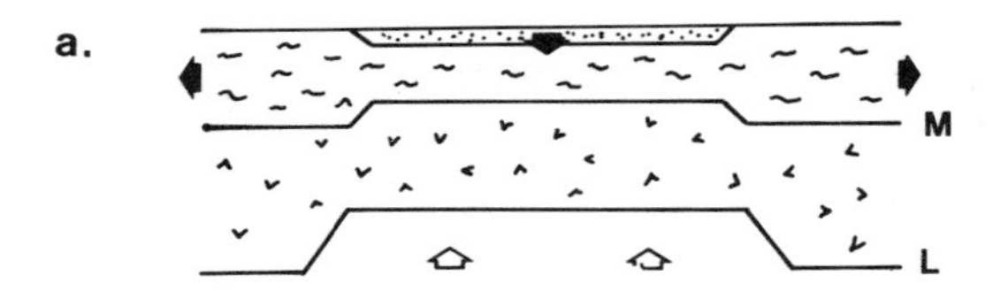

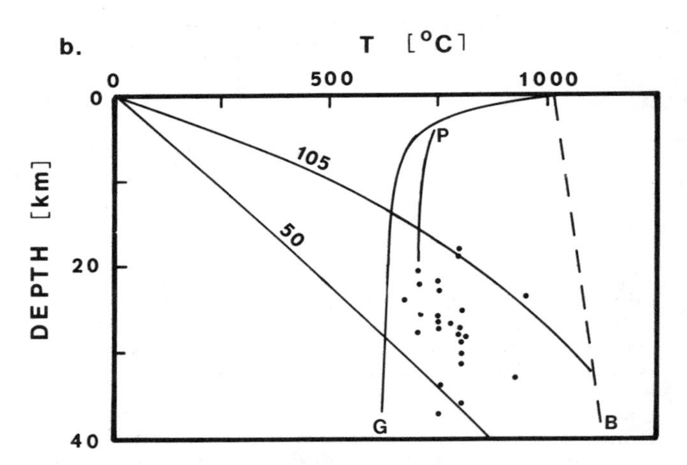

Fig. 2. (a) Model of extension of the crust without magmatism. (b) Possible geothermal gradients for simple extension based on models for the Basin and Range [Lachenbruch and Sass, 1978] calculated for asthenospheric heat flow of 33.5 mW/m^2. The two curves correspond to gradients for reduced heat flow values of 50 and 105 mW/m^2 that require extension rates of 0.6 and 5.6%/m.y., respectively. Also shown is the solidus curve for basalt (B) from Lachenbruch et al. [1985]. Other curves and data points are the same as in Figure 1.

stretching is depth dependent. Temperatures in the crust rise because high temperature asthenosphere is transported upward. The thermal state of the crust also depends upon the amount of sedimentation [e.g., McKenzie, 1978; Lachenbruch et al., 1985].

Examples of the various thermal gradients calculated for stretching without sedimentation and subsidence are given in the treatment of Lachenbruch and Sass [1978], an analysis of possible heat transfer mechanisms in the Basin and Range province in the western United States. The simplest of these models invokes stretching of the lithosphere accompanied by accretion of crystalline material at the base of the crust. Calculated thermal gradients for this model are presented in Figure 2b where each curve corresponds to a calculation for a constant reduced heat flow and, therefore, must correspond to a different extension rate. In this case, reduced heat flow increases with an increase in extension rate.

Keen [1987] points out that these simple stretching models may not be appropriate because they are limited in their ability to explain the geological history of rifts. In particular, simple stretching models do not easily account for the complex history of vertical motions of rifts, uplift of rift shoulders and the variable degree of volcanism observed during rifting [Keen, 1987].

Extension with Distributed Magmatism

Lachenbruch and Sass [1978] and Lachenbruch et al. [1985] investigated the thermal effects of the situation where emplacement of magmas into the crust accompanies extension. These models require even lower extension rates for a given reduced heat flow than the previous model because magmas provide heat to the crust in addition to that added by simple extension.

To explain the thermal evolution of the Salton trough in southern California, Lachenbruch et al. [1985] developed a model in which lithospheric extension was accompanied by sedimentation and distributed dike and sill emplacement (Figure 3a). Magma injection allows heat transfer from the mantle to the lower crust and elevates the temperatures in that regime. Subsidence and sedimentation decrease temperatures in the sedimentary section. A series of gradients calculated for different extension rates (Figure 3b) exhibit a characteristic sigmoidal shape that, Chapman [1986] argues, minimizes the problem of the excessively high extrapolated temperatures predicted by the one-dimensional conductive geothermal gradients. The dimensional axes in Figure 3b were recalculated from Lachenbruch et al. [1985] for a 30 km thick crust as opposed to the much thinner Salton trough crust used in their paper. These figures demonstrate that low thermal gradients can result for low extension rates (<10%/m.y.) thus indicating that lower crust in extensional terrains can be relatively cool in such regions. An example of gradients for the case of no sedimentation or subsidence computed for the Basin and Range province model [Lachenbruch and Sass, 1978] is also shown in Figure 3b.

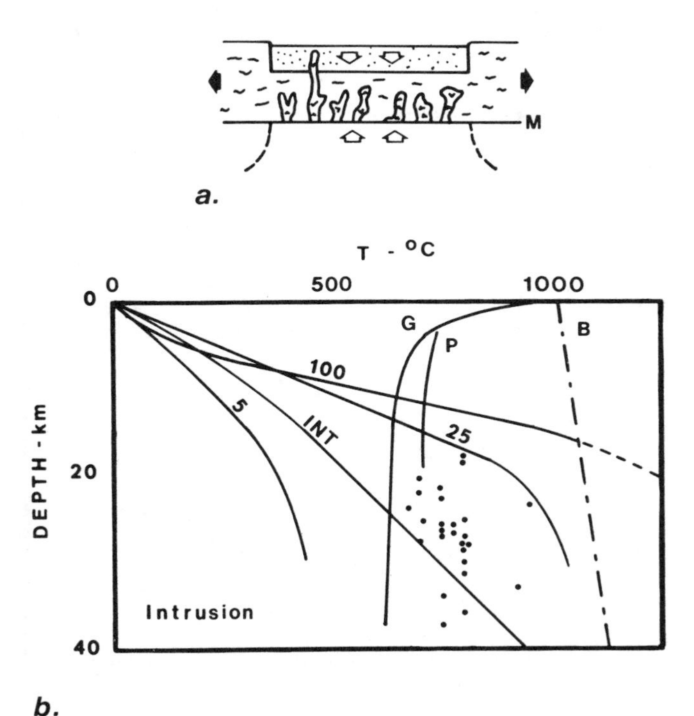

Fig. 3. (a) Distributed intrusion model of Lachenbruch et al. [1985]. (b) Possible geothermal gradients for the distributed intrusion model [Lachenbruch et al., 1985] at a constant sedimentation rate for extension rates of 5, 25 and 100%/m.y. Curve labeled INT corresponds to a model of distributed intrusion without sedimentation for the Basin and Range province [Lachenbruch and Sass, 1978] for asthenospheric heat flow of 33.5 mW/m^2. Other curves and data points are the same as in Figures 1 and 2.

Extension with Underplating

Recently, several investigators realized that the upwelling mantle material under an extending continental crust may partially melt due to decompression and that the resultant magma may accrete to the base of the crust [Dixon et al., 1981; Foucher et al., 1982; McKenzie, 1984a; Wickham and Oxburgh, 1985; Furlong and Fountain, 1986; Keen, 1987; White et al., 1987]. This accretionary process, magmatic underplating, has long been viewed as a possible mechanism to generate lower continental crust during the Precambrian [e.g., Fyfe, 1973; Fyfe and Leonardos, 1973; Wells, 1980; Lambert, 1983] but, until recently, has not received much attention for its potential role in the manufacture of Phanerozoic lower continental crust. Underplating occurs as a consequence of rapid, nearly adiabatic diapiric upwelling of mantle material during extension and resultant decompression partial melting as the diapir crosses the peridotite solidus. The basaltic liquid product separates from the mantle (see Ribe [1987] for review of melt segregation) and ponds at the base of the crust because of the density contrast between basalt liquids and typical crustal rocks [e.g., Herzberg et al., 1983].

The thermal consequences of magmatic underplating are illustrated by models developed by Lachenbruch and Sass [1978] for the Basin and Range province and Lachenbruch et al. [1985] for the Salton trough. Examples of thermal gradients for the case where there is no sedimentation and subsidence (Basin and Range) are presented in Figure 4b. Also shown are gradients for the case where sedimentation and subsidence are included (Salton trough).

Although computed under the assumption that the lithosphere undergoes steady state extension, these thermal gradients (Figures 2 through 4) can be regarded as transient gradients when considered in the context of duration of extension relative to geologic time. As such, rocks resident in the lower crust during extension experience peak thermal conditions during the extensional event and eventually cool to conditions typical of stable regions. This suggests that P-T conditions of metamorphism, as determined by geobarometry and geothermometry, represent transient thermal events and should not be taken as guides to steady state crustal geotherms in the geologic past.

Consequences of Extension and Underplating Lower Crust

In the previous section I reviewed a model of extension that required the addition of basaltic magmas to the base of the continental crust to account for the surface heat flow of extensional terrains. As illustrated by the models of Lachenbruch and Sass [1978], Lachenbruch et al. [1985], Keen [1987] and White et al. [1987], upwelling of asthenosphere and consequent decompression melting may occur under extensional terrains thereby causing the lower crust to grow in response to underplating. The magnitude of volcanism associated with continental extension coupled with the mass balance calculations that require additions to the crust of regions such as the Basin and Range [Gans, 1987] leads to the suggestion that underplating may be an important crustal growth mechanism during extension [e.g., Lister et al., 1986; Okaya and Thompson, 1986; Gans, 1987; White et al., 1987; Keen, 1987]. Below, I review a few of the important aspects of underplating and discuss the geological and geophysical consequences of the process.

Volume of Underplated Material

There have been several attempts to quantify the amount of melt generated during adiabatic rise of asthenospheric material for a variety of tectonic scenarios [Cawthorn, 1975; Ahern and Turcotte, 1979; Sleep and Windley, 1982; McKenzie, 1984b; Furlong and Fountain, 1986; Keen, 1987; White et al., 1987]. If

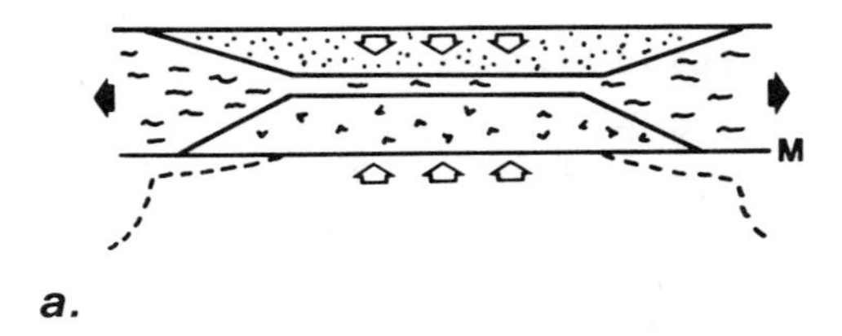

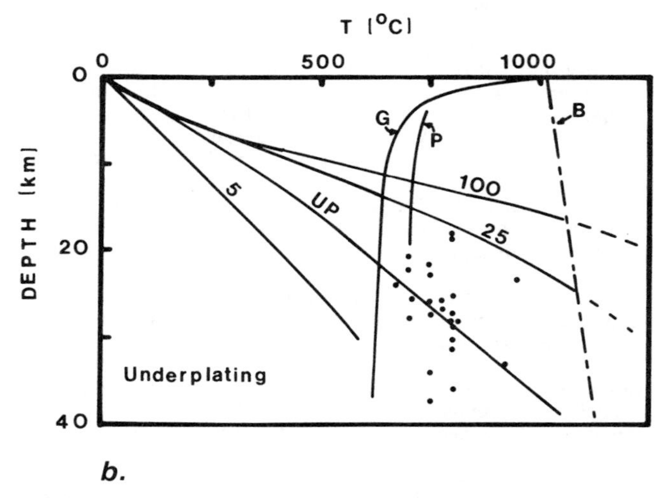

Fig. 4. (a) Extension with underplating model [Lachenbruch et al., 1985]. (b) Possible geothermal gradients for underplating model [Lachenbruch et al., 1985] at a constant sedimentation rate for extension rates of 5, 25 and 100%/m.y. Curve labeled UP corresponds to a model of extension with underplating without sedimentation for the Basin and Range province [Lachenbruch and Sass, 1978] for asthenospheric heat flow of 33.5 mW/m^2. Other curves and data points are the same as in Figures 1 and 2.

the dimensions of the ascending peridotite body are large [e.g., Furlong and Fountain, 1986] and if the ascent rate is fast [e.g., Foucher et al., 1982; White et al, 1987; Keen, 1987], the material follows the upper mantle adiabat. At some depth, the upwelling peridotitic mass crosses the peridotite solidus, partial melting begins and basalt liquids are produced. McKenzie [1984b] and Furlong and Fountain [1986] demonstrated that the amount of melt generated will depend upon the initial temperature of the peridotite diapir relative to the peridotite solidus and the final ascent depth reached by the upwelling mantle. For example, if upwelling asthenosphere at an initial temperature close to the solidus temperature at depths of 125 km ascends to Moho depths, 15 to 25% of the material will be liquid [Furlong and Fountain, 1986]. When integrated over the thickness of the ascending column, these melt volumes may potentially form layers in excess of 10 km thick (see Figure 5 in Furlong and Fountain, 1986] if all the melt separates and ponds at the base of the crust. This simple calculation assumes all lithosphere mantle is removed and asthenosphere rises to the Moho. In extension, the entire column of lithospheric mantle is probably not removed and the asthenosphere rises to some depth far removed from the continental Moho.

To quantify this situation, White et al. [1987] consider the role of the potential temperature of the asthenosphere which is the temperature asthenospheric material would have if it moved to the surface adiabatically without melting. Because the degree of partial melting depends on the potential temperature and the ultimate ascent position of the upwelling mantle, the degree of partial melting can be recast in terms of the amount of lithospheric stretching (ß). White et al. [1987] present new estimates of the variation of the thickness of partial melt produced by decompression melting of upwelling asthenosphere as a function of lithospheric stretching (ß) for the case of uniform stretching (Figure 5). Their calculations allow for variation in potential temperature and of initial lithospheric thickness. Large amounts of melt will accumulate as ß increases for high potential temperatures. If potential temperature is low, small amounts of melt are generated even for large values of ß. Thus, magmatic underplating may be unimportant in rift environments characterized by cool asthenosphere or low amounts of extension. Figure 5 also shows the range of ß estimates for continental extensional terrains and, thereby, emphasizes that significant melt thicknesses will be generated for relatively high potential temperatures. Similarly, Keen [1987] examined underplating during extension from a temporal point a view. She calculated ß as a function of time for a constant extension rate and, using an approach similar to White et al. [1987], converted ß into a melt layer thickness (see Figure 3 in Keen [1987]).

Fundamentally, this process is directly analogous to that postulated to operate under the mid-ocean ridges [e.g., Ahern and Turcotte, 1979; Sleep and Windley, 1982]. The solidified melt layer, in that case, becomes the oceanic crust and the density barrier to further upward ascent of magma is sea water. In detail, there are many complexities in the growth of the oceanic crust, but the first order process appears to be separation of basaltic melt from ascending mantle with melt volumes

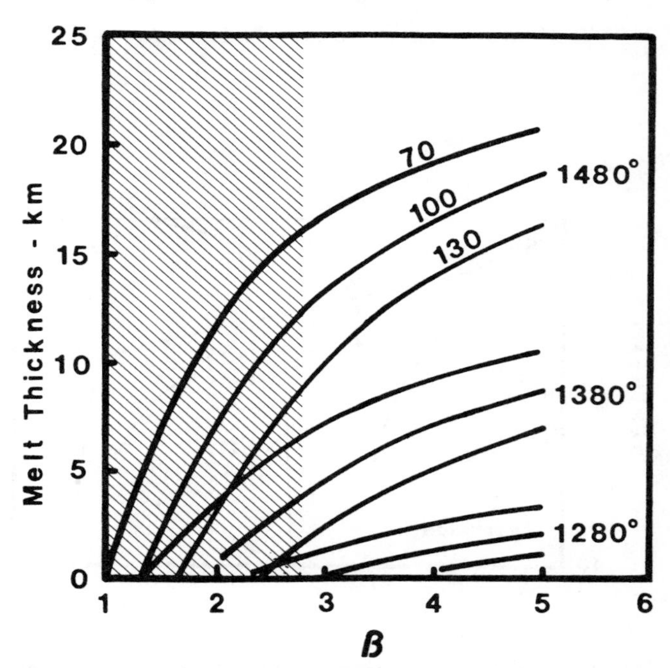

Fig. 5. Thickness of partial melt generated by decompression of passively upwelling asthenosphere for different amounts of lithosphere thinning assuming uniform stretching (from White et al. [1987]). Curves correspond to asthenosphere potential temperatures of 1280°, 1380°, and 1480°C and an intitial lithosphere thickness of 100 km. Shaded area indicates approximate range of typical stretching factors (ß) for regions of continental extension [Kusznir and Park, 1987].

of sufficient magnitude necessary to form the cumulative thickness of layers 2 and 3.

Petrological Nature of Underplated Material

The composition of the magma generated by partial melting of upwelling asthenosphere depends strongly on the initial depth (or potential temperature) of melting and the extent of melting during ascent. Recently, Klein and Langmuir [1987] estimated the major element compositon of magmas produced by decompression melting of asthenosphere under mid-ocean ridges. Not surprisingly, these theoretical results indicate that basaltic liquids will be produced, even for large percentages of partial melting and that, for the mid-ocean ridge case, more silicic melts are produced if the ascending asthenosphere intersects the solidus at lower pressures (i.e., lower potential temperature) than for higher pressure solidus intersection conditions (see Table 3 in Klein and Langmuir [1987]). Because magmatic underplating during continental extension is analogous to the processes that form the oceanic crust at mid-ocean ridges, the same analysis applies to magmas accreted to the base of continents. The results of Klein and Langmuir [1987] indicate that, in general, magmas low in SiO_2, Al_2O_3 and Na_2O and high in MgO, FeO and CaO will be produced from upwelling material with high potential temperatures. Formation of layered mafic-ultramafic igneous complexes should occur thereby developing lower crustal layers of anorthosite, pyroxenite, peridotite and gabbroic rocks.

Once these melts are accreted to the base of or intruded into the continental crust, crystallization occurs in the pyroxene granulite-garnet granulite-eclogite stability fields (Figure 6). There are two extreme situations to consider. The accreted material may isobarically cool to some ambient geothermal gradient following emplacement and the resulting mineral assemblage will reflect the equilibrium conditions along that geotherm. If isobaric cooling were to progress through geologic time (Figure 6) and reactions occurred, the mineral assemblages would tend toward the eclogite field and, for greater depths, cross into the eclogite field [Griffin and O'Reilly, 1986]. Alternatively, the reactions may be arrested as soon as the material solidifies and the mineral assemblage formed under those conditions would remain through isobaric cooling. The resulting rocks would exhibit disequilibrium assemblages. Assemblages between these two end member cases may result as well. Any basaltic magma added to the lower crust in a high temperature regime, such as an extensional terrain, should crystallize as a pyroxene granulite, garnet granulite or an eclogite depending upon depth and cooling history provided conditions are anhydrous.

Metamorphism in the Lower Crust

The previous sections outlined mechanisms by which the lower continental crust of extensional terrains might grow by addition of magmas. Once these magmas crystallize, they will likely exhibit a granulite (or, possibly, eclogite) facies mineralogy. Implicit in the preceding discussion is the notion that heat is added to the lower crust as a consequence of magma emplacement and that the resultant high temperatures can drive high grade metamorphism. The idea that high grade metamorphism occurs in zones of active extension recently received increased attention [e.g., Wickham and Oxburgh, 1985; Sandiford and Powell, 1986; Okaya and Thompson, 1986] and several authors proposed that granulite facies metamorphism occurs in the lower crust in extensional tectonic settings [Sandiford and Powell, 1986; Windley and Tarney, 1986; Chapman, 1986; Frost and Frost, 1987].

To illustrate the importance of lower crustal high grade metamorphism in extended lower crust, geobarometric and geothermometric data for granulite facies rocks compiled by

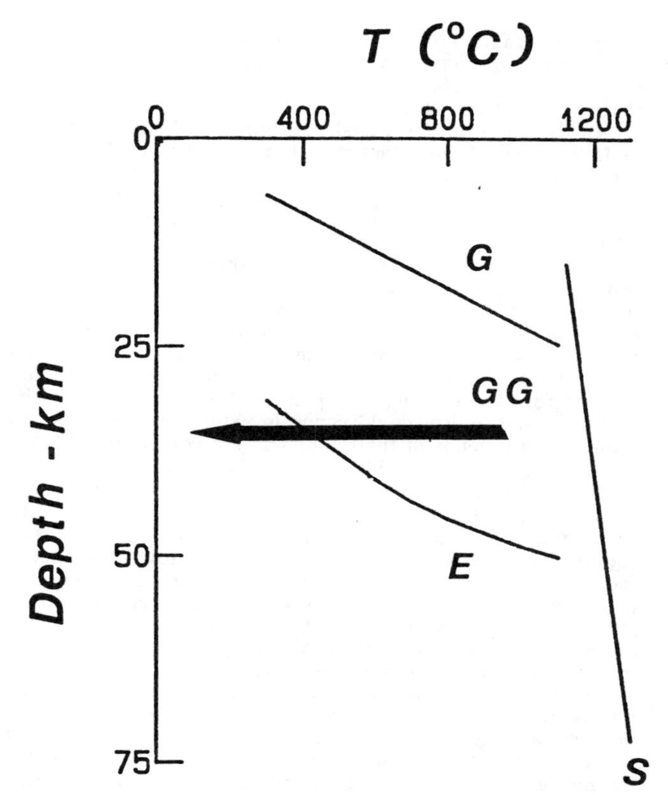

Fig. 6. Pyroxene granulite (G)-garnet granulite (GG)-eclogite (E) phase diagram for olivine gabbro compositions from Wood [1984]. Also shown is the dry basalt solidus (line S). The arrow indicates the changes in metamorphic facies expected with isobaric cooling for the case where metamorphic reactions proceed to equilibrium conditions.

Bohlen [1987] from various sources in the literature are shown in Figures 1 through 4. Clearly, there is considerable overlap between the field of granulite facies P-T estimates and the gradients predicted by the various thermal-mechanical models of extensional terrains, in particular those that require rapid uniform extension, distributed intrusion or magmatic underplating. This comparison clearly demonstrates that the high temperatures required for granulite facies metamorphism are clearly met at lower crustal depths under extensional terrains characterized by high extension rates, large amounts of extension and high asthenospheric potential temperature.

Until recently, granulite facies metamorphism was postulated to occur exclusively in regions of crustal overthrust [Newton and Perkins, 1982; Newton, 1983]. Bohlen [1987], however, presented cogent arguments that high temperatures are the common important ingredient for granulite facies metamorphism and proposed that granulite facies metamorphism occurs in the lower crust of magmatic arcs. It is important to recognize that the high temperatures required for the arc model of granulite genesis are also obtained in the lower crust of certain extensional terrains (Figures 2 through 4). This suggests that there is no common tectonic environment for granulite metamorphism and an understanding of the causes of granulite facies in a high grade terrain requires a complete analysis of its thermal and tectonic history. I hasten to point out that the P-T data displayed in Figures 1 through 4 represent available data for granulite facies rocks. There is no intention to imply that all these rocks experienced metamorphism in extensional regimes but, rather, to emphasize the consistency of the P-T data with thermal conditions expected in extensional lower crust.

An argument for the crustal doubling hypothesis of granulite facies metamorphism is that granulite facies terrains commonly show compressional structures. Recently, however, detailed structural analyses in a few granulite facies terrains revealed evidence of extensional strain. Specifically, Brodie and Rutter [1987] identified high temperature shear zones in the Ivrea zone granulite facies rocks in the Valle d'Ossola region whose geometry suggests extension. New field work in the Kapuskasing granulite domain in Ontario [Moser, 1988] also revealed extensional shear zones. In both cases, the timing of extension appears to be closely related to the time of peak metamorphism although much more work remains to verify this relationship.

Sandiford and Powell [1986] presented the case that granulite facies metamorphism could occur in extensional terrains. An important argument in their case was the contention that the retrograde paths exhibited by granulite facies rocks should show evidence of isobaric cooling; this would distinguish an extensional cooling history from that of the crustal doubling hypothesis (see Bohlen [1987]). Bohlen [1987] argues that isobaric cooling paths should be associated with cooling of granulite facies rocks after a magmatic heating event, a history evident in many granulite terrains (see Figure 3 in Sandiford and Powell [1986] and Figure 6 in Bohlen [1987]). These retrograde paths can be interpreted as cooling paths following magmatic heating in either an arc or extensional environment.

The possible addition of basaltic magmas to the lower crust during underplating associated with extension has another implication in terms of crustal metamorphism. Recently, Frost and Frost [1987] advocated a model for granulite facies metamorphism in which magmas intruded into the lower crust provide CO_2 to surrounding rocks. The CO_2 flux serves to dehydrate neighboring rocks thereby driving granulite facies metamorphism. In this model, melts not only provide additional heat but also high volumes of CO_2 that encourage granulite facies metamorphism. Frost and Frost [1987] argue that this process can lead to formation of charnockites and is likely to occur in extensional regimes.

An interesting implication of synmetamorphic deformation associated with extension is that rocks at specific crustal levels will not be fixed in position. As extension progresses, deep level lower crustal rocks could be juxtaposed against shallower lower crustal rocks. If a metamorphic terrain formed in the manner envisioned above were to be exposed at the Earth's surface, unusual patterns in the spatial distribution of pressure and temperature estimates might be observed as neighboring packages of rocks could exhibit very different P-T data as derived from geobarometric and geothermometric analyses.

Magma Genesis in Extensional Lower Crust

If metamorphic temperatures are high enough to drive granulite facies metamorphism, they certainly are high enough to cause melting or partial melting of rocks in the lower crust as postulated by Sandiford and Powell [1986]. Figures 1 through 4 also show melting curves for H_20-saturated granite and pelites [e.g., Wyllie, 1977; Grant, 1985]. Conditions for partial melting of these rock types are clearly achieved in the lower crust of extensional terrains, especially those in which rapid uniform stretching, magmatic intrusion or magmatic underplating occurs. Consequently, silicic magmas should be generated in the lower crust, migmatization should be common at depth, and restites should be important constituents for extensional lower crust. Hildreth [1981], from a petrological point of view, develops the case that rhyolitic magmas formed in extensional terrains are products of lower crustal partial melting caused by the addition of heat by the ponding of mantle-derived basalt magmas. Gans [1987] relates late Tertiary Basin and Range volcanism to the addition of mantle-derived magmas to the lower crust of that region and Wickham and Oxburgh [1985] attribute the genesis of Hercynian plutons in the Pyrenees to a similar mechanism.

Rheology and Deformation of Extensional Lower Crust

Recently, there has been considerable attention paid to the rheological nature of the lower crust. The numerous models proposed [e.g., Brace and Kohlstedt, 1980; Meissner and Strehlau, 1982; Smith and Bruhn, 1984; Kusznir and Park, 1986, 1987; Ranalli and Murphy, 1987] generally assume an extrapolation of Byerlee's law for the upper crust and apply steady state creep laws to the lower crust. Commonly, the models use Maryland diabase to simulate the lower crust or assume that a dominant mineral (such as quartz or feldspar) will control the lower crustal rheology. Another important ingredient in these modelling efforts is the necessary assumption of a geothermal gradient to calculate the flow strength. Models for the Basin and Range [Smith and Bruhn, 1984] assume a mafic lower crust (Maryland diabase) and employ a one-dimensional, conductive geotherm for the 90 mW/m^2 surface heat flow case [Pollack and Chapman, 1977]. In this case, the lower crust is relatively strong in its upper portions and weakens with depth. Another strength boundary occurs at the Moho where an olivine-dominated rheological behavior controls the flow strength.

As discussed earlier, the assumption of a one-dimensional, conductive geotherm may not be an appropriate choice for an actively extending terrain where dynamic processes of stretching and magmatic accretion occur. What would be the strength profile of extensional lower crust if we consider the elevated geothermal gradients predicted in models such as those presented by Lachenbruch and Sass [1978] or Lachenbruch et al. [1985]? In general these models predict temperatures greater than 750° C (granulite facies conditions) at typical lower crustal depths. Some models (Figures 2 through 4) predict that 750°C can be attained at depths as shallow as 15 to 20 km. Figure 7

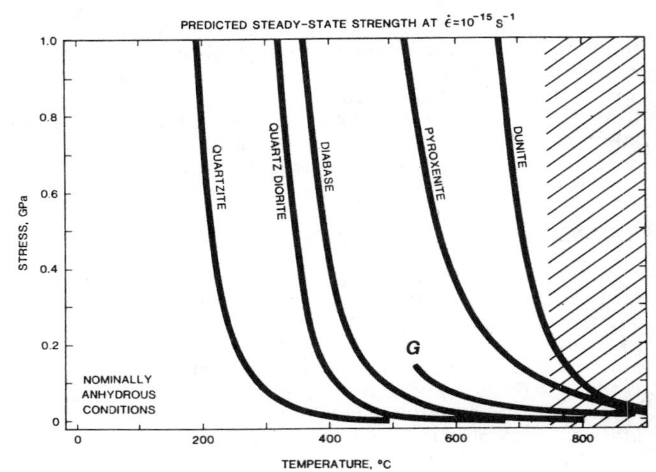

Fig. 7. Steady state flow stress (at a strain rate of 10^{-15} s^{-1}) as a function of temperature for various dry crustal rocks and minerals with a dunite included for comparison (from Kirby [1985]) and new data for a mafic granulite from Carter and Wilks [1987]. The shaded area corresponds to temperatures associated with the granulite facies metamorphism conditions likely to prevail to depths as shallow as 15 to 20 km in extensional lower crust where magmatic underplating is a major process.

shows this temperature field superimposed on a diagram from Kirby [1985] that shows flow strength for various dry crustal minerals and rocks as a function of temperature. Added to this diagram are new data from Carter and Wilks [1987] for a mafic granulite from the Pikwitonei granulite province in Manitoba. All crustal rocks and minerals in Figure 7 have flow stresses much less than 100 MPa under these high temperatures associated with magmatic underplating and granulite facies metamorphism. Ultramafic rocks are also weak at mantle temperatures anticipated by the various models in Figures 2 through 4. This suggests that the flow stress contrast between the crust and upper mantle is small in actively extending terrains characterized by rapid or large extension and/or magmatic accretion. Extensional terrains with low extension rates, small ß or no significant underplating should have higher strength lower crust and, therefore, not exhibit the ductility of high temperature extensional lower crust. This is supported by observations of lower crustal earthquakes in the East African rift [Shudofsky et al., 1987].

Unfortunately, this approach provides little insight into the issue of whether strain is localized or distributed in extensional lower crust. Wernicke [1981] theorizes that crustal-penetrating shear zones accommodate extensional strain and, consequently, deform the lower crust. Hamilton [1982] advanced the idea that extensional strain in the lower crust is achieved along ductile shear zones that anastomose around relatively undeformed lenses of rock. In contrast to models that argue for localized deformation, alternative theories [Stewart, 1971; Eaton, 1979; Miller et al., 1983; Smith and Bruhn, 1984] contend that extended lower crust deforms by pervasive ductile flow and magmatism below the seismogenic upper crust. Hamilton [1987] further modified his approach to suggest that deeper crustal levels are deformed by crustal flattening. Allmendinger et al. [1987] find that COCORP data in Nevada do not provide unique support for any one of these models although there is some evidence for the ductile flow model in the western Basin and Range and little evidence for the crustal-penetrating shear zone model.

Perhaps the most compelling evidence for the style of lower crustal deformation is derived from observations of deformation in exposed deep crustal terrains. Brodie and Rutter [1987] identified mylonite zones in the Ivrea zone that exhibit extensional strain along localized ductile shear zones to depths as great as about 25 km. Moser [1988] identified shear zones in the upper amphibolite rocks structurally above the Kapuskasing granulite terrain and interpreted these as extensional shear zones.

Seismic Properties of Extensional Lower Crust

The various geological consequences of extension in the lower continental crust must have a significant impact on the seismic character of the deep crust. The effects should be particularly pronounced in regions where extension rates were high, total extension was large and potential temperature of the asthenosphere was high thus encouraging additions of magma to the lower crust. These seismic characteristics can be anticipated from the previous discussion and have been discussed in some depth in the seismological literature [e.g., Klemperer et al., 1986; Drummond and Collins, 1987; Valasek et al., 1987; Warner, 1987].

The accretion of large thicknesses of basaltic liquid to the base of the crust will have significant effects on the seismic velocity structure of the deep crust. Using recalculated phase boundaries for the pyroxene granulite-garnet granulite-eclogite facies from Wood [1984] and estimates of volumetric abundances of the phases through the phase transitions [Green and Ringwood, 1967], Furlong and Fountain [1986] estimated the mineralogy of the solidified melt layer for the equilibrium and disequilibrium cases. From these estimates, they were able to calculate the expected variation of compressional wave velocity with depth in a layer of accreted material. The calculated velocities agree well with laboratory velocities for pyroxene granulites, garnet granulites and eclogites [e.g., Manghnani et al., 1974; Christensen and Fountain, 1975; Fountain, 1976; Jackson and Arculus, 1984].

Importantly, the calculated velocities correspond to those commonly reported in the lower crust and uppermost mantle in extensional terrains [Furlong and Fountain, 1986]. Of particular note are the high velocities (7.2-7.5 km/s) observed in the lower 3 km of the central Basin and Range [Valasek et al., 1987] and the high velocities under old rifts such as the Mississippi embayment [Mooney et al., 1983]. White et al. [1987] and the LASE Study Group [1986] report thick lower crustal prism-shaped bodies with velocities in excess of 7 km/s under rifted continental margins bordering the Atlantic Ocean. Both groups interpret these bodies as masses of mafic rocks formed by underplating during extension. High velocities in the same range (7.0-7.5 km/s) are reported under other continental margins [e.g., Keen et al., 1975; Talwani et al., 1979; Falvey and Middleton, 1981; Weigel et al., 1982]. Surprisingly, similarly high compressional wave velocities are reported under cratonic basins such as the Williston basin [Morel-a-l'Huissier et al., 1987] and the large basins in Australia [Drummond and Collins, 1986]. This similarity of velocities does not provide a unique test of the underplating model, but does indicate that magmatic underplating should be given serious consideration as a lower crustal growth mechanism in extensional terrains.

In addition to the high velocities of crystallized underplated material, the seismic characteristics of some pre-existing rocks can be changed by effects of high grade metamorphism and anatexis in the lower crust. For instance, data presented by Fountain [1976] and Burke [1987] demonstrate that granulite facies meta-pelites have higher velocities than lower grade equivalents. This is particularly pronounced when the granulite facies meta-pelites have experienced a partial melting event. Not all lithologies will be affected this way, but there will be a tendency for compressional wave velocities for several rock types to be high in extensional lower crust because of the effects of anatexis and high grade metamorphism. These increases in velocity of some lithologies with respect to others serve to enhance acoustic contrasts in the deep crust.

Recently, much attention has focused on the high reflectivity of the lower crust and the reflection characteristics of the Moho. These features seem to be prominent under Phanerozoic extensional terrains such as the Basin and Range province [e.g., Allmendinger et al., 1987; Valasek et al., 1987], the British Isles [e.g., Cheadle et al., 1987] and eastern North America [e.g., Hutchinson et al., 1986] and, according to Allmendinger et al. [1987], less prominent for older extensional lower crust. Several authors contend that the high reflectivity of the lower crust is related to the relative youth of the latest tectonothermal event in a region [Meissner, 1984; Klemperer, 1987; Cheadle et al., 1987]. Many attribute this reflectivity to layered mafic rocks emplaced in the crust during extension and underplating [e.g., Cheadle et al., 1987; Allmendinger et al., 1987] whereas others emphasize the importance of extensional strain as a prime agent in creating near-horizontal structures in the lower crust [e.g., Hamilton, 1987; Valasek et al., 1987].

An Example of Extensional Lower Crust

Another method to investigate the evolution of the lower continental crust in zones of extension is to examine exposed examples of lower crust [e.g., Fountain and Salisbury, 1981] that exhibit an extensional geologic history. Unfortunately, the evolution of high grade metamorphic terrains is rarely interpreted in the light of an extensional tectonic model. One important case to the contrary is represented by recent investigations of the Ivrea-Verbano terrain of northern Italy. Although there are still many outstanding questions awaiting resolution concerning the geologic evolution of this region, there is some evidence that this terrain experienced extension [Hodges and Fountain, 1984; Handy, 1986, 1987; Schmid et al., 1987; Brodie and Rutter, 1987] while it resided at depths between 18 to, possibly, 30 km [Schmid and Wood, 1976; Sills, 1984]. Below, I review the evidence for extension and interpret some of the main aspects of the zone in terms of the extensional model.

The Ivrea-Verbano terrain (Figure 8) consists of upper amphibolite to granulite facies carbonates, metapelites, mafic gneisses and ultramafic rocks (see Zingg [1983] for review). It is bounded to the east by the Strona-Ceneri zone, a terrain dominated by various amphibolite facies paragneisses and orthogneisses. Much of the length of the contact between the Ivrea-Verbano and Strona-Ceneri zones is marked by the Pogallo fault zone (PFZ). Recently, Hodges and Fountain [1984] proposed that the PFZ was a pre-Alpine low-angle normal fault situated at intermediate crustal levels that was rotated into a near-vertical position during the Alpine orogeny.

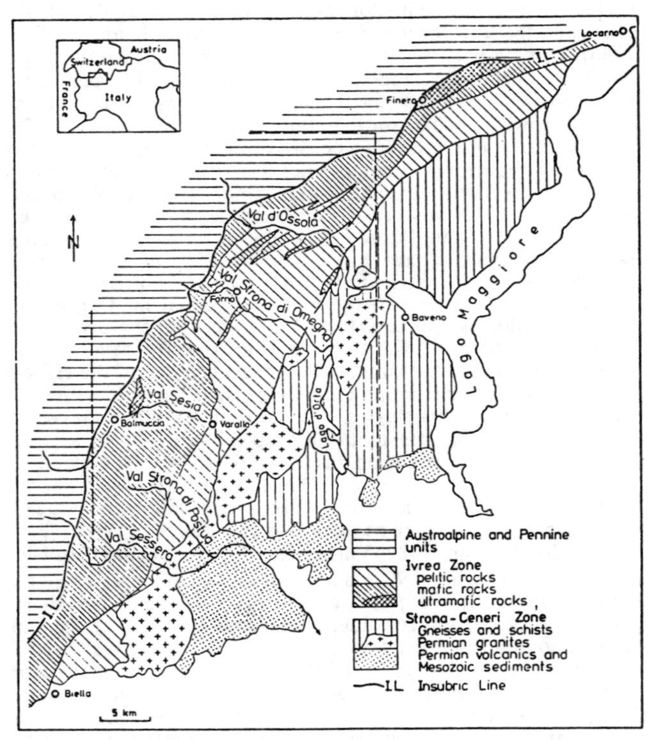

Fig. 8. Generalized geologic map of the Ivrea-Verbano and Strona-Ceneri zones, northern Italy (from Zingg [1983]).

The PFZ [Boriani, 1970; Handy, 1986, 1987; Schmid et al., 1987] is a kilometer wide zone of ductile and subordinate brittle tectonites with a strike oblique to the trend of structures in the Strona-Ceneri zone. Metamorphism in the zone is a retrograde greenschist to amphibolite facies overprint of the metamorphic assemblages of the adjacent Ivrea-Verbano zone. The fabrics in the PFZ change systematically along strike indicating that deformation mechanisms change from low temperature to higher temperature mechanisms toward the northeast. This indicates that deeper portions of the fault zone are exposed along the northeastern trend of the fault as suggested by Hodges and Fountain [1984]. Sense of shear indicators yield a sinistral sense of motion in present coordinates. Movement on the fault zone presumably post-dates Permo-Carboniferous (295-310 Ma) diorite agmatites in the Valle d'Ossola. The zone displaces Permian granites, and temperature-time paths based on metamorphic and isotopic data suggest that movement along the PFZ occurred between 230 and 180 Ma [Fountain and Hodges, 1984; Schmid et al., 1987].

More important evidence for extension in the lower crust, as represented by the Ivrea-Verbano zone, comes from studies of the high temperature shear zones within the Ivrea zone in the Valle d'Ossola area [Brodie, 1981; Brodie and Rutter, 1987]. These zones are less than 50 meters wide and, more commonly, less than 10 meters wide. Zones of grain size reduction occur in mafic rocks (pyriclasites) where the plagioclase grains are significantly reduced in size and pyroxene, amphibole and garnet behaved passively during deformation. Brodie [1981] and Brodie and Rutter [1987] find no evidence of hydration or retrogression in the shear zones but, instead, report evidence for prograde conditions during deformation. Brodie and Rutter [1987] were able to reconstruct the geometry of the high temperature shear zones by removing the effects of Alpine deformation and present a revised lower crustal cross section based on the structure of the Valle d'Ossola section (Figure 9). The deduced stretching direction of the shear zones is similar to the direction of motion determined for the PFZ by Handy [1986, 1987].

If the low-angle normal fault hypothesis for the Pogallo line is valid, the various mylonitic zones in the Ivrea section could

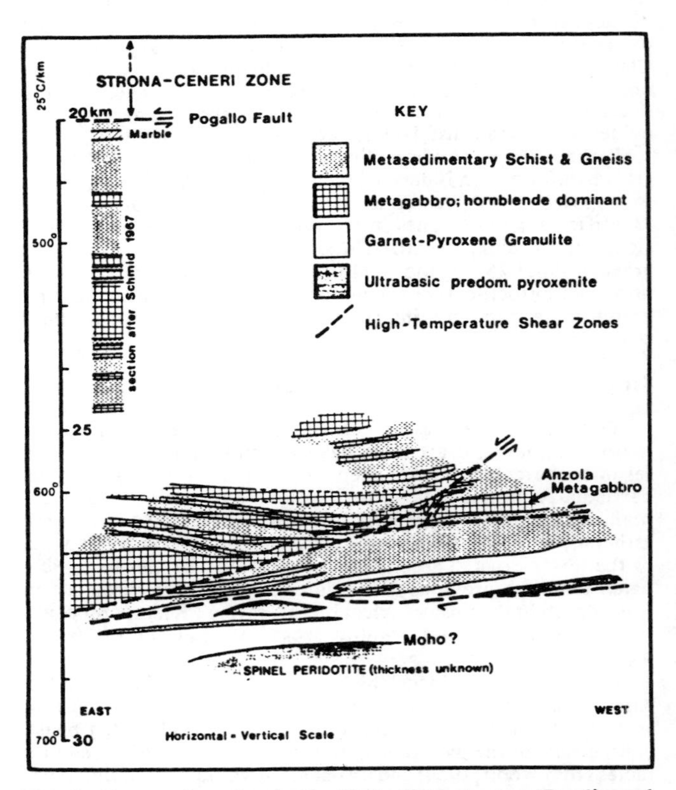

Fig. 9. Restored section in the Valle d'Ossola area [Brodie and Rutter, 1987] after removal of effects of Alpine folding and faulting. The temperatures indicated are minima required by the data.

be regarded as models of mylonite zones in the deep crust under extensional terrains. Handy [1986, 1987] postulated that the PFZ formed along the continental margin of Italy during the Triassic-Jurassic rifting stage associated with development of the Tethys before the Alpine collision. Several years ago, Shervais [1979], based on the high temperatures determined for the emplacement of the Ivrea ultramafic bodies (e.g., Balmuccia), suggested that metamorphism in the Ivrea zone may have occurred under thermal conditions commonly associated with the Basin and Range province. The Ivrea mylonite zones perhaps provide a view into deep crustal deformation of intra-continental rifts and rifted margins.

Although many structural features of the Ivrea-Verbano zone can be interpreted in terms of an extensional strain history, the timing of this deformation relative to other key events in the zone remains enigmatic. Brodie and Rutter [1987] present evidence for syn-metamorphic development of extensional shear zones in the Valle d'Ossola region. Based on a Rb-Sr whole rock isochron for samples collected over a wide area in the zone, Hunziker and Zingg [1980] argue for an Ordovician date for peak metamorphic conditions. Pin [1986], however, supports a late Hercynian age (285 Ma) of metamorphism based on U-Pb zircons from a gabbro-diorite complex, a view shared by Köppel [1974]. Unfortunately, these dates pre-date the inferred age of faulting on the PFZ [Hodges and Fountain, 1984; Handy, 1986, 1987; Schmid et al., 1987] perhaps precluding a kinematic relationship between movement on the PFZ and the high temperature shear zones within the Ivrea granulite facies rocks. The high temperature shear zones may be older than the PFZ and, therefore, related to earlier events in the complex.

Despite the problems concerning timing of metamorphism, several major geological features of the Ivrea-Verbano zone are similar to those expected in extensional terrains. The most important of these is the pervasive granulite facies metamorphism that reflects high temperature metamorphic conditions during peak metamorphism. P-T data for metabasic rocks in the Val Sesia yield temperatures of 750-800° C at 800 MPa whereas metasedimentary rocks from the Val Strona give values of 750° C at 600 MPa [Sills, 1984]. Although the rocks apparently originated at different depths in the crust, they experienced similar high temperature conditions. Schmid and Wood [1976] report high temperatures from the Valle d'Ossola area where garnet granulite facies mafic rocks are exposed [Schmid, 1967]. Wood [1983], furthermore, argues that high temperatures drove the granulite facies metamorphism in the Ivrea zone. Sills [1984] presents evidence that isobaric cooling followed high temperature conditions in the mafic rocks.

In addition to the high grade metamorphism, other features of this area can be interpreted in the light of an extensional tectonic model. Thermo-mechanical models incorporating extension with underplating or distributed intrusion, predict that garnet granulites should be prevalent in the lower crust and overly an ultramafic residua from which they were extracted as melts. Garnet granulite facies rocks are common in the Ivrea zone and are spatially associated with large ultramafic rocks that are residua from melting episodes [Ottonello et al., 1984]. A few of these bodies are in contact with layered mafic complexes that equilibrated at high temperatures [e.g., Shervais, 1979; Rivalenti et al., 1981] at lower crustal depths [Sills, 1984]. Portions of these layered complexes developed from melts derived from the ultramafic masses during upwelling [e.g., Shervais, 1979; Sinigoi et al., 1980]. Although these features could be interpreted in the light of the extension with an underplating model, recent Nd isotopic data suggest that the partial melting or emplacement of the peridotites occurred about 600 Ma [Pin and Sills, 1986; Voshage et al., 1987]. Clearly, these dates are older than the postulated time of movement on the PFZ and are even older than the postulated metamorphic dates.

Finally, higher structural levels in the Ivrea-Verbano and Strona-Ceneri zones exhibit features that can be interpreted as consequences of an extensional history, although the relative timing of these features may not fit with events summarized above. For instance, migmatites are common in the transition between the granulite and amphibolite facies and their development was probably associated with the high grade metamorphism [Mehnert, 1975]. High level Hercynian granites are found in the Strona-Ceneri zone. Some workers [Pin, 1986; Pin and Sills, 1986; Fountain, 1986] have associated the genesis of these plutons with an anatectic event evident in the meta-sedimentary granulite facies rocks of the Ivrea zone, although there is no clear evidence for this in the isotope geochemistry or geochronology.

Compressional wave velocities measured at high pressures for rocks from the Ivrea zone [Fountain, 1976; Burke, 1987], permit construction of realistic seismic models of this lower crustal section. Hale and Thompson [1982] developed a one-dimensional reflection model based on the Valle d'Ossola geology. Fountain [1986] expanded this approach with a generalized two-dimensional synthetic reflection section for the entire complex. Hurich and Smithson [1987] presented a more detailed two-dimensional model based on the Valle d'Ossola section. None of these studies benefitted from the new data of Burke [1987] and all are based on geometries that include the effects of Alpine deformation. Burke [1987] used the reconstructions of Brodie and Rutter [1987] with the new velocity data as the basis of a more refined reflection model of the lower crust (Figure 10). This synthetic seismogram section shows

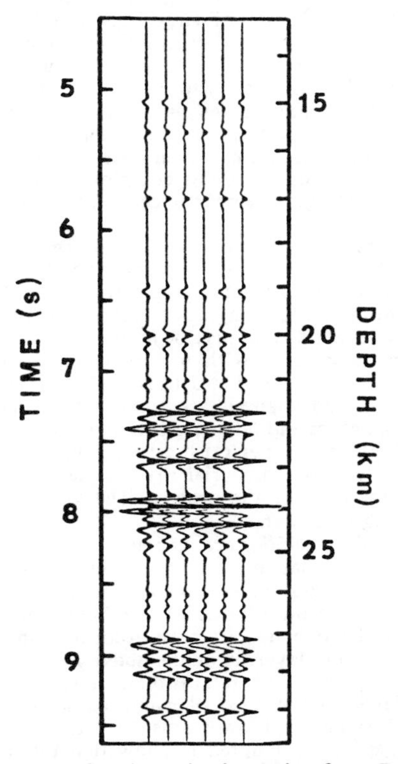

Fig. 10. Synthetic reflection seismic section form Burke [1987] based on the geometry presented in Figure 9 and laboratory determined seismic velocities [Fountain, 1976; Burke, 1987].

many of the features evident in earlier models. Of particular importance is that the velocity contrasts in this complexly lithologically layered sequence of rocks are of appropriate magnitude to generate strong reflections of the type reported in extensional terrains [e.g., Allmendinger et al., 1987; Cheadle et al., 1987]. The geometry of this section is a consequence of multiple strain events (extensional?) superimposed on a layered sequence of rocks.

Discussion and conclusions

Several parameters control the effects extension will have on the growth and development of the continental lower crust. Analysis of advective heat transfer mechanisms indicates that extension rate is a major control of the temperatures in the lower crust with high temperatures favored by high extension rates. In certain situations, basaltic liquids can accrete to the base of the crust. Magmatic underplating requires high potential temperatures of the asthenosphere and/or large values of ß. Thus, lower crust in rifts characterized by low extension rates, low asthenosphere potential temperatures or small stretching factors may not be appreciably changed during the rifting process. Geological and geophysical characteristics of such lower crust may differ little from those of the adjacent, undisturbed lower crust. Its strength may remain high during rifting as suggested by reports of lower crustal earthquakes in East Africa (Shudofsky et al., 1987).

The most significant changes in extensional lower continental crust are likely to occur in regions where extension rates are high, asthenosphere potential temperatures are elevated and extension amounts are large. These conditions favor magmatic underplating where the most significant effect on the lower crust will be the addition of large bodies of basaltic magmas. These magmas can differentiate into layered igneous complexes composed of gabbroic, anorthositic and ultramafic layers. Importantly, the magmas will crystallize in the pyroxene granulite, garnet granulite and, in special cases, eclogite facies. This metamorphic pattern is superimposed on the igneous layering as is evident in the layered series in the Valle Sesia area of the Ivrea zone or the Giles complex within the granulite domain of the Musgrave Range in central Australia [e.g., Goode and Moore, 1975].

High temperatures associated with magmatic underplating provide the heat necessary to drive upper amphibolite to granulite facies metamorphism of pre-existing rocks. CO_2 liberated from intruding basaltic magmas enhances the probability that granulite facies conditions may be attained. Because of the high temperatures associated with underplating, these high grade conditions may be reached at depths as shallow as 15 to 20 km. Migmatization and anatexis are further consequences of this process. Migmatites and restites can be expected to be important constituents in the deeper crust whereas upper to mid-crustal granitoid plutons and subaerial rhyolitic lavas are the shallow level manifestations of the process. The lower crust is highly ductile during extension.

High seismic velocities are an important characteristic of a lower crust that has grown by underplating as pyroxene granulites, garnet granulites, anorthosites and ultramafic rocks will be the dominant lithologies. High velocity restites will contribute to the high velocity crust. Acoustic contrasts are large because of the interlayering of the above rock types with pre-existing lithologies that retain their lower velocities. This lithologic layering the most plausible cause of the reflective character of extensional lower crust.

In this paper, I attempted to show that the lower crust in extensional regimes will grow and modify to varying degrees during extension. This paper has focused on the extreme case where the lower crust is greatly modified during extension because crust of this type may be easiest to recognize in geophysical data. However, I emphasize that there should exist many variations on this theme and many different types of lower crust between this end member and that of the virtually unmodified lower crust expected to be present in zones of low extension rates, low ß and cool asthenosphere. Thus, not only is it difficult to geophysically distinguish lower crust formed in different tectonic environments but lower crust formed and modified within similar tectonic regimes may also have very different characteristics because of the interplay of thermal-mechanical parameters. In particular, extended lower crust exhibits great diversity because of the sensitivity of crustal evolution to important parameters of extension rate, asthenosphere potential temperature and extension amount.

Acknowledgements. This paper is an outgrowth of research associated with N.S.F. Grant EAR-8418350. The manuscript benefitted from discussions with R. Frost and K. Furlong.

References

Ahern, J.L., and D.L. Turcotte, Magma migration beneath an ocean ridge, Earth Planet. Sci. Lett., 45, 115-122, 1979.

Allmendinger, R.W., J.W. Sharp, D. Von Tish, L.Serpa, L. Brown, S. Kaufman, J. Oliver, and R.B. Smith, Cenozoic and Mesozoic structure of the eastern Great Basin province, Utah, from COCORP seismic-reflection data, Geology, 11, 532-536, 1983.

Allmendinger, R.W., T.A. Hague, E.C. Hauser, C.J. Potter, S.L. Klemperer, K.D. Nelson, and P. Knuepfer, Overview of the COCORP 40° N transect, western United States: The fabric of an orogenic belt, Geol. Soc. Am. Bull., 98, 308-319, 1987a.

Allmendinger, R.W., T.A. Hague, E.C. Hauser, C.J. Potter' and J. Oliver, Tectonic heredity and the layered lower crust in the Basin and Range Province, western United States, Continental Extensional Tectonics, Geol. Soc. Spec. Publ., vol. 28, edited by M.P. Coward, J.F. Dewey, and P.L. Hancock, Blackwell, Oxford, 223-246, 1987b.

Beaumont, C., C.E. Keen, and R. Boutilier, A comparison of foreland and rift margin sedimentary basins, Phil. Trans. R. Soc. Lond., A305, 295-317, 1982.

Bohlen, S.R., Pressure-temperature-time paths and a tectonic model for the evolution of granulites, J. Geol., 95, 617-632, 1987.

Boriani, A., The Pogallo Line and its connection with the metamorphic and anatectic phases of "Massiccio dei Laghi" between the Ossola Valley and Lake Maggiore (Northern Italy), Boll. Soc. Geol. Ital., 89, 415-433, 1970.

Boriani, A., and R. Sacchi, Geology of the junction between the Ivrea-Verbano and Strona-Ceneri zones, Mem. Ist. Geol. Mineral. Padova, 28, 1-36, 1973.

Boriani, A., and E. Giobbi Origoni, High-grade regional metamorphism, degranitisation and origin of granites: an example from the South-Alpine basement, Proc. 27th Int. Geol. Congress, v. 9, Petrology (Igneous and Metamorphic Rocks), VNU Press, Utrecht, 41-66, 1984.

Brace, W.F., and D.L. Kohlstedt, Limits on lithospheric stress imposed by laboratory experiments, J. Geophys. Res., 85, 6248-6522, 1980.

Brodie, K.H., Variation in amphibole and plagioclase composition with deformation, Tectonophysics, 78, 385-402, 1981.

Brodie, K.H., and E.H Rutter, Deep crustal extensional faulting in the Ivrea Zone of northern Italy, Tectonophysics, 140, 193-212, 1987.

Burke, M., Compressional wave velocities in rocks from the Ivrea-Verbano and Strona-Ceneri zones, Southern Alps. Northern Italy: Implications for models of crustal structure, M.S. Thesis, Univ. of Wyoming, Laramie, 78p., 1987.

Carter, N.L., and K.R. Wilks, Rheology of the lower continental

crust and mechanical contrast at Moho (abstract), Int. Un. Geod. Geophys. XIX Gen. Assem. Abstracts, 1, 65, 1987.
Cawthorn, R.G., Degrees of melting in mantle diapirs and the origin of ultrabasic liquids, Earth Planet. Sci. Lett., 27, 113-120, 1975.
Chapman, D.S., Thermal gradients in the continental crust, The Nature of the Lower Continental Crust, Geol. Soc. Spec. Publ., vol. 24, edited by J.B. Dawson, D.A. Carswell, J. Hall, and K.H. Wedepohl, Blackwell, Oxford, 63-70, 1986.
Cheadle, M.J., S. McGeary, M.R. Warner, and D.H. Matthews, Extensional structure on the western UK continental shelf: a review of evidence from deep seismic profiling, Continental Extensional Tectonics, Geol. Soc. Spec. Publ., vol. 28, edited by M.P. Coward, J.F. Dewey, and P.L. Hancock, 445-465, 1987.
Chenet, P., L. Montadert, H. Gairaud, and D. Roberts, Extension ratio measurements on the Galicia, Portugal and North Biscay continental margins; Implications for evolutionary models of passive continental margins, Studies in Continental Margin Geology, Am. Assoc. Petrol. Geol. Mem., vol. 34, edited by J.S. Watkins and C.L. Drake, 703-715, 1982.
Christensen, N.I., and D.M. Fountain, Constitution of the lower continental crust based on experimental studies of seismic velocities in granulite, Geol. Soc. Am. Bull., 86, 227-236, 1975.
Dewey, J.F., Diversity in the lower continental crust, The Nature of the Lower Continental Crust, Geol. Soc. Spec. Publ., vol. 24, edited by J.B. Dawson, D.A. Carswell, J. Hall, and K.H. Wedepohl, Blackwell, Oxford, 71-78, 1986.
Dixon, J.E., J.G. Fitton, and R.T.C. Frost, The tectonic significance of post-Carboniferous igneous activity in the North Sea basin, Petroleum Geology of the Continental Shelf of North-West Europe, Institute of Petroleum, London, 121-137, 1981.
Drummond, B.J., and C.D.N. Collins, Seismic evidence for underplating of the lower continental crust of Australia, Earth Planet. Sci. Lett., 79, 361-372, 1986.
Eaton, G.P., Regional geophysics, Cenozoic tectonics, and geologic resources of the Basin and Range Province and adjoining regions, Basin and Range Symposium, Rocky Mountain Assoc. Geologists 1979 Symposium, edited by G.W. Newman, and H.D. Goode, 11-39, 1979.
Falvey, D.A., and M.F. Middleton, Passive continental margins: Evidence for a pre-breakup deep crustal metamorphic subsidence mechanism, Colloquium on Geology of Continental Margins (C3), Oceanologica Acta, 4 (Supplement), 103-114, 1981.
Foucher, J.P., X. LePichon, and J.G. Sibuet, The ocean-continent transition in the uniform lithospheric stretching model; role of partial melting in the mantle, Phil. Trans. R. Soc. London, A305, 27-43, 1982.
Fountain, D.M., The Ivrea-Verbano and Strona-Ceneri Zones, northern Italy: A cross-section of the continental crust - New evidence from seismic velocities, Tectonophysics, 33, 145-165, 1976.
Fountain, D.M., Implications of deep crustal evolution for seismic reflection interpretation, Reflection Seismology: The Continental Crust, Am. Geophys. Un. Geodyn. Ser., v. 14, edited by M. Barazangi, and L. Brown, 1-7, 1986.
Fountain, D.M., and Salisbury, M.H., Exposed cross-sections through the continental crust: implications for crustal structure, petrology and evolution, Earth Planet. Sci. Lett., 56, 263-277, 1981.
Frost, B.R., and C.D. Frost, CO_2, melts and granulite metamorphism, Nature, 327, 503-506, 1987.
Furlong, K.P., and D.M. Fountain, Continental crustal underplating; Thermal considerations and seismic-petrologic consequences, J. Geophys. Res., 91, 8285-8294, 1986.
Fyfe, W.S., The granulite facies, partial melting and the Archean crust, Phil. Trans. R. Soc. London, A273, 457-461, 1973.
Fyfe, W.S., and O.H Leonardos, Ancient metamorphic-migmatite belts of the Brazilian African coasts, Nature, 224, 501-502, 1973.
Gans, P.B., An open-system, two-layer crustal stretching model for the eastern Great Basin, Tectonics, 6, 1-12, 1987.
Goode, A.D.T., and A.C. Moore, High pressure crystallization of the Ewarara, Kalka and Gosse Pile intrusions, Giles Complex, central Australia, Contrib. Mineral. Petrol., 51, 77-97, 1975.
Grant, J.A., Phase equilibria in partial melting of pelitic rocks, Migmatites, edited by J.R. Ashworth, Blackie and Son, Glasgow, 85-144, 1985.
Green, D.H., and A.E. Ringwood, An experimental investigation of the gabbro to eclogite transformation and its petrological applications, Geochim. Cosmochim. Acta, 31, 767-833, 1967.
Griffin, W.L., and S.Y. O'Reilly, Is the continental Moho the crust-mantle boundary, Geology, 15, 241-244, 1987.
Hale, L.D., and G.A. Thompson, The seismic reflection character of the Mohorovičić discontinuity, J. Geophys. Res., 87, 4625-4635, 1982.
Hamilton, W., Crustal evolution by arc magmatism, Phil. Trans. R. Soc. Lond., A301, 279-291, 1981.
Hamilton, W., Structural evolution of the Big Maria Mountains, northeastern Riverside County, southeastern California, Mesozoic-Cenozoic Tectonic Evolution of the Colorado River Region California, Arizona, and Nevada, edited by E.G. Frost, and D.L. Martin, 1-27, 1982.
Hamilton, W., Crustal extension in the Basin and Range Province, southwestern United States, Continental Extensional Tectonics, Geol. Soc. Spec. Publ., vol. 28, edited by M.P. Coward, J.F. Dewey, and P.L. Hancock, Blackwell, Oxford, 155-176, 1987.
Handy, M.R., The structure and rheological evolution of the Pogallo fault zone, A deep crustal dislocation in the southern Alps of northwestern Italy (Prov. Novara), PhD. Diss., Geologisch-Paläontologisches Institut, Basel, 327p., 1986.
Handy, M.R., The structure, age and kinematics of the Pogallo fault zone; Southern Alps, northwestern Italy, Eclogae Helv., 80, 593-632, 1987.
Herzberg, C.T., W.S. Fyfe, and M.J. Carr, Density constraints on the formation of the continental Moho and crust, Contrib. Mineral. Petrol., 84, 1-5, 1983.
Hildreth, W., Gradients in silicic magma chambers: Implications for lithospheric magmatism, J. Geophys. Res., 86, 10153-10192, 1981.
Hodges, K.V., and D.M. Fountain, Pogallo Line, South Alps, northern Italy: An intermediate crustal level, low angle normal fault?, Geology, 12, 151-155, 1984.
Hunziker, J.C., and A. Zingg, Lower Paleozoic amphibolite to granulite facies metamorphism in the Ivrea zone, southern Alps, northern Italy, Schweiz. Min. Petr. Mitt., 60, 181-213, 1980.
Hurich, C.A., and S.B. Smithson, Compositional variation and the origin of deep crustal reflections, Earth Planet. Sci. Lett., 85, 416-426, 1987.
Hutchinson, D.R., J.A. Grow, and K.D. Klitgord, Crustal reflections from the Long Island platform of the U.S. Atlantic continental margin, Reflection Seismology: the Continental Crust, Am. Geophys. Un. Geodyn. Ser., vol. 14, edited by M. Barazangi, and L.D. Brown, 173-188, 1986.
Jackson, I., and R.J. Arculus, Laboratory wave velocity measurements on lower crustal xenoliths from Calcutteroo, South Australia, Tectonophysics, 101, 185-197, 1984.
Kay, R.M., and S.M. Kay, Petrology and geochemistry of the lower continental crust: an overview, The Nature of the Lower Continental Crust, Geol Soc. Spec. Publ., vol. 24,

edited by J.B. Dawson, D.A. Carswell, J. Hall, and K.H. Wedepohl, Blackwell, Oxford, 147-159, 1986.

Keen, C.E., Some important consequences of lithospheric extension, Continental Extensional Tectonics, Geol. Soc. Spec. Publ., vol. 28, edited by M.P. Coward, J.F. Dewey, and P.L. Hancock, 67-73, 1987.

Keen, C.E., M.J. Keen, D.L. Barrett, and D.E. Heffler, Some aspects of the ocean-continent transition at the continental margin of eastern North America, Offshore Geology of Eastern Canada, Geol. Surv. Can. Pap., vol. 74-30, edited by W.J.M. van den Linden, and J.A. Wade, 189-197, 1975.

Kirby, S.H., Rock mechanics observations pertinent to the rheology of the continental lithosphere and the localization of strain along shear zones, Tectonophysics, 119, 1-27, 1985.

Klein, E.M., and C.H. Langmuir, Global correlations of ocean ridge basalt chemistry with axial depth and crustal thickness, J. Geophys. Res., 92, 8089-8115, 1987.

Klemperer, S.L., A relation between continental heat flow and the seismic reflectivity of the lower crust, J. Geophys., 61, 1-11, 1987.

Klemperer, S.L., T.A. Hauge, E.C. Hauser, J.E. Oliver, and C.J. Potter, The Moho in the northern Basin and Range province, Nevada, along the COCORP 40° N seismic reflection transect, Geol. Soc. Am. Bull., 97, 603-618, 1986.

Kligfield, R., J. Crespi, J. Naruk, and G.H. Davis, Displacement and strain patterns of extensional orogens, Tectonics, 3, 557-609, 1984.

Köppel, V., Isotopic U-Pb ages of monazites and zircons from the crust-mantle transition and adjacent units of the Ivrea and Ceneri zones (southern Alps, Italy), Contrib. Miner. Petrol., 43, 55-70, 1974.

Kusznir, N.J., and R.G. Park, Continental lithosphere strength: the critical role of lower crustal deformation, The Nature of the Lower Continental Crust, Geol. Soc. Spec. Publ., vol. 24, edited by J.B. Dawson, D.A. Carswell, J. Hall, and K.H. Wedepohl, Blackwell, Oxford, 79-94, 1986.

Kusznir, N.J., and R.G. Park, The extensional strength of the continental lithosphere: its dependence on geothermal gradient, and crustal composition and thickness, Continental Extensional Tectonics, Geol. Soc. Spec. Publ., vol. 28, edited by M.P. Coward, J.F. Dewey, and P.L. Hancock, 35-52, 1987.

Lachenbruch, A.H., and J.H. Sass, Models of an extending lithosphere and heat flow in the Basin and Range province, Geol. Soc. Am. Mem., vol. 152, 209-250, 1978.

Lachenbruch, A.H., J.H. Sass, and S.P. Galanis, Jr., Heat flow in southernmost California and the origin of the Salton Trough, J. Geophys. Res., 90, 6709-6739, 1985.

Lambert, R. St. J., Metamorphism and thermal gradients in the Proterozoic continental crust, Geol. Soc. Am. Mem., vol. 161, 155-165, 1983.

LASE Study Group, Deep structure of the US East Coast passive margin from large aperture seismic experiments (LASE), Mar. Petrol. Geol., 3, 234-242, 1986.

Lister, G.S., M.A. Etheridge, and P.A. Symonds, Detachment faulting and the evolution of passive continental margins, Geology, 14, 246-250, 1986.

Manghnani, M.H., R. Ramananantoandro, and S.P. Clark, Jr., Compressional and shear wave velocities in granulite facies rocks and eclogites to 10 kbar, J. Geophys. Res., 70, 5427-5446, 1974.

McGeary, S., Nontypical BIRPS on the margin of the northern North Sea: The SHET survey, Geophys. J. R. astr. Soc., 89, 231-238, 1987.

McKenzie, D., Some remarks on the development of sedimentary basins, Earth Planet. Sci. Lett., 40, 25-32, 1978.

McKenzie, D., A possible mechanism for epeirogenic uplift, Nature, 307, 616-618, 1984a.

McKenzie, D., The generation and compaction of partially molten rock, J. Petrol., 25, 713-165, 1984b.

Mehnert, K., The Ivrea zone, a model of the deep crust, N. Jb. Mineral. Abh., 125, 156-199, 1975.

Meissner, R., The continental crust in central Europe as based on data from reflection seismology, Int. Symp. Deep Structure of the Continental Crust, 57-58, Cornell University, 1984.

Meissner, R., and J. Strehlau, Limits of stresses in the continental crust and their relation to the depth-frequency distribution of shallow earthquakes, Tectonics, 1, 73-89, 1982.

Miller, E.L., P.B. Gans, and J. Garing, The Snake Range decollement: An exhumed mid-Tertiary ductile-brittle transition, Tectonics, 2, 239-263, 1983.

Mooney, W.D., M.C. Andrews, A. Ginzburg, D.A. Peters, and R.M. Hamilton, Crustal structure of the northern Mississippi embayment and a comparison with other continental rift zones, Tectonophysics, 94, 327-348, 1983.

Morel-a-l'Huissier, A.G. Green, and C.J. Pike, Crustal refraction surveys across the Trans-Hudson orogen/Williston basin of South Central Canada, J. Geophys. Res., 92, 6403-6420, 1987.

Moser, D., Structure of the Wawa gneiss terrane near Chapleau, Ontario, Current Research, Part C, Geol. Surv. Canada Pap., 88-1C, 93-99, 1988.

Newton, R.C., Geobarometry of high-grade metamorphic rocks, Am. J. Sci., 283A, 1-28, 1983.

Newton, R.C., and D. Perkins, III, Thermodynamic calibration of geobarometers and basic granulites based on the assemblages garnet-plagioclase-orthopyroxene (clinopyroxene)-quartz with applications to high grade metamorphism, Am. Mineral., 67, 203-222, 1982.

Okaya, D.A., and G.A. Thompson, Involvement of deep crust in extension of the Basin and Range province, Geol. Soc. Am. Mem., vol. 208, 15-22, 1986.

Ottonello, G., W.G. Ernst, and J.L. Joron, Rare earth and 3d transition element geochemistry of peridotitic rocks: I. Peridotites from the Western Alps, J. Petrol., 25, 343-372, 1984.

Pakiser, L.C., Seismic exploration of the crust and upper mantle of the Basin and Range province, Geol. Soc. Am. Cent. Spec. Vol., vol. 1, 453-469, 1985.

Pin, C., Datation U-Pb sur zircons a 285 M.a. du complexe gabbro-dioritique du Val Sesia-Val Mastallone et age tardi-hercynien du metamorphsime granulitique de la zone Ivrea-Verbano (Italie), C.R. Acad. Sc. Paris, vol. 303, Ser. II, no. 9, 827-830, 1986.

Pin, C., and J. Sills, Petrogenesis of layered gabbros and ultramafic rocks from Val Sesia, the Ivrea zone, northwest Italy: trace element and isotope geochemistry, The Nature of the Lower Continental Crust, Geol. Soc. Spec. Publ., vol. 24, edited by J.B. Dawson, D.A. Carswell, J.Hall, and K.H. Wedepohl, Blackwell, Oxford, 231-249, 1986.

Pollack, H.N., and D.S. Chapman, On the regional variation of heat flow, geotherms and lithosphere thickness, Tectonophysics, 38, 279-296, 1977.

Ranalli, G., and D.C. Murphy, Rheological stratification of the lithosphere, Tectonophysics, 132, 281-295, 1987.

Ribe, N.M., Theory of melt segregation - A review, J. Volcan. Geoth. Res., 33, 241-253, 1987.

Rivalenti, G., G. Garuti, A. Rossi, F. Siena, and S. Sinigoi, Existence of different peridotite types and of a layered igneous complex in the Ivrea zone of the western Alps, J. Petrol., 22, 127-153, 1981.

Royden, L., and C.E. Keen, Rifting process and thermal evolution of the continental margin of eastern Canada determined from subsidence curves, Earth Planet. Sci. Lett., 51, 343-361, 1980.

Sandiford, M., and R. Powell, Deep crustal metamorphism during continental extension: modern and ancient examples, Earth Planet Sci. Lett., 79, 151-158, 1986.

Schmid, R., Zur Petrographie und Struktur der Zone Ivrea-Verbano zwischen Valle D'Ossola und Val Grande (Prov. Novara, Italien), Schweiz. Min. Petr. Mitt., 47, 935-1117, 1967.
Schmid, R., Are the metapelites in the Ivrea Zone restites?, Mem. Ist. Geol. Mineral. Univ. Padova, 33, 67-69, 1978/1979.
Schmid, R., and B.J. Wood, Phase relationships in granulite meta-pelites from the Ivrea-Verbano zone (Northern Italy), Contrib. Miner. Petrol., 54, 255-279, 1976.
Schmid, S.M., A. Zingg, and M. Handy, The kinematics of movements along the Insubric Line and the emplacement of the Ivrea Zone, Tectonophysics, 135, 47-66, 1987.
Sclater, J.G., L. Royden, F. Horvath, B.C. Burchfiel, S. Semken, and L. Stegena, Subsidence and thermal evolution of the intra-Carpathian Basins, Earth Planet. Sci. Lett., 51, 139-162, 1980.
Shervais, J.W., Thermal emplacement model for the Alpine lherzolite massif at Balmuccia, Italy, J. Petrol., 20, 795-820, 1979.
Shudofsky, G.N., S. Cloetingh, S. Stein, and R. Wortel, Unusually deep earthquakes in East Africa: Constraints on the thermo-mechanical structure of a continental rift system, Geophys. Res. Lett., 14, 741-744, 1987.
Sills, J.D., Granulite facies metamorphism in the Ivrea Zone, N.W. Italy, Schweiz. Min. Petr. Mitt., 64, 169-191, 1984.
Sinigoi, S., P. Comin-Chiaromonti, and A.A. Alberti, Phase relations in the partial melting of the Baldissero spinel-lherzolite (Ivrea-Verbano Zone, Western Alps, Italy), Contrib. Miner. Petrol., 75, 111-121, 1980.
Sleep, N.H., and B.F. Windley, Archean plate tectonics: Constraints and inferences, J. Geol., 90, 363-379, 1982.
Smith, R.B., Seismicity, crustal structure, and intraplate tectonics of the interior of the western Cordillera, Cenozoic Tectonics and Regional Geophysics of the western Cordillera, Geol. Soc. Am. Mem., vol. 152, edited by R.B. Smith, and G.P. Eaton, 111-144, 1978.
Smith, R.B., and R.L. Bruhn, Intraplate extensional tectonics of the eastern Basin-Range: Inferences on structural style from seismic-reflection data, tectonics, thermal-mechanical models of brittle-ductile deformation, J. Geophys. Res., 89, 5733-5762, 1984.
Steckler, M.S., and A.B. Watts, Subsidence history and tectonic evolution of Atlantic-type continental margins, Dynamics of Passive Margins, Am. Geophys. Un. Geodyn. Ser., vol. 6, edited by R.A. Scrutton, 184-196, 1982.
Stewart, J.H., Basin and Range structure: A system of horsts and grabens produced by deep-seated extension, Geol. Soc. Am. Bull., 82, 1019-1044, 1971.
Talwani, M., J.C. Mutter, R. Houtz, and M. Konig, The crustal structure and evolution of the area underlying the magnetic quiet zone on the margin of South Austalia, Geological and Geophysical Investigations of Continental Margins, Am. Assoc. Petrol. Geol. Mem., vol. 29, edited by J.S. Watkins, L. Montadert, and P.W. Dickerson, 151-176, 1979.
Valasek, P.A., R.B. Hawman, R.A. Johnson, and S.B. Smithson, Nature of the lower crust and Moho in eastern Nevada from "wide-angle" reflection measurements, Geophys. Res. Lett., 14, 1111-1114, 1987.
Voshage, H., J.C. Hunziker, A.W. Hoffmann, and A. Zingg, A Nd and Sr isotopic study of the Ivrea zone, Southern Alps, N. Italy, Conrib. Miner. Petrol., 97, 31-42, 1987.
Warner, M.R., Continental growth by mafic intrusion during lithospheric extension (abstract), Int. Un. Geod. Geophys. XIX Gen. Assem. Abstracts, 1, 66, 1987.
Wass, S.Y., and J.D. Hollis, Crustal growth in south-eastern Australia - evidence from lower crustal eclogitic and granulitic xenoliths, J. Metamorph. Geol., 1, 25-45, 1983.
Weigel, W., G. Wissmann, and P. Goldflam, Deep seismic structure (Mauritania and Central Morocco), Geology of the Northwest African Continental Margin, edited by U. vanRad, K. Hinz, M. Sarntheim, and E. Seibold, Springer-Verlag, Berlin, Heidelberg, 132-159, 1982.
Wells, P.R.A., Thermal models for the magmatic accretion and subsequent metamorphism of continental crust, Earth Planet. Sci. Lett., 46, 253-265, 1980.
Wernicke, B., Low-angle normal faults in the Basin and Range Province: Nappe tectonics in an extending orogen, Nature, 291, 645-648, 1981.
White, R.S., G.D. Spence, S.R. Fowler, D.P. McKenzie, G.K. Westbrook, and A.N. Bowen, Magmatism at rifted continental margins, Nature, 330, 439-444, 1987.
Wickham, S.M., and E.R. Oxburgh, Continental rifts as a setting for regional metamorphism, Nature, 318, 330-333, 1985.
Windley, B.F., and J. Tarney, The structural evolution of the lower crust of orogenic belts, present and past, The Nature of the Lower Continental Crust, Geol. Soc. Spec. Publ., vol. 24, edited by J.B. Dawson, D.A. Carswell, J. Hall, and K.H. Wedepohl, Blackwell, Oxford, 221-230, 1986.
Wood, B.J., Petrological constraints on the constitution of the lower crust, Geol. Soc. Am. Abstr. with Prog., 15, 722, 1983.
Wood, B.J., Mineralogic constitutions of granulites under lower crustal conditions, Eos Trans. Am. Geophys. Un., 65, 288, 1984.
Wyllie, P.J., Crustal anatexis: An experimental review, Tectonophysics, 43, 41-71, 1977.
Zingg, A., Regional metamorphism in the Ivrea zone (Southern Alps, Northern Italy)-a review, Schweiz. Min. Petr. Mitt., 63, 361-392, 1983.

GRANULITE TERRANES AND THE LOWER CRUST OF THE SUPERIOR PROVINCE

John A. Percival

Lithosphere and Canadian Shield Division
Geological Survey of Canada, Ottawa, Ontario

Abstract. Granulites are a common component of the lower crust but the study of surface granulites can provide information on the contemporary lower crust only in certain rare examples. For example, granulite terranes of the Archean Superior Province formed in three distinct tectonic environments which governed their burial and uplift history: 1) ancient terranes, such as the Minnesota River Valley gneiss terrane, metamorphosed at modest pressure (4.5-6.5 kbar) in the granulite facies, possibly representing a continental collision environment; 2) greenstone-granite terranes, with granulitic (7.5-9 kbar) lower crust, that formed in magmatic arc environments; and 3) metasedimentary belts, with small to large, 4.5-6.5 kbar granulite occurrences, that represent accretionary wedges, tectonically thickened and subsequently metamorphosed during uplift. Collisional and accretionary settings lead to the rapid exposure of high-grade metamorphic rocks through tectonic thickening, heating and consequent erosion; however, thickening through magmatism rarely results in exhumation of deep crustal rocks. The Kapuskasing and Pikwitonei terranes of the Superior Province represent basal magmatic arcs, exposed late in their metamorphic history by Proterozoic faults, based on characteristics attributed to exposed crustal cross-sections, including association with faults, gradational metamorphic zonation, high maximum metamorphic pressures and retarded cooling rates.

Granulite formation is an integral part of the process of production of stable continental crust in the Superior Province. Information on the nature of the contemporary lower crust can be inferred only from faulted granulite terranes although similar petrogenetic processes are recognized in all terrane types.

Introduction

Rocks of the granulite facies constitute large parts of the lower continental crust, based on evidence from geophysical interpretation (Smithson and Brown, 1977), studies of xenoliths from the deep crust (Kay and Kay, 1981) and exposed crustal cross-sections (Giese, 1968; Berckhemer, 1969; Fountain and Salisbury, 1981; Percival and Berry, 1987). Granulites also make up a significant part of the contemporary upper crust, particularly of Precambrian shield areas. A common approach to understanding the character and evolution of the generally inaccessible lower crust is to use exposed granulite terranes as analogues. However, an important first requirement is to establish the regional context and evolution of the terrane in order to determine whether or not it represents the present day lower crust.

The process of establishing the regional context of a given granulite terrane involves lithological and structural correlation of high and lower grade regions, as well as estimation of the relative timing of events and quantification of metamorphic conditions, particularly depth. Regional correlation can be accomplished only through field mapping, in conjunction with precise geochronology. Many U-Pb studies of Precambrian terranes have demonstrated short but complex evolution of volcanic, intrusive and tectonic events (e.g. Hoffman and Bowring, 1984), necessitating the precision of modern U-Pb zircon techniques (e.g. Krogh, 1982). Similarly, reliable techniques for estimation of depths of granulite metamorphism (e.g. Newton and Perkins, 1982; Bohlen and Lindsley, 1987) are required to establish relationships in the third dimension. Together, these tools are used to re-assess the significance of granulite terranes.

Granulites and the Lower Crust

Granulite terranes are made up of high grade metamorphic rocks (700-900°C, 4-12 kbar; Newton and Perkins, 1982) containing orthopyroxene in appropriate bulk compositions. Supracrustal rocks are common, if minor, components. Criteria for recognition of granulite terranes that are representative of the contemporary lower crust must be based on an understanding of the origin of specific terranes. For example, Bohlen (1987) used P-T path information from several granulite terranes, with particular regard to the Adirondacks, to infer magmatic arc origins. In this environment, granulite metamorphism occurs in the middle and lower parts of a magmatically thickened crust; the exposed middle part may not be representative of the underlying lower crust. Conversely, granulite terranes at the base of exposed crustal cross-sections such as the Ivrea zone of the southern Alps (Schmid, 1967; Berckhemer, 1969;

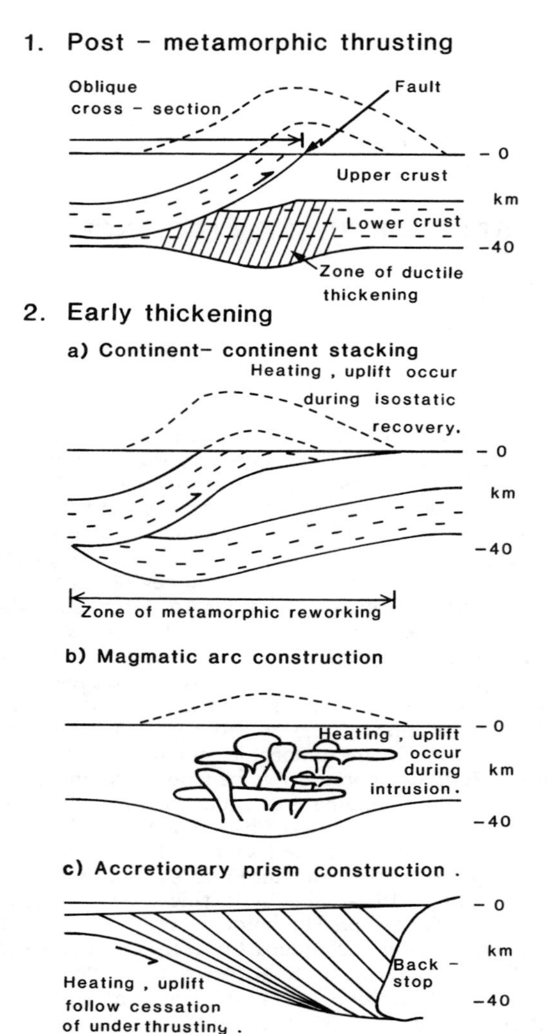

Fig. 1. Tectonic environments responsible for formation and/or emplacement of granulite terranes. 1) Post-metamorphic uplift of granulite terranes representative of contemporary lower crust. 2) Environments of early crustal thickening leading to metamorphism and uplift; a) continental collision setting, leading to synmetamorphic removal of the upper plate; b) magmatic arc setting, leading to thickening without necessarily significant uplift; c) accretionary wedge setting, involving synmetamorphic removal of the overthickened crustal volume.

Zingg, 1980; Fountain and Salisbury, 1981) were thrust into position at some time after establishment of the gross crustal structure; with the exception of structures imposed during uplift, the features of these terranes can be used to derive information about the contemporary deep crust. Similar rock types may form at identical P-T conditions in several different environments along diverse burial and uplift paths, making additional criteria necessary for distinction.

The two examples cited above represent fundamentally different modes of granulite emplacement (Figure 1): 1) slivers of granulite from the deep crust exhumed late in their metamorphic history; and 2) large, regional terranes metamorphosed and subsequently exposed as a consequence of crustal thickening early in the tectonic history. Mechanisms for thickening of 2) are grouped into three classes based on tectonic setting (Figure 1), including a) continent-continent stacking, b) magmatic arc zones, and c) accretionary prisms. If surface granulites are representative of the basal component of normal contemporary crust, there must also be tectonic processes by which supracrustal rocks are buried and metamorphosed to high grade, but not returned to the surface.

Two additional mechanisms of granulite formation have been considered to be important in the literature. First, high grade metamorphism may occur in the lower crust in response to mafic intrusion and high heat flow in extensional settings (e.g. Sandiford and Powell, 1986; Fountain, 1987). For example, xenolith studies, in conjunction with heat flow considerations, indicate that granulites may be currently forming beneath the Rio Grande rift (Padovani and Carter, 1977). Mafic sills in the high-grade Ivrea zone are interpreted as rift features (Sills and Tarney, 1984), later exposed by continental obduction (Fountain and Salisbury, 1981). A second process involves syn-metamorphic CO_2 flushing, and dehydration by fluids derived from mantle degassing or devolatilization of subducted carbonates (e.g. Newton et al., 1980). Recent work on the southern Indian localities of classic carbonic metamorphism suggests that the process may be minor in importance (Stahle et al., 1987).

A variety of granulite terranes and terrane types is present in the late Archean Superior Province, from which examples will be drawn to illustrate the diversity within granulite terranes that results from differences in tectonic setting during formation and/or emplacement.

Granulite Terranes of the Superior Province

Metamorphic grade in the Superior Province ranges from sub-greenschist facies (Jolly, 1978) to granulite facies. In general, the east-striking greenstone belts (Figure 2) of the southern part of the Province are characterized by greenschist and amphibolite facies (Ermanovics and Froese, 1978), whereas the intervening metasedimentary gneiss belts are at upper amphibolite and locally granulite facies (Breaks et al., 1978; Pirie and Mackasey, 1978; Fig. 2). The east-striking belts are transected by the Kapuskasing uplift which exposes granulites over about 400 km of strike length (Percival and McGrath, 1986). Smaller granulite terranes occur on the northwestern (Pikwitonei region) and southwestern (Minnesota River Valley) margins of the Province. Relatively little is known of two large granulite terranes in the northeastern part of the Province, the Minto and Ashuanipi subprovinces (Card and Ciesielski, 1986), mapped only at reconnaissance scale (Eade, 1966; Stevenson, 1968; Herd, 1978). Each terrane will be reviewed in terms of its setting, age, metamorphic conditions and origin, arbitrarily working upwards in scale.

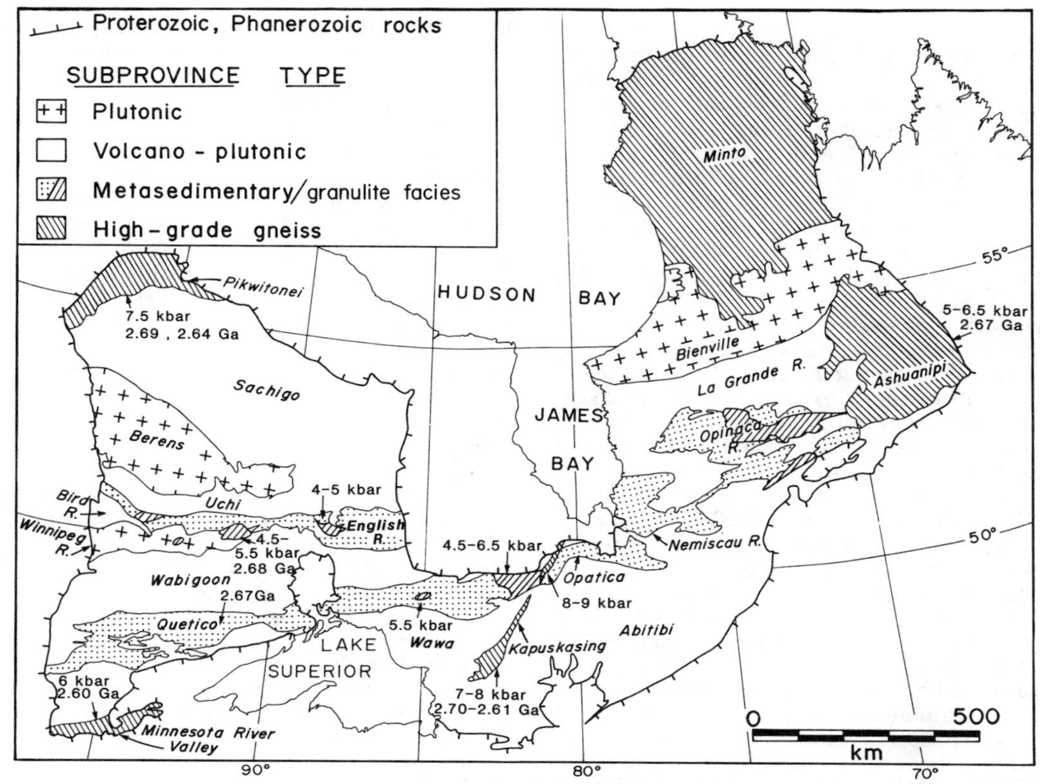

Fig. 2. Tectonic map of Superior Province, showing distribution of subprovinces (after Card and Ciesielski, 1986) and granulite terranes. Sources of data for pressure (kbar) and time (Ga) of metamorphism in individual terranes are mentioned in the text.

English River Subprovince

Several small granulite nodes (Figure 2) occur in the metasedimentary English River and adjacent plutonic Winnipeg River subprovince (Card and Ciesielski, 1986). Most of the English River subprovince is made up of steeply dipping units of metasedimentary migmatite, foliated and gneissic tonalite and peraluminous granite (Breaks et al., 1978), whereas the Winnipeg River subprovince consists mainly of plutonic rocks of 2.7-3.1 Ga age (Beakhouse, 1985; Krogh et al., 1976). Metamorphic grade in the migmatites ranges from upper amphibolite to granulite facies (Thurston and Breaks, 1978; Perkins and Chipera, 1985), with the boundary defined either by an orthopyroxene isograd, or by the reaction:

biotite + sillimanite + quartz = garnet + cordierite + melt.

Pressure estimates for orthopyroxene-, garnet-bearing migmatites range from 4.5 to 5.5 kbar (Perkins and Chipera, 1985), indicating removal of 15-20 km of cover. Metamorphism is dated by U-Pb on zircon from leucosome at 2680 Ma (Krogh et al., 1976). In the Winnipeg River subprovince to the south, a pod of granulite occurs in metaplutonic rocks and another straddles the metasedimentary-metaplutonic boundary (Breaks et al., 1978). In the eastern continuation of the metasedimentary belts in Quebec (Nemiscau River subprovince), several patches of granulite facies, of slightly larger dimension than the western occurrences, have been reported (Eade, 1966; Herd, 1978).

Beneath the English River subprovince, the crust appears to be thinner, by about 5 km, than that of the adjacent regions, whereas the Conrad discontinuity is deeper, at about 15 km (Hall and Hajnal, 1973). Gravity modeling suggests continuity of surface units to depths of about 10 km (Gupta and Barlow, 1984).

Several tectonic models have been suggested to explain the high-grade metamorphism in the English River and Winnipeg River subprovinces. Langford and Morin (1976) proposed a fore-arc or back-arc setting, based on the high proportion of sedimentary rock types, whereas Breaks et al. (1978), noting abundant intrusions, favoured a magmatically heated environment. Perkins and Chipera (1985) suggested collisional processes, aided by input of magmatic heat. Further discussion of the origin of these granulites in their regional context follows description of other metasedimentary gneiss belts.

Quetico Subprovince

The Quetico, like the English River subprovince, consists dominantly of complexly deformed metasedimentary rocks, with concordant bodies of peraluminous and other granites (Pirie and Mackasey, 1978; Percival, in press). The belt is fault bounded against low-grade rocks of metavolcanic origin to the north and south. Metamorphic grade increases systematically from greenschist facies at the belt margins, to migmatite grade, with abundant intrusive granite, in the centre (Pirie and Mackasey, 1978; Percival et al., 1985; Sawyer, 1987). Assemblages of garnet-cordierite-andalusite-staurolite ± sillimanite in the central part of the belt in the west (Percival et al., 1985), similar to those in metasediments of the Wabigoon volcano-plutonic subprovince to the north (Poulsen, 1984), indicate pressures of 3.5-4 kbar and suggest similar erosion levels in adjacent metasedimentary and greenstone-granite subprovinces. Orthopyroxene occurs within the migmatite zone in the east at two localities (Percival, 1985; Percival and McGrath, 1986). Metamorphic pressure recorded by the garnet-orthopyroxene-plagioclase-quartz barometer (Newton and Perkins, 1982), is in the 4.5-6.5 kbar range, slightly higher than that inferred for the central part of the belt to the west. Although the granulite metamorphism has not been dated directly, regional metamorphism in the Quetico belt was dated indirectly at 2670 Ma by U-Pb on monazite from peraluminous granite (Percival and Sullivan, 1985), which appears to be both a product of deep anatexis and a heat source at the exposed level. Based on structural, stratigraphic and geochronological considerations, Percival and Williams (in press) proposed that the Quetico sediments were deposited in an accretionary prism, shortly after 2700 Ma ago (Percival and Sullivan, in press).

Minnesota River Valley Gneiss Terrane (MRVGT)

This poorly exposed terrane is well known for its ancient rocks, in excess of 3.3 Ga, in contrast to the common 3.1-2.65 Ga ages of the bulk of the Superior Province. The oldest rocks are amphibolite facies orthogneiss and migmatite, with 3.3 and 3.05 Ga components (Goldich et al., 1980). Paragneiss and mafic gneiss contain garnet-orthopyroxene assemblages (Himmelberg and Phinney, 1967) which yield P-T estimates of 6 kbar, 700°C (Moecher et al., 1986; Perkins and Chipera, 1985). High grade metamorphism and intrusion by late granites are dated at approximately 2.6 Ga (Goldich et al., 1980).

The relationship of the MRVGT to the northern Superior Province is unclear. It may have been an ancient nucleus, to which the late Archean belts were accreted, or alternatively, it may have itself been an exotic block, accreted to the Superior Province (Stockwell, 1982). A COCORP line across the terrane (Gibbs et al., 1984) indicates north-dipping reflectors, which do not resolve this question.

Pikwitonei Region

The Pikwitonei region of Archean granulite facies rocks occurs at the northwestern edge of the Superior Province adjacent to the Proterozoic Churchill Province. It is separated from lithologically and structurally contiguous lower grade granite-greenstone terrane to the east by an orthopyroxene isograd (Hubretgse, 1980) and from younger rocks to the northwest by a zone of retrogression as well as oblique thrust and sinistral transcurrent faults (Bleeker, in press). Rock types of the Pikwitonei region include steeply dipping units of tonalite, granodiorite and anorthosite, with less abundant mafic gneiss, paragneiss, calc-silicates and ultramafic bodies. Metamorphic conditions based on coexisting sapphirine and quartz were estimated at 9 kbar, 980°C (Arima and Barnett, 1984); Paktunc and Baer (1986) found evidence for an east-west pressure gradient of 6-12 kbar at 700-800°C. However, re-evaluation by Mezger et al. (1986), using more reliable geobarometers, suggests consistent 7-8 kbar values across the terrane. A zone of partial to complete amphibolite facies retrogression of Proterozoic age occurs adjacent to the Churchill Province (Weber and Scoates, 1978). Two high grade metamorphic events were recognized on the basis of cross-cutting, orthopyroxene-bearing leucosomes (Hubretgse, 1980); zircon geochronology indicates growth of metamorphic zircon at 2690 and 2637 Ma (Heaman et al., 1986). The younger age is anomalous with respect to the age of regional metamorphism of greenstone belts of the Superior Province, generally in the range 2670-2700 Ma (Krogh et al., in Ayres and Thurston, 1985). Heat production in the granulites is lower than that in the corresponding lithologies at lower metamorphic grade, suggesting some geochemical depletion as the result of high grade metamorphism (Fountain et al., 1987).

Several models for the tectonic setting of the Pikwitonei region have been proposed. Fountain and Salisbury (1981) considered it to be an exposed crustal cross-section that suffered Proterozoic effects during uplift whereas Paktunc and Baer (1986) proposed an Archean continent-collision zone model. Bohlen (1987) regarded the terrane as an example of a deeply eroded magmatic arc. Bleeker (in press) suggested that the Pikwitonei region is the upwarped foreland to an early Proterozoic collisional orogen (see also Hoffman, 1988).

Kapuskasing Uplift

Geophysical anomalies associated with the northeast-striking Kapuskasing uplift transect the easterly subprovince structure of the Superior Province over a distance of almost 500 km (Figure 2). Interpretations of the origin of the structure have come full circle from Garland's (1950) "thinning of the granitic upper crustal layer", through models of a Proterozoic rift-related structure (Innes, 1961; Bennett et al., 1967; Burke and Dewey, 1973), to a transcurrent fault zone of early Proterozoic age (Watson, 1980), and more recently, an intracratonic, "basement-uplift" style thrust, exposing an oblique crustal cross-section (Percival and Card, 1983 & 1985; Percival and McGrath, 1986). This latter interpretation has significant implications for the properties and origin of the contemporary lower crust beneath the Superior Province and is undergoing rigorous tests as part of the Canadian LITHOPROBE program (Geis, et al., 1988; Percival and Green, in press).

High-grade rocks of the Kapuskasing uplift comprise

three geologically and geophysically distinct segments, from south to north, the Chapleau, Groundhog River and Fraserdale-Moosonee blocks (Percival and McGrath, 1986). In the south, a continuous transition is exposed from the greenschist facies Michipicoten greenstone belt of the Wawa subprovince, through amphibolite facies tonalitic orthogneiss of the Wawa gneiss terrane, to a heterogeneous gneiss sequence in the upper amphibolite and granulite facies in the Chapleau block of the Kapuskasing uplift. Across this lithological and metamorphic transition, metamorphic pressure increases from 2-3 kbar in greenschists (Studemeister, 1983), to 5-6 kbar in amphibolites, based on hornblende geobarometry (Hammarstrom and Zen, 1986) of tonalites, to 7-8 kbar in granulites, based on garnet-pyroxene-plagioclase-quartz barometry (Percival, 1983). A moderately dipping sequence of tonalitic, dioritic, mafic rocks, paragneiss and anorthosite, interlayered on a km scale, makes up the Chapleau block (Thurston et al., 1977; Percival and Card, 1985). It is separated from low grade rocks of the Abitibi belt to the east by a brittle, northwest dipping thrust, the Ivanhoe Lake fault (Percival and Card, 1983; Cook, 1985; Geis et al., 1988). Over part of its length, the western boundary is a complex lithological, structural and metamorphic transition which may represent an exposed example of the Conrad mid-crustal seismic discontinuity (Percival, 1986), based on density and seismic velocity contrasts (Percival and Fountain, in press). In a zone several tens of km west of and structurally above this boundary, Moser (1988) reported evidence of subhorizontal, synmagmatic extensional structures; recent work indicates similar structural style throughout the high-grade Kapuskasing zone (Bursnall, 1988; D. Moser, pers. comm., 1988). Although interpretation of the exact timing and regional significance of these structures depends upon geochronology in progress, they could be analogues of the subhorizontal, discontinuous reflectors present at mid to deep crustal levels on many seismic reflection profiles.

The present 53+ km crust beneath parts of the Chapleau block is thicker by 10-15 km than that of the adjacent regions (Northey and West, 1986; Boland et al., 1987). This Moho downwarp may be considered as an isostatic crustal root supporting the anomalously dense slab of granulite, out of position near the surface.

In an attempt to define depth-related changes in crustal composition, Truscott and Shaw (1986) determined weighted average compositions for the middle and lower parts of the crustal profile. In comparison with published estimates of the composition of the lower continental crust, the Chapleau block is more aluminous and siliceous, undoubtedly reflecting the presence of the large Shawmere anorthosite complex. Ashwal et al. (1987) reported variable, but generally low values of heat production from all levels of the crustal section, which they attributed to the mafic character of the greenstone belt crust.

Profiles of radiometric age through the crustal section show decreasing apparent age with increasing paleopressure and crustal level (Percival et al., 1988), in the U-Pb zircon (2.7-2.61 Ga) and sphene systems (2.69-2.49 Ga) (Percival and Krogh, 1983; Krogh et al., 1986 & 1988), the K-Ar and $^{40}Ar/^{39}Ar$ hornblende (2.69-2.48 Ga) and biotite (2.50-2.0 Ga) systems (Hunt and Roddick, 1987; Archibald et al., 1986; Lopez-Martinez and York, 1986) and the Rb-Sr biotite system (2.50-1.93 Ga) (Z.E. Peterman, pers. comm., 1985). Furthermore, the age gap between systems with different blocking temperatures widens with depth. The age-depth pattern can be interpreted in terms of progressively slower cooling with depth, down to temperatures in the 300°C range, because uplift of rocks above 300°C would have caused instantaneous cooling, with a resultant plateau in the distribution of biotite ages. This concept of uplift of a cold slab is consistent with the brittle nature of the exposed faults that are inferred to have carried the crustal section toward the surface.

The metamorphic environment of the basal granulites of the section (25-30 km; 700-800°C; Percival, 1983) could have been produced at depth within a magmatic arc, based on the high proportion of igneous rocks throughout the section and evidence for prolonged residence in the lower crust. The thermal effects of the long magmatic history may be reflected in the metamorphism and growth of new zircon for about 100 Ma after major magmatism (Krogh et al., 1988), in the manner modeled by Wells (1981). Percival and Card (1985) proposed a tectonic model whereby supracrustal rocks of the Kapuskasing zone were progressively buried to lower crustal depths by volcanic accumulations and subsequent intraplated tonalitic intrusions. Many authors have considered Archean greenstone belts to be analogous to contemporary island arcs, however the 40 ± 5 km average crustal thickness of greenstone belts is anomalous with respect to modern 20 km thick intra-oceanic arcs. Either accretion of arcs causes structural and magmatic thickening and subsequent intracrustal melting, or thicker, more felsic arcs were produced in the Archean owing to subduction of warmer oceanic crust of average younger age than at present (Abbott and Hoffman, 1984; Martin, 1986).

North of the Chapleau block, the Groundhog River and Fraserdale-Moosonee blocks are bounded by faults both to the east and west. Although the eastern faults are considered to represent the northward continuation of the Ivanhoe Lake fault, the western faults appear to have variable west-side down normal displacement (Percival and McGrath, 1986; Leclair and Nagerl, 1988). The Groundhog River block differs from the Chapleau block in its lack of anorthosite, widespread two-pyroxene assemblages, slightly lower metamorphic pressures, more positive aeromagnetic anomalies and contained Matachewan diabase dykes, present regionally but absent from the Chapleau block (Thurston et al., 1977). The Fraserdale-Moosonee block, which transects the Quetico-Opatica metasedimentary gneiss belt, consists dominantly of metasedimentary granulites with 8-9 kbar metamorphic signatures (Percival and McGrath, 1986), as well as minor tonalite and mafic gneiss (Card, 1982; Percival, 1985). The two northern blocks of the Kapuskasing uplift therefore expose deep crust representative of the lower parts of the Superior Province, however the faulting history appears to have left the blocks stranded.

Measurement of heat flow and heat generation in the

area of the Groundhog River block indicated low values for the highest grade sampled (Cermak and Jessop, 1971), although no granulites were intersected owing to non-coincidence of geophysical anomalies and granulites. In a test of whether the lower crust, known to be electrically conductive at normal depth (20-40 km), is also conductive where it reaches the surface, Woods and Allard (1986) measured conductivity on a regional scale in the Kapuskasing area. The lack of an anomaly associated with the structure led Woods and Allard to suggest that lower crustal conductivity is a result of the presence of deep fluids which escaped from the Kapuskasing rocks during uplift.

Ashuanipi Complex

This 300 x 300 km zone of high grade metamorphic rocks occurs at the western end of the easterly subprovince structure of the Superior Province. It is known from regional reconnaissance (Eade, 1966; Herd, 1978) as well as more detailed work on the eastern side (Percival, 1987; Percival and Girard, 1988), in the vicinity of minor gold showings (Lapointe, 1986).

The complex consists mainly of intrusive bodies of orthopyroxene-bearing granodiorite (Eade, 1966) which contain remnants of paragneiss, iron formation and orthogneiss and are cut by late granitoid plutons, pegmatite and syenite. Orthopyroxene, garnet and biotite are ubiquitous constituents of felsic rock types; cordierite and sillimanite are rare. Two-pyroxene, hornblende assemblages characterize the rare mafic and ultramafic units. Paragneiss, the oldest unit recognized, is interlayered with sills of variably deformed and recrystallized orthopyroxene-bearing tonalite, as well as local pyroxenite. Following high grade metamorphism and deformation that produced moderate to steep, northwesterly gneissic layering, the sequence was intruded by voluminous orthopyroxene $\pm$ garnet-bearing granodiorite, similar in composition to, and probably derived from, paragneiss. Equilibration of garnet-orthopyroxene in both paragneiss and granodiorite occurred at 700-800°C, 5-6.5 kbar (Percival and Girard, 1988). This sequence was then involved in open, upright, second-phase folds and cut by syn-tectonic pegmatites before intrusion of late plutons. U-Pb zircon geochronology defines a sequence of events beginning with deposition of sedimentary rocks slightly after 2700 Ma, followed by intrusion of early tonalite at about 2690 Ma, and by major orthopyroxene granodiorite intrusions at 2690-2670 Ma and later pegmatite injection at about 2652 Ma (Mortensen and Percival, 1987; Percival et al., 1988). Monazite U-Pb ages, recording cooling to the 650-700°C level (Parrish, 1988), occupy the 2668-2660 Ma bracket, suggesting rapid cooling from the metamorphic peak, probably during unroofing of the complex. The granulites were exposed by the time of deposition of the Kaniapiskau Supergroup of the Labrador Trough about 2150 Ma ago. Patchy amphibolite retrogression (Herd, 1978) is dated by growth of new metamorphic zircon at 2642 $\pm$ 3 Ma (Mortensen and Percival, 1987).

Based on lithological, metamorphic and geochronological similarity and on-strike position, Mortensen and Percival (1987) correlated the Ashuanipi complex with metasedimentary gneiss belts of the western Superior Province, such as the English River, Quetico and Nemiscau, implying a common origin. Hence, rather than considering the Ashuanipi complex to be an isolated block of granulite, it is part of a belt of metasedimentary rocks of at least 2000 km strike length, which contains several nodes of low-pressure granulite. The scale of the belt resembles that of components of Phanerozoic mountain belts; geological features such as sedimentary character with early tonalitic intrusions and low-P metamorphic zonation are similar to those of some accretionary wedges, for example the Chugach terrane of Alaska (Hudson and Plafker, 1982; Sisson and Hollister, 1988). A model for the origin of the Ashuanipi complex and western Superior Province metasedimentary belts involves deposition and accretion of a thick sedimentary wedge, slightly after 2700 Ma ago. Heating of the base of the overthickened crust, by intrusions and thermal relaxation following cessation of subduction, produced crustal melts which rose and crystallized, transporting heat and volatile components upward. If this model is valid, the exposed Ashuanipi granulites are not representative of the 35 km thick granulite column beneath, which would be relatively refractory.

Minto Subprovince

This 500 x 500 km zone of high-grade metamorphic rocks forms a distinct northern part of the Superior Province, characterized by northerly structural and aeromagnetic trends (Stevenson, 1968; Card and Ciesielski, 1986). The major lithological components are similar to those of the Ashuanipi complex, comprising an early sequence of paragneiss, iron formation, orthogneiss and small ultramafic bodies, intruded by large orthopyroxene-bearing plutons and late granitoid bodies (Stevenson, 1968). Herd (1978) noted widespread granulite facies metamorphism, overprinted by extensive amphibolite facies retrogression.

Based on the discordance in aeromagnetic and structural trends, Card and Ciesielski (1986) suggested that the Minto subprovince could be an older nucleus onto which the easterly belts of the southern Superior Province were accreted. Sparse geochronological data on granodiorites from the eastern Minto subprovince, indicating plutons of 2720 Ma age (Machado et al., 1987), do not support this interpretation. Correlating the Ashuanipi and Minto subprovinces on the basis of lithological similarity and broadly similar metamorphic history would make the structural discordance difficult to explain. Clearly, more work is needed to provide constraints on regional tectonic interpretation.

Lower Crust?

Granulites are commonly considered to have formed in the lower part of a thickened continental crust, to account for evidence of 6-8 kbar metamorphism at a surface underlain by normal crustal thickness (e.g. Newton and Perkins, 1982). If the crust is thickened tectonically, a consequence is rapid isostatic uplift (England and Richardson, 1977; England and Thompson, 1984), revealed in decompressional P-T paths (St-Onge, 1987) and rapidly cooling isotopic systems (Parrish, 1983;

Tucker et al., 1987). Evidence for rapid, regional scale, uplift in the Ashuanipi complex suggests such an early overthickening event, possibly related to accretion. Granulite terranes that were thickened early in their tectonic history do not represent the contemporary lower crust, although similar processes occurred at equivalent depths.

Granulite terranes which experienced prolonged residence at depth and incidental tectonic uplift, such as the Kapuskasing uplift and the Pikwitonei region, are inferred to be samples of the contemporary lower crust. The southern Kapuskasing and Pikwitonei regions probably originated in magmatic arc environments, beneath greenstone belts, whereas the northern Kapuskasing structure exposes the lower crust of the Quetico metasedimentary belt. Late tectonic events are required to transport lower crust formed in these environments to the surface (cf. Ellis, 1987). Proterozoic faults are implicated in the uplift of both the Pikwitonei and Kapuskasing terranes. Prolonged magmatic underplating was suggested by Bohlen (1987) as a mechanism for substantial crustal thickening and uplift of mid-crustal rocks. The nature of the uplift process should be distinguishable by the duration of the uplift history, which would be brief and follow immediately on intrusion and metamorphism in Bohlen's (1987) model.

Granulite terranes produced by continent-continent stacking are rare in the Superior Province. A possible example is the Minnesota River Valley terrane, where distinctly pre-metamorphic crustal components have been identified. An older (3.30-3.05 Ga) terrane was reworked during a granulite facies event at about 2.6 Ga. A similar history characterizes parts of the Winnipeg River subprovince, where 3.1 Ga rocks were metamorphosed and intruded at about 2.68 Ga. Both of the older terranes are relatively small and could have been microcontinental fragments involved in accretionary processes, rather than the true continents necessary to drive metamorphism (England and Thompson, 1984) in collisional settings.

Conclusions

Granulite terranes of the Superior Province formed in three distinct tectonic environments which controlled their burial and uplift history. Two settings, accretionary wedges and magmatic arcs, accounting for most granulites of the Superior Province, yield high-grade metamorphic rocks as an integral part of formation and stabilization of the continental crust. Continent-continent stacking, presumed to be the major environment for production of granulite terranes that reach the surface in younger belts, may be represented by terranes such as the Minnesota River Valley gneiss terrane and the Winnipeg River subprovince. In these areas, older, sialic rocks were involved in later Archean high-grade metamorphism, but the limited extent of the terranes may not qualify them as true continents and thus this setting is apparently less significant than the others.

Uplift magnitude and timing can be related to the amount of tectonic thickening. Both accretionary wedges and continental collisions may lead to transitory states of overthickened crust which responds isostatically to remove excess overburden. Syn-metamorphic heating and uplift of up to 20 km produce familiar clockwise P-T loops and a 35 ± 5 km granulitic crust beneath the present exposure level. Direct information on the contemporary lower crust in these environments is not available from surface studies. Magmatic arc environments, represented by greenstone-granite belts in the Superior Province, generally show minor uplift to expose greenschist or amphibolite facies rocks. Where uplifted along later structures, the dominantly magmatic lower crust shows a complex history of intrusion, metamorphism and retarded cooling. Information on the nature of the contemporary lower crust can be derived from such environments, provided that uplift was relatively passive, because deep crustal characteristics were frozen in at the time of cooling.

Insights into lower crustal structure and evolution beneath an Archean greenstone-granite terrane have been derived from the Kapuskasing uplift. These include documentation of continuous high-grade metamorphic activity for approximately 100 Ma after crustal stabilization (Krogh et al., 1988) and recognition of spaced, subhorizontal, extensional, high-strain zones (Moser, 1988) which could be analogues of deep crustal reflections observed on many crustal-scale seismic profiles.

Acknowledgements. Discussion and a review by K.D. Card are appreciated. This is Geological Survey of Canada Contribution No. 21588.

References

Abbott, D.H., and S.E. Hoffman, Archean plate tectonics revisited 1: Heat flow, spreading rate, and the age of subducting oceanic lithosphere and their effects on the origin and evolution of continents, Tectonics, 3, 420-448, 1984.

Archibald, D.A., E. Farrar and J.A. Hanes, An $^{40}Ar/^{39}Ar$ study of the eastern boundary of the Kapuskasing structural zone: Ivanhoe Lake cataclastic zone and contiguous Abitibi greenstone belt, Geol. Assoc. Can. Prog. Abstr., 11, 41-42, 1986.

Arima, M., and R.L. Barnett, Sapphirine bearing granulites from the Sipiwesk Lake area of the late Archean Pikwitonei granulite terrain, Manitoba, Canada, Contrib. Mineral. Petrol., 88, 102-112, 1984.

Ashwal, L.D., P. Morgan, S. Kelley, and J.A. Percival, Heat production in an Archean crustal profile and implications for heat flow and mobilization of heat producing elements, Earth Planet. Sci. Lett., 85, 439-450, 1987.

Ayres, L.D., and P.C. Thurston, Archean supracrustal sequences in the Canadian Shield: an overview, in L.D. Ayres, P.C. Thurston, K.D. Card, and W. Weber, eds., Evolution of Supracrustal Sequences, Geol. Assoc. Can. Sp. Pap., 28, 343-380, 1985.

Beakhouse, G.P., The relationship of supracrustal sequences to a basement complex in the western English River subprovince; in L.D. Ayres, P.C. Thurston, K.D. Card, and W. Weber, eds.,

Evolution of Supracrustal Sequences, Geol. Assoc. Can. Sp. Pap., 28, 169-178, 1985.

Bennett, G., D.D. Brown, P.T. George, and E.J. Leahy, Operation Kapuskasing, Ontario Dept. Mines, Misc. Pap., 10, 98 p., 1967.

Berckhemer, H., Direct evidence for the composition of the lower crust and the Moho, Tectonophysics, 8, 97-105, 1969.

Bleeker, W., New structural-metamorphic constraints on Early Proterozoic oblique collision along the Thompson nickel belt, northern Manitoba, Canada, in J.F. Lewry and M.R. Stauffer, eds., The Early Proterozoic Trans-Hudson Orogen, Geol. Assoc. Can. Sp. Pap., (in press).

Bohlen, S.R., Pressure-temperature-time paths and a tectonic model for the evolution of granulite; J. Geol., 95, 617-632, 1987.

Bohlen, S.R., and D.H. Lindsley, Thermometry and barometry of igneous and metamorphic rocks, Ann. Rev. Earth Planet. Sci., 15, 397-420, 1987.

Boland, A.V., R.M. Ellis, G.F. West, and D.J. Northey, A crustal scale seismic refraction experiment over the Kapuskasing structural zone, northern Ontario, (Abstr.) Inter. Un. Geodesy Geophys. XIX, 1, 311, 1987.

Breaks, F.W., W.D. Bond, and D. Stone, Preliminary geological synthesis of the English River subprovince, northwestern Ontario, and its bearing upon mineral exploration, Ontario Geol. Surv. Misc. Pap., 72, 55 p., 1978.

Burke, K., and J.F. Dewey, Plume-generated triple junctions: key indicators in applying plate tectonics to old rocks, J. Geol., 81, 406-433, 1973.

Bursnall, J.T., Deformation sequence from the southeastern part of the Kapuskasing structural zone and its boundary with the Abitibi subprovince in the vicinity of Ivanhoe Lake, in 1988 KSZ Lithoprobe Wkshp Univ. Toronto, 75-84, 1988.

Card, K.D., Progress report on regional geological synthesis, central Superior Province, Geol. Surv. Can. Pap., 82-1A, 23-28, 1982.

Card, K.D., and A. Ciesielski, Subdivisions of the Superior Province of the Canadian Shield, Geosci. Can., 13, 5-13, 1986.

Cermak, V., and A.M. Jessop, Heat flow, heat generation and crustal temperature in the Kapuskasing area of the Canadian Shield, Tectonophysics, 11, 287-303, 1971.

Cook, F.A., Geometry of the Kapuskasing structure from a Lithoprobe pilot reflection survey, Geology, 13, 368-371, 1985.

Eade, K.E., Fort George River and Kaniapiskau River (west half) map-areas, New Quebec, Geol. Surv. Can. Mem., 339, 84 p., 1966.

Ellis, D.J., Origin and evolution of granulites in normal and thickened crusts, Geology, 15, 167-170, 1987.

England, P.C., and S.W. Richardson, The influence of erosion upon the mineral facies of rocks from different metamorphic environments, J. Geol. Soc. Lond., 134, 201-213, 1977.

England, P.C., and A.B. Thompson, Pressure-temperature-time paths of regional metamorphism I. Heat transfer during the evolution of regions of thickened continental crust, J. Petrol., 25, 894-928, 1984.

Ermanovics, I.F., and E. Froese, Metamorphism of the Superior Province in Manitoba, Geol. Surv. Can. Pap., 78-10, 17-24, 1978.

Fountain, D.M., Growth and modification of lower continental crust in extended terranes - a review (Abstr.), Inter. Un. Geodesy Geophys. XIX, 1, 56, 1987.

Fountain, D.M., and M.H. Salisbury, Exposed cross-sections through the continental crust: Implications for crustal structure, petrology, and evolution, Earth Planet. Sci. Lett., 56, 263-277, 1981.

Fountain, D.M., M.H. Salisbury, and K.P. Furlong, Heat production and thermal conductivity of rocks from the Pikwitonei-Sachigo continental cross-section, central Manitoba: implications for the thermal structure of Archean crust, Can. J. Earth Sci., 24, 1583-1594, 1987.

Garland, G.D., Interpretations of gravimetric and magnetic anomalies on traverses in the Canadian Shield in northern Ontario, Dom. Observ., Ottawa Publ., 16, 57 p., 1950.

Geis, W.T., A.G. Green, and F.A. Cook, Preliminary results from the high resolution seismic reflection profile: Chapleau block, Kapuskasing structural zone, in 1988 KSZ Lithoprobe Wkshp, Univ. Toronto, 155-165, 1988.

Gibbs, A.K., B. Payne, T. Setzer, L.D. Brown, J.E. Oliver, and S. Kaufman, Seismic-reflection study of the Precambrian crust of central Minnesota, Geol. Soc. Am. Bull., 95, 280-294, 1984.

Giese, P., Die Struktur der Erdkruste im Bereich der Ivrea Zone. Ein Vergleich verschiedener, seismischer Interpretationen und der Versuch einer petrographisch-geologischen Deutung. Schweiz. Mineral. Petrogr. Mitt., 48, 261-284, 1968.

Goldich, S.S., C.E. Hedge, T.W. Stern, J.L. Wooden, J.B. Bodkin, and R.M. North, Archean rocks of the Granite Falls area, southwestern Minnesota, Geol. Soc. Am. Sp. Pap., 182, 19-43, 1980.

Gupta, V.K., and R.B. Barlow, A detailed gravity profile across the English River subprovince, northwestern Ontario, Can. J. Earth Sci., 21, 145-151, 1984.

Hall, D.H., and Z. Hajnal, Deep seismic crustal studies in Manitoba, Seis. Soc. Am. Bull., 63, 885-910, 1973.

Hammarstrom, J.M., and E-an Zen, Aluminum in hornblende: An empirical igneous geobarometer, Amer. Mineral., 71, 1297-1313, 1986.

Heaman, L., N. Machado, T. Krogh, and W. Weber, Preliminary U-Pb zircon results from the Pikwitonei granulite terrain, Manitoba, Geol. Assoc. Can. Prog. Abstr., 11, 79, 1986.

Herd, R.K., Notes on metamorphism in New Quebec, Geol. Surv. Can. Pap., 78-10, 79-83, 1978.

Himmelberg, G.R., and W.C. Phinney, Granulite-facies metamorphism, Granite Falls-Montevideo area, Minnesota, J. Petrol., 8, 325-348, 1967.

Hoffman, P.F., United plates of America, the birth of a craton: Early Proterozoic assembly and growth of Laurentia, Ann. Rev. Earth Planet. Sci., 16, 543-603, 1988.

Hoffman, P.F., and S.A. Bowring, Short-lived 1.9 Ga continental margin and its destruction, Wopmay orogen, northwest Canada, Geology, 12, 68-72, 1984.

Hubretgse, J.J.M.W., The Archean Pikwitonei granulite domain and its position at the margin of the northwestern Superior Province (Central Manitoba), Manitoba Mineral. Res. Pub., GP80-3, 16 p., 1980.

Hudson, T., and G. Plafker, Paleogene metamorphism of an accretionary flysch terrane, eastern Alaska, Geol. Soc. Am. Bull., 93, 1280-1290, 1982.
Hunt, P.A., and J.C. Roddick, A compilation of K-Ar ages, Geol. Surv. Can. Pap., 87-2, 143-210, 1987.
Innes, M.J.S., Gravity and isostasy in northern Ontario and Manitoba, Dom. Observ. Ottawa Publ., 21, 265 p., 1961.
Jolly, W.T., Metamorphic history of the Archean Abitibi belt; in Metamorphism in the Canadian Shield, Geol. Surv. Can. Pap., 78-10, 63-78, 1978.
Kay, R.W., and S.M. Kay, The nature of the lower continental crust: Inferences from geophysics, surface geology and crustal xenoliths, Rev. Geophys. Space Phys., 19, 271-297, 1981.
Krogh, T.E., Improved accuracy of U-Pb zircon ages by the creation of more concordant systems using an air abrasion technique, Geochim. Cosmochim. Acta, 46, 637-649, 1982.
Krogh, T.E., N.B.W. Harris, and G.L. Davis, Archean rocks from the eastern Lac Seul region of the English River gneiss belt, northwestern Ontario, part 2. Geochronology, Can. J. Earth Sci., 13, 1212-1215, 1976.
Krogh, T.E., F. Corfu, and J.A. Percival, U-Pb zircon and sphene ages from the Kapuskasing zone in the Chapleau-Agawa Bay region, Geol. Assoc. Can. Prog. Abstr., 11, 91, 1986.
Krogh, T.E., L.M. Heaman, and N. Machado, Detailed U-Pb chronology of successive stages of zircon growth at medium and deep levels using parts of single zircon and titanite grains, in 1988 KSZ Lithoprobe Wkshp, 243, 1988.
Langford, F.F., and J.A. Morin, The development of the Superior Province of northwestern Ontario by merging island arcs, Amer. J. Sci., 276, 1023-1034, 1976.
Lapointe, B., Reconnaissance géologique de la région du lac Paillerault-Territoire-du Nouveau-Quebec, Min. Ener. Ress. Quebec, MB 85-73, 10 p., 1986.
Leclair, A., and P. Nagerl, Geology of the Chapleau, Groundhog River and Val Rita blocks, Kapuskasing area, Ontario, Geol. Surv. Can. Pap. 88-1C, 83-91, 1988.
Lopez-Martinez, M., and D. York, A $^{40}Ar/^{39}Ar$ age study of the Kapuskasing structural zone, Geol. Assoc. Can. Prog. Abstr., 11, 96, 1986.
Machado, N., N. Goulet, and C. Gariepy, Evolution of the northern Labrador Trough basement: evidence from U-Pb geochronology, Geol. Assoc. Can. Prog. Abstr., 12, 67, 1987.
Martin, H., Effect of steeper Archean geothermal gradients on geochemistry of subduction-zone magmas, Geology, 14, 753-756, 1986.
Mezger, K., G.N. Hanson, and S.R. Bohlen, U-Pb systematics of garnets in amphibolite to granulite grade rocks, Pikwitonei domain, Manitoba, Canada (Abstr.), EOS, 67, 1248, 1986.
Moecher, D.P., D. Perkins, P.J. Leier-Englehardt, and L.G. Medaris Jr., Metamorphic conditions of late Archean high-grade gneisses, Minnesota River valley, U.S.A., Can. J. Earth Sci., 23, 633-645, 1986.
Mortensen, J.K., and J.A. Percival, Reconnaissance U-Pb zircon and monazite geochronology of the Lac Clairambault area, Ashuanipi complex, Quebec, Geol. Surv. Can. Pap., 87-2, 135-142, 1987.
Moser, D., Structure and chronology of tonalitic gneisses in the Chapleau area, Ontario, Geol. Surv. Can. Pap., 88-1C, 93-99, 1988.
Newton, R.C., and D. Perkins, Thermodynamic calibration of geobarometers based on the assemblages garnet-plagioclase-orthopyroxene (clinopyroxene)-quartz, Amer. Mineral., 67, 203-222, 1982.
Newton, R.C., J.V. Smith, and B.F. Windley, Carbonic metamorphism, granulites and crustal growth, Nature, 288, 45-50, 1980.
Northey, D.J., and G.F. West, A crustal scale seismic refraction experiment over the Kapuskasing structural zone, Geol. Assoc. Can., Prog. Abstr., 11, 108, 1986.
Padovani, E., and J. Carter, Aspects of the deep crustal evolution beneath south central New Mexico, in J.G. Heacock, ed., The Earth's crust: Its nature and physical properties, Amer. Geophys. Un. Geophys. Monogr. Ser., 20, 19-55, 1977.
Paktunc, A.D., and A.J. Baer, Geothermobarometry of the northwestern margin of the Superior Province: Implications for its tectonic evolution, J. Geol., 94, 381-394, 1986.
Parrish, R.R., Cenozoic thermal evolution and tectonics of the Coast Mountains of British Columbia 1, Fission track dating, apparent uplift rates, and patterns of uplift, Tectonics, 2, 601-631, 1983.
Parrish, R.R., U-Pb systematics of monazite and a preliminary estimate of its closure temperature based on natural examples, Geol. Assoc. Can. Prog. Abstr. 13, A94, 1988.
Percival, J.A., High-grade metamorphism in the Chapleau-Foleyet area, Ontario, Amer. Mineral., 68, 667-686, 1983.
Percival, J.A., The Kapuskasing structure in the Kapuskasing-Fraserdale area, Ontario, Geol. Surv. Can. Pap., 85-1A, 1-5, 1985.
Percival, J.A., A possible exposed Conrad discontinuity in the Kapuskasing uplift, Ontario; in M. Barazangi and L.D. Brown eds., Reflection Seismology: The Continental Crust, Amer. Geophys. Un. Geodynam. Ser. 14, 135-141, 1986.
Percival, J.A., Geology of the Ashuanipi granulite complex in the Schefferville area, Geol. Surv. Can. Pap., 87-1A, 1-10, 1987.
Percival, J.A., A regional perspective of the Quetico metasedimentary belt, Superior Province, Canada, Can. J. Earth Sci., in press.
Percival, J.A., and M.J. Berry, The lower crust of the continents, in K. Fuchs, and C. Froidevaux, eds., Composition, structure and dynamics of the lithosphere-asthenosphere system, Amer. Geophys. Un. Geodynam. Ser., 16, 33-59, 1987.
Percival, J.A., and K.D. Card, Archean crust as revealed in the Kapuskasing uplift, Superior Province, Canada, Geology, 11, 323-326, 1983.
Percival, J.A., and K.D. Card, Structure and evolution of Archean crust in central Superior Province, Canada, in L.D. Ayres, P.C. Thurston, K.D. Card and W. Weber eds., Evolution of Archean Supracrustal Sequences, Geol. Assoc. Can. Sp. Pap., 28, 179-192, 1985.
Percival, J.A., and D.M. Fountain, Metamorphism and melting at an exposed example of the Conrad discontinuity, Kapuskasing uplift, Canada, in D. Bridgwater, ed., Fluid Movements, Element Transport

and the Composition of the Crust. NATO Adv. Study Inst., Reidel, in press.
Percival, J.A., and R. Girard, Structural character and history of the Ashuanipi complex in the Schefferville area, Quebec-Labrador, Geol. Surv. Can. Pap., 88-1C, 51-60, 1988.
Percival, J.A., and A.G. Green, Lithoprobe seismic profiles of exposed lower crust in the Kapuskasing uplift, Superior Province, Canada, Geol. Soc. Amer. Abstr. Prog., in press.
Percival, J.A., and T.E. Krogh, U-Pb zircon geochronology of the Kapuskasing structural zone and vicinity in the Chapleau-Foleyet area, Ontario, Can. J. Earth Sci., 20, 830-843, 1983.
Percival, J.A., and P.H. McGrath, Deep crustal structure and tectonic history of the northern Kapuskasing uplift of Ontario: an integrated petrological-geophysical study, Tectonics, 5, 553-572, 1986.
Percival, J.A., and R.W. Sullivan, Age constraints on the evolution of the Quetico belt, Superior Province, Ontario: in Tectonic Evolution of Greenstone Belts, Lunar Planet. Inst. Tech. Rep., 86-10, 167-169, 1985.
Percival, J.A., and R.W. Sullivan, Age constraints on the evolution of the Quetico belt, Superior Province, Canada, in Geol. Surv. Can. Pap. 88-2, in press.
Percival, J.A., and H.R. Williams, The late Archean Quetico accretionary complex, Superior Province, Canada, Geology (in press).
Percival, J.A., J.K. Mortensen, and J.C.M. Roddick, The Ashuanipi granulite complex: conventional and ion probe U-Pb data, Geol. Assoc. Can. Prog. Abstr. 13, A97, 1988.
Percival, J.A., R.R. Parrish, T.E. Krogh, and Z.E. Peterman, When did the Kapuskasing zone come up? in 1988 KSZ Lithoprobe Wkshp, Univ. Toronto, 43-47, 1988.
Percival, J.A., R.A. Stern, and M.R. Digel, Regional geological synthesis of the western Superior Province, Ontario, Geol. Surv. Can. Pap., 85-1A, 385-397, 1985.
Perkins, D., and S.J. Chipera, Garnet-orthopyroxene-plagioclase-quartz barometry: Refinement and application to the English River subprovince and the Minnesota River Valley, Contrib. Mineral. Petrol., 89, 69-80, 1985.
Pirie, J.A., and W.O. Mackasey, Preliminary examination of regional metamorphism in parts of Quetico metasedimentary belt, Superior Province, Ontario, in Geol. Surv. Can. Pap., 78-10, 37-48, 1978.
Poulsen, K.H., The geological setting of mineralization in the Mine Centre-Fort Frances area, District of Rainy River, Ontario Geol. Surv., Open File Rep., 5512, 126 p., 1984.
St-Onge, M.R., Zoned poikiloblastic garnets: P-T paths and syn-metamorphic uplift through 30 km of structural depth, Wopmay Orogen, Canada, J. Petrol., 28, 1-21, 1987.
Sandiford, M., and R. Powell, Deep crustal metamorphism during continental extension: modern and ancient examples, Earth Planet. Sci. Lett., 79, 151-158, 1986.
Sawyer, E.W., The role of partial melting and fractional crystallization in determining discordant migmatite leucosome compositions, J. Petrol., 28, 445-473, 1987.
Schmid, R., Zur Petrographie und Struktur der Zone Ivrea-Verbano zwischen Valle d'Ossola und Val Grande (Prov. Novara, Italien). Schweiz. Mineral. Petrogr. Mitt., 47, 935-1117, 1967.
Sills, J.D., and J. Tarney, Petrogenesis and tectonic significance of amphibolites interlayered with metasedimentary gneisses in the Ivrea zone, southern Alps, NW Italy, Tectonophysics, 107, 187-206, 1984.
Sisson, V.B., and L.S. Hollister, Low-pressure series metamorphism in an accretionary sedimentary prism, southern Alaska, Geology, 4, 358-361, 1988.
Smithson, S.B., and S.K. Brown, A model for lower continental crust, Earth Planet. Sci. Lett., 35, 134-144, 1977.
Stahle, H.J., M. Raith, S. Hoernes, and A. Delfs, Element mobility during incipient granulite formation at Kabbaldurga, southern India, J. Petrol., 28, 803-834, 1987.
Stevenson, I.M., A geological reconnaissance of Leaf River map-area, New Quebec and Northwest Territories, Geol. Surv. Can. Mem., 356, 112 p., 1968.
Stockwell, C.H., Proposals for time classification and correlation of Precambrian rocks and events in Canada and adjacent areas of the Canadian Shield, Part 1: A time classification of Precambrian rocks and events, Geol. Surv. Can. Pap., 80-19, 135 p., 1982.
Studemeister, P.A., The greenschist facies of an Archean assemblage near Wawa, Ontario, Can. J. Earth Sci., 20, 1409-1420, 1983.
Thurston, P.C., and F.W. Breaks, Metamorphic and tectonic evolution of the Uchi-English River subprovince, Geol. Surv. Can. Pap., 78-10, 49-62, 1978.
Thurston, P.C., G.N. Siragusa, and R.P. Sage, Geology of the Chapleau area, Districts of Algoma, Sudbury and Cochrane, Ontario Div. Mines Geol. Rep., 157, 293 p., 1977.
Truscott, M.G., and D.M. Shaw, Preliminary estimate of lower continental crust composition in the Kapuskasing structural zone, Geol. Assoc. Can. Prog. Abstr., 11, 138, 1986.
Tucker, R.D., A. Raheim, T.E. Krogh, and F. Corfu, Uranium-lead zircon and titanite ages from the northern portion of the Western Gneiss Region, south-central Norway, Earth Planet. Sci. Lett., 81, 203-211, 1987.
Watson, J., The origin and history of the Kapuskasing structural zone, Ontario, Canada, Can. J. Earth Sci., 17, 866-876, 1980.
Weber, W., and R.F.J. Scoates, Archean and Proterozoic metamorphism in the northwestern Superior Province and along the Churchill-Superior boundary, Manitoba, Geol. Surv. Can. Pap., 78-10, 5-16, 1978.
Wells, P.R.A., Thermal models for the magmatic accretion and subsequent metamorphism of continental crust, Earth Planet. Sci. Lett., 46, 253-265, 1981.
Woods, D.V., and M. Allard, Reconnaissance electromagnetic induction study of the Kapuskasing structural zone: implications for lower crustal conductivity, Phys. Earth Planet. Inter., 42, 135-142, 1986.
Zingg, A., Regional metamorphism in the Ivrea zone (Southern Alps, N. Italy): field and microscopic investigations, Schweiz. Mineral. Petrogr. Mitt., 60, 153-179, 1980.

PHASE TRANSFORMATIONS IN THE LOWER CONTINENTAL CRUST AND ITS SEISMIC STRUCTURE

Stephan V. Sobolev and Andrey Yu. Babeyko

Institute of Physics of the Earth, B. Gruzinskaya, 10, Moscow, D-242, USSR

Abstract. Data on lower crustal xenoliths and high grade metamorphic terrains suggest that the lower crust of tectonically stable continental regions can consist of anhydrous mafic rocks with equilibration temperatures higher than 700°C. These rocks are chemically unstable in the cool continental crust and survive only due to the extremely low reaction rates at T<700°C.

The method of calculation of density-depth and seismic wave velocity-depth relations in the mafic lower crust is developed. The calculations are made in two steps. Firstly, we determine chemically stable mineral assemblages in pressure-temperature diagram using Gibbs energy minimization technique for multicomponent systems and compute corresponding densities and seismic wave velocities of the rocks. Secondly, we estimate the deviations of densities and seismic wave velocities from those at equilibrium using a linearized kinetic model of solid state phase reactions in rocks and models of thermal evolution of the lithosphere.

The method suggested is applied to a number of problems pertaining to the seismic structure of the lower crust. It is shown that quartz tholeiite chemical composition of the lower crust is consistent with seismic data if kinetics of solid state phase transformations is considered. The emplacement of thin mafic intrusions into the mafic lower crust produces seismic layering even if the intrusions have the same chemical composition as a bulk of the lower crust. Phase transformations in the mafic lower crust coupled with the mechanical instability at the crust-mantle boundary limit the thickness of the crust. The "extra" crustal material goes into the mantle and participates in the process of recycling.

Introduction

New information on the detailed structure of the continental lower crust actually makes the theoretical prediction of mineralogical composition, density and seismic wave velocities in the lower crust of given chemical composition. The solution of this direct problem is a necessary step of the fundamental inverse problem, i.e. determination of chemical and mineralogical composition of the lower crust based on geophysical data. Furthermore the evaluation of density in the lower crust is very important for construction of geodynamic models of the continental lithosphere, particularly its isostatic movements caused by temperature variations in the lithosphere.

In order to calculate the seismic parameters in the lower crust we should have an idea about the chemical composition and thermodynamic state of lower crustal rocks. This problem is briefly discussed in the next section.

Then we suggest the method of evaluation of the seismic parameters in the crust of given composition which involves thermodynamic and kinetic calculations.

Finally, we apply this method to various problems pertaining to the seismic structure of the lower crust.

Composition and Thermodynamic State of the Lower Crustal Rocks

An important source of data on composition and thermodynamics of the lower crustal rocks are xenoliths carried by kimberlites and basalts. They often are anhydrous mafic garnet granulites with equilibration temperatures between 700-900°C. For example, the lower crustal xenoliths from the Udachnaya kimberlite pipe in Yakutia, USSR (Stenina and Shatsky, 1985) contain plagioclase, garnet and clinopyroxene (see Figure 1) and have olivine gabbro chemical composition. Their equilibration temperature is about 800°C and the few estimates of pressure give 10 kbars.

The Udachnaya pipe was emplaced in the typical shield area. Consequently the temperature in the crust there was likely much lower than 800°C at the time of emplacement. Thus these rocks were most likely chemically unstable in the lower crust. This is strongly confirmed by the variation of

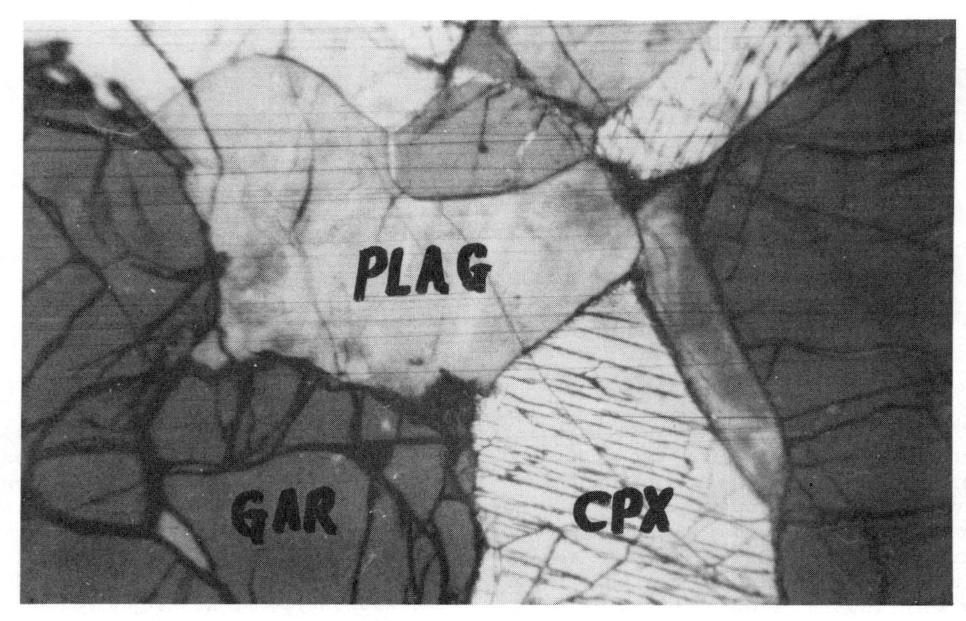

Fig. 1. Thin section of xenolith from Udachnaya kimberlite pipe, Yakutia, USSR. GAR-garnet, CPX-clinopyroxene, PL-plagioclase. With kind permission by Dr. Shatsky.

chemical composition of pyroxene within the range of a thin section (V. Shatsky, personal communication, 1987). "Freezing" of mineral assemblages at temperatures of about 700-900°C and pressures of the lower crust is a very common phenomenon in granulite terrains, see for example (Newton and Perkins, 1982). It is reasonable to suggest the same also for the lower crust which is not exposed.

Thus the data on xenoliths and metamorphic rocks suggest that the lower crust of cool, tectonically stable continental areas could consist of anhydrous mafic rocks with a mineralogical composition that was "frozen" at a temperature higher than 700°C during the cooling of the crust after the previous epoch of its strong heating. This idea was suggested in the papers (Sobolev, 1976, 1978) where it was applied to seismic models of the continental lower crust.

The absence of chemical equilibrium indicates very low reaction rates at temperatures less than 700-900°C. This is consistent with estimations of diffusion rates in mafic rocks (Ahrens and Shubert, 1975; Sobolev, 1976, 1978) but seems to contradict the equilibration times in experiments with mafic rocks which are about two weeks at 800-900°C (Ito and Kennedy, 1971). The contradiction, however, disappears if the difference between experimental and natural conditions is considered. Indeed, it is easy to show that the difference in grain size of natural and experimental specimens, difference in strain rates, etc. result in much more rapid diffusion-controlled reactions in experiments than in natural conditions (Sobolev, 1978).

If chemical equilibrium is achieved, mineralogical composition of a rock is determined by current temperature and pressure values. If not, mineralogical composition also becomes a function of the temperature and pressure history of the rocks. That is why the method of theoretical prediction of physical properties of the lower crust of given chemical composition should consist of two main blocks; 1) determination of equilibrium phase diagrams with isolines of the physical properties of interest; 2) determination

Table 1. Adjusted thermodynamic data

Value[1]	Saxena & Eriksson [1983]	This work
S_{Py}	88.00	87.75
S_{An}	205.43	201.43
W^s_{PyGr}	6.26	4.00
W^s_{GrPy}	6.26	4.00
W^p_{GrPy}	0.058	0.00
W^h_{PyGr}	4.182	5.682
W^h_{GrPy}	16.927	19.427

[1]Entropies and W^s are in J/K•cat, W^p is in J/bar•cat and W^h are in kJ/cat.

Table 2. Accepted densities and elastic moduli of species

Phase	Component	ρ^1, g/cm^3	K^1, kbar	μ^1, kbar	Reference
Spinel	$MgAl_2O_4$	3.58	1933	769	Belikov et al., [1970]
Spinel	$FeAl_2O_4$	4.27	1845	508	estimation[2]
Olivine	$MgSi_{0.5}O_2$	3.21	1281	810	v_p and v_s data by Kumazawa and Anderson [1969]
Olivine	$FeSi_{0.5}O_2$	4.39	1223	535	v_p and v_s data by Chung [1970]
Orthopyroxene	$MgSiO_3$	3.21	1156	793	v_p and v_s data by Liebermann [1974]
Orthopyroxene	$(CaMg)_{0.5}SiO_3$	2.93	1013	666	estimation, v_p and v_s for diopside were used
Orthopyroxene	$FeSiO_3$	4.00	1163	552	v_p and v_s data by Liebermann [1974]
Orthopyroxene	$AlAlO_3$	3.72	1288	846	estimation, v_p and v_s for diopside were used
Clinopyroxene	$MgSiO_3$	3.21	1156	793	estimation, v_p and v_s for enstatite were used
Clinopyroxene	$(CaMg)_{0.5}SiO_3$	3.28	1135	746	v_p and v_s data by Liebermann and Maison [1976]
Clinopyroxene	$FeSiO_3$	4.00	1163	552	estimation, v_p and v_s for ferrosilite were used
Clinopyroxene	$AlAlO_3$	3.89	1348	886	estimation, v_p and v_s for diopside were used
Clinopyroxene	$(NaAl)_{0.5}SiO_3$	3.35	1426	949	v_p and v_s data by Hughes and Nishitake [1963]
Garnet	$MgAl_{2/3}SiO_4$	3.56	1750	900	Leitner et al. [1980]
Garnet	$FeAl_{2/3}SiO_4$	4.33	1760	980	Leitner et al. [1980]
Garnet	$CaAl_{2/3}SiO_4$	3.59	1690	1040	Leitner et al. [1980]
Plagioclase	$CaAl_2Si_2O_8$	2.61	545	282	Belikov et al., [1970]
Plagioclase	$NaAlSi_3O_8$	2.76	860	391	Belikov et al., [1970]
Quartz	SiO_2	2.65	383	446	Belikov et al., [1970]
Kyanite	Al_2SiO_5	3.68	1913	966	v_p, v_s data by Belikov et al. [1970]
Rutile	TiO_2	4.25	2160	1136	Belikov et al., [1970]
Hematite	Fe_2O_3	5.27	982	928	Belikov et al., [1970]

All densities are recalculated from molar volumes by Saxena and Eriksson [1983] except for Rutile and Hematite which are from Clark [1966].
[1]Values are at 25 C, 1 bar.
[2]$M_{Fe-Sp} = M_{Mg-Sp} \cdot (M_{Fe-Ol}/M_{Mg-Ol})$, $M = K, \mu$

of deviations of mineralogical composition and physical properties of rocks from those at equilibrium based on kinetic models of the solid phase transformations and models of thermal evolution of the crust.

Equilibrium Phase Diagrams

Unfortunately there are only few experimental phase diagrams for mafic rocks at the pressures and temperatures of consideration (Ringwood and Green, 1966; Ito and Kennedy, 1971). Furthermore, there are only very poor data on density and elastic wave velocities in mafic rocks at high pressures and temperatures. That is why we tried to add thermodynamic computations to experimental data. We modified the program by Eriksson (1975) for computation of the equilibrium mineral composition of rocks at different P and T. Plagioclase, garnet, pyroxenes, olivine, and spinel were considered as solid solutions. In this work the thermodynamic data of Saxena and Eriksson (1983) were used which seem to be good for ultramafic system. This data set was the only one which was completed enough and was available until 1987. Very recently Wood (1987) has published a new, more sophisticated model. His data we still have in work. We slightly transformed Saxena and Eriksson's (1983) data in order to fit the experimental results on $CaO-MgO-Al_2O_3-SiO_2$ system by Perkins and Newton (1980). The changes are given in Table 1. We neglect changes in mineral volumes due to mixing in the calculations of rock density and use so-called Voigt-Reuss-Hill (VRH) average in calculations of elastic wave velocities of the multiphase system. Accepted densities and elastic moduli of species are given in Table 2. Computed phase diagrams for quartz tholeiite and olivine

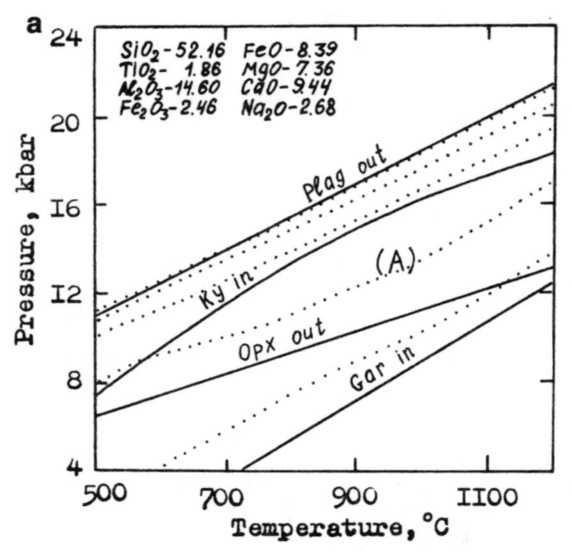

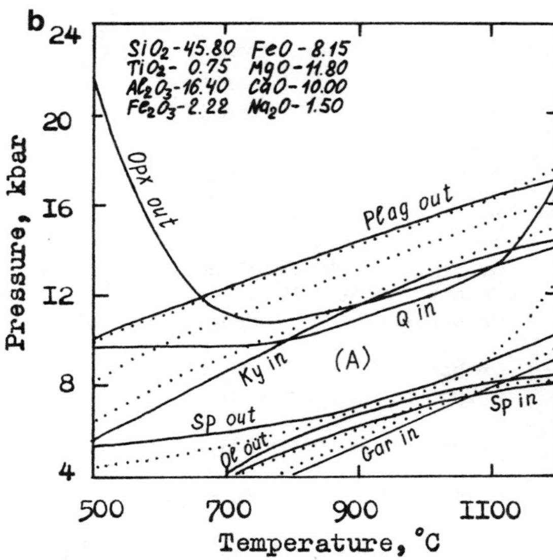

Fig. 2. Computed phase diagrams with isolines of compressional velocity, Poisson ratio and density reduced to 25°C, 1 bar. (a) quartz tholeiite, A = Pl+Gar+Cpx+Q+Rutile+Hem, 1 - 7.1 km/s, 0.258, 3.14 g/cm³; 2 - 7.3, 0.248, 3.24; 3 - 7.5, 0.240, 3.32; 4 - 7.7, 0.231, 3.38; 5 - 7.9, 0.222, 3.44; (b) olivine gabbro, A = Pl+Gar+Opx+Cpx+Rutile+Hem, 1 - 7.3 km/s, 0.272, 3.13 g/cm³, 2 - 7.5, 0.268, 3.22, 3 - 7.7, 0.263, 3.33, 4 - 7.9, 0.260, 3.42, 5 - 8.1, 0.253, 3.48, 6 - 8.3, 0.247, 3.54. Triangles represent experimental data by Ringwood and Green (1966).

gabbro are shown in Figure 2. Densities, compressional wave velocities and Poisson ratios computed for corresponding mineral assemblages at T=25°C, P=1 bar are shown by isolines. It is seen that computed plagioclase-out curve for quartz tholeiite fit well the experimental data by Ringwood and Green (1966) at 1000-1200°C, but the angles of curves differ considerably. This produces considerable difference in mineralogical compositions at low temperatures. Computed garnet -in curve lies considerably lower than experimental points. This curve can be easily elevated by decreasing the entropy of almandines (S_{Alm}), but we need too low S_{Alm}=91.48 J/K.cat in order to fit experimental data within the framework of Saxena and Eriksson's model. Nevertheless we use both types of quartz tholeiite phase diagram in our calculations of lower crustal seismic structure. As it is shown below the difference in the lower crustal models produced by the elevation of garnet-in curve is not large.

Figure 3 shows compressional wave velocity-pressure relations at constant temperatures. The typical two-step shape of these functions is consistent with experimental data by Ito and Kennedy (1971). There is also an impression that VRH average somewhat underestimates compressional wave velocities for quartz tholeiite.

Kinetic Model

There are no doubts that the problem of construction of a precise kinetic model for solid phase transformations in mafic rocks is extraordinarily complicated. We are, however, interested in approximate values of bulk physical characteristics of rocks as density and elastic wave velocities and not in the fine structure of rocks. This makes us feel optimistic.

The approaches based on linear irreversible thermodynamics and solid state diffusion can be used for construction of approximate kinetic models of phase transformations in mafic rocks. Assume that a finite number of inner variables ξ_1, ξ_2, ..., ξ_n should be added to the outer ones (P,T) in order to describe the state of a closed system when chemical equilibrium is not achieved. The technique of linear irreversible thermodynamics (de Groot, 1951) gives the following system of kinetic equations for ξ_1, ..., ξ_n:

$$\frac{d\xi_1}{dt} = - \sum_{j=1}^{n} \frac{\xi_1 - \xi_{ei}(P,T)}{\tau_{ij}(P,T)} \tag{1}$$

where τ^{-1}_{ij} is the matrix of inversed relaxation times, ξ_{ei} is the equilibrium value of ε_1. Diagonalization of matrix gives a spectrum of relaxation times that corresponds to the spectrum of kinetic processes. Now assume that the value of some physical parameter θ, which can be expressed as a linear function of ξ_1, ... ξ_n is controlled by a single process, i.e. the growth of garnet grains characterized by the relaxation time τ.

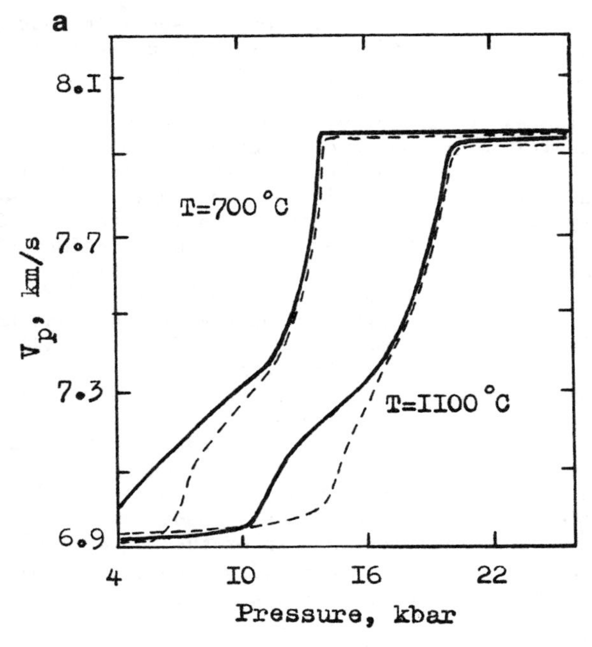

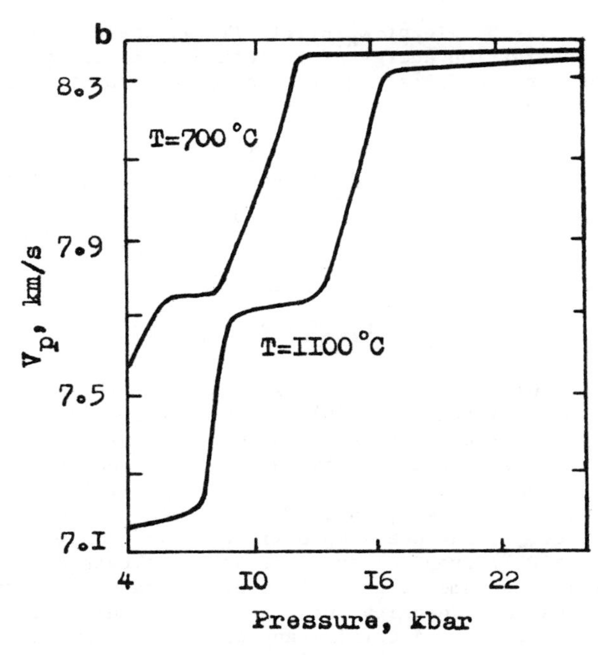

Fig. 3. Computed compressional velocity-pressure relations at constant temperatures for: (a) quartz tholeiite, solid curves are for diagram shown in Figure 2a, dashed lines are for diagram with elevated garnet-in curve; (b) olivine gabbro.

Then we obtain the following approximate kinetic equation for θ:

$$-\frac{d\theta}{dt} = \frac{\theta - \theta_e(P,T)}{\tau(P,T)} \qquad (2)$$

where $\theta_e(P,T)$ is the equilibrium value of θ at given P and T. The function $\tau(P,T)$ cannot be found within the framework of thermodynamic approach.

The rate of solid phase transformations involving the diffusion of species in most of the cases is limited by the rate of diffusion (Fisher, 1978). Suppose that spherical garnet grains grow in a pyroxene-plagioclase matrix. Let Al^{3+} be the slowest among the diffusion species. Now suppose that the growth of garnet grains is controlled by quasistatic diffusion of Al^{3+} in the matrix. The Al^{3+} flux (J_{Al}) is given by the equation:

$$J_{Al} = -\bar{L}_{Al}\frac{\partial\mu_{Al}}{\partial r} \qquad (3)$$

where μ_{Al} is the chemical potential of Al^{3+} and L_{Al} is the effective transport coefficient of Al^{3+} in the matrix and r is the distance from the center of garnet grain. There are two sorts of garnet grains. The first one consists of grains growing at the site of plagioclase and the second of those growing at the site of pyroxene. Growing grains of the first sort accept Mg^{2+}, Fe^{2+} cations and produce Al^{3+}, Ca^{2+} cations. Grains of the second sort act in an opposite way. In the mean field approximation the interaction between the grains is considered by the quasistatic mean value of μ_{Al} achieved far away from the garnet grains. The solution of the corresponding problem of diffusion gives the following kinetic equation:

$$\frac{d\alpha}{dt} = \frac{9}{2}\alpha^{1/3}\bar{L}_{Al}\,Vg\,\frac{(\mu_{Al}-\mu_{gAl})}{r^2_{ec}} \qquad (4)$$

$$\alpha = C/C_{ec},\ C \to C_e,\ \alpha \to \alpha_e = C_e/C_{ec}$$

where C and C_e are the current and equilibrium volumetric concentrations of garnet, respectively, C_{ec} is the volumetric concentration of garnet in eclogite, r_{ec} is the radius of garnet grains in eclogite, Vg is the molar volume of garnet, μ_{gAl} is the chemical potential of Al^{3+} in garnet. Assuming $\mu_{Al}-\mu_{gAl} = \Delta\mu(\alpha_e-\alpha)$ where $\Delta\mu$ is a constant, we obtain a nonlinear kinetic equation. Its linearization gives the expression for τ in eq. 2:

$$\tau = \frac{2}{9\,\alpha_e^{1/3}}\,\frac{r^2_{ec}}{D_{Al}\,\Delta\mu}\,\frac{\partial\mu_{Al}}{\partial n_{Al}}\,\frac{1}{Vg} \qquad (5)$$

where D_{Al} is the coefficient of Al^{3+} diffusion in the matrix, n_{Al} is the number of gramm-atoms of Al^{3+} in the volume unit of matrix. Substitution of $D(T) = D(T_o)\exp(-E_a/R(1/T-1/T_o))$, where E_a is the activation energy of diffusion, R is the gas constant and T_o is some reference temperature, in (5) gives:

$$\tau = \tau_o \exp(\frac{E_a}{R}(\frac{1}{T} - \frac{1}{T_o})) \quad (6)$$

$$\tau_o = \frac{2}{9\alpha_e^{1/3}} \frac{r_{ec}^2}{D_{Al}(T_o)\,\Delta\mu} \frac{\partial\mu_{Al}}{\partial n_{Al}} \frac{1}{Vg}$$

Calculations show that vairations of the pre-exponential factor in (6) with temperature are much less important than that of exponential one. Furthermore, nonlinearity in (4) is so weak that its linearization does not affect the results very much. That is why, in the first linear approximation, we can use simple kinetic equation (2) with $\tau(P,T)$ from (6) and τ_o=const to obtain any physical parameter controlled by the garnet growth, for instance, density and elastic wave velocities. Now in order to determine any of these parameters, say compressional wave velocity (V_p), at some P and T we should solve the kinetic equation:

$$\frac{dv_p^o}{dt} = \frac{v_{pe}^o(P,T)-v_p^o}{\tau_o} \exp(\frac{E_a}{R}(\frac{1}{T_o} - \frac{1}{T})) \quad (7)$$

and consider the "pure" (with constant mineralogical composition) influence of P and T on v_p:

$$v_p = v_p^o + \left.\frac{\partial v_p}{\partial T}\right|_{\xi,P} (T-298K) + \left.\frac{\partial v_p}{\partial P}\right|_{\xi,T} (P-1bar) \quad (8)$$

where $v_{pe}^o(P,T)$ is compressional wave velocity of mineral assemblage chemically stable at given P and T measured at P=1 bar and T=298K.

Thermal Models

The one-dimensional heat transfer equation in the lithosphere where phase transformations occur is as follows:

$$(C_p + \Delta C_p(\bar{\xi}))(\frac{\partial T}{\partial t} + V_z\frac{\partial T}{\partial z}) =$$

$$\frac{\partial}{\partial z}[(\lambda + \Delta\lambda(\bar{\xi}))\frac{\partial T}{\partial z}] + W_r + \sum_i \left.\frac{\partial h}{\partial \xi_i}\right|_{P,T} \frac{d\xi_i}{dt} \quad (9)$$

$$+ \sigma_{ij}\dot{\varepsilon}_{ij}$$

where C_p and λ are the volumetric specific heat and the conductivity of the initial phase assemblage, ΔC_p and $\Delta\lambda$ are their variations due to phase transformations, $\bar{\xi}$ is the vector of inner variables, v_z is the vertical displacement velocity, h is the specific enthalpy, σ_{ij}, $\dot{\varepsilon}_{ij}$ are the stress and strain rate tensors, respectively, and W_r is the radiogenic heat generation. It is easy to show that the last dissipative term in (9) is negligible. The convective term, transformation heat generation and terms with ΔC_p and $\Delta\lambda$ are of the same order of magnitude. Calculations using the technique of perturbation theory show that they may be neglected too in the case of solid phase transformations in mafic rocks.

Thus, we see that the heat transfer equation is separated from kinetic equation in the first approximation. That is why we can solve the heat transfer equation separately and then integrate the kinetic equation for a given function T(t). Different initial and boundary conditions for eq. (9) produce different models of thermal evolution of the lithosphere. We shall use here two simple one-dimensional models simulating cooling of the strongly heated crust and emplacement of hot thin mafic intrusion into the lower crust.

Lower Crust Models

Different chemical compositions (different phase diagrams) and different scenarios of thermal evolution of the lithosphere produce many different physical-petrological models of the lower crust. We discuss two models here that have implications concerning the problems of lower crust composition, thickness of the continental crust and nature of seismic layering.

Many regions of the continental crust were once subjected to a strong heating and maybe even magmatic underplating. Chemical equilibrium is set very rapidly at high temperature, so the lower crustal rocks should be in the equilibrium state during the epoch of strong heating. Subsequent cooling of the crust results in "freezing" of mineral assemblages and hence in "freezing" of the density and elastic wave velocities in the lower crust. Figure 4 shows the computation results for this case.

We have assumed that mafic rocks equilibrated at solidus temperature (1200°C) were kept for 50 Ma in strongly heated crust with temperature profile shown in Figure 4a (t=0-50). Then the cooling of the crust occurred which was simulated by a thermal model identical to the cooling plate model by Parsons and Sclater (1977). Solution of the thermal problem is shown in Figure 4a. The activation energy in kinetic equation was taken to be 60 kcal/mol and the main kinetic parameter $\tau_o=\tau(800^oC)$ was varied in wide range. The typical values of $\partial v_p/\partial P=1.5\cdot10^{-2}$ km/s·kbar and $\partial v_p/\partial T=-5\cdot10^{-4}$ km/s·K were used. The calculations were made for quartz tholeiite (Figure 4b,c) and olivine gabbro (Figure 4d) lower crust compositions.

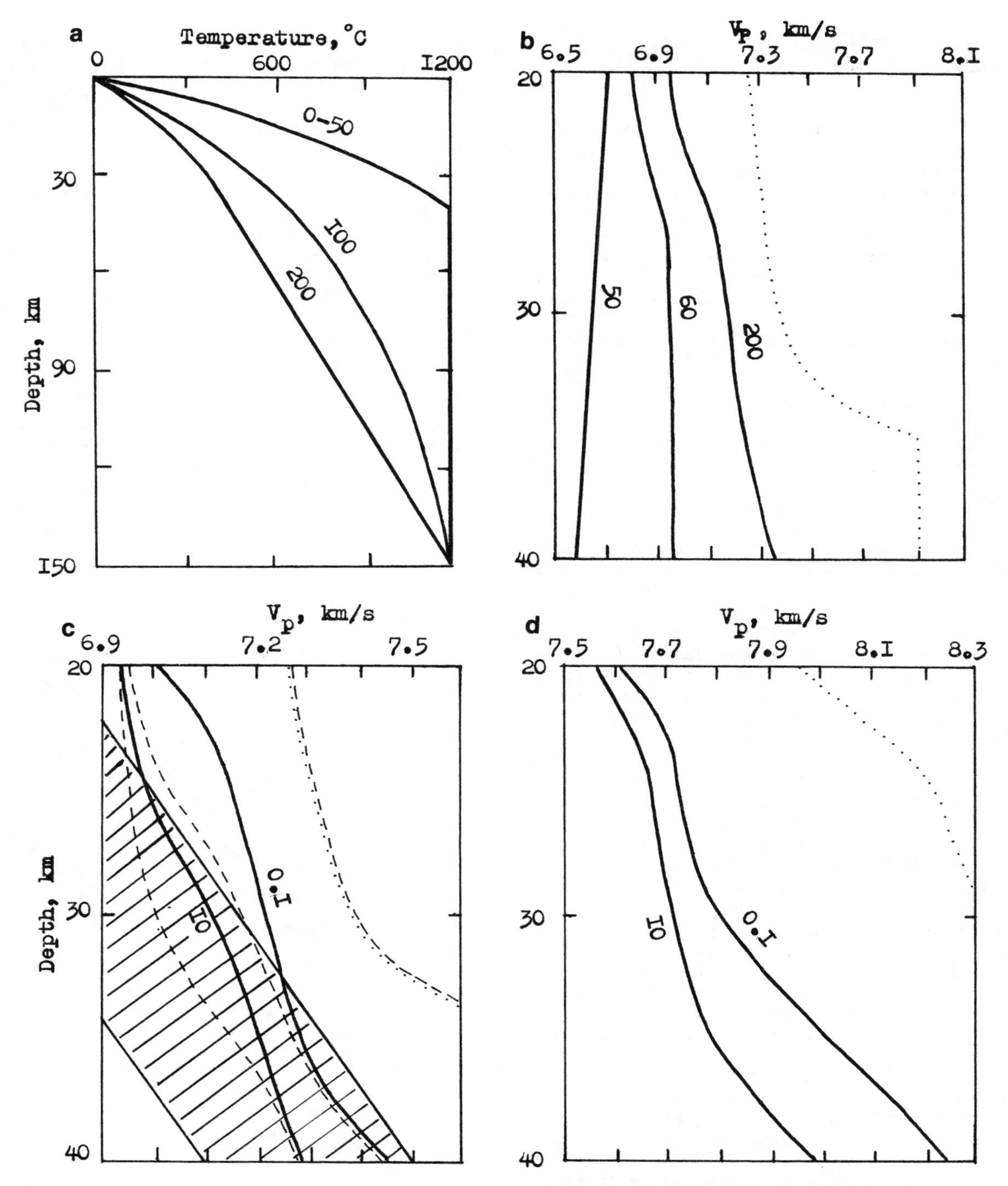

Fig. 4. Computed seismic and thermal models of the lower crust cooling after a strong heating event: (a) thermal model, figures are time in Ma; (b) seismic model for quartz tholeiite, figures are time in Ma, τ_o = 1 Ma; also shown are seismic models for the cooled quartz tholeiite (c) and olivine gabbro (d) lower crust. Figures denote τ_o in Ma. Dotted curves present compressional wave velocities for equilibrium mineral assemblages in cooled crust. Dashed curves refer to the case of phase diagram of quartz tholeiite with elevated garnet-in curve. Dashed area presents the typical seismic data for stable continental crust by Kosminskaya and Pavlenkova (1980).

The calculations lead us to the following conclusions.

1. Anhydrous quartz tholeiite composition of the bulk of the lower crust is consistent with typical seismic data on tectonically stable continental crust if the equilibration time at 800°C, τ_o, is greater than 0.1-1 Ma (see Figure 4c). These values correspond to an effective diffusional coefficient greater than 10^{-15}-10^{-16} cm²/s at 800°C and provide equilibration temperatures of rocks about 650-750°C. For Ringwood and Green's (1966) phase diagram for quartz tholeiite, which produces the eclogite stability in the cool lower crust, the calculations are consistent with seismic data at τ_o greater than 1-10 Ma (Artyushkov and Sobolev, 1982).
2. Rocks with olivine gabbro composition have too high seismic velocities to be the major component of the lower crust (see Figure 4d), but they may form thin intrusions or parts of intrusions in the crust.
3. "Freezing" temperature of rocks only slightly depends on depth and varies within the range of less than 100 degrees. That is why the first approximation of mineralogical composition-depth relation in the cool crust should be given by an isothermal cross-section of P-T diagram. The typical two-step shape of $V_p(z)$ and $\rho(z)$ functions is much more pronounced in this case than in the case of equilibrium state when the mineralogical composition, $V_p(z)$ and $\rho(z)$ functions are given by geothermal cross-section of P-T diagram.

The lamination of the lower crust is often explained by mafic magma intrusions. This quite natural idea presents the question of why intrusions are localized mainly in the lower crust and are not so common in the upper crust. The most effective mechanism for magma transportation through the lithosphere is likely to be magma fracturing (Artyushkov and Sobolev, 1984; Spence and Turcotte, 1985). The upper crust is usually characterized by deviatoric compression (Magnitsky and Artyushkov, 1978) and low density of rocks. These two phenomena may lock a magmatic crack in the lower crust (Sobolev and Siplivets, 1988) and result in the formation of horizontal intrusions in the lower ductile crust where the deviatoric stresses are relaxed.

Let us see what would happen if a thin intrusion is emplaced into the lower crust. Assume that the lower crust has quartz tholeiite composition. The horizontal intrusion 100 m in thickness is emplaced at the 35 km depth and its temperature is 1200°C. The temperature of rocks before the emplacement is 700°C and τ_o value is 1 Ma. Consider two cases. In the first case the intrusion is isochemical with crustal rocks and for the second case the intrusion has olivine gabbro composition. After the emplacement, the intrusion rapidly cools. This rapid cooling produces mineral assemblages with the temperature of last equilibrium higher than that in the bulk of the lower crust. Consequently the compressional wave velocity is lower in the "frozen" quartz tholeiite intrusion than in surrounding rocks by 0.2-0.3 km/s. Thus even the emplacement of the nonlayered intrusion isochemical with respect to the sufficiently cool lower crust results in seismic layering. New heating event (i.e. tectonic activation) leads to reequilibration of rocks and hence erases mineralogical and seismic difference between intrusion and surrounding rocks.

The compressional velocity in the "frozen" olivine gabbro intrusion is higher than that in surrounding rocks by 0.4-0.5 km/s. Further heating of the crust up to the temperature of 800-900°C makes this difference greater. Intrusions of olivine gabbro composition can act as strong reflectors. Their magmatic layering, which is likely a common process, increases their reflective properties even more. Series of intrusions of different chemical composition (within the range of mafic rocks) may produce a complicated laminated seismic structure which could change with time due to heating or cooling of the crust. It is possible that xenoliths from the Udachnaya

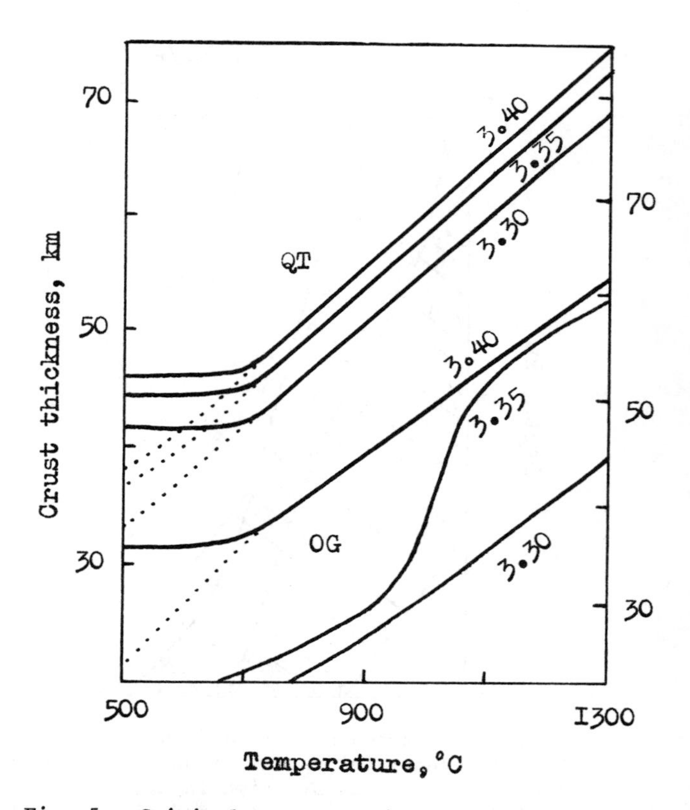

Fig. 5. Critical crustal thickness for the Earth (left) and Venus (right) as a function of temperature at the base of the crust. Figures denote mantle density in g/cm³. QT is quartz tholeiite, OG is olivine gabbro.

kimberlite pipe are samples from seismic lamellae of the lower crust.

High pressure modifications of dry mafic rocks, i.e. eclogites, are denser than mantle ultramafic rocks. That is why we may introduce the critical thickness of the crust H_c which is defined as the depth where the density of lower crustal mafic rocks equals the density of the mantle. If the thickness of the crust is greater than H_c, its lowermost layer becomes mechanically unstable and tends to sink into the mantle.

Ringwood and Green (1967) suggested that basalts which were intruded into the crust crystallized as eclogites and sank through the crust into the mantle. However, they did not associate this process with the thickness of the crust. Our main point is that the sinking of the lower crustal rocks into the mantle may take place if the thickness of the crust exceeds some critical value which is the function of the lower crustal chemical composition and temperature.

Figure 5 shows the dependence of H_c value on temperature at the base of the crust for different suppositions about the density of the mantle. The maximal thickness of the crust which can be formed by the underplating process (Furlong and Fountain, 1986) or doubling of the crust in collision zones is about 75 km for the Earth in the case of quartz tholeiite lower crust composition. The "extra" crustal material goes into the mantle and participates in the process of recycling. Cooling

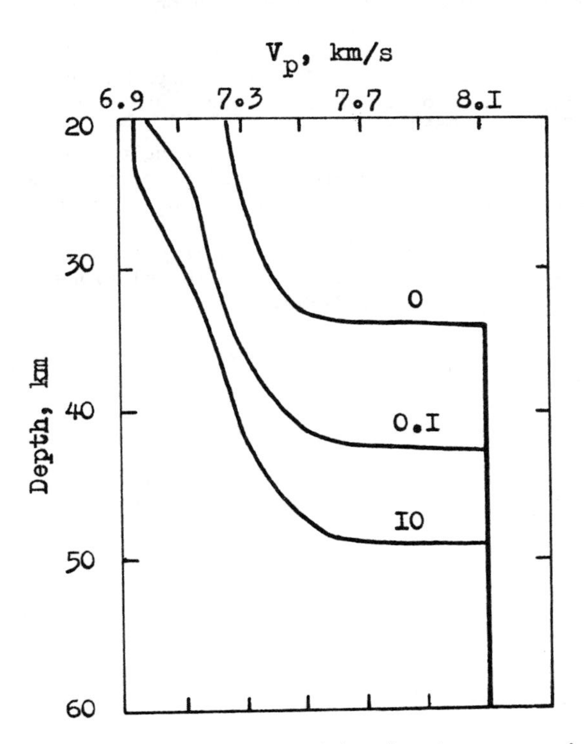

Fig. 6. Compressional velocity in the crust with critical thickness. Figures denote τ_0 in Ma.

of the crust from the solidus temperature leads to the decrease of H_c values. At T<700-800°C, H_c becomes independent of temperature because of "freezing" of phase transformations. Figure 6 shows the compressional wave velocity profiles in the cooled quartz tholeiitic continental crust with thickness H_c. The density is continuous at the crust-mantle boundary according to the H_c definition but the seismic wave velocity jumps from 7.6 to 8.1 km/s. Reasonable predicted values of crustal thickness (Figure 5) and of crustal seismic structure (Figure 6) suggest that phase transformations in the mafic lower crust coupled with mechanical instability at the crust-mantle boundary could play an important role in formation of the continental crust on the earth.

The idea of critical crustal thickness can be easily applied to every planet with mafic lower crust and ultramafic upper mantle; the H_c values simply change in inverse proportion to the gravity. Figure 5 (right vertical axes) shows H_c for Venus as a function of the temperature at the base of the crust. We may predict that the maximal thickness of the Venus crust should be between 45-85 km if its chemical composition is between olivine gabbro and quartz tholeiite, and the solidus temperature is achieved at the base of the crust.

Acknowledgements. We thank V.S. Shatsky who kindly provided some data on lower crustal xenoliths from Udachnaya kimberlite pipe and R.C. Liebermann for useful advice in search of data on elastic constants of minerals. We also thank D.M. Fountain, B.J. Wood and B.R. Frost for constructive reviews.

References

Artyushkov, E.V., and S.V. Sobolev, Mechanism of passive margins and inland seas formation, AAPG Memoir, 34, 689-701, 1982.

Artyushkov, E.V., and S.V. Sobolev, Physics of the kimberlite magmatism, in: Kimberlites and related rocks, Kornprobst J. (ed.), Elsevier, 309-322, 1984.

Ahrens, T.J., and G. Shubert, Gabbro-eclogite reaction rate and its geophysical significance, Rev. Geophys. Space Phys., 13, 383-400, 1975.

Belikov, B.P., K.S. Alexandrov, and T.V. Ryzhova, Elastic properties of rock-forming minerals and rocks, Publ. Office "Nauka", Moscow, 1970.

Chung, D.H., Effects of Iron/Magnesium ratio on P- and S-wave velocities in olivine, J.Geophys. Res., 75, 7353-7361, 1970.

Clark, S.P.Jr. (ed.) Handbook of physical constants, Mem. Geol.Soc.Am., 97, 1966.

de Groot, S.R., Thermodynamics of irreversible processes, North.-Holland, Amsterdam, 1951.

Erikkson, G.,Thermodynamic studies of high temperature equilibria. XII. SOLGASMIX, Computer program for calculation of equilibrium compositions in multiphase systems, Chem.Scr., 8, 100-103, 1975.

Fisher, G.W., Rate laws of metamorphism, Geochim. Cosmochim.Acta., 42, 1035-1051, 1978.

Furlong, K.P., and D.M. Fountain, Continental crustal underplating, thermal consideration and seismic-petrological consequences, J.Geophys. Res., 91, 8285-8294, 1986.

Hughes, D.S., and T. Nishitake, Measurements of elastic wave velocity in Armco iron and jadeite under high pressures and high temperatures, in: Geophys. Papers Dedicated to Professor Kehzo Sasso, Kyoto Univ. Geophys. Institute, 379-385, 1963.

Ito, K., and G.C. Kennedy, An experimental study of basalt-garnet granulite-eclogite transformation, in: The structure and physical properties of the Earth's crust, Geophys.Monograph.Ser., 14, 303-314, 1971.

Kosminskaya, I.P., and N.I. Pavlenkova, A generalized seismic model of the continental crust, in: Seismic models of the lithosphere for the major geostructures, Publ. Office "Nauka", Moscow, 141-152, 1980 (in Russian).

Kumazawa, M., and O.L. Anderson, Elastic moduli, pressure derivatives of single-crystal olivine and single-crystal forsterite, J.Geophys.Res. 74, 5961-5972, 1969.

Leitner, B.J., D.J. Weidner, and R.C. Liebermann, Elasticity of single crystal pyrope and implications for garnet solid solution series, Phys.Earth Planet. Interiors, 22, 111-121, 1980.

Liebermann, R.C., Elasticity of pyroxene - garnet and pyroxene-ilmenite phase transformations in germanates, Phys.Earth Planet. Interiors, 8, 361-374, 1974.

Liebermann, R.C., and D.J. Mayson, Elastic properties of polycrystalline diopside ($CaMgSi_2O_6$), Phys.Earth Planet.Interiors, 11, 1-4, 1976.

Magnitsky, V.A., and E.V. Artyushkov, Some general problems of the Earth's dynamics, in: Tectonosphere of the Earth, Publ. Office "Nauka", Moscow, 487-525, 1978 (in Russian).

Newton, R.C., and D. Perkins, III, Thermodynamic calibration of geobarometers based on the assemblages garnet-plagioclase-orthopyroxene (clinopyroxene)-quartz, Amer.Mineral., 67, 203-232, 1982.

Parsons, B., and J. Sclater, Analysis of variations of ocean floor bathymetry and heat flow with age, J.Geophys.Res., 82, 803-827, 1977.

Perkins, D., and R.C. Newton, The composition of coexisting pyroxenes and garnet in the system $CaO-MgO-Al_2O_3-SiO_2$ at 900-1100°C and high pressures, Contr.Miner. Petrol., 75, 291-300, 1980.

Ringwood, A.E., and D.H. Green, An experimental investigation of the gabbro-eclogite transformation and some geophysical implications, Tectonophysics, 3, 383-427, 1966.

Saxena, S.K., and G. Eriksson, Theoretical computation of mineral assemblages in pyrolite and lherzolite, J.Petrol., 24, Part 4, 538-555, 1983.

Sobolev, S.V., Seismic models of the Moho in the case of a phase transformation, in: Materialy XIV Vsesoyusnoy studentcheskoy conferentsii, ser. Geologia, Novosibirsk Univ. Publ. 62-69, 1976 (in Russian).

Sobolev, S.V., Models of the lower crust on continents with consideration of the gabbro-eclogite transformation, in: Petrological problems of the Earth's crust and mantle, Publ. Office "Nauka", Novosibirsk, 347-355, 1978 (in Russian).

Sobolev, S.V. and S.G. Siplivets, Influence of magma chamber shape and melting on the dimensions of critical magma fracture, Doklady Akademii Nauk SSSR, in press, 1988.

Spence, D.A., and D.L. Turcotte, Magma driven propagation of cracks, J.Geophys.Res. 90, 575-580, 1985.

Stenina, N.G., and V.S. Shatsky, Exsolution structure in clinopyroxenes of eclogite-like rocks, Geology and Geophysics, No. 3, 51-64, 1985 (in Russian).

Wood, B.J., Thermodynamics of multicomponent systems containing several solid solutions, Rev. Mineral., 17, 71-95, 1987.

GEOPHYSICAL PROCESSES INFLUENCING THE LOWER CONTINENTAL CRUST

D. L. Turcotte

Department of Geological Sciences, Cornell University, Ithaca, New York

Abstract. The primary mechanisms for the addition of material from the mantle to the continental crust at the present time are island arc volcanics and continental rift and hot spot volcanics. We propose as a working hypothesis that these mechanisms have also been responsible for the addition of material to the continental crust in the past. However, since these volcanics have near basaltic compositions, an essential question is how the continental crust develops a silicic composition. Partial melting of the lower continental crust can create magmas with a silicic composition; this volcanism can lead to an internal crustal fractionation with a more silicic upper crust and a more mafic lower crust. In order to achieve a bulk silicic composition for the continental crust it is necessary to remove the mafic lower crust. We propose that this is done by the delamination of substantial portions of the continental lithosphere including the lower continental crust. We suggest that the mean rate of continental delamination is near 0.1 km^2/yr. This is about 4% of the present rate of subduction.

Introduction

The geophysical mechanisms associated with the creation of the oceanic crust at ocean ridges and its subduction at ocean trenches are quite well understood. Much less is known about the geophysical processes associated with the creation (and destruction) of the continental crust. A fundamental question is how the continental crust is formed. Can currently observed processes lead to the formation of continental crust; examples are subduction related volcanism and continental rift or hot spot volcanism. Or was the continental crust primarily formed by processes in the Archean that are no longer active. A related question is whether the formation of the continental crust is continuous or episodic.

One basic question concerning the continental crust is its silicic composition. It is generally agreed that the crust has a mean composition that is more silicic than magmas being produced in the mantle today. One hypothesis for the formation of the continental crust is that silicic magmas were generated in the mantle in the Archean and that these magmas were responsible for the silicic composition of the continental crust.

Following the suggestions of Kay and Kay (1988), this paper examines a general hypothesis for the generation of the continental crust. This hypothesis is broken into three parts:

1) That basaltic volcanism from the mantle associated with island arc volcanics and continental rifts and hot spots are responsible for the formation of the continental crust.
2) That intracrustal melting and high temperature metamorphism is responsible for differentiation of the crust so that the upper crust becomes more silicic and the lower crust becomes more mafic.
3) That delamination of substantial quantities of continental lithosphere including the mantle and lower crust return a substantial fraction of the more mafic lower crust to the mantle reservoir. The residuum, composed primarily of upper crust, thus becomes more silicic.

Mechanisms of Crustal Formation

A variety of processes can lead to the formation of continental crust. The relative importance of these processes is a matter of considerable controversy.

Island Arc Processes

Island arcs can contribute to the formation of continental crust in two ways. If an island arc stands on oceanic crust it generates the formation of thick crust. If this island arc subsequently collides with a continent it can add material in the form of exotic terranes. If a subduction zone is adjacent to a continent then the subduction zone can add mantle derived magmas directly to the crust. This is happening today in the Andes. The importance of island arc processes in forming continental crust has been discussed in detail by Taylor (1967, 1977), Taylor and White (1965),

Jakeŝ and White (1971) and by Jakeŝ and Taylor (1974).

Hot Spot Volcanism

The association between flood basalts, continental rifts, and mantle plumes is still a subject of debate (Burke and Dewey, 1973; Turcotte and Emerman, 1983). However, flood and rift volcanics have added a considerable volume of mantle derived magmas to the continental crust. McKenzie (1984) argues that the continental crust is being continuously underplated by extensive intrusive volcanics.

Summary

Our working hypothesis is that island arc and flood basalt processes are the primary means of formation of continental crust. Since these mantle derived magmas have near basaltic compositions our working hypothesis is also that material added to the continents has a basic composition (Kay and Kay, 1985, 1986).

This hypothesis is certainly controversial, particularly in terms of Archean processes. Brown (1977) suggests the direct addition of silicic magmas from the mantle. Tarney and Windley (1977) conclude that Archean processes were primarily responsible for the formation of the silicic continental crust. We suggest, in this paper, that ad hoc Archean processes are not necessary to explain the origin and evolution of the continental crust and suggest an alternative involving the delamination of the lower continental crust.

Mechanisms of Reprocessing

Reprocessing of the continental crust can result in fractionation and systematic differences between the upper and lower continental crust.

Erosion

Erosion physically removes young mountain belts. The resulting sediments are deposited either in low-lying regions of the continental crust or on the sea floor. At subduction zones a fraction of the sediments deposited on the sea floor form accretionary prisms. In some cases, i.e. the central valley of California, these accretional prisms are sufficiently thick to form the entire thickness of the continental crust. Thus erosion and accretion can transfer material from the upper continental crust to the lower continental crust.

Crustal Melting and High Temperature Metamorphism

The origin of granites is another subject of long standing controversy. It is another hypothesis of this paper that intracrustal differentiation processes associated with melting and high temperature metamorphism lead to the formation of a silicic upper crust. Detailed justifications for this hypothesis have been given by Fyfe (1973, 1978) and Wyllie (1977). The melting can be caused by massive intrusions of basaltic, mantle derived magmas in either an Andean or hot spot environment, or by increased crustal temperatures in continental collisions.

Mechanisms of Recycling

Several mechanisms have been proposed for returning continental crust to the upper mantle reservoir. Crustal recycling can reduce the mean age of the continental crust and can significantly alter its bulk chemical composition.

Alteration of the Oceanic Crust

Hydrothermal circulations of sea water through the oceanic crust can deposit significant quantities of continental crustal elements that are soluble (Wolery and Sleep, 1976). Examples include magnesium, calcium, and uranium. When the sea floor is subducted these elements may be returned to the mantle. We can estimate the magnitude of this process in changing the bulk composition of the continents by estimating the fraction of the volume of the continental crust that can be recycled. The mean rate of subduction of ocean crust over the last 500 Myrs has been 25 km^3/yr (Turcotte and Schubert, 1982, pp. 163-167). If 1% of this was deposited continental crust, 2% of the continental crust would be recycled in 500 Myrs. The conclusion is that it is difficult to explain variations in the bulk composition of the continental crust using this mechanism.

Sediment Subduction

Although a fraction of the sediments entering an oceanic trench are scraped off to form an accretionary prism, a substantial fraction may be subducted. The absence of either an accretionary prism or accumulated sediments in some trenches is taken as evidence that sediments are subducted. In many trenches the upper oceanic crust is broken into a series of horsts and grabens when the oceanic lithosphere bends as it enters the trench. This rough morphology provides a means of entraining substantial quantities of sediments in the subduction process. Dewey and Windley (1981) estimated the volume of sediments approaching subduction zones, subtracted the volume accumulating in accretionary prisms, and concluded that about 200 m of sediments are subducted. Kay (1980) and Karig and Kay (1981) estimate that about 50 m of sediments must be subducted in order to account for the potassium budget in the Marianas. If 200 m of sediments on a world-wide basis are subducted with a mean density of 2,350 kg/m^3, this would represent the subduction of about 5% of the continental crust in 500 Myrs.

Again it appears that this process is not adequate to explain changes in the bulk composition of the continental crust. Also, sediment subduction would preferentially recycle the upper continental crust. Thus the process would not be expected to make the continental crust more silicic.

It should be emphasized that alteration of the oceanic crust and sediment subduction recycle sediments into the mantle reservoir only if they escape being returned to the continental crust in island arc volcanism. There is direct observational evidence that subducted sediments form a fraction of island arc volcanics. But the fraction of the subducted continental crustal material that is removed by this volcanism is unknown.

Crustal Erosion at Subduction Zones

At a continental margin the subducting lithosphere may erode and entrain the overlying wedge of continental crust (Karig, 1974). This is a mechanism for recycling the lower continental crust but again the entrained material must escape being returned to the continental crust in island arc volcanics. It is very difficult to make realistic estimates of the importance of crustal erosion as a mechanism for returning continental crust to the mantle; but it is doubtful that it is volumetrically significant.

Delamination

In subduction, the entire oceanic lithosphere sinks into the earth's interior. In delamination, a fraction of the continental lithosphere sinks into the earth's interior. If this fraction includes some continental crust, then delamination leads to the recycling of continental crust into the mantle reservoir. Continental delamination was proposed and studied by Bird (1979) and by Bird and Baumgardner (1981). Direct evidence for continental delamination does not exist. However, there is considerable indirect evidence that delamination of the continental lithosphere occurs quite often (see e.g. Mueller, 1977). There is a number of continental areas in which the mantle lithosphere is absent. The best example is the western United States. Plateau uplifts such as the Altiplano in Peru are also associated with the absence of mantle lithosphere. Delamination is an efficient mechanism for the removal of continental lithosphere. Alternative mechanisms for thinning the lithosphere include heat transfer from an impinging plume and heat transport by magmas. However, the former is very slow (Emerman and Turcotte, 1983) and the latter requires very large volumes of magma (Lachenbruch and Sass, 1978). McKenzie and O'Nions (1983) suggest that the mantle lithosphere delaminates in subduction zones. An important question is whether delamination includes the lower continental crust. It is well documented that an intracrustal weak zone exists in orogenic zones (Hadley and Kanamori, 1977; Eaton, 1980; Yeats, 1981; Mueller, 1982, Turcotte et al., 1984). The presence of a soft layer at an intermediate depth in the crust can be attributed to the presence of quartz (Kirby and McCormick, 1979). Delamination at this intracrustal weak zone can explain intracrustal decollements in the Alps (Oxburgh, 1972) and in the southern Appalachians (Cook et al., 1979). A quantitative evaluation of crustal delamination will be given in a later section.

Crustal Composition

Although it is relatively easy to estimate the composition of the upper continental crust it is difficult to estimate the composition of the crust as a whole. Direct evidence for the composition of the lower continental crust comes from surface exposures of high grade metamorphic rocks and lower crustal xenoliths transported to the surface in diatremes and magma flows. Indirect evidence for the composition of the lower crust comes from comparisons between in situ seismic velocities and laboratory measurements of seismic velocities for relevant minerals. Several estimates of the bulk composition of the continental crust are given in Table 1. In Table 2 the average composition from Table 1 is compared with a typical basalt composition. Also included in Table 2 are the mean compositions of Archean and post-Archean clastic sediments. Estimates for the mean composition of the continental crust are clearly more basic than the composition of the upper continental crust but do not approach a basaltic composition. The fact that the upper continental crust is more silicic than the lower continental crust is consistent with the remelting hypothesis for the origin of granites. However, if the only mantle melt responsible for forming the continental crust is basalt, then the mean composition of the continental crust should be basaltic. This is clearly not the case. There is also some evidence that the continental crust has become more silicic with time (Ronov, 1972). This change is supported by the comparison between Archean and post-Archean sediments given in Table 2.

Isotopic Constraints

Isotope systematics can be used to determine the mean age of the continental crust. The age in this context is the time when the crustal rock was removed from the mantle and is not affected by reprocessing. Turcotte and Kellogg (1986) used isotope systematics for the samarium-neodymium and rubidium-strontium systems to conclude that the mean age of the continental crust is 2.1 ± 0.7 Gyrs. Similar results were obtained by Allegre et al. (1983) and others. Without crustal recycling the size of the continental crust would be one-half its present value at that time. This requirement does not appear to be consistent with sea level data. It appears reasonable to

TABLE 1. Various Estimates for the Composition of the Bulk Continental Crust

	1	2	3	4	5	6	7	8	Aver-age
SiO_2	61.9	63.9	57.8	61.9	62.5	63.8	63.2	57.3	61.7
TiO_2	1.1	0.8	1.2	0.8	0.7	0.7	0.6	0.9	0.8
Al_2O_3	16.7	15.4	15.2	15.6	15.6	16.0	16.1	15.9	15.8
FeO	6.9	6.1	7.6	6.2	5.5	5.3	4.9	9.1	6.4
MgO	3.5	3.1	5.6	3.1	3.2	2.8	2.8	5.3	3.6
CaO	3.4	4.2	7.5	5.7	6.0	4.7	4.7	7.4	5.4
Na_2O	2.2	3.4	3.0	3.1	3.4	4.0	4.2	3.1	3.3
K_2O	4.2	3.0	2.0	2.9	2.3	2.7	2.1	1.1	2.5

1. Goldschmidt (1933), 2. Vinogradov (1962), 3. Pakiser and Robinson (1966), 4. Ronov and Yaroshevsky (1969), 5. Holland and Lambert (1972), 6. Smithson (1978), 7. Weaver and Tarney (1984), 8. Taylor and McLennan (1985).

hypothesize that the volume of water in the oceans has remained nearly constant for at least 3 Ga. If the volume of the continents had increased substantially during this time the displaced water would have caused sea level to increase. If this had occurred the stable cratons would be covered with sediments, but they are not. In fact, observational evidence argues that sea level has fallen relative to the stable cratons (Ambrose, 1964; Wise, 1972, 1974; Watson, 1976; Windley, 1977; Hallam, 1977; Dewey and Windley, 1981; and McLennan and Taylor, 1983). Some fall in sea level can be attributed to a decrease in ridge volume but this is not consistent with a 50% increase in crustal volume since 2 Gyrs ago (Reymer and Schubert, 1984, 1986; Schubert and Reymer, 1985).

Isotopic studies of the earliest available rocks can be interpreted as inferring that the volume of the continental crust was a significant fraction of the present value 3 Gyrs ago (Armstrong, 1968, 1981; Armstrong and Hein, 1973; DePaolo, 1983; and Nelson and DePaolo, 1985). A large continental volume in the Archean and a young mean age implies substantial recycling of the continental crust into the mantle. A number of authors have studied the implications of isotope data on the volume of the continental crust (McCulloch and Wasserburg, 1978; O'Nions et al., 1979; DePaolo, 1979, 1980; Allègre, 1982; Hamilton et. al., 1983; and Allègre and Rousseau, 1984). The general conclusion of these studies is that substantial volumes of continental crust have been recycled. However, this conclusion is certainly not universally accepted. For example, Hofmann et al. (1986) argue against significant recycling on the basis of studies of Nb and Pb in oceanic basalts.

TABLE 2. Average Composition of Basalts (Nockolds, 1954), the Mean Composition of the Continental Crust from Table 1, and the Mean Compositions of Archean and Post-Archean Clastic Sediments (Taylor and McLennan, 1985, p. 99)

	Basalt	Average Continental Crust	Archean Clastic Sediments	Post-Archean Clastic Sediments
SiO_2	50.8	61.7	65.9	70.4
TiO_2	2.0	0.8	0.6	0.7
Al_2O_3	14.1	15.8	14.9	14.3
FeO	9.0	6.4	6.4	5.3
MgO	6.3	3.6	3.6	2.3
CaO	10.4	5.4	3.3	2.0
Na_2O	2.2	3.3	2.9	1.8
K_2O	0.8	2.5	2.2	3.0

Crustal Production and Recycling

It is of interest to consider a simple model for crustal production and recycling. It is based on the hypothesis of uniformitarianism, that processes in the past were similar to those today, and follows the approach used by Allègre and Jaupart (1985) and by Gurnis and Davies (1985, 1986). We assume that the change in mass of the continental crust is given by

$$\frac{dM_c}{dt} = F_o \exp\left[\frac{\tau_e - t}{\tau_o}\right] - \frac{1}{\tau_r} M_c \qquad (1)$$

where F_o is the present rate of mass addition to the continental crust. This rate was higher in the past because of more vigorous convection in the mantle due to larger concentrations of heat producing elements. Time t is measured forward from the time of initial crustal growth τ_e. We associate τ_o with the time variation of heat production in the mantle and take $\tau_o = 3.6$ Gyrs (Turcotte and Schubert, 1982, pp. 139-142). We further hypothesize that the rate of crustal recycling is proportional to the mass of the continental crust with the characteristic time τ_r determining the rate. The solution of (1) must satisfy the conditions $M_c = 0$ at $t = 0$ and $M_c = M_{co}$ at $t = \tau_e$ where M_{co} is the present mass of the continental crust. The required solution is

$$M_c = M_{co}\left[\frac{e^{-t/\tau_o} - e^{-t/\tau_r}}{e^{-\tau_e/\tau_o} - e^{-\tau_e/\tau_r}}\right] \qquad (2)$$

with

$$F_o = \frac{M_{co}\ (1/\tau_r - 1/\tau_o)}{[1 - \exp(\tau_e/\tau_o - \tau_e/\tau_r)]} \qquad (3)$$

The mean age of the continental crust $\bar{\tau}_c$ is given by

$$\bar{\tau}_c = \frac{1}{(1/\tau_r - 1/\tau_o)} + \frac{\tau_e}{[1 - \exp(\tau_e/\tau_r - \tau_e/\tau_o)]} \qquad (4)$$

If the mean age $\bar{\tau}_c$ is specified the recycling time τ_r can be determined.

As a specific example we take $\tau_c = 4$ Gyrs, $\tau_o = 3.6$ Gyrs, and $\bar{\tau}_c = 2.1$ Gyrs; from (4) we find that $\tau_r = 4.94$ Gyrs. The corresponding dependence of M_c/M_{co} on τ from (2) is given in Figure 1. From (3) the present rate of crustal production is $F_o = 1.58$ km^3/yr and the present rate of crustal recycling is $F_r = M_{co}/\tau_r = 1.49$ km^3/yr. Thus the net rate of crustal growth is 0.09 km^3/yr.

Phanerozoic rates of crustal production and recycling have been studied by Reymer and Schubert (1984); their preferred values are $F_o = 1.65$ km^3/yr and $F_r = 0.59$ km^3/yr. Based on their geochemical studies Armstrong (1981) gives $F_r = 2 \pm 1$ km^3/yr and DePaolo (1983) gives $F_r = 2.5 \pm 1.2$ km^3/yr.

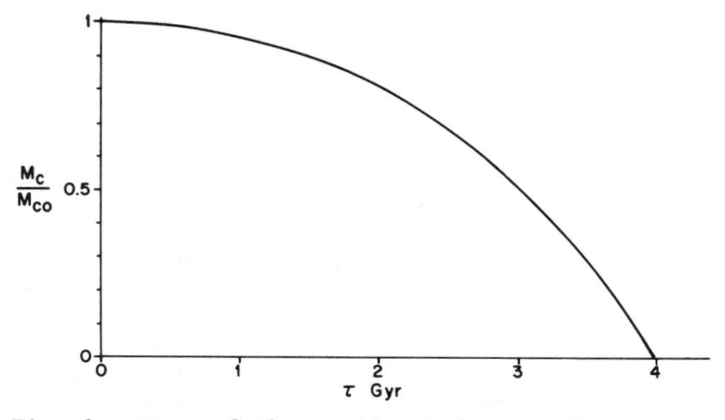

Fig. 1. Mass of the continental crust M_c as a fraction of the present value M_{co} given as a function of time τ measured backwards from the present.

Delamination

The estimate of crustal recycling rates given in the previous section can be used to estimate the amount of delamination that might be occurring. If 15 km of lower continental crust is included in the delaminated lithosphere, then 10^7 km^2 of continental lithosphere must delaminate every 100 Ma in order to provide a recycling flux of 1.5 km^3/yr. This corresponds to an area of 3,000 km by 3,000 km every 100 Myr. The mean rate of required delamination would be 0.1 km^2/yr; this compares with the present rate of subduction that is 2.8 km^2/yr. Thus the required rate of delamination is about 4% of the rate of subduction. Although it is not possible to give direct observational evidence for this amount of crustal delamination, it does not appear to be unreasonably large.

The continental lithosphere is gravitationally buoyant because of the buoyancy of the continental crust. This is why the continental lithosphere does not subduct with the entire continental crust. If the continental mantle lithosphere has the same mean composition as the upper mantle it will be negatively buoyant due to thermal contraction. If the continental mantle is substantially depleted of the basaltic component, it will have some chemical buoyancy (Oxburgh and Parmentier, 1978). Jordan (1981) has also argued in favor of significant chemical buoyancy. Since it is not possible to place a constraint on the chemical buoyancy of the continental mantle lithosphere we will neglect this effect. The buoyancy of the continental lithosphere below a depth y is given by (Turcotte and Schubert, 1982, pp. 181-183)

$$\delta M = \int_{y}^{y_c} (\rho_{mo} - \rho_{co}) dy$$

$$- \frac{\rho_{mo}\alpha(T_m - T_o)y_L}{1.16} \int_{1.16y/y_L}^{\infty} \text{erfc } \eta \, d\eta \qquad (5)$$

where the first term represents the positive buoyancy of the continental crust and the second term the negative buoyancy of the continental lithosphere due to thermal contraction.

As a specific example we take ρ_{mo} = 3,300 kg/m^3, ρ_{co} = 2,700 kg/m^3 for 0 < y < 20 km and ρ_{co} = 2,800 kg/m^3 for 20 < y < 35 km, y_c = 35 km, α = 3.2 x 10^{-5} °K^{-1}, T_m - T_o = 1300°K, and y_L = 180 km (Turcotte and McAdoo, 1979). The buoyancy δM for the continental lithosphere deeper than a depth y is given as a function of y in Figure 2. Beneath a depth of about 16 km the lithosphere has negative buoyancy and can be delaminated.

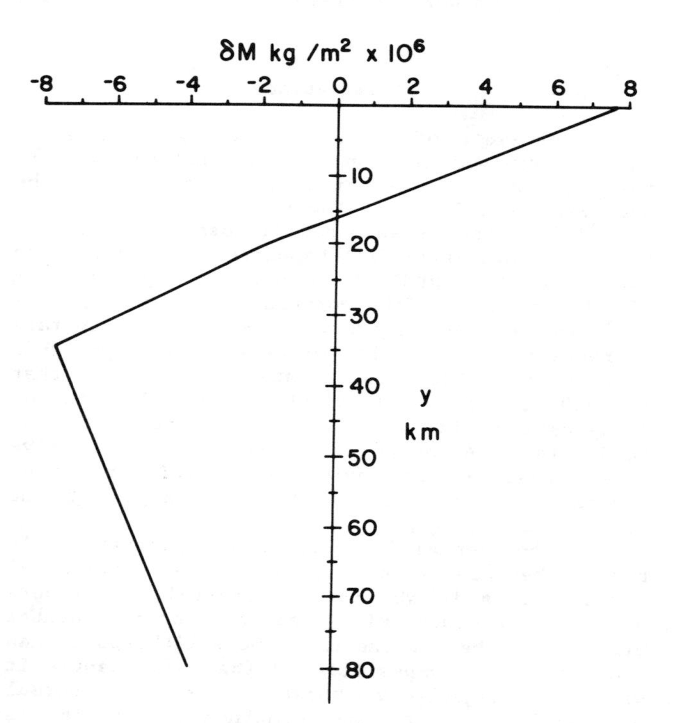

Fig. 2. Buoyancy per unit area of the continental lithosphere δM below a depth y as a function of that depth.

Bird (1979) in his studies of delamination hypothesized that delamination occurred in a manner similar to subduction. The lithosphere developed a flexure beneath the decollement similar to the flexure of the oceanic lithosphere seaward of an ocean trench. However, this type of delamination would result in large surface gravity signatures that are not observed. Another approach to delamination has been given by Houseman et al. (1981), Houseman and England (1986), and England and Houseman (1986). These authors treat the continental lithosphere as a viscous fluid which thickens in a collision zone. The thick, cold lithosphere is gravitationally unstable and sinks or delaminates. The essential question is whether it is appropriate to treat the cold, rigid lithosphere as a viscous fluid. This led Turcotte (1983) to propose an alternative mechanism of lithospheric stoping. This mechanism is illustrated in Figure 3. Soft mantle rock penetrates the continental crust in a zone of volcanism; possible sites would be the volcanic lines associated with subduction zones adjacent to continents and continental rifts; this penetration occurs at mid-crustal levels where the rocks have the softest rheology (Figure 3a). Eventually the continental lithosphere fails along a pre-existing zone of weakness (e.g. a fault). The decoupled block of continental lithosphere (including the lower crust) sinks into the mantle and is replaced by hot asthenosphere mantle rock. This lithosphere stoping process continues away from the initial zone of volcanism.

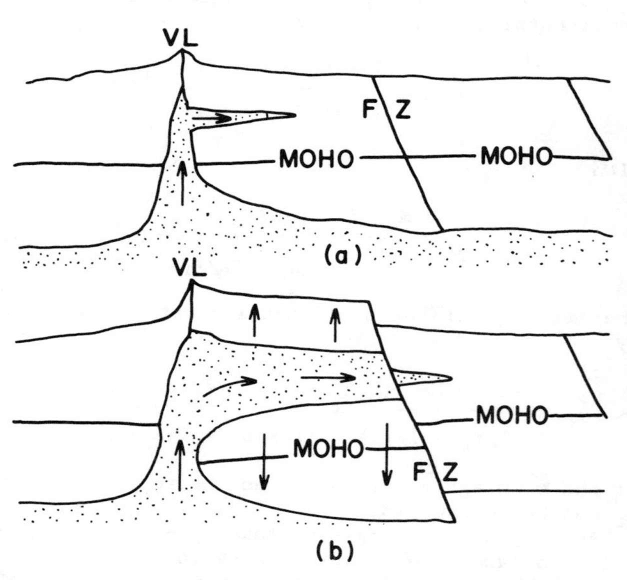

Fig. 3. Illustration of the delamination of the continental crust by a lithospheric stoping process. (a) The asthenosphere penetrates into the continental crust along a volcanic line (VL) associated with a subduction zone. It then splits the crust behind the volcanic line along an intracrustal (horizontal) zone of weakness. (b) The lower continental crust and mantle lithosphere beneath the penetrating asthenosphere breaks away along a pre-existing fault (FZ) and delaminates.

Conclusions

It is generally accepted that the lower continental crust is more mafic than the upper continental crust and that the mean continental crust is considerably more silicic than a typical basaltic magma. Material added to the continental crust in the recent past due to island arc volcanics and continental rift and hot spot volcanics has a near basaltic composition.

As a working hypothesis we propose that these mechanisms were also responsible for the addition of material to the continental crust throughout geologic time. We further hypothesize that partial melting in the lower crust creates silicic melts that ascend and fractionate the crust. If the continental crust is created with a near basaltic composition the internal fractionation would make a more mafic lower crust complementary to the silicic upper crust. Both direct and indirect evidence suggests that this is not the case. Accepting the above, it is necessary to provide a mechanism for removing mafic material from the lower crust and returning it to the mantle. This mechanism is crustal delamination.

The basic cycle responsible for the evolution of the continental crust is illustrated in Figure 4. The oceanic crust is created at mid-ocean ridges. Selected elements from the continental crust are added to the oceanic crust by solution and deposition. Sediments derived from the continents coat the oceanic crust. Some fraction of the sediments and the altered oceanic crust are subducted at ocean trenches. The remainder of the sediments are returned to the continental crust in accretionary prisms.

The subducted oceanic crust and entrained sediments are partially melted beneath island arc volcanoes; this partial melting induces partial melting in the overlying mantle wedge. The result is a near basaltic composition with rare earth and isotopic compositions contaminated with the signature of the altered oceanic crust and entrained continental sediments. The resulting island arcs are believed to be a primary source of new continental crust. If the subduction zone lies adjacent to a continent the volcanics add material directly to the continent. Hot spot and continental rift volcanism also adds material with a basaltic composition to the continents.

Delamination of the continental lithosphere including the lower continental crust can return substantial volumes of continental crust to the mantle. Our calculations suggest that the mean rate of continental delamination is near 0.1 km^2/yr. This is about 4% of the present rate of subduction.

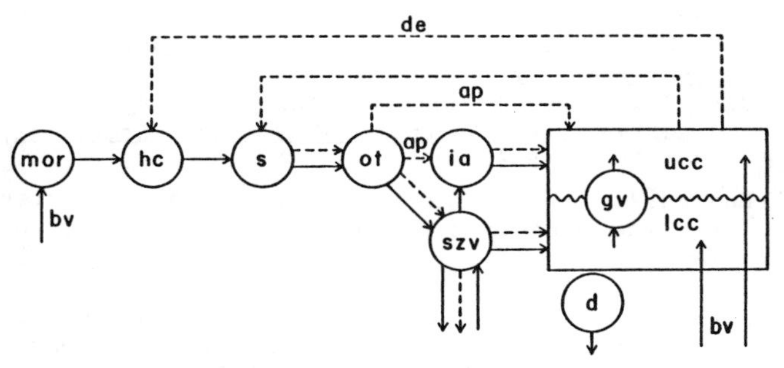

Fig. 4. Flow diagram for the creation and destruction of continental crust: bv - basaltic volcanism, mor - mid-ocean ridge, hc-hydrothermal circulations of sea water, de-desolved elements from continental crust in sea water, s - sediments, ot -ocean trench, ap-accretionary prism, szv - subduction zone volcanics, ia - island arc, ucc - upper continental crust, lcc - lower continental crust, gv - granitic volcanism, d - delamination.

Acknowledgments. The author acknowledges many stimulating discussions of this topic with R.W. Kay and S.M. Kay. This research has been supported in part by the Division of Earth Sciences, National Science Foundation under grant EAR-8721172.

References

Allègre, C.J., Chemical geodynamics, Tectonophys., 81, 109-132, 1982.

Allègre, C.J., S.R. Hart, and J.F. Minster, Chemical structure and evolution of the mantle and continents determined by inversion of Nd and Sr isotopic data, II. Numerical experiments and discussion, Earth Planet. Sci. Let., 66, 191-213, 1983.

Allègre, C.J. and C. Jaupart, Continental tectonics and continental kinetics, Earth Planet. Sci. Let., 74, 171-186, 1985.

Allègre, C.J. and D. Rousseau, The growth of the continent through geological time studied by Nd isotope analysis of shales, Earth Planet. Sci. Let., 67, 19-34, 1984.

Ambrose, J.W., Exhumed paleoplains of the Precambrian Shield of North America, Am. J. Sci., 262, 817-857, 1964.

Armstrong, R.L., A model for the evolution of strontium and lead isotopes in a dynamic Earth, Rev. Geophys., 6, 175-200, 1968.

Armstrong, R.L., Radiogenic isotopes: the case for crustal recycling on a near-steady-state no-continental-growth Earth, Phil. Trans. Roy. Soc. London, 301A, 443-472, 1981.

Armstrong, R.L. and S.M. Hein, Computer simulation of Pb and Sr isotope evolution of the Earth's crust and upper mantle, Geochim. Cosmochim. Acta, 37, 1-18, 1973.

Bird, P., Continental delamination and the Colorado Plateau, J. Geophys. Res., 84, 7561-7571, 1979.

Bird, P. and J. Baumgardner, Steady propagation of delamination events, J. Geophys. Res., 86, 4891-4903, 1981.

Brown, G.C., Mantle origin of Cordilleran granites, Nature, 265, 21-24, 1977.
Burke, K. and J.F. Dewey, Plume-generated triple junctions: key indicators in applying plate tectonics to old rocks, J. Geol., 81, 406-433, 1973.
Cook, F.A., D.S. Albaugh, L.D. Brown, S. Kaufman, J.E. Oliver, R.D. Hatcher, Thin-skinned tectonics in the crystalline southern Appalachians: COCORP seismic-reflection profiling of the Blue Ridge and Piedmont, Geology, 7, 563-567, 1979.
DePaolo, D.J., Implications of correlated Nd and Sr isotopic variations for the chemical evolution of the crust and mantle, Earth Planet. Sci. Let., 43, 201-211, 1979.
DePaolo, D.J., Crustal growth and mantle evolution: inferences from models of element transport and Nd and Sr isotopes, Geochim. Cosmochim. Acta, 44, 1185-1196, 1980.
DePaolo, D.J., The mean life of continents: estimates of continent recycling rates from Nd and Hf isotopic data and implications for mantle structure, Geophys. Res. Let., 10, 705-708, 1983.
Dewey, J.F. and B.F. Windley, Growth and differentiation of the continental crust, Phil. Trans. Roy. Soc. London, 301A, 189-206, 1981.
Eaton, G.P., Geophysical characteristics of the crust of the Basin and Range Province, Continental Tectonics, eds. B.C. Burchfiel, J.E. Oliver, and L.T. Silver, pp. 96-113, National Academy of Sciences, Washington, D.C., 1980.
Emerman, S.H. and D.L. Turcotte, Stagnation flow with a temperature-dependent viscosity, J. Fluid Mech., 127, 507-517, 1983.
England, P., and G. Houseman, Finite strain calculations of continental deformation 2. Comparison with the India-Asia collision zone, J. Geophys. Res., 91, 3664-3676, 1986.
Fyfe, W.S., The generation of batholiths, Tectonophys., 17, 273-283, 1973.
Fyfe, W.S., The evolution of the Earth's crust: Modern plate tectonics to ancient hot spot tectonics? Chem. Geol., 23, 89-114, 1978.
Goldschmidt, V.M., Grundlagen der quantitativen Geochemie, Fortschr. Min. Krist. Petrog., 17, 112-156, 1933.
Gurnis, M. and G.F. Davies, Simple parametric models of crustal growth, J. Geodynamics, 3, 105-135, 1985.
Gurnis, M. and G.F. Davies, Apparent episodic crustal growth arising from a smoothly evolving mantle, Geology, 14, 396-399, 1986.
Hadley, D. and K. Kanamori, Seismic structure of the Transverse Ranges, California, Geol. Soc. Am. Bull., 88, 1469-1478, 1977.
Hallam, A., Secular changes in marine inundation of USSR and North America through the Phanerozoic, Nature, 269, 769-772, 1977.
Hamilton, P.J., R.K. O'Nions, D. Bridgwater, and A. Nutman, Sm-Nd studies of Archaean metasediments and metavolcanics from west Greenland and their implications for the Earth's early history, Earth Planet. Sci. Let., 62, 263-272, 1983.
Hofmann, A.W., K.P. Jochum, M. Seufert, and W.M. White, Nb and Pb in oceanic basalts: new constraints on mantle evolution, Earth Planet. Sci. Let., 79, 33-45, 1986.
Holland, J.G. and R.St.J. Lambert, Major element chemical composition of shields and the continental crust, Geochim. Cosmochim. Acta, 36, 673-683, 1972.
Houseman, G., and P. England, Finite strain calculations of continental deformation 1. Method and general results for convergent zones, J. Geophys. Res., 91, 3651-3663, 1986.
Houseman, G.A., D.P. McKenzie, and P. Molnar, Convective instability of a thickened boundary layer and its relevance for the thermal evolution of continental convergence belts, J. Geophys. Res, 86, 6115-6132, 1981.
Jakeŝ, P. and S.R. Taylor, Excess europium content in Precambrian sedimentary rocks and continental evolution, Geochim. Cosmochim. Acta, 38, 739-745, 1974.
Jakeŝ, P. and A.J.R. White, Composition of island arcs and continental growth, Earth Planet. Sci. Let., 12, 224-230, 1971.
Jordan, T.H., Continents as a chemical boundary layer, Phil. Trans. Roy. Soc. London, 301A, 359-373, 1981.
Karig, D.E., Tectonic erosion at trenches, Earth Planet. Sci. Let., 21, 209-212, 1974.
Karig, D.E. and R.W. Kay, Fate of sediments on the descending plate at convergent margins, Phil. Trans. Roy. Soc. London, 301A, 233-251, 1981.
Kay, R.W., Volcanic arc magmas: Implications of a melting-mixing model for element recycling in the crust-upper mantle system, J. Geology, 88, 497-522, 1980.
Kay, R.W. and S.M. Kay, Petrology and geochemistry of the lower continental crust: an overview, in The Nature of the Lower Continental Crust, J.B. Dawson, D.A. Carswell, J. Hall, and K.H. Wadepohl, eds., pp. 147-159, Geol. Soc. London, Spec. Publ., 24, 1986.
Kay, R.W. and S.M. Kay, Crustal recycling and the Aleutian arc, Geochim. Cosmochim. Acta, 52, 1351-1359, 1988.
Kay, S.M. and R.W. Kay, Role of crystal cumulates and the oceanic crust in the formation of the lower crust of the Aleutian arc, Geology, 13, 461-464, 1985.
Kirby, S.H. and J.W. McCormick, Creep of hydrolytically weakened synthetic quartz crystals oriented to promote 2110-0001 slip: a brief summary of work to date, Bull. Mineral., 102, 124-137, 1979.
Lachenbruch, A.H. and J.H. Sass, Models of an extended lithosphere and heat flow in the Basin and Range province, Geol. Soc. Am. Mem., 152, 209-250, 1978.
McCulloch, M.T. and G.J. Wasserburg, Sm-Nd and Rb-Sr chronology of continental crust formation, Science, 200, 1003-1011, 1978.

McKenzie, D., A possible mechanism for epeirogenic uplift, Nature, 307, 616-618, 1984.
McKenzie, D. and R.K. O'Nions, Mantle reservoirs and ocean island basalts, Nature, 301, 229-231, 1983.
McLennan, S.M. and S.R. Taylor, Continental freeboard, sedimentation rates and growth of continental crust, Nature, 306, 169-172, 1983.
Mueller, S., A new model of the continental crust, Am. Geophys. Un. Mono., 20, 289-317, 1977.
Mueller, S., Deep structure and recent dynamics in the Alps, in Mountain Building Processes, K.J. Hsü, ed., pp. 181-199, Academic Press, London, 1982.
Nelson, B.K. and D.J. DePaolo, Rapid production of continental crust 1.7 to 1.9 b.y. ago: Nd isotopic evidence from the basement of the North American mid-continent, Geol. Soc. Am. Bull., 96, 746-754, 1985.
Nockolds, S.R., Average chemical compositions of some igneous rocks, Geol. Soc. Am. Bull., 65, 1007-1032, 1954.
O'Nions, R.K., N.M. Evensen, and P.J. Hamilton, Geochemical modeling of mantle differentiation and crustal growth, J. Geophys. Res., 84, 6091-6101, 1979.
Oxburgh, E.R., Flake tectonics and continental collision, Nature, 239, 202-204, 1972.
Oxburgh, E.R. and E.M. Parmentier, Thermal processes in the formation of continental lithosphere, Phil. Trans. Roy. Soc. London, A288, 415-429, 1978.
Pakiser, L.C. and R. Robinson, Composition of the continental crust as estimated from seismic observations, Am. Geophys. Un. Mono., 10, 620-626, 1966.
Reymer, A. and G. Schubert, Phanerozoic addition rates to the continental crust and crustal growth, Tectonics, 3, 63-77, 1984.
Reymer, A. and G. Schubert, Rapid growth of some major segments of continental crust, Geology, 14, 299-302, 1986.
Ronov, A.B., Evolution of rock composition and geochemical processes in the sedimentary shell of the Earth, Sedimentology, 19, 157-172, 1972.
Ronov, A.B., and A.A. Yaroshevsky, Chemical composition of the Earth's crust, Am. Geophys. Un. Mono., 13, 37-57, 1969.
Schubert, G. and A.P.S. Reymer, Continental volume and freeboard through geological time, Nature, 316, 336-339, 1985.
Smithson, S.B., Modelling continental crust: structural and chemical constraints, Geophys. Res. Let., 5, 749-753, 1978.
Tarney, J. and B.F. Windley, Chemistry, thermal gradients, and evolution of the lower continental crust, J. Geol. Soc. London, 134, 153-172, 1977.
Taylor, S.R., The origin and growth of continents, Tectonophys., 4, 17-34, 1967.
Taylor, S.R., Island arc models and the composition of the continental crust, in Island Arcs, Deep Sea Trenches, and Back Arc Basins, M. Talwani and W.C. Pitman, eds., Maurice Ewing Series 1, pp. 325-335, American Geophysical Union, Washington, D.C., 1977.
Taylor, S.R. and S.M. McLennan, The Continental Crust: Its Composition and Evolution, (Blackwell Scientific Publications, Oxford, 1985) 312 p.
Taylor, S.R. and A.J.R. White, Geochemistry of andesites and the growth of continents, Nature, 208, 271-273, 1965.
Turcotte, D.L., Mechanisms of crustal deformation, J. Geol. Soc. London, 140, 701-724, 1983.
Turcotte, D.L. and S.H. Emerman, Mechanisms of active and passive rifting, Tectonophys., 94, 39-50, 1983.
Turcotte, D.L. and L.H. Kellogg, Isotopic modeling of the evolution of the mantle and crust, Rev. Geophys., 24, 311-328, 1986.
Turcotte, D.L., J.Y. Liu, and F.H. Kulhawy, The role of an intracrustal asthenosphere on the behavior of major strike-slip faults, J. Geophys. Res., 89, 5801-5816, 1984.
Turcotte, D.L. and D.C. McAdoo, Geoid anomalies and the thickness of the lithosphere, J. Geophys. Res., 84, 2381-2387, 1979.
Turcotte, D.L. and G. Schubert, Geodynamics, (John Wiley and Sons, New York, 1982), p. 450.
Vinogradov, A.P., Average contents of chemical elements in the principal types of igneous rocks of the Earth's crust, Geochem., 7, 641-664, 1962.
Watson, J.V., Vertical movements in Proterozoic structural provinces, Phil. Trans. Roy. Soc. London, A280, 629-640, 1976.
Weaver, B.L. and J. Tarney, Empirical approach to estimating the composition of the continental crust, Nature, 310, 575-577, 1984.
Windley, B.F., Timing of continental growth and emergence, Nature, 270, 426-428, 1977.
Wise, D.U., Freeboard of continents through time, Geol. Soc. Am. Memoir, 132, 87-100, 1972.
Wise, D.U., Continental margins, freeboard and the volumes of continents and oceans through time, in The Geology of Continental Margins, C.A. Burk and C.L. Drake, eds., pp. 45-56, Springer-Verlag, New Nork, 1974.
Wolery, T.J. and N.H. Sleep, Hydrothermal circulation and geochemical flux at mid-ocean ridges, J. Geol., 84, 249-275, 1976.
Wyllie, P.J., Crustal anatexis: An experimental review, Tectonophys., 43, 41-71, 1977.
Yeats, R.S., Quaternary flake tectonics of the California Transverse Ranges, Geology, 9, 18-20, 1981.

GEOCHEMISTRY OF Rb, Sr AND REE IN NIUTOUSHAN BASALTS IN THE COASTAL AREA OF FUJIAN PROVINCE, CHINA

Yu Xueyuan

Institute of Geochemistry, Academia Sinaca, Guiyang, P. R. China

Abstract. Analyses of major elements, REE, Rb, Sr and Rb-, Sr-isotopes show that the Niutoushan basalts of Fujian, China belong to the subalkaline volcanic rock series of active continental margins, Predominated by subalkaline basalts in association with minor amounts of alkaline basalts, this volcanic cone is well differentiated. The tholeiite in this area contains abundant mantle-source xenoliths and features a high Mg/Fe ratio. In addition, its Rb/Sr(0.0358),87Rb/86Sr (0.101) and 87Sr/86Sr(0.7038) ratios are very close to those typical of pyrolite. It is thus believed to be the parent magma of the basaltic rocks in this area. The subalkaline and alkaline basalts have similar 87Sr/86Sr ratios, the initial 87Sr/86Sr ratio is 0.7039±0.004. Batch Melting Model calculations on trace elements indicate that the olivine tholeiitic magma was derived from 7 % partial melting of garent lherzolite from the upper mantle. From a simulated calculation on REE it is inferred that garnet was the residual phase during the upper mantle melting. Calculation of a Rb, Sr fractional crystallization model indicates that the alkaline basalt was the residual melt formed after the fractional crystallization of either 67-42% plagioclase and pyroxene cumulates or 43-25% plagioclase cumulates.

Introduction

Like other modern subalkaline volcanic suites in the world, the Niutoushan basalt belt occurs in the active marginal zone of plate junctures in the lithosphere. In this belt the Niutoushan volcanic cone is better differentiated, and is predominated by tholeiite, associated with minor amounts of alkaline basalt. The latter is considered to be the product of fractional crystallization under high pressures (Yu Xueyuan, 1985). This paper is intended to further discuss the possibility for the alkaline and subalkaline basaltic magmas to form as a result of fractional crystallization under high pressures.

Geological Setting

Belonging to the Tertiary Futan Group, the basaltic rocks in this area overlie Cretaceous granites and migmatites. K-Ar dating has assigned the tholeiite a whole rock age of 13.2±0.6 Ma.

The Niutoushan volcanic cone, exposed over an area of about 0.5 Km^2, bears the feature of multicyclic eruptions. The early stage facies is predominantly flood-phase subalkaline basalt and the late stage facies mainly sheet-like alkaline basalt (Fig.1). In addition, in the volcanic cone there extensively occur various types of inclusions, including mantle-source xenoliths of spinel lherzolite and spinel olivine websterite, cogenetic inclusions of pyroxene anorthosite gabbro and websterite, as well as Al-enriched pyroxene megacrysts. It is worth noting that mantle-source lherzolite was obsereved in olivine tholeiite as a rare occurrence of mantle-source xenoliths.

Analytical Results and Interpretation

Major Elements

Given in Table 1 are the results of both major elements (by AAS) and normative minerals of the basalts in this area.

The Niutoushan basalts can be divided into subalkaline and alkaline series. According to the characteristics of the normative minerals, the subalkaline series is subdivided into quartz-tholeiite (containing Q) and olivine tholeiite (containing Ol), and the alkaline series into olivine basalt (containing Ol with low Hy content) and alkaline olivine basalt(containing Ol and Ne with Hy absent).

The M value ($M=Mg/(Mg+Fe_T)x100$), an indicator of petrogenesis, tends to decrease from subalkaline to alkaline series, concordant with the rule that M value declines during the late stage differentiation of basic magmas. Bultitude & Green(1971) stated that basaltic magma in equilibrium with mantle or residual peridotite should be relatively enriched in magmesium with M values varying from 0.68 to 0.72. On the other hand, the differentiation of

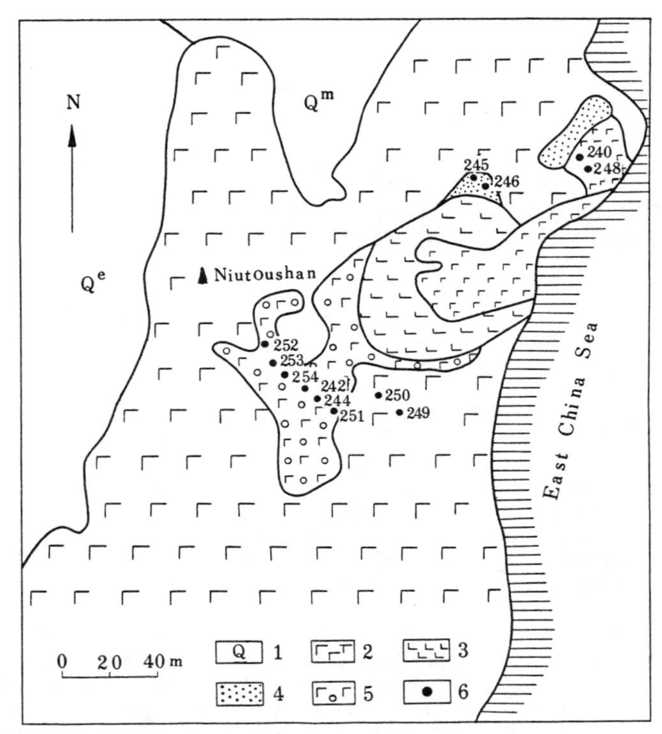

Fig.1. Generalized Geological Map of Niutoushan Basalts (After the No.3 Regional Geological Survey Team of Fujian Province, 1977).
1) Oceanic deposits and residual slide deposits;
2) Tholeiite;
3) Tholeiite with well developed columnar joints;
4) Olivine thoeliite;
5) Alkaline basalt;
6) Sampling locality.

basaltic magma in a closed system, which underwent the precipitation of pyroxene, olivine and garnet, could cause the decrease of M values. Therefore, it can be assumed that the olivine tholeiite with high M values (M= 61.68-63.96) was the parent magma of this area.

The subalkaline basalt has relatively low Na and K contents(Na_2O + K_2O = 2.79-3.99%) with a Na/K ratio of 6-9, which falls between those pertinent to continental and island-arc basalts, thus belonging to the soda-basalt series of active continental margins. From the subalkaline to the alkaline series both M values and Na/K ratios decrease but the total alkali (ecpecially K) amount increases. These facts suggest that the alkaline basalt is the product of fractional crystallization. In the Ol'-Ne'-Qu' diagram (Fig.2) such a differentiation trend is quite noticeable, the projecting points of the basalts running across the Si-saturation division surface (Ab-Opx) and the low-pressure thermal barries (Ab-Ol'). It appears that the differentiation proceeded toward the increasing of Ne and decrease Hy in the magma. The fact that the alkaline basalt

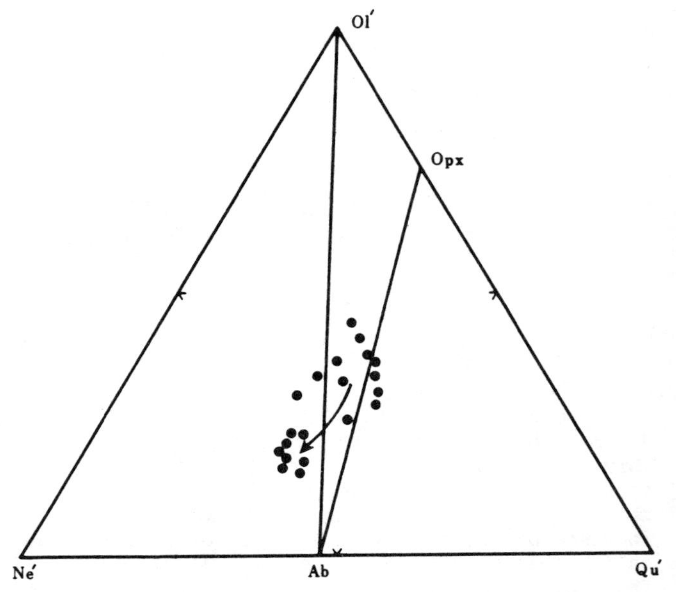

Fig.2. Ol'-Ne'-Qu' Diagram of Niutoushan Basalts. (Arrow indicates the evolution trend of magma.)

resulting from fractional crystallization is projected across the low-pressure thermal barrier strongly suggests that the fractional crystallization should occur under high pressure conditions.

REE

The major REE contents in various types of basalts in this area given in Table 2 (by XRF).

Olivine tholeiite and quartz tholeiite are similar in their REE abundances and distribution patterns, conforming to the REE characteristics of continental tholeiite---the enrichment of LREE. The alkaline basalt has a higher REE abundance with LREE highly concentrated, the $(La/Yb)_N$ ratio being up to 18.54, and features a highly fractionated REE distribution pattern.

The mantle-source spinel lherzolite xenoliths exhibit an LREE-enriched pattern(Fig.3), in which La and Ce are 3 times and the rest 1.5 times higher in content than the respective REE in chondrite. This indicates that there might exist LREE-enriched upper mantle in the area studied.

Rb, Sr and Their Isotopes

Rb and Sr abundances(by XRF) and their isotopic compositions for various types of Niutoushan basalts are shown in Table 3.

The Rb and Sr abundances in the subalkaline basalt are fairly close to those in continental tholeiite. For example, in the olivine tholeiite of Columbia River continental tholeiite suite Rb= 2.3-49 ppm and Sr = 234-383 ppm(Gary et al.,1981), both being higher than those typical of island-arc

TABLE 1. Major Element and Normative Mineral Contents in Niutoushan Basalts

Rock	Subalkaline Baslt							Alkaline Basalt					
	Quartz tholeiite				Olivine tholeiite			Olivine basalt			Alkaline olivine basalt		
No.	240	248	249	250	237	245	246	252	253	254	242	244	251
SiO_2	51.17	51.74	51.00	46.89	49.60	49.92	49.36	49.23	48.99	49.10	48.88	48.96	49.21
TiO_2	1.03	0.96	1.04	2.49	1.33	1.50	1.50	2.14	2.12	2.11	2.08	2.14	2.16
Al_2O_3	17.73	16.91	17.46	17.29	14.91	16.24	16.24	14.86	14.82	14.44	14.63	13.94	14.83
Fe_2O_3	1.26	1.72	1.26	4.32	1.71	3.51	2.88	5.76	4.51	5.73	4.53	4.88	5.14
FeO	8.48	7.60	8.44	4.83	9.01	6.65	6.77	4.92	5.91	4.71	5.75	5.73	5.14
MnO	0.02	0.03	0.02	0.12	0.22	0.16	0.14	0.10	0.10	0.10	0.10	0.11	0.09
MgO	6.30	6.16	6.20	5.73	7.16	7.36	7.86	6.20	6.18	5.90	6.43	6.38	6.13
CaO	9.21	8.53	9.43	7.63	10.17	8.95	9.09	8.06	7.91	8.36	8.37	8.26	8.39
Na_2O	2.54	2.66	2.49	2.79	2.90	2.90	3.17	3.38	3.35	3.40	3.60	3.67	3.74
K_2O	0.34	0.36	0.30	1.29	0.37	0.73	0.78	2.35	2.32	2.30	2.28	2.34	2.28
P_2O_5	0.17	0.10	0.18	0.40	0.19	0.26	0.28	0.61	0.59	0.59	0.56	0.54	0.56
M	55.62	56.97	55.34	60.36	58.35	61.68	63.69	59.77	58.35	59.30	59.78	59.17	59.69
Op	0.35	0.21	0.37	0.82	0.39	0.63	0.68	1.48	1.44	1.44	1.36	1.32	1.36
Il	1.96	1.82	1.98	4.73	2.53	2.90	2.90	4.16	4.16	4.16	4.06	4.19	4.20
Or	2.04	2.13	1.77	7.62	2.19	4.40	4.70	14.22	14.16	14.04	13.86	14.20	13.79
Ab	21.48	22.50	21.06	23.60	24.53	25.01	27.35	29.29	29.27	29.73	30.23	30.84	31.56
An	35.98	33.14	35.58	30.85	26.58	29.70	28.32	18.88	19.16	17.93	17.51	15.11	17.34
Mt	1.83	2.49	1.83	6.26	2.48	5.19	4.26	8.55	6.75	8.59	6.75	7.30	7.73
Di	7.01	6.96	8.23	3.44	18.51	11.12	12.71	14.09	13.85	16.38	17.03	18.65	16.95
Hy	25.07	22.91	24.14	14.05	14.17	20.01	12.00	7.27	6.12	6.73	-	-	-
Q	2.49	4.57	2.81	2.30	-	-	-	-	-	-	-	-	-
Ol	-	-	-	-	6.60	0.44	7.09	2.06	5.09	1.03	8.60	7.70	6.73
Ne	-	-	-	-	-	-	-	-	-	-	0.60	0.64	0.45

TABLE 2. Major REE Contents in Various Types of Basalts and Mantle-Source Xenoliths in Niutoushan Area(ppm)

	Alkaline olivine basalt	Olivine tholeiite	Tholeiite	Spinel lherzolite
Ce	82.60	26.23	26.0	2.37
Sm	9.19	3.38	3.9	0.31
Eu	3.22	1.12	1.2	0.14
Yb	1.70	1.37	1.1	0.32

and mid-ocean-ridge tholeiites. In this area, the alkaline basalt has higher Rb and Sr contents than the subalkaline basalt with Rb far more concentrated than Sr.

The enrichment of K, LREE,Rb and Sr in alkaline basaltic magma resulting from fractional crystallization is related to their geochemical properties. These incompatible elements commonly possess a rather small solid-liquid partition coefficient; and, with respect to basaltic melt, their total partition coefficients are much less than unity. Therefore, they tend to be concentrated in the residual melt produced from fractional crystallization. Some strongly incompatible elements, such as K, Ce and Rb, have a total partition coefficient of much less than 1 (D 1). So it would be easier for them to enter the melt phase as compared with Sm, Eu and Sr. This feature is responsible for the enrichment of LREE, K and Rb, the high-degree fractionation of REE and the increase of Rb/Sr ratio in the residual melt (the alkaline basalt).

As seen from Fig.4, in terms of Rb-, Sr-isotopic ratios the basalts of this area are plotted along a straight line roughly parallel to the abscissa; i.e., from subalkaline to alkaline basalts, 87Rb/86Sr ratio gradually increases while 87Sr/86Sr ratio shows a trivial variation only. The linear relationship between the two ratios can be expressed as follows.

$$(87Sr/86Sr) = 0.7037 + 0.001(87Rb/86Sr).$$

The initial 87Sr/86Sr ratio (0.7037±0.0004) is very close to the mean 87Sr/86Sr ratio of pyrolite---0.704±0.002(Faure, 1977). The Rb/Sr and 87Rb/86Sr ratios of the olivine tholeiite, the parent magma of this area, are also similar to those pertinent to the modern earth as a whole---Rb/Sr=0.029±0.003 and $(87Rb/86Sr)_{UR}$=0.0839(Depaolo and Wasserburg, 1976). All these facts suggest that the Niutoushan basalts were derived from the upper mantle.

Briqueu and Lancelot(1979) pointed out that during fractional crystallization of magma with homogeneous isotopic compositions the Sr-isotopic composition should not change with the degree of differentiation, and that the correlation between

TABLE 3. Abundances and Isotopic Composition of Rb and Sr in Niutoushan Basalts

No.	Rock	Rb ppm	Sr ppm	Rb/Sr	87Rb μml/g	86Sr μml/g	87Rb/ 86Sr	87Sr/ 86Sr
248	Quartz tholeiite	7.3	274	0.0266	0.0234	0.3146	0.0743	0.7036
250		21,8	706	0.0309	0.0699	0.8100	0.0863	0.7041
240		8.5	246	0.0346	0.0273	0.2825	0.0966	0.7039
	Average	12.5	409	0.0307	0.0402	0.4690	0.0857	0.7039
237	Olivine tholeiite	6.0	310	0.0194	0.0192	0.3559	0.0540	0.7037
245		14.5	308	0.0471	0.0477	0.3529	0.1352	0.7039
246		12.8	313	0.0409	0.0409	0.3592	0.1140	0.7037
	Average	11.1	310	0.0358	0.0359	0.3560	0.1010	0.7038
242	Alkaline olivine	63.1	565	0.1117	0.2021	0.6473	0.3122	0.7042
244	basalt	63.6	941	0.0676	0.2035	1.0790	0.1885	0.7039
251		62.0	725	0.0855	0.1984	0.8307	0.2388	0.7039
252		64.4	710	0.0907	0.2062	0.8139	0.2533	0.7040
253		64.5	666	0.0968	0.2064	0.7638	0.2702	0.7039
254		64.1	731	0.0875	0.2053	0.8377	0.2451	0.7038
	Average	63.6	723	0.0899	0.2036	0.8287	0.2513	0.7039

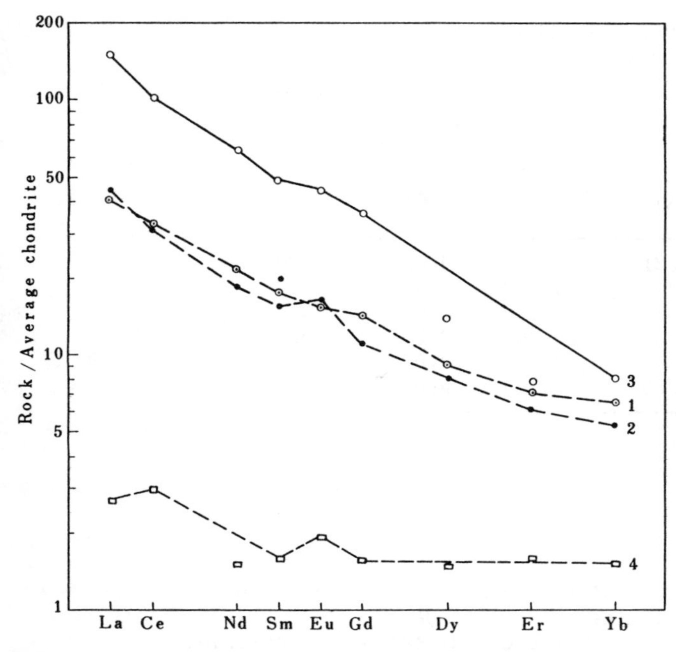

Fig.3. REE Distribution Patterns of Niutoushan Basalts. 1) Olivine tholeiite; 2) Quartz tholeiite; 3) Alkaline olivine basalt; 4) Spinel lherzolite xenolith.

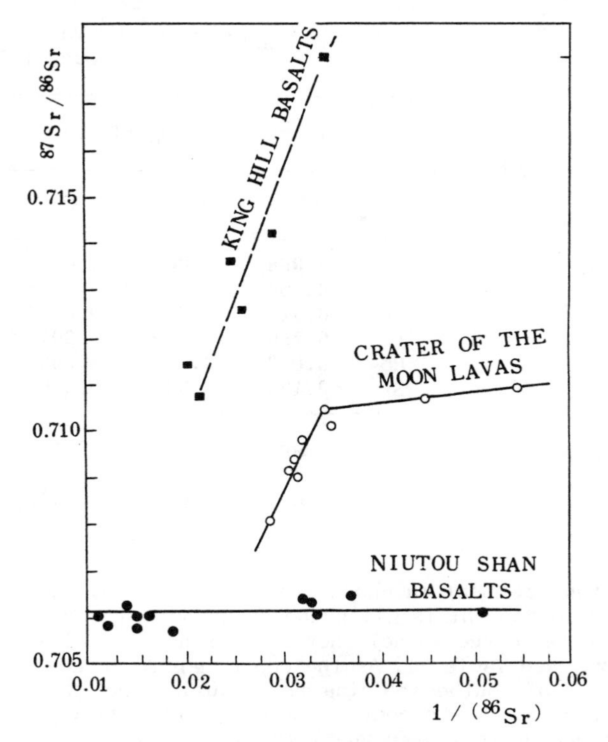

Fig.5. Relationship between 87Sr/86Sr and 1/(86Sr) for Niutoushan Basalts (For King Hill Basalts and Crater of the Moon Lavas, after Briqueu and Lancelot, 1979).

87Sr/86Sr and 1/(86Sr) could be used as an indicator of contamination. In the correlation diagram the non-contaminated fractions are plotted along a horizontal line parallel to the abscissa of 1/(86Sr) while the contaminated fractions along a declined with a positive slope (see Fig.5, the King Hill Basalt and the Crater of the Moon Lavas, Briqueu and Lancelot, 1979).

As can be seen from Fig.5, the data points of the various basalts are plotted along a horizontal line parallel to the abscissa, suggesting that they resulted from differentiation of cogenetic magmas and that both the parent magma and the products of fractional crystallization were not contaminated by radiogenic Sr from the crust, probably because the basaltic magma penetrated a relatively thin continental crust while ascending from the mantle---the crust thickness in coastal areas of East China is about 31 Km (Song Zhong and Tan Chenye, 1965).

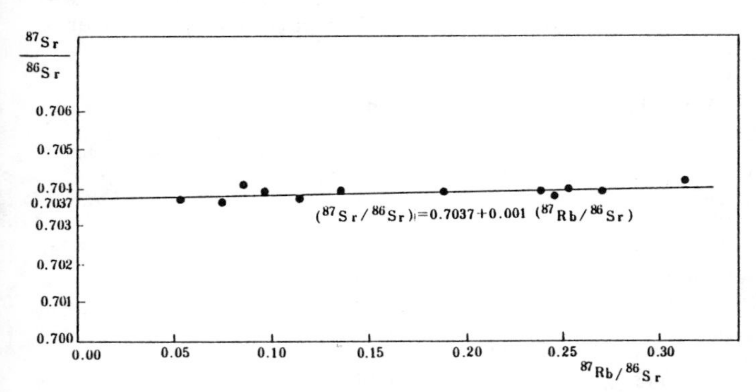

Fig.4. Variation of 87Sr/86Sr and 87Rb/86Sr Ratios of Niutoushan Basalts.

Genesis

Origin of Olivine Tholeiitic Magma

As mentioned above, the olivine tholeiite in this area contains abundant mantle-source spinel lherzolite xenoliths, features high Mg/Fe ratios (M =61.68-63.69) and possesses Rb/Sr, 87Rb/86Sr and 87Sr/86Sr ratios similar to those typical of pyrolite. These facts indicate that olivine tholeiite was the parent magma of the basalts in this area. And it is believed that this parent magma was derived from the partial melting of garnet lherzolite in upper mantle. Based on the Batch Melting Model(Hanson, 1980) we will calculate the degree of partial melting. Table 4 gives the Kd's used in the calculation.

The Sm, Eu, Yb, Rb and Sr contents (C_0) in the garnet lherzolite from upper mantle are each assigned a value of 2 times the abundance of the re-

TABLE 4. Kd Values of Mineral-Silicate Melts Used in Calculations

	Melting of Mantle[1]				Fractional Crystallization of Basaltic Magma[2]		
	CPX	OPX	OL	GAR	CPX	OPX	PLA
Ce	0.098	0.003	0.0005	0.021	0.070	0.024	0.120
Sm	0.260	0.010	0.0013	0.217	0.180	0.054	0.067
Eu	0.310	0.013	0.0016	0.320	0.180	0.054	0.340
Yb	0.280	0.049	0.0015	4.030	0.160	0.340	0.067
Rb[3]	0.003	0.003	0.0001	0.010	0.003	0.003	0.050
Sr[4]	0.120	0.020	0.0001	0.080	0.120	0.020	1.830

1 - Hanson G. N. (1980)
2 - Arth J.G. (1976)
3,4 - Gary E.L. et al.(1981)

levant element in chondrite (Hanson, 1980), and the Ce content is given the value obtained for mantle-source spinel lherzolite xenoliths (2.37 ppm, see Table 2). As the upper mantle in coastal areas of Southeast China may plausibly be LREE-enriched, its Ce content is taken for 3 times that of chondrite (Yu Xueyuan, 1985).

In Table 5, the F_1 values were calculated by assuming the residues of partial melting as harzburgite(Ol:Opx:Cpx=75:20:5). The six sets of data differ significantly, the F_1 value of Yb group being up to 0.285. It is generally considered that alkaline basaltic magma and tholeiite were the products respectively of 5% ± 1% and 15-20% ± 5% partial melting of pyrolite(Gary et al., 1981).The partial melting degree of olivine tholeiite magma is estimated to be within the range of 5-15%. Thus the F_1 values could not be used. The second set of data(F_2) were obtained by assuming residues to be garnet lherzolite (Ol:Opx:Cpx:Gar=65:18:12:5). The five sets of data for REE groups are fairly consistent, the F values ranging between 7.2-7.8%, while the data of Rb/Sr group seem somewhat lower.

Shown in Fig.6 is the variation trend of Rb/Sr ratios during the melting process of the two different residues. The general trend is the increase of the ratio with decreasing partial melting degrees. But great difference is observed in the variation rate of Rb/Sr ratios. This is because garnet, the principal HREE carrier, was left in the mantle thereby causing the fractionation of REE; in other words, LREE entered the melt and HREE were left in the residual solid phase (Fig.6A). If garnet had not been left in the residual phase, the REE in the melt would not have been fractionated (Fig. 6B).

According to a simulated batch melting calculation on REE, Rb and Sr, the olivine tholeiitic magma of this area should be the product of 7% partial melting of garnet lherzolite from upper mantle, during which garnet might be left as a residual phase.

Formation of Alkaline Olivine Basalt

The occurrence, the evolution trend of major elements, REE, Rb, Sr and Rb-, Sr-isotopes, and the mineralogical characteristics of Niutoushan basalts all suggest that the alkaline olivine basalt in this area is the product of fractional crystallization of the olivine tholeiitic parent magma under

TABLE 5. Results of Batch Melting Calculation on Gernet Lherzolite

	C_0 (ppm)	C_L (ppm)	D_1	F_1	D_2	F_2
Ce	2.37	26.23	0.0059	0.056	0.0137	0.078
Sm	0.384	3.38	0.0162	0.099	0.0447	0.075
Eu	0.1444	1.12	0.0197	0.112	0.0566	0.072
Yb	0.416	1.37	0.0256	0.285	0.2449	0.077
Ce/Yb	5.70	19.15		0.003		0.078
Rb/Sr	0.0296	0.0409		0.049		0.051

higher pressures (Yu Xueyuan, 1985; Huang Wankang, 1983).

In order to discuss the origin of the alkaline olivine basaltic magma based on the fractional differentiation model, it is imperative to first determine the mineral composition and the fraction of cumulates during fractional crystallization. As described above, in the olivine tholeiite there are considerable amounts of cogenetic inclusions of websterite, gabbro and pyroxene anorthosite, as well as pyroxene megacryts which are considered to be the cumulates formed during fractional crystallization under high pressures. Three types of cumulates, namely anorthosite (PLA=100%), gabbro (PLA:OPX:CPX=50:30:20) and websterite (OPX:CPX=50:50), were chosen for a simulated calculation on fractional crystallization. The results obtained are given in Fig.7.

Comparing the three fractional crystallization curves, one may yield the following understandings. If the cumulates were anorthosite, Rb and Sr would be fractionated to a greatest extent. Gabbro is next and websterite still next to anorthosite in causing Rb-Sr fractionation (Fig.7). In other words, the degree of Rb-Sr fractionation is related to the content of plagioclase in the cumulates. The higher the plagioclase content, the greater the extent of Rb-Sr fractionation. This is because plagioclase has an excessively high Sr partition coefficient (Kd_{Sr}=1.83).

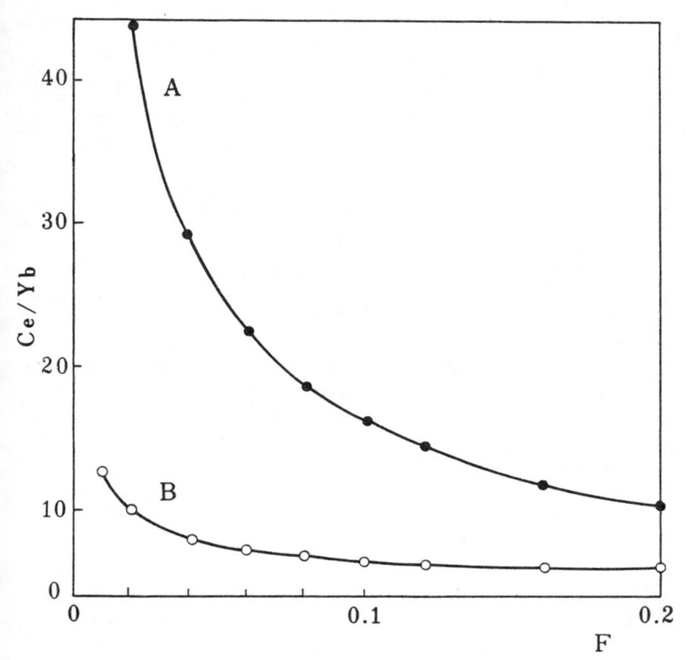

Fig.6. Variation of Rb/Sr Ratios during the Partial Melting of Garnet Lherzolite.
(A) Garnet lherzolite as residual phase;
(B) Harzburgite as residual phase.

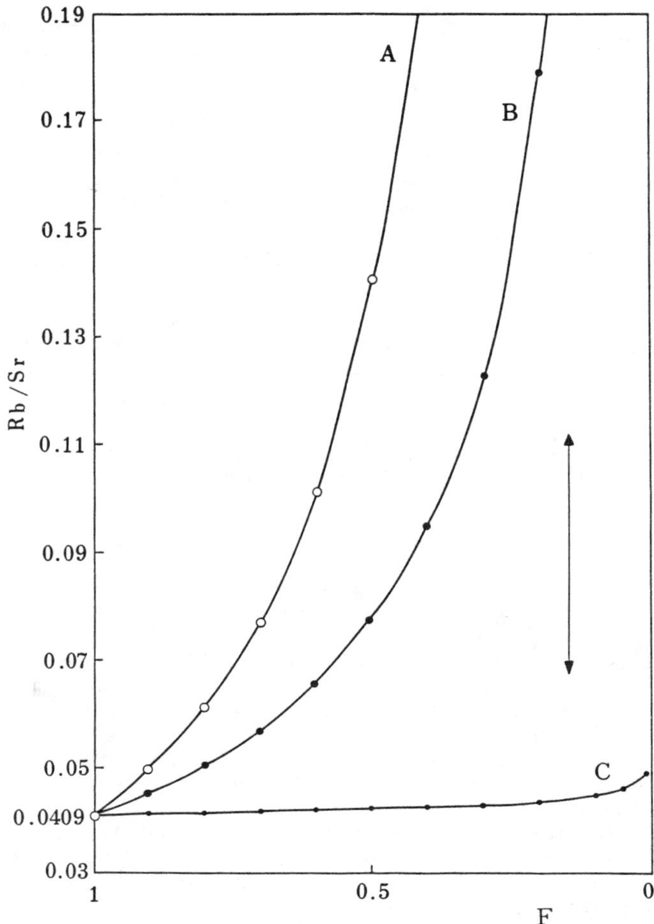

Fig.7. Variation of Rb/Sr Ratios during the Fractional Crystallization of Olivine Tholeiitic Magma. (A) Anorthosite cumulates; (B) Gabbro cumulates; (C) Websterite cumulates. (The Rb/Sr ratio of the parent magma is 0.0409 as seen from Table 3, sample No.246. Arrow indicates the variation range for Niutoushan alkaline basalt.)

Given in Table 6 are the percentages of the alkaline olivine basaltic magma formed after the separation respectively of anorthosite and gabbro from the olivine tholeiitic magma (F_A and F_B). If the cumulates are composed entirely of anorthosite, then the fraction of the residual melt from fractional crystallization is F_A=57-75%. In the case of gabbro cumulates the fraction is rather small: F_B=33-58%. Owing to the difficulty in accurately evaluating the ratio between the two cumulates, the F values can only be given a rough range with the F_A values being the upper limit.

The simulated calculation on fractional crystallization based on Rb/Sr ratios has confirmed that the parent magma in this area (the olivine tholeiitic magma) could give rise to alkaline olivine ba-

TABLE 6. Calculated Results on the Fractional Crystallization of Olivine Tholeiitic Magma

No.	Rock	Rb/Sr	F_A	F_B
245	Olivine tholeiite	0.0471	0.9237	0.8575
242	Alkaline olivine basalt	0.1117	0.5687	0.3349
244	Ditto	0.0676	0.7540	0.5786
251	Ditto	0.0855	0.6608	0.4480
252	Ditto	0.0907	0.6392	0.4201
253	Ditto	0.0968	0.6163	0.3914
254	Ditto	0.0899	0.6523	0.4369

saltic magma through the fractional crystallization of anorthosite and gabbro cumulates.

Conclusion

1. The compositional characteristics of major elements K and Na, trace elements REE, Rb and Sr, and Rb- Sr-isotopes suggest that the basalts in this area belong to soda-basalt series of active continental margins.
2. From the subalkaline to the alkaline basalts, various elements vary regularly; i.e., M value decreases, incompatible elements K, Ce, Rb and Sr increase in content, and LREE-HREE and Rb-Sr are considerably fractionated.
3. The initial 87Sr/86Sr ratio of the basalts in this area is 0.7037 and the Rb/Sr, 87Rb/86Sr ratios of the subalkaline basalts are similar to those of the present-day earth, indicating that they were derived from the upper mantle.
4. The correlation analysis between 87Sr/86Sr and 1/86Sr indicates that the parent magma and the products of its fractional crystallization were not contaminated by radiogenetic Sr from the crust.
5. The olivine tholeiite was the parent magma of the basalts in this area. It was formed from 7 % partial melting of garnet lherzolite from the upper mantle. Simulated quantitative calculation on REE suggests that garnet was the residual phase from the partial melting of upper mantle.
6. The alkaline olivine basaltic magma is the residual melt formed by the separation of 42-67% gabbro and 25-43% anorthosite cumulates from the olivine tholeiitic magma.

Acknowledgements. The author wishes to thank Miss Jiang Chengzhong for major element analysis; Mr Qian Zhixin, Mr Feng Liangyuan, Miss Li Ruolin and Miss Zhang Yawen for REE analysis; and Mr Xue Xiaofeng for Rb, Sr and Rb-, Sr-isotopes analysis.

References

Arth, J. G., Behavior of Trace Elements During Magmatic Processes - A Summary of Theoretical Models and Their Applications, J. Res. US. Geol. Surv., 4, 41-47, 1976.

Briqueu, L. and Lancelot, J. R., Rb-Sr Systematics and Crustal Contamination Models for Calc-Alkaline Igneous Rocks, Earth and Planet. Sci. Letters, 43, 385-396, 1979.

Bultitude, R. J. and Green, D. H., Experimental Study of Crystal-Liquid Relationships at High Pressure in Olivine Nephelinite and Basanite comositions, J. Petrol., 12(1), 121-147, 1971.

DePaolo, D. J. and Wasserburg, G. J., Inferences about Magma Sources and Mantle Structure from Variations of 143Nd/144Nd, Geophys. Res. Letters, 3(12), 743-746, 1976.

Engel, A. E. J., Engel, C. G. and Havens, G., Chemical Characteristics of Oceanic Basalts and the Upper Mantle, Bull. Geol. Soc. Am.,76, 719-734, 1965.

Faure, G., Principles of Isotope Geology, John Eiley & Sons, Inc., 107-138, 1977.

Gary, E. L. et al., Petrology and Chemistry of Terrestrial, Lunar and Meteoritic Basalts, in Basaltic Volcanism on the Terrestrial Planets, Edited by W. M. Kaula, Pergamon Press, 78-160, 1981.

Hanson, G. N., Rare Earth Elements in Petrogenetic Studies of Igneous Systems, Ann. Rev. Earth & Planet. Sci., 8, 371-406, 1980.

Huang Wankang et al., Mineralogical Study of Spinel-Lherzolite Inclusions from basalts in Southern and Eastern China, Geochemistry, 2(4), 361-376, 1983.

Song Zhong and Tan Chenye, Journal of Geophysics (in Chinese), 14(1), 1965.

Yu Xueyuan, The Origin of Basaltic Rocks in Niutoushan Area - High Pressure Differentiation, Chinese Journal of Geochemistry, 4(2), 150-158, 1985.